Theoretical Chemical Engineering

Christo Boyadjiev

Theoretical Chemical Engineering

Modeling and Simulation

Springer

PD Dr. Christo Boyadjiev
Institute of Chemical Engineering (IChE)
Bulgarian Academy of Sciences (BAS)
Acad. G. Bonchev Str., Bl. 103
1113 Sofia, Bulgaria
email:chboyadj@bas.bg

ISBN 978-3-642-10777-1 e-ISBN 978-3-642-10778-8

DOI 10.1007/978-3-642-10778-8

Springer Heidelberg Dordrecht London New York

Library of Congress Control Number: 2010934858

Cover design: WMXDesign GmbH

Printed on acid-free paper

Springer is part of Springer Science+Business Media (www.springer.com)

Motto

Experimenters *are the striking force of science. The* ***experiment*** *is a question which science puts to nature. The* ***measurement*** *is the registration of nature's answer. But* **before** *the question is put to nature, it* ***must be formulated****. Before the measurement result is used, it* ***must be explained****, i.e., the answer must be understood correctly. These* ***two problems*** *are obligations of the* **theoreticians**.

Max Planck

This book is dedicated to my parents.

Christo Boyanov Boyadjiev

Abstract

The theoretical methods of chemical engineering for modeling and simulation of industrial processes are surveyed in this book. On this basis it is possible to formulate correct experimental conditions and to understand correctly the experimental results.

The continuous media approach is used for modeling simple processes such as hydrodynamic processes, mass transfer processes, and heat transfer processes. The theory of scalar, vector, and tensor fields permits one to create the basic equations and boundary conditions. Problems of rheology, turbulence, turbulent diffusion, and turbulent mass transfer are examined too.

The chemical processes and adsorption models and especially the stoichiometry, reaction mechanism, reaction route, kinetics of simple and complex chemical reactions, physical and chemical adsorption, and heterogeneous reactions are discussed.

Different types of complex process models are presented depending on the process mechanism. The relation between the mechanism and the mathematical description is shown in the case of physical absorption. Characteristic scales, generalized variables, and dimensionless parameters are used for analysis of the process mechanism. Full information about this mechanism permits the creation of theoretical models. Mass transfer in film flows is an example of such models, where the effects of a chemical reaction and gas motion and absorption of slightly and highly soluble gases are considered.

The very complicated hydrodynamic behavior in column apparatuses is a reason for using diffusion-type models in the cases of mass transfer with a chemical reaction and interphase mass transfer. An average concentration model of an airlift reactor is presented.

Similarity theory models are demonstrated in the case of absorption in packed-bed columns. Generalized (dimensionless) variables and generalized individual cases are used for formulation of the similarity conditions and similarity criteria. The dimension analysis, mathematical structure of the models, and some errors in criteria models are discussed.

Regression models are preferred when there is complete absence of information about the process mechanism and the least-squares method is used for parameter identification.

A theoretical analysis of models of the mass transfer theories is presented in the cases of linear and nonlinear mass transfer. The model theories, boundary layer theory, mass transfer in countercurrent flows, influence of the intensive mass transfer on the hydrodynamics, boundary conditions of the nonlinear mass transfer problem, nonlinear mass transfer in the boundary layer, and the Marangoni effect are examined.

A qualitative theoretical analysis is presented as a generalized analysis. The use of generalized variables permits the analysis of the models of mass transfer with a chemical reaction, nonstationary processes, and stationary processes and the effect of the chemical reaction rate.

The generalized analysis permits the analysis of the mechanism of gas–liquid chemical reactions in the cases of irreversible chemical reactions, homogenous catalytic reactions, and reversible chemical reactions and the relationships between the chemical and physical equilibria during absorption.

A comparative qualitative analysis for process mechanism identification is presented in the cases of different nonlinear effects, nonstationary absorption mechanisms, and nonstationary evaporation kinetics.

A quantitative theoretical analysis is presented for solution of the scale-up problems and statistical analysis of the models. The similarity and scale-up, scale effect and scale effect modeling, scale-up theory and hydrodynamic modeling, and scale effect and scale-up of column apparatuses are discussed. The statistical analysis ranges over basic terms, statistical treatment of experimental data, testing of hypotheses, significance of parameters, and model adequacy of different types of models.

The stability analysis of the models examines the general theory of stability (evolution equations, bifurcation theory), hydrodynamic stability (fundamental equations, power theory, linear theory, stability, bifurcations, and turbulence), the Orr–Sommerfeld equation (parallel flows, almost parallel flows, linear stability, and nonlinear mass transfer), and self-organizing dissipative structures (interphase heat and mass transfer between gas–liquid immovable layers, Oberbeck–Boussinesq equations, gas absorption, and liquid evaporation).

The calculation problems in chemical engineering theory are related to the solutions of differential equations and identification of the model parameters (estimation). Different analytical methods, such as the similarity variables method, Green's functions, Laplace transforms, the Sturm–Liouville problem, the eigenvalue problem, and perturbation methods, are presented. Numerical methods (finite differences method, finite elements method) are examined on the basis of commercial software. Iterative solution methods are considered too.

Parameter estimation methods are discussed in the case of incorrect (ill-posed) inverse problems. An iterative method for parameter identification is presented for solution of correct, incorrect, and essentially incorrect problems. The optimization methods are examined as a basis of the least squares function minimization.

Models of chemical plant systems are presented as a set of process models and the relations between them. An algorithm for simulation of chemical plants is proposed. The methods of optimal synthesis of chemical plants are considered in the case of optimal synthesis of heat recuperation systems. The renovation of chemical plants is formulated as a mathematical model. The main problems are the renovation by optimal synthesis, renovation by introduction of new equipment, and renovation by introduction of new processes.

Examples from the author's investigations are presented at the end of all chapters.

Christo B. Boyadjiev

Preface

The role of *theory* in science was formulated very brilliantly by **Max Planck**:

> ***Experimenters*** *are the striking force of science.* The ***experiment*** *is a question which science puts to nature. The* ***measurement*** *is the registration of nature's answer. But* ***before*** *the question is put to nature,* it ***must be formulated****. Before the measurement result is used,* it***must be explained****, i.e., the answer must be understood correctly. These* ***two problems*** *are obligations of the* ***theoreticians****.*

Chemical engineering is an experimental science, but theory permits us to formulate correct experimental conditions and to understand correctly the experimental results. The theoretical methods of chemical engineering for modeling and simulation of industrial processes are surveyed in this book.

Theoretical chemical engineering solves the problems that spring up from the necessity for a quantitative description of the processes in the chemical industry. They are quite different at the different stages of the quantitative description, i.e., a wide circle of theoretical methods are required for their solutions.

Modeling and simulation are a united approach to obtain a quantitative description of the processes and systems in chemical engineering and chemical technology, which is necessary to clarify the process mechanism or for optimal process design, process control, and plant renovation.

Modeling is the creation of the mathematical model, i.e., construction of the mathematical description (on the basis of the process mechanism), calculation of the model parameters (using experimental data), and statistical analysis of the model adequacy.

Simulation is a quantitative description of the processes by means of algorithms and software for the solution of the model equations and numerical (mathematical) experiments.

The processes in chemical engineering are composed of many simple processes, such as hydrodynamic, diffusion, heat conduction, and chemical processes. The models are created in the approximations of continuous media mechanics.

The complex process model is constructed on the basis of the physical mechanism hypothesis. In cases where full information is available, it is possible to create a theoretical type of model. If the information is insufficient (it is not possible to formulate the hydrodynamic influence on the heat and mass transfer), the model is pattern theory, diffusion type or similarity criteria type. The absence of information leads to the regression model.

The theoretical analysis of the models solves qualitative, quantitative, and stability problems. The qualitative analysis clarifies the process mechanism or similarity conditions. The quantitative analysis solves the problems related to the scale-up and model adequacy. The stability analysis explains the increase of the process efficiency as a result of self-organizing dissipative structures.

All theoretical methods are related to calculation problems. The solutions of the model equations use analytical and numerical methods. The identification (estimation) of the model parameters leads to the solutions of the inverse problems, but very often they are incorrect (ill-posed) and need the application of regularization methods, using a variational or an iterative approach. The solutions of many chemical engineering problems (especially parameter identification) use minimization methods.

The book ideology briefly described above addresses the theoretical foundation of chemical engineering modeling and simulations. It is concerned with building, developing, and applying the mathematical models that can be applied successfully for the solution of chemical engineering problems. Our emphasis is on the description and evaluation of models and simulations. The theory selected reflects our own interests and the needs of models employed in chemical and process engineering. We hope that the problems covered in this book will provide the readers (Ph.D. students, researchers, and teachers) with the tools to permit the solution of various problems in modern chemical engineering, applied science, and other fields through modeling and simulations.

The solutions of the theoretical problems of modeling and simulations employ a number of mathematical methods (exact, asymptotic, numerical, etc.) whose adoption by engineers will permit the optimal process design, process control, and plant renovation.

The modeling and the simulations of chemical systems and plants can be achieved very often through a hierarchical modeling. This approach uses the structural analysis of the process systems. The result of the structural analysis is a quantitative description allowing further optimal process design, process control, and plant renovation. The effectiveness of the optimal solutions can be enhanced if they are combined with suitable methods of optimal synthesis. The latter is a methodical basis and a guide for process system renovations.

The book incorporates a lot of fundamental knowledge, but it is assumed that the readers are familiar with the mathematics at engineering level of usual university courses.

The above comments are the main reasons determining the structure of this book.

Part 1 concerns model construction problems. The mechanics of the continuum approach is used for modeling hydrodynamic, diffusion, and heat conduction processes as basic (elementary) processes in chemical engineering. The modeling of complex processes in chemical engineering is presented on the basis of the relation between the process mechanism and the mathematical description. The models are classified in accordance with the knowledge available concerning the process mechanisms. This means a situation when a theoretical model is available, i.e., sufficient knowledge of the process mechanism as well as the opposite situation of knowledge deficiency, which leads to regression models. Theoretical diffusion, dimensionless, and regression types of models are illustrated. The linear, nonlinear, and pattern mass transfer theories are considered too.

Part 2 focuses on theoretical analysis of chemical engineering process models. The qualitative analysis uses generalized (dimensionless) variables and shows the degree to which the different physical effects participate in a complex process. On this basis, similarity criteria and physical modeling conditions are shown. The quantitative analysis concerns the scale-up problems and statistical analysis of the models. The stability analysis of the models permits the nonlinear mass transfer effects to be obtained and the creation of the self-organizing dissipative structures with very intensive mass transfer.

Part 3 addresses the calculation problems in modeling and simulation. Different analytical and numerical methods for the solution of differential equations are considered. The estimation of the model parameters is related to the solutions of the ill-posed inverse problems. An iterative method for incorrect problem solutions is presented. Different methods for function minimization are shown for the purposes of process optimization and model parameter identification.

Part 4 examines modeling and simulation of the chemical plant systems. The simulation of the systems on the basis of structure system analysis is presented. The optimal synthesis of chemical plants is considered in the case of the optimal synthesis of heat recuperation systems.

This book can be used as a basis for theoretical and experimental investigations in the field of the chemical engineering. The methods and analyses presented permit theoretical problems to be solved, the experimental conditions to be correctly formulated, and the experimental results to be interpreted correctly.

The fundamental suggestion in this book is the necessity for full correspondence (direct and inverse) between the separated physical effect in the process and the mathematical (differential) operator in the model equation.

The main part of this book has a monographic character and the examples are from the author's papers. The book uses the author's lectures "Course of modeling and optimization" (subject chemical cybernetics in the Faculty of Chemistry of Sofia University "St. Kliment Ohridski"), "Course of modeling and simulation of chemical plant systems" (Bourgas University "Prof. Asen Zlatarov"), and "Master's classes of theoretical chemical engineering" (Bourgas University "Prof. Asen Zlatarov"). That is why, as a whole, it is possible for it to be used as teaching material for modeling and simulation. This book proposes an exact formulation and the correct solution of quantitatively described problems in chemical engineering. It may be useful for scientists, Ph.D. students, and undergraduate students.

Some of the results presented in the book were obtained with financial support from the National Fund "Scientific Researches" of the Republic of Bulgaria (contracts no. TH-154/87, TH-162/87, TH-89/91, TH-127/91, TH-508/95, TH-4/99, TH-1001/00, TH-1506/05).

The author would like to thank Assoc. Prof. PhD Jordan Hristov, Assoc. Prof. PhD Natasha Vaklieva-Bancheva, Assoc. Prof. PhD Boyan Ivanov, Assist. Prof. PhD Maria Doichinova, Assist. Prof. Petya Popova, Assist. Prof. Elisaveta Shopova and Dipl. Eng., M.Sc. Boyan Boyadjiev for their help in the preparation of this book.

Christo Boyanov Boyadjiev

Contents

Part I Model Construction Problems

Introduction

1 Quantitative Description

Modeling and simulation are the principal approaches employed for quantitative descriptions of processes and process systems in chemical and process engineering for the solution of scientific and engineering problems.

From a scientific point of view, the quantitative process description addresses the process mechanism. The creation of a hypothesis about the process mechanism is followed by a mathematical model and proof of its adequacy, which in fact is a proof of the hypothesis.

From an engineering aspect, the quantitative description of a process (and of process systems too) forms the basis of the engineering optimization of new chemical plants and control and renovation of older ones as well.

The recovery of quantitative information concerning processes and process systems through modeling and simulations has some advantages with respect to physical experiments. The most important of these advantages are:

- Reduction of required material resources.
- Simulations of extreme (or dangerous) conditions give an opportunity to avoid or eliminate the risks.
- Short time simulations of long technological cycles.

In all cases the quantitative description is oriented towards the kinetics of the processes and the systems. The rates of nonequilibrium processes (in accordance with the Onsager approach) depend on the deviations from their equilibrium states. Thus, the quantitative description needs knowledge concerning the process statics (thermodynamics).

Besides the variety of the problems mentioned, both the modeling and the simulation follow almost unified approaches including several stages (see Table 1). In several particular cases, some stages can be reduced or eliminated.

2 Modeling and Simulation

The Association for the Advancement of Modeling and Simulation Techniques in Enterprises (AMSE) defines the purpose of **modeling** as *a schematic description* of the processes and the systems, whereas **the simulations** are*employments of the models for process investigations or process optimizations* **without experiments with real systems**.

Table 1 Modeling and simulation stages

Quantitative description
Modeling
Determination of mathematical description structure
Identification of parameter estimates in the mathematical description using physical experiments
Verification of the statistical significance of the parameters and model adequacy
Simulation
Creation of an algorithm for solution of the model equations
Computer realization of the solution algorithm
Quantitative description obtained by means of computing (mathematical) experiments

Obviously, it is easy to realize that modeling addresses the first three stages (see Table 1), whereas the simulations utilize the last three stages, where the final step means performance of numerical experiments.

On the other hand, the concept of *the mathematical modeling* as a **unified method** employs the assumption that the model building and the simulations are steps of it. This approach considers the *mathematical description* as *a model* after the creation of computer-oriented algorithms and codes. From this point of view, the difference between the physical and the mathematical models disappears. The physical modeling replaces the mathematical analogy, whereas the physical experiment corresponds to the numerical experiment.

The present book addresses **process modeling** *as a technology concerning selection* (or creation) of *mathematical structures* (the model equations), *parameter identification* (on the basis of data obtained through physical experiments), and *a check of the model adequacy*.

According to the definition of the operator of the direct and inverse problem solutions, if A and A^{-1} are the direct and inverse operators, the simulation is a direct problem solution, namely,

$$y = Ax, \tag{2.1}$$

This implies obtaining the target (object) function y of the real process (the big apparatus on the book cover) if the model parameter values x are known. Obviously the direct problem solution has an experimental equivalent (the target function can be obtained experimentally).

The modeling is an inverse problem solution, i.e.,

$$x = A^{-1} y_e, \tag{2.2}$$

and addresses obtaining the model parameter value x of the modeled process (the small apparatus on the book cover). The latter implies that the experimental values of the target function y_e are known; in this case finding the inverse problem solution is possible by means of calculation only.

3 Chemical Engineering and Chemical Technology

The employment of modeling and simulation during the quantitative description of processes and systems needs clear determination of the **chemical engineering** and the **chemical technology** problems.

Chemical engineering is a branch of the chemical sciences concerning implementation of physical chemistry processes at an industrial scale in chemical plants. Chemical engineering employs modeling or simulation of processes as the principal tool based on an amalgam of fundamentals of chemistry, physics, and mathematics. The main target of such investigations is the quantitative description of the process mechanisms and kinetics under industrial conditions. These descriptions are the first steps of the optimal process design, process control, and plant renovation.

The industrial implementation of physical chemistry processes considers purely chemical and physical phenomena whose performance is affected mainly by the scale of the contacting devices used.

Chemical technology is a science oriented towards the creation of technological schemes including the consequences of processes. The flowsheet synthesis provides all the relationships between the processes at the flowsheet sublevels that need a systematic approach to be employed. In this context, *the objects of the investigations* of chemical technology are *technological systems*. The analyses of such systems with the methods of the *system techniques* are in the domain of **process system engineering**. Such an approach is not trivial and it is based on some common suggestions, among them.

Separate chemical and physical (hydrodynamic, diffusion, thermal, adsorption, etc.) processes of chemical technologies are the subjects of **chemistry** and **physics**. The simultaneous occurrence of these processes in industrial devices is the subject of **chemical engineering**. In this context, the complex interrelations between the processes of technological systems are a subject of **chemical technology**. It should be noted that the system synthesis and optimization are also branches of **process system engineering**. These standpoints allow chemical technology to be described as a **chemical engineering system technique** employing both modeling and simulations as working methods based on the developments of **chemical engineering** and **applied mathematics**. The main goal of studies employing these methods is the quantitative descriptions of the systems and the subsequent optimal process design, process control, and plant renovation.

It should be noted that optimal reconstruction (retrofit) could be defined as a **renovation** from a more generalized point of view. Thus, the development of the system towards a better and more efficient (economically) state with respect to the existing situation is the main purpose of the renovation.

4 Theoretical Problems and Methods

Chemical engineering processes are combinations of basic processes (hydrodynamic, diffusion, thermal, chemical, etc.) that occur simultaneously (or in sequence) under conditions imposed by the complex geometry of the industrial contactors. The ordered sequence of separate stages (elementary processes) is the mechanism of the chemical engineering process. From this point of view, the structure of the mathematical description of the process depends on the mathematical description of the elementary processes involved and the interrelations between them. Therefore, the creation of the mathematical description of the elementary processes involved is the first step towards the modeling of the entire chemical engineering process.

The analysis of the separate stages of the modeled process (see Table 1) shows that the main step is the creation of the mathematical structures of the model utilizing submodels of elementary process and the interrelation mechanisms. Obviously, these structures (submodels) depend on the knowledge available about the process mechanisms.

A very important stage of model development is parameter identification (on the basis of experimental data) through inverse problem solutions. The main difficulties are related to the incorrectness of the inverse problem.

The *mathematical structures* developed (structures with identified parameters) *become models* after the evaluation of both the *parameter significance* and the *model adequacy*.

The model created (before the simulations) can be used for a qualitative analysis of the relation between the process mechanism (and kinetics) and the values of the model parameters. This allows some levels of a hierarchical modeling to be defined as well as the scale-up effects of the processes. All these steps facilitate the subsequent simulation and *very often they are required preliminary steps*.

The modeling and simulation of both the processes and the systems are related to a lot of calculation problems. Moreover, there are various approaches of the optimal process design, process control, and plant renovation.

Many of the calculation problems are related to the solution of differential equations. Different analytical (*similarity variables method*, *Green's functions*, *Laplace transforms*, *Sturm–Liouville problem*, *eigenvalue problem*, *perturbation method*) and numerical (*finite differences method, finite elements method, iterative solution methods*) methods have to be used.

The calculation problems of the identification of the model parameters need the use of sets, metric spaces, functionals, variational and iterative methods, as well as different methods for function minimization.

The modeling and simulation of chemical plant systems deal with quite specific calculation problems that should be solved with the help of graphs, matrixes, nonlinear, integer, and heuristic programming, etc.

5 Physical Fundamentals of Theoretical Chemical Engineering

Theoretical chemical engineering uses physical approximations of the mechanics of continua. Simply, this implies that the material point of the medium corresponds to a volume sufficiently small with respect to the entire volume under consideration but at the same time sufficiently large with respect to the intermolecular distances of the medium. Modeling in chemical engineering utilizes mathematical structures (descriptions) provided by the mechanics of the continua

The principal problem in theoretical chemical engineering is the necessity for full adequacy of the mathematical operators in the mathematical models of the physical effects described. This is very important in the theoretical and diffusion types of models and the mass transfer theory is a very important example of such problems.

The theoretical analyses employing similarity theory models demonstrate that if the mathematical methods are very simple, the formalistic use of the theory could result in wrong results. This note addresses, for example, incorrect formulation of similarity conditions, independent and dependent dimensionless parameters, etc. Physical ideas form the basis of the similarity theory and its mathematical methods. In this context, the correct physical approach might help in the use of the similarity theory as a powerful method of quantitative investigations because the correct understanding of the process physics is of primary importance

Part I
Model Construction Problems

Chemical reaction rates are functions of the concentrations of the reagents, whereas the kinetic constants are functions of the temperature. In industrial conditions these processes are realized in moving fluids, where the concentrations of the reagents and the temperature are the results of diffusion, heat and mass transfer, and convective transfer. As a result, chemical engineering processes are complex systems of simple processes that interact with each other in a manner defined by the mechanism of the industrial process. Thus, the model of a particular chemical engineering process can be represented as a suitable combination of the models of the simple processes.

Simple Process Models

Simple processes in chemical engineering concern hydrodynamic, diffusion, heat conduction, adsorption, and chemical processes. These are typical nonequilibrium processes and the relevant mathematical descriptions concern quantitatively their kinetics. This gives a ground to utilize the laws of irreversible thermodynamics as mathematical structures building the models of the simple processes [1].

The quantitative description of irreversible processes depends on the level of the process description. From such a point of view, one can define three basic levels of description—*thermodynamic*, *hydrodynamic*, and *Boltzmann levels*. These different levels of process description form a natural hierarchy. Thus, going up from one level to the next, the description becomes richer, i.e., more detailed. This approach allows the kinetic parameters defined at a lower level to be described through relevant kinetic parameters at an upper level.

The thermodynamic level utilizes quantitative descriptions through *extensive variables* (internal energy, volume, and mass). If there is a distributed space, the volume must be represented as a set of unit cells, where the variables are the same but have different values in different cells.

The hydrodynamics is the next level, where a new extensive variable participates in the processes. This variable is the momentum. Therefore, the hydrodynamic level of description can be considered as a generalization of the lower, thermodynamic level. Here, the extensive variables (taking into account their distribution in the space) are mass density, momentum, and energy. In the isolated systems they are conserved and the conservation laws of mass, momentum, and energy are used.

The Boltzmann level is the next upper level of description and concerns only the mass density as a function of the distribution of the molecules in space and their momenta.

The kinetics of irreversible processes employs mathematical structures following from *Onsanger's linear principle* [1]. According to them, the mean values of the time derivatives of the extensive variables and the mean deviations of their adjoined intensive variables from the equilibrium are expressed through linear

C. Boyadjiev, *Theoretical Chemical Engineering*,
DOI: 10.1007/978-3-642-10778-8_1, © Springer-Verlag Berlin Heidelberg 2010

relationships. The principle is valid close to the equilibrium and the coefficients of the proportionality are the kinetic constants. When the process takes place far from equilibrium (highly intensive processes), the kinetic constants become kinetic complexes depending on the corresponding *intensive variables* (in the case of fusion of two identical systems, the extensive variables double, whereas the intensive variables remain the same).

The hydrodynamic level is widely applicable in the mechanics of continua. Here, the *material point* corresponds to a sufficient volume of the medium that is simultaneously *sufficiently small* with respect to the entire volume under consideration and at the same time *sufficiently large* with respect to the intermolecular distances of the medium.

Modeling in chemical engineering utilizes *mathematical structures* (descriptions) provided by the mechanics of the continua. The principal reason for this is the fact that these structures sufficiently well describe the phenomena in detail. Moreover, they employ physically well defined models with a low number of experimentally defined parameters.

Modeling the fundamental processes by the concepts of the continuum mechanics follows the first three stages defined in Table 1. Levels 2 and 3 (see Page XXVI) will be especially discussed further, so the present chapter focuses on the mathematical description of the simple processes. The discussion developed employs a mathematical description following from Onsanger's principle and the field theory widely applicable in the mechanics of continua.

1 Mechanics of Continuous Media

Simple chemical engineering processes in liquids and gases are of macroscopic type and must be considered in the continuum approximation [2], i.e., the size of such an elementary volume of liquid (gas) is sufficiently large with respect to the intermolecular distances (the elementary volume consists of many molecules).

The simple processes of continua are related to variations of basic physical quantities, such as *density* (ρ), *velocity* ($\boldsymbol{u}$), *pressure* (p), *concentration* (c), and *temperature* (t):

$$\rho = \rho(x, y, z, \tau), \boldsymbol{u} = \boldsymbol{u}(x, y, z, \tau), p = p(x, y, z, \tau), c = c(x, y, z, \tau), t = t(x, y, z, \tau), \tag{1.1}$$

where x, y, and z are Cartesian coordinates and τ is the time. The quantities ρ, p, c, and t are *scalars,* whereas velocity $\boldsymbol{u}$ is a *vector*. The scalars are defined by numbers, and the vector is defined by a number and a direction.

1.1 Scalar and Vector Fields

The mechanics of continua assumes that functions (1.1) are *defined* and *continuous* over a certain area of the space considered as a *field*. Therefore, depending on the type of the physical quantity, we have *scalar fields* or a *vector field* [3, 4, 11]. Each point of the scalar field is characterized by one number, i.e., the value of the scalar magnitude. The vector field is characterized by three arranged numbers, i.e., the magnitudes of the vector projections on the Cartesian coordinates.

The *scalar field* is represented by the space-distributed scalar function φ:

$$\varphi = \varphi(x, y, z). \tag{1.2}$$

All the points where $\varphi = \varphi_{0.}$ = const. form a *surface*:

$$\varphi(x, y, z) = \varphi_0. \tag{1.3}$$

This surface is characterized by the *property* that sliding on it, the value of φ remains unchanged (constant φ_0). In all other directions φ is inconstant and its change is maximal (minimal) in the direction of the surface normal. That is why at each point M (x, y, z) of the scalar field, the biggest change of φ is a vector, directed to the interface normal, i.e., its projections are

$$\frac{\partial \varphi}{\partial x}, \frac{\partial \varphi}{\partial y}, \frac{\partial \varphi}{\partial z}, \tag{1.4}$$

where as its magnitude is equal to the rate of the growth of the function at point M. This vector is called a scalar field *gradient*:

$$\textbf{\textit{grad}}\, \varphi = \frac{\partial \varphi}{}\boldsymbol{i} + \frac{\partial \varphi}{\partial y}\boldsymbol{j} + \frac{\partial \varphi}{\partial z}\boldsymbol{k}, \tag{1.5}$$

where $\boldsymbol{i}, \boldsymbol{j}$, and $\boldsymbol{k}$ are the unit vectors of the coordinate axes x, y, and z, respectively.

The vector of the gradient (or antigradient) determines the *direction* of the faster (slower) function growth, at a particular point, with respect to the space coordinates. The variations of the function along any other direction can be determined through projection of the particular direction vector considered on the gradient.

The gradient of a scalar field can also be represented through the differential operator ∇ ("*nabla*"):

$$\nabla = \frac{\partial}{\partial x}\boldsymbol{i} + \frac{\partial}{\partial y}\boldsymbol{j} + \frac{\partial}{\partial z}\boldsymbol{k}, \tag{1.6}$$

or

$$\nabla\varphi = \mathbf{grad}\ \varphi. \tag{1.7}$$

The gradient forms a *vector field*.

Processes in media with variable properties (density, concentration, or temperature) are very common in chemical engineering. Thus, the function φ in (1.7) can be replaced by a particular scalar quantity (see 1.1).

The *vector field* represents the space distribution of a *vector function*:

$$\boldsymbol{\varphi} = \varphi_{\mathrm{x}}\boldsymbol{i} + \varphi_{\mathrm{y}}\boldsymbol{j} + \varphi_{\mathrm{z}}\boldsymbol{k}, \tag{1.8}$$

where φ_{x}, φ_{y} and φ_{z} are *scalars* and they are the *projections* of the vector $\boldsymbol{\varphi}$ on the coordinate axes (a Cartesian coordinate system in this particular case). For example, the *velocity vector* (1.1) may be expressed through its *components* (projections at the coordinate axes) as

$$\boldsymbol{u} = u_x\boldsymbol{i} + u_y\boldsymbol{j} + u_z\boldsymbol{k}. \tag{1.9}$$

The vector fields have some basic properties expressed through elementary operations on them such as a *scalar product* and a *divergence*.

The *scalar product* of two vectors $\boldsymbol{a}$ and $\boldsymbol{b}$ is a *scalar*:

$$\boldsymbol{ab} = a_x b_x + a_y b_y + a_z b_z = ab\cos\alpha, \tag{1.10}$$

where a_i, b_i ($i = x, y, z$) are the *vector components,* i.e., the scalar product (1.10) is a product of the vector *magnitudesa* and b and the cosine function of the *angle*α between them.

The *vectorflux* through a plane surface is the product of the *cross-sectional area* and the *vector projection* on the normal vector of the surface at the point where the flux crosses the surface. If we have an arbitrary surface S and the flux J of the vector $\boldsymbol{\varphi}$ crosses it, the *flux density* at an elementary area ds is:

$$j = \varphi\,\boldsymbol{n}\ \mathrm{d}s = (\varphi\cos\alpha)\mathrm{d}s. \tag{1.11}$$

The integration of j over the all surface S leads to

$$J = \iint_S \varphi\boldsymbol{n}\ \mathrm{d}s = \iint_S \varphi\cos\alpha\ \mathrm{d}s. \tag{1.12}$$

The product $\boldsymbol{\varphi}.\boldsymbol{n}$ is a scalar product of the *normal vector*$\boldsymbol{n}$ of the surface ds and *vector* $\boldsymbol{\phi}$. It represents the vector magnitude along the direction defined by the normal vector $\boldsymbol{n}$. The same product expressed through the *vector magnitude*φ and the *angle* between the vector and the normal to the surface is the second integral of (1.12). Here it is assumed that magnitude of $\boldsymbol{n}$ is 1, i.e., $\boldsymbol{n}$ in the first integral of (1.12) is the *unit vector* of the *surface normal.*

If the surface S encloses a volume v, the flux of the vector $\boldsymbol{\phi}$ through the surface S when $v \to 0$, i.e., the limit $\lim\limits_{v \to 0} \boldsymbol{\varphi}$, is the vector *divergence:*

$$div\ \boldsymbol{\varphi} = \lim_{v \to 0}\left(\frac{J}{v}\right) = \frac{\partial \varphi_x}{\partial x} + \frac{\partial \varphi_y}{\partial y} + \frac{\partial \varphi_y}{\partial z}. \tag{1.13}$$

Therefore, the divergence of a *vector field* is a *scalar field.* Formally, the divergence can be represented by a *scalar product* of the vector $\boldsymbol{\varphi}$ and the *symbolic vector* ∇ (1.6):

$$div\ \boldsymbol{\varphi} = \nabla \cdot \boldsymbol{\varphi} = \left(\frac{\partial}{\partial x}\boldsymbol{i} + \frac{\partial}{\partial y}\boldsymbol{j} + \frac{\partial}{\partial z}\boldsymbol{k}\right)\left(\varphi_x \boldsymbol{i} + \varphi_y \boldsymbol{j} + \varphi_z \boldsymbol{k}\right). \tag{1.14}$$

The vector $\boldsymbol{\varphi}$ could be the fluid velocity $\boldsymbol{u}$. Thus, the flux of $\boldsymbol{u}$ through a surface S is the quantity of fluid passing through this surface per unit time. If the surface S encloses a volume v, the flow rate (input or output flow *per unit volume)* is the *divergence* of the vector $\boldsymbol{u}$.

1.2 Stress Tensor and Tensor Field

Developing the description of continua through *scalar* and *vector* fields and continuing, we come to *tensor* fields. The type of the tensors is directly related to the physical quantities described by them. The simplest and the friendliest is the stress tensor [3, 4], so we will describe it as an example since it is a basic tool in continuum mechanics.

Following the Newton law, the product of the mass per unit volume (*density*) (ρ) and its *acceleration* ($\boldsymbol{a}$) equals the sum of both the surface ($\boldsymbol{P}$) and the volume ($\boldsymbol{K}$) forces acting on it:

$$\rho \boldsymbol{a} = \boldsymbol{K} + \boldsymbol{P}. \tag{1.15}$$

Here $\boldsymbol{a}$ is the *substantial* acceleration, i.e., the velocity change with respect to the *space coordinates* and the *time*.

The volumetric (mass) forces are an effect of the influence of *external* forces on the volume (mass) of the medium considered. The latter means that acting on a mass of the particular volume, they are produced by external physical fields such as the gravitational force (the body force of gravity), electrical forces (Coulomb forces between charged bodies), centrifugal forces, and buoyant forces. The volume forces are *proportional* to the mass on which they act and are described by the coordinates of the volume of that mass. Obviously, they form vector fields and usually the forces are in proportion to the gradient of the particular scalar field under consideration.

In the dynamics of continua, the volume forces are usually expressed through the density of their distribution (***k***) or, in other words, through the force per unit mass:

$$\delta \boldsymbol{K} = \rho \boldsymbol{k}(\delta v), \tag{1.16}$$

where (δv) is a sufficiently small volume. If (Δv) contains the material point M (x, y, z), the unit vector ***k*** is defined as

$$\boldsymbol{k} = \lim_{\Delta m \to 0} \frac{\Delta \boldsymbol{K}}{\Delta m} = \frac{1}{\rho} \lim_{\Delta v \to 0} \frac{\Delta \boldsymbol{K}}{\Delta v}, \tag{1.17}$$

where Δm is the mass of the volume Δv.

The density of the force distribution (see 1.17) is in fact the acceleration, produced by the action on the mass of the medium. For example, when ***K*** is the gravitational force, the density ***k*** is the gravity acceleration ***g***.

The surface forces differ principally from the body forces. They also act on a particular volume of the medium, but *are not proportional* to its mass. They are *internally* determined forces, i.e., they represent *the reaction of the medium* against the application of the external (to the volume) forces. Thus, they represent the *stressed state* of the medium.

Let us consider an elementary volume under the action of surface forces and let one of the volume dimensions vanish to zero (i.e., apply the limit approach). As a result, the volume becomes a surface with surface forces applied on it. The latter depend on the surface orientation (the orientation of the surface normal vector). Examples of such forces are *pressure* and the *forces of the internal friction*.

The surface forces **P** are described by the *stress vector****p***:

$$\delta \boldsymbol{P} = \boldsymbol{p}\, ds, \tag{1.18}$$

where ds is an *infinitely small* surface area. If $\Delta \boldsymbol{s}$ contains point M (x, y, z), the definition of ***p*** is

$$\boldsymbol{p} = \lim_{\Delta s \to 0} \frac{\Delta P}{\Delta s}, \tag{1.19}$$

where $\Delta \boldsymbol{s}$ is a small surface area and contains point M. In the above equations, ***p*** is the force acting on a *unit surface* of a plane containing point M (x, y, z) and characterized by its *normal vector****n***, i.e., **p** depends on the normal vector components.

The general expression concerning the surface forces per unit volume of a deformable medium can be derived if a small unit volume of it is considered. If this volume is a cube, the three normal vectors of the walls characterize the *stressed state* of the medium, i.e., nine scalar quantities. Let this volume be managed in accordance with the Cartesian coordinate system (Fig. 1). The elementary volume is $\delta v = \delta x \delta y \delta z$.

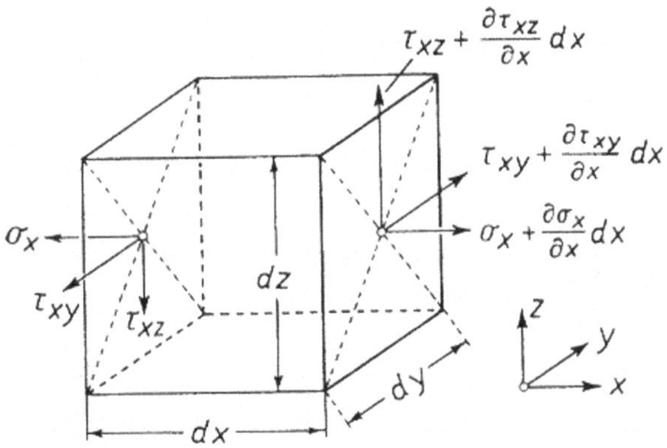

Fig. 1 Stresses on a volume element Δv

The stresses acting on both surfaces perpendicular to the x-axis are

$$-\boldsymbol{p}_{\mathrm{x}} \text{ and } \boldsymbol{p}_{\mathrm{x}} + \frac{\partial \boldsymbol{p}_{\mathrm{x}}}{\partial x}\delta x. \tag{1.20}$$

The corresponding surface forces can be determined through multiplication of the stresses and the surface of the elementary volume (where they are applied) $\delta s_{\mathrm{x}} = \delta y \delta z$ as follows:

$$-\boldsymbol{p}_{\mathrm{x}}\delta s_{\mathrm{x}} \text{ and } -\boldsymbol{p}_{x}\delta s_{x} + \frac{\partial \boldsymbol{p}_{x}}{\partial x}\delta v. \tag{1.21}$$

For the other forces acting on one surface perpendicular to the axes y and z we have in a similar way that

$$-\boldsymbol{p}_{\mathrm{y}}\delta s_{\mathrm{y}} \text{ and } -\boldsymbol{p}_{y}\delta s_{y} + \frac{\partial \boldsymbol{p}_{y}}{\partial y}\delta v, \quad -\boldsymbol{p}_{\mathrm{z}}\delta s_{\mathrm{z}} \text{ and } -\boldsymbol{p}_{\mathrm{z}}\delta s_{z} + \frac{\partial \boldsymbol{p}_{z}}{\partial z}\delta v. \tag{1.22}$$

The integral of the surface force **P** per unit volume follows from the summation of (1.21) and (1.22) along the coordinate axes and is expressed as the sum per unit volume (i.e., dividing by δv after the summation):

$$\boldsymbol{P} = \frac{\partial \boldsymbol{p}_{x}}{\partial x} + \frac{\partial \boldsymbol{p}_{y}}{\partial y} + \frac{\partial \boldsymbol{p}_{z}}{\partial z}. \tag{1.23}$$

The vectors $\boldsymbol{p}_{\mathrm{x}}$, $\boldsymbol{p}_{\mathrm{y}}$, and $\boldsymbol{p}_{\mathrm{z}}$ can be expressed through their components along the coordinate axes.

One of these components is normal to the wall. In the present case (see Fig. 1) it is parallel to the axis to which the wall is perpendicular. The other two components are in the plane of the wall and are parallel to the other coordinate axes. The former is termed a normal stress (σ), whereas the latter two are shear stresses (τ). As a result,

$$\boldsymbol{p}_x = \sigma_x \boldsymbol{i} + \tau_{xy} \boldsymbol{j} + \tau_{xz} \boldsymbol{k},$$

$$\boldsymbol{p}_y = \tau_{yx} \boldsymbol{i} + \sigma_y \boldsymbol{j} + \tau_{yz} \boldsymbol{k},$$

$$\boldsymbol{p}_z = \tau_{zx} \boldsymbol{i} + \tau_{zy} \boldsymbol{j} + \sigma_z \boldsymbol{k}. \tag{1.24}$$

The form of (1.24) shows that the stressed state of the medium is determined by nine scalar quantities forming the so-called *stress tensor*:

$$\Pi = \begin{pmatrix} \sigma_x & \tau_{xy} & \tau_{xz} \\ \tau_{yx} & \sigma_y & \tau_{yz} \\ \tau_{zx} & \tau_{zy} & \sigma_z \end{pmatrix}. \tag{1.25}$$

In the general case, the stress tensor is nine numbers arranged as a matrix, which characterize the stress vector at every point of the field.

The stress tensor depends on the coordinates of point M only and forms a *tensor field*. The form of (1.25) indicates that Π is a tensor of second rank (the vectors are tensors of rank 1, whereas the scalars are tensors of rank 0).

The vector of the stress **p** can be determined if we consider a small polygon ABC (see Fig. 1) with a normal vector $\boldsymbol{n}$:

$$\boldsymbol{n} = n_x \boldsymbol{i} + n_y \boldsymbol{j} + n_z \boldsymbol{k}. \tag{1.26}$$

The vector $\boldsymbol{p}$ represents a stress produced by the surface force $\boldsymbol{P}$ acting on the polygon ABC with a surface δs. The balance of the surface forces acting on the volume MABC is

$$\boldsymbol{p}\delta s = \boldsymbol{p}_x \delta s_x + \boldsymbol{p}_y \delta s_y + \boldsymbol{p}_z \delta s_z. \tag{1.27}$$

The surface of the volume MABC can be determined as the sum of their elements. A more convenient way is to express them as projections of the wall ABC on the coordinate planes formed by the coordinate axes. This requires knowing the angles between the normal vector of the plane ABC and the coordinate axes. Thus, we have

$$\delta s_x = \delta s \cos(n, x) = n_x \delta s,$$

$$\delta s_y = \delta s \cos(n, y) = n_y \delta s,$$

$$\delta s_z = \delta s \cos(n, z) = n_z \delta s. \tag{1.28}$$

In this way, with the help of (1.23) and (1.24) we can obtained the stress vector as

$$\boldsymbol{p} = n_x \boldsymbol{p}_x + n_y \boldsymbol{p}_y + n_z \boldsymbol{p}_z. \tag{1.29}$$

The projections along the coordinate axes through (1.24) and (1.26) are

$$p_x = n_x\sigma_x + n_y\tau_{xy} + n_z\tau_{xz},$$
$$p_y = n_x\tau_{xy} + n_y\sigma_y + n_z\tau_{zy},$$
$$p_z = n_x\tau_{xz} + n_y\tau_{yz} + n_z\sigma_z. \quad (1.30)$$

The expressions developed above, i.e., (1.25), (1.26), and (1.30), permit us to represent the stress vector (1.29) with components (1.30) as an internal product of the normal vector (1.26) and stress tensor (1.25), i.e.

$$\boldsymbol{p} = \boldsymbol{n}\Pi. \quad (1.31)$$

This result allows us to represent the stress vector only through one vector depending on the surface (plane) orientation and one tensor of second rank determined by the coordinates of the plane.

The form of (1.30) shows that the components of the stress vector depend simultaneously on the coordinates of point M and the orientation of the plane (defined by ABC) where this point lies. Therefore, the stress vector is defined by nine numbers and does not form a vector field.

2 Hydrodynamic Processes

The hydrodynamic processes concerning the flow problems of fluids (liquids or gases) [2, 4] are characterized by five variables—three projections of the velocity vector u_x, u_y, u_z (Cartesian coordinate system), fluid density ρ, and the pressure p. Their determinations employ a set of:

- The continuity equation (mass conservation equation)
- Three equations of motion (*momentum conservation* equations)
- One equation describing the thermodynamic state (a relationship between the density and the pressure)

In the case of nonisothermal processes, all the fluid properties depend on the temperature, and an additional equation for the *energy conservation* is needed.

The intensive mass transfer in moving fluids as a result of large concentration gradients induces secondary flows [5] and we must add the convection–diffusion equation.

2.1 Basic Equations

The basic equations of hydrodynamic processes will be obtained for a small (control) volume Δv containing point M (x, y, z). This volume belongs to the bulk of the medium and does not interact with the macroscopic boundaries of the flow.

Under the assumption of a constant density (the control volume is very small) the mass of the control volume is

$$\delta m = \rho \delta v. \tag{2.1}$$

The velocity of the liquid particle described by the velocity vector (1.9) has components representing the variations of the coordinates of point M with time:

$$u_{\mathrm{x}} = \frac{\mathrm{d}x}{\mathrm{d}t}, \quad u_{\mathrm{y}} = \frac{\mathrm{d}y}{\mathrm{d}t}, \quad u_{\mathrm{z}} = \frac{\mathrm{d}z}{\mathrm{d}t}. \tag{2.2}$$

The forces acting on the volume change the control volume $\delta v = \delta x \delta y \delta z$ and the rate of its variation with time is

$$\frac{\mathrm{d}}{\mathrm{d}t}(\delta v) = \left[\frac{1}{\delta x}\frac{\mathrm{d}}{\mathrm{d}t}(\delta x) + \frac{1}{\delta y}\frac{\mathrm{d}}{\mathrm{d}t}(\delta y) + \frac{1}{\delta z}\frac{\mathrm{d}}{\mathrm{d}t}(\delta z)\right]\delta v. \tag{2.3}$$

If the control volume δv is that described in Fig. 1, its walls are parallel to the coordinate plane. Concerning the lines of intersection of each neighboring wall, we could decide that they are parallel to the coordinate axes. Such a line of intersection forms the edge of the control volume (specific for the Cartesian system chosen here). An elementary increment of the length of such an edge with time is the elementary increment of the velocity in the direction of the corresponding coordinate axis. Therefore, *the deformation with time* is the basic definition of a *flow*. Thus, the result is

$$\frac{\mathrm{d}}{\mathrm{d}t}(\delta x) = \frac{\mathrm{d}}{\mathrm{d}t}(x - \bar{x}) = u_{\mathrm{x}} - \bar{u}_{\mathrm{x}} = \delta u_{\mathrm{x}} = \frac{\partial u_{\mathrm{x}}}{\partial x}\delta x. \tag{2.4}$$

The derivations of the velocities along the other two coordinate axes proceed in a similar way and the use of (1.3) leads to

$$\frac{\mathrm{d}}{\mathrm{d}t}(\delta v) = \delta v \left(\frac{\partial u_{\mathrm{x}}}{\partial x} + \frac{\partial u_{\mathrm{y}}}{\partial y} + \frac{z}{\partial z}\right) = \delta v \, div \, \boldsymbol{u}. \tag{2.5}$$

The mass conservation equation follows from (2.1) in the form

$$\frac{\mathrm{d}}{\mathrm{d}t}(\delta m) = \frac{\mathrm{d}}{\mathrm{d}t}(\rho \delta v) = 0. \tag{2.6}$$

The differentiation of (2.6) and the subsequent use of (2.5) gives

$$\frac{\mathrm{d}\rho}{\mathrm{d}t}\delta v + \rho\frac{\mathrm{d}}{\mathrm{d}t}\delta v = \left(\frac{\mathrm{d}\rho}{\mathrm{d}t} + \rho div \, \boldsymbol{u}\right)\delta v = 0, \tag{2.7}$$

which can be expressed in the most common form of the *continuity equation*:

$$\frac{\mathrm{d}\rho}{\mathrm{d}t} + \rho div \, \boldsymbol{u} = 0. \tag{2.8}$$

Following the rules for differentiation, we have (with the help of 1.5 and 1.9):

$$\frac{\mathrm{d}\rho}{\mathrm{d}t} = \frac{\partial \rho}{\partial t} + \frac{\partial \rho}{\partial x}\frac{dx}{dt} + \frac{\partial \rho}{\partial y}\frac{dy}{dt} + \frac{\partial \rho}{\partial z}\frac{dz}{dt} = \frac{\partial \rho}{\partial t} + u_{\mathrm{x}}\frac{\partial \rho}{\partial x} + u_{\mathrm{y}}\frac{\partial \rho}{\partial y} + u_{\mathrm{z}}\frac{\partial \rho}{\partial z} = \frac{\partial \rho}{\partial t} + \boldsymbol{u}\nabla\rho. \tag{2.9}$$

The application of the definition of the divergence,

$$div(\rho \boldsymbol{u}) = \boldsymbol{u} div\rho + \ \boldsymbol{u}, \tag{2.10}$$

allows (2.9) to be expressed as

$$\frac{\partial \rho}{\partial t} + div(\rho \boldsymbol{u}) = \frac{\partial \rho}{\partial t} + \frac{\partial}{\partial x}(\rho u_{\mathrm{x}}) + \frac{\partial}{\partial y}(\rho u_{\mathrm{y}}) + \frac{\partial}{\partial z}(\rho u_{\mathrm{z}}) = 0. \tag{2.11}$$

For the steady-state processes, the density is *time-independent*, i.e.,

$$div(\rho \boldsymbol{u}) = 0. \tag{2.12}$$

When the density does not vary, $\rho =$ const., i.e.,

$$div\, u = 0. \tag{2.13}$$

As mentioned earlier (see 1.15), the integral action of the surface and the body forces accelerates the control volume δv with acceleration **a**:

$$\begin{aligned}\mathbf{a} &= \frac{\mathrm{d}\mathbf{u}}{\mathrm{d}t} = \frac{\partial \mathbf{u}}{\partial t} + \frac{\partial \mathbf{u}}{\partial x}\frac{\mathrm{d}x}{\mathrm{d}t} + \frac{\partial \mathbf{u}}{}\frac{\mathrm{d}y}{\mathrm{d}t} + \frac{\partial \mathbf{u}}{\partial z}\frac{\mathrm{d}z}{\mathrm{d}t} = \frac{\partial \mathbf{u}}{\partial t} + u_{\mathrm{x}}\frac{\partial \mathbf{u}}{\partial x} + u_{\mathrm{y}}\frac{\partial \mathbf{u}}{\partial y} + u_{\mathrm{z}}\frac{\partial \mathbf{u}}{\partial z} \\ &= \frac{\partial \mathbf{u}}{\partial t} + (\mathbf{u}\cdot\nabla)\mathbf{u},\end{aligned} \tag{2.14}$$

where $(\boldsymbol{u}\cdot\nabla)$ is the scalar product of the vectors $\boldsymbol{u}$ and ∇ (see 1.6 and 1.9).

The definitions of the acceleration expressed by (1.15) and (1.17) show that the principal equation of fluid dynamics for a control volume δv is

$$\delta ma = (\rho\delta v)a = \rho g\delta v + P\delta v \tag{2.15}$$

or

$$\rho\,\mathbf{a} = \rho\mathbf{g} + \mathbf{P}, \tag{2.16}$$

where **g** is the acceleration due to the body forces and **P** is the integral of the surface force (see 1.23).

Therefore, we have (from 2.14, 2.16, and 1.23)

$$\begin{aligned}&\rho\left(\frac{\partial \mathbf{u}}{\partial t}\right) + u_{\mathrm{x}}\frac{\partial \mathbf{u}}{\partial x} + u_{\mathrm{y}}\frac{\partial \mathbf{u}}{\partial y} + u_{\mathrm{z}}\frac{\partial \mathbf{u}}{\partial z} \\ &= \rho\mathbf{g} + \left(\frac{\partial \sigma_{\mathrm{x}}}{\partial x} + \frac{\partial \tau_{\mathrm{yx}}}{\partial y} + \frac{\partial \tau_{\mathrm{zx}}}{\partial z}\right)\boldsymbol{i} + \left(\frac{\partial \tau_{\mathrm{xy}}}{\partial x} + \frac{\partial \sigma_{\mathrm{y}}}{\partial y} + + \frac{\partial \tau_{\mathrm{zy}}}{\partial z}\right)\boldsymbol{j} + \left(\frac{\partial \tau_{\mathrm{zx}}}{\partial x} + \frac{\partial \tau_{\mathrm{zy}}}{\partial y} + \frac{\partial \sigma_{\mathrm{z}}}{\partial z}\right)\boldsymbol{k}.\end{aligned} \tag{2.17}$$

The introduction of the pressure through the normal stresses only,

$$\sigma_x = -p + \sigma'_x, \quad \sigma_y = -p + \sigma'_y, \quad \sigma_x = -p + \sigma'_z, \tag{2.18}$$

permits us to express (2.17) in the form

$$\frac{\partial \boldsymbol{u}}{\partial t} + (\boldsymbol{u}\nabla)\boldsymbol{u} = \mathbf{g} - \frac{1}{\rho}\nabla p$$
$$+ \frac{1}{\rho}\left[\left(\frac{\partial \sigma'_x}{\partial x} + \frac{\partial \tau_{yx}}{\partial y} + \frac{\partial \tau_{zx}}{\partial z}\right)\boldsymbol{i} + \left(\frac{\partial \tau_{xy}}{\partial x} + \frac{\partial \sigma'_y}{\partial y} + + \frac{\partial \tau_{zy}}{\partial z}\right)\boldsymbol{j} + \left(\frac{\partial \tau_{zx}}{\partial x} + \frac{\partial \tau_{zy}}{\partial y} + \frac{\partial \sigma'_z}{\partial z}\right)\boldsymbol{k}\right]. \tag{2.19}$$

For many gases and liquids, there are linear relationships (*experimentally derived laws*) between the stresses and the shear rates (the deformation rate). The Stokes postulation [3, 4] employs the idea of the Newton law (Onsanger's principle [1]). In a more general form, including all the components of the stress tensor, the Stokes postulate gives

$$\sigma'_x = \mu\left(2\frac{x}{\partial x} - \frac{2}{3} div\, \boldsymbol{u}\right), \quad \tau_{xy} = \mu\left(\frac{\partial u_x}{\partial y} + \frac{\partial u_y}{\partial x}\right),$$

$$\sigma'_y = \mu\left(2\frac{\partial u_y}{\partial y} - \frac{2}{3} div\, \boldsymbol{u}\right), \quad \tau_{yz} = \mu\left(\frac{\partial u_y}{\partial z} + \frac{\partial u_z}{\partial y}\right),$$

$$\sigma'_z = \mu\left(2\frac{\partial u_z}{\partial z} - \frac{2}{3} div\, \boldsymbol{u}\right), \quad \tau_{xz} = \mu\left(\frac{\partial u_z}{\partial x} + \frac{\partial u_x}{\partial z}\right). \tag{2.20}$$

The substitution of (2.20) into (2.19) and the subsequent projection along the coordinate axes gives the so-called Navier–Stokes equations:

$$\rho\left(\frac{\partial u_x}{\partial t} + u_x\frac{\partial u_x}{\partial x} + u_y\frac{\partial u_x}{\partial y} + u_z\frac{\partial u_x}{\partial z}\right) = \rho g_x - \frac{\partial p}{\partial x}$$
$$+ \frac{\partial}{\partial x}\left[\mu\left(2\frac{\partial u_x}{\partial x} - \frac{2}{3} div\, \boldsymbol{u}\right)\right] + \frac{\partial}{\partial y}\left[\mu\left(\frac{\partial u_x}{\partial y} + \frac{\partial u_y}{\partial x}\right)\right] + \frac{\partial}{\partial z}\left[\mu\left(\frac{\partial u_z}{\partial x} + \frac{\partial u_x}{\partial z}\right)\right],$$

$$\rho\left(\frac{\partial u_y}{\partial t} + u_x\frac{\partial u_y}{\partial x} + u_y\frac{\partial u_y}{\partial y} + u_z\frac{\partial u_y}{\partial z}\right) = \rho g_y - \frac{\partial p}{\partial y}$$
$$+ \frac{\partial}{\partial y}\left[\mu\left(2\frac{\partial u_y}{\partial y} - \frac{2}{3} div\, \boldsymbol{u}\right)\right] + \frac{\partial}{\partial z}\left[\mu\left(\frac{\partial u_y}{\partial z} + \frac{\partial u_z}{\partial y}\right)\right] + \frac{\partial}{\partial x}\left[\mu\left(\frac{\partial u_x}{\partial y} + \frac{\partial u_y}{\partial x}\right)\right],$$

$$\rho\left(\frac{\partial u_z}{\partial t} + u_x\frac{\partial u_z}{\partial x} + u_y\frac{\partial u_z}{\partial y} + u_z\frac{\partial u_z}{\partial z}\right) = \rho g_z - \frac{\partial p}{\partial z}$$
$$+ \frac{\partial}{\partial z}\left[\mu\left(2\frac{\partial u_z}{\partial z} - \frac{2}{3} div\, \boldsymbol{u}\right)\right] + \frac{\partial}{\partial x}\left[\mu\left(\frac{\partial u_z}{\partial x} + \frac{\partial u_x}{\partial z}\right)\right] + \frac{\partial}{\partial y}\left[\mu\left(\frac{\partial u_y}{\partial z} + \frac{\partial u_z}{\partial y}\right)\right], \tag{2.21}$$

where g_x, g_y, g_z are the components of the acceleration vector $\mathbf{g}$. If the gravitational field generates the body force (i.e., the force of weight) and the x-axis is oriented vertically downwards, we have

$$g_x = g, \quad g_y = g_z = 0, \tag{2.22}$$

where g is the gravitational acceleration.

Gases can be considered as *incompressible* if their velocities are *subsonic* (i.e., less than the velocity of sound). At higher velocities it is necessary to utilize the relationship between the density and the pressure:

$$p - \rho RT = 0, \tag{2.23}$$

where R is the gas constant and T is the absolute temperature.

Incompressible liquids have constant densities, so in the case of a gravity body force ($g = g_x$) and if $div\boldsymbol{u} = 0$ (2.13), the Navier–Stokes equations become

$$\begin{aligned}
&\frac{\partial u_x}{\partial t} + u_x\frac{\partial u_x}{\partial x} + u_y\frac{\partial u_x}{\partial y} + u_z\frac{\partial u_x}{\partial z} = g - \frac{1}{\rho}\frac{\partial p}{\partial x} + \nu\left(\frac{\partial^2 u_x}{\partial x^2} + \frac{\partial^2 u_x}{\partial y^2} + \frac{\partial^2 u_x}{\partial z^2}\right),\\
&\frac{\partial u_y}{\partial t} + u_x\frac{\partial u_y}{\partial x} + u_y\frac{\partial u_y}{\partial y} + u_z\frac{\partial u_y}{\partial z} = - \frac{1}{\rho}\frac{\partial p}{\partial y} + \nu\left(\frac{\partial^2 u_y}{\partial x^2} + \frac{\partial^2 u_y}{\partial y^2} + \frac{\partial^2 u_y}{\partial z^2}\right),\\
&\frac{\partial u_z}{\partial t} + u_x\frac{\partial u_z}{\partial x} + u_y\frac{\partial u_z}{\partial y} + u_z\frac{\partial u_z}{\partial z} = - \frac{1}{\rho}\frac{\partial p}{\partial z} + \nu\left(\frac{\partial^2 u_z}{\partial x^2} + \frac{\partial^2 u_z}{\partial y^2} + \frac{\partial^2 u_z}{\partial z^2}\right),
\end{aligned} \tag{2.24}$$

where $\nu = \frac{\mu}{\rho}$ is the fluid kinematic viscosity.

The vector form of (2.24) is

$$\frac{\partial \boldsymbol{u}}{\partial t} + (\boldsymbol{u} \cdot \nabla)\boldsymbol{u} = \boldsymbol{g} - \frac{1}{\rho}\nabla p + \nu\nabla^2\boldsymbol{u}, \tag{2.25}$$

where ∇^2 is the Laplacian operator:

$$\nabla^2 = \frac{\partial^2}{\partial x^2} + \frac{\partial^2}{\partial y^2} + \frac{\partial^2}{\partial z^2}. \tag{2.26}$$

2.2 Cylindrical Coordinates

In cylindrical coordinates r, φ, z the velocity vector $\boldsymbol{u} = (u_r, u_\varphi, u_z)$ must satisfy the continuity equation, momentum equations, and viscous stresses [4]:

$$\frac{\partial \rho}{\partial t} + \frac{1}{r}\frac{\partial(\rho r u_r)}{\partial r} + \frac{1}{r}\frac{\partial(\rho u_\varphi)}{\partial \varphi} + \frac{\partial(\rho u_z)}{\partial z} = 0;$$

$$\rho\left(\frac{\partial u_r}{\partial t}+u_r\frac{\partial u_r}{\partial r}+\frac{u_\varphi \partial u_r}{r\,\partial \varphi}-\frac{u_\varphi^2}{r}+u_z\frac{\partial u_r}{\partial z}\right)=f_r-\frac{\partial p}{\partial r}+\frac{1}{r}\frac{\partial(r\tau_{rr})}{\partial r}+\frac{1}{r}\frac{\partial(r\tau_{r\varphi})}{\partial \varphi}+\frac{\partial \tau_{rz}}{\partial z}-\frac{\tau_{\varphi\varphi}}{r},$$

$$\rho\left(\frac{\partial u_\varphi}{\partial t}+u_r\frac{\partial u_\varphi}{\partial r}+\frac{u_\varphi \partial u_\varphi}{r\,\partial \varphi}-\frac{u_r u_\varphi}{r}+u_z\frac{\partial u_\varphi}{\partial z}\right)$$
$$=f_\varphi-\frac{1}{r}\frac{\partial p}{\partial \varphi}+\frac{1}{r^2}\frac{\partial}{\partial r}\left(r^2\tau_{r\varphi}\right)+\frac{1}{r}\frac{\partial \tau_{\varphi\varphi}}{\partial \varphi}+\frac{\partial \tau_{\varphi z}}{\partial z},$$

$$\rho\left(\frac{\partial u_z}{\partial t}+u_r\frac{\partial u_z}{\partial r}+\frac{u_\varphi \partial u_z}{r\,\partial \varphi}+u_z\frac{\partial u_z}{\partial z}\right)=f_z-\frac{\partial p}{\partial z}+\frac{1}{r}\frac{\partial(r\tau_{rz})}{\partial r}+\frac{1}{r}\frac{\partial \tau_{\varphi z}}{\partial \varphi}+\frac{\partial \tau_{zz}}{\partial z};;$$

$$\tau_{rr}=\mu\left(2\frac{\partial u_r}{\partial r}-\frac{2}{3}div\,\boldsymbol{u}\right),\ \tau_{\varphi\varphi}=\mu\left[2\left(\frac{1}{r}\frac{\partial u_\varphi}{\partial \varphi}+\frac{u_r}{r}\right)-\frac{2}{3}div\,\boldsymbol{u}\right],$$

$$\tau_{zz}=\mu\left(2\frac{\partial u_z}{\partial z}-\frac{2}{3}div\,\boldsymbol{u}\right),\ \tau_{r\varphi}=\mu\left[r\frac{\partial}{\partial r}\left(\frac{u_\varphi}{r}\right)+\frac{1}{r}\frac{\partial u_r}{\partial \varphi}\right],$$

$$\tau_{rz}=\mu\left(2\frac{\partial u_z}{\partial r}+\frac{\partial u_r}{\partial z}\right),\ \tau_{\varphi z}=\mu\left(\frac{\partial u_\varphi}{\partial z}+\frac{1}{r}\frac{\partial u_z}{\partial \varphi}\right). \tag{2.27}$$

2.3 Boundary Conditions

The *Navier–Stokes equations* (2.21) together with the *continuity equation* (2.11) and the *equation of state* (2.23) form a *closed set* of *five* differential equations allowing us to determine *five variables*: three velocity components (u_x, u_y, u_z), the density (ρ), and the pressure (p).

The boundaries of the macroscopic system should affect the solution through the *boundary conditions.* The type of a particular boundary condition depends on the physical situation at the boundary between the fluid and the surroundings.

Despite the variety of conditions existing at the boundaries, some of them are common for a large number of hydrodynamic problems. For example, if we have a phase boundary, commonly represented as a boundary between a flowing fluid and a confining wall, several types of boundary conditions are possible. If u_n and u_τ are the normal and tangential velocity components, nonslip conditions must be used:

$$u_\tau=0,\quad u_n=0. \tag{2.28}$$

At the mobile (gas–liquid, liquid–liquid) interphase these conditions are

$$u_\tau^{(1)}=u_\tau^{(2)},\quad u_n^{(1)}=u_n^{(2)}=0, \tag{2.29}$$

where the last condition expresses that the fluids are immiscible (interphase impermeability condition).

The interaction between two mobile phases must be expressed by the equality of the tangential (P_τ) and normal (P_n) components of the stress vector at the interphase:

$$P_\tau^{(1)} = P_\tau^{(2)}, P_n^{(1)} = P_n^{(2)} + P_\sigma, \tag{2.30}$$

where P_σ is capillary pressure.

The conditions for the *interface impermeability* (2.30) are not valid if there is an intensive mass transfer across it.

2.4 Laminar Boundary Layer

Many processes in chemical engineering are realized in two-dimensional flows. For theoretical analysis of these flows let us consider a semi-infinite solid interface with coordinates $x \in [0, \infty)$ and $y = 0$, streamlined by a potential flow with velocity u_∞. At the fluid–solid interface *a nonslip condition* exists and the velocity of the fluid is zero. Far from the interface the velocity is equal to u_∞. Under these conditions we will analyze a *steady-state* and *two-dimensional flow*. The flow in a horizontal plane with coordinates *x*, *y*, i.e., the Navier–Stokes equations (2.24), has the form

$$u_x \frac{\partial u_x}{\partial x} + u_y \frac{\partial u_x}{\partial y} = -\frac{1}{\rho}\frac{\partial p}{\partial x} + \nu\left(\frac{\partial^2 u_x}{\partial x^2} + \frac{\partial^2 u_x}{\partial y^2}\right),$$

$$u_x \frac{\partial u_y}{\partial x} + u_y \frac{\partial u_y}{\partial y} = -\frac{1}{\rho}\frac{\partial p}{\partial y} + \nu\left(\frac{\partial^2 u_y}{\partial x^2} + \frac{\partial^2 u_y}{\partial y^2}\right),$$

$$\frac{\partial u_x}{\partial x} + \frac{\partial u_y}{\partial y} = 0, \tag{2.31}$$

with boundary conditions

$$x = 0, \quad u_x = u_\infty, \quad \frac{\partial u_x}{\partial x} = 0; \quad u_y = 0, \quad \frac{\partial u_y}{\partial x} = 0;$$

$$y = 0, \quad u_x = u_y = 0; \quad y \to \infty, \quad u_x = u_\infty, \quad u_y = 0. \tag{2.32}$$

A *generalized analysis* (see Chap. 4.1) permits the qualitative investigation of the process, using *generalized (dimensionless) variables*.

Let us consider an area of the fluid defined by

$$0 \le x \le l, \quad 0 \le y \le \delta, \tag{2.33}$$

where the main variations of the fluid velocity ($0 \leq u_x \leq u_\infty$) occur. The upper limit δ means that at $y \to \delta$ the velocity $u_x \to u_\infty$. This allows us to employ l, δ, and u_∞ as naturally defined *scales* of the process. The dimensionless variables defined through these scales are

$$X = \frac{x}{l}, \quad Y = \frac{y}{\delta}, \quad U_x = \frac{u_x}{u_\infty}, \quad U_y = \frac{l u_y}{u_\infty \delta}, \quad P = \frac{p}{\rho u_\infty^2}. \tag{2.34}$$

As a result, the order of magnitude of the dimensionless (generalized) variables and its derivates is 1.

The dimensionless form of equations (2.31) and the boundary conditions (2.32) are

$$U_x \frac{\partial U_x}{\partial X} + U_y \frac{\partial U_x}{\partial Y} = -\frac{\partial P}{\partial X} + \left(\frac{\nu l}{u_\infty \delta^2}\right)\left(\frac{\delta^2}{l^2}\frac{\partial^2 U_x}{\partial X^2} + \frac{\partial^2 U_x}{\partial Y^2}\right),$$

$$\left(\frac{\delta^2}{l^2}\right)\left(U_x \frac{\partial U_y}{\partial X} + U_y \frac{\partial U_y}{\partial Y}\right) = -\frac{\partial P}{\partial Y} + \left(\frac{}{u_\infty \delta^2}\right)\left(\frac{\delta^2}{l^2}\right)\left(\frac{\delta^2}{l^2}\frac{\partial^2 U_y}{\partial X^2} + \frac{\partial^2 U_y}{\partial Y^2}\right),$$

$$\frac{\partial U_x}{\partial X} + \frac{\partial U_y}{\partial Y} = 0;$$

$$X = 0, \quad U_x = 1; \quad \frac{\partial U_x}{\partial X} = 0; \quad U_y = 0, \quad \frac{\partial U_y}{\partial X} = 0;$$

$$Y = 0, \quad U_x = U_y = 0; \quad Y \to \infty, \quad U_x = 1, \quad U_y = 0, \tag{2.35}$$

where the orders of magnitude of the physical effects are determined by the orders of magnitude of the dimensionless parameters.

The dimensionless equations (2.35) defined above indicate, that inertia effects (left-hand side of the first equation in 2.35) are balanced by the viscous (right-hand side of the same equation) forces if the order of the dimensionless parameter $\left(\frac{\nu l}{u_\infty \delta^2}\right)$ is 1, i.e.,

$$\frac{\nu l}{u_\infty \delta} = 1, \quad \delta = \sqrt{\frac{\nu l}{u_\infty}} = \frac{l}{\sqrt{Re}}, \quad Re = \frac{u_\infty l}{\nu}, \quad \frac{\delta}{l} = \frac{1}{\sqrt{Re}}, \tag{2.36}$$

where Re is the Reynolds number in these conditions.

The results (2.36) indicate that at large Reynolds numbers $\left(\frac{\delta^2}{l^2} = \frac{1}{Re} \leq 10^{-2}\right)$ the longitudinal effect of the viscous forces is negligible,

$$\frac{\delta^2}{l^2}\frac{\partial^2 U}{\partial X^2} \leq 10^{-2}, \quad \frac{\partial^2 U}{\partial X^2} \sim 1, \tag{2.37}$$

because an effect less than 1% is impossible to register experimentally.

From (2.36) it follows that a velocity change from 0 to u_∞ is realized in the thin layer δ (laminar boundary layer [4]) if $Re \geq 10^2$. In this case a boundary layer approximation can be used:

$$\frac{\delta^2}{l^2} = \frac{1}{Re} = 0. \tag{2.38}$$

If we introduce (2.36) and (2.38) into (2.35), the boundary layer approximation of the problem (Prandtl equations) has the form

$$U_x \frac{\partial U_x}{\partial X} + U_y \frac{\partial U_x}{\partial Y} = -\frac{\partial P}{\partial X} + \frac{\partial^2 U_x}{\partial Y^2}, \quad \frac{\partial P}{\partial Y} = 0, \quad \frac{\partial U_x}{\partial X} + \frac{\partial U_y}{\partial Y} = 0;$$

$$X = 0, \quad U_x = 1; \quad Y = 0, \quad U_x = U_y = 0; \quad Y \to \infty, \quad U_x = 1. \tag{2.39}$$

If $u_\infty = \text{const.}$, $\frac{\partial P}{\partial X} = 0$, i.e., $P = \text{const.}$ [4]. This approximation is widely applied in chemical engineering for processes occurring in thin layers near the interphase surface. The scaling approach demonstrated here will be utilized further in other problems in the book for generalized (qualitative) analysis of the models.

2.5 Two-Phase Boundary Layers

Interphase heat and mass transfer in gas–liquid and liquid–liquid systems has a large range of application in science and technology. These two flow processes are characterized by a moving interphase surface [6]. The shape of the surface may be flat [7] or wavy [8] and depends on many interface effects, such as a two-phase interaction [7], surfactants [7], nonlinear heat and mass transfer effects [5], and surface instability [8].

Despite the significant number of works devoted to these issues, there is no single answer to the problems for the mechanism of the appearance of waves and the approximations of the equations in which the shape of the moving interface has to be determined.

Let us consider a two-dimensional two-phase co-current flow. The mathematical description of this flow is given by the Navier–Stokes equations (2.31) for two phases,

$$u_i \frac{\partial u_i}{\partial x} + v_i \frac{\partial u_i}{\partial y} = \upsilon_i \left(\frac{\partial^2 u_i}{\partial x^2} + \frac{\partial^2 u_i}{\partial y^2} \right) - \frac{1}{\rho_i} \frac{\partial p_i}{\partial x},$$

$$u_i \frac{\partial v_i}{\partial x} + v_i \frac{\partial v_i}{\partial y} = \upsilon_i \left(\frac{\partial^2 v_i}{\partial x^2} + \frac{\partial^2 v_i}{\partial y^2} \right) - \frac{1}{\rho_i} \frac{\partial p_i}{\partial y},$$

$$\frac{\partial u_i}{\partial x} + \frac{\partial v_i}{\partial y} = 0, \tag{2.40}$$

with boundary conditions assuming the contact of the potential flows with velocities u_i^∞, i = 1,2 and continuity of the components of the velocities and stress vector at the interface:

$$y = h(x); x = 0, \quad u_i = u_i^\infty, \quad v_i = 0; \quad x \to \infty, \quad \frac{\partial u_i}{\partial x} = 0, \quad v_i = 0;$$

$$y \to \infty, \quad u_1 = u_1^\infty, \quad v_1 = 0; \quad y \to -\infty, \quad u_2 = u_2^\infty, \quad v_2 = 0;$$

$$y = h(x), \quad u_1 = u_2, \quad v_1 = v_2,$$

$$p_1 + 2\mu_1 \frac{1 - h'^2}{1 + h'^2}\frac{\partial u_1}{\partial x} + 2\mu_1 \frac{h'}{1 + h'^2}\left(\frac{\partial u_1}{\partial y} + \frac{\partial v_1}{\partial x}\right)$$
$$= p_2 + 2\mu_2 \frac{1 - h'^2}{1 + h'^2}\frac{\partial u_2}{\partial x} + 2\mu_2 \frac{h'}{1 + h'^2}\left(\frac{\partial u_2}{\partial y} + \frac{\partial v_2}{\partial x}\right),$$

$$\mu_1 \frac{1 - h'^2}{1 + h'^2}\left(\frac{\partial u_1}{\partial y} + \frac{\partial v_1}{\partial x}\right) - 4\mu_1 \frac{h'}{1 + h'^2}\frac{\partial u_1}{\partial x}$$
$$= \mu_2 \frac{1 - h'^2}{1 + h'^2}\left(\frac{\partial u_2}{\partial y} + \frac{\partial v_2}{\partial x}\right) - 4\mu_2 \frac{h'}{1 + h'^2}\frac{\partial u_2}{\partial x}, \quad h' = \frac{dh}{dx}, \quad h(0) = 0. \tag{2.41}$$

If we use characteristic scales similar to (2.34), the dimensionless variables have the form

$$x = lX, \quad y = \delta_i Y_i, \quad u_i = u_i^\infty U_i, \quad v_i = u_i^\infty \frac{\delta_i}{l} V_i, \quad p_i = \rho_i u_i^{\infty 2} P_i,$$
$$h(x) = \delta_i H_i(X),$$

$$h' = \sqrt{\varepsilon_i} H_i', \quad \varepsilon_i = \left(\frac{\delta_i}{l}\right)^2. \tag{2.42}$$

As a result, the problem in dimensionless variables has the form

$$U_i \frac{\partial U_i}{\partial X} + V_i \frac{\partial U_i}{\partial Y_i} = \frac{\upsilon_i l}{u_i^\infty \delta_i^2}\left[\varepsilon_i \frac{\partial^2 U_i}{\partial X^2} + \frac{\partial^2 U_i}{\partial Y_i^2} - \frac{\partial P_i}{\partial X}\right],$$

$$\varepsilon_i \left(U_i \frac{\partial V_i}{\partial X} + V_i \frac{\partial V_i}{\partial Y_i}\right) = \frac{\upsilon_i l}{u_i^\infty \delta_i^2}\left[\varepsilon_i^2 \frac{\partial V_i}{\partial X} + \varepsilon_i \frac{\partial V_i}{\partial Y_i} - \frac{\partial P_i}{\partial Y_i}\right],$$

$$\frac{\partial U_i}{\partial X} + \frac{\partial V_i}{\partial Y_i} = 0;$$

$$X = 0, \quad U_i = 1, \quad V_i = 0; \quad x \to \infty, \frac{\partial U_i}{\partial X} = 0, \quad V_i = 0;$$

$$Y_1 \to \infty, \quad U_1 = 1, \quad V_1 = 0; \quad Y_2 \to -\infty, \quad U_2 = 1, V_2 = 0;$$

$$Y_1 = H_1(X), \quad Y_2 = H_2(X), \quad U_1 = \theta_1 U_2, \quad V_1 = \theta_2 V_2,$$

$$
\begin{gathered}
P_1 + 2\varepsilon_1 \frac{1-\varepsilon_1 H_1'^2}{1+\varepsilon_1 H_1'^2}\frac{\partial U_1}{\partial X} + 2\varepsilon_1 \frac{H_1'}{1+\varepsilon_1 H_1'^2}\left(\frac{\partial U_1}{\partial Y_1} + \varepsilon_1 \frac{\partial V_1}{\partial X}\right) \\
= \theta_3 \left[P_2 + 2\varepsilon_2 \frac{1-\varepsilon_2 H_2'^2}{1+\varepsilon_2 H_2'^2}\frac{\partial U_2}{\partial X} + 2\varepsilon_2 \frac{H_2'}{1+\varepsilon_2 H_2'^2}\left(\frac{\partial U_2}{\partial Y_2} + \varepsilon_2 \frac{\partial V_2}{\partial X}\right)\right], \\
\frac{1-\varepsilon_1 H_1'^2}{1+\varepsilon_1 H_1'^2}\left(\frac{\partial U_1}{\partial Y_1} + \varepsilon_1 \frac{\partial V_1}{\partial X}\right) - \varepsilon_1 \frac{H_1'}{1+\varepsilon_1 H_1'^2}\frac{\partial U_1}{\partial X} \\
= \theta_4 \left[\frac{1-\varepsilon_2 H_2'^2}{1+\varepsilon_2 H_2'^2}\left(\frac{\partial U_2}{\partial Y_2} + \varepsilon_2 \frac{\partial V_2}{\partial X}\right) - \varepsilon_2 \frac{H_2'}{1+\varepsilon_2 H_2'^2}\frac{\partial U_2}{\partial X}\right], \quad i = 1,2.
\end{gathered} \tag{2.43}
$$

The balance between inertia and surface forces effects leads to

$$
\frac{\upsilon_i l}{u_i^\infty \delta_i^2} = 1, \quad i = 1,2; \tag{2.44}
$$

and the parameters in (2.43) have the forms

$$
\delta_i = \sqrt{\frac{\upsilon_i l}{u_i^\infty}}, \quad \varepsilon_i = \left(\frac{\delta_i}{l}\right)^2 = Re_i^{-1}, \quad Re_i = \frac{u_i^\infty l}{\upsilon_i},
$$

$$
\theta_1 = \frac{u_2^\infty}{u_1^\infty}, \quad \theta_2 = \sqrt{\frac{u_2^\infty \upsilon_2}{u_1^\infty \upsilon_1}}, \quad \theta_3 = \frac{\rho_2 u_2^\infty}{\rho_1 u_1^\infty}, \quad \theta_4 = \theta_1^{3/2}\left(\frac{\upsilon_2^\infty}{\upsilon_1^\infty}\right)^{\frac{1}{2}}, \quad i = 1,2. \tag{2.45}
$$

In the approximation of the boundary layer [4], $10^{-2} > \varepsilon_i = 0$, $i = 1, 2$ and problem (2.43) has the form

$$
\begin{gathered}
U_i \frac{\partial U_i}{\partial X} + V_i \frac{\partial U_i}{\partial Y_i} = -\frac{\partial P_i}{\partial X} + \frac{\partial^2 U_i}{\partial Y_i^2}, \quad \frac{\partial U_i}{\partial X} + \frac{\partial V_i}{\partial Y_i} = 0, \quad \frac{\partial P_i}{\partial Y_i} = 0; \\
X = 0, \quad U_i = 1; \quad Y_1 \to \infty, \quad U_1 = 1; \quad Y_2 \to -\infty, \quad U_2 = 1; \\
Y_1 = H_1, \quad Y_2 = H_2, \quad U_1 = \theta_1 U_2, \quad V_1 = \theta_2 V_2, \quad P_1 = \theta_3 P_2, \quad \frac{\partial U_1}{\partial Y_1} = \theta_4 \frac{\partial U_2}{\partial Y_2}, \\
i = 1,2.
\end{gathered} \tag{2.46}
$$

From (2.43) and (2.45) it follows that increase of the Reynolds numbers Re_i, ($i = 1, 2$) does not affect the conditions for the existence of boundary layers (in practice $Re \geq 10^2$).

The shape of the interphase surface can be obtained using the condition that this surface is impermeable, i.e.,

$$
u_{n1} = \frac{v_1(x,h) - h' u_1(x,h)}{\sqrt{1 + (h')^2}} = 0, \tag{2.47}
$$

where u_{n1} is the normal component of the velocity at the interface $y = h(x)$. The boundary layer problem (2.46) is solved [25] using the perturbation method [21, 22]. The results obtained show that in this approximation $v(x, h) \equiv 0$. In the boundary layer approximation ($\varepsilon_1 = 0$) from (2.47) and (2.41) it follows that

$$h'(x) \equiv 0, \quad h \equiv 0, \tag{2.48}$$

i.e., the moving interface is flat.

The results obtained show that the solution of the boundary layer problem (2.46) with a flat interface is a partial solution of (2.43) and exists when the two-phase systems are stable. The interface may not be flat if the systems lose their stability.

The appearance of random periodic disturbances leads to an increase of their amplitudes to a stable state. Methods for nonlinear stability analysis have to be used to obtain the amplitude's value [9].

2.6 Particular Processes

In some specific cases, the fluid motion equations can be simplified significantly. In some cases, this is due to certain general assumptions, whereas in other situations adequate analysis leads to simpler equations. The fluid flow within the laminar boundary layer discussed earlier is an example of such an adequate analysis.

Let the second approach be applied again to the case of a steady-state (non-wavy) gravity-driven falling-film flow, and let us consider a falling film flowing over a smooth solid surface [7]. The situation allows us to consider a two-dimensional flow of incompressible fluid. Therefore, these general considerations lead to $u_z \equiv 0$, $\rho = const.$, and that u_x and u_y are time-independent. Assuming that the x-axis is oriented along the flow, i.e., vertically downwards, we have through (2.1) and (2.24)

$$u_x \frac{\partial u_x}{\partial x} + u_y \frac{\partial u_x}{\partial y} = g - \frac{1}{\rho}\frac{\partial p}{\partial x} + \nu\left(\frac{\partial^2 u_x}{\partial x^2} + \frac{\partial^2 u_x}{\partial y^2}\right),$$

$$u_x \frac{\partial u_y}{\partial x} + u_y \frac{\partial u_y}{\partial y} = -\frac{1}{\rho}\frac{\partial p}{\partial y} + \nu\left(\frac{\partial^2 u_y}{\partial x^2} + \frac{\partial^2 u_y}{\partial y^2}\right),$$

$$\frac{\partial u_x}{\partial x} + \frac{\partial u_y}{\partial y} = 0, \quad \frac{\partial}{\partial x}\int\limits_0^{h(x)} u\,dy = h' u_x - u_y = 0, \tag{2.49}$$

where the last equation expresses the macroscopic mass balance in the film flow. This equation permits us to obtain the film thickness and follows from the boundary condition $u_n = 0$ too.

The boundary conditions (2.27–2.30) assume *no slip* at the solid surface of the plate at $y = 0$, whereas at the liquid–air interface $y = h(x)$ the assumption is of equal tangential velocities of both the phase and the stress vector components. Along the normal vector of the film surface, no flow of both phases is assumed, so the surface should be considered as *impermeable.* These physical conditions can be expressed as

$$y = 0, \quad u_\tau = u_x = 0, \quad u_n = u_y = 0;$$

$$y = h(x), \quad u_\tau = \frac{u_x + h' u_y}{\sqrt{1 + h'^2}} = u_{\tau g}, \quad u_n = \frac{u_y - h' u_x}{\sqrt{1 + h'^2}} = u_{ng} = 0,$$

$$P_n = P_\sigma + P_{ng}, \quad P_\tau = P_{\tau g}, \tag{2.50}$$

where P_n, P_τ, P_{ng}, and $P_{\tau g}$ are the stress vector components (2.41) in the liquid and in the gas, respectively. In the next solution it will be assumed that the gas is immobile, i.e., gas velocity and stress tensor components are equal to zero.

The components of the liquid stress vector and the capillary pressure P_σ are

$$P_n = -p - 2\mu \frac{1 - h'^2}{1 + h'^2} \frac{\partial u_x}{\partial x} - 2\mu \frac{h'^2}{1 + h'^2} \left(\frac{\partial u_x}{\partial y} + \frac{y}{\partial x} \right), \quad P_\sigma = \frac{\sigma h''}{[1 + h'^2]^{\frac{3}{2}}},$$

$$P_\tau = \mu \frac{1 - h'^2}{1 + h'^2} \left(\frac{\partial u_x}{\partial y} + \frac{\partial u_y}{\partial x} \right) - 4\mu \frac{h'}{1 + h'^2} \frac{\partial u_x}{\partial x}. \tag{2.51}$$

The boundary conditions (2.50) of the set of equations (2.49) need four additional boundary conditions for u_x and u_y (see the comments later).

2.7 Generalized Variables

A further simplification of the mathematical description can be performed if the mathematical model of a falling film is expressed though *generalized variables* (see Chap. 4). These dimensionless variables are obtained using characteristic scales. The main condition for the scales is that generalized variables *must have an order of magnitude of unity*:

$$x = lX, \quad y = h_0 Y, \quad u_x = \bar{u} U_x, \quad u_y = \varepsilon_0 \bar{u} U_y, \quad p = \rho \bar{u}^2 P, \quad h = h_0 H, \quad \varepsilon_0 = \frac{h_0}{l}, \tag{2.52}$$

where h_0 is the film thickness at $x \to \infty$ and $\bar{u}$ is the average velocity of the film flow.

The introduction of the new variables (2.52) into (2.49) yields

$$\varepsilon_0\left(U_x\frac{\partial U_x}{\partial X}+U_y\frac{\partial U_x}{\partial Y}\right)=-\varepsilon_0\frac{\partial P}{\partial X}+\frac{1}{Fr}+\frac{1}{Re}\left(\varepsilon_0^2\frac{\partial^2 U_x}{\partial X^2}+\frac{\partial^2 U_x}{\partial Y^2}\right),$$

$$\varepsilon_0^2\left(U_x\frac{\partial U_x}{\partial X}+U_y\frac{\partial U_y}{\partial Y}\right)=-\frac{\partial P}{\partial Y}+\frac{\varepsilon_0}{\mathrm{Re}}\left(\varepsilon_0^2\frac{\partial^2 U_y}{\partial X^2}+\frac{\partial^2 U_y}{\partial Y^2}\right),$$

$$\varepsilon_0\frac{\partial U_x}{\partial X}+\frac{\partial U_y}{\partial Y}=0, \qquad (2.53)$$

where

$$Re=\frac{\bar{u}h_0}{\nu},\qquad Fr=\frac{\bar{u}^2}{gh_0} \qquad (2.54)$$

are the Reynolds and Froude numbers, respectively.

Here and to the end of the book, the dimensionless parameters (obtained after introducing the generalized variables) express the order of magnitude of the physical effects, expressed by means of the mathematical operators (in generalized variables) related to them. The neglecting of the physical effect (mathematical operator) when its parameter α is very small ($0 = \alpha \leq 10^{-2}$) may be made because the relative influence of this effect is less than 1% and practically there are no experimental techniques for its registration.

In fact, the film thickness is negligible with respect to its length, so practically $\varepsilon_0 \ll 1$. In the cases when $\varepsilon_0 \leq 10^{-2}$, we can use $\varepsilon_0 = 0$ in (2.53) and the equations defining the velocity profiles and the pressure of the falling liquid flow under the approximation of a *film flow* ($\varepsilon_0 = 0$) are

$$\frac{\partial^2 U_x}{\partial Y^2}=-\frac{Re}{Fr},\quad \frac{\partial P}{\partial Y}=0,\quad \frac{\partial U_y}{\partial Y}=0. \qquad (2.55)$$

The equation defining u_y follows from (2.50) and (2.55):

$$\frac{\partial u_y}{\partial y}=0;\quad y=0,\quad u_y=0. \qquad (2.56)$$

Therefore, we have

$$u_y\equiv 0. \qquad (2.57)$$

The introduction of (2.57) into the expression for u_n (2.50) yields

$$h'u_x=0. \qquad (2.58)$$

Hence,

$$h'=0;\quad x\to\infty,\quad h=h_0, \qquad (2.59)$$

i.e.,

$$h \equiv h_0. \tag{2.60}$$

The boundary conditions for U_x in (2.55) follow directly from (2.50)–(2.52). It should be taken into account that the continuity equation in (2.49) and (2.57) is an inherent result of the fact that $\frac{\partial U_x}{\partial X} = 0$. Therefore, if the effect of the second phase in (2.50) is neglected (i.e., $p_{\tau g} = 0$), we have from (2.49)–(2.51) that the simplified equations defining u_x are

$$\nu \frac{\partial^2 u_x}{\partial y^2} + g = 0, \quad \frac{\partial u_x}{\partial x} + \frac{\partial u_y}{\partial y} = 0; \quad y = 0, \quad u_x = 0, \quad u_y = 0; \quad y = h_0, \quad \frac{\partial u_x}{\partial y} = 0. \tag{2.61}$$

Thus, the result is

$$u_x = \frac{g}{2\nu}\left(2h_0 y - y^2\right), \quad u_y = 0; \quad \bar{u} = \int\limits_0^{h_0} u_x dy = \frac{g h_0^2}{3\nu}, \quad u_s = \frac{g h_0^2}{2\nu}, \tag{2.62}$$

where u_s is the interface velocity of the film flow.

The second equation in (2.60) gives $p =$ const. The constant is defined by a condition requiring equality of the normal components of the stress vector:

$$P = -P_{ng}, \tag{2.63}$$

i.e., the pressure in the liquid is constant and equals the pressure at the liquid–gas interphase.

The result just derived shows that flow of a nonwavy falling film is stratified ($u_y \equiv 0$), the thickness is constant ($h = h_0$), and the film velocity does not vary along its longitudinal axis ($\partial u_x / \partial x \equiv 0$). The latter fact tells us that the boundary conditions along the x-axis are not required (see 2.50).

The example discussed illustrates a general approach permitting us to simplify the mathematical structures describing particular processes using the introduction of characteristic scales and evaluation of the significance of the terms of the equations. The latter means that through the evaluation of the order of magnitude of the terms of the equation, we in fact evaluate the contributions of elementary physical effects in the process and their weights in the balance equations.

2.8 Basic Parameters

Equations (2.21) describe a broad range of hydrodynamic processes. The model built suggests a linear relationship between the stress and the shear rate (2.20). Therefore, the models incorporate the Newtonian rheological law (see the further discussion). When the relationship is non-linear, the fluid is non-Newtonian [3]. The rheological laws will be commented on especially in the next section.

The dynamic viscosity coefficient of Newtonian liquids is a basic parameter of the mathematical models describing their flows. It is a value *experimentally determined* by utilization of a simple hydrodynamic model (e.g., Couette flow) [3]. However, this is not a general principle and not every simple hydrodynamic model can be used for the determination of the dynamic viscosity coefficient. The problem just discussed (2.61) allows us to determine analytically the velocity profile (2.62) and the volumetric flow rate (by integration over the film cross-sectional area). The volumetric flow rate per unit length of the film width is

$$Q = \int_0^{h_0} u_x dy = \frac{g}{3\nu} h_0^3. \tag{2.64}$$

From (2.64) it follows directly that

$$\mu = \frac{g h_0^3 \rho}{3Q}. \tag{2.65}$$

Therefore, the dynamic viscosity can be determined through measurements of the macroscopic flow parameters Q and h_0. Unfortunately, this equation is inapplicable in practice since the small film thickness cannot be measured accurately. Usually, Hagen–Poiseuille flow in a cylindrical cross-sectional tube of radius R and length L is a more convenient experimental situation. For a pressure drop across the tube section over the distance L, we have for the volumetric flow rate

$$Q = \frac{\pi R^4}{8\mu L} \Delta P. \tag{2.66}$$

Thus,

$$\mu = \frac{\pi R^4}{8LQ} \Delta P, \tag{2.67}$$

where R, L, ΔP, and Q are easily and accurately determinable quantities.

Obviously this is valid for fluids exhibiting linear relationships between the shear stresses and the sheer rate, i.e., for Newtonian fluids or in other words for fluids where the Navier–Stokes equation is valid.

2.9 Rheology

Equations (2.20) briefly commented on already show [4] a linear relationship between the shear stress and the velocity gradient (the Newton law):

$$\tau = -\mu \frac{\partial u}{\partial n}. \tag{2.68}$$

The basic assumption of the Newton law is that *the flow is laminar* and the shear stress between two elementary layers is proportional to the velocity gradient in a transverse direction ($\partial u/\partial n$). Here, n is the normal vector of the surface separating two adjacent fluid layers. The factor (i.e., the coefficient of proportion) μ is defined as a *dynamic viscosity coefficient.*

In fact, if the fluid is incompressible, relationship (2.68) can be expressed [3] as

$$\tau = -\left(\frac{\mu}{\rho}\right)\frac{\partial}{\partial n}(\rho u) = -v\frac{\partial}{\partial n}(\rho u), \tag{2.69}$$

where $v = \mu/\rho$ is the *fluid kinematic viscosity coefficient*.

This form of the Newton law is more fundamental and relevant to Onsanger's linear principle. Thus, the *flux* of the momentum (the shear stresses τ) and the momentum gradient [$\partial(\rho u)/\partial n$] are related through a linear diffusive-type (see Fick's law later) law. The kinematic viscosity v is the *diffusion coefficient* of the momentum ρu under the assumption of a laminar flow.

Equation (2.68) can be defined also as a special case (linear case) of the more general relationship between the *stress tensor*Π and the *tensor* of deformation rates S. If the fluid is isotropic and viscous, the law can be expressed as

$$\Pi = aS + b\varepsilon, \tag{2.70}$$

where ε is the unit tensor, and a and b are scalars.

In an arbitrary coordinate system,

$$\varepsilon_{\mathrm{ij}} = \left\{\begin{matrix} 0, & if\ \mathrm{i} \neq \mathrm{j} \\ 1, & if\ \mathrm{i} = \mathrm{j} \end{matrix}\right\} \mathrm{i,j} = 1, 2, 3. \tag{2.71}$$

The coordinate axes 1, 2, 3 correspond to x, y, z in the case of Cartesian coordinates.

The factor a in (2.70) is a *constant* equal to 2μ (independent of the components of Π and S).

If the fluid structure is not isotropic (Newtonian fluids are isotropic), the factor a is a tensor characterizing the medium anisotropy. The scalar b in (2.70) can be related linearly to the tensors Π and S. In the case of an *isotropic fluid*, these linear relationships are *linear scalar combinations* of the tensor components, i.e., of their *linear invariants*.

The linear invariant of Π is

$$p_{11} + p_{22} + p_{33}, \tag{2.72}$$

and the linear invariant of S is

$$S_{11} + S_{22} + S_{ss} = \frac{\partial V_1}{1} + \frac{\partial V_2}{\partial x_2} + \frac{\partial V_3}{\partial x_3} = div\boldsymbol{V} = 0. \tag{2.73}$$

The scalar b can be obtained if one introduces the linear invariant components of the stress tensor into (2.69) and the sum of all three equations, then the result is

$$p_{11} + p_{22} + p_{33} = 3b = 3p. \tag{2.74}$$

Here $p = -p_{11} = -p_{22} = -p_{33}$ is the pressure at an arbitrary point of the fluid. The final form of (2.69) can be expressed as

$$\Pi = 2\mu S - \varepsilon p. \tag{2.75}$$

The component form of (2.75) is

$$p_{\mathrm{ij}} = \begin{cases} \mu\left(\dfrac{\partial V_{\mathrm{i}}}{\partial x_{\mathrm{j}}} + \dfrac{\partial V_{\mathrm{j}}}{\partial x_{\mathrm{i}}}\right), & at\ \mathrm{i} \neq \mathrm{j}; \\ -p + 2\mu\dfrac{\partial V_{\mathrm{i}}}{\partial x_{\mathrm{i}}}, & at\ \mathrm{i} = \mathrm{j}. \end{cases} \tag{2.76}$$

The result is equal to the results in (2.18) and (2.20).

The hydrodynamic definition of the pressure p in (2.75) is an additional *hypothesis* in the generalized form of the Newton law. It cannot be proved because of pure hydrodynamic assumptions. The problem arises from the fact that pressure p is an invariant scalar quantity equal to the same thermodynamic characteristics defined by the ideal gas law.

The rheological equation (2.76) describes the behavior of all gases and a wide class of liquids. In many cases (fine suspensions, dyes, coating liquids, etc.) this linear law is not valid. In these practically important liquids, the viscosity coefficients depend on the shear rate and the time. All liquids that do not satisfy the Newton rheology law are called *non-Newtonian.*

An important, and widely applicable, rheological model is that describing the flow of viscous-plastic liquids, known as Bingham liquids:

$$\tau = \tau_0 + \mu'\frac{\partial u}{\partial n}, \tag{2.77}$$

where τ_0 is the yield stress. If $\tau < \tau_0$, the fluid does not flow, which means $\frac{\partial u}{\partial n} = 0$.

The plastic coefficient μ' (the *structural viscosity*) depends on the shear rate. Aqueous clay solutions, concretes, some dyes, and pastes are such liquids. Relationship (2.77) indicates that if $\tau < \tau_0$, in the fluid body there exists a structure that resists the external deformation and is subsequently broken at $\tau = \tau_0$.

Another important rheological model describing the "pseudo-plastic" fluid flow concerns a law without a yield stress (i.e., $\tau_0 = 0$):

$$\tau = k\left(\frac{\partial u}{\partial l}\right)^n, \tag{2.78}$$

where k (the *consistency index*) and $n < 1$ (if $n \to 1$ the liquid approaches Newtonian behavior) are constants that do not vary over a wide range of shear rates. *The apparent viscosity* coefficients can be expressed as $k\left(\frac{\partial n}{\partial l}\right)^{n-1}$. The fluid flow decreases with increase of the shear rate and thus exhibits a shear thinning behavior.

When the power-law model (2.78) applied to a non-Newtonian liquid yields values of $n > 1$, the flow behavior is that of a *dilatant*, which is very often exhibited by concentrated suspensions.

Viscoplastic media (epoxy resins, very viscous liquids) exhibit simultaneously *elastic* and *viscous* properties, so the additive Voigt model [3] adequately describes their rheological behavior:

$$\tau = \tau_1 + \mu \frac{\partial u}{\partial l}, \tag{2.79}$$

where τ_1 is the elastic stress.

The above examples do not cover all the rheological models available for describing of non-Newtonian liquid behavior. However, they give the basic idea of how the shear stresses and the shear rate are related, thus forming the additional momentum equations in a way already described in detail for Newtonian liquids. Obviously, the incorporation of the non-Newtonian rheological law into the stress divergence terms (the viscous Stokes terms in the case of Navier–Stokes equations) causes mathematical difficulties.

2.10 Turbulence

At high Reynolds numbers, the flow character differs significantly from that discussed in the previous sections.

Beyond a certain Reynolds number a stationary laminar fluid becomes a nonstationary chaotic flow and the motions along a particular direction should be considered on the basis of mean characteristics only. A nonstationary flow regime with a velocity oscillating around a certain mean value is called a *turbulent flow*.

Simple experiments on the onset of turbulence indicate that it is related to instabilities of the preceding laminar flow regime emerging at *Re* numbers greater than a critical value Re_{cr} [2, 10, 18].

The main assumption is that infinitely small perturbations can be superposed on the main fluid flow. At $Re < Re_{cr}$ the perturbations occurring in the fluid attenuate very fast, whereas at $Re > Re_{cr}$ they superimpose on each other and grow rapidly and produce a turbulent flow. In these conditions the imposition of pulsations without any deterministic behavior on the main flow results in velocity pulsation amplitudes that at $Re \gg Re_{cr}$ are comparable with the *mean* flow velocity.

Let us imagine that is possible to follow the motion of a particular small fluid mass. The observations would demonstrate a complex and irreproducible motion of the chosen material points. The average flow, however, could be considered as more oriented. The trajectories of the particular fluid masses resemble the motions of gas molecules. Therefore, the unpredictable fluid motion of the turbulent flow should be *stochastic* in nature.

Let us consider as a first problem the common qualitative characteristics of the turbulent flow at $Re \gg Re_{cr}$ [2, 10, 18]. This motion is very often called a *fully developed turbulence*.

Small perturbations with various amplitudes are imposed on the main flow having a velocity U. The turbulent pulsations should be characterized not only by a mean velocity but also by a certain distance over which they propagate. Such a distance is often called the *length scale* of the motion. The very fast pulsations have greater length scales. Their velocities are of order of magnitude of

$$v' \sim \Delta U, \tag{2.80}$$

where ΔU is the variation of the mean velocity over a scale of length l. For example, in a pipe flow, the maximum length scale of the turbulent pulsations l_{max} equals the pipe diameter, whereas the velocity scale will be the maximum value at the pipe axis.

Most of the kinetic energy of the turbulent flow is transported by motions of such large-scale pulsations (*large eddies*). The Reynolds numbers of these large eddies, defined as $Re_l = \frac{\Delta U l}{\nu}$, are of the order of magnitude of the mean Reynolds number defined through the mean flow velocity.

Another part of the kinetic energy is transported by *eddies* with length scales of $\lambda < l$ and velocities $v_\lambda < v'$ and Reynolds number $Re_\lambda = \frac{v_\lambda \lambda}{\nu}$. The comparison of the Reynolds numbers of the large-scale and small-scale turbulent pulsations shows $Re_l \gg Re_\lambda$. In the case of $\lambda = \lambda_0$, when $Re_{\lambda_0} \leq 1$,, the big part of the energy dissipation (heat generation) is a result of the small-scale (λ_0) pulsations. The influence of the viscous forces of the large-scale pulsations ($Re_l \gg 1$) is very small, i.e., energy dissipation is negligible. The superposition of the *large eddies* provokes the *small eddies* with Reynolds numbers vanishing as fast as the corresponding length scales decrease.

Furthermore, the small eddies can be considered as a mechanism allowing the kinetic energy of the large eddies to be transformed into heat. If the flow was under conditions without large variations, the energy transfer would have a stationary character. Therefore, despite the fact that turbulence occurs at very high Reynolds numbers, it is associated with high energy dissipation. From a macroscopic point of view, the momentum transport of the turbulent flow could be associated with an "apparent" viscosity allowing joining of the energy dissipation (per unit fluid volume) and the macroscopic flow characteristics:

$$\varepsilon = -\frac{dE}{dt} = \mu_{turb}\left(\frac{\Delta U}{l}\right)^2. \tag{2.81}$$

The order of magnitude of the turbulent viscosity can be estimated from an analysis of the flow similitude. The quantity $\varepsilon = -\frac{dE}{dt}$ does not depend on the length scale of the flow and can be assumed to be a characteristic constant (of the flow itself). For example, this energy equals the energy dissipation of the largest eddies used for the creation of motions at lower length scales (lower eddies). This physically described process occurs at high Reynolds numbers and is practically unaffected by the molecular viscosity of the fluid μ. Therefore, the value of ε should be defined by quantities characterizing the macroscopic flow, i.e., the large eddy motions. Such quantities are only the velocity ΔU, the length scale l, and the fluid density ρ (these quantities together with the viscosity form all the variables of the hydrodynamic equations). The use of ΔU, l, and ρ allows us to form a common quantity with dimension $[\varepsilon] = \frac{erg}{m^3 s}$ that can be expressed as

$$\varepsilon \approx \rho \frac{(\Delta U)^3}{l}. \tag{2.82}$$

From (2.81) and (2.82) it can be derived that the apparent turbulent dynamic viscosity is

$$\mu_{\text{turb}} = \rho l(\Delta U), \tag{2.83}$$

whereas the corresponding kinematic viscosity is

$$\nu = \frac{\mu_{\text{turb}}}{\rho} = \Delta U l. \tag{2.84}$$

The formula just derived can be obtained from a different point of view based on the analogy between the turbulent motions and the chaotic motions of the gas molecules. If we assume that both phenomena are similar, the length scale l should be considered as analogous to the free path of the gas molecules, whereas the velocity pulsations should correspond to the mean velocity of the gas molecules.

Assuming the above physical similarity, we can derive (2.83) and (2.84) directly from the kinetic theory of gases. Developing the analogy, we find that the pulsation velocity ΔU is

$$\Delta U \approx l \frac{\partial U}{\partial l}. \tag{2.85}$$

The effective turbulent viscosity is significantly greater that the turbulent viscosities exhibited by fluids under normal laminar conditions. If we create the ratio $\frac{\nu}{\nu_{\text{turb}}}$, we find that

$$\frac{\nu}{\nu_{\text{turb}}} = \frac{\nu}{\Delta U l} \sim \frac{1}{Re} << 1. \tag{2.86}$$

The friction forces acting on a unit solid surface (e.g., 1 cm^2), i.e., the fluid shear stresses at the boundary interface, can be expressed as

$$\tau = \frac{F}{S} \approx \mu_{\text{turb}} \left(\frac{\Delta U}{l}\right) \approx \mu_{\text{turb}} \left(\frac{\partial U}{\partial l}\right) \approx \rho l^2 \left(\frac{\partial U}{\partial l}\right)^2 = \alpha \rho l^2 \left(\frac{\partial U}{\partial l}\right)^2, \tag{2.87}$$

where α is an unknown factor of proportionality.

The further discussion will consider two special cases of turbulent motions: (1) turbulence with a scale length $\lambda \ll l$, i.e., the small eddy turbulence, far from the solid surfaces, and (2) turbulent motion near solid surfaces.

Let us consider the small eddy turbulence with $\lambda \ll l$ in the fluid bulk. The first assumption that we could make is that $\lambda \gg \lambda_0$, meaning that any viscous effects are ignored. Now we look for the turbulent pulsation velocity v_λ with a length scale λ. In other words we try to define the gradient of the velocity v_λ over a distance equal to λ. The value of v_λ can depend only on ρ, λ, and the factor of proportionality ε, i.e., on quantities characterizing the fluid motion at any scale. At length scales $\lambda \gg \lambda_0$ the velocity is independent of the viscosity v. Moreover, the value of v is independent of the macroscopic scales of the flow—the characteristic length l and the velocity U—since $\lambda \ll l$.

The unique combination of ρ, λ, and ε, available to define a quantity with a dimension of a velocity, is $\left(\frac{\varepsilon\lambda}{\rho}\right)^{\frac{1}{3}}$, i.e.,

$$v_\lambda \sim \left(\frac{\varepsilon\lambda}{\rho}\right)^{\frac{1}{3}}. \tag{2.88}$$

Expressing ε via ΔU with the help of (2.82), we find that

$$v_\lambda \sim \Delta U\left(\frac{\lambda}{l}\right)^{\frac{1}{3}}. \tag{2.89}$$

Therefore, the scale of the pulsation velocity of motions, defined by the length scale λ by (2.89), is less than ΔU because $\left(\frac{\lambda}{l}\right)^{\frac{1}{3}} < 1$.

The decrease of both the velocity and the scale corresponds to a decrease of the Reynolds numbers expressed through the length scale λ as

$$Re_\lambda = \frac{v_\lambda \lambda}{v} = \frac{\Delta U \lambda^{\frac{4}{3}}}{v l^{\frac{1}{3}}} = Re\left(\frac{\lambda}{l}\right)^{\frac{4}{3}}. \tag{2.90}$$

At a certain value of the length scale λ_0, very often termed the *internal scale* of the turbulence, the Reynolds number Re_{λ_0} becomes of an order of unity. Obviously, λ_0 has an order of magnitude defined as

$$\lambda_0 \sim \left(\frac{1}{Re}\right)^{\frac{3}{4}} \sim \left(\frac{v^3\rho}{\varepsilon}\right)^{\frac{1}{4}}. \tag{2.91}$$

Starting with this value of the length scale and decreasing its contribution to the viscous forces on the flow growth, it exhibits a viscous *behavior.* The turbulent pulsations with length scales $\lambda \leq \lambda_0$ do not disappear immediately, but attenuate gradually, damped by the viscosity.

The turbulence near a solid surface will be discussed below under conditions allowing the flow to be considered as one along an infinite semiplane ($y = 0$).

Let the mean flow velocity along the x-axis be $\bar{v}_x = U$. The velocity depends on the distance from the surface; thus, we have $U = U(y)$. The turbulent pulsation superposes on the mean flow along the x-axis.

Let us determine the profile $U = U(y)$. Employing (2.87), we have

$$\frac{\partial U}{\partial y} = \left(\frac{1}{l}\right)\sqrt{\frac{\tau}{\rho\alpha}}. \tag{2.92}$$

All the points at the solid surface ($y = 0$) are equivalent and the shear stress can be assumed to be homogeneously distributed along the x-axis. This should be interpreted physically in a simple manner. The value of τ represents the momentum flux transferred from the fluid to the wall. In the fluid, flowing along the solid surface, there is a continuous transfer of momentum from the *far-field* points (large distances from the wall) towards the *near-field* points of the layers almost at the wall.

Taking into account that the momentum transport equation is in fact a conservation law, we can suppose that along the x-axis the momentum remains unchanged or U is independent of x. This assumption does not consider the energy dissipation due to the molecular viscosity. Therefore, under the assumption of $\tau = \text{const.}$, (2.92) becomes

$$U = \sqrt{\frac{\tau}{\rho\alpha}}\int\frac{dy}{l} + \text{const.} = \frac{v_0}{\sqrt{\alpha}}\int\frac{dy}{l(y)} + C_1, \tag{2.93}$$

where C_1 is a constant of integration, and

$$v_0 = \sqrt{\frac{\tau}{\rho}}. \tag{2.94}$$

The integration of (2.93) requires the profile $l(y)$, i.e., the variation of the length scale with the distance to the wall. However, the flow problem discussed does not consider either the size of the plate streamlined by the fluid or any other macroscopic dimension that should be utilized as a length scale of the macroscopic turbulent pulsations l. Intuitively, we could suggest that

$$l(y) \sim y. \tag{2.95}$$

The suggestion (it could be assumed as a condition too) (2.95) indicates that the scale of the pulsations grows as the point of interest moves from the wall towards the bulk of the fluid flow. The physical suggestion of that flow description assumes the solid surface to be a damper, so the pulsations should attenuate as the point of interest becomes closer to the wall.

The scaling suggestion (2.95) allows utilization of (2.93) to express the mean velocity profile as

$$U = \frac{v_0}{\sqrt{\alpha}}\ln y + C_1. \tag{2.96}$$

The physical sense of v_0 needs some additional explanations. Remember that the pulsation velocity v' (2.80) is

$$v' \approx \Delta U \approx U(y+l) - U(y) \approx v_0. \tag{2.97}$$

Thus, v_0 is a characteristic *velocity scale* for the turbulent pulsations. The determination of the constant C_1 should account for the fact that approaching the wall ($y \to 0$), the pulsations attenuate. Thus, the Reynolds number

$$Re = \frac{v_0 l(y)}{v} \tag{2.98}$$

will decrease as $y \to 0$.

At a certain $l = \delta_0$ the Reynolds number becomes of the order of unity. The area defined by $y < \delta_0$ is called a *viscous sublayer* since the viscous flow behavior dominates. The thickness of the viscous sublayer can be defined from the condition

$$\frac{v_0 \delta_0}{v} \approx 1, \tag{2.99}$$

which gives

$$\delta_0 = a\frac{v}{v_0}. \tag{2.100}$$

Here a is a factor of proportionality. The constant in (2.96) must satisfy the condition that at $l \sim \delta$ the mean flow velocity is negligible with respect to the characteristic velocity scale of the turbulent pulsations v_0.

Under these conditions the *logarithmic velocity profile*

$$U = \frac{v_0}{\sqrt{\alpha}} \ln\left(\frac{v_0 y}{a v}\right) \tag{2.101}$$

represents the mean velocity distribution across the viscous sublayer.

Expressing v_0 through the shear stresses τ (2.94), we can define that

$$U = \sqrt{\frac{\tau}{\rho\alpha}} \ln \frac{y}{a v}\sqrt{\frac{\tau}{\rho}}. \tag{2.102}$$

Thus, the apparent turbulent kinematic viscosity can be expressed as

$$v_{\mathrm{turb}} \sim v_0 l \sim v_0 y \sim v\frac{y}{\delta_0}. \tag{2.103}$$

The logarithmic velocity profile (2.102) contains two unknown constants, a and α. They should be defined on the basis of experimental data of the velocity distribution near the wall. Very often, it is convenient to introduce a dimensionless variable:

$$y_* = \frac{v_0 y}{v} \sim \frac{y}{\delta_0}. \tag{2.104}$$

The experimental data (see, e.g., [10]) indicate that the simple logarithmic profile matches the real velocity distribution at $y_* \geq 30$ and the corresponding value of α is about 0.17. The determination of a directly from the experimental profiles has no a physical sense since this constant (according to the definition given above) represents the situation at $Re = y_* \sim 1$, i.e., the area of the viscous sublayer.

The analysis performed above introduced the viscous sublayer as special zone of the flow. There are several hypotheses about the velocity profiles across it. The well-known Prandtl hypothesis suggests that at $y < \delta_0$ the flow behavior is *purely laminar*. The physical basis of Prandtl's assumption is the fact that at $y < \delta_0$ the Reynolds number becomes *lower that unity*.

Within the laminar sublayer the shear stress τ_0 satisfies the Newton law:

$$\tau_0 = \rho \nu \frac{dU}{dy}. \tag{2.105}$$

Therefore, the velocity profile is expressed through a linear relationship:

$$U = \frac{\tau_0}{\rho \nu} y + C. \tag{2.106}$$

The constant of integration C must be zero because of the nonslip conditions at the fluid–solid interface. Thus, at $y < \delta_0$ we have

$$U = \frac{\tau_0}{\rho \nu} y. \tag{2.107}$$

The "stitch up" of the linear and the logarithmic velocity profiles did not satisfy well the experimental data [2, 10], so a new hypothesis was developed by Karman.

According to Karman, the turbulent flow near the wall is separated into three sublayers: (1) turbulent flow, (2) "damping" sublayer, and (3) laminar boundary layer. The "damping" (very often called a *buffer* sublayer) attenuates the turbulent pulsation according to the scaling hypothesis $l(y) \sim y$ (2.95), which is also valid for the turbulent flow developed. However, within the fame of the buffer sublayer there is a *viscous effect* that leads to integration constants of the logarithmic profile of the mean velocity different from those defined for the turbulent flow.

Another hypothesis was conceived by Landau and Levich [2, 10]. It suggests that the turbulent motion within the viscous sublayer *does not disappear sharply*, but *attenuates gradually* towards the wall. However, the *attenuation law* for the turbulent pulsations, or in other words the scaling relationship $l = l(y)$ for the viscous sublayer, cannot be defined on the basis of scale analysis as was performed in the case of a developed turbulence. With the viscous sublayer all the quantities *can depend on the viscosity* ν and the distance to the wall *is not a unique length scale* of the flow. The scaling of the turbulent pulsation attenuation within the viscous sublayer can be developed on the basis of several assumptions.

Let us assume that velocity profile across the viscous sublayer is like that for the laminar flow, i.e.,

$$v_x \sim y. \tag{2.108}$$

Within the viscous sublayer there are no turbulent pulsations, but they approach it from the top, i.e., when $y > \delta_0$. The velocities of turbulent motion within the viscous sublayer have the same scale as the mean velocities. Therefore, we can scale as

$$v'_x \sim y. \tag{2.109}$$

Taking into account the continuity equation,

$$\frac{\partial v'_x}{\partial x} + \frac{\partial v'_y}{\partial y} = 0, \tag{2.110}$$

the normal velocity component is

$$v'_y = -\int \frac{\partial v'_x}{\partial x} dy \sim y^2. \tag{2.111}$$

The factor of proportionality in the relationship for v'_y may be expressed from the condition that at $y \sim \delta_0$, i.e., at the boundary of the viscous sublayer, the pulsation velocity v'_y has the same order of magnitude as the characteristic velocity scale of the turbulent flow v_0. Thus,

$$v'_y = v_0 \left(\frac{y}{\delta_0}\right)^2. \tag{2.112}$$

Therefore, the distributions of both the tangential and the normal components of both the *mean* and the *pulsation* velocities across the viscous sublayer are like in the laminar boundary layer. However, this is the *only similarity* between the *viscous sublayer* and the *laminar boundary layer* concepts.

The coefficient of apparent turbulent viscosity of the viscous sublayer could be defined if the length scale of the turbulent motion were known. Looking for the information required, we could suggest that in the viscous sublayer the Reynolds number is lower than unity. The latter means that only linear terms of the Navier–Stokes equation are significant, whereas second-order terms can be neglected. Thus, the velocity profile of the viscous sublayer can be determined by linearized equations. If some turbulent pulsations, however, penetrate the viscous sublayer, they are suppressed rapidly. The motions could be considered as *independent periodic motions* whose periods T remain unchanged over the entire viscous sublayer.

Under such considerations, the periods of turbulent motions inside the viscous sublayer do not depend on the distance to the wall y. Thus, the length scale of the motion directed along the y-axis (towards the wall) is

$$l \sim v'_y T. \tag{2.113}$$

On the other hand, taking into account that the period T is independent of y, we can scale as

$$l \sim y^2. \tag{2.114}$$

At $y \sim \delta_0$ the length scale of the motion should coincide with that of the turbulent boundary layer, i.e., $l \sim \delta_0$. Therefore, the length can be normalized by a scale of δ_0, which leads to

$$l = \left(\frac{1}{\delta_0}\right) y^2. \tag{2.115}$$

The scale of the turbulent motions of the viscous sublayer decreases (as a function of the distance y) faster than the corresponding scale of the turbulent boundary layer.

Following the definition of the momentum flux transported by the turbulent pulsations, we have

$$\tau_{turb} = \rho \nu_{turb} \frac{dU}{dy} = \left(\rho v'_y l\right) \frac{dU}{dy} \sim \left(\frac{\rho v_0 y^4}{\delta_0^3}\right) \frac{dU}{dy}. \tag{2.116}$$

Thus, the kinematic coefficient of the turbulent viscosity is

$$\nu_{turb} \sim \frac{v_0 y^4}{\delta_0^3} \sim \nu \left(\frac{y}{\delta_0}\right)^4. \tag{2.117}$$

At $y < \delta_0$ the momentum flux transported by the turbulent pulsations is lower than the momentum fluxes transported by a molecular mechanism and $\nu_{turb} < \nu$. As a result, $\tau \sim \tau_0$ and the velocity profile can be obtained using (2.107).

The turbulent pulsations exist practically in the vicinity of the wall. They contribute mainly to the mass and heat transfer from the fluid to the solid surface. The hypothesis of the viscous sublayer and the turbulent pulsation behavior within it allows us to avoid the problem with the conjunction of the velocity profiles as commented on above [10]. There is other hypothesis about the damping of turbulent pulsations in the viscous sublayer. It considers the length scale of the turbulent pulsations defined by (2.95) (i.e., it assumes that $l \sim y$). In other words, the length scale for the viscous sublayer and that in the region of $y > \delta_0$ are the same. This assumption leads to a turbulent viscosity expressed as

$$\nu_{turb} \sim \nu \left(\frac{y}{\delta_0}\right)^3. \tag{2.118}$$

The choice of a hypothesis depends on the assumption concerning the *mechanism of damping* of the turbulent pulsations. The detection of the turbulent pulsations and the exact definition of the law of their attenuation can be performed only experimentally. The problem is not easy, but some satisfactory results can be derived through studies concerning the diffusion of species dissolved in a turbulent liquid flow. The experimental data [10] confirm the hypothesis explaining the turbulent pulsation damping according to the scaling equation (2.115). Therefore, the further discussion will utilize the length scale defined by (2.115) and the relevant expression for the turbulent kinematic viscosity ν_{turb} (2.117) for the velocity profile determination. The velocity profiles within the conjunction zone can be defined because of the assumption that in that area the turbulent pulsation is the *only mechanism* of the momentum exchange with small effects of the molecular (Newtonian) viscosity. These assumptions mean that the scaling of the turbulent pulsation suppression (2.103) has no sense in the area of the conjunction of the velocity profiles. Obviously, the turbulent viscosity for the area of velocity profile conjunction can be interpolated in a manner that satisfies both scaling relationships (2.103, 2.117), e.g.,

$$\nu_{\text{turb}} = b\nu\left(\frac{y}{\delta_0}\right)^2. \tag{2.119}$$

The latter defines the shear stresses as

$$\tau = (\nu + \nu_{\text{turb}})\frac{dU}{dy} = \rho\left[\nu + b\nu\left(\frac{y}{\delta_0}\right)^2\right]\frac{dU}{dy}, \tag{2.120}$$

where b is a constant.

The integration of (2.120) gives the profile of the mean velocity over the area of conjunction:

$$U = \frac{v_0}{\sqrt{b}}\,\text{arctg}\sqrt{b}\left(\frac{y}{\delta_0}\right) + C. \tag{2.121}$$

Assuming that at $y_* \sim 5$ the velocity distribution is practically equal to that expressed by (2.107) and at $y_* \sim 30$ the logarithmic suitable profile matches well the data, we can define the unknown constants b and C.

Finally, the profiles of the mean velocities can be expressed as

$$\frac{U}{v_0} = y_*, \quad 0 \le y_* \le 5,$$

$$\frac{U}{v_0} = 10\text{arctg}(0.1y_*) + 1.2, \quad 5 \le y_* \le 30,$$

$$\frac{u_0}{v_0} = 5.5 + 2.5\ln y_*, \quad y_* > 30, \tag{2.122}$$

fitting satisfactorily the experimental measurements.

The boundary layers in turbulent flows near semi-infinite solid interface are shown in [10].

3 Mass and Heat Transfer Processes

The sets of equations derived in (1.2) are valid for pure gases (liquids) or homogeneous gas (liquid) mixtures. If the composition of the mixture is variable, the equations change their form. We will consider the cases when the concentration of the one of components is variable.

3.1 Basic Equations

Let us consider the mass transport of component with a local concentration c with a nonuniform distribution over the flowing fluid [2]. The common definition of the *concentration* is the mass of the component per *unit mass of the fluid*. Concerning an elementary control volume δv, we have

$$\delta m_1 = \rho c \delta v. \tag{3.1}$$

The variation of the concentration of the component concerned may be a result of both mass transfer mechanisms defined later.

The physical model described above can be expressed directly from the continuity equation (2.11). The simple substitution of the mass of the elementary volume ρ in (2.11) by the mass (ρc) of the component *with a nonuniform concentration field* leads to

$$\frac{\partial}{\partial t}(\rho c) + div(\rho c\, \boldsymbol{u}) = 0. \tag{3.2}$$

Let us consider the surface δs which encloses the volume δv. The mass transfer across this surface will be a result of the molecular diffusion (chaotic motion of the nonuniform distributed molecules) and convection (arranged motion of these molecules with the flow).

As a result of the concentration distribution nonuniformity, the density of the diffusion mass flux across surface δs is

$$\boldsymbol{q}_c = -Dgrad(\rho c) = -D\nabla(\rho c), \tag{3.3}$$

where the linearity of (3.3) follows from Onsanger's linear principle [1] for small $\nabla(\rho c)$ and D is the *diffusivity* (molecular diffusion coefficient). Equation (3.3) is Fick's first law, and for small concentrations $D = \text{const}$.

The diffusion mass flux (3.3) takes part in the mass balance in the elementary volume δv as $div\boldsymbol{q}_c$:

$$\frac{\partial}{\partial t}(\rho c) + div(\rho c\, \boldsymbol{u}) + div\, \boldsymbol{q}_c = 0. \tag{3.4}$$

The combination of (3.3) and (3.4) gives the *convection–diffusion equation*:

$$\frac{\partial}{\partial t}(\rho c) + div(\rho c\, \boldsymbol{u}) = div[D\nabla(\rho c)]. \tag{3.5}$$

Assuming D and ρ are *constants* and the *Cartesian* coordinate system as a *default*, we have

$$\frac{\partial c}{\partial t} + u_x \frac{\partial c}{\partial x} + u_y \frac{\partial c}{\partial y} + u_z \frac{\partial c}{\partial z} = D \left[\frac{\partial^2 c}{\partial x^2} + \frac{\partial^2 c}{\partial y^2} + \frac{\partial^2 c}{\partial z^2} \right] \tag{3.6}$$

or

$$\frac{\partial c}{\partial t} + (\boldsymbol{u}\nabla)c = D\nabla^2 c. \tag{3.7}$$

Here the *Laplacian* follows directly from the *gradient* [11], since

$$\nabla^2 = div(\nabla). \tag{3.8}$$

If the medium does not flow (i.e., a stagnant fluid), (3.7) reduces to the molecular diffusion equation (Fick's second law):

$$\frac{\partial c}{\partial t} = D\nabla^2 c. \tag{3.9}$$

A similar result could be obtained for the heat transfer if in the mass conservation law (3.5) we replace the mass (ρc) with the heat, i.e., internal energy ($\rho\varepsilon$) of the mass, where

$$\varepsilon = c_p T + \varepsilon_0. \tag{3.10}$$

In (3.10) c_p is the specific heat of the substance at pressure p, T is the absolute temperature, and ε_0 is a nonessential constant.

By analogy to the mass transfer, the heat transfer is a result of conduction and convection. In chemical engineering processes, radiation heat transfer effects are not so big because these effects are the result of high temperature.

In many cases the relation between the conductive (molecular) heat flux $\boldsymbol{q}_T$ and the temperature gradient is linear (the first Fourier law):

$$\boldsymbol{q}_T = -\lambda \boldsymbol{grad}\, T = -\lambda\nabla T, \tag{3.11}$$

where λ is thermal conductivity.

The convective heat transfer in moving fluids with variable temperature can be obtained from (3.2) by replacing the mass concentration with the internal energy concentration:

$$\frac{\partial}{\partial t}(\rho\varepsilon) + div(\rho\varepsilon\boldsymbol{u}) = 0. \tag{3.12}$$

The conductive heat transfer may supplement the convective one and the convection–conduction equation has the form:

$$\frac{\partial}{\partial t}(\rho\varepsilon) + div(\rho\varepsilon\boldsymbol{u}) + div\boldsymbol{q}_T = 0. \tag{3.13}$$

If ρ, c_p, and λ are constants, from (3.11) and (3.12) it follows that

$$\frac{\partial T}{\partial t} + div(\boldsymbol{u}T) = a\nabla^2 T, \quad a = \frac{\lambda}{\rho c_p}, \tag{3.14}$$

where a is the thermal diffusivity. In the Cartesian coordinate system the convection–conduction equation (3.14) has the form

$$\frac{\partial T}{\partial t} + u_x\frac{\partial T}{\partial x} + u_y\frac{\partial T}{\partial y} + u_z\frac{\partial T}{\partial z} = a\left(\frac{\partial^2 T}{\partial x^2} + \frac{\partial^2 T}{\partial y^2} + \frac{\partial^2 T}{\partial z^2}\right) \tag{3.15}$$

or

$$\frac{\partial T}{\partial t} + (\boldsymbol{u}\nabla)T = a\nabla^2 T. \tag{3.16}$$

For an immobile medium the convective heat transfer is equal to zero and from (3.16) the Fourier's second law follows:

$$\frac{\partial T}{\partial t} = a\nabla^2 T, \tag{3.17}$$

i.e., the nonstationary conduction equation.

3.2 Boundary Conditions

The boundary conditions leading to uniqueness of the solutions of the differential equations depend on the physical conditions imposed at the boundaries of the systems modeled.

The uniqueness conditions of the convection–diffusion equation (3.6) can be of the first, second, or third kind.

Very often at the phase boundary there exists a thermodynamic equilibrium and the equilibrium of the concentration is known beforehand (e.g., c = const.). In these cases the boundary conditions are of the first kind and contain the concentration only (Dirichlet condition).

If the interphase mass transfer through the interphase surface occurs, the boundary conditions are of the second kind and contain the concentration derivatives only (Neumann conditions). In these cases the boundary condition specifies the derivative of the concentration along the normal vector to the interface. If the mass flux though the interface is absent, from Fick's first law it follows that the derivative is equal to zero.

In the cases of chemical reaction at the interphase surface, the boundary conditions are of the third kind and contain the concentration and the concentration derivative.

The boundary conditions of the convection–conduction equation are usually of the first kind (the temperature at the boundaries is specified) or of the second kind (the heat flux at the boundaries is specified).

3.3 Transfer Processes Rate

The diffusion equation has closed-form analytical solutions in different cases. For example, if we consider one-dimensional liquid vapor diffusion through a layer of inert gas (consider a long vertical tube with a liquid at the bottom and a gas layer above, very often known as a *Stephan tube*), the governing equation (see 3.9) is

$$\frac{\partial c}{\partial t} = D\frac{\partial^2 c}{\partial x^2}. \tag{3.18}$$

The *boundary conditions* consider a *constant* (equilibrium) concentration c^* at the liquid surface ($x = 0$) and a null concentration far from the interphase surface ($x \to \infty$):

$$x = 0, \quad c = c^*; \quad x \to \infty \quad c = 0. \tag{3.19}$$

The initial condition defining the onset of the process is $t = 0$, $c = 0$.

The solution of (3.18) with the boundary and initial conditions defined above is well known [10]:

$$c = c^*\left(1 - \frac{2}{\sqrt{\pi}}\int_0^{\frac{x}{2\sqrt{Dt}}} e^{-n^2}dn\right) = c^* erfc\frac{x}{2\sqrt{Dt}}. \tag{3.20}$$

Solution (3.20) defines the *diffusion rate* (mass flux per unit area and per unit time) as

$$q_c = -D\left(\frac{\partial c}{\partial x}\right)_{x=0} = c^*\sqrt{\frac{D}{\pi t}}. \tag{3.21}$$

On the other hand, the *total quantity* of the liquid evaporated over time t_0 is

$$Q_c = \int_0^{t_0} q_c dt = 2c^*\sqrt{\frac{t_0 D}{\pi}}. \tag{3.22}$$

The modeling of the basic diffusion processes is related mainly to the determination of the diffusion coefficient. This needs experimental data concerning Q_c for a period of time defined by t_0. Thus, the solution of (3.22) with respect to D gives

$$D = \frac{Q_c^2\pi}{4(c^*)^2 t_0}. \tag{3.23}$$

The power law with respect to Q_c requires high precision of the experimental data (owing to the exponent 2 of Q_c).

The convection–conduction equation can be solved for different cases. For example, the heat transfer in a long rod with the ends at fixed temperatures follows from (3.17):

$$\frac{\partial T}{\partial \tau} = a\frac{\partial^2 T}{\partial x^2}, \tag{3.24}$$

with the initial and boundary conditions

$$t = 0, T = T_1; \quad x = 0, T = T_1; \quad x \to \infty, T = T_\infty. \tag{3.25}$$

In (3.24) and (3.25) it is supposed that the surface of the rod is heat-isolated. The temperature distribution in the rod is obtained by the analogy with (3.20):

$$T = T_2 + (T_1 - T_2)\, erfc\frac{x}{2\sqrt{at}}. \tag{3.26}$$

From (3.26) we can obtain the heat transfer rate,

$$q_T = -\lambda\left(\frac{\partial T}{\partial x}\right)_{x=0} = \lambda\frac{T_2 - T_1}{\sqrt{\pi a\, t}}, \tag{3.27}$$

and the heat quantity for the time t_0,

$$Q_T = \int_0^{t_0} q_T dt = 2\lambda(T_2 - T_1)\sqrt{\frac{t_0}{\pi a}}. \tag{3.28}$$

The modeling of the heat transfer processes is related to the identification of the conductivity (obtaining ρ and c_p is very simple). From (3.28) it follows that

$$\lambda = \frac{\pi\, Q_T^2}{4\,(T_2 - T_1)\rho\, c_p t_0}, \tag{3.29}$$

where Q_T must be measured very precisely.

3.4 Diffusion Boundary Layer

In many interphase processes the mass transfer is realized through thin layers near the interphase surface.

In the case of a semi-infinite plate streamlined by a fluid flow, the *convection–diffusion equation* takes the form

$$u\frac{\partial c}{\partial x} + v\frac{\partial c}{\partial y} = D\left(\frac{\partial^2 c}{\partial x^2} + \frac{\partial^2 c}{\partial y^2}\right) \tag{3.30}$$

In the case of crystallization, the boundary conditions assume a constancy of the concentration over the volume of the phase ($c = c_0$) and equilibrium concentration ($c = c^*$) at the solid surface ($y = 0$):

$$x = 0, \quad c = c_0; \quad y = 0, \quad c = c^*, \quad y \to \infty, \quad c = c_0. \tag{3.31}$$

The velocity components in (3.30) satisfy equations (2.31).

For generalized analysis (see Sect. 4.1) we will be use natural scales of the mass transfer process:

$$x = lX, \quad y = \delta_{\text{D}} Y, \quad c = c_0 + (c^* - c_0)C,$$

$$u = u_\infty U, \quad v = u_\infty \frac{\delta_D}{l} V, \tag{3.32}$$

where δ_{D} is the order of magnitude of the *diffusion boundary layer thickness*.

This transforms (3.30) into

$$U\frac{\partial C}{\partial X} + V\frac{\partial C}{\partial X} = \frac{Dl}{u_\infty \delta_{\text{D}}^2}\left(\frac{\delta_{\text{D}}^2}{l^2}\frac{\partial^2 C}{\partial X^2} + \frac{\partial^2 C}{\partial Y^2}\right). \tag{3.33}$$

Equation (3.33) indicates that the convective transfer (the left-hand side) is balanced by diffusive transfer (the right-hand side) if the order of the dimensionless parameter $\left(\frac{Dl}{u_\infty \delta_{\text{D}}^2}\right)$ is 1:

$$\frac{Dl}{u_\infty \delta_{\text{D}}^2} = 1, \quad \delta_{\text{D}} = \sqrt{\frac{Dl}{u_\infty}} = \frac{l}{\sqrt{Pe}}, \quad Pe = \frac{u_\infty l}{D}, \quad \frac{\delta_{\text{D}}}{l} = \frac{1}{\sqrt{Pe}}, \tag{3.34}$$

where Pe is the Péclet number.

For great values of the Péclet number ($Pe > 10^2$) we can use the approximation of the diffusion boundary layer ($Pe^{-1} = 0$), i.e.,

$$U\frac{\partial C}{\partial X} + V\frac{\partial C}{\partial Y} = \frac{\partial^2 C}{\partial Y^2};$$

$$X = 0, \quad C = 0; \quad Y = 0, \quad C = 1, \quad Y \to \infty, \quad C = 0 \tag{3.35}$$

is *the diffusion boundary layer equation* [10].

In this approximation $\left[\left(\frac{\delta_{\text{D}}}{l}\right)^2 = 0\right]$ the first equation in (2.39) at $u_\infty = \text{const.}$ has the form

$$U\frac{\partial U}{\partial X} + V\frac{\partial U}{\partial Y} = Sc\frac{\partial^2 U}{\partial Y^2}, \tag{3.36}$$

where $Sc = \frac{\nu}{D}$ is the Schmidt number.

3.5 Turbulent Diffusion

The mass transfer mechanism in turbulent flow is very different in the fluid volume and near the solid interfaces. That is why these two cases will be analyzed separately [2, 10].

The turbulent fluid flow in the phase (gas or liquid) volume can be represented as a combination of a stationary flow (large-scale turbulent pulsations averaged in time) and a nonstationary flow (chaotic motion of the small-scale turbulent pulsations). As a result, only the mass transfer is convective, i.e., the diffusive (by a molecular mechanism) transfer is negligible.

The mass transfer in turbulent flow is a combination of two components. The first is a convective mass transfer as a result of the stationary flow:

$$J = \rho c\, \boldsymbol{u}, \tag{3.37}$$

where $\boldsymbol{u}$ is the velocity of the large-scale turbulent pulsations.

The second is the convective mass transfer as a result of the chaotic motion of the small-scale turbulent pulsations and is similar to the molecular diffusion (caused by the chaotic motion of the molecules). It is named *turbulent diffusion* and Fick's first law can be used for the mathematical description:

$$\boldsymbol{j}_{\text{turb}} = -D_{\text{turb}}\mathbf{grad}\rho c, \tag{3.38}$$

where D_{turb} is the *turbulent diffusion coefficient.* If we introduce (3.37) and (3.38) into (3.5), the equation of turbulent convection–diffusion has the form

$$\frac{\partial}{\partial t}(\rho c) + \text{div}(\rho c\mathbf{u}) = \text{div}[D_{\text{turb}}\mathbf{grad}(\rho c)]. \tag{3.39}$$

Near the solid interface the velocity decreases to zero and the scale of the turbulent pulsations decreases too. This damping of the turbulent pulsations near solid interface influences the turbulent diffusivity, i.e., D_{turb} depends on the distance to the solid interface.

The turbulent diffusivity is related to the characteristic quantities of the turbulent flow (ρ, l, ΔU). The unique combination between these quantities with dimensions of meters squared per second is

$$D_{\text{turb}} \approx \Delta U l, \tag{3.40}$$

where l is the length of the free path and ΔU is the average velocity of the molecules This result is similar to the expression in the kinetic theory of gases,.

From (3.40) it follows directly that

$$\Delta U \approx \frac{\partial U}{\partial l} l, \quad D_{\text{turb}} \approx l^2 \frac{\partial U}{\partial l}. \tag{3.41}$$

From (2.103) and (3.40) it follows that

$$\nu_{\text{turb}} \approx \Delta U l, \tag{3.42}$$

i.e., ν_{turb} and D_{turb} have the same order of magnitude (this is from analogy with the diffusivity and viscosity in the kinetic theory of an ideal gas).

D_{turb} is proportional to the macroscopic scale of the turbulent pulsations and in developed turbulent flow is many times greater than the molecular diffusivity. As a result, the concentration in the volume is constant. This rule is breached in the boundary layer near the solid interface.

If we introduce (2.95) into (3.41), the expression for the turbulent viscosity has the form

$$D_{turb} \approx y^2 \frac{\partial U}{\partial y}, \tag{3.43}$$

i.e., the dependence of D_{turb} on y is determined by the dependence of U on y in the turbulent boundary layer (2.101).

For mass flux (turbulent mass transfer in the concentration boundary layer), we obtain

$$j_{turb} = D_{turb} \frac{\partial c}{\partial y} = \beta_0 y^2 \frac{\partial U}{\partial y} \frac{\partial c}{\partial y}, \quad \beta_0 = \text{const.} \tag{3.44}$$

Let us consider a flat solid interface $y = 0$. From (2.101), we can obtain U and substitute it into (3.44). As a result,

$$j_{turb} = \beta \nu_0 y \frac{\partial c}{\partial y}, \quad \beta = \frac{\beta_0}{\sqrt{\alpha}}. \tag{3.45}$$

In (3.45) $j_{turb} = 0$ (according to the mass conservation law) and integration permits us to obtain the concentration distribution in the turbulent boundary layer:

$$c_{11} = \frac{j_{turb}}{\beta \nu_0} \ln y + a_1. \tag{3.46}$$

The concentration distribution (3.46) is valid in the layer $\beta_0 < y < d$, where the viscous sublayer thickness is β_0 and d is the turbulent boundary layer thickness [10]:

$$d \approx \frac{\nu_0 x}{U_0}, \tag{3.47}$$

where U_0 is the velocity of the boundary layer border.

Outside the turbulent boundary layer ($y \geq d$), i.e., in the nucleus of the turbulent flow, the velocity U_0 and the concentration c_0 do not depend on the distance to the solid interface. Using the boundary condition $y = d$, $c_{11} = c_0$, and (3.46) leads to

$$c_{11} = \frac{j_{turb}}{\beta \nu_0} \ln \frac{y}{d} + c_0. \tag{3.48}$$

3.6 Turbulent Mass Transfer

The turbulent mass transfer in the phase volume is described by (3.39) if the velocity is variable in the space, i.e., the difference from laminar flow is only the diffusivities ($D_{\text{turb}} \gg D$).

Of special interest is the mass transfer near the solid interface, where the mass transfer rate is limited by the mass transfer in the diffusion sublayer [10]. This rate depends on the law (mechanism) of the turbulent pulsations fading in the viscous sublayer. There are different hypotheses for this mechanism:

- Full fading in the viscous sublayer (Prandtl–Taylor [10, 14–16])
- Gradual fading near the solid interface (Landau–Levich [2, 10])

According to the *Prandtl–Taylor hypothesis* the turbulent pulsations fade fully at $y < \delta_0$ and the mass transfer is a result of the molecular diffusion, i.e., its rate is the diffusion flux:

$$j = D\frac{\partial c}{\partial y}, \tag{3.49}$$

where $j =$ const. and the maximum concentration value $c = c_0$is in the main turbulent flow $y \geq d$. Having in mind the boundary condition at the solid interface $y = 0$, $c = 0$, we obtain from (3.49)

$$c_{111} = \frac{j}{D}y. \tag{3.50}$$

At the boundary $y = \delta_0$, $c_{11} = c_{111}$, and from (2.100), (3.48), and (3.50) we obtain

$$j = \frac{Dc_0}{\delta_0\left(1 - \frac{1}{a\beta Pr}\ln\frac{\delta_0}{d}\right)}, \qquad Pr = \frac{\nu}{D}. \tag{3.51}$$

For liquids Pr $\approx 10^3$, i.e.,

$$j = \frac{Dc_0}{\delta_0} \approx \frac{c_0 v_0}{Pr}. \tag{3.52}$$

The main results of the Prandtl–Taylor hypothesis are:

- $y > d$, $c = c_1 = c_0 =$ const., developed turbulence area
- $\delta_0 < y < d$, $c = c_{11}$, see (3.48), turbulent boundary layer
- $0 < y < \delta_0$, $c = c_{111}$, see (3.50)–(3.52), viscous sublayer

A change to this hypothesis was proposed by Karman [16], especially the existence of a buffer sublayer between the turbulent and viscous sublayers. For Pr $\gg 1$ the results coincide with Prandtl's result.

The *Landau–Levich hypothesis* for gradual fading of the turbulent pulsations near the solid interface leads to a complicated structure of the concentration boundary layer, where the turbulent boundary layer is divided into two sublayers—viscous ($\delta < y < \delta_0$) and diffusion ($0 < y < \delta$).

In the viscous sublayer ($\delta < y < \delta_0$) the mass transfer is a result of the turbulent diffusion:

$$j = D_{\text{turb}} \frac{\partial c}{\partial y}. \tag{3.53}$$

D_{turb} can be obtained from (3.41), and v_0 is obtained from (2.100):

$$D_{\text{turb}} \approx \frac{v_0 y^4}{\delta_0^3}. \tag{3.54}$$

If we introduce (3.54) into (3.53), we obtain the following expressions for the mass transfer rate and the concentration distribution in the viscous sublayer:

$$j = \gamma v_0 \frac{y^4 \partial c}{\delta^3 \partial y}, \quad \gamma = \text{const.}, \quad c_{111} = \frac{j\delta_0^3}{3vv_0 y^3}\frac{1}{} + a_2, \quad a_2 = \text{const.} \tag{3.55}$$

From (3.54) is seen that D_{turb} decreases very fast in the viscous boundary layer and at $y = \delta$ is equal to the molecular diffusion coefficient:

$$D_{\text{turb}}(\delta) = \frac{\gamma v_0 \delta^4}{\delta_0^3} = D, \tag{3.56}$$

where δ is the diffusion sublayer thickness,

$$\delta = \sqrt{4}\frac{D\delta_0^3}{\gamma v_0}. \tag{3.57}$$

The expressions for the mass transfer rate and the concentration distribution are obtained immediately:

$$j = D\frac{\partial c}{\partial y}, \quad c_{1111} = \frac{j}{D} y. \tag{3.58}$$

The elimination of the concentrations in (3.55) and (3.58), i.e., $c_{111} = c_{1111}$ for $y = \delta$, permits us to obtain the constant a, the mass transfer rate, and the diffusion sublayer thickness:

$$a_2 = \frac{j\delta_0^3}{3vv_0\delta^3}\frac{1}{} + \frac{j\delta}{D}, \quad j = \frac{Dc_0}{\frac{D}{\beta v_0}\ln\frac{\delta_0}{d} + \frac{\delta_0^3 D}{3\gamma v_0}\left(\frac{1}{\delta^3} - \frac{1}{\delta_0^3}\right) + \delta}, \quad \delta = \frac{\delta_0}{\sqrt{4}10\gamma Pr}. \tag{3.59}$$

The main results of the Landau–Levich hypothesis are:

- $y > d$, $c = c_1 = c_0 =$ const., developed turbulence area
- $\delta_0 < y < d$, $c = c_{11}$, see (3.48), turbulent boundary layer

- $\delta < y < \delta_0$, $c = c_{111}$, see (3.55), viscous sublayer
- $0 < y < \delta$, $c = c_{1111}$, see (3.58) and (3.59), diffusion sublayer

The results presented show that the fundamental problem of turbulent mass transfer is the law of the turbulent pulsations fading in the viscous sublayer. A solution of this problem is possible on the basis of experimental data of heat or mass transfer in turbulent conditions. Different theoretical results in this field are shown in [10].

4 Chemical Processes and Adsorption

Chemical processes and adsorption as hydrodynamic, diffusion, and heat processes are the most important simple processes in process system engineering and chemical technology. In practice, these simple processes are mutually connected and depend on each other, but here they will be represented individually, i.e., in conditions of constant temperature, a static medium, and in the absence of concentration gradients. In this way, the basis of these processes becomes homogeneous and heterogeneous chemical reactions.

4.1 Stoichiometry

The mathematical description of a chemical reaction is a stoichiometric equation:

$$\sum_{i=1}^{n} a_i A_i = 0, \tag{4.1}$$

which expresses that $a_1,\ldots,a_{n_0}$ molecules of substances $A_1,\ldots,A_{n_0}$ ($a_i > 0$, $i = 1,\ldots, n_0$) react chemically and as a result $a_{n_0+1},\ldots,a_n$ molecules of substances $A_{n_0+1},\ldots,A_n$ ($a_i < 0$, $i = n_0 + 1,\ldots, n$) are obtained. In (4.1) $a_i (i = 1,\ldots, n)$ are the stoichiometric coefficients and the processes can be convertible or inconvertible. Chemical reactions are simple if they are realized in one stage and $n_0 \leq 2$. Complex chemical reactions are realized in several stages, in which intermediate substances take part. They consist of simultaneous and (or) consecutive realization of some simple chemical reactions. Their stoichiometric equations are obtained through summation of the stoichiometric equations of simple chemical reactions.

4.2 Mechanism and Reaction Route

The combination of stages (simple reactions) in a chemical reaction represents its *mechanism*. The stoichiometric equations of different stages can be multiplied by proper numbers, so after the summation, the intermediate substances in the equation of a complex chemical reaction disappear. This set of stoichiometric numbers determines the *route of the chemical reaction* [13]. It represents an order of numbers, arranged in sequence, i.e., it has vector properties. If the mechanism is as given, it could be realized by more than one linear independent route. Such vectors cannot be represented by a linear combination of the other ones. All of the linear independent routes form the basis of the routes from which an arbitrary number of linear dependent routes can be realized [13].

The considerations mentioned above can be shown with the complex reaction of vinyl chloride production. It is carried out in four stages, according to one probable mechanism:

1.	$HgCl_2{\cdot}HCl + C_2H_2 \Leftrightarrow HgCl_2{\cdot}C_2H_2{\cdot}HCl$	1	0	1
2.	$HgCl_2{\cdot}HCl + HCl \Leftrightarrow HgCl_2{\cdot}2HCl$	0	1	1
3	$HgCl_2{\cdot}C_2H_2{\cdot}HCl + HCl \Rightarrow C_2H_3Cl + HgCl_2{\cdot}HCl$	1	0	1
4.	$HgCl_2{\cdot}2HCl + C_2H_2 \Rightarrow C_2H_3Cl + HgCl_2{\cdot}HCl$	0	1	1
	$C_2H_2 + HCl = C_2H_3Cl$			

The vectors of different routes are represented on the right side. The first and second vectors are linearly independent and form the basis of the routes. The third vector is linearly dependent and is determined by the summation of the first and second vectors.

4.3 Kinetics of Simple Chemical Reactions

The main problem in modeling chemical processes is the determination of the kinetic model for simple chemical processes, which is further used in the model description of complex reactions and processes. Since hydrodynamic, diffusion, and heat processes are associated with chemical reactions, it can be done in two different ways:

1. The rate of the complex process determines the rate of the chemical process, accounting for the hydrodynamics and mass transfer effects.
2. The rate of mass transfer, accounting for chemical reaction effects, is considered.

In all cases the *chemical reaction kinetics model* has to be determined.

Chemical reactions can be realized in the phase volume or at the boundary of two phases. In the first case, they are called "homogeneous" and their rate is determined by the amount of substance reacted in a unit volume for a unit time. In the second case, they are heterogeneous ones and the rate is determined by the amount of substance reacted on a unit surface for a unit time. This difference is why they are considered individually.

The equation for the homogeneous *chemical reaction rate* represents the law of mass action. In the case of the simplest chemical reaction,

$$A_1 + A_2 = A_3, \tag{4.2}$$

it is as follows:

$$r = kc_1c_2, \tag{4.3}$$

where k is a rate constant and c_1 and c_2 are volume concentrations of substances A_1 and A_2. In (4.3) it is assumed that the number of reacted chemical molecules of substances A_1 and A_2 is proportional to the number of collisions between them, i.e., of the probability that a molecule of A_1 and a molecule of A_2 will be together at one point. The coefficient of proportionality is

$$k = k_0 \exp\left(-\frac{E}{RT}\right). \tag{4.4}$$

This equation expresses the condition that a chemical linkage is formed between two molecules (as a result of the collision between them) if the collision energy is not smaller than the activation energy E (T is absolute temperature, R is the gas constant). From (4.4) it can be seen that not each collision with sufficient energy is an effective one when $k_0 < 1$.This can be observed in the case of big molecules, where it is necessary for the active centers of the molecules to meet and not just the molecules themselves. In this case, k_0 represents a spherical factor and it gives the probability of collision between active centers of the molecules. The probability defines the relationship of the surface of the active center over all the surface of the molecule.

The chemical reaction shown in (4.2) is bimolecular. The monomolecular reaction

$$A_1 \rightarrow A_2 \tag{4.5}$$

has an analogous kinetic equation:

$$r = kc_1, \tag{4.6}$$

where the probability of transformation of substance A_1 into substance A_2 is proportional to its concentration.

The probability of three of molecules being located at one point is negligible. For this reason, monomolecular and bimolecular reactions are considered as simple and complex ones (4.1), where $n_0 > 2$ represents simultaneous and (or) consecutive realization of some monomolecular and bimolecular reactions. The rate of homogeneous reactions expresses the change of volume concentration over time. Taking into consideration that the concentrations of substances A_1 and A_2 decrease with the time as a result of the chemical reaction for reaction (4.2), we can write

$$r_1 = -\frac{dc_1}{dt}, \quad r_2 = -\frac{dc_2}{dt}, \quad r_3 = \frac{dc_3}{dt}, \quad r_1 = r_2 = r_3 = r = kc_1c_2. \tag{4.7}$$

Analogously, we obtain for (4.5)

$$r_1 = -\frac{dc_1}{dt}, \quad r_2 = \frac{dc_2}{dt}, \quad r_1 = r_2 = r = kc_1. \tag{4.8}$$

If we add the required initial conditions, the results obtained allow formulation of the mathematical description of the kinetics of simple monomolecular and bimolecular reactions:

$$\frac{dc_1}{dt} = -kc_1, \quad c_2 = c_{10} - c_1; \quad t = 0, \quad c_1 = c_{10}. \tag{4.9}$$

$$\frac{dc_1}{dt} = -kc_1c_2, \quad c_2 = c_{20} - c_{10} + c_1, \quad c_3 = c_{10} - c_1; \quad t = 0, \quad c_1 = c_{10}, \tag{4.10}$$

where the concentrations are expressed as moles for a unit volume, c_{10} and c_{20} are initial concentrations, and k is as in (4.4).

The equations above solve as

$$c_1 = c_{10}\exp(-kt), \quad c_2 = c_{10} - c_{10}\exp(-kt). \tag{4.11}$$

$$\begin{aligned} c_1 &= \frac{\Delta c \frac{c_{10}}{c_{20}} \exp(-\Delta ckt)}{1 - \frac{c_{10}}{c_{20}} \exp(-\Delta ckt)}, \quad \Delta c = c_{20} - c_{10}, \quad c_2 = \frac{\Delta c}{1 - \frac{c_{10}}{c_{20}} \exp(-\Delta ckt)}, \\ c_3 &= c_{10} \frac{1 - \exp(-\Delta ckt)}{1 - \frac{c_{10}}{c_{20}} \exp(-\Delta ckt)}. \end{aligned} \tag{4.12}$$

Equations (4.1.4), (4.1.9), and (4.1.10) show that mathematical models of simple reactions contain two parameters—activation energy (E) and a prefactor (k_0), which have to be determined from experimental data. The experimental dependence of the concentration on temperature and time or the relationship between the reaction rate and concentrations of the reagents is used.

4.4 Kinetics of Complex Reactions

Complex chemical reactions are realized in many stages, and in each of them a simple reaction takes place. The combination of stages represents the mechanism of the chemical reaction. The stoichiometric equations of the different stages of chemical reactions are summed. As a result the overall stoichiometric equation is obtained. If these equations are multiplied by given stoichiometric numbers, the overall equations in different routes are obtained. Each route of one complex reaction has an individual set of stoichiometric equations for the different stages and leads to a different set of kinetic equations.

The real rate of reaction determines the reaction rate on the basis of the routes, using the rule of summation of vectors. The mathematical descriptions using complex reactions as a combination of the mathematical description of the simple reactions results in multiparametric models, since each simple reaction has own activation energy and prefactor. In such a way, the probable mechanism of the reaction can be determined. This description is useful when the process rate is limited by the chemical kinetics and all additional processes lead to unessential additional effects.

Simulation of the stoichiometric equations at the different stages, accounting for different possible routes, results in the overall stoichiometric equation looks like (4.1). In this case, the chemical reaction rate can be used in the following equation:

$$r_i = \frac{dc_i}{dt} = a_i k \prod_{j=1}^{n_0} c_j^{\alpha_j}, \quad i = 1, \ldots, n_0, \tag{4.13}$$

where parameters α_j ($j = 1, \ldots, n_0$) are *reaction orders* in respect of different reagents. Using them, we can obtain the overall order of the reaction:

$$\alpha = \sum_{j=1}^{n_0} \alpha_j. \tag{4.14}$$

For some comparatively simple reactions the order could be 1 or 2, showing that the reaction is monomolecular or bimolecular, but in more cases this is not true and the order is a fraction. In the case of complex reactions with a kinetics equation of type (4.13), it is necessary to determine the activation energy, prefactor, and the reactions' orders as well. The logarithm of (4.13) leads to linear equations concerning parameters in the mathematical description of chemical kinetics.

4.5 Adsorption Processes

Different distances between the molecules in different phases lead to different strengths of the interactions between them. Because of this, there are layers with no equilibration strength interactions on the phase boundaries between gas, liquid, and the solid surface. Equilibration can be attained by changing the concentrations of the substances on the interphase surface. This results in solid and liquid surfaces whose molecules combine physically or chemically by physical (van der Waals's) or chemical (valence) forces with molecules of the gas or liquid phase of the volume contacting with them. This process is called "adsorption." It is a physical or a chemical process and depends on the strengths of the interactions that combine adsorbed molecules on the surface. The rate of the adsorption process determines analogously the rate of the chemical reactions, as law of mass action is changed by the law of surface action.

4.6 Physical Adsorption

In the case of physical bonds on the interphase surface, the adsorption rate is proportional to the free surface and the volume concentration of the adsorbed substance:

$$r_1 = k_1 c\left(1 - \frac{\Gamma}{\Gamma_\infty}\right), \tag{4.15}$$

where k_1 is the rate constant, c is the volume concentration of the adsorbed substance, and Γ and Γ_∞ are its surface concentrations. In this case the process is convertible and the desorption rate can be obtained by an analogous consideration:

$$r_2 = k_2 \Gamma, \tag{4.16}$$

where k_2 is the rate constant, which depends on temperature by analogy with (4.4). The process rate is

$$r = r_1 - r_2 = k_1 c\left(1 - \frac{\Gamma}{\Gamma_\infty}\right) - k_2 \Gamma. \tag{4.17}$$

In the case of physical equilibrium $r = 0$, i.e.,

$$\Gamma = \frac{kc}{1 + \frac{kc}{\Gamma_\infty}}, \quad k = \frac{k_1}{k_2}. \tag{4.18}$$

Equation (4.18) is the *Langmuir isotherm* and k is the equilibrium constant.

4.7 Chemical Adsorption

The presence of chemical bonds at the interphase surface leads to the next expression for the *adsorption rate*:

$$r = k_0 \exp(-E/RT) \prod_{i=1}^{m} z_i^{\alpha_i} \prod_{j=1}^{n} p_j^{\beta_j} z_0^{\left(m_t - \sum_{i=1}^{m} \alpha_i\right)}, \tag{4.19}$$

where z_i are the parts of the interphase surface occupied by the molecules of substances A_i (i = 1,..., m), p_j are partial pressures (volume concentrations) of substances B_j (j = 1,..., n) in the gas (liquid), z_0 is the part of the free surface which is able to form physical bonds with the molecules of substances

B_j (j = 1,..., n), and m_t is the number of active places at the interface which form the physical bonds.

In (4.19) it is assumed that the molecules A_i (i = 1,..., m) forming the interphase surface react chemically with some of the molecules B_j (j = 1,..., n), whereas the other molecules form physical bonds with the active places m_t. The heterogeneous reaction rate and the reactions orders are $r, \alpha_i (i = 1, \ldots, m)$ and $\beta_j (j = 1, \ldots, n)$.

The adsorption leads to a decrease of the activation energy in (4.19) and the chemical reaction rate increases (heterogeneous catalytic reaction). Analogous effects are possible in the cases of homogeneous chemical reactions, but they are the result of the substances, which change the reaction route and as a result the general activation energy decreases too.

In the case of reversible heterogeneous chemical reactions, the equation for the adsorption rate follows from (4.18) and (4.19):

$$r = \frac{k \prod_{j=1}^{n} c_j}{\left(1 + \sum_{j=1}^{n} k_{pj} c_j\right)}, \tag{4.20}$$

where k_{pj} are the equilibrium constants of the reagents B_j(j = 1,...,n).

All equations for the adsorption kinetics are based on the ideal adsorption layer model. Practically, most of the adsorption processes and heterogeneous catalytic reactions are related to real adsorption layers, i.e., the catalytic surfaces are non-homogeneous as a result of the changes of the solid-phase structure.

4.8 Heterogeneous Reactions

Heterogeneous reactions can be considered in a broader sense, where “disappearance” (“appearance”) of the substance on the gas–solid, liquid–solid, or gas–liquid phase boundary results not only from a chemical reaction. It could be a result of other processes, since other kinetics laws could hold, which are generalized as heterogeneous kinetics [12].

If one substance transfers on the phase boundary through convection and diffusion, it could disappear on the phase boundary as a result of different processes:

- Chemical reactions, for example, gas absorption, associated with a fast chemical reaction with a component of the liquid phase
- Electrochemical reactions, for example, electrochemical crystallization
- Physical or chemical adsorption, for example, ion exchange in synthetic anionites
- Mass transfer between two phases—in these processes the substance transferred through the phase boundary towards the second phase “disappears” from the first phase

Analogous cases could be considered in the case of the substance "created" on the phase boundary and its transfer into the phase volume by convection and diffusion.

In this case the rate of the processes is determined by heterogeneous reactions kinetics, which depends on the diffusion kinetics (convection–diffusion process) and the kinetics of heterogeneous processes (chemical, electrochemical, adsorption, diffusion). In all of the cases, the rate of the common process is limited by the slow process, i.e., the process with the smallest rate constant. It has not to be forgotten that mass transfer in the phase and reaction on the phase boundary are two consecutive processes and their rates (mol m^{-2} s^{-1}) are equal, but they differ in their rate constants. A number of concrete cases will be considered next.

5 Examples

5.1 Dissolution of a Solid Particle

The problem of solid particle dissolution is very interesting [17], because the process is rather complicated as a result of the particle radius decreasing (dissolution of a solid particle) and secondary flow at the interface (effect of a large concentration gradient).

5.1.1 Particle Radius is Constant ($r_0 = r_0^{(0)} = \text{const.}$)

Let us consider a neutrally buoyant spherical particle of radius r_0 suspended in an unbounded viscous fluid. The particle medium is soluble in the fluid or it contains a soluble admixture. For a short time the particle radius is constant and the concentration distribution near the sphere satisfies the following equations and boundary conditions:

$$\frac{\partial c}{\partial t} = D\left(\frac{\partial^2 c}{\partial r^2} + \frac{2}{r}\frac{\partial c}{\partial r}\right); t = 0, c = c_0; \quad r = r_0^{(0)}, c = c^*; \quad r \to \infty, c = c_0; \quad (c^* > c_0). \tag{5.1}$$

The solution of the problem uses the next dimensionless (generalized) variables:

$$t = t_0 T, \quad r = r^{(0)} + lR, \quad c = c_0 + (c * - c_0)C, \quad l = \sqrt{Dt_0}. \tag{5.2}$$

As a result, we obtain

$$\frac{\partial C}{\partial T} = \frac{\partial^2 C}{\partial R^2} + \frac{2}{\alpha_0 + R}\frac{\partial C}{\partial R}; \quad T = 0, C = 0; \quad R = 0, \quad C = 1; \quad R \to \infty, \quad C = 0; \left(\alpha_0 = \frac{r_0^{(0)}}{l}\right). \tag{5.3}$$

The solution of (5.3) is

$$C = \frac{\alpha_0}{\alpha_0 + R} erfc \frac{R}{2\sqrt{T}}. \tag{5.4}$$

This result permits us to obtain the mass transfer rate J,

$$J = -D\left(\frac{\partial C}{\partial r}\right)_{r=r(t)}, \tag{5.5}$$

and the velocity of the decrease of the radius of the solid particle v,

$$v = -\frac{J}{\rho}, \tag{5.6}$$

where ρ is the density of the solid phase.

The expression for the change of the particle radius follows directly from these results:

$$r_0(t) = r_0^{(0)} + \int_0^t v(t)\, dt \tag{5.7}$$

and its dimensionless form is

$$\begin{aligned} \alpha = \alpha(t) &= \frac{r_0(t)}{l} = \alpha(t_0 T) = A_0(T) \\ &= \alpha_0 + \frac{c^* - c_0}{\rho} \int_0^T \left(\frac{\partial C}{}\right)_{R=0} dT = \alpha_0 - \frac{c^* - c_0}{\rho}\left(\frac{T}{\alpha_0} + \frac{2\sqrt{T}}{\sqrt{\pi}}\right), \end{aligned} \tag{5.8}$$

$$A_0'(T) = \frac{dA_0}{dT} = \frac{c^* - c_0}{\rho}\left(\frac{\partial C}{\partial R}\right)_{R=0} - \frac{c* - c_0}{\rho}\left(\frac{1}{\alpha_0} + \frac{1}{\sqrt{\pi T}}\right).$$

5.1.2 Particle Radius Decreases ($r_0 = r_0(t)$)

In the case when the dissolution time is not very short, the decrease of the particle radius and the dimensionless variables are

$$t = t_0 T, \quad r = r_0(t) + lR, \quad c = c_0 + (c * - c_0)C \tag{5.9}$$

and the problem has the form

$$\frac{\partial C}{\partial T} = \frac{\partial^2 C}{\partial R^2} + \left(A' + \frac{2}{A+R}\right)\frac{\partial C}{\partial R}; \quad T = 0, \quad C = 0; \quad R = 0, \quad C = 1; \quad R \to \infty, \quad C = 0. \tag{5.10}$$

The solution of (5.10) uses an iterative approach and solves the following problem at every step:

$$\frac{\partial C_i}{\partial T}=\frac{\partial^2 C_i}{\partial R^2}+\left(A'_{i-1}+\frac{2}{A_{i-1}+R}\right)\frac{\partial C_i}{\partial R};\quad T=0,\quad C_i=0;\quad R=0,\quad C_i=1;$$
$$R\to\infty, C_i=0, \tag{5.11}$$

where for A_0 and A'_0 we must use (5.8).

The iteration stop criterion is $\varepsilon < 10^{-3}$, where

$$\varepsilon=\frac{\int\limits_0^1\left[\left(\frac{\partial C_i}{\partial R}\right)_{R=0}-\left(\frac{\partial C_{i-1}}{\partial R}\right)_{R=0}\right]^2 dT}{\int\limits_0^1\left[\left(\frac{\partial C_i}{\partial R}\right)_{R=0}\right]^2 dT}. \tag{5.12}$$

The solution of (5.10) is shown in Fig. 2.

5.2 Contemporary Approach of Turbulence Modeling

The development of the turbulence theory is difficult owing to the complexity of the phenomena and the use of semiempirical theory methods. The increase in the performance of the central processing units of computers permits direct numerical modeling, i.e., the numerical integration of the nonstationary Navier–Stokes equations without using additional empirical models and constants. The three-dimensional turbulent flow parameters obtained permit us to calculate all average flow characteristics.

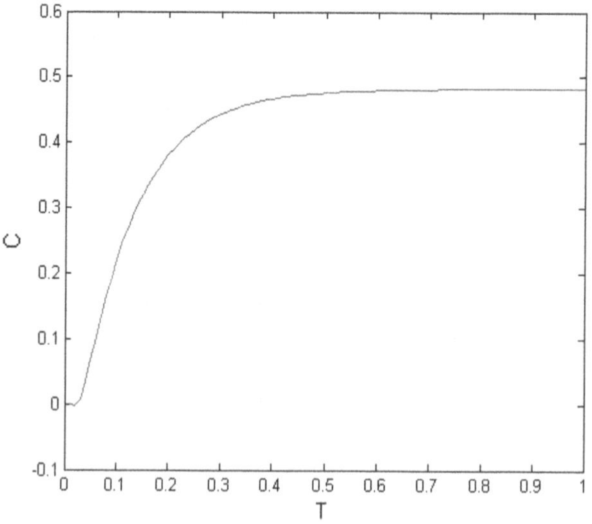

Fig. 2 Concentration-time distribution for $R = 0.5$, $\alpha = 30$, $(c^* - c_0)/\rho = 5 \times 10^{-2}$

The direct numerical simulation of a three-dimensional turbulent flow is a numerical experiment based on an effective model, an exact method for the solution of differential equations, and a multiprocessor computing system.

The theoretical analysis of the turbulent flow as a stochastic process is not a very good approach because it is a pseudo-chance phenomenon. The transition from laminar to turbulent flow is accompanied by a loss of stability, but the deviation of the flow parameters is not so large as a result of the viscous forces. This three-dimensional process is nonstationary but the average values of the flow parameters approach those of a stationary process.

The direct solution of the full turbulent flow problem needs a multiprocessor computing system with an effective performance 10^{12}–10^{14} floating point operations per second. At the present time this is not possible, but obtaining solutions with an error of 10^{-2}–10^{-4}% is possible.

The direct numerical simulation method permits us to obtain different interesting results, such as the modeling of the homogeneous isotropic turbulence at moderate values of the Reynolds numbers and modeling of the inertia subdomain of the spectrum of two- and three-dimensional turbulence. As the result it was obtained that in the case of two-dimensional turbulence the energy transfer is from a small-scale to a large-scale vortex (pulsations), whereas in the case of three-dimensional turbulence the energy transfer is in the opposite direction (from the large-scale to the small-scale vortexes).

The limitations of the direct numerical simulation method is the reason for the development of modeling of large-scale vortexes—large eddy simulation. This approach is used for turbulent flow modeling in channels [19] and at the atmospheric boundary layer [20].

The basis of large eddy simulation is the hypothesis concerning the independence of the large-scale vortexes from the molecular viscosity. As the result, it can create a numerical model of the large-scale turbulent pulsations without calculation of the small vortexes (the calculations are independent of the Reynolds number). The method is used for solution of different interesting problems [21, 22].

The direct numerical simulation and large eddy simulation methods need very advanced computing resources. That is why a new approach has been presented [23] for modeling separated flows, named detached eddy simulation. The great potential advantage of detached eddy simulation over methods and the simplicity of the realization has led to the introduction of detached eddy simulation in computational fluid dynamics products such as CFX, COBALT, FLUENT, and STAR CD.

The detached eddy simulation method is very efficient [23, 24] and very good agreement between computed and experimental data has been obtained.

References

1. Keizer J (1987) Statistical thermodynamics of non-equilibrium processes. Springer, New York
2. Landau LD, Lifshitz EM (1989) Fluid mechanics, 2nd edn. Pergamon, Oxford

3. Loitsianskii LG (1966) Mechanics of liquids and gases, 1st edn. Pergamon, Oxford
4. Schlichting H, Gerstein K (2000) Boundary layer theory, 8th edn. Springer, Berlin
5. Boyadjiev CB, Babak VN (2000) Non-linear mass transfer and hydrodynamic stability. Elsevier, Amsterdam
6. Boyadjiev C, Doichinova M (2004) In: Proceedings, Borovets'2004, 9th workshop on transport phenomena in two-phase flow, Bulgaria, 27 August–1 September, p 159
7. Chr Boyadjiev, Beschkov V (1984) Mass transfer in liquid film flows. Bulgarian Academy of Sciences, Sofia
8. Alekseenko SV, Nakoryakov VE, Pokusaev BG (1994) Wave flow of liquid films. CRC, London
9. Joseph DD (1976) Stability of fluid motion. Springer, New York
10. Levich VG (1962) Physicochemical hydrodynamics. Prentice Hall, New York
11. Korn GA, Korn TM (1968) Mathematical handbook for scientists and engineers, 2nd edn. McGraw-Hill, New York
12. Franc-Kamenetskii VA (1969) Diffusion and heat transfer in chemical kinetics. Plenum, New York
13. Butt J (1980) Reaction kinetics and reactor design. Wiley, New York
14. Prandtl L (1910) Eine Beziehung zwischen. Warmeaustausch und Stromungswider stand der Fliissigkeiten. Phys Z 11:1072
15. Prandtl L (1928) Bemerkung iiberdenW ~ rmeiibergangimRohr. Phys Z 29:487
16. Karman T (1934) In: Proceedings of the 4th international congress for applied mechanics, Cambridge, p 77
17. Doichinova M, Lavrenteva O, Boyadjiev C (2005) In: Proceedings, Sunny Beach'2004, 10th workshop on transport phenomena in two-phase flow, Bulgaria, 10–15 September, p 189
18. Konkov OA, Nikolaev NA (2008) Mathematical models of turbulence. ZAO "Novoe znanie", Kazan (in Russian)
19. Deardorff JW (1970) J Fluid Mech 41:453
20. Deardorff JW (1973) J Fluid Mech 91:429
21. Spalart PR (2000) Strategies for turbulence modeling and simulations. J Heat Fluid Flow 21:252
22. Rodi W, Ferziger JH, Breuer M, Pourquie M (1997) Status of large eddy simulation: results of a workshop. J Fluids Eng 119(2):248
23. Spalart PR, Jou W-H, Strelets M, Allmaras SR (1997) In: Proceedings of 1st AFOSR international conference on DNS/LES, Ruston, LA, August, 4–8. Greyden, Columbus, p 137
24. Mazaev A (2003) Comparison of DES, RANS and LES for the separated flow around a flat plate at high incidence. Int J Numer Methods Fluids 41:357
25. Chr Boyadjiev (1971) Int Chem Eng 11(3):464

Complex Process Models

Chemical engineering processes are aggregates of simple chemical and physical processes having mathematical descriptions based on fundamental chemical and physical laws. Establishment of a mathematical model of a certain complex process needs information about the interactions between the simple processes involved. The latter simply means that such complex mechanisms have to permit mathematical descriptions consisting of mathematical operators corresponding to the physical and chemical effects contributing to the processes. In accordance with this standpoint, the basic *theoretical models* of mass transfer, for instance, contain equations of hydrodynamics, convection–diffusion, and chemical reactions.

In many cases, the model construction becomes more complicated owing to the complex hydrodynamic situation (turbulence, fluid motion as drops, bubbles, or through a bed of particles, etc.). Using physical analogies (e.g., between turbulent and molecular diffusion) or average velocity and concentration models, one can solve the modeling problems.

Otherwise, if there is a lack of information about the process mechanism, then *similarity theory models can be employed.* In cases when the information about the process mechanism is completely missing, *regression models* allow one to create functional relationships.

1 Mechanism and Mathematical Description

The significance of the models for science and practice is mainly related to their exactness, which in general addresses the correspondence between the separate mathematical operators in the model and the physical and the chemical effects in the process. That is why the principal step in the process of model building is the correct theoretical analysis of the process mechanism being considered. This will be exemplified by model building concerning a physical absorption process [1].

C. Boyadjiev, *Theoretical Chemical Engineering*,
DOI: 10.1007/978-3-642-10778-8_2, © Springer-Verlag Berlin Heidelberg 2010

1.1 Mechanism of Physical Absorption

Let us consider a two-dimensional gas–liquid co-current flow under laminar conditions in a horizontal channel with a flat interphase surface $y = 0$. The area of gas motion is $0 \le x \le l$, $0 \le y \le \delta_1$, whereas that of the liquid phase is $0 \le x \le l$, $-\delta_2 \le y \le 0$.

The inlet ($x = 0$) velocities $(\bar{u}_1, \bar{u}_2)$ and the concentrations $(\bar{c}_1, \bar{c}_2)$ in the phases are constant and unaffected in the phase volumes far from the interface $y = 0$. If the concentration of the gas component is sufficiently high, i.e., $\bar{c}_1 - \chi\bar{c}_2 > 0$ (χ is Henry's constant), then absorption of this component is possible. The mechanism of this process has four stages:

1. Diffusion of the substance from the gas volume to the interphase surface.
2. Adsorption of the substance at the interphase surface.
3. Desorption of the substance in the liquid.
4. Diffusion of the substance from the interphase surface to the liquid volume.

The experimental data analysis [2] shows that the adsorption–desorption processes are very fast (in comparison with the diffusion processes) and there exists a thermodynamic equilibrium at the interphase surface. The later implies that the concentrations in both the gas and the liquid at the interface obey Henry's law.

1.2 Mathematical Description

The process mechanism presented reveals that the mathematical description consists of dimension equations of hydrodynamics and convection–diffusion for the gas ($i = 1$) and liquid ($i = 2$) phases, respectively:

$$\begin{aligned}
&u_x^{(i)}\frac{\partial u_x^{(i)}}{\partial x} + u_y^{(i)}\frac{\partial u_x^{(i)}}{\partial y} = -\frac{1}{\rho_i}\frac{\partial p^{(i)}}{\partial x} + \nu_i\left[\frac{\partial^2 u_x^{(i)}}{\partial x^2} + \frac{\partial^2 u_x^{(i)}}{\partial y^2}\right],\\
&u_x^{(i)}\frac{\partial u_y^{(i)}}{\partial x} + u_y^{(i)}\frac{\partial u_y^{(i)}}{\partial y} = -\frac{1}{\rho_i}\frac{\partial p^{(i)}}{\partial y} + \nu_i\left[\frac{\partial^2 u_y^{(i)}}{\partial x^2} + \frac{\partial^2 u_y^{(i)}}{\partial y^2}\right],\\
&\frac{\partial u_x^{(i)}}{\partial x} + \frac{\partial u_y^{(i)}}{\partial y} = 0, \quad u_x^{(i)}\frac{\partial c_i}{\partial x} + u_y^{(i)}\frac{\partial c_i}{\partial y} = D_i\left[\frac{\partial^2 c_i}{\partial x^2} + \frac{\partial^2 c_i}{\partial y^2}\right], \quad i = 1,2.
\end{aligned} \tag{1.1}$$

The boundary conditions at the inlet of two phase flows are

$$x=0,\ u_x^{(i)}=\bar{u}_i,\ \frac{\partial u_x^{(i)}}{\partial x}=0,\ u_y^{(i)}=0,\ \frac{\partial u_y^{(i)}}{\partial x}=0,\ p_i=\bar{p}_i,\ c_i=\bar{c}_i,\ i=1,2. \tag{1.2}$$

At the interphase surface ($y = 0$) the boundary conditions express the interphase nonslip and the impermeability conditions, the continuity of both the stress tensor and the mass flux, and the thermodynamic equilibrium (Henry's law):

$$y = 0, \quad u_x^{(1)} = u_x^{(2)}, u_y^{(i)} = 0, \quad i = 1,2, \quad p^{(1)} + 2\mu_1 \frac{\partial u_x^{(1)}}{\partial x} = p^{(2)} + 2\mu_2 \frac{\partial u_x^{(2)}}{\partial x},$$
$$\mu_1 \left[\frac{\partial u_x^{(1)}}{\partial y} + \frac{\partial u_y^{(1)}}{\partial x} \right] = \mu_2 \left[\frac{\partial u_x^{(2)}}{\partial y} + \frac{\partial u_y^{(2)}}{\partial x} \right], \quad D_1 \frac{\partial c_1}{\partial y} = D_2 \frac{\partial c_2}{\partial y}, \quad c_1 = \chi c_2. \tag{1.3}$$

At the frontiers of the area the following boundary conditions are valid:

$$y = (-1)^{i-1} \delta_i, \quad u_x^{(i)} = \bar{u}_i, \quad u_y^{(i)} = 0, \quad c_i = \bar{c}_i. \tag{1.4}$$

1.3 Generalized Variables and Characteristic Scales

The mathematical description of the absorption process contains many parameters, but their number can be reduced if dimensionless parameters are used. Suitably obtaining these parameters permits us to make a qualitative (generalized) analysis of the proposed process mechanism concerning order of magnitude analysis of separate parameters and relevant physical effects. For this purpose, dimensionless (generalized) variables should be introduced using characteristic (inherent) scales of the process. These scales have to be selected in a manner allowing the dimensionless (generalized) variables to be of the order of magnitude of unity, i.e., $O(1)$. Consequently, the order of magnitude of the separate terms in the model equations (separate physical effects) will be equal to the order of magnitude of their dimensionless parameters.

In the case of the absorption process considered, the following dimensionless variables will be used:

$$x = lX, \quad y = \delta_i Y_i = \delta_{Di} Y_{Di}, \quad u_x^{(i)}(x,y) = \bar{u}_i U_x^{(i)}(X, Y_i), \quad u_y^{(i)}(x,y) = \tilde{u}_i U_y^{(i)}(X, Y_i),$$
$$p^{(i)}(x,y) = \bar{p}_i P^{(i)}(X, Y_i), \quad c_i(x,y) = \bar{c}_i - (-\chi)^{1-i}(\bar{c}_1 - \chi\bar{c}_2) C_i(X, Y_{Di}), \quad i = 1,2, \tag{1.5}$$

where l, $\bar{u}_i$ and $(\bar{c}_1 - \chi\bar{c}_2)$ are known scales, and δ_i, δ_{Di}, $\tilde{u}_i$, and $\bar{p}_i$ $(i = 1,2)$ are unknown scales, which should be obtained as combinations of the known ones.

If by means of (1.5) the continuity equation is transformed, its dimensionless form is

$$\frac{\partial U_x^{(i)}}{\partial X} + \frac{\tilde{u}_i l}{\bar{u}_i \delta_i} \frac{\partial U_y^{(i)}}{\partial Y_i} = 0, \quad i = 1,2. \tag{1.6}$$

In the case of

$$\frac{\tilde{u}_i l}{\bar{u}_i \delta_i} \sim 1, \quad \tilde{u}_i = \frac{\bar{u}_i \delta_i}{l}, \quad i = 1, 2, \tag{1.7}$$

the flow is stratified ($u_{\mathrm{y}}^{(i)} = 0$), *i.e.*,

$$\frac{\tilde{u}_i l}{\bar{u}_i \delta_i} \ll 0, \quad \frac{\partial U_{\mathrm{x}}^{(i)}}{\partial X} = 0, \quad i = 1, 2, \tag{1.8}$$

and $\tilde{u}_i$ $(i = 1, 2)$ cannot be obtained.

1.4 Dimensionless Parameters and Process Mechanism

Introduction of (1.5) into (1.1) leads to a dimensionless (generalized) form of the model:

$$\begin{gathered}
U_{\mathrm{x}}^{(i)} \frac{\partial U_{\mathrm{x}}^{(i)}}{\partial X} + U_{\mathrm{y}}^{(i)} \frac{\partial U_{\mathrm{x}}^{(i)}}{\partial Y_i} = -\frac{\bar{p}_i}{\rho_i \bar{u}_i^2} \frac{\partial P^{(i)}}{\partial X} + \frac{1}{\varepsilon_i Re_i} \left[\varepsilon_i^2 \frac{\partial^2 U_{\mathrm{x}}^{(i)}}{\partial X^2} + \frac{\partial^2 U_{\mathrm{x}}^{(i)}}{\partial Y_i^2} \right], \\
\varepsilon_i^2 \left[U_{\mathrm{x}}^{(i)} \frac{\partial U_{\mathrm{y}}^{(i)}}{\partial X} + U_{\mathrm{y}}^{(i)} \frac{\partial U_{\mathrm{y}}^{(i)}}{\partial Y_i} \right] = -\frac{\bar{p}_i}{\rho_i \bar{u}_i^2} \frac{\partial P^{(i)}}{\partial Y} + \frac{\varepsilon_i}{Re_i} \left[\varepsilon_i^2 \frac{\partial^2 U_{\mathrm{y}}^{(i)}}{\partial X^2} + \frac{\partial^2 U_{\mathrm{y}}^{(i)}}{\partial Y_i^2} \right], \\
\frac{\partial U_{\mathrm{x}}^{(i)}}{\partial X} + \frac{\partial U_{\mathrm{y}}^{(i)}}{\partial Y_i} = 0, \quad U_{\mathrm{x}}^{(i)} \frac{\partial C_i}{\partial X} + U_{\mathrm{y}}^{(i)} \frac{\partial C_i}{\partial Y_{\mathrm{Di}}} = \frac{1}{\varepsilon_{\mathrm{Di}} Pe_i} \left[\varepsilon_{\mathrm{Di}}^2 \frac{\partial^2 C_i}{\partial X^2} + \frac{\partial^2 C_i}{\partial Y_{\mathrm{Di}}^2} \right], \quad i = 1, 2,
\end{gathered} \tag{1.9}$$

where

$$\varepsilon_i = \frac{\delta_i}{l}, \quad \varepsilon_{\mathrm{Di}} = \frac{\delta_{\mathrm{Di}}}{l}, \quad Re_i = \frac{\bar{u}_i \delta_i}{\nu_i}, \quad Pe_i = \frac{\bar{u}_i \delta_{\mathrm{Di}}}{D_i}, \quad i = 1, 2. \tag{1.10}$$

In (1.10) *Re* and *Pe* are Reynolds and Péclet numbers, respectively.

The order of magnitude of the functions and their derivatives in (1.9) is 1 and all physical effects are equivalent to each other if the following conditions are satisfied:

$$\frac{p_i}{\rho_i \bar{u}_i^2} \sim 1, \quad \varepsilon_i Re_i \sim 1, \quad \varepsilon_{\mathrm{Di}} Pe_i \sim 1, \quad i = 1, 2. \tag{1.11}$$

This indicates that the unknown scales can be defined as

$$\bar{p}_i = \rho_i \bar{u}_i^2, \quad \delta_i = \sqrt{\frac{\nu_i l}{\bar{u}_i}}, \quad \delta_{\mathrm{Di}} = \sqrt{\frac{D_i l}{\bar{u}_i}}, \quad i = 1, 2. \tag{1.12}$$

Practically, the boundary layer approximations are valid if

$$0 = \varepsilon_i^2 = Re_i^{-1} < 10^{-2}, \quad 0 = \varepsilon_{Di}^2 = Pe_i^{-1} < 10^{-2}. \tag{1.13}$$

In gases ($i = 1$) $\nu_1 \sim D_1$, whereas in liquids ($i = 2$) $\nu_2 \gg D_2$, i.e., $\delta_1 \sim \delta_{D1}$, $\delta_2 \gg \delta_{D2}$. Hence, using δ_{Di} ($i = 1, 2$) as linear scales in mass transfer problems is very convenient. As a result, in the diffusion boundary layer approximation of problem (1.9) we have

$$\begin{aligned} &U_x^{(i)}\frac{\partial U_x^{(i)}}{\partial X} + U_y^{(i)}\frac{\partial U_x^{(i)}}{\partial Y_{Di}} = -\frac{\partial P^{(i)}}{\partial X} + Sc_i\frac{\partial^2 U_x^{(i)}}{\partial Y_{Di}^2}, \quad \frac{\partial P^{(i)}}{\partial Y_{Di}} = 0, \quad Sc_i = \frac{\nu_i}{D_i}, \\ &\frac{\partial U_x^{(i)}}{\partial X} + \frac{\partial U_y^{(i)}}{\partial Y_{Di}} = 0, \quad U_x^{(i)}\frac{\partial C_i}{\partial X} + U_y^{(i)}\frac{\partial C_i}{\partial Y_{Di}} = \frac{\partial^2 C_i}{\partial Y_{Di}^2}, \end{aligned} \tag{1.14}$$

where Sc is the Schmidt number.

From the second equation in (1.14) it follows that the pressure in the boundary layers is equal to the pressure outside them. Moreover, the flows are nonviscous (potential flows): with $|y| > \delta_i$, $u_x^{(i)} = \bar{u}_i$, $u_y^{(i)} = 0$. As a result, from the first equation in (1.1) it follows that

$$\bar{u}_i\frac{\partial \bar{u}_i}{\partial x} = -\frac{1}{\rho_i}\frac{\partial p^{(i)}}{\partial x}, \quad \bar{u}_i = \text{const.}, \quad -\frac{\partial p^{(i)}}{\partial x} = 0, \quad i = 1, 2. \tag{1.15}$$

The next results will illustrate the case $\bar{u}_i = \text{const.}$

The boundary conditions corresponding to the new approximation are

$$\begin{aligned} &X = 0, \quad U_x^{(i)} = 1, \quad C_i = 0; \quad Y_{Di} = (-1)^{i-1}, \quad U_x^{(i)} = 1, \quad C_i = 0; \\ &Y_i = 0, \quad U_x^{(1)} = \theta_1 U_x^{(2)}, \quad \theta_1 = \frac{\bar{u}_2}{\bar{u}_1}, \quad U_y^{(i)} = 0, \quad \theta_2\frac{\partial U_x^{(1)}}{\partial Y_{D1}} = \frac{\partial U_x^{(2)}}{\partial Y_{D2}}, \quad \theta_2 = \frac{\mu_1\bar{u}_1\delta_2}{\mu_2\bar{u}_2\delta_1}, \\ &C_1 + C_2 = 1, \quad \frac{\partial C_1}{\partial Y_{D1}} = -\varepsilon\frac{\partial C_2}{\partial Y_{D2}}, \quad \varepsilon = \frac{1}{\chi}\sqrt{\frac{\bar{u}_2 D_2}{\bar{u}_1 D_1}}, \quad i = 1, 2, \end{aligned} \tag{1.16}$$

where the unnecessary conditions are omitted.

For the development of analytical solutions, the boundary condition outside the boundary layers is very conveniently expressed as

$$Y_{Di} \to (-1)^{i-1}\infty, \quad U_x^{(i)} = 1, \quad C_i = 0. \tag{1.17}$$

This analysis simply shows that the mathematical description of the process was possible because the process mechanism was known. Moreover, the mathematical description permits us to specify the process mechanism and to omit insignificant physical effects represented by small parameter values.

1.5 Boundary Conditions and Mechanism

The condition $0 = \theta_1 < 10^{-2}$ leads to the boundary condition $U_x^{(1)}(X,0) = 0$, i.e., the gas phase moves over an immobile (solid) interface and the hydrodynamic equations can be solved consecutively and independently.

The condition $0 = \varepsilon < 10^{-2}$ leads to

$$Y_1 = 0, \quad \frac{\partial C_1}{\partial Y_{D1}} = 0, \quad C_1 \equiv 0, \tag{1.18}$$

i.e., the mass transfer rate in the gas phase is very large and as a result the concentration in the gas phase is uniformly distributed. This is the case when the interphase mass transfer is limited by the mass transfer in the liquid phase. In the opposite case ($0 = \varepsilon^{-1} < 10^{-2}$), the diffusion resistance is determined by the gas phase.

1.6 Kinetics and Mechanism

The theoretical analysis of the interphase mass transfer kinetics requires the interphase mass transfer coefficient (k) to be obtained. In the case of the process considered, the process rate (J) relates to the concentration difference as

$$J = k(\bar{c}_1 - \chi \bar{c}_2). \tag{1.19}$$

If $c_i^*(i = 1, 2)$ are equilibrium concentrations at the interphase surface ($c_1^* = \chi c_2^*$), then the interphase mass transfer rate can be represented using the mass transfer coefficients in the separated phases, namely,

$$J = k_1(\bar{c}_1 - c_1^*) = k_2(c_2^* - \bar{c}_2), \tag{1.20}$$

where k_1 and k_2 are the mass transfer coefficients in the gas and liquid phases.

From (1.19) and (1.20) it follows that

$$\bar{c}_1 - \chi \bar{c}_2 = (\bar{c}_1 - c_1^*) + \chi(c_2^* - \bar{c}_2) = \frac{J}{k} = \frac{J}{k_1} + \chi \frac{J}{k_2}. \tag{1.21}$$

In accordance with the law of diffusion resistance arranged in a series, we have

$$\frac{1}{k} = \frac{1}{k_1} + \frac{\chi}{k_2}. \tag{1.22}$$

When highly soluble gases are used, $0 = \chi < 10^{-2}$ and $k = k_1$, then the interphase mass transfer is limited by the mass transfer in the gas phase and the model only allows the equation to be solved for the gas phase.

Another version of the process rate is

$$J = \tilde{k}\left(\frac{\bar{c}_1}{\chi} - \bar{c}_2\right), \quad \frac{\bar{c}_1}{\chi} - \bar{c}_2 = \frac{1}{\chi}\left(\bar{c}_1 - c_1^*\right) + \left(c_2^* - \bar{c}_2\right) \tag{1.23}$$

and consequently the result is

$$\frac{1}{\tilde{k}} = \frac{1}{\chi k_1} + \frac{1}{k_2}, \tag{1.24}$$

with slightly soluble gases, $0 = \chi^{-1} \leq 10^{-2}$ and $\tilde{k} = k_2$, the interphase mass transfer is limited by the mass transfer in the liquid phase and consequently only the equation for the liquid phase has to be solved.

The results obtained above are correct with respect to the film mass transfer theory assumptions. The ε criterion comes from the diffusion boundary layer theory and leads to the same result (in comparison with the χ criterion) but it is more precise (see Chap. 3) as a tool.

The mass transfer rate can be represented using the diffusion flux

$$I = -D_i\left(\frac{\partial c_i}{\partial y}\right)_{y=0}, \quad i = 1, 2 \tag{1.25}$$

and the average mass transfer rate for length l as

$$J = \frac{1}{l}\int_0^l I dx. \tag{1.26}$$

From (1.25) and (1.26) and replacing J from (1.19), (1.20), and (4.23), we can express the Sherwood number (Sh) as

$$\begin{aligned}
Sh_i &= \frac{kl}{D_i} = -\frac{1}{\bar{c}_1 - \chi\bar{c}_2}\int_0^l \left(\frac{\partial c_i}{\partial y}\right)_{y=0} dx = (-\chi)^{1-i}\sqrt{Pe_i}\int_0^1 \left(\frac{\partial C_i}{\partial Y_{Di}}\right) dX, \quad i = 1, 2, \\
\widehat{Sh}_1 &= \frac{k_1 l}{D_1} = -\frac{1}{\bar{c}_1 - c_1^*}\int_0^l \left(\frac{\partial c_1}{\partial y}\right)_{y=0} dx = \frac{\bar{c}_1 - \chi\bar{c}_2}{\bar{c}_1 - c_1^*}\sqrt{Pe_1}\int_0^1 \left(\frac{\partial C_1}{\partial Y_{D1}}\right) dX, \\
\widehat{Sh}_2 &= \frac{k_2 l}{D_2} = -\frac{1}{c_2^* - \bar{c}_2}\int_0^l \left(\frac{\partial c_2}{\partial y}\right)_{y=0} dx = -\frac{1}{\chi}\frac{\bar{c}_1 - \chi\bar{c}_2}{c_2^* - \bar{c}_2}\sqrt{Pe_2}\int_0^1 \left(\frac{\partial C_2}{\partial Y_{D2}}\right) dX.
\end{aligned} \tag{1.27}$$

Here, $Sh_i (i = 1, 2)$ are the Sherwood numbers of the interphase mass transfer, and $\widehat{Sh}_i$ $(i = 1, 2)$ are the Sherwood numbers of the mass transfer in the gas and liquid phases, respectively.

The analysis of the absorption (desorption) process shows that after the formulation of the mathematical model of the process mechanism it is possible to perform a qualitative theoretical analysis of the model equations through nondimensionalization and an order of magnitude analysis. Such a generalized analysis permits us to specify the order of magnitudes of the different physical effects and to eliminate insignificant ones, which results in model reduction.

2 Theoretical Models: Mass Transfer in Film Flow

Many processes in chemical engineering are carried out in gas (vapor)–liquid film flows. In such flows, the effects of different volume (i.e., velocity distributions, chemical reactions) and interface (gas motion, surfactants, capillary waves) effects on the mass transfer in liquid film flows [2, 3] have to taken into account, as exemplified next.

2.1 Film with a Free Interface

Let us consider absorption of a slightly soluble gas in a laminar liquid film, flowing over a flat vertical surface ($y = 0$). In Sect. 1.2 the velocity distribution (1.62) was obtained, and permits us to represent the convection–diffusion equation as

$$\frac{g}{2\nu}\left(2h_0 y - y^2\right)\frac{\partial c}{\partial x} = D\left(\frac{\partial^2 c}{\partial x^2} + \frac{\partial^2 c}{\partial y^2}\right), \tag{2.1}$$

The relevant boundary conditions are

$$\begin{aligned} &x = 0, \quad c = c_0; \qquad x \to \infty, \quad c = c^*; \qquad y = 0, \quad \frac{\partial c}{\partial y} = 0; \\ &y = h_0, \quad c = c^*, \end{aligned} \tag{2.2}$$

A thermodynamic equilibrium exists at the film interface ($y = h_0$) and c^* denotes the equilibrium concentration. The solid surface ($y = 0$) is impenetrable for the diffusing substance with inlet concentration $c_0 < c^*$ (absorption), and a film of length l will be considered as an example. The diffusion boundary layer thickness δ (see page 44 (3.34)) is less than the liquid film thickness h_0 that permits the diffusion boundary layer approximation to be applied. As a consequence of this approach, the next generalized variables can be introduced:

$$x = lX, \quad y = h_0 - \delta Y, \quad c = c_0 + (c^* - c_0)C, \tag{2.3}$$

where $\delta \ll h_0$ and $h_0 \ll l$.

The introduction of (2.3) into (2.1) yields

$$\left(1 + \frac{\delta^2}{h_0^2}Y^2\right)\frac{\partial C}{\partial X} = \frac{Dl}{u_{av}\delta^2}\left(\frac{\delta^2}{l^2}\frac{\partial^2 C}{\partial X^2} + \frac{\partial^2 C}{\partial Y^2}\right), \tag{2.4}$$

where

$$u_{\mathrm{av}} = \frac{gh_0^2}{3\nu}, \quad \delta = \sqrt{\frac{Dl}{u_{\mathrm{av}}}}, \quad \frac{\delta^2}{h_0^2} = \frac{Dl}{u_{\mathrm{av}} h_0^2} = Fo < 10^{-1}, \quad \frac{\delta^2}{l^2} = \frac{D}{u_{\mathrm{av}} l} = Pe^{-1} < 10^{-2}. \tag{2.5}$$

In (2.5) u_{av} is the average velocity of the film flow, Fo is a small parameter (like the Fourier number), and Pe is the Péclet number. Under these conditions the problem was solved [4, 5] in the diffusion boundary layer approximation ($10^{-2} > Pe^{-1} = 0$), namely,

$$\left(1 + FoY^2\right)\frac{\partial C}{\partial X} = \frac{\partial^2 C}{\partial Y^2}; \quad X = 0, \quad C = 0; \quad Y = 0, \quad C = 1; \quad Y \to \infty, \quad C = 0. \tag{2.6}$$

The *mass transfer rate* (J) in a liquid film flow with length l is the average value of the local mass flux through the interphase surface ($y = h_0$). On the other hand, this rate can also be represented using the mass transfer coefficient k. Hence, we have

$$J = \frac{D}{l}\int_0^l \left(\frac{\partial c}{\partial y}\right)_{y=h_0} dx = k(c^* - c_0). \tag{2.7}$$

The adimensionalization of (2.6) by (2.3) allows the Sherwood number (Sh) to be expressed as

$$Sh = \frac{kl}{D} = -\sqrt{Pe}\int_0^1 \left(\frac{\partial C}{\partial Y}\right)_{Y=0} dX, \tag{2.8}$$

In (2.8) the concentration profile $C(X, Y)$ is the solution of (2.6) [4, 5] developed by a perturbation method [6, 7]. This approach uses an expression of the concentration profile as a series in ascending powers of the small parameter Fo and results in the following dimensionless expression of the Sherwood number:

$$Sh = \sqrt{\frac{6Pe}{\pi}}\left(1 - \frac{Fo}{6} - \frac{19Fo^2}{120}\right). \tag{2.9}$$

The expression (2.9) shows the effect of the velocity distribution on the concentration boundary layer and the mass transfer rate. In the approximation of Higbie's penetration theory [8], for example, the velocity distribution is a constant and $Fo = 0$.

2.2 Effect of a Chemical Reaction

Homogeneous chemical reactions can be represented as source (sink) terms of the substances, with intensities equal to the chemical reaction rates. Hence, the mass balance in the convection–diffusion equation can be expressed as

$$u_x \frac{\partial c}{\partial x} + u_y \frac{\partial c}{\partial y} + u_z \frac{\partial c}{\partial z} = D\left(\frac{\partial^2 c}{\partial x^2} + \frac{\partial^2 c}{\partial y^2} + \frac{\partial^2 c}{\partial z^2}\right) \pm Q(c), \tag{2.10}$$

The term $Q(c)$ denotes the volumetric chemical reaction rate: the plus sign means of the creation (substance source), whereas the minus sign indicates disappearance (substance sink) of the transferred substance.

In the case of absorption, the transferred substance reacts with a component in the liquid. If the concentration of this component is very large, then the chemical reaction is of first order: $Q = k_1 c$, where k_1 is the chemical reaction rate constant. Hence, from (2.10) it follows that

$$\frac{g}{2\nu}\left(2h_0 y - y^2\right)\frac{\partial c}{\partial x} = D\left(\frac{\partial^2 c}{\partial x^2} + \frac{\partial^2 c}{\partial y^2}\right) - k_1 c. \tag{2.11}$$

The solution of (2.11) is obtained similarly to that of (2.1) using the same boundary conditions. The perturbation method and Green's functions permit us to express the Sherwood number [9] as

$$Sh = \sqrt{\frac{6Pe}{\pi}}\left\{\begin{array}{l} -\frac{1}{2}\exp(-K_1) + \left(\frac{3}{4}\sqrt{\frac{\pi}{K_1}} - \frac{1}{2}\sqrt{\pi K_1}\right)\operatorname{erf}\sqrt{K_1} - Fo\left[\begin{array}{l} -\frac{1}{4K_1}\exp(-K_1) + \\ \frac{1}{8K_1}\sqrt{\frac{\pi}{K_1}}\operatorname{erf}\sqrt{K_1} \end{array}\right] - \\ -Fo^2\frac{19}{120}\exp(-K_1) \end{array}\right\},$$

$$K_1 = \frac{k_1 l}{u_{av}}, \quad \operatorname{erf}\sqrt{K_1} = \frac{2}{\sqrt{\pi}}\int_0^{\sqrt{K_1}} \exp(-s^2)\,ds. \tag{2.12}$$

2.3 Effect of Gas Motion

Numerous processes in chemical engineering are carried out in gas–liquid film flows [2, 3]. In a co-current vertical liquid film and a laminar gas flow, the mathematical problem in the boundary layer approximation has the form

$$\begin{aligned} &\frac{\partial^2 u_x}{\partial y^2} + \frac{g}{\nu} = 0, \quad \frac{\partial u_x}{\partial x} + \frac{\partial u_y}{\partial y} = 0; \\ &v_x \frac{\partial v_x}{\partial x} + v_y \frac{\partial v_x}{\partial y} = \nu_g \frac{\partial^2 v_x}{\partial y^2}, \quad \frac{\partial v_x}{\partial x} + \frac{\partial v_y}{\partial y} = 0. \end{aligned} \tag{2.13}$$

The boundary conditions of (2.13) represent the mutual influence of the phases: *the equality of the velocity and the stress tensor at the interphase surface*. These conditions can be obtained by (1.2.41) in the boundary layer approximation, namely,

$$
\begin{aligned}
&x=0, \quad v_x=v_{av}; \quad y=0, \quad u_x=u_y=0; \\
&y=h(x), \quad u_x=v_x, \quad \mu\frac{\partial u_x}{\partial y}=\mu_g\frac{\partial v_x}{\partial y}; \quad y\to\infty, \quad v_x=v_{av}.
\end{aligned} \tag{2.14}
$$

The film thickness $h(x)$ is a variable dependent on the gas motion, and the *macroscopic balance* over the liquid film yields

$$
\frac{\partial h}{\partial t}+\frac{\partial}{\partial x}\int\limits_0^h u_x dy=0. \tag{2.15}
$$

If no waves are considered in the film $\left(\frac{\partial h}{\partial t}=0\right)$, then we have

$$
h' u_x(x,h)-u_y(x,h)=0, \quad h'=\frac{dh}{dx}. \tag{2.16}
$$

For large value of $x(\varepsilon_0<10^{-2})$, we have $h(x)\equiv h_0$, which is, in fact, the boundary condition for (2.16).

The solution of the problem needs the following dimensionless (generalized) variables to be used:

$$
\begin{aligned}
&x=lX, \quad y=h_0Y=h(x)+\delta_gY_g, \quad \delta_g=\sqrt{\frac{v_g l}{v_{av}}}, \quad h(x)=h_0H(X), \\
&\varepsilon_0=\frac{h_0}{l}, \quad \varepsilon_g=\frac{\delta_g}{l}, u_x=u_{av}U_x, \quad u_y=\varepsilon_0u_{av}U_y, \quad u_{av}=\frac{gh_0^2}{3v}, \quad v_x=v_{av}V_x, \\
&v_y=\varepsilon_g v_{av}V_y.
\end{aligned} \tag{2.17}
$$

The dimensionless form of the governing equations is

$$
\begin{aligned}
&\frac{\partial^2U_x}{\partial Y^2}=-3, \quad \frac{\partial U_x}{\partial X}+\frac{\partial U_y}{\partial Y}=0; \quad H'U_x(X,H)-U_y(X,H)=0, \quad H'=\frac{dH}{dX}; \\
&V_x\frac{\partial V_x}{\partial X}+V_y\frac{\partial V_x}{\partial Y_g}=\frac{\partial^2V_x}{\partial Y_g^2}, \quad \frac{\partial V_x}{\partial X}+\frac{\partial V_y}{\partial Y_g}=0; \quad X=0, \quad V_x=0; \quad X\to\infty, \\
&H=1; \quad Y=0, \quad U_x=U_y=0; \quad Y_g\to\infty, \quad V_x=1; \\
&Y=H(X), \quad Y_g=0, \quad V_x=\theta_1U_x, \quad \frac{\partial U_x}{\partial Y}=\theta_2\frac{\partial V_x}{\partial Y_g}, \quad V_y=0,
\end{aligned} \tag{2.18}
$$

where θ_1 and θ_2 are small parameters in the gas–liquid systems,

$$\theta_1 = \frac{u_{\mathrm{av}}}{v_{\mathrm{av}}}, \quad \theta_2 = \frac{\mu_{\mathrm{g}} v_{\mathrm{av}} h_0}{\mu u_{\mathrm{av}} \delta_{\mathrm{g}}}. \tag{2.19}$$

The solution of (2.18) was obtained in [2, 3, 10] by perturbation methods [6, 7], expressing the solution as a series in ascending powers of the small parameters θ_1 and θ_2. In (2.18) the approximation $0 = \varepsilon_1 = \frac{h_0}{\delta_{\mathrm{g}}} < 10^{-2}$ for gas–liquid film flow systems is used.

From (2.18) it can be seen that the velocity distribution in the gas flow depends on the interface velocity of the liquid film flow (a kinematic condition):

$$Y = H(X), \quad Y_{\mathrm{g}} = 0, \quad V_{\mathrm{x}}(X,0) = \theta_1 U_{\mathrm{x}}(X,H) = \theta_1 A(X), \tag{2.20}$$

The function *A*(*X*) can be obtained from the velocity distribution in the liquid film. On the other hand, the velocity distribution in the film flow depends on the friction force with the gas flow (a dynamic condition):

$$Y = H(X), \quad Y_{\mathrm{g}} = 0, \quad \frac{\partial U_{\mathrm{x}}}{\partial Y} = \theta_2 \frac{\partial V_{\mathrm{x}}}{\partial Y_{\mathrm{g}}} = \theta_2 B(X). \tag{2.21}$$

The function *B*(*X*) can be obtained from the velocity distribution in the gas flow.

Introducing the new boundary conditions (2.20) and (2.21) into (2.18), we can represent the problem as three separate subproblems:

$$\begin{gathered}
\frac{\partial^2 U_{\mathrm{x}}}{\partial Y^2} = -3, \quad \frac{\partial U_{\mathrm{x}}}{\partial X} + \frac{\partial U_{\mathrm{y}}}{\partial Y} = 0; \qquad Y = 0, \quad U_{\mathrm{x}} = U_{\mathrm{y}} = 0; \\
Y = H(X), \quad \frac{\partial U_{\mathrm{x}}}{\partial Y} = \theta_2 B(X).
\end{gathered} \tag{2.22}$$

$$H' U_{\mathrm{x}}(X,H) - U_{\mathrm{y}}(X,H) = 0; \quad X \to \infty, \quad H = 1. \tag{2.23}$$

$$\begin{gathered}
V_{\mathrm{x}} \frac{\partial V_{\mathrm{x}}}{\partial X} + V_{\mathrm{y}} \frac{\partial V_{\mathrm{x}}}{\partial Y_{\mathrm{g}}} = \frac{\partial^2 V_{\mathrm{x}}}{\partial Y_{\mathrm{g}}^2}, \frac{\partial V_{\mathrm{x}}}{\partial X} + \frac{\partial V_{\mathrm{y}}}{\partial Y_{\mathrm{g}}} = 0; \\
X = 0, \quad V_{\mathrm{x}} = 0; \qquad Y_{\mathrm{g}} \to \infty, \quad V_{\mathrm{x}} = 1; \qquad Y_{\mathrm{g}} = 0, \quad V_{\mathrm{x}}(X,0) = \theta_1 A(X), \quad V_{\mathrm{y}} = 0.
\end{gathered} \tag{2.24}$$

The separate problems (2.22–2.24) have to be solved for arbitrary functions *A*(*X*) and *B*(*X*), which, in fact, can be obtained subsequently from (2.20) and (2.21).

From (2.22–2.24) it follows that the solutions depend on the small parameters θ_1 and θ_2:

$$\begin{gathered}
U_{\mathrm{x}} = U_{\mathrm{x}}(X, Y, \theta_1, \theta_2), \quad U_{\mathrm{y}} = U_{\mathrm{y}}(X, Y, \theta_1, \theta_2), \quad V_{\mathrm{x}} = V_{\mathrm{x}}(X, Y_g, \theta_1, \theta_2), \\
V_{\mathrm{y}} = V_{\mathrm{y}}(X, Y_g, \theta_1, \theta_2), \\
H = H(X, \theta_1, \theta_2), \quad A = A(X, \theta_1, \theta_2), \quad B = B(X, \theta_1, \theta_2),
\end{gathered} \tag{2.25}$$

This permits us to look for solutions expressed as a series in ascending powers of the small parameters θ_1 and θ_2, namely,

$$F = F^0 + \theta_1 F^1 + \theta_2 F^2 + \theta_1^2 F^{11} + \theta_2^2 F^{22} + \theta_1 \theta_2 F^{12} + \cdots, \qquad (2.26)$$

where

$$F = \left\| U_x, U_y, V_x, V_y, H, A, B \right\|. \qquad (2.27)$$

In (2.26) F^0 is the zeroth-order approximation, F^1 and F^2 are first-order approximations, and F^{11}, F^2, and F^{12} are second-order approximations of the solution F. It is necessary for (2.26) to be introduced into (2.22–2.24) for the equations of the separate approximations to be developed. This implies that the terms of the separate approximations have to be unified as separate expressions and then equated to zero.

The zeroth-order approximations follow from (2.22) to (2.24) with $\theta_1 = \theta_2 = 0$:

$$\begin{aligned} &\frac{\partial^2 U_x^0}{\partial Y^2} = -3, \quad \frac{\partial U_x^0}{\partial X} + \frac{\partial U_y^0}{\partial Y} = 0; \qquad Y = 0, \quad U_x^0 = U_y^0 = 0; \\ &Y = H^0(X), \quad \frac{\partial U_x^0}{\partial Y} = 0. \end{aligned} \qquad (2.28)$$

$$H^{0\prime} U_x^0 - U_y^0 = 0; \quad X \to \infty, \quad H^0 = 1. \qquad (2.29)$$

$$\begin{aligned} &V_x^0 \frac{\partial V_x^0}{\partial X} + V_y^0 \frac{\partial V_x^0}{\partial Y_g} = \frac{\partial^2 V_x^0}{\partial Y_g^2}, \\ &X = 0, \quad V_x^0 = 0; \quad Y_g \to \infty, \quad V_x^0 = 1; \quad Y_g = 0, \quad V_x^0(X,0) = 0, \quad V_y^0 = 0. \end{aligned} \qquad (2.30)$$

Problems (2.28) and (2.29) describe the case of a film flow with a free interface (1.2.62). The solutions in generalized variables (1.5.17) are

$$U_x^0 = 3Y - \frac{3}{2}Y^2, \quad U_y^0 \equiv 0, \quad H^0 \equiv 1. \qquad (2.31)$$

Equations (2.30) describe a laminar boundary layer problem [11] with a solution expressed with similarity variables:

$$V_x^0 = f_0', \quad V_y^0 = \frac{1}{2\sqrt{X}}\left(\eta f_0' - f_0\right), \quad f_0 = f_0(\eta), \quad \eta = \frac{Y_g}{\sqrt{X}}. \qquad (2.32)$$

Here $f_0(\eta)$ is the Blasius function, obtained as a solution of the problem

$$2f_0''' + f_0 f_0'' = 0, \quad f_0(0) = f_0'(0) = 0, \quad f_0'(\infty) = 1. \qquad (2.33)$$

The numerical solution of (2.33) is tabulated in [12], where $f_0''(0) = \alpha = 0.33205$.

The approximations of the velocity in the liquid flow are obtained from (2.22) as

$$\frac{\partial^2 U_x^i}{\partial Y^2}=0,\quad \frac{\partial U_x^i}{\partial X}+\frac{\partial U_y^i}{\partial Y}=0;\qquad Y=0,\quad U_x^i=U_y^i=0;\qquad i=1,2,11,22,12;$$
$$Y=H^0=1,\quad \frac{\partial U_x^1}{\partial Y}=0,\quad \frac{\partial U_x^2}{\partial Y}=B^0,\quad \frac{\partial U_x^{11}}{\partial Y}=0,\quad \frac{\partial U_x^{22}}{\partial Y}=B^2,\quad \frac{\partial U_x^{12}}{\partial Y}=B^1. \tag{2.34}$$

The solutions of problems (2.34) are

$$U_x^1=U_y^1\equiv 0;\quad U_x^2=B^0Y,\quad U_y^2=-\frac{1}{2}\left(B^0\right)'Y^2;\qquad U_x^{11}=U_y^{11}\equiv 0;$$
$$U_x^{22}=B^2Y,\quad U_y^{22}=-\frac{1}{2}\left(B^2\right)'Y^2;\qquad U_x^{12}=B^1Y,\quad U_y^{12}=-\frac{1}{2}\left(B^1\right)'Y^2. \tag{2.35}$$

The approximations of the film thickness are obtained from (2.23), namely,

$$\left(H^1\right)'=0, H^1(\infty)=0;\quad \left(H^2\right)'=-\frac{1}{3}\left(B^0\right)',\quad H^2(\infty)=0;\quad \left(H^{11}\right)'=0,\quad H^{11}(\infty)=0;$$
$$\left(H^{22}\right)'=\frac{2}{9}B^0\left(B^0\right)'-\frac{1}{2}\left(B^2\right)',\quad H^{22}(\infty)=0;\quad \left(H^{12}\right)'=-\frac{1}{3}\left(B^1\right)',\quad H^{12}(\infty)=0. \tag{2.36}$$

The solutions of (2.36) are the following:

$$H^1\equiv 0,\quad H^2=-\frac{1}{3}B^0,\quad H^{11}\equiv 0,\quad H^{22}=\frac{1}{9}\left(B^0\right)^2-\frac{1}{2}B^2,\quad H^{12}=-\frac{1}{3}B^1. \tag{2.37}$$

The approximations of the function $A(X)=U_x(X, H)$, using (2.31) and (2.35) for $H=H^0=1$, are

$$A^0\equiv\frac{3}{2},\quad A^1\equiv 0,\quad A^2=B^0,\quad A^{11}\equiv 0,\quad A^{22}=B^2,\quad A^{12}=B^1. \tag{2.38}$$

The approximations of the velocity in the gas are obtained from (2.24) as

$$V_x^0\frac{\partial V_x^i}{\partial X}+V_y^0\frac{\partial V_x^i}{\partial Y_g}+V_x^i\frac{\partial V_x^0}{\partial X}+V_y^i\frac{\partial V_x^0}{\partial Y_g}=\frac{\partial^2 V_x^i}{\partial Y_g^2},\quad \frac{\partial V_x^i}{\partial X}+\frac{\partial V_y^i}{\partial Y_g}=0;$$
$$X=0,\quad V_x^i=0;\quad Y_g=0,\quad V_y^i=0;\quad Y_g\to\infty,\quad V_x^i=0;\quad i=1,2,22,12;$$
$$Y_g=0,\quad V_x^1=\frac{3}{2},\quad V_x^2=V_x^{22}=V_x^{12}=0. \tag{2.39}$$

$$V_{\mathrm{x}}^{0}\frac{\partial V_{\mathrm{x}}^{11}}{\partial X}+V_{\mathrm{y}}^{0}\frac{\partial V_{\mathrm{x}}^{11}}{\partial Y_{\mathrm{g}}}+V_{\mathrm{x}}^{11}\frac{\partial V_{\mathrm{x}}^{0}}{\partial X}+V_{\mathrm{y}}^{11}\frac{\partial V_{\mathrm{x}}^{0}}{\partial Y_{\mathrm{g}}}+V_{x}^{1}\frac{\partial V_{\mathrm{x}}^{1}}{\partial X}+V_{\mathrm{y}}^{1}\frac{\partial V_{\mathrm{x}}^{1}}{\partial Y_{\mathrm{g}}}=\frac{\partial^{2}V_{\mathrm{x}}^{11}}{\partial Y_{\mathrm{g}}^{2}},\quad \frac{\partial V_{\mathrm{x}}^{11}}{\partial X}+\frac{\partial V_{\mathrm{y}}^{11}}{\partial Y_{\mathrm{g}}}=0;$$
$$X=0,\quad V_{\mathrm{x}}^{11}=0;\quad Y_{\mathrm{g}}=0,\quad V_{\mathrm{x}}^{11}=V_{\mathrm{y}}^{11}=0;\quad Y_{\mathrm{g}}\to\infty,\quad V_{\mathrm{x}}^{11}=0. \tag{2.40}$$

The solution (2.39) needs the following similarity variables to be employed:

$$V_{\mathrm{x}}^{i}=f_{i}^{\prime},\quad V_{\mathrm{y}}^{i}=\frac{1}{2\sqrt{X}}\left(\eta f_{i}^{\prime}-f_{i}\right),\quad f_{i}=f_{i}(\eta),\quad \eta=\frac{Y_{\mathrm{g}}}{\sqrt{X}},\quad i=1,2,22,12. \tag{2.41}$$

As a result, (2.39) has the form

$$\begin{gathered}2f_{\mathrm{i}}^{\prime\prime\prime}+f_{0}f_{\mathrm{i}}^{\prime\prime}+f_{0}^{\prime\prime}f_{\mathrm{i}}=0;\quad f_{\mathrm{i}}(0)=0,\quad f_{\mathrm{i}}^{\prime}(\infty)=0;\quad \mathrm{i}=1,2,22,12;\\ f_{1}^{\prime}(0)=\frac{3}{2},\quad f_{2}^{\prime}(0)=f_{22}^{\prime}(0)=f_{12}^{\prime}(0)=0.\end{gathered} \tag{2.42}$$

The solutions of (2.42) are

$$f_{1}=\frac{3}{2\alpha}f_{0}^{\prime},\quad f_{2}=f_{22}=f_{12}\equiv 0. \tag{2.43}$$

For V_{x}^{11} and V_{y}^{11} the following similarity variables [10] are used:

$$V_{\mathrm{x}}^{11}=\frac{9}{4}\ f_{11}^{\prime},V_{\mathrm{y}}^{11}=\frac{9}{8\sqrt{X}}(\eta f_{11}^{\prime}-f_{11}),\quad f_{11}=f_{11}(\eta),\quad \eta=\frac{Y_{\mathrm{g}}}{\sqrt{X}}. \tag{2.44}$$

As a result [10], we have

$$2f_{11}^{\prime\prime\prime}+f_{0}f_{11}^{\prime\prime}+f_{0}^{\prime\prime}f_{11}=-\frac{1}{\alpha^{2}}f_{0}^{\prime}f_{0}^{\prime\prime};\quad f_{11}(0)=f_{11}^{\prime}(0)=0,\quad f_{11}^{\prime}(\infty)=0. \tag{2.45}$$

The numerical solution of (2.45) is tabulated in [12] with $f_{11}^{\prime\prime}(0)=\beta=-0.54470$.

From (2.21) it follows that

$$B^{i}(X)=\left(\frac{\partial V_{\mathrm{x}}^{i}}{\partial Y_{\mathrm{g}}}\right)_{Y_{\mathrm{g}}=0}=\frac{f_{i}^{\prime\prime}}{\sqrt{X}},\quad i=0,1,2. \tag{2.46}$$

Solutions (2.32), (2.37), (2.38), and (2.43) permit us to obtain the unknown functions, namely,

$$B^{0}=\frac{\alpha}{\sqrt{X}},\quad B^{1}=B^{2}\equiv 0,\quad A^{2}=\frac{\beta}{\sqrt{X}},\quad A^{22}=A^{12}\equiv 0,\quad H^{2}=-\frac{\alpha}{3\sqrt{X}}. \tag{2.47}$$

From these results, the velocity distributions in dimension form [10] follows:

$$u_x = \frac{g}{2\nu}\left(2h_0y - y^2 + \frac{2\alpha v_{av}}{g\rho}\sqrt{\frac{\mu_g\rho_g v_{av}}{x}}y\right), \quad u_y = \frac{\alpha v_{av}}{4\mu x}\sqrt{\frac{\mu_g\rho_g v_{av}}{x}}y^2,$$
$$h = h_0 - \frac{\alpha v_{av}}{g\rho}\sqrt{\frac{\mu_g\rho_g v_{av}}{x}}. \tag{2.48}$$

$$v_x = v_{av}f_0' + \frac{3u_{av}}{2\alpha}f_0'' + \frac{9u_{av}^2}{4v_{av}}f_{11}',$$
$$v_y = \sqrt{\frac{v_{av}v_g}{4x}}\left[\eta f_0' - f + \frac{3u_{av}}{2\alpha v_{av}}(\eta f_0'' - f_0') + \frac{9u_{av}^2}{4v_{av}^2}(\eta f_{110}' - f_{11})\right];$$
$$2f_0''' + f_0f_0'' = 0, \quad 2f_{11}''' + f_0f_{11}'' + f_0''f_{11} = -\frac{1}{\alpha^2}f_0'f_0''; \quad f_0 = f_0(\eta), \quad f_{11} = f_{11}(\eta),$$
$$\eta = (y-h)\sqrt{\frac{v_{av}}{v_gx}}; \quad f_0(0) = f_0'(0) = 0, \quad f_0''(0) = \alpha, \quad f_{11}(0) = f_{11}'(0) = 0,$$
$$f_{11}''(0) = \beta, \alpha = 0.33205, \beta = -0.54470. \tag{2.49}$$

In (2.49) ρ and ρ_g are the liquid and gas densities, respectively.

This theoretical analysis reveals the effect of the liquid film velocity on the velocity distribution in the gas phase (the effect of parameter θ_1) and the friction force with the effect of the gas phase on the velocity distribution in the liquid phase (the effect of parameter θ_2).

2.4 Absorption of Slightly Soluble Gas

The absorption rate of a slightly soluble gas is limited by the mass transfer in the liquid phase.

For short liquid films ($l < l_0$) the diffusion boundary layer thickness (δ_c) is less than the film thickness:

$$\delta_c = \sqrt{\frac{Dl_0}{u_{av}}} < h_0. \tag{2.50}$$

In this case, the concentration distribution of the absorbed substance can be obtained by the convection–diffusion equation in the boundary layer approximation:

$$u_x\frac{\partial c}{\partial x} + u_y\frac{\partial c}{\partial y} = D\frac{\partial^2 c}{\partial y^2}; \quad x = 0, \quad c = c_0; \quad y = 0, \quad c = c_0; \quad y = h(x), \quad c = c^*, \tag{2.51}$$

Here the velocity components u_x and u_y are represented by (2.48) and the boundary conditions are similar to (2.2). To solve the problem, the following generalized variables of the diffusion boundary layer are used [13]:

$$x = lX, \quad y = h + \delta_c Y_c, \quad u_x(x,y) = u_{av} U_x(X, Y_c), \quad u_y(x,y) = u_{av}\frac{\delta_c}{l} U_y(X, Y_c),$$
$$c(x,y) = c_0 + (c^* - c_0) C(X, Y_c). \tag{2.52}$$

With the new variables, problem (2.51) has the form

$$\left(\frac{3}{2} - \varepsilon_1 Y_c^2 + \varepsilon_2 \frac{3}{\sqrt{X}}\right)\frac{\partial C}{\partial X} + \varepsilon_2 \frac{3Y_c}{2X\sqrt{X}}\frac{\partial C}{\partial Y_c} = \frac{\partial^2 C}{\partial Y_c^2};$$
$$X = 0, \quad C = 0; \quad Y_c \to -\infty, \quad C = 0; \quad Y_c = 0, \quad C = 1, \tag{2.53}$$

where $\varepsilon_1$and ε_2 are small parameters, namely,

$$\varepsilon_1 = \frac{3Dl}{2u_{av}h_0^2} \sim 10^{-1}, \quad \varepsilon_2 = \frac{\alpha v_{av}}{\rho g h_0}\sqrt{\frac{\mu_g \rho_g v_{av}}{l}} \sim 10^{-2}. \tag{2.54}$$

Having in mind the orders of magnitude of the parameters ε_1 and ε_2, we can express the solution as [13]

$$C = C^0 + \varepsilon_1 C^1 + \varepsilon_2 C^2 + \varepsilon_1^2 C^{11}, \tag{2.55}$$

ensuring accuracy of about 1%.

The zeroth approximation is

$$\frac{3}{2}\frac{\partial C^0}{\partial X} = \frac{\partial^2 C^0}{\partial Y_c^2}; \quad X = 0, \quad C^0 = 0; \quad Y_c \to -\infty, \quad C^0 = 0; \quad Y_c = 0, \quad C^0 = 1; \tag{2.56}$$

The solution of (2.56) was developed by similarity variables [13]:

$$C^0(X, Y_c) = \operatorname{erfc}\eta_c = 1 - \frac{2}{\sqrt{\pi}}\int_0^{\eta_c} \exp(-s^2)ds, \quad \eta_c = -\sqrt{\frac{3}{8}}\frac{Y_c}{\sqrt{X}}. \tag{2.57}$$

The next approximation is

$$\frac{3}{2}\frac{\partial C^1}{\partial X} = \frac{\partial^2 C^1}{\partial Y_c^2} + Y_c^2 \frac{\partial C^0}{\partial X}; \quad X = 0, \quad C^1 = 0; \quad Y_c \to -\infty, \quad C^1 = 0; \quad Y_c = 0,$$
$$C^1 = 0. \tag{2.58}$$

The solution of (2.58) can be obtained by Green's functions as

$$C^1(X, Y_c) = \frac{1}{3\sqrt{\pi}}\left(Y_c\sqrt{\frac{2X}{3}} + \frac{Y_c^3}{\sqrt{6X}}\right)\exp\left(-\frac{3Y_c^2}{8X}\right). \tag{2.59}$$

The problem for C^2 is analogous to that expressed by (2.56) with zeroth-order boundary conditions:

$$C^2(X, Y_c) \equiv 0. \tag{2.60}$$

The last boundary value problem is analogous to (2.58):

$$\frac{3}{2}\frac{\partial C^{11}}{\partial X} = \frac{\partial^2 C^{11}}{\partial Y_c^2} + Y_c^2\frac{\partial C^1}{\partial X}; \quad X = 0, \quad C^{11} = 0; \quad Y_c \to -\infty, C^{11} = 0; \quad Y_c = 0, \quad C^{11} = 0; \tag{2.61}$$

and the solution is

$$C^{11}(X, Y_c) = \frac{1}{3\sqrt{6\pi}}\left(\frac{38}{27}Y_c X\sqrt{X} - \frac{19}{18}Y_c^3\sqrt{X} + \frac{4}{15}\frac{Y_c^5}{\sqrt{X}} + \frac{1}{24}\frac{Y_c^7}{X\sqrt{X}}\right)\exp\left(-\frac{3Y_c^2}{8X}\right). \tag{2.62}$$

The solution of problem (2.53) permits us to obtain the Sherwood number (2.8) and the mass transfer rate (2.7), namely,

$$Sh = \frac{kl}{D} = Pe\int_0^1\left(\frac{\partial C}{\partial Y_c}\right)_{Y_c=0} dX = \sqrt{\frac{6Pe}{\pi}}\left(1 - \frac{2\varepsilon_1}{27} - \frac{38\varepsilon_1^2}{1443}\right). \tag{2.63}$$

In (2.63) $\varepsilon_1 = \frac{9}{4}Fo$, i.e., (2.63) coincides with (2.9). This result shows that the velocity distribution in the liquid film, affected by the gas motion, is independent of the mass transfer rate in the liquid phase (the effect is less than 1%).

2.5 Absorption of Highly Soluble Gas

The absorption rate of a highly soluble gas is limited by the mass transfer in the gas phase. For short liquid films ($l < l_0$) the diffusion boundary layer thickness $\left(\tilde{\delta}_g\right)$ is

$$\tilde{\delta}_g = \sqrt{\frac{D_g l_0}{v_{av}}} \sim \delta_g, \quad \left(D_g \sim \nu_g\right). \tag{2.64}$$

In this case the concentration distribution of the absorbed substance can be obtained from the convection–diffusion equation in the boundary layer approximation:

$$v_x\frac{\partial c_g}{\partial x}+v_y\frac{\partial c_g}{\partial y}=D_g\frac{\partial^2 c_g}{\partial y^2};\quad x=0,\quad c=c_{g0};\quad y\to\infty,\quad c_g=c_{g0};\quad y=h(x),\quad c=c_g^*,\tag{2.65}$$

where the velocity components v_x and v_y are represented in (2.49). For solution of the problem, the generalized variables of the diffusion boundary layer are used [13]:

$$x=lX,\quad y=h+\tilde{\delta}_g\tilde{Y}_g,\quad v_x(x,y)=v_{av}V_x\left(X,\tilde{Y}_g\right),\quad v_y(x,y)=v_{av}\frac{\tilde{\delta}_g}{l}V_y\left(X,\tilde{Y}_g\right),\quad c_g=c_g^*+\left(c_{g0}-c_g^*\right)C_g.\tag{2.65a}$$

In new variables, the problem (2.65) has the form

$$\left(f_0'+\frac{\theta_1}{\alpha}f_0''+\theta_1^2f_{11}'\right)\frac{\partial C_g}{\partial X}+\frac{\sqrt{Sc_g}}{2\sqrt{X}}\left[\eta f_0'-f_0+\frac{\theta_1}{\alpha}\left(\eta f_0''-f_0'\right)+\theta_1^2\left(\eta f_{11}'-f_{11}\right)\right]\frac{\partial C_g}{\partial \tilde{Y}_g}=\frac{\partial^2 C_g}{\partial \tilde{Y}_g^2};\ X=0,\ C_g=1;\ \tilde{Y}_g=0,\ C_g=0;\ \tilde{Y}_g\to\infty,\ C_g=1,\tag{2.66}$$

where

$$\eta=\frac{\tilde{Y}_g}{\sqrt{Sc_gX}},\qquad Sc_g=\frac{v_g}{D_g}.\tag{2.67}$$

Having in mind the orders of magnitude of the parameter $\theta_1\sim 10^{-1}$, we obtain the solution of (2.66) in the form [13]

$$C_g=C_g^0+\theta_1C_g^1+\theta_1^2C_g^{11},\tag{2.68}$$

with an accuracy of about 1%.

The zeroth approximation leads to the following boundary value problem:

$$f_0'\frac{\partial C_g^0}{\partial X}+\frac{\sqrt{Sc_g}}{2\sqrt{X}}\left(\eta f_0'-f_0\right)\frac{\partial C_g^0}{\partial \tilde{Y}_g}=\frac{\partial^2 C_g^0}{\partial \tilde{Y}_g^2};\quad X=0,\quad C_g^0=1;\quad \tilde{Y}_g=0,\quad C_g^0=0;\quad \tilde{Y}_g\to\infty,\quad C_g^0=1.\tag{2.69}$$

With the similarity variable (2.67) we have

$$C_g^0\left(X,\tilde{Y}_g\right)=F_0(\eta)=\beta_0\int_0^\eta \exp\left[\frac{Sc_g}{2}\int_0^s f_0(t)dt\right]ds,\tag{2.70}$$

where

$$\beta_0 = \frac{1}{\varphi_0(Sc_g)}, \quad \varphi_0(Sc_g) = \int_0^\infty \exp\left[\frac{Sc_g}{2}\int_0^s f_0(t)dt\right] ds. \tag{2.71}$$

The numerical integration [5] of (2.71) yields

$$\varphi_0(Sc_g) = 3.01 Sc_g^{-0.35}. \tag{2.72}$$

The first approximation to (2.68) is obtained by solving the boundary value problem [13]:

$$f_0' \frac{\partial C_g^1}{\partial X} + \frac{\sqrt{Sc_g}}{2\sqrt{X}}(\eta f_0' - f_0)\frac{\partial C_g^1}{\partial \tilde{Y}_g} = \frac{\partial^2 C_g^1}{\partial \tilde{Y}_g^2} - \frac{1}{\alpha} f_0'' \frac{\partial C_g^0}{\partial X} - \frac{\sqrt{Sc_g}}{2\alpha\sqrt{X}}(\eta f_0'' - f_0')\frac{\partial C_g^0}{\partial \tilde{Y}_g};$$
$$X = 0, \quad C_g^1 = 0; \quad \tilde{Y}_g = 0, \quad C_g^1 = 0; \quad \tilde{Y}_g \to \infty, \quad C_g^1 = 0. \tag{2.73}$$

The solution of (2.73) is found similarly to that of (2.69), namely,

$$C_g^1(X, \tilde{Y}_g) = F_1(\eta) = \int_0^\eta \left[\frac{\beta_0^2}{\alpha} - \frac{\beta_0 Sc_g}{2\alpha} f_0(s)\right] \exp\left[\frac{Sc_g}{2}\int_0^s f_0(t)dt\right] ds. \tag{2.74}$$

To determine C_g^{11} the following boundary value problem has to be solved:

$$f_0' \frac{\partial C_g^{11}}{\partial X} + \frac{\sqrt{Sc_g}}{2\sqrt{X}}(\eta f_0' - f_0)\frac{\partial C_g^{11}}{\partial \tilde{Y}_g} = \frac{\partial^2 C_g^{11}}{\partial \tilde{Y}_g^2} - \frac{1}{\alpha} f_0'' \frac{\partial C_g^1}{\partial X} - \frac{\sqrt{Sc_g}}{2\alpha\sqrt{X}}(\eta f_0'' - f_0')\frac{\partial C_g^1}{\partial \tilde{Y}_g}$$
$$- f_{11}' \frac{\partial C_g^0}{\partial X} + \frac{\sqrt{Sc_g}}{2\sqrt{X}}(\eta f_{11}' - f_{11})\frac{\partial C_g^0}{\partial \tilde{Y}_g};$$
$$X = 0, \quad C_g^{11} = 0; \quad \tilde{Y}_g = 0, \quad C_g^{11} = 0; \quad \tilde{Y}_g \to \infty, \quad C_g^{11} = 0. \tag{2.75}$$

The solution of (2.73) is found similarly to that of (2.69), namely,

$$C_g^{11}(X, \tilde{Y}_g) = F_{11}(\eta) = \int_0^\eta \left[\beta_1 - \frac{\beta_0^2 Sc_g}{2\alpha^2} f_0(s) + \frac{\beta_0 Sc_g^2}{8\alpha^2} f_0^2(s) - \frac{\beta_0 Sc_g}{2}\int_0^s f_{11}(t)dt\right]$$
$$\times \exp\left[\frac{Sc_g}{2}\int_0^s f_0(t)dt\right] ds; \quad \beta_1 = \frac{\beta_0^3}{\alpha} - \frac{\beta_0^2 Sc_g^2}{8\alpha^2}\varphi_1(Sc_g) + \frac{\beta_0^2 Sc_g}{2}\varphi_2(Sc_g),$$

$$\varphi_1(Sc_g) = \int_0^\infty f_0^2(s) \times \exp\left[\frac{Sc_g}{2}\int_0^s f_0(t)dt\right] ds,$$

$$\varphi_2(Sc_g) = \int_0^\infty \left[\int_0^s f_{11}(t)dt\right] \times \exp\left[\frac{Sc_g}{2}\int_0^s f_0(t)dt\right] ds. \tag{2.76}$$

These results permit the Sherwood number to be obtained [13]:

$$Sh_{g} = \frac{k_{g} l}{D_{g}} = 2\sqrt{Re_{g}}\varphi(Sc_{g}), Re_{g} = \frac{v_{av} l}{\nu_{g}},$$
$$\varphi = \frac{1}{\varphi_0} + \frac{\theta_1}{\alpha\varphi_0^2} + \theta_1^2\left(\frac{1}{\alpha\varphi_0^3} - \frac{\varphi_1 Sc_g^2}{8\alpha^2\varphi_0^2} + \frac{\varphi_2 Sc_g}{2\varphi_0^2}\right). \tag{2.77}$$

The analysis of the absorption rate

$$J_{g} = \frac{D_{g}}{l}\int_{0}^{l}\left(\frac{\partial c_{g}}{\partial y}\right)_{y=h} dx = k_{g}\left(c_{g0} - c_{g}^{*}\right), \tag{2.78}$$

of highly soluble gas in a laminar liquid film shows the effect on the surface film velocity (θ_1).

3 Diffusion-Type Models

The analysis of the theoretical models indicates that the main difficulties are related to the hydrodynamic equations of the models. The first problem results from the nonlinearity of the equations, whereas the second one originates from the very complicated or unidentified shape of the interface separating the phases. The first problem can be avoided with the help of advanced numerical codes. The second one is a common problem in columnar devices for performing chemical reactions in gas–liquid and liquid–liquid systems such as airlifts and bubble columns. In all these cases, the formulation of the boundary conditions at the interphase surface is practically impossible and as a result the velocity distribution in these devices cannot be obtained. The solution of this problem in model theories (see Sect. 3.1) results in large discrepancies between the theoretical predictions and the performance of the real processes. A way to overcome these difficulties and to develop solutions of existing problems is the use of diffusion-type models.

3.1 Mass Transfer with a Chemical Reaction

Many mass transfer processes in columnar devices may be described by convection–diffusion equations with volumetric reaction terms [14], among them are gas absorption in bubble columns and packed beds performing homogeneous or heterogeneous reactions [15–18] and airlifts for chemical, biochemical, or photochemical reactions [19–23].

The convective transfer in columns results from laminar or turbulent (large-scale pulsations) flows. The diffusive transfer is molecular or turbulent (small-scale pulsations). The volumetric reaction is a mass source (or sink) as a result of chemical reactions or interphase mass transfer.

Let us consider a liquid motion in a column with a homogeneous chemical reaction [14] between two liquid components. If the difference between the component concentrations is very large, then the chemical reaction will be first order. Further, if liquid circulation takes place, the process is nonstationary. If we define the velocity and concentration distributions in the column as

$$u = u(r,z), \quad v = v(r,z), \quad c = c(t,r,z), \tag{3.1}$$

the model can be expressed as

$$\begin{aligned}
&\frac{\partial c}{\partial t} + u\frac{\partial c}{\partial z} + v\frac{\partial c}{\partial r} = D\left(\frac{\partial^2 c}{\partial z^2} + \frac{1}{r}\frac{\partial c}{\partial r} + \frac{\partial^2 c}{\partial r^2}\right) - kc, \quad \frac{\partial u}{\partial z} + \frac{\partial v}{\partial r} + \frac{v}{r} = 0;\\
&t = 0, \quad c = c_0; r = 0, \quad \frac{\partial c}{\partial r} = 0; \quad r = R, \quad \frac{\partial c}{\partial r} = 0;\\
&z = 0, \quad c(t,r,0) = \bar{c}(t,l), \quad \bar{u}(0)\,\bar{c}(t,1) = uc - D\frac{\partial c}{\partial z},
\end{aligned} \tag{3.2}$$

where u and v are the velocities, c is the concentration of the reagent (with small concentration), k is the chemical reaction rate constant, t is the time, r and z are radial and axial coordinates, D is the diffusivity, c_0 is the initial concentration, $\bar{c}(t,l)$ is the average concentration, $\bar{u}(0)$ is the velocity at the inlet (outlet) of the column, and R and l are the column radius and height.

In the cases of heterogeneous reactions performed at the surfaces of catalysts, the chemical reaction rate (mol m^{-2} s^{-1}) should be multiplied by the specific catalytic surface (m^2 m^{-3}) and inserted into (3.2) as a volumetric mass source (sink).

3.2 Interphase Mass Transfer

In the case of interphase mass transfer in gas–liquid or liquid–liquid systems, model (3.2) has to contain convection–diffusion equations for the two phases. Moreover, the chemical reaction rate has to be replaced by the interphase mass transfer rate:

$$k(c_1 - \chi c_2), \tag{3.3}$$

Here, k is interphase mass transfer coefficient, c_1 is the concentration of the transferred substance in the gas (liquid) phase, c_2 is the concentration of the transferred substance in the liquid phase, and χ is Henry's constant (the liquid–liquid mass distribution coefficient).

As a result of all these assumptions, the diffusion model for interphase mass transfer in a column apparatus becomes

$$\varepsilon_i u_i \frac{\partial c_i}{\partial z} = \varepsilon_i D_i\left(\frac{\partial^2 c_i}{\partial z^2} + \frac{1}{r}\frac{\partial c_i}{\partial r} + \frac{\partial^2 c_i}{\partial r^2}\right) - (-1)^{i+1}k(c_1 - \chi.c_2), \tag{3.4}$$

where D_i and $\varepsilon_i (i = 1, 2)$ are diffusivities and holdup coefficients ($\varepsilon_1 + \varepsilon_2 = 1$). The boundary conditions for (3.4) are similar to those in (3.2), but differences depending on the conditions for contact between the two phases are possible.

Let us consider a countercurrent gas–liquid bubble column with column height l, where $c_1(z_1, r)$ and $c_2(z_2, r)$ are the concentrations of the absorbed substance in the gas and liquid phases ($z_1 = l - z_2$). The boundary conditions of (3.4) have the form

$$
\begin{gathered}
z_1 = 0, \quad c_1(0,r) = c_1^0, \quad \bar{u}_1(0)c_1^0 = u_1(0,r)c_1(0,r) - D_1\left(\frac{\partial c_1}{\partial z_1}\right)_{z_1=0}; \\
z_2 = 0, \quad c_2(0,r) = c_2^0, \quad \bar{u}_2(0)c_2^0 = u_2(0,r)c_2(0,r) - D_2\left(\frac{\partial c_2}{\partial z_2}\right)_{z_2=0}; \\
r = 0, \quad \frac{\partial c_1}{\partial r} = \frac{\partial c_2}{\partial r} = 0; \quad r = R, \quad \frac{\partial c_1}{\partial r} = \frac{\partial c_2}{\partial r} = 0,
\end{gathered}
\tag{3.5}
$$

where $\bar{u}_i(0), i = 1, 2$, are the inlet average velocities in the gas and liquid phases.

The extraction processes commonly performed in columnar devices are examples of application of models (3.5).

3.3 Average Concentration Models

Generally, in the diffusion-type models (3.2) and (3.5), the velocity distributions in the phases cannot be obtained. The problem can be avoided if the average values of the velocity and the concentration over the cross-sectional area of the column are used. This approach is more adequate with respect to the experimental data commonly used for the purpose of parameter identification, because measurements of average values (velocity or concentration) are simpler with respect to local (point) measurements.

Let us consider a cylinder with radius $R = R\ (\varphi)$, where φ is the angle in cylindrical coordinates (z, r, φ). The average value of the function $f(z, r, \varphi)$ for the cross-sectional area of the cylinder is

$$
\bar{f}(z) = \frac{\int_{(S)} f(z, r, \varphi)dS}{S}, \tag{3.6}
$$

where

$$
S = \int_0^{2\pi} \frac{[R(\varphi)]^2}{2} d\varphi, \quad \iint_{(S)} f(z, r, \varphi)\, dS = \int_0^{2\pi} \left[\int_0^{R(\varphi)} r f(z, r, \varphi)\, dr \right] d\varphi. \tag{3.7}
$$

For a circular cylinder with R = const. and from (3.6) and (3.7) we have

$$S = \pi R^2, \quad \iint_{(S)} f(z,r)\,dS = 2\pi \int_0^R rf(z,r)\,dr, \quad \bar{f}(z) = \frac{2}{R^2}\int_0^R rf(z,r)\,dr. \quad (3.8)$$

For the average values of the velocity and concentration for the column cross-sectional area from (3.8) it follows that

$$\bar{u}(z) = \frac{2}{R^2}\int_0^R ru(r,z)dr, \quad \bar{v}(z) = \frac{2}{R^2}\int_0^R rv(r,z)dr, \quad \bar{c}(t,z) = \frac{2}{R^2}\int_0^R rc(t,r,z)dr. \quad (3.9)$$

Functions (3.1) can be represented with the help of the average functions (3.9):

$$u(r,z) = \bar{u}(z)\tilde{u}(r,z), \quad v(r,z) = \bar{v}(z)\tilde{v}(r,z), \quad c(t,r,z) = \bar{c}(t,z)\,\tilde{c}(r,z). \quad (3.10)$$

Here $\tilde{u}(r,z)$, $\tilde{v}(r,z)$, and $\tilde{c}(r,z)$ represent the radial nonuniformity of both the velocity and the concentration distributions satisfying the conditions

$$\frac{2}{R^2}\int_0^R r\tilde{u}(r,z)dr = 1, \quad \frac{2}{R^2}\int_0^R r\tilde{v}(r,z)dr = 1, \quad \frac{2}{R^2}\int_0^R r\,\tilde{c}(r,z)dr = 1. \quad (3.11)$$

The average concentration model may be obtained if we put (3.10) into (3.2) and then multiply by r and integrate with respect to r over the interval [0, R]. The result is

$$\frac{\partial\bar{c}}{\partial t} + \alpha(R,z)\bar{u}\frac{\partial\bar{c}}{\partial z} + \frac{\partial\alpha}{\partial z}\bar{u}\bar{c} + a\bar{v}\bar{c} = D\frac{\partial^2\bar{c}}{\partial z^2} - k\bar{c};$$

$$t = 0, \quad \bar{c}(0,z) = c_0;$$

$$t = 0, \quad \bar{c}(0,z) = c_0; \quad z = 0, \quad \bar{c}(t,0) = \bar{c}(t,l), \quad \bar{u}(0)\bar{c}(t,l) = \alpha(R,z)\,\bar{u}\,\bar{c} - D\frac{\partial\bar{c}}{\partial z}, \quad (3.12)$$

where

$$\alpha(R,z) = \frac{2}{R^2}\int_0^R r\tilde{u}\tilde{c}dr, \quad a(R,z) = \frac{2}{R^2}\int_0^R r\tilde{v}\frac{\partial\tilde{c}}{\partial z}dr. \quad (3.13)$$

The average radial velocity component $\bar{v}$ may be obtained from the continuity equation in (3.2) if it is multiplied by r^2 and then integrated with respect to r over the interval [0, R]:

$$\bar{v} = b\frac{\partial\bar{u}}{\partial z} + \frac{\partial b}{\partial z}\bar{u}, \quad b = \frac{2}{R^2}\int_0^R r^2\tilde{u}dr. \quad (3.14)$$

If we put (3.14) into (3.12), the diffusion model has the form

$$\frac{\partial \bar{c}}{\partial t} + \alpha(R,z)\bar{u}\frac{\partial \bar{c}}{\partial z} + \beta\,\bar{u}\bar{c} + \gamma\,\bar{c}\frac{\partial \bar{u}}{\partial z} = D\frac{\partial^2 \bar{c}}{\partial z^2} - k\bar{c}, \tag{3.15}$$

where

$$\beta = \frac{\partial \alpha}{\partial z} + a\frac{\partial b}{\partial z}, \quad \gamma = ab. \tag{3.16}$$

In model (3.15), $\bar{u}$ is the average velocity of the laminar or turbulent flow in the column and D is the diffusivity or turbulent diffusivity (as a result of the small-scale pulsations). The model parameters α, β, and γ are related to the radial nonuniformities and show the effect of the column radius on the mass transfer kinetics.

The parameter k in model (2.13.15) may be obtained in advance by chemical kinetics modeling. If the velocity and the concentration radial nonuniformities are independent of the axial coordinate z, then the parameters α and γ are related to the column radius only. A constant average velocity in these conditions leads to the simplest model $\left(\frac{\partial \bar{u}}{\partial z} = 0\right)$.

The parameters in the diffusion model (3.15) show the scale-up effect of increase of column radius on the mass transfer kinetics if there exists a radial nonuniformity in the velocity distribution.

The identification of the parameters α, β, γ, and D may be performed by means of experimental data concerning average velocities and concentrations, obtained at a laboratory scale. In the case of scale-up, only α (R, z), β (R, z), and γ (R, z) need be obtained because the values of D and k are the same for both the laboratory model and the real-scale column.

The average concentration of the model in the case of interphase mass transfer may be obtained by the analogy with (3.15), namely,

$$\alpha_{\mathrm{i}}(R,z_{\mathrm{i}})\bar{u}_{\mathrm{i}}\frac{\partial \bar{c}_{\mathrm{i}}}{\partial z_{\mathrm{i}}} + \beta_{\mathrm{i}}(R,z_{\mathrm{i}})\bar{u}_{\mathrm{i}}\bar{c}_{\mathrm{i}} + \gamma_{\mathrm{i}}(R,z_{\mathrm{i}})\bar{c}_{\mathrm{i}}\frac{\partial \bar{u}_{\mathrm{i}}}{\partial z_{\mathrm{i}}} = D_{\mathrm{i}}\frac{\partial^2 \bar{c}_{\mathrm{i}}}{\partial z_{\mathrm{i}}^2} - (-1)^{\mathrm{i}+1}\frac{k}{\varepsilon_{\mathrm{i}}}(\bar{c}_1 - \chi\bar{c}_2). \tag{3.17}$$

where α_i, $\beta_i$and $\gamma_i(i = 1, 2)$ are similar to α, β and γ in (3.13), (3.14), and (3.16).

The boundary conditions of (3.17) are

$$\begin{aligned} z_1 &= 0, \quad \bar{c}_1(0) = c_1^0, \quad \bar{u}_1(0)c_1^0 = \alpha_1(R,z_1)\bar{u}_1(0)\bar{c}_1(0) - D_1\left(\frac{\partial \bar{c}_1}{\partial z_1}\right)_{z_1=0}; \\ z_2 &= 0, \quad \bar{c}_2(0) = c_2^0, \quad \bar{u}_2(0)c_2^0 = \alpha_2(R,z_2)\bar{u}_2(0)\bar{c}_2(0) - D_2\left(\frac{\partial \bar{c}_2}{\partial z_2}\right)_{z_2=0}. \end{aligned} \tag{3.18}$$

The radial nonuniformity of the velocity is the main cause of the scale-up effect manifesting itself by reduced process efficiency with increasing column diameter.

The results developed demonstrate that the diffusion models in columnar devices allow us to replace the radial distributions of both the velocity and the concentration by the average values over the column cross-sectional area.

These new models permit us to identify the model parameters using a hierarchical approach. As a first step, the chemical reaction (or mass transfer) rate constant should be defined. The next step addresses the identification of the parametersα, β, γ, and D using experimental data, obtained with real liquids. The scale-up refers to specification of the parameter α, β, and γ for the real-scale device. As a reasonable outcome, the suggested mathematical model may be used for the simulation of real-scale processes carried out in columnar devices.

3.4 Airlift Reactor

The approach developed for modeling mass transfer and chemical processes in columnar devices permits us to model complex processes such as chemical, photochemical, and biochemical reactions in airlift reactors.

The hydrodynamics of the gas and liquid flows in airlift reactors is very complicated. Investigations of airlift reactors have shown [19–23] that the convection–diffusion equation with a volumetric reaction may be used as a mathematical structure for the model. Under these conditions the convective transfer, the diffusive transfer, and the volumetric reactions are carried out simultaneously. The convective transfer is due to laminar or turbulent (large-scale pulsations) flows, whereas the diffusive transport is of a molecular of nature or is promoted by small-scale turbulent pulsations. The volumetric reactions are mass sources (sinks) as a result of chemical (photochemical or biochemical) reactions and interphase mass transfer.

Let us consider a chemical reaction carried out in an airlift reactor [23] with cross-sectional area F_0 of the riser and F_1 of the downcomer. The length of both working zones is l. The volumetric flow rates are Q_0 and Q_1 for the gas phase and liquid phase, respectively. The gas and liquid holdups in the riser are ε and $(1 - \varepsilon)$, respectively.

The concentration of the active gas component in the gas phase is $c(x, r, t)$, whereas that in the liquid phase is $c_0(x, r, t)$ for the riser and $c_1(x_1, r, t)$ for the downcomer, where $x_1 = l - x$. The concentration of the active liquid component in the downcomer is $c_2(x_1, r, t)$, whereas that in the riser is $c_3(x, r, t)$.

The interphase mass transfer rate in the riser is

$$R = k(c - \chi c_0). \tag{3.19}$$

The average velocities in the gas and liquid phases at the inlet (outlet) of the column are

$$\bar{\bar{u}}_0 = \frac{Q_0}{F_0}, \quad \bar{\bar{u}}_1 = \frac{Q_1}{F_0}, \quad \bar{\bar{u}} = \frac{Q_1}{F_1}. \tag{3.20}$$

The chemical reaction rates in the riser and in the downcomer are

$$R_1 = k_0 c_0^{\alpha_1} c_3^{\alpha_2}, \quad R_2 = k_0 c_1^{\alpha_1} c_2^{\alpha_2}. \tag{3.21}$$

The building of the mathematical model of chemical processes in an airlift reactor will be performed on the basis of the differential mass balance in the reactor volume [14], which means employment of convection–diffusion equations with volumetric reactions.

The equations for the active gas component concentration distributions in the gas and liquid phases in the riser are

$$\begin{aligned}
&\varepsilon\frac{\partial c}{\partial t} + \varepsilon u_0 \frac{\partial c}{\partial x} + \varepsilon v_0 \frac{\partial c}{\partial r} = \varepsilon D\left(\frac{\partial^2 c}{\partial x^2} + \frac{1}{r}\frac{\partial c}{\partial r} + \frac{\partial^2 c}{\partial r^2}\right) - k(c - \chi c_0),\\
&\frac{\partial u_0}{\partial x} + \frac{\partial v_0}{\partial r} + \frac{v_0}{r} = 0,\\
&(1-\varepsilon)\frac{\partial c_0}{\partial t} + (1-\varepsilon)\left(u_1 \frac{\partial c_0}{\partial r} + v_1 \frac{\partial c_0}{\partial x}\right) = (1-\varepsilon) D_0 \left(\frac{\partial^2 c_0}{\partial x^2} + \frac{1}{r}\frac{\partial c_0}{\partial r} + \frac{\partial^2 c_0}{\partial r^2}\right)\\
&k(c - \chi c_0) - k_0 c_0^{\alpha_1} c_3^{\alpha_2}, \frac{\partial u_1}{\partial x} + \frac{\partial v_1}{\partial r} + \frac{v_1}{r} = 0.
\end{aligned} \tag{3.22}$$

It is possible to suppose that $\varepsilon = \text{const}$.

The equations for the concentration distributions for the active liquid and gas components in the liquid phase in the downcomer are

$$\begin{aligned}
&\frac{\partial c_1}{\partial t} + u\frac{\partial c_1}{\partial x_1} + v\frac{\partial c_1}{\partial r} = D_1\left(\frac{\partial^2 c_1}{\partial x_1^2} + \frac{1}{r}\frac{\partial c_1}{\partial r} + \frac{\partial^2 c_1}{\partial r^2}\right) - k_0 c_1^{\alpha_1} c_2^{\alpha_2},\\
&\frac{\partial c_2}{\partial t} + u\frac{\partial c_2}{\partial x_1} + v\frac{\partial c_2}{\partial r} = D_2\left(\frac{\partial^2 c_2}{\partial x_1^2} + \frac{1}{r}\frac{\partial c_2}{\partial r} + \frac{\partial^2 c_2}{\partial r^2}\right) - k_0 c_1^{\alpha_1} c_2^{\alpha_2}.
\end{aligned} \tag{3.23}$$

The equation for a concentration distribution for the active liquid component in the riser is

$$\begin{aligned}
(1-\varepsilon)\frac{\partial c_3}{\partial t} + (1-\varepsilon)\left(u_1 \frac{\partial c_3}{\partial r} + v_1 \frac{\partial c_3}{\partial x}\right) &= (1-\varepsilon) D_3 \left(\frac{\partial^2 c_3}{\partial x^2} + \frac{1}{r}\frac{\partial c_3}{\partial r} + \frac{\partial^2 c_3}{\partial r^2}\right)\\
&\quad - k_0 c_0^{\alpha_1} c_3^{\alpha_2}.
\end{aligned} \tag{3.24}$$

The initial conditions will be formulated for this case, when at $t = 0$ the process starts with the beginning of gas motion:

$$\begin{aligned}
&t = 0,\ c(x,r,0) = c^{(0)},\ c_0(x,r,0) = 0,\ c_1(x_1,r,0) = 0,\ c_2(x_1,r,0)\\
&= c_2^{(0)},\ c_3(x,r,0) = c_2^{(0)},
\end{aligned} \tag{3.25}$$

where $c^{(0)}$ and $c_2^{(0)}$ are the initial concentrations of the reagents in the two phases.

The boundary conditions are, in fact, equalities of the concentrations and the mass fluxes at both ends of the working zones $x = 0(x_1 = l)$ and $x = l(x_1 = 0)$. The boundary conditions for $c(x, r, t)$ and $c_0(x, r, t)$ in Eqs. (3.22) are

$$x=0,\quad c=c^{(0)},\quad \bar{\bar{u}}_0 c^{(0)} = u_0 c(0,r,t) - D\left(\frac{\partial c}{\partial x}\right)_{x=0},\quad c_0(0,r,t)=\bar{c}_1(l,t),$$
$$\bar{\bar{u}}\bar{c}_1(l,t) = u_1 c_0(0,r,t) - D_0\left(\frac{\partial c_0}{\partial x}\right)_{x=0};\quad r=0, \frac{\partial c}{\partial r}=\frac{\partial c_0}{\partial r}=0;\quad r=r_0, \frac{\partial c}{\partial r}=\frac{\partial c_0}{\partial r}=0. \tag{3.26}$$

The boundary conditions for $c_1(x_1, r, t)$, $c_2(x_1, r, t)$ and $c_3(x_1, r, t)$ in Eqs. (3.23) and (3.24) are

$$\begin{aligned}
&x_1=0,\quad c_1(0,r,t)=\bar{c}_0(l,t),\quad \bar{\bar{u}}_1\bar{c}_0(l,t)=uc_1(0,r,t)-D_1\left(\frac{\partial c_1}{\partial x_1}\right)_{x_1=0};\\
&r=r_0,\quad \frac{\partial c_1}{\partial r}=0;\quad r=R_0,\quad \frac{\partial c_1}{\partial r}=0;\\
&x_1=0,\quad c_2(0,r,t)=\bar{c}_0(l,t),\quad \bar{\bar{u}}_1\bar{c}_0(l,t)=uc_2(0,r,t)-D_2\left(\frac{\partial c_2}{\partial x_1}\right)_{x_1=0};\\
&r=r_0,\quad \frac{\partial c_2}{\partial r}=0;\quad r=R_0,\quad \frac{\partial c_{21}}{\partial r}=0;\\
&x=0,\quad c_3(0,r,t)=\bar{c}_2(l,t),\quad \bar{\bar{u}}\bar{c}_2(l,t)=u_1c_3(0,r,t)-D_2\left(\frac{\partial c_3}{\partial x}\right)_{x=0};\\
&r=r_0,\quad \frac{\partial c_2}{\partial r}=0;\quad r=R_0,\quad \frac{\partial c_{21}}{\partial r}=0;
\end{aligned} \tag{3.27}$$

The development of the average concentration model starts with the first equation in (3.22) by analogy with (3.17) for $i = 1$ using the average velocities and concentration:

$$\bar{u}_0(x)=\frac{2}{r_0^2}\int_0^{r_0} ru_0(x,r)dr,\quad \bar{v}_0(x)=\frac{2}{r_0^2}\int_0^{r_0} rv_0(x,r)dr, \bar{c}(x,t)=\frac{2}{r_0^2}\int_0^{r_0} rc(x,r,t)dr, \tag{3.28}$$

where $\bar{u}_0(0)=\bar{\bar{u}}_0$. As a result, the following forms are developed:

$$\frac{\partial \bar{c}}{\partial t}+A(r_0,x)\,\bar{u}_0\frac{\partial \bar{c}}{\partial x}+B(r_0,x)\bar{u}_0\bar{c}+G(r_0,x)\,\bar{c}\frac{\partial \bar{u}_0}{\partial x}=D\frac{\partial^2 \bar{c}}{\partial x^2}-\frac{k}{\varepsilon}(\bar{c}-\chi\bar{c}_0), \tag{3.29}$$

where

$$\begin{aligned}
A(r_0,x)&=\frac{2}{r_0^2}\int_0^{r_0} r\,\tilde{u}_0\tilde{c}dr,\quad g(r_0,x)=\frac{2}{r_0^2}\int_0^{r_0} r\,\tilde{v}_0\frac{\partial \tilde{c}}{\partial r}dr,\\
h(r_0,x)&=\frac{2}{r_0^2}\int_0^{r_0} r^2\tilde{u}_0dr, B(r_0,x)=\frac{\partial A}{\partial x}+g\frac{}{\partial x},\quad G(r_0,x)=gh.
\end{aligned} \tag{3.30}$$

The boundary conditions of (3.30) are

$$t = 0, \quad \bar{c}(x,0) = c^{(0)}; \quad x = 0, \quad \bar{c}(0,t) = c^{(0)},$$
$$\bar{\bar{u}}_o c^{(0)} = A(r_0,x)\bar{u}_0(0)\bar{c}(0,t) - D\left(\frac{\partial \bar{c}}{\partial x}\right)_{x=0}; \tag{3.31}$$

The parameters in model (3.29) are of two types—*specific model parameters* (D, k, ε, χ) and *scale model parameters* (A, B, G). The scale parameters are functions of the column radius r_o. They result from the radial nonuniformity in the velocity and the concentration fields, and show the effect of the scale-up on the model equations. The parameter χ may be obtained by thermodynamic measurements in advance.

From (3.29) it follows that the average radial velocity component affects the transfer process in the cases when $\partial \bar{u}_0/\partial x \neq 0$. This simply means that the gas holdup is not constant along the column height. For many practical interesting cases it is possible to assume $\varepsilon =$ const., which yields $\partial \bar{u}_0/\partial x = 0$ and the radial velocity component ($\bar{v}_0 = 0$) can be omitted.

The holdup ε can be obtained from the balance relationship:

$$\varepsilon = \frac{(l - l_0)(F_0 + F_1)}{(l - l_0)(F_0 + F_1) + F_0 l_0}, \tag{3.32}$$

where l and l_o are liquid levels in the riser with and without gas motion.

The values of the parameters D, k, A, B, and G should be obtained by means if experimental data concerning $\bar{c}\,(x,t)$ measured at the laboratory scale. In the case of scale-up, only A, B, and G have to be specified when using a column with a large diameter since D and k do not change with increase of the scale of the device.

The same technology may be applied to the equations of the model, and the result is

$$\frac{\partial \bar{c}_0}{\partial t} + A_0(r_0,x)\,\bar{u}_1\frac{\partial \bar{c}_0}{\partial x} + B_0(r_0,x)\bar{u}_1\bar{c}_0 - G_0(r_0,x)\bar{c}_0\frac{\partial \bar{u}_1}{\partial x}$$
$$= D_0\frac{\partial^2 \bar{c}_0}{\partial x^2} + \frac{k}{1-\varepsilon}(\bar{c} - \chi\bar{c}_0) - M_0(r_0,x)\frac{k_0}{1-\varepsilon}\bar{c}_0^{\alpha_1}\bar{c}_3^{\alpha_2};$$
$$M_0(r_0,x) = \frac{2}{r_0^2}\int_0^{r_o} r\,\tilde{c}_0^{\alpha_1}\tilde{c}_3^{\alpha_2}\,dr; \tag{3.33}$$
$$t = 0, \quad \bar{c}_0\,(x,0) = 0; \quad x = 0, \quad \bar{c}_0\,(0,t) = \bar{c}_1(l,t),$$
$$\bar{c}_1(l,t)\bar{\bar{u}} = A_0(r_0,x)\bar{u}_1(0)\bar{c}_0(0,t) - D_0\left(\frac{\partial \bar{c}_0}{\partial x}\right)_{x=0},$$

where A_0, B_0, and G_0 are obtained by analogy with A, B, and G. The actual expressions for A, B, and G are not of interest because their values should be obtained by means of experimental data.

The average concentration model for $\bar{c}_1$ is obtained by analogy with $\bar{c}$:

$$\begin{gathered}\frac{\partial \bar{c}_1}{\partial t} + A_1(r_0, R_0, x_1)\bar{u}\frac{\partial \bar{c}_1}{\partial x_1} + B_1(r_0, R_0, x_1)\overline{uc}_1 + G_1(r_0, R_0, x_1)\bar{c}_1\frac{\partial \bar{u}}{\partial x_1} \\ = D_1\frac{\partial^2 \bar{c}_1}{\partial x_1^2} - k_0 M(r_0, R_0, x_1)\bar{c}_1^{\alpha_1}\bar{c}_2^{\alpha_2} \\ t = 0, \quad \bar{c}_1(x_1, 0) = 0; \quad x_1 = 0, \quad \bar{c}_1(0, t) = \bar{c}_0(l, t), \\ \bar{c}_0(l, t)\bar{\bar{u}}_1 = A_1(r_0, R_0, x_1)\bar{u}(0)\bar{c}_1(0, t) - D_1\left(\frac{\partial \bar{c}_1}{\partial x_1}\right)_{x_1 = 0},\end{gathered} \tag{3.34}$$

where

$$\begin{gathered}\bar{u}(x_1) = \frac{2}{R_0^2 - r_0^2}\int_{r_0}^{R_0} r u(x_1, r)dr, \quad \bar{c}_1(x_1, t) = \frac{2}{R_0^2 - r_0^2}\int_{r_0}^{R_0} r c_1(x_1, r, t)dr; \\ M(r_0, R_0, x_1) = \frac{2}{R_0^2 - r_0^2}\int_{r_0}^{R_0} r\,\tilde{c}_1^{\alpha_1}\tilde{c}_2^{\alpha_2}dr\end{gathered} \tag{3.35}$$

and A_1, B_1, G_1 are obtained by analogy with A, B, and G, but the limits of the integrals are $[r_0, R_0]$.

The average concentration model for $\bar{c}_2$ is

$$\begin{gathered}\frac{\partial \bar{c}_2}{\partial t} + A_2(r_0, R_0, x_1)\bar{u}\frac{\partial \bar{c}_2}{\partial x_1} + B_2(r_0, R_0, x_1)\bar{u}\bar{c}_2 + G_2(r_0, R_0, x_1)\bar{c}_2\frac{\partial \bar{u}}{\partial x_1} \\ = D_2\frac{\partial^2 \bar{c}_2}{\partial x_1^2} - M(r_0, R_0, x_1)k_0\bar{c}_1^{\alpha_1}\bar{c}_2^{\alpha_2}; \\ t = 0, \quad \bar{c}_2 = c_2^{(0)}; \quad x_1 = 0, \quad \bar{c}_2(0, t) = \bar{c}_3(l, t), \\ \bar{\bar{u}}_1\bar{c}_3(l, t) = A_2(r_0, R_0, x_1)\bar{u}(0)\bar{c}_2(0, t) - D_2\left(\frac{\partial \bar{c}_2}{\partial x_1}\right)_{x_1 = 0},\end{gathered} \tag{3.36}$$

where

$$\bar{c}_2 = \frac{2}{R_0^2 - r_0^2}\int_{r_0}^{R_0} r c_2(x_1, r, t)dr, \quad \bar{c}_3 = \frac{2}{r_0^2}\int_0^{r_0} r c_3(x, r, t)dr. \tag{3.37}$$

The integration of (3.24) leads to the equation for $\bar{c}_3$, namely,

$$\frac{\partial \bar{c}_3}{\partial t}+A_3(r_0,x)\bar{u}_1\frac{\partial \bar{c}_3}{\partial x}+B_3(r_0,x)\bar{u}_1\bar{c}_3+G_3(r_0,x)\bar{c}_3\frac{\partial \bar{u}_1}{\partial x}=D_3\frac{\partial^2 \bar{c}_3}{\partial x^2}-M_0(r_0,x)\frac{k_o}{1-\varepsilon}\bar{c}_0^{\alpha_1}\bar{c}_3^{\alpha_2};$$

$$t=0,\ \bar{c}_3(x,0)=c_2^{(0)};\ \ x=0,\ \bar{c}_3(0,t)=\bar{c}_2(l,t),$$

$$\bar{\bar{u}}\bar{c}_2(l,t)=A_3(r_0,x)\bar{u}_1(0)\bar{c}_3(0,t)-D_3\left(\frac{\partial \bar{c}_3}{\partial x}\right)_{x=0}, \tag{3.38}$$

where A_3, B_3, and G_3 are obtained by analogy with A, B, and G.

For many practically interesting cases the specific volume ($m^3\ m^{-3}$) of the catalytic particles or gas holdup are almost constant along the column, i.e.,

$$\frac{\partial \bar{u}}{\partial x}=\frac{\partial \bar{u}_0}{\partial x}=\frac{\partial \bar{u}_1}{\partial x}=0, \quad \bar{v}=\bar{v}_0=\bar{v}_1=0, \tag{3.39}$$

and the number of model parameters decreases, i.e., $G=G_0=G_1=G_2=G_3=0$.

Problems (3.29), (3.33), (3.34), (3.36), and (3.38) are the mathematical model of an airlift chemical reactor applicable to either homogeneous or heterogeneous reactions. The model parameters are of five types:

1. Known beforehand ($c^{(0)}$, $c_2^{(0)}$, R_0, r_0).
2. Obtained beforehand (ε, χ, α_{1z}, α_2, k_0).
3. Obtained without a chemical reaction (k, D, D_0, A, B, A_0, B_0, G, G_0).
4. Obtained with a chemical reaction (D_1, D_2, D_3, M, M_0).
5. Obtained in the modeling and specified in the scale-up (A, A_0, A_1, A_2, A_3, B, B_0, B_1, B_2, B_3, G, G_0, G_1, G_2, G_3, M, M_0).

These results show the possibility to create models for an airlift reactor using average velocities and concentrations only instead of the corresponding radial distribution profiles. This approach permits us to solve the scale-up problem caused by radial nonuniformities in the velocity and concentration fields. The model parameter identification is based on average values of experimental data concentration, which results in advantages with respect to local concentration measurements.

4 Similarity Theory Models

Physical modeling is another possibility for quantitative description of the processes in chemical engineering. Obtaining experimental data from a process in a model apparatus and using scale coefficients, we can find a quantitative description of a given process in a real device, if these two processes are similar, of course. The similarity theory formulates similarity conditions on the basis of a penetrating physical analysis. This theory uses very simple mathematical methods and many applications are encountered. However, in many cases when the physical basis of the similarity theory is ignored, the results of this theory are unreasonable and the *theory of similarity becomes a similarity of theory.* However, we follow the correct approach, which be exemplified by the gas absorption problem developed next.

4.1 Absorption in a Packed-Bed Column

Let us consider absorption of a slightly soluble gas in a packed-bed column [1]. The packed bed is ordered (structured packing) and forms vertical canals of width d and height l. Both the gas and the liquid move in the canals in gas–liquid film co-current flow mode. In such a case (see Sect. 2.1 for details) the mathematical description has the form

$$
\begin{aligned}
&u_x^{(g)}\frac{\partial u_x^{(g)}}{\partial x}+u_y^{(g)}\frac{\partial u_x^{(g)}}{\partial y}=\nu_g\frac{\partial^2 u_x^{(g)}}{\partial y^2},\quad \frac{\partial u_x^{(g)}}{\partial x}+\frac{\partial u_y^{(g)}}{\partial y}=0;\\
&\nu\frac{\partial^2 u_x}{\partial y^2}+g=0,\quad \frac{\partial u_x}{\partial x}+\frac{\partial u_y}{\partial y}=0;\ u_x\frac{\partial c}{\partial x}+u_y\frac{\partial c}{\partial y}=D\frac{\partial^2 c}{\partial y^2};\\
&x=0,\ u_x^{(g)}=u_0^{(g)};\ y=0,\ u_x=0,\ u_y=0;\\
&y=h,\ u_x^{(g)}=u_x, u_y^{(g)}=0,\ \mu_g\frac{\partial u_x^{(g)}}{\partial y}=\mu\frac{\partial u_x}{\partial y},\ c=c^*;\\
&y=0,\ \frac{\partial c}{\partial y}=0; y=\frac{d}{2},\ \frac{\partial u_x^{(g)}}{\partial y}=0,
\end{aligned}
\tag{4.1}
$$

where $x = 0$ is the canal inlet, $y = 0$ is the solid interface, h is the film thickness (obtained from liquid flow rate Q; see 1.2.64), and $c^*$is the equilibrium concentration.

4.2 Generalized (Dimensionless) Variables

The generalized variables are defined through the characteristic process scales as

$$
\begin{aligned}
&X=\frac{x}{l},\quad Y_g=\frac{y-h}{d},\quad Y=\frac{y}{h},\quad \varepsilon_g=\frac{d}{l},\quad \varepsilon=\frac{h}{l},\\
&U_x^{(g)}=\frac{u_x^{(g)}}{u_0^{(g)}},\quad U_y^{(g)}=\frac{u_y^{(g)}}{\varepsilon_g u_0^{(g)}},\quad U_x=\frac{u_x}{u_0},\quad U_y=\frac{u_y}{\varepsilon u_0},\quad C=\frac{c-c^*}{c_0-c^*}.
\end{aligned}
\tag{4.2}
$$

If (4.2) is introduced into (4.1), then the mathematical description has the form

$$U_x^{(g)}\frac{\partial U_x^{(g)}}{\partial X}+U_y^{(g)}\frac{\partial U_x^{(g)}}{\partial Y_g}=\frac{1}{\varepsilon_g Re_g}\frac{\partial^2 U_x^{(g)}}{\partial Y_g^2},\qquad \frac{\partial U_x^{(g)}}{\partial X}+\frac{\partial U_x^{(g)}}{\partial Y}=0,$$

$$\frac{\partial^2 U_x}{\partial Y^2}+\frac{Re}{Fr}=0,\qquad \frac{\partial U_x}{\partial X}+\frac{\partial U_y}{\partial Y}=0,\qquad U_x\frac{\partial C}{\partial X}+U_y\frac{\partial C}{\partial Y}=\frac{1}{\varepsilon Pe}\frac{\partial^2 C}{\partial Y^2};$$

$$X=0,\quad U_x^{(g)}=1,\quad U_x=1,\quad C=1;\quad Y=0,\quad U_x=0,\quad U_y=0,\quad \frac{\partial C}{\partial Y}=0;$$

$$Y=1,\quad Y_g=0,\quad U_x^{(g)}=\frac{u_0}{u_0^{(g)}}U_x,\quad U_y^{(g)}=0,\quad \frac{\partial U_x}{\partial Y}=\frac{\mu_g h u_0^{(g)}}{\mu d u_0}\frac{\partial U_x^{(g)}}{\partial Y_g},\qquad C=0;$$

$$Y_g=\frac{1}{2}-\frac{h}{d},\quad \frac{\partial U_x^{(g)}}{\partial Y_g}=0, \tag{4.3}$$

where

$$Re_g=\frac{u_0^{(g)}h}{\nu_g},\quad Re=\frac{u_0 h}{\nu},\quad Fr=\frac{u_0^2}{gh},\quad Pe=\frac{u_0 h}{D}, \tag{4.4}$$

where Re, Fr, and Pe are Reynolds, Froude, and Péclet numbers, respectively.

From (4.3) it can be seen that the mathematical description of the process contains six dimensionless (generalized) parameters:

$$A_1=\left(\varepsilon_g Re_g\right)^{-1},\quad A_2=\frac{Re}{Fr},\quad A_3=(\varepsilon Pe)^{-1},\quad A_4=\frac{\mu_g h u_0^{(g)}}{\mu d u_0},\quad A_5=\frac{u_0}{u_0^{(g)}},$$
$$A_6=\frac{h}{d}, \tag{4.5}$$

The parameters A_1, …, A_4 are of complex type (complex of different dimension quantity), whereas A_5, A_6 are of simplex type (ratio of quantities with the same dimensions).

The absorption (desorption) rate can be presented as

$$J=k(c^*-c_0)=\frac{D(c^*-c_0)}{h}\int_0^1\left(\frac{\partial C}{\partial Y}\right)_{Y=1}dX, \tag{4.6}$$

from which the Sherwood number follows:

$$Sh=\frac{kh}{D}=\int_0^1\left(\frac{\partial C}{\partial Y}\right)_{Y=1}dX. \tag{4.7}$$

4.3 Generalized Individual Case and Similarity

Problem (4.3) is a mathematical description of a *generalized case*, which represents an unlimited set of processes. For particular (individual) values of the parameters $A_i = A_i^0$, $i = 1, \ldots, 6$, there exists a subset, which will be named [24] *generalized individual case*. All processes in the generalized individual case are similar to each other, i.e., each process in the generalized individual case could be considered as a physical model of the others. The equality of the separate dimensionless parameters in the mathematical description by means of generalized variables of the two processes considered is the *similarity criterion* between them.

From (4.3) it can be seen that the dimensionless concentration distribution depends only on the independent variables and parameters:

$$C = F_1(X, Y, A_1, \ldots, A_6). \tag{4.8}$$

Hence, we have

$$\begin{aligned} &\left(\frac{\partial C}{\partial Y}\right)_{Y=1} = F_2(X, A_1, \ldots, A_6), \quad \int_0^1 F_2(X, A_1, \ldots, A_6)dX = F(A_1, \ldots, A_6), \\ &Sh = F(A_1, \ldots, A_6), \end{aligned} \tag{4.9}$$

where the function F can be obtained after solution of the problem (4.3).

The particular experimental conditions of a real process (absorption) allow us to assess the values of the parameters A_i^0, $i = 1, \ldots, 6$. If the particular experimental conditions of the model process assess the same parameter values, then an experimental determination of $Sh = Sh^0$ for the model permits us to determines the Sherwood number for the real-scale process, because the relation

$$Sh^0 = F\left(A_1^0, \ldots, A_6^0\right) \tag{4.10}$$

is valid for both the model and the real process.

Let us denote the model quantities with subscript 1. The absorption rate in the real process J can be obtained from experimental data taken from the model (J_1), namely,

$$k_1 = \frac{J_1}{c^* - c_0}, \quad Sh^0 = \frac{k_1 h_1}{D_1}, \quad k = Sh^0 \frac{D}{h}, \quad J = k(c^* - c_0). \tag{4.11}$$

These results show that the experimental determination of the absorption rate in the model determines the absorption rate in the real process. In other words the experimental results obtained with the model are the same as those obtained with the real process. The experimental approach for the Sherwood number determination without solution of problem (4.3) is, in fact, an *experimental solution of a boundary valueproblem.*

This analysis clearly reveals that the similarity condition means the equality of the similarity criteria in both the model and the real process. However, each dimensionless parameter is not a similarity criterion. From (1.4.10) it can be seen that the equality of Sh^0 of the model with that of the real process is not a similarity criterion, but only a result of the process similarity.

4.4 Mathematical Structure of the Models

From (4.10) a similarity criteria model can be obtained if a suitable mathematical structure of the function F is available and experimental data to obtain Sh^0 for different values of the parameters A_i^0, i = 1, ..., 6 are used.

The function F describes a physical process, i.e., its mathematical structure must be invariant with respect to similarity transformations [24].

Let us consider the model equation

$$f(x_1, \ldots, x_m) = 0, \tag{4.12}$$

which is invariant regarding similarity transformations,

$$\bar{x}_i = k_i x_i, \quad i = 1, \ldots, m, \tag{4.13}$$

if f is a homogenous function, i.e.,

$$f(k_1 x_1, \ldots, k_m x_m) = \varphi(k_1, \ldots, k_m) f(x_1, \ldots, x_m) = 0. \tag{4.14}$$

A brief expression of (4.14) is

$$f[\bar{x}_i] = \varphi[k_i] f[x_i]. \tag{4.15}$$

The problem is expressed in a form allowing us to obtain the function f satisfying equation (4.15).

Differentiation of (4.15) concerning k_1 leads to

$$\frac{\partial f[\bar{x}_i]}{\partial k_1} = \frac{\partial \varphi}{\partial k_1} f(x_i). \tag{4.16}$$

On the other hand, we have

$$\frac{\partial f[\bar{x}_i]}{\partial k_1} = \frac{\partial f[\bar{x}_i]}{\partial \bar{x}_1} \frac{\partial \bar{x}_1}{\partial k_1} = \frac{\partial f[\bar{x}_i]}{\partial \bar{x}_1} x_1. \tag{4.17}$$

Equation (4.17) is valid for different values of k_i, including $k_i = 1 (i = 1, \ldots, m)$. As a result, we get $\bar{x}_i = x_i (i = 1, \ldots, m)$ and from (4.16) and (4.17) it follows that

$$\frac{\partial f[\bar{x}_i]}{\partial \bar{x}_1} x_1 = b_1 f[x_i], \tag{4.18}$$

where

$$b_1 = \left(\frac{\partial \varphi}{\partial k_1} \right)_{k_i = 1}. \quad (4.19)$$

From (4.18) it follows that

$$\frac{1}{f} \frac{\partial f}{\partial x_1} = \frac{b_1}{x_1}, \quad (4.20)$$

That is,

$$f = c_1 x_1^{b_1}. \quad (4.21)$$

If we reiterate the upper operations for $x_2, \ldots, x_m$, then the homogenous function f takes the form

$$f = c x_1^{b_1}, \ldots, x_m^{b_m}, \quad (4.22)$$

which means that the *function f is homogenous if it is represented as a product of a power–law complex.* This result is invariant regarding similarity (metric) transformations.

The result (4.22) defines the use of power–law-function complexes as elements of mathematical structures of the similarity theory models. Similarly, relations between primary and secondary quantities can be proved through the so-called *dimension analysis*. In this context, the power–law-function complexes are quite convenient as kinetic models of complicated chemical reactions (introducing chemical reaction order higher than 1).

From (4.10) and (4.22) it follows that the models of similarity theory (*criteria models*) have the power–law-function complexes expressed as

$$Sh^0 = b_0 \prod_{i=1}^{6} A_i^{0 b_i}. \quad (4.23)$$

If the logarithm of (4.23) is applied, then the result is a linear-regression-type model, namely,

$$\lg Sh^0 = \lg b_0 + \sum_{i=6}^{6} b_i \lg A_i^0, \quad (4.24)$$

The model parameters b_i, $i = 1, \ldots, 6$ can be obtained (like in the regression models; see Sect. 2.5) using the Sherwood number values Sh_j^0, $j = 1, \ldots, N$, $N \geq 7$, calculated from experimental data under N different experimental conditions. The calculated similarity criteria values are A_{ij}^0, $i = 1, \ldots, 6$, $j = 1, \ldots, N$, respectively. The determination of b_0 after applying an antilogarithm to (4.24) is very inexact as an approach because b_0 has to be calculated after determination of b_i, $i = 1, \ldots, 6$:

$$b_0 = \frac{1}{N}\sum_{j=1}^{N} \frac{Sh_j^0}{\prod_{i=1}^{6}\left(A_{ij}^0\right)^{b_i}}. \tag{4.25}$$

The theoretical analysis of the process similarity [24] shows that the similarity criteria models contain two types of dimensionless parameters—*determinant* or *independent dimensionless variables* determining the similarity conditions and *determined variables* (*dependent dimensionless variables*), whose values are determined by the similarity conditions imposed. In model (4.23) the determinant (independent) parameters playing the roles of similarity criteria are $A_1^0, \ldots, A_6^0$, whereas Sh^0 is a determined (dependent) parameter (process variable). Obviously, the specification of the dimensionless model parameters is very important when building the criteria model. For example, from (2.6) it can be seen that Fo is a determinant (independent) parameter, whereas the determined parameter is the ratio $Sh\Big/\sqrt{\frac{6Pe}{\pi}}$. Hence, the form of the criteria model is

$$Sh\sqrt{\frac{\pi}{6Pe}} = \left(1 - \frac{Fo}{6} - \frac{19Fo^2}{120}\right). \tag{4.26}$$

The determinant parameters obtained by scaling the differential equations are ratios of physical effects represented by the terms (differential operators). More precisely, A_1 for instance, expresses the ratio between the force driving the fluid representing the convective motion and the surface (viscosity) force representing the diffusive transport of momentum. In this context, A_2 is the ratio of the surface (viscosity) forces and the body forces (gravitational ones in this particular example), whereas A_3 represents the ratio of the convective and diffusive mass transfer:

$$\begin{aligned} A_1 &= \left(\varepsilon_g Re_g\right)^{-1} \sim \frac{\left(u_x^{(g)}\frac{\partial u_x^{(g)}}{\partial x} + u_y^{(g)}\frac{\partial u_x^{(g)}}{\partial y}\right)}{\nu_g \frac{\partial^2 u_x^{(g)}}{\partial y^2}}, A_2 = \frac{Re}{Fr} \sim \frac{\nu\frac{\partial^2 u_x}{\partial y^2}}{g}, A_3 \\ &= (\varepsilon Pe)^{-1} \sim \frac{\left(u_x\frac{\partial c}{\partial x} + u_y\frac{\partial c}{\partial y}\right)}{D\frac{\partial^2 c}{\partial y^2}}. \end{aligned} \tag{4.27}$$

The determinant parameters defined by the boundary conditions are ratios of physical scales. For example, A_4 is the ratio of the tangential components of the stress tensor in two phases (shear forces), whereas A_5 and A_6 are ratios of velocity and linear scales, respectively:

$$A_4 = \frac{\mu_g h u_0^{(g)}}{\mu d u_0} \sim \frac{\mu_g \frac{\partial u_x^{(g)}}{\partial y}}{\mu \frac{\partial u_x}{\partial y}}, \quad A_5 = \frac{u_0}{u_o^{(g)}} \sim \frac{u_x^{(g)}}{u_x}, \quad A_6 = \frac{h}{d}. \tag{4.28}$$

The determined parameters are usually dimensionless target (subject) functions such as mass transfer rate (J) and pressure drop (Δp). In the context of the above comments we have

$$Sh = \frac{kh}{D} \sim \frac{k(c^* - c_0)}{\frac{D}{l}\int\limits_0^l \left(\frac{\partial c}{\partial y}\right)_{y=h} dx}, \qquad Eu = \frac{\Delta p}{\rho u^2}, \tag{4.29}$$

where Eu is the Euler number.

The presence of N values of $Sh_j^0, j = 1, \ldots, N, N \geq 7$ indicates the existence of experimental data obtained from N models and characterized by their own parameter values A_{ij}^0, $i = 1, \ldots, 6$, $j = 1, \ldots, N$. Model (4.23) permits us to simulate all processes with the determinant parameters within the intervals defined by its minimum and maximum values:

$$A_{ij}^{0\min} \leq A_{ij}^0 \leq A_{ij}^{0\max}, \quad i = 1, \ldots, 6, j = 1, \ldots, N. \tag{4.30}$$

Hence, (4.23) can be used only for interpolation.

The presentation of the similarity theory models (criteria models) shows that they are mathematical models through physical modeling of experimental data.

4.5 Dimension Analysis

A possibility for similarity criteria formulation in the case of the absence of the mathematical description of the process is the *dimension analysis* approach. It is applicable if the complete (exact) combinations of the physical quantities involved in the process are well known, which affects the form of the target (subject) function of the process.

Let us suppose that the process target function z depends on n physical quantities:

$$z = f(x_1, \ldots, x_{\mathrm{m}}; y_1, \ldots, y_{\mathrm{r}}), \quad \mathrm{m} + \mathrm{r} = \mathrm{n}, \tag{4.31}$$

where $x_i (i = 1, \ldots, \mathrm{m})$ are primary quantities, i.e., their values are the result of direct measurement (length, time, mass), whereas $y_j (j = 1, \ldots, \mathrm{r})$ are secondary quantities, i.e., their values are the result of the combinations of primary quantity values:

$$y_j = f_j(x_1, \ldots, x_{\mathrm{m}}), j = 1, \ldots, \mathrm{r}. \tag{4.32}$$

These functions must be invariant regarding similarity (metric) transformations, i.e., to *represent a power–law-function complex (product)*:

$$y_j = x_1^{A_{1j}}, \ldots, x_{\mathrm{m}}^{A_{\mathrm{m}j}}, j = 1, \ldots, \mathrm{r}. \tag{4.33}$$

The physical quantities have symbols and dimensions. If the symbols of the primary quantities (length, time, mass) are L, T, M and their dimensions are denoted by $[L]$, $[T]$, $[M]$, then we may suggest that they are equal, i.e.,

$$[L] = L, \quad [T] = T, \quad [M] = M. \tag{4.34}$$

The dimensions of the secondary quantities can be determined by power–law functions. For example, for the velocity and the force involved in a certain process, we have

$$[V] = LT^{-1}, \quad [F] = MLT^{-2}. \tag{4.35}$$

The main problem of dimension analysis is how to obtain relations between the physical quantities in a quantitative description of the process. The problem is solved by employment of the condition of *equality of thedimensions* of both sides of the physical equations, i.e., *both sides of the equation should be dimensionally homogeneous.*

If the dimensions of the physical quantities in (4.31) are

$$[x_i] = k_i, i = 1, \ldots, m; \quad [y_j] = K_j, j = 1, \ldots, r, \tag{4.36}$$

then from (4.33) it follows that

$$K_j = k_1^{A_{1j}} \ldots k_m^{A_{mj}}. \tag{4.37}$$

Let us suppose that the function (4.31) has the form of a product of power–law functions, namely,

$$\pi = x_1^{\alpha_1} \ldots x_m^{\alpha_m} \cdot y_1^{\beta_1} \ldots y_r^{\beta_r}, \tag{4.38}$$

Then, the dimension equation is

$$k_\pi = k_1^{\alpha_1} \ldots k_m^{\alpha_m} \cdot K_1^{\beta_1} \ldots K_r^{\beta_r}, \tag{4.39}$$

and k_i, $i = 1, \ldots, m$; K_j, $j = 1, \ldots, r$, have to be obtained from the condition of dimensional homogeneity of both sides of (4.39).

If π is dimensionless, $k_\pi = 1$ and from (4.39) it follows that

$$\prod_{i=1}^{m} k_i^{\gamma_i} = 1, \tag{4.40}$$

where

$$\gamma_i = \alpha_i + \sum_{j=1}^{r} A_{ij}\beta_j = 0, \quad i = 1, \ldots, m. \tag{4.41}$$

Let us consider a solution of (4.41), $\alpha_1^{(0)}$, $\beta_j^{(0)}$, $i = 1, \ldots, m$, $j = 1, \ldots, r$, i.e.,

$$\alpha_i^{(0)} + \sum_{j=1}^{r} A_{ij}\beta_j^{(0)} = 0, \quad i = 1, \ldots, m \tag{4.42}$$

and the power–law complex $x_1^{\alpha_1^{(0)}} \ldots x_m^{\alpha_m^{(0)}} \cdot y_1^{\beta_1^{(0)}} \ldots y_r^{\beta_r^{(0)}}$ is dimensionless. The set of linear equations (4.42) consists of m equations, n = m + r unknowns, and r = n − m independent solutions, i.e., n different physical quantities. m are primaries, and form r = n − m dimensionless complexes (the "π-theorem" of Buckingham).

The dimension analysis approach has restricted application owing to some limitations, among them are:

- Incorrect combinations of physical quantities lead to wrong results.
- It is impossible to specify similarity criteria and determined parameters.

4.6 Some Errors in Criteria Models

The establishment of the criteria model is very simple, but the absence of the physical analysis leads to incorrect results.

Let us consider heat transfer between solid particles and a column wall in a fluidized-bed column [25]. There are enough experimental data about the heat transfer coefficient (k) for particles of various diameters (d) and a variety of gas velocities (u). With these data the Reynolds and the Sherwood numbers using the particle diameter as a length scale can be obtained, namely,

$$Re = \frac{ud\rho}{\mu}, \quad Nu = \frac{kd}{\lambda}, \tag{4.43}$$

where ρ, μ, λ are the density, viscosity, and thermal conductivity of the gas. From these data, the criteria model has the form

$$Nu = 0.13Re^{0.79}. \tag{4.44}$$

A comparison of this model with experimental data is shown in Fig. 1. The correlation is not very good although the exponent of Re is close to 0.8 (occurring very often in many empirical heat transfer/fluid flow correlations published in the literature). It is reasonable to suggest that some devices did work properly.

Another possibility for the process modeling is to use the Weber and the Stanton numbers in gas–liquid bubbly flows because the existence of bubbles is natural:

$$We = \frac{u^2 d\rho}{\sigma}, \quad St = \frac{k}{\rho u c_p}, \tag{4.45}$$

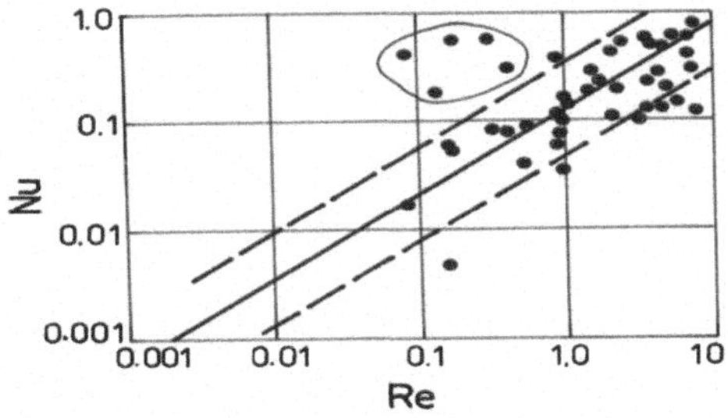

Fig. 1 Nusselt number–Reynolds number relation

In gas-fluidized beds both the gas bypasses through the bubbles together with the promoted mixing affect the heat transfer. Having in mind that bubbles have a surface energy associated with the gas–liquid interface, we can reasonably to suggest that the Weber number may characterize the bed behavior. In (4.45) σ and c_p are the liquid surface tension and the specific heat capacity, respectively. Further, the convective transfer mainly contributes to the heat transfer, which allows the Nusselt number to be replaced by the Stanton number. Hence, under these conditions, the model is

$$St = 0,2We^{-0,5}. \tag{4.46}$$

A comparison with the experimental data is shown in Fig. 2.

Other possibilities for the creation of criteria models using the same experimental data are presented in [25], for example,

$$\begin{aligned} &St = 2Re^{-1,12}, \quad Nu = f_1(Fr), \quad Nu = f_2(Gr), \quad Nu = \frac{d}{D} = 0,26Gr^{0,82}, \\ &St = f_3(Fr.We), Fr = \frac{u^2}{gd}, Gr = \frac{d^3\rho^2 g\beta\Delta\vartheta}{\mu^2}, \end{aligned} \tag{4.47}$$

Here, Fr and Gr are the Froude and Grashof numbers, respectively, g the gravity acceleration, D the column diameter, β the thermal expansion coefficient, and $\Delta\vartheta$ is the temperature difference. All these models are in good agreement with the experimental data [25].

In reality, the foregoing example is fictitious. The numbers ascribed to the quantities u, d, and k from which the various dimensionless groups were created

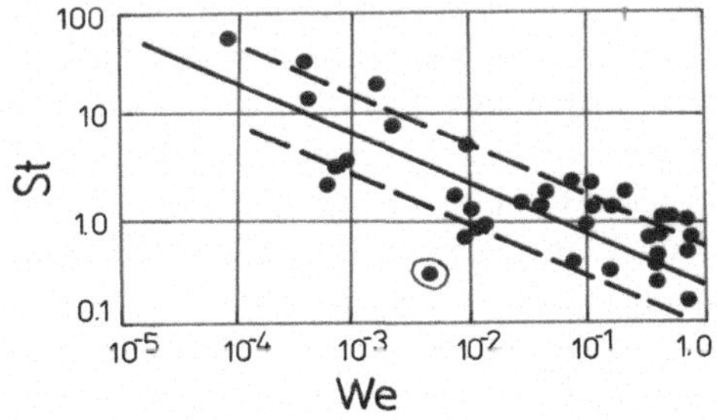

Fig. 2 Stanton number–Weber number relation

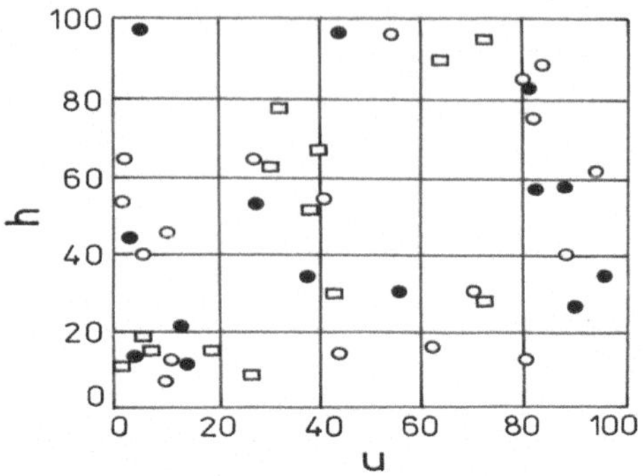

Fig. 3 Heat transfer coefficient (kcal $m^{-1}s^{-1o}$ C^{-1})—velocity relation (*m/s*)

were taken from adjacent columns in a table of random numbers varying from 0 to 99. Hence, a correlation between the experimental data is completely missing. The original data are plotted in Fig. 3 on a linear scale as *k* against *u* for "large", "medium," and "small" values of *d*.

The curious fact mentioned above shows that flagrant errors can be produced in the creation of the criteria models. This especially refers to the *selection of the determinant and determined dimensionless complexes (groups). The cause of this curious fact is the presence of a joint (common) quantity contributing simultaneously to both the determinant and the determined dimensionless complexes*

In the above models, the next correlations are used:

$$\begin{aligned}&\frac{kd}{\lambda}\to\frac{ud\rho}{\mu},\quad \frac{k}{u\rho c_{\mathrm{p}}}\to\frac{ud\rho}{\mu},\quad \frac{k}{u\rho c_{\mathrm{p}}}\to\frac{u^2d\rho}{\sigma},\quad \frac{k}{u\rho c_{\mathrm{p}}}\to\frac{u^2}{gd},\\&\frac{kd}{\lambda}\to\frac{d^3\rho^2g\rho\Delta\vartheta}{\mu^2},\quad \frac{kd}{\lambda}\frac{d}{D}\to\frac{d^3\rho^2g\rho\Delta\vartheta}{\mu^2},\quad \frac{k}{u\rho c_{\mathrm{p}}}\to\frac{u^2}{gd}\frac{u^2d\rho}{\sigma},\end{aligned}\tag{4.48}$$

Having in mind that in (4.48) the variables are *k*, *u*, and *d* only, the correlations in (4.48) are very simple:

$$kd\to ud,\quad \frac{k}{u}\to ud,\quad \frac{k}{u}\to u^2d,\quad \frac{k}{u}\to\frac{u^2}{d},\quad kd\to d^3,\quad kd^2\to d^3,\quad \frac{k}{u}\to u^4.\tag{4.49}$$

From (4.49) it can be seen that if a correlation between *k*, *d*, and *u* does not exist, *kd* and *ud* are correlated because they are dependent on *d*. A similar situation exists between *k*/*u* and *ud*, where this "correlating" effect is caused by *u*. It is evident that this *correlating effect* increases if the exponent of the *correlating quantity* increases (see other cases in 4.49).

The examples presented show [25, 49] that the absence of serious physical analysis in the process of using the similarity theory and the similarity criteria models leads to unusable results.

The theoretical analysis of the similarity theory demonstrates that the usage of mathematical methods is very simple, but a formal utilization of this theory can lead to a wrong result (e.g., as a result of incorrect formulation of the similarity conditions and the determinant and determined dimensionless parameters). Learning the physical ideas, which are the basis of the similarity theory and its mathematical methods, may lead to the correct understanding of this theory as an approach for quantitative investigations. Application of the similarity theory requires first of all good understanding of its physical background. *Modeling without understanding (realization) the similarity may lead to reduction of the similarity theory to a similarity of the theory.*

5 Regression Models

The experimental data of a concrete process can be represented as a discrete function:

$$y_{\mathrm{n}}^{\mathrm{exp}} = \varphi^{\mathrm{exp}}(x_{1\mathrm{n}}, \ldots, x_{\mathrm{mn}}), \quad \mathrm{n} = 1, \ldots, \mathrm{N}, \tag{5.1}$$

where y is the process subject (target) function, x is an independent variable of the process, and N is the number of experiments.

The model of the process represents a continuous function:

$$y = \varphi(x_1, \ldots, x_{\mathrm{m}}; b_1, \ldots, b_{\mathrm{k}}), \tag{5.2}$$

where b is a model parameter, which must be obtained using the experimental data (5.1). In the case of full absence of information about the process mechanism, regression models can be used. The form of function (5.2) is unknown, but we can suppose that this function is continuous (its derivates too) in the vicinity of point $x_0 = x_{10}, \ldots, x_{\mathrm{m}0}$ and can be represented as a Taylor series:

$$y(x) = y(x_0) + \sum_{\mathrm{i}=1}^{\mathrm{m}} \left. \frac{\partial y(x)}{\partial x_{\mathrm{i}}} \right|_{x=x_0} (x_{\mathrm{i}} - x_{\mathrm{i}0}) + \sum_{\mathrm{i}=1}^{\mathrm{m}-1} \sum_{\mathrm{j}=\mathrm{i}+1}^{\mathrm{m}} \left. \frac{1}{2!} \frac{\partial^2 y(x)}{\partial x_{\mathrm{i}} \partial x_j} \right|_{x=x_0} (x_1 - x_{10})(x_j - x_{\mathrm{j}10})$$
$$+ \ldots + \sum_{\mathrm{i}=1}^{\mathrm{m}} \left. \frac{1}{2!} \frac{\partial^2 y(x)}{\partial x_{\mathrm{i}}^2} \right|_{x=x_0} (x_{\mathrm{i}} - x_{\mathrm{i}10})^2 + \ldots \tag{5.3}$$

The form of the derivates in (5.3) is unknown, but (5.3) can be represented as

$$y(x) = b_0 + \sum_{i=1}^{\mathrm{m}} b_1 x_1 + \sum_{i=1}^{\mathrm{m}-1} \sum_{j=i+1}^{\mathrm{m}} b_{ij} x_i x_j + \sum_{i=1}^{\mathrm{m}} b_{\mathrm{ii}} x_i^2 + \cdots, \tag{5.4}$$

where b is a parameter in the regression model (5.4) and will be obtained using the experimental data (5.1).

5.1 Regression Equations

The polynomial models (5.4) are applicable to a wide range of problems, but in the regression models, more complicated functions can be used:

$$y(x) = \sum_{i=1}^{k} b_i f_i(x). \tag{5.5}$$

The choice of the regression model starts with a linear model,

$$y = \sum_{i=1}^{k} b_i x_i, \tag{5.6}$$

and after adequate verification more complicated (nonlinear) models can be used.

5.2 Parameter Identification

The main problem of regression models is parameter identification (estimation).

Let us consider model (5.5), where $x = x_1, \ldots, x_m$. If we put experimental data (5.1) into (5.5), a set of N equations will be used to obtain the model parameters b_i, $i = 1, \ldots, k$, $k \leq N$:

$$\begin{aligned} b_1 f_1(x_{11}, \ldots, x_{m1}) + \cdots + b_k f_k(x_{11}, \ldots, x_{m1}) &= y_1^{exp}, \\ b_1 f_1(x_{1n}, \ldots, x_{mn}) + \cdots + b_k f_k(x_{1n}, \ldots, x_{mn}) &= y_n^{exp}, \\ b_1 f_1(x_{1N}, \ldots, x_{mN}) + \cdots + b_k f_k(x_{1N}, \ldots, x_{mN}) &= y_N^{exp}. \end{aligned} \tag{5.7}$$

The set of equations does not have an exact solution and there exists a difference between both sides of equations (5.7):

$$b_1 f_1(x_{1n}, \ldots, x_{mn}) + \cdots + b_k f_k(x_{1n}, \ldots, x_{mn}) - y_n^{exp} = \varepsilon_n, \quad n = 1, \ldots, N. \tag{5.8}$$

The solution of the parameter identification problem can be obtained as a minimization of the function

$$F(b_1, \ldots, b_k) = \sum_{n=1}^{N} \varepsilon_n^2 = \sum_{n=1}^{N} \left[b_1 f_1(x_{1n}, \ldots, x_{mn}) + \cdots + b_k f_k(x_{1n}, \ldots, x_{mn}) - y_n^{exp} \right]. \tag{5.9}$$

5.3 Least-Squares Method

Identification of model parameters uses the *least-squares method*, i.e., the minimization of the *least-squares function* (5.9), where the parameter values obtained must minimize the difference between calculated and experimental values of the objective function.

The conditions for a minimum of function (5.9) are

$$\frac{\partial F}{\partial b_i} = 0, \quad i = 1, \ldots, \mathrm{k}, \tag{5.10}$$

i.e.,

$$\sum_{\mathrm{n}=1}^{\mathrm{N}} 2f_i(x_{1\mathrm{n}}, \ldots, x_{\mathrm{mn}}) \left[b_1 f_1(x_{1\mathrm{n}}, \ldots, x_{\mathrm{mn}}) + \cdots + b_{\mathrm{k}} f_{\mathrm{k}}(x_{1\mathrm{n}}, \ldots, x_{\mathrm{mn}}) - y_{\mathrm{n}}^{\mathrm{exp}} \right] = 0,$$
$$\mathrm{i} = 1, \ldots, \mathrm{k}. \tag{5.11}$$

The linear set of equations (5.11) is named a *normal set* and its solution provides the parameter values in the regression model. This procedure is used in commercial mathematical software. The mathematical problems of model parameter identifications are described in the next section.

6 Examples

6.1 Effect of Surfactants

The influence of surfactants on laminar liquid film flows is reduced to two main effects—damping of the ripples on the film surface and alteration of the velocity profile. The next theoretical results are related to the effect of soluble surfactants on the velocity distribution in a waveless liquid film.

Let us consider a vertical film flow containing *soluble surfactants* with constant inlet concentration $c = c_0$. The concentration in the film is variable because of the adsorption on the surface $y = h(x)$. The bulk concentration is denoted by $c^* = c(x, h)$. The surface concentration Γ and surface tension σ are variable along the gas–liquid interphase surface $y = h(x)$. It has been shown that in these conditions a surface tension gradient gradσ must exist on the flowing liquid surface [26].

For a mathematical description of the film flow, (1.2.49) and (1.2.50) can be used, where

$$y = h(x), \quad P_{\mathrm{ng}} = P_{\tau\mathrm{g}} = 0, \quad P_\tau = \frac{\partial \sigma}{\partial \tau} = \frac{\frac{\partial \sigma}{\partial x} + h' \frac{\partial \sigma}{\partial y}}{\sqrt{1 + h'^2}} \tag{6.1}$$

and $\frac{\partial \sigma}{\partial \tau}$ is a tangential derivate.

In the *film flow approximation* $\left[10^{-2} > \varepsilon_0 = \frac{h_0}{l} = 0\right]$ the problem has the form (1.2.61), where

$$y = h, \quad \mu \frac{\partial u_{\mathrm{x}}}{\partial y} = \frac{\partial \sigma}{\partial x}. \tag{6.2}$$

If we use the generalized variables (1.2.52), from (1.2.61), (1.2.62), and (6.2) it follows that

$$\frac{\partial^2 U_x}{\partial Y^2} = -3, \quad \frac{\partial U_x}{\partial X} + \frac{\partial U_y}{\partial Y} = 0; \quad Y = 0, \quad U_x = U_y = 0; \quad Y = H, \\ \frac{\partial U_x}{\partial Y} = -\alpha F(X), \tag{6.3}$$

where

$$\alpha = \frac{\sigma_0 h_0}{\mu \bar{u} l}, \quad F(X) = \frac{\partial \bar{\sigma}}{\partial X}, \quad \bar{\sigma} = \frac{\sigma}{\sigma_0} \tag{6.4}$$

and σ_0 is the surface tension of the pure liquid (water).

In the case $\alpha \sim 1$ problem (6.3) cannot be solved if $F(X)$ is an arbitrary function (the interval $0 < l < l_0$ in Fig. 4).

From the boundary conditions in (6.3) it can be seen that for $\alpha < 10^{-2}$ the surfactants do not have an influence on the velocity distribution in the film flow and in this condition ($\alpha = 0$) the solution coincides with (1.2.62) (see the interval $l_1 < l < l_2$ in Fig. 4).

If $\alpha$is a small parameter ($10^{-2} < \alpha < 1$), the solution can be obtained in the interval

$$\frac{\sigma_0 h_0}{\mu \bar{u} l} = l_0 < l < l_1 = \frac{\sigma_0 h_0}{\mu \bar{u} l} 10^2 \tag{6.5}$$

using the perturbation method:

$$U_x = U_x^0 + \alpha U_x^1, \quad V_x = \alpha V_x^1, \quad H = 1 + \alpha H^1, \tag{6.6}$$

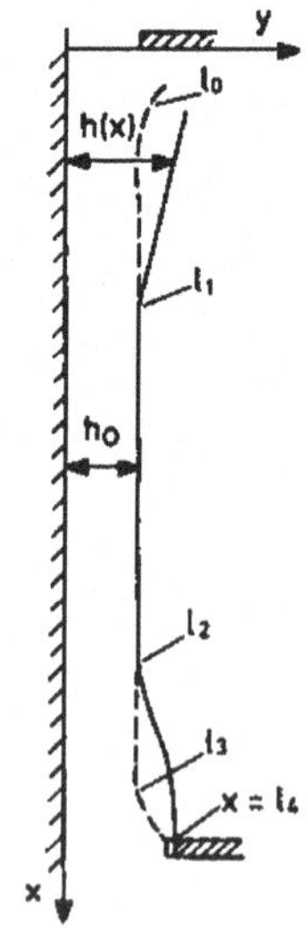

Fig. 4 The film thickness profile

where U_x^0 is the solution of problem (1.2.55). As a result, the following is obtained [2]:

$$U_x = 3Y - \frac{3}{2}Y^2 + \alpha F(X)Y, \quad U_y = -\frac{\alpha}{2}F'(X)Y^2, \quad H = 1 + \frac{\alpha}{3}F(X), \tag{6.7}$$

where the arbitrary function $F(X)$ represents the effect of the surfactants (see 6.4),

$$F(X) = \frac{\partial \bar{\sigma}}{\partial X}. \tag{6.8}$$

The surface tension σ is related to the surface concentration Γ of the surfactants, i.e.,

$$y = h, \quad \frac{\partial \sigma}{\partial x} = \frac{\partial \sigma}{\partial \Gamma}\frac{\partial \Gamma}{\partial x}, \tag{6.9}$$

where $\frac{\partial \sigma}{\partial \Gamma}$ can be obtained from the state equation for a two-dimensional gas [2].

The surface concentration Γ can be obtained using the law of mass conservation at the film surface [26]. For the steady-state case it has the form

$$div_s(j_{conv} - j_{diff}) + j = 0, \tag{6.10}$$

where div_s is the surface divergence and j_{conv} and j_{diff} are the convective and the diffusive mass flux, respectively:

$$j_{conv} = \Gamma u_{s,} j_{diff} = D_s \, grad \, \Gamma, \tag{6.11}$$

where $\boldsymbol{j}$ is the mass flux from the bulk. The surface velocity and surface diffusivity of the surfactants are denoted by $\boldsymbol{u}_s$ and D_s. It is usually assumed that $\boldsymbol{j}_{diff} \approx 0$, because the convective flux is much greater than the diffusive one. This simplification can be made when the surfactants are soluble.

The *mass flux from the bulk is determined by the rate of diffusion and adsorption*. Further, we will consider the limiting cases when the rate of the transport of the surfactant from the bulk towards the film surface is controlled by the bulk diffusion (when the thermodynamic equilibrium is rapidly attained) or by the adsorption associated with a certain energetic barrier.

The first case of mass transport presumes the rapid establishment of an equilibrium and Γ and the volume concentration at the interface c^* are related by the Langmuir isotherm:

$$\Gamma = \frac{kc^*}{1 + \frac{kc^*}{\Gamma_\infty}}, \tag{6.12}$$

where Γ_∞ is the limiting surface concentration, corresponding to the dense monomolecular layer. In this case the diffusion is slow and the mass flux is determined by the volume concentration:

$$j = -D\left(\frac{\partial c}{\partial n}\right)_{y=h}, \tag{6.13}$$

where D is the surfactant bulk diffusivity and $\partial c/\partial c\partial n.\partial n$ is the normal derivative. In the film flow approximation ($\varepsilon_0 = 0$) and from (6.10), (6.11), and (6.13) it follows that

$$y = h, \quad u_s \frac{\partial \Gamma}{\partial x} = D \frac{\partial c}{\partial y}, \tag{6.14}$$

where c is the solution of the convection–diffusion equation:

$$u \frac{\partial c}{\partial x} = D \frac{\partial^2 c}{\partial y^2}. \tag{6.15}$$

The boundary conditions of (6.15) will be discussed later.

In the case when the transport process is limited by adsorption, the diffusion is sufficiently rapid and the bulk concentration is practically constant and equal to the inlet one, $c = c_0$, because of the small capacity of the surface layer. Then j is the difference of the rates of desorption P and adsorption Q:

$$j = P(\Gamma) - Q(c_0, \Gamma), \tag{6.16}$$

where

$$P = \alpha_0 \Gamma, \quad Q = \beta_0 c_0 \left(1 - \frac{\Gamma}{\Gamma_\infty}\right). \tag{6.17}$$

Here α_0 and β_0 are rate constants of the processes. From (6.10), (6.11), and (6.16) it follows that

$$y = h, \quad u_s \frac{\partial \Gamma}{\partial x} = \alpha_0 \Gamma_\infty \left[\frac{kc_0}{\Gamma_\infty} - \left(1 + \frac{kc_0}{\Gamma_\infty}\right)\right], \tag{6.18}$$

where $k = \beta_0/\beta_0\alpha_0.\alpha_0$ is the equilibrium constant in the Langmuir isotherm.

$F(X)$ in (6.7) is determined in a different way, according to the controlling stage of the mass transfer of the surfactant from the volume to the interface. For the case of a *diffusion-controlled process*, the relation is

$$F(X) = \frac{\partial \bar{\sigma}}{\partial X} = \left(\frac{\partial \bar{\sigma}}{\partial C}\right)_{C=C^*} \left(\frac{\partial C}{\partial X}\right)_{Y=1}, \tag{6.19}$$

where

$$C = \frac{c}{c_0}, \quad C^* = \frac{c^*}{c_0}. \tag{6.20}$$

From (6.7) it can be seen that $F(X)$ does not depend on α, i.e., the concentration $C(X, Y)$ must be determined by (6.15), which is written in the zeroth approximation of α:

$$U_x^0 \frac{\partial C}{\partial X} = \theta^2 \frac{\partial^2 C}{\partial Y^2}, \quad \theta = \sqrt{\frac{lD}{h_0^2 u_{av}}}. \tag{6.21}$$

The boundary conditions for solving the convection–diffusion equation (1.6.21) express the constant concentration at the film inlet, the impermeability of the solid wall down which the film is flowing, and the convection–diffusion equation at the interface:

$$X = 0, \quad C = 1; \quad Y = 0, \quad \frac{\partial C}{\partial Y} = 0; \quad Y = 1, \quad U^0_x \frac{\partial \bar{\Gamma}}{\partial X} = -\theta^2 \frac{c_0 h_0}{\Gamma_\infty} \frac{\partial C}{\partial Y}. \quad (6.22)$$

The problem (6.21, 6.22) is solved by means of the perturbation method [2]. Then retaining the terms with an order of magnitude θ^2, we obtain the following relationship for the function $F(X)$:

$$F(X) = \frac{2Re}{We}\left(\frac{\partial \bar{\sigma}}{\partial C}\right)_{C=C^*}\left[\frac{\theta h_0}{k}\sqrt{\frac{3}{2\pi X}} - \left(\frac{\theta h_0}{k}\right)^2\left(1 - \frac{4kc_0}{\pi \Gamma_\infty}\right)\right]. \quad (6.23)$$

This function can be determined in the case of *mass transfer under adsorption control*. For this purpose, the zeroth approximation with respect to α in (6.9) and (6.18) is used,

$$Y = 1, \quad \frac{\partial \bar{\sigma}}{\partial X} = \left(\frac{\partial \bar{\sigma}}{\partial \bar{\Gamma}}\right)_{\bar{\Gamma}=\bar{\Gamma}(X,1)} \frac{\partial \bar{\Gamma}}{\partial X}; \quad (6.24)$$

$$\frac{\partial \bar{\Gamma}}{\partial X} = \frac{2\alpha_0 l}{3u_{av}}\left[\frac{kc_0}{\Gamma_\infty} - \left(1 + \frac{kc_0}{\Gamma_\infty}\right)\bar{\Gamma}\right] \quad (6.25)$$

with the reasonable boundary condition

$$X = 0, \quad \bar{\Gamma} = 0. \quad (6.26)$$

As a result, for F we obtain [2]

$$F(X) = \frac{Re}{We}\left(\frac{\partial \bar{\sigma}}{\partial \bar{\Gamma}}\right)_{\bar{\Gamma}=\bar{\Gamma}(X,1)} \frac{\alpha_0 l}{u_{av}} \frac{kc_0}{\Gamma_\infty} \exp\left[-\frac{\alpha_0 l}{u_{av}}\left(1 + \frac{kc_0}{\Gamma_\infty}\right)X\right]. \quad (6.27)$$

The solutions obtained are valid in the interval $l_0 < l < l_1$ (see 6.5). For long films ($l_1 < l < l_2$, see Fig. 4) the asymptotic solution (1.2.62) is valid.

The presence of surfactants in a liquid leads to the onset of some interesting effects at the film exit. At some point near the film exit the surface velocity becomes zero. This stagnant point could be treated as an apparent barrier at point $x = l_4$ (see Fig. 4). That is why an accumulation of surfactant molecules near this point takes place and in the interval $l_3 \le l \le l_4$ on the surface $y = h$, $\Gamma = \Gamma_\infty$, $u_s = 0$.

The solution of the problem in the interval $l = l_3 - l_4$ is performed [2] in a new coordinate system, where $0 \le x \le l$. At the two ends of this interval the following conditions can be specified:

$$\begin{aligned} &x = 0, \quad u = \frac{g}{2v}(2h_0 y - y^2), \quad v = 0, \quad h = h_0, \quad c = c_0, \quad \Gamma = \Gamma_0; \\ &x = l, \quad u = \frac{g}{2v}(2h_1 y - y^2), \quad v = 0, \quad h = h_1 = h_0\sqrt{[3]4}, \quad \Gamma = \Gamma_\infty. \end{aligned} \quad (6.28)$$

The solution of (6.3) in the interval $0 \le x \le l$ can be obtained by assuming the film thickness increases linearly with the length:

$$h = h_0 + \theta_0 x, \quad \theta_0 = \frac{h_1 - h_0}{l}. \tag{6.29}$$

The velocity distribution can be found in the following form:

$$u_x = f(x)y - \frac{g}{2v}y^2, \quad u_y = -\frac{1}{2}f'(x)y^2, \tag{6.30}$$

where the function $f(x)$ is determined according to the transport mechanism, assuming that θ_0 is a small parameter ($10^{-2} < \theta_0 < 1$).

In the case of *diffusion-controlled mass transfer* the solution is [2]

$$f(x) = \frac{g}{2v} - \theta_0 f_1(x), \tag{6.31}$$

where

$$f_1(x) = 2A\left[\exp\left(\frac{x}{Ah_0}\right) - 1\right] - 2\frac{x}{h_0} + \alpha_1\sqrt{A}\left[\left(3A - \frac{2x}{h_0}\right)\exp\left(\frac{x}{Ah_0}\right) erf\sqrt{\frac{x}{Ah_0}} - 6\sqrt{\frac{Ax}{\pi h_0}}\right]$$

$$A = -\frac{2c_0}{\rho g h_0^2}\left(1 + \frac{kc_0}{\Gamma_\infty}\right)\left(\frac{\partial\bar{\sigma}}{\partial C}\right)_{C=C^*}, \alpha_1 = \sqrt{\frac{2vD}{gh_0k^2}}\left(1 + \frac{kc_0}{\Gamma_\infty}\right)^2. \tag{6.32}$$

The *adsorption-controlled process* leads to

$$f(x) = f_2(x) + \frac{gh}{v} - \theta_0^2\frac{gK}{\mu}, \quad K = -\frac{\Gamma_0}{\alpha_0}\frac{\partial\sigma}{\partial\Gamma}. \tag{6.33}$$

The function $f_2(x)$ is a solution of the boundary-value problem,

$$hf_2'' + 2f_2' + \frac{\mu}{K\theta_0^2}f_2 = 0; \quad x = 0, \quad f_2 = \frac{gK\theta_0^2}{\mu v}; \quad x = l, \quad f_2 = \frac{gK\theta_0^2}{\mu v} - \frac{gh_1}{2v}, \tag{6.34}$$

obtained in terms of the Bessel functions of the first order [2].

These theoretical results are in a good agreement with the experimental data (see Fig. 5). The film thickness profile is observed [27] in films 10^{-7} M aqueous solutions of sodium heptadecyl sulfate. The dotted line in Fig. 5 denotes the case with a free interface.

The experimental film thickness values fit well equation (6.7), combined with (6.37) in the case of adsorption-controlled transport, where the constants are $\alpha_0 = 0.095\ \mathrm{s}^{-1}$ and $(\partial\sigma/\partial\Gamma)\Gamma_0 = -41 \times 10^{-3}\ \mathrm{Nm}^{-1}$. These values are in good agreement with the experimental data [29] for the adsorption kinetics of sodium

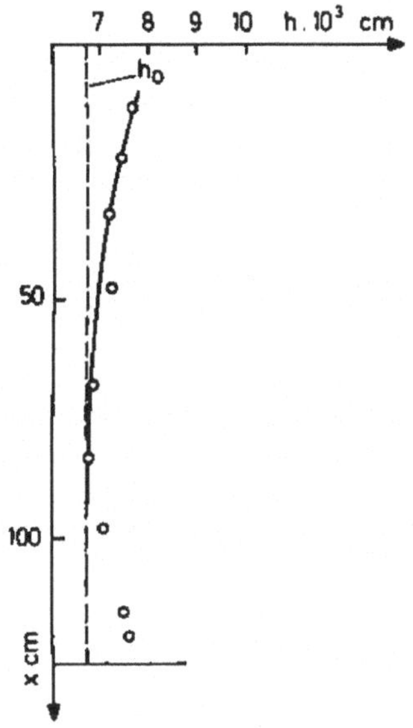

Fig. 5 Comparison of the theoretical and experimental values for the film thickness

alkyl sulfates at an air–water interface. The experimental determination of the surface velocity in a film flow in the presence of surfactants [28] provides results that fit well the computed ones (see Fig. 6) according to equations (6.7) and (6.37).

For the parameters in (6.37), calculation using these experimental data results in $(\partial\sigma/\partial\Gamma)\Gamma_0 = -\ 2.1 \times 10^{-3}\ \mathrm{Nm}^{-1}$ and $\alpha_0 = 5.8\ \mathrm{s}^{-1}$, whereas $(\partial\sigma/\partial\Gamma)\Gamma_0 = -2.5 \times 10^{-3}\ \mathrm{Nm}^{-1}$ is computed [28] by means of the numerical solution of the Navier–Stokes equation. Different theoretical and experimental investigations are presented in [28, 30, 31].

The presence of surfactants in the liquid of the film may affect the mass transfer rate [2, 32]. The effect differs depending on whether the surfactants form a third phase at the interface or not.

We will assume that the surfactants are soluble, even though the case of insoluble substances forming dense surface films is also of great importance [33, 34].

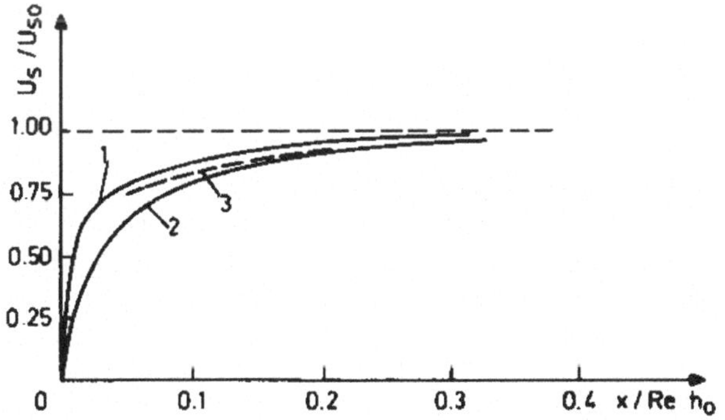

Fig. 6 Comparison of the theoretical [27] (*line 3*) and experimental [28] values of the surface velocity u_s in an inclined film flow. *Line 1* clean water surface, *line 2* aqueous solution of heptanoic acid, $u_{s0} = g \sin \gamma h_0^2/2\nu$

The quantitative explanation of the hydrodynamic effect of surfactants on the mass transfer rate is initiated from the solution of the convection–diffusion equation by introducing a velocity profile modified by their action. Within the region $0(l_0) \leq x \leq l_1$ the velocity profile (1.5.86) should be used and the following equation for the Sherwood number is obtained:

$$Sh = Sh_0 - \sqrt{\frac{2Pe}{3\pi}}\left\{\int_0^1 F(X)dX - \lim_{X\to 0}\left[\frac{\int F(X)dX}{\sqrt{X}}\right]\right\}, \tag{6.35}$$

where Sh_0 is determined from (2.9) and $F(X)$ is determined from (6.23) or (6.27) depending on the mechanism of transport assumed.

The mass transfer kinetics for $l_1 \leq x \leq l_2$ can be determined from (2.9).

For the region $l_2 \leq x \leq l_3$ the velocity profiles (1.5.109) permit a computation of Sh [2]:

$$Sh = Sh_0 - 4\left(\sqrt[3]{4} - 1\right)\sqrt{\frac{\theta_0}{\pi}}\left(A_0\int_0^{A_0^{-0.5}} e^{-s^2}ds - A_0 - \frac{1}{3}\right), \quad A_0 = \varepsilon_0 A, \tag{6.36}$$

$$\theta_0 = \frac{\sqrt[3]{4} - 1}{l_3 - l_2}h_0.$$

Experimental data for oxygen desorption from films of aqueous solutions of ethylene glycol in the presence of sodium heptadecyl sulfate are presented in [2] for different film lengths.

6.2 Effect of Interface Waves

There have been many theoretical investigations of the wavy film flow [2, 35]. Here we will present an *integral method of moments* for calculation of the wave's parameters and the velocity profiles in wavy films.

Theoretical analysis [36] shows that the application of the boundary layer equations to the film flow is justified only in a very small range of the Reynolds number:

$$\left(g\nu^4\right)^{1/3}(\rho/\sigma)Re^{5/3} \ll 1, \quad \left(g\nu^4\right)^{1/12}(\rho/\sigma)^{1/4}Re^{-1/12} \ll 1. \tag{6.37}$$

This range appears to be $1 \leq Re \leq 20$ for water and $1 \leq Re \leq 7$ for ethanol. Later, the analysis of the wavy film flow is proposed [36] without any simplifications in the two-dimensional equations of motion:

$$\frac{\partial u_x}{\partial t} + u_x\frac{\partial u_x}{\partial x} + u_y\frac{\partial u_x}{\partial y} = -\frac{1}{\rho}\frac{\partial p}{\partial x} + \nu\left(\frac{\partial^2 u_x}{\partial x^2} + \frac{\partial^2 u_x}{\partial y^2}\right) + g, \tag{6.38}$$

$$\frac{\partial u_y}{\partial t} + u_x\frac{\partial u_y}{\partial x} + u_y\frac{\partial u_y}{\partial y} = -\frac{1}{\rho}\frac{\partial p}{\partial y} + \nu\left(\frac{\partial^2 u_y}{\partial x^2} + \frac{\partial^2 u_y}{\partial y^2}\right), \tag{6.39}$$

$$\frac{\partial u_x}{\partial x} + \frac{\partial u_y}{\partial y} = 0, \tag{6.40}$$

with the boundary conditions at the free surface $y = h(x, t)$, obtained from (1.2.50) and (1.2.51) in the case $P_{ng} = P_{\tau g} = 0$,

$$p + \frac{\sigma h''}{(1 + h'^2)^{3/2}} + 2\mu\frac{1 + h'^2}{1 - h'^2}\frac{\partial u_x}{\partial x} = 0, \tag{6.41}$$

$$\frac{4h'^2}{1 - h'^2}\frac{\partial u_x}{\partial x} - \left(\frac{\partial u_x}{\partial y} + \frac{\partial u_y}{\partial x}\right) = 0. \tag{6.42}$$

At the sold surface $y = 0$:

$$u_x = u_y = 0. \tag{6.43}$$

The solution of this problem must be represented by the following power series:

$$u_x = \sum_{m=0}^{M} a_m(\xi)y^m, \quad u_y = -\sum_{m=0}^{M}\frac{a'_m(\xi)}{m+1}y^{m+1}, \quad M = 2,$$
$$a_m = \sum_{k=-\infty}^{+\infty} a_{mk}\exp\left(ik\frac{2\pi}{\lambda}\xi\right), \quad h(\xi) = \sum_{k=-\infty}^{+\infty} h_k\exp\left(ik\frac{2\pi}{\lambda}\xi\right), \quad \xi = \frac{x - \alpha u_0 t}{h_0}, \quad \alpha = \frac{c}{u_0}, \tag{6.44}$$

where c is the phase velocity, u_0 is the mean film velocity defined by

$$u_0 = \frac{1}{\lambda h_0}\int_0^{\lambda}\int_0^{h} u(\xi, y)d\xi dy, \tag{6.45}$$

and λ is the wavelength.

The functions $a_n(\xi)$ and $h(\xi)$ are determined [36] from Eqs. (6.38–6.43) as follows:

1. The pressure is eliminated by integrating (6.39) over y, using the boundary condition (6.41). After differentiation of x, it is introduced into (6.38) to determine $a_n(\xi)$.
2 Expressions (6.44) are introduced into (6.38). Then we can determine u_x and u_y with the desired accuracy, depending on the maximum power in the series in (6.44). After multiplication of (6.38) consequently by 1, y, y^2, ..., y^{N-1} and the

following integration over y on the whole cross section, we get a set of equations using (6.42) and the macroscopic balance equation:

$$\frac{\partial h}{\partial t} + \frac{\partial}{\partial x}\int u_x dy = 0. \tag{6.46}$$

3. Since the solution of this set is extremely hard to obtained, it is more convenient to use the Fourier series in (6.44). As a result, a set of algebraic nonlinear equations will be obtained [36] for the determination of the numerical coefficients.

The numerical analysis of the problem [36] shows that at $Re \leq 20$ we can use $N = 2$.

The determination of the wave amplitude $A = (h_{max} - h_0)/h_0$ requires an additional condition to be introduced. The latter expresses the Reynolds principle of flow stability, i.e., the kinetic energy E_k of the flow should not grow in time:

$$\frac{dE_k}{dt} = \frac{\rho}{2}\int_0^{\lambda}\int_0^{h}\frac{\partial}{\partial t}(\mathbf{v}^2)dydx \leq 0, \tag{6.47}$$

where $\boldsymbol{v}$ is the velocity vector (with components u_x and u_y).

The method presented [36] permits us to calculate the wave number $n = 2\pi h_0/\lambda$, wave amplitude A, and phase velocity α as functions of the Weber number:

$$We = \frac{\rho h_0 u_0^2}{\sigma}. \tag{6.48}$$

A comparison [36] of the computed wave characteristics with the aid of the method presented with the experimental data is shown in Figs. 7, 8 and 9.

The solution for the velocity distribution in the film flow permits us to obtain the instant streamlines ψ for some fixed moment of time,

$$u_x = \frac{\partial \psi}{\partial y}, \quad u_y = -\frac{\partial \psi}{\partial x}, \tag{6.49}$$

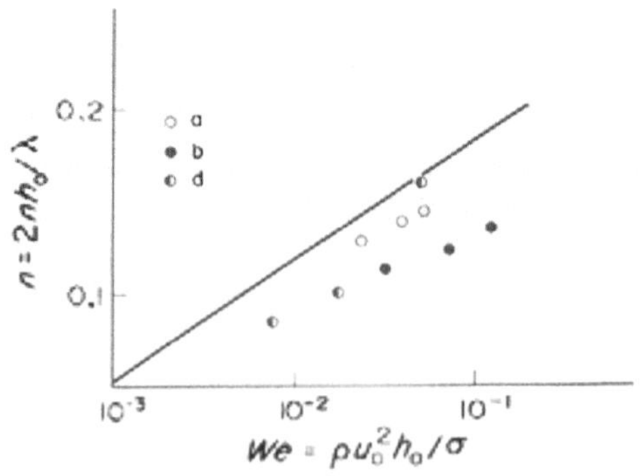

Fig. 7 Theoretical [36] and experimental [37–39] data for the wave number (**a** from [37], **b** from [38], **d** from [39])

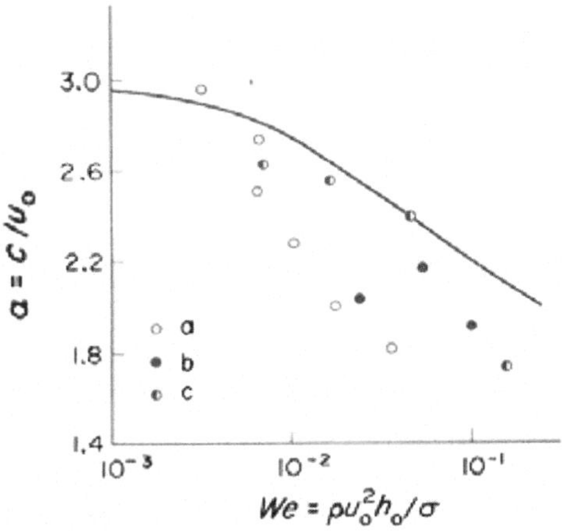

Fig. 8 Theoretical [36] and experimental [37–39] [37], data for the phase velocity (**a** from [37], **b** from [38], **c** from [39])

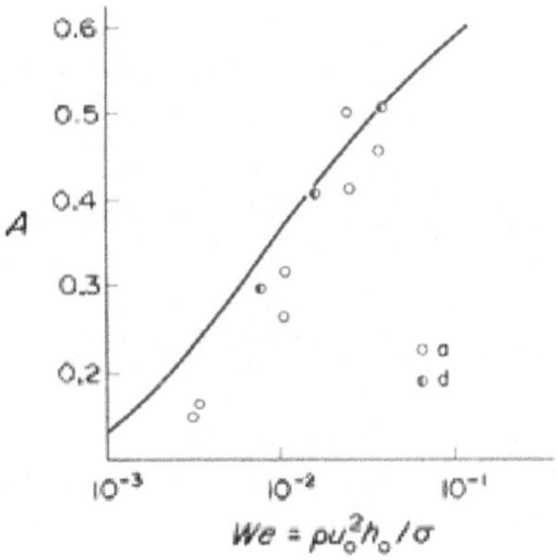

Fig. 9 Comparison of the theoretical [36] and experimental [37, 39] data for the wave amplitude. (**a** from [37], **d** from [39])

and the liquid particle trajectories in coordinates translating along the film length with phase velocity $c = \alpha u_0$ with respect to the solid wall,

$$\frac{1}{h_0}\frac{dy}{d\xi} = -\frac{u_y}{c - u_x}. \tag{6.50}$$

In the cases of stationary flows these two type of lines are equivalent, but in wave film flows they are very different. In Fig. 10 these lines for $\alpha = 1.98$ are shown. These results demonstrate the absence of the film surface renewal and the mass transfer in the wave film flow is a result of the velocity distribution only [36]. A numerical solution of the nonstationary convection–diffusion equation in the case of wavy film flows is presented in [40].

Fig. 10 Instant streamline particle trajectories (*dotted lines*)

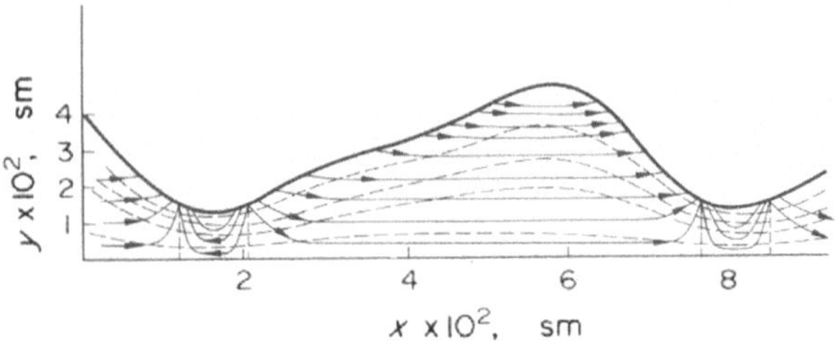

6.3 Photobioreactor Model

Photobioprocesses include dissolution of an active gas component (CO_2, O_2) in a liquid and its reaction with a photoactive material (cells). These two processes may take place in different systems, such as mixed bioreactors; bubble columns, and airlift photobioreactors [41–44]. The comparison of these systems shows apparent advantages in the use of airlift photobioreactors, because of the possibility of manipulation of the light-darkness history of the photosynthetic cells [45–47].

The hydrodynamic behavior of the gas and the liquid in airlift reactors is very complicated, but in all cases the process includes convective transport, diffusive transport, and volume reactions. That is why the convection–diffusion equation with a volume reaction may be used as a mathematical structure of the model. Using the average velocities permits us to solve the mass transfer problem without solution of the hydrodynamic equations. Introducing the average concentrations is the basis for the scale-up problem solution.

Let us consider an airlift reactor with a horizontal cross-sectional area F_0 for the riser zone and F_1 for the downcomer zone. The length of the working zones is l. The gas flow rate is Q_0 and the liquid flow rate (water) is Q_1. The gas and the liquid holdup in the riser are ε and $(1 - \varepsilon)$. The concentrations of the active gas component (CO_2) are $c(x, r, t)$ in the gas phase, and $c_0(x, r, t)$ for the riser and $c_1(x_1, r, t)$ for the downcomer and in the liquid phase, where $x_1 = l - x$. The concentration of the photoactive substance in the downcomer is $c_2(x_1, r, t)$ and in the riser's section is $c_3(x_1, r, t)$.

The average velocities in the gas and liquid phases are

$$\bar{\bar{u}}_0 = \frac{Q_0}{F_0}, \quad \bar{\bar{u}}_1 = \frac{Q_1}{F_0}, \quad \bar{\bar{u}} = \frac{Q_1}{F_1}. \tag{6.51}$$

The interphase mass transfer rate in the riser is

$$I_0 = k(c - \chi c_0), \tag{6.52}$$

where I_0 is a volume source in the convection–diffusion equations (see 6.61, 6.62), k *is the* interphase mass transfer coefficient, and χ is Henry's constant.

The photoreaction rates in the downcomer and the riser (for small values of c_0, c_1, J, J_1) are taken, respectively, as

$$I = k_0 c_1 c_2 J, \quad I_1 = k_0 c_0 c_3 J_1, \tag{6.53}$$

instead of the more general formulation, presented in (6.94). This is done to avoid mathematical complexity that is not essential in many cases. The photon flux densities $J\,(x_1, r, t)$ and $J_1\,(x, r, t)$ are functions of the radial coordinate r, as will be shown below.

Let us consider a cylindrical surface with radius R_0 and length 1 m, which is regularly illuminated with a photon flux density J_0. The photon flux densities over a cylindrical surface with $r < R_0$ is

$$i(r) = \frac{R_0 J_0}{r}. \tag{6.54}$$

The increase of the photon flux density between r and $r - \Delta r$ is

$$\Delta J_1 = \frac{J_0 R_0}{r - \Delta r} - \frac{J_0 R_0}{r} = \frac{J_0 R_0 \Delta r}{r(r - \Delta r)}. \tag{6.55}$$

The volume between the cylindrical surfaces (m^3 liquid/m^2 surface) with radiuses r and $r - \Delta r$ is

$$V = \Delta r\left(1 - \frac{\Delta r}{2r}\right), \tag{6.56}$$

and the decrease of the photon flux density as a result of the light absorption (from the photoactive cells) in this volume is

$$\Delta J_2 = J(x_{1.}, r, t)\beta c_2 \Delta r\left(1 - \frac{\Delta r}{2r}\right), \tag{6.57}$$

where $c_2 = c_2(x_1, r, t)$ is the concentration of the photoactive cells in the downcomer.

The difference between photon flux densities for r and $r - \Delta r$ is

$$\Delta J = \Delta J_1 - \Delta J_2 = \frac{J_0 R_0 \Delta r}{r(r - \Delta r)} - J\beta c_2 \Delta r\left(1 - \frac{\Delta r}{2r}\right). \tag{6.58}$$

As a result,

$$\lim_{\Delta r \to 0} \frac{\Delta J}{\Delta r} = \frac{\partial J}{\partial r} = \frac{R_0 J_0}{r^2} - \beta c_2 J, \tag{6.59}$$

where $J\,(R_0) = J_0$. The solution of (6.59) for $c_2 = c_2\,(x_1, r, t)$ is

$$J(x_1,r,t)=\exp\left(\beta\int_r^{R_0} c_2(x_1,\rho,t)d\rho\right)\left\{J_0 - R_0 J_0 \int_r^{R_0} \frac{1}{\rho^2}\exp\left[-\beta\int_\rho^{R_0} c_2(x_1,\eta,t)d\eta\right]d\rho\right\}. \tag{6.60}$$

The mathematical model of the process in the airlift photobioreactor will be built on the basis of the differential mass balances in the reactor volume [14–18]. The convection–diffusion equation with a volume reaction will be used, where convective transfer will be the result of the laminar flow or large-scale turbulent pulsations. The diffusivity is taken as being molecular or turbulent (as a result of the small-scale turbulent pulsations) and the sources (volume reactions) are interphase mass transfer and photochemical reaction.

The equations for the distribution of the active gas component (CO_2) in the gas and liquid phases in the riser are

$$\begin{aligned} &\varepsilon\frac{\partial c}{\partial t}+\varepsilon u_0\frac{\partial c}{\partial x}+\varepsilon v_0\frac{\partial c}{\partial r}=\varepsilon D\left(\frac{\partial^2 c}{\partial x^2}+\frac{1}{r}\frac{\partial c}{\partial r}+\frac{\partial^2 c}{\partial r^2}\right)-k(c-\chi c_0),\\ &\frac{\partial u_0}{\partial x}+\frac{\partial v_0}{\partial r}+\frac{v_0}{r}=0, \end{aligned} \tag{6.61}$$

$$\begin{aligned} &(1-\varepsilon)\frac{\partial c_0}{\partial t}+(1-\varepsilon)\left(u_1\frac{\partial c_0}{\partial r}+v_1\frac{\partial c_0}{\partial x}\right)=(1-\varepsilon)D_0\left(\frac{\partial^2 c_0}{\partial x^2}+\frac{1}{r}\frac{\partial c_0}{\partial r}+\frac{\partial^2 c_0}{\partial r^2}\right)\\ &+k(c-\chi c_0)-\alpha k_0 c_0 c_3 J_1,\\ &\frac{\partial u_1}{\partial x}+\frac{\partial v_1}{\partial r}+\frac{v_1}{r}=0. \end{aligned} \tag{6.62}$$

It will be assumed, that $\varepsilon = \text{const}$.

The equations for the distribution of the active gas component (CO_2) and photoactive substance (cells) in the downcomer are

$$\frac{\partial c_1}{\partial t}+u\frac{\partial c_1}{\partial x_1}+v\frac{\partial c_1}{\partial r}=D_1\left(\frac{\partial^2 c_1}{\partial x^2}+\frac{1}{r}\frac{\partial c_1}{\partial r}+\frac{\partial^2 c_1}{\partial r^2}\right)-\alpha k_0 c_1 c_2 J, \tag{6.63}$$

$$\begin{aligned} &\frac{\partial c_2}{\partial t}+u\frac{\partial c_2}{\partial x_1}+v\frac{\partial c_2}{\partial r}=D_2\left(\frac{\partial^2 c_2}{\partial x^2}+\frac{1}{r}\frac{\partial c_2}{\partial r}+\frac{\partial^2 c_2}{\partial r^2}\right)-k_0 c_1 c_2 J,\\ &\frac{\partial u}{\partial x_1}+\frac{\partial v}{\partial r}+\frac{v}{r}=0, \end{aligned} \tag{6.64}$$

where $x_1 = 1 - x$.

A photochemical reaction may take place in riser too, and the equation for the cell concentration is

$$\begin{aligned} &(1-\varepsilon)\frac{\partial c_3}{\partial t}+(1-\varepsilon)\left(u_1\frac{\partial c_3}{\partial r}+v_1\frac{\partial c_3}{\partial x}\right)\\ &=(1-\varepsilon)D_3\left(\frac{\partial^2 c_3}{\partial x^2}+\frac{1}{r}\frac{\partial c_3}{\partial r}+\frac{\partial^2 c_3}{\partial r^2}\right)-k_0 c_0 c_3 J_1, \end{aligned} \tag{6.65}$$

where $J_1 = J_1\ (x, r, t)$ is photon flux density in the riser

$$\frac{\partial J_1}{\partial r} = \frac{r_0 J(x_1, r_0, t)}{r^2} - \beta\, c_3 J_1, \quad r = r_0, \quad J_1 = J(x_1, r_0, t), \quad x_1 = l - x \quad (6.66)$$

and $c_3 = c_3\ (x, r, t)$ is the concentration of the photoactive substance in the riser.

The initial conditions will be formulated for the case of thermodynamic equilibrium between gas and liquid phases, i.e., a full liquid saturation with the active gas component and the process starts on starting the illumination:

$$\begin{gathered} t = 0, \quad c(x, r, 0) = c^{(0)}, \quad c_0(x, r, 0) = \frac{c^{(0)}}{\chi}, \\ c_1(x_1, r, 0) = \frac{c^{(0)}}{\chi}, \quad c_2(x_1, r, 0) = c_2^{(0)}, \quad c_3(x, r, 0) = c_2^{(0)}, \end{gathered} \quad (6.67)$$

where $c^{(0)}$ and $c_2^{(0)}$ are the initial concentrations of the active gas component in gas phase and the photoactive substance in the liquid phase.

The boundary conditions are equalities of the concentrations and mass fluxes at both ends of the working zones $x = 0 (x_1 = l)$ and $x = l\ (x_1 = 0)$.

The boundary conditions for $c(x, r, t)$ and $c_0(x, r, t)$ in (6.61) and (6.62) are

$$\begin{aligned} &x = 0, \quad c(0, r, t) = c^{(0)}, \quad \bar{\bar{u}}_0 c^{(0)} = u_0 c(0, r, t) - D\left(\frac{\partial c}{\partial x}\right)_{x=0}; \\ &x = 0, \quad c_0(0, r, t) = \bar{c}_1(l, t), \quad \bar{c}_1(l, t)\bar{\bar{u}} = c_0(0, r, t) u_1 - D_0\left(\frac{\partial c_0}{\partial x}\right)_{x=0}; \\ &r = 0, \quad \frac{\partial c}{\partial r} = \frac{\partial c_0}{\partial r} = 0; \quad r = r_0, \quad \frac{\partial c}{\partial r} = \frac{\partial c_0}{\partial r} = 0. \end{aligned} \quad (6.68)$$

The boundary conditions for $c_1(x_1, r, t)$, $c_2(x_1, r, t)$ and $c_3(x, r, t)$ are

$$\begin{aligned} &x_1 = 0, \quad c_1(0, r, t) = \bar{c}_0(l, t), \quad \bar{c}_0(l, t)\bar{\bar{u}}_1 = c_1(0, r, t) u - D_1\left(\frac{\partial c_1}{\partial x_1}\right)_{x_1=0}; \\ &r = r_0, \quad \frac{\partial c_1}{\partial r} = 0; \quad r = R_0, \quad \frac{\partial c_1}{\partial r} = 0; \\ &x_1 = 0, \quad c_2(0, r, t) = \bar{c}_3(l, t), \quad \bar{c}_3(l, t)\bar{\bar{u}} = c_2(0, r, t) u - D_2\left(\frac{\partial c_2}{\partial x_1}\right)_{x_1=0}; \\ &r = r_0, \quad \frac{\partial c_2}{\partial r} = 0; \quad r = R_0, \quad \frac{\partial c_2}{\partial r} = 0; \\ &x = 0, \quad c_3(0, r, t) = \bar{c}_2(l, t), \quad \bar{c}_2(l, t)\bar{\bar{u}}_1 = c_3(0, r, t) u_1 - D_3\left(\frac{\partial c_3}{\partial x}\right)_{x=0}; \\ &r = 0, \quad \frac{\partial c_3}{\partial r} = 0; \quad r = r_0, \quad \frac{\partial c_3}{\partial r} = 0. \end{aligned}$$

(6.69)

In (6.68) and (6.69) $\bar{c}_0, \bar{c}_1, \bar{c}_2, \bar{c}_3$ are the average concentrations for the cross-sectional area at the ends of the column. In these boundary conditions (Danckwerts conditions) a balance between convective and diffusive transfer is used (like convection–diffusion equation).

The column scale effect (decreasing process efficiency with increasing column diameter) in the column scale-up is result of the radial nonuniformity of the velocity only. In the specific case of photoreactions, an additional factor is the local variations of light availability. Here the average velocity and concentration in any cross-sectional area are used. This approach has a sensible advantage in the collection of experimental data for the parameter identification, because measurements of the average concentrations are very simple in comparison with local concentration measurements.

Let us consider equation (6.61). The velocity $u_0(x, r)$ and concentration $c\ (x, r, t)$ in cylindrical coordinates practically do not depend on the angular coordinate since symmetry is assumed. In this case the average velocities and concentration in any cross-sectional area are

$$\bar{u}_0(x) = \frac{2}{r_0^2}\int_0^{r_0} r u_0(x,r)dr, \quad \bar{v}_0(x) = \frac{2}{r_0^2}\int_0^{r_0} r v_0(x,r)dr,$$
$$\bar{c}(x,t) = \frac{2}{r_0^2}\int_0^{r_0} r c(x,r,t)dr. \tag{6.70}$$

The velocity (concentration) distribution can be represented using the average functions:

$$u_0(x,r) = \bar{u}_0(x)\tilde{u}_0(x,r), \quad v_0(x,r) = \bar{v}_0(x)\tilde{v}_0(x,r), \quad c(x,r,t) = \bar{c}(x,t)\tilde{c}(x,r), \tag{6.71}$$

where the velocity (concentration) nonuniformity is the ratio between the velocity (concentration) distribution and its average value.

In (6.71) the velocity (concentration) distribution practically is not a function of time. If we multiply (6.71) by r and integrate over r in the interval $[0, r_0]$, the following properties [14] of the average functions are valid:

$$\frac{2}{r_0^2}\int_0^{r_0} r\tilde{u}_0(x,r)dr = 1, \quad \frac{2}{r_0^2}\int_0^{r_0} r\tilde{v}_0(x,r)dr = 1, \quad \frac{2}{r_0^2}\int_0^{r_0} r\tilde{c}(x,r)dr = 1. \tag{6.72}$$

To introduce (6.70) into (6.61) we must put (6.71) into (6.61), multiply by r and integrate over r in the interval $[0, r_o]$. As a result, the following equation for $\bar{c}$ is obtained:

$$\frac{\partial \bar{c}}{\partial t} + A(r_0,x)\bar{u}_0\frac{\partial \bar{c}}{\partial x} + \frac{\partial A}{\partial x}\bar{u}_0\bar{c} + g(r_0,x)\bar{v}_0\bar{c} = D\frac{\partial^2 \bar{c}}{\partial x^2} - \frac{k}{\varepsilon}(\bar{c} - \chi\bar{c}_0), \tag{6.73}$$

where

$$A(r_0,x)=\frac{2}{r_0^2}\int_0^{r_0} r\,\tilde{u}_0\tilde{c}dr,\quad g(r_0,x)=\frac{2}{r_0^2}\int_0^{r_0} r\,\tilde{v}_0\frac{\partial\tilde{c}}{\partial r}dr,\quad \left(\frac{\partial\tilde{c}}{\partial r}\right)_{r=r_0}=0. \tag{6.74}$$

Introduction of (6.72) into the second equation in (6.61), multiplication by r^2, and integration over r in the interval $[0,\ r_0]$, leads to

$$\bar{v}_0=h(r_0,x)\frac{\partial\bar{u}_0}{\partial x}+\frac{\partial h}{\partial x}\bar{u}_0, \tag{6.75}$$

where

$$h(r_0,x)=\frac{2}{r_0^2}\int_0^{r_0} r^2\tilde{u}_0dr. \tag{6.76}$$

Introducing (6.75) into (6.73), the final form of the model is

$$\frac{\partial\bar{c}}{\partial t}+A(r_0,x)\bar{u}_0\frac{\partial\bar{c}}{\partial x}+B(r_0,x)\bar{u}_0\bar{c}+G(r_0,x)\bar{c}\frac{\partial\bar{u}_0}{\partial x}=D\frac{\partial^2\bar{c}}{\partial x^2}-\frac{k}{\varepsilon}(\bar{c}-\chi\bar{c}_0), \tag{6.77}$$

where

$$B(r_0,x)=\frac{\partial A}{\partial x}+g\frac{\partial h}{\partial x},\qquad G(r_0,x)=gh. \tag{6.78}$$

The boundary condition of (6.77) has thc form

$$\begin{aligned}
&t=0,\quad \bar{c}(x,0)=c^{(0)};\\
&x=0,\quad \bar{\bar{u}}_0c^{(0)}=A(r_0,x)\bar{u}_0(0)\bar{c}(0,t)-D\left(\frac{\partial\bar{c}}{\partial x}\right)_{x=o};\\
&x=l,\quad \bar{c}(l,t)=\chi\bar{c}_0(l,t).
\end{aligned} \tag{6.79}$$

The holdup ε can be obtained using

$$\varepsilon=\frac{(l-l_0)\,(F_0+F_1)}{(l-l_0)\,(F_0+F_1)+F_0l_0}, \tag{6.80}$$

where l and l_o are liquid levels in the riser with and without gas motion.

The parameters in the model (6.77, 6.79) are of two types: specific model parameters (D, k, ε, χ) and model scale parameters (A, B, G). The scale parameters are functions of the column radius r_0. They are the result of the radial nonuniformity of the velocity and the concentration and show the influence of the scale-up on the model equations. The parameter χ may be obtained beforehand as a result of thermodynamic measurements.

From (6.73, 6.75, 6.77) it follows that the radial velocity component influences the transfer process in cases, where $\partial\bar{u}_0/\partial x \neq 0$, i.e., when the gas holdup is not constant along the column height. For many cases of practical interest $\varepsilon = \text{const.}$ and $\bar{v}_0 = 0$. As a result, $\frac{\partial\bar{u}_0}{\partial x} = 0, B = \frac{\partial A}{\partial x}$ and model (6.77) has the form

$$\frac{\partial\bar{c}}{\partial t} + A(r_0,x)\bar{u}_0\frac{\partial\bar{c}}{\partial x} + \frac{\partial A}{\partial x}\bar{u}_0\bar{c} = D\frac{\partial^2\bar{c}}{\partial x^2} - \frac{k}{\varepsilon}(\bar{c} - \chi\bar{c}_0), \tag{6.81}$$

where $\bar{u}_0 = \bar{\bar{u}}_0$.

The values of the parameters D, k, and A must be obtained using experimental data for the average velocity and concentration, measured on the laboratory column. In the case of scale-up, only A must be specified because it is a function of the column radius and the radial nonuniformity of the velocity and concentration (D and k do not change at scale-up).

The same procedure may be used for (6.62) and (6.68) and as a result

$$\begin{aligned}
&\frac{\partial\bar{c}_0}{\partial t} + A_0(r_0,x)\bar{u}_1\frac{\partial\bar{c}_0}{\partial x} + B_0(r_0,x)\bar{u}_0\bar{c} + G_0(r_0,x)\bar{c}_0\frac{\partial\bar{u}_1}{\partial x} \\
&\quad = D_0\frac{\partial^2\bar{c}_0}{\partial x^2} + \frac{k}{1-\varepsilon}(\bar{c} - \chi\bar{c}_0) - \frac{\alpha k_0}{1-\varepsilon}M_3(r_0,x)\bar{c}_0\bar{c}_3\bar{J}_1; \\
&t = 0, \quad \bar{c}_0(x,0) = \frac{c^{(0)}}{\chi}; \quad x = 0, \quad \bar{c}_0(0,t) = \bar{\bar{c}}_1(l,t), \\
&\bar{\bar{c}}_1(l,t)\bar{\bar{u}} = A(r_0,x)\bar{u}_1(0)\bar{c}_0(0,t) - D_0\left(\frac{\partial\bar{c}_0}{\partial x}\right)_{x=0},
\end{aligned} \tag{6.82}$$

where

$$\bar{J}_1 = \frac{2}{r_0^2}\int_0^{r_0} rJ_1(x,r,t)dr, \quad M_3(r_0,x) = \frac{2}{r_0^2}\int_0^{r_0} r\tilde{c}_0\tilde{c}_3\tilde{J}_1 dr. \tag{6.83}$$

A_0, B_0 and G_0 are obtained in the same way as A, B and G (see 6.74, 6.76, 6.78). The concrete expressions of A_0, B_0 and G_0 are not very important because those values must be obtained using experimental data in all cases. In the practically interesting cases $\varepsilon = \text{const.}$ and $\frac{\partial\bar{u}_1}{\partial x} = 0,\ B_0 = \frac{\partial A_0}{\partial x},\ \bar{u}_1 = \bar{\bar{u}}_1$.

In (6.63) and (6.64) with boundary conditions (6.69) we must put the average velocity, concentrations, and photon flux density:

$$\begin{aligned}
u(x_1,r) &= \bar{u}(x_1)\tilde{u}(x_1,r), \quad c_1(x_1,r,t) = \bar{c}_1(x_1,t)\tilde{c}_1(x_1,r), \\
c_2(x_1,r,t) &= \bar{c}_2(x_1,t)\tilde{c}_2(x_1,r), \quad J(x_1,r,t) = \bar{J}(x_1,t)\tilde{J}(x_1,r),
\end{aligned} \tag{6.84}$$

where

$$\bar{u}(x_1) = \frac{2}{R_0^2 - r_0^2} \int_{r_0}^{R_0} ru(x_1, r)dr, \quad \bar{c}_1(x_1, \mathrm{t}) = \frac{2}{R_0^2 - r_0^2} \int_{r_0}^{R_0} rc_1(x_1, r, t)dr,$$

$$\bar{c}_2(x_1, t) = \frac{2}{R_0^2 - r_0^2} \int_{r_0}^{R_0} rc_2(x_1, r, t)dr, \quad \bar{J}(x_1, t) = \frac{2}{R_0^2 - r_0^2} \int_{r_0}^{R_0} rJ(x_1, r, t)dr. \tag{6.85}$$

Using the same procedure, after integration of (6.63) and (6.64) over r in the interval $[r_0, R_0]$, the problem has the form

$$\begin{aligned}
&\frac{\partial \bar{c}_1}{\partial t} + A_1(r_0, R_0, x_1)\bar{u}\frac{\partial \bar{c}_1}{\partial x_1} + B_1(r_0, R_0, x_1)\bar{u}\bar{c}_1 + G_1(r_0, R_0, x_1)\bar{c}_1\frac{\partial \bar{u}}{\partial x_1}\\
&= D_1\frac{\partial^2 \bar{c}_1}{\partial x_1^2} - \alpha k_0 M(r_0, R_0, x_1)\bar{c}_1\bar{c}_2\bar{J};\\
&t = 0, \quad \bar{c}_1(x_1, 0) = \frac{c^{(0)}}{\chi}; \quad x_1 = 0, \quad \bar{c}_1(0, t) = \bar{c}_0(l, t),\\
&\bar{c}_0(l, t)\bar{\bar{u}}_1 = A_1(r_0, R_0, x_1)\bar{u}(0)\bar{c}_1(0, t) - D_1\left(\frac{\partial \bar{c}_1}{\partial x_1}\right)_{x_1=0},
\end{aligned} \tag{6.86}$$

$$\begin{aligned}
&\frac{\partial \bar{c}_2}{\partial t} + A_2(r_0, R_0, x_1)\bar{u}\frac{\partial \bar{c}_2}{\partial x_1} + B_2(r_0, R_0, x_1)\bar{u}\bar{c}_2 + G_2(r_0, R_0, x_1)\bar{c}_2\frac{\partial \bar{u}}{\partial x_1}\\
&= D_2\frac{\partial^2 \bar{c}_2}{\partial x_1^2} + k_0 M(r_0, R_0, x_1)\bar{c}_1\bar{c}_2\bar{J};\\
&t = 0, \quad \bar{c}_2 = c_2^{(0)}; \quad x_1 = 0, \quad \bar{c}_2(0, t) = \bar{c}_3(l, t),\\
&\bar{c}_3(l, t)\bar{\bar{u}}_1 = A_2(r_0, R_0, x_1)\bar{u}(0)\bar{c}_2(0, t) - D_2\left(\frac{\partial \bar{c}_2}{\partial x_1}\right)_{x_1=0},
\end{aligned} \tag{6.87}$$

where

$$M(r_0, R_0, x_1) = \frac{2}{R_0^2 - r_0^2} \int_{r_0}^{R_0} r\tilde{c}_1\tilde{c}_2\tilde{J}dr \tag{6.88}$$

and A_1, A_2, B_1, B_2, G_1, G_2 are obtained in a similar way as A, B, G (see 6.74, 6.76, 6.78), but taking into account that the limits of the integrals are $[r_0, R_0]$. For $\varepsilon = \text{const.}$, $\bar{u} = \bar{\bar{u}}, B_1 = \frac{\partial A_1}{\partial x}$ and $B_2 = \frac{\partial A_2}{\partial x}$.

$\bar{J}$ may be obtained by introducing (6.84) into (6.59), multiplying (6.59) by r^3, and integrating over r in the interval $[r_0, R_0]$. As a result,

$$\bar{J} - \frac{1}{N_1 + \beta N_2 \bar{c}_2}, \tag{6.89}$$

where

$$N_1(r_0, R_0, x_1) = \frac{2}{R_0 J_0 (R_0^2 - r_0^2)} \int_{r_0}^{R_0} r^3 \frac{\partial \tilde{J}}{\partial r} dr,$$
$$N_2(r_0, R_0, x_1) = \frac{2}{R_0 J_0 (R_0^2 - r_0^2)} \int_{r_0}^{R_0} r^3 \tilde{c}_2 \tilde{J} dr. \tag{6.90}$$

The same procedure for (6.65) and (6.69) leads to the equation for $\bar{c}_3$:

$$\frac{\partial \bar{c}_3}{\partial t} + A_3(r_0, x)\bar{u}_1 \frac{\partial \bar{c}_3}{\partial x} + B_3(r_0, x)\bar{u}_1 \bar{c}_3 - G_3(r_0, x)\bar{c}_3 \frac{\partial \bar{u}_1}{\partial x} = D_3 \frac{\partial^2 \bar{c}_3}{\partial x^2} + \frac{k_0}{(1-\varepsilon)} M_3(r_0, x)\bar{c}_0 \bar{c}_3 \bar{J}_1;$$
$$t = 0, \quad \bar{c}_3(x, 0) = c_2^{(0)}; \quad x = 0, \quad \bar{c}_3(0, t) = \bar{c}_2(l, t),$$
$$\bar{c}_2(l, t)\bar{\bar{u}} = A_3(r_0, x)\bar{u}_1(0)\bar{c}_3(0, t) - D_3 \left(\frac{\partial \bar{c}_3}{\partial x} \right)_{x=0}, \tag{6.91}$$

where

$$\bar{c}_3(x, t) = \frac{2}{r_0^2} \int_0^{r_0} r c_3(x, r, t) dr, \quad M_3(r_0, x) = \frac{2}{r_0^2} \int_0^{r_0} r \tilde{c}_0 \tilde{c}_3 \tilde{J}_1 dr \tag{6.92}$$

and A_3, B_3, and G_3 are obtained in a way similar to A, B, and G.

$\bar{J}_1$ may be obtained from (6.66), by analogy with (6.89):

$$\bar{J}_1(x, t) = \frac{\bar{J}(x_1, t)}{P_1 - \beta P_2 \bar{c}_3}, \tag{6.93}$$

where

$$P_1(r_0, x) = \frac{2}{r_0^3} \int_0^{r_0} r^3 \frac{\partial \tilde{J}_1}{\partial r} dr, \quad P_2(r_0, x) = \frac{2}{r_0^3} \int_0^{r_0} r^3 \tilde{c}_3 \tilde{J}_1 dr. \tag{6.94}$$

For many cases of practical interest $\varepsilon = $ const., i.e.,

$$\frac{\partial \bar{u}}{\partial x} = \frac{\partial \bar{u}_1}{\partial x} = 0, \quad \bar{u} = \bar{\bar{u}}, \quad \bar{u}_1 = \bar{\bar{u}}_1, \quad B_1 = \frac{\partial A_1}{\partial x}, \quad B_2 = \frac{\partial A_2}{\partial x}, \quad B_3 = \frac{\partial A_3}{\partial x}, \tag{6.95}$$

and the number of parameters in the model decrease.

The photochemical reaction rate equations shown in (6.53) are acceptable when J and c_1 are very small. A more general form of the photochemical reaction rate equation [48] (written here for the downcomer) is:

$$I = \frac{I_{\max} J}{k_j + J + J^2/k_{\text{inhb}}} \frac{c_1 c_2}{k_c + c_1}. \tag{6.96}$$

Another possible form for equations (6.53) could be

$$I = \bar{k}_0 c_1^{\gamma_1} c_2^{\gamma_2} J^{\gamma}, \quad I_1 = \bar{k}_0 c_0^{\gamma_1} c_3^{\gamma_2} J_1^{\gamma}, \tag{6.97}$$

where the kinetic parameters $\bar{k}_0, \gamma, \gamma_1, \gamma_2$ must be obtained using experimental data. Applying expressions (6.97) in equations (6.86–6.88) and (6.91), the photochemical reaction rate equations has the form

$$\begin{gathered} -\alpha k_0 M(r_0, R_0, x_1) \bar{c}_1^{\gamma_1} \bar{c}_2^{\gamma_2} \bar{J}^{\gamma}; \quad +k_0 M(r_0, R_0, x_1) \bar{c}_1^{\gamma_1} \bar{c}_2^{\gamma_2} \bar{J}^{\gamma}; \\ +\frac{k_0}{1-\varepsilon} M_3(r_0, x) \bar{c}_0^{\gamma_1} \bar{c}_3^{\gamma_2} \bar{J}_1^{\gamma}, \end{gathered} \tag{6.98}$$

where

$$M(r_0, R_0, x_1) = \frac{2}{R_0^2 - r_0^2} \int_{r_0}^{R_0} r \bar{c}_1^{\gamma_1} \bar{c}_2^{\gamma_2} \bar{J}^{\gamma} dr, \quad M_3(r_0, x) = \frac{2}{r_0^2} \int_0^{r_0} r \bar{c}_0^{\gamma_1} \bar{c}_3^{\gamma_2} \bar{J}_1^{\gamma} dr. \tag{6.99}$$

Problems (6.81), (6.82), and (6.86–6.88) are mathematical model of an airlift photobioreactor. The model parameters are of five types:

1. Known beforehand ($c^{(0)}$, $c_2^{(0)}$, R_0, J_0, r_0).
2. Obtained beforehand (ε, χ, α, β, k_o, γ, γ_1, γ_2).
3. Obtained without a photobioreaction (k, D, D_0, A, A_0).
4. Obtained with a photobioreaction (D_1, D_2, D_3), because diffusion of the gas and the photoactive substance is the result of the photobioreaction.
5. Obtained in the modeling and specified in the scale-up (A, A_0, A_1, A_2, A_3, M, M_3, P_1, P_2), because they are functions of the column radius and radial nonuniformity of the velocity concentration.

The parameters $c^{(0)}$, $c_2^{(0)}$, J_0, ε, χ, α, β, k_0, γ, γ_1, γ_2, k, D, D_0, D_1, D_2, D_3 are related to the process (gas absorption with a photobioreaction in the liquid phase), but the parameters R_0, r_0, A, A_0, A_1, A_2, A_3, M, M_3, P_1, and P_2 are related to the apparatus (column radius and radial nonuniformity of the velocities and concentrations).

References

1. Boyadjiev Chr (1993) Fundamentals of modeling and simulation in chemical engineering and technology. Bulgarian Academy of Sciences, Sofia (in Bulgarian)
2. Boyadjiev Chr, Beschkov V (1984) Mass transfer in liquid film flows. Bulgarian Academy of Sciences, Sofia
3. Boyadjiev Chr, Beschkov V (1984) Mass transfer in following liquid films. Mir, Moscow (in Russian)

4. Boyadjiev Chr, Levich V, Krylov V (1968) Int Chem Eng 8:393
5. Boyadjiev Chr (1971) Int Chem Eng 11:459
6. van Dyke M (1964) Perturbation methods in fluid mechanics. Academic, New York
7. Nayfen AH (1973) Perturbation methods. Wiley, New York
8. Higbie R (1935) Trans AIChE 31:365
9. Boyadjiev Chr (1972) C R Bulg Acad Sci 25(2):205
10. Boyadjiev Chr (1971). Int Chem Eng 11:470
11. Schlichting H, Gerstein K (2000) Boundary layer theory, 8th edn. Springer, Berlin
12. Boyadjiev Chr, Piperova M (1971) Int Chem Eng 11:479
13. Boyadjiev Chr (1971) Int Chem Eng 11:464
14. Boyadjiev Chr (2006) Int J Heat Mass Transf 49:796
15. McMullen AK, Miyauchi T, Vermenlen T (1958) UCRI—3911. US Atomic Energy Commission
16. Miyauchi T, Vermenlen T (1963) Ind Eng Chem Fundam 2:113
17. Sleicher CA Jr (1959) AIChE J 5:145
18. Boyadzhiev L, Boyadjiev Chr (1973) Chem Eng J 6:107
19. Al Masry WA, Abasaeed AE (1998) Chem Eng Sci 53(24):4085
20. Heijnen JJ, Hols J, van der Lans RGJM, van Leeuwen HLJM, Weltevrede R (1997) Chem Eng Sci 52(15):2527
21. Abashar ME, Narsingh U, Rouillard AE, Judd R (1998) Ind Eng Chem Res 37(4):1251
22. Camarasa E, Meleiro LAC, Carvalho E, Domingues A, Maciel Filho R, Wild G, Poncin S, Midoux N, Bouillard J (2001) Comput Chem Eng 25(4–6):577
23. Boyadjiev Chr (2006) Int J Heat Mass Transf 49:2053
24. Guhman AA (1973) Introduction in the similarity theory. High School, Moscow (in Russian)
25. Rowe PN (1974) CHEMTECH 14(1):9
26. Levich VG (1962) Physicochemical hydrodynamics. Prentice Hall, New York
27. Beschkov V, Boyadjiev Chr (1977) Chem Eng J 14:1
28. Cerro RL, Whitaker S (1971) J Colloid Interface Sci 37:33
29. Matuura R, Kimizuka H, Miyamoto S, Yatsunami K (1959) Bull Chem Soc Jpn 32:405
30. Strobel WJ, Whitaker S (1969) AIChE J 15:527
31. Ruckenstein E (1974) Chem Eng Sci 20:853
32. Davies JT (1960) Trans Inst Chem Eng 38:289
33. Sada E, Himmelblau DM (1967) AIChE J 13:860
34. Princen HM, Overbeek JG, Mason SG (1967) J Colloid Interface Sci 24:125
35. Alekseenko SV, Nakoryakov VE, Pokusaev BG (1994) Wave flow of liquid films. CRC, London
36. Penev V, Krylov VS, Boyadjiev Chr, Vorotilin VP (1972) Int J Heat Mass Transf 15:1395
37. Kapitza PL, Kapitza SP (1949) Zh Eksp Teor Fiz 19:105
38. Massot C, Irani F, Lightfoot EN (1966) AIChE J 12:443
39. Olevsky VM (1969) Dissertation. Moskovskii Khimiko-Tekhnologicheskii Institut, Moscow
40. Beschkov V, Boyadjiev Chr (1983) Chem Eng Commun 20:173
41. Lee JK, Low GS (1992) Biotechnol Bioeng 40:1119
42. Frohlich BT, Webster IA, Ataai MM, Shuler ML (1983) Biotechnol Bioeng Symp 13:331
43. Ogbonna JC, Yada H, Masu H, Tanaka H (1966) J Ferment Bioeng 82:61
44. Prokop A, Erickson LE (1994) In: Asenjo JA, Merchuk JC (eds) Bioreactor system design. Dekker, New York
45. Merchuk JC, Ladwa JC, Bulmer M (1993) In: Ninehow A (ed) Bioprocesses and bioreactor fluid dynamics. BHRA–Elsevier, Amsterdam, p 61
46. Merchuk JC, Gluz M (1999) In: Flickinger MC, Drew SW (eds) Encyclopedia of bioprocess technology. Wiley, New York, p 320
47. Schlotelburg C, Gluz M, Popovic M, Merchuk JC (1999) Can J Chem Eng 77:804
48. Aibo S (1982) Adv Biochem Biochem Eng 23:85
49. Krug RR (1978) Ind Eng Chem Fundam 17:306

Mass Transfer Theories

The contemporary development of the chemical, power, biotechnology, oil processing, and food processing industries, for example, requires the creation of devices with high throughput and is associated with mass transfer rate problems in both theoretical model building and the background mass transfer theory. In this context, particularly when small concentration gradients control the process of interest, linear mass transfer models are widely applicable.

1 Linear Mass Transfer Theory

The balance of the convective and the diffusive transfer mechanisms determines the overall mass transfer in a moving fluid. If both the velocity distribution and the concentration field are denoted as $\boldsymbol{u}(x, y, z)$ and $c(x, y, z)$, then the mass flux ($\boldsymbol{j}$) per unit surface of a given elementary volume is the sum of the convective and the diffusive fluxes, namely,

$$\boldsymbol{j} = \boldsymbol{u}c - D\boldsymbol{grad}\, c, \tag{1.1}$$

where D is the molecular diffusivity.

When a stationary process takes place and a substance volumetric source (sink) is missing, the mass balance of the substance in an elementary volume is, in fact, an integration of the flow over the whole surface of this volume, namely,

$$\operatorname{div} \boldsymbol{j} = 0. \tag{1.2}$$

From (1.1) and (1.2) we have

$$\boldsymbol{u}\ \boldsymbol{grad}\, c = D\nabla^2 c. \tag{1.3}$$

Further, using (1.3), one can formulate a two-dimensional mass transfer problem within an area with dimensions L and h. Here, $y = 0$ is the surface (interphase

DOI: 10.1007/978-3-642-10778-8_3, © Springer-Verlag Berlin Heidelberg 2010

surface), through which the mass transfer towards another phase (solid, liquid, and gas) takes place:

$$\begin{aligned} & u\frac{\partial c}{\partial x}+v\frac{\partial c}{\partial y}=D\left(\frac{\partial^2 c}{\partial x^2}+\frac{\partial^2 c}{\partial y^2}\right); \\ & x=0, \quad y>0, \quad c=c_0, \quad x=L, \quad 0\le y<h, = c^*; \\ & y=0, \quad 0\le x<L, \quad c=c^*; \quad y=h, \quad c=c_0. \end{aligned} \tag{1.4}$$

From (1.4) the mass transfer rate (J) can be defined through the mass transfer coefficient (k) and the local mass flux (i), namely,

$$J=k(c^*-c_0)=\frac{h}{L}\int_0^L i dx, \quad i=-D\left(\frac{\partial c}{\partial y}\right)_{y=0}. \tag{1.5}$$

In a dimensionless form, this leads to the definition of the Sherwood number:

$$Sh=\frac{kL}{D}=-\frac{1}{c^*-c_0}\int_0^L\left(\frac{\partial c}{\partial y}\right)_{y=0}dx. \tag{1.6}$$

Expressions (1.5) and (1.6) reveal that the determination of the mass transfer rate requires the mass transfer coefficient k or the Sherwood number Sh to be known, i.e., problem (1.4) has to be solved. The principal problem emerging in this solution is the determination of the velocity field since the Navier–Stokes equations are strongly nonlinear [1]. The problem can be avoided by application of some model theories of the mass transfer, as outlined next.

1.1 Model Theories

The basic difficulties in the mathematical description of mass transfer processes are related to the problem of modeling complicated hydrodynamic conditions forming the process background. In many mass transfer theories, the accepted simplifications replace the real conditions by to some extent unjustified hydrodynamic models.

The first mass transfer theory, conceived by Nernst, was the so-called *film theory* [2]. Similar theories concerning gas–liquid and liquid–liquid systems are those of Langmuir [5] and Lewis and Whitman [6]. The film theory states that the mass transfer is performed by steady-state diffusion processes through an immovable fluid film with thickness h. The film theory, in fact, is an approximation of the linear mass transfer theory (1.4) if the film theory conditions are accepted, namely,

$$u=v=0, \quad \text{const.}=h\ll L. \tag{1.7}$$

In this way, from (1.4) it follows that.

$$\frac{\partial^2 c}{\partial y^2} = 0; \quad y = 0, \quad c = c^*; \quad y = h, \quad c = c_0, \tag{1.8}$$

i.e.,

$$c = \frac{c_0 - c^*}{h} y + c^*, \quad k = \frac{D}{h}. \tag{1.9}$$

The principal disadvantages of the film theory are due to the linear relationship of k and the molecular diffusivity D, a fact that does not match the experimental results. In this context, the unknown film thickness h avoids the need for the mass transfer coefficient to be predicted theoretically. However, to some extent, some ideas and outcomes of this theory are still valid and have a more fundamental background. Precisely, the assumption that the mass transfer takes place in a thin layer near the phase boundary and there is thermodynamic equilibrium at the interphase surface are principal outcomes of the film theory, together with the assumption that the diffusional resistances are connected in series (additivity of the diffusion resistances) [3, 4]. In the cases of mass transfer complicated with a fast chemical reaction, where the mass transfer rate does not depend on the velocity of the flow, the outcomes of this theory are valid.

Higbie's *penetration theory* [7] is another approach to approximate linearly the mass transfer process assuming the process is nonstationary in a coordinate system moving with interphase velocity u_0:

$$\frac{\partial c}{\partial \tau} = D \frac{\partial^2 c}{\partial y^2}, \quad \tau = \frac{x}{u_0}. \tag{1.10}$$

This case, in fact, is equivalent to mass transfer in a thin layer of thickness δ flowing with a constant fluid velocity u_0:

$$u = u_0, \quad v = 0, \quad \delta \ll h < L. \tag{1.11}$$

As a result of this assumption, from (1.4) one can obtain.

$$u_0 \frac{\partial c}{\partial x} = D \frac{\partial^2 c}{\partial y^2}; \quad x = 0, \quad c = c_0; \quad y = 0, \quad c = c^*; \quad y \to \infty, \quad c = c_0. \tag{1.12}$$

The solution of (1.12) has been obtained by means of Green's functions [8, 9] in the form.

$$c = c_0 + (c^* - c_0) erfc\, y \sqrt{\frac{u_0}{4Dx}}, \tag{1.13}$$

which yields.

$$Sh = \frac{2}{\sqrt{\pi}} Pe^{\frac{1}{2}}, \quad Pe = \frac{u_0 L}{D}. \tag{1.14}$$

In fact, the velocity in the thin layer varies [3, 4, 10, 11, 77] and (1.14) is the zeroth approximation in the solution of the problem, assuming a homogeneous velocity profile.

The first attempts to account for the real hydrodynamic conditions near fluid–solid interfaces were made by Prandtl [12] and Taylor [13]. They supposed that in turbulent conditions a laminar flow (like Couette flow) exists near the solid interface.

On the other hand, the *surface renewal theory of* Kishinevsky [14, 15] and Danckwerts [16], looks more deeply at the processes occurring near the interface. The main idea of this theory is there is a permanent replacement of the fluid elements contacting the solid surface. The motion of the fluid elements approaching the surface, contacting it for a certain time, and then going back to the bulk of the fluid phase is promoted by the turbulence. The contact time $\Delta\tau$ between a given fluid element and the solid surface is either constant as was assumed by Kishinevsky or spans a range of values as stated by Danckwerts. The contact time $\Delta\tau$ cannot be predicted theoretically and its determination needs experimental data to be processed. The turbulent pulsations vanish near the solid surface within the viscous sublayer, and this forms the basis of the turbulent mass transfer theory. During the time of contact $\Delta\tau$, the mass transfer in the fluid element is only due to transient diffusion.

There exist versions of the penetration and the renewal theories, such as the film-penetration model of Toor and Marchelo [17] and the development of Ruckenstein [18–20]. However, the introduction of the transient (nonstationary) diffusion mechanism in the model theories has no clear physical basis. The parameters calculated on the basis of experimental data are in good agreement with the experimental results but do not allow prediction of the process behavior under new conditions.

The theoretical analysis of the turbulent mass transfer (see also Sect. 1.3) shows that the calculation of the mass transfer rate is possible if the turbulent pulsation fading law in the viscous sublayer ($D_{turb} \sim y^n$) is known. Different values of n ($n = 2$ [21], $n = 3$ [22–25], $n = 4$ [26–30], $n = 5$) have been suggested in the literature [31]. Obviously additional experimental data are necessary.

Several other model theories of linear mass transfer exist but they do not differ significantly from those mentioned above and have similar disadvantages of insufficient physical reasoning of their basic assumptions. In the context of these critical comments, the mass transfer in the boundary layer approximation has the best physical reasoning.

1.2 Boundary Layer Theory

The interface mass transfer in gas (liquid)–solid systems [1, 32, 33] takes place through an immobile phase boundary. In this context, a potential flow with a constant velocity u_0 on a semi-infinite flat plate will be discussed next. Under such assumptions, from (1.2.39) and (1.3.35) one can obtain directly.

$$u\frac{\partial u}{\partial x}+v\frac{\partial u}{\partial y}=v\frac{\partial^2 u}{\partial y^2},\quad \frac{\partial u}{\partial x}+\frac{\partial v}{\partial y}=0,\quad u\frac{\partial c}{\partial x}+v\frac{\partial c}{\partial y}=D\frac{\partial^2 c}{\partial y^2};$$
$$x=0,\quad u=u_0,\quad c=c_0;\quad y=0,\quad u=0,\quad v=0,\quad c=c^*;y\to\infty,$$
$$u=u_0,\quad c=c_0. \tag{1.15}$$

Here the boundary conditions state there is a thermodynamic equilibrium at the phase boundary.

($y = 0$). Depending on the sign of the difference ($c^*- c_0$), a process of dissolving or crystallization takes place. Problem (1.15) has a solution if the following similarity variables are used:

$$u=0,5u_0\varepsilon\varphi',\quad v=0,5\left(\frac{u_0 v}{x}\right)^{0,5}(\eta\varphi'-\varphi),c=c_0+(c^*-c_0)\psi,\quad y=\eta\left(\frac{u_0}{4Dx}\right)^{-0,5},$$
$$\varepsilon=Sc^{0,5},\quad \varphi=\varphi(\eta),\quad \psi=\psi(\eta). \tag{1.16}$$

The introduction of (1.16) into (1.15) leads to.

$$\begin{aligned}&\varphi'''+\varepsilon^{-1}\varphi\varphi''=0,\quad \psi''+\varepsilon\varphi\psi'=0,\\ &\varphi(0)=0,\quad \varphi'(0)=0,\quad \psi(0)=1,\quad \varphi'(\infty)=2\varepsilon^{-1},\quad \psi(\infty)=0.\end{aligned} \tag{1.17}$$

The solution of (1.17) is obtained [32] through the Blasius function $f(z)$:

$$\varphi(\eta)=f(z),\quad z=\frac{2}{c}\eta,\quad \psi(\eta)=1-\frac{1}{\varphi}\int_0^z E(\varepsilon,p)dp,$$
$$E(\varepsilon,p)=\exp\left[-\frac{\varepsilon^2}{2}\int_0^p f(s)\right]ds, \tag{1.18}$$
$$\varphi=\int_0^\infty E(\varepsilon,p)dp\approx\begin{cases}3,01\,Sc^{-0,35}\text{—for gases}\\ 3,12\,Sc^{-0,34}\text{—for liquids}\end{cases}.$$

The Blasius function is the solution of the problem.

$$2f'''+ff''=0,\quad f(0)=0,\quad f'(0)=0,\quad f''(0)=0,33205 \tag{1.19}$$

and its values are given elsewhere [34].

The introduction of (1.18) into (1.6) determines the Sherwood number:

$$Sh=\frac{kL}{D}=-Pe^{0,5}\psi'(0)\approx\frac{2}{3}\sqrt{Re}\sqrt{[3]Sc}. \tag{1.20}$$

1.3 Two-Phase Boundary Layers

The mass transfer in gas–liquid and liquid–liquid systems occurs at a moving phase boundary. In the boundary layer approximation, the problem is expressed as.

$$u_j \frac{\partial u_j}{\partial x} + v_j \frac{\partial u_j}{\partial y} = v_j \frac{\partial^2 u_j}{\partial y^2}, \quad \frac{\partial u_j}{\partial x} + \frac{\partial v_j}{\partial sy} = 0,$$
$$u_j \frac{\partial c_j}{\partial x} + v_j \frac{\partial c_j}{\partial y} = D_j \frac{\partial^2 c_j}{\partial y^2}, \quad j = 1, 2. \tag{1.21}$$

The boundary conditions take into account the continuity of the velocity, stress tensor, and mass flux at the phase boundary:

$$x = 0, \quad u_j = u_{j0}, \quad c_j = c_{j0}, \quad j = 1, 2; \quad y = 0, \quad u_1 = u_2, \quad \mu_1 \frac{\partial u_1}{\partial y} = \mu_2 \frac{\partial u_2}{\partial y},$$
$$c_1 = \chi c_2, \quad D_1 \frac{\partial c_1}{\partial y} = D_2 \frac{\partial c_2}{\partial y}, \quad v_j = 0, \quad j = 1, 2;$$
$$y \to \infty, \quad u_1 = u_{10}, \quad c_1 = c_{10}; y \to -\infty, \quad u_2 = u_{20}, \quad c_2 = c_{20}. \tag{1.22}$$

The subscripts are j = 1 for the first phase (gas or liquid) and j = 2 the second one (liquid). At the phase boundary, a phase equilibrium is assumed and χ is the distribution coefficient: the Henry constant in the case of gas–liquid systems or the coefficient of separation of liquid–liquid counterparts.

The average rate of mass transfer between the phases is determined in a similar way by integration and averaging of the local mass fluxes, namely,

$$J = K_1(c_{10} - \chi c_{20}) = \frac{1}{L}\int_0^L I_1 dx = k_1(c_{10} - c_1^*)$$
$$= K_2\left(\frac{c_{10}}{\chi} - c_{20}\right) = \frac{1}{L}\int_0^L I_2 dx = k_2(c_2^* - c_{20}), \tag{1.23}$$
$$c_1^* = \chi c_2^*.$$

Here K_j (j = 1, 2) are the interphase mass transfer coefficients, k_j (j = 1, 2) are mass transfer coefficients, and c_1^* and c_2^* are the concentrations of both phases at the interphase surface (y = 0). The local mass fluxes (after solution of 1.21, 1.22) are.

$$I_j = -D_j\left(\frac{\partial c_j}{\partial y}\right)_{y=0}, \quad j = 1, 2. \tag{1.24}$$

Then, from (1.23) and (1.24) it follows that the Sherwood numbers are.

$$Sh_j = \frac{K_j L}{D_j} = \frac{\chi^{j-1}}{c_{10} - \chi c_{20}} \int_0^L \left(\frac{\partial c_j}{\partial y}\right)_{y=0} dx, \quad j = 1,2. \tag{1.25}$$

From (1.23) the law of additivity of the diffusion resistances follows directly, namely,

$$K_1^{-1} = k_1^{-1} + \chi k_2^{-1}; \quad \chi = 0, \quad K_1 = k_1; \quad K_2^{-1} = (\chi k_1)^{-1} + k_2^{-1}; \quad \chi \to \infty, \\ K_2 = k_2. \tag{1.26}$$

Expressions (1.26) reveal that when the interphase mass transfer rate is limited by the diffusion resistance in one of the phases, then the interphase mass transfer coefficient equals the mass transfer coefficient in the same phase.

The problem (1.21, 1.22) has a solution after introducing the following similarity variables:

$$\begin{aligned} &u_j = 0,5 j u_{j0} \varepsilon_j \varphi_j', \quad v_j = (-1)^{j-1} 0,5 j \left(\frac{u_{j0} v_j}{x}\right)^{0,5} \left(\eta_j \varphi_j' - \varphi_j\right), \\ &c_j = c_{j0} - (-\chi)^{1-j} (c_{10} - \chi c_{20}) \psi_j, \quad \varphi_j = \varphi_j(\eta_j), \quad \psi_j = \psi_j(\eta_j), \\ &\eta_j = (-1)^{j-1} y \left(\frac{u_{j0}}{4 D_j x}\right)^{0,5}, \quad \varepsilon_j = Sc_j^{0,5}, \quad Sc_j = \frac{v_j}{D_j}, \quad j = 1,2. \end{aligned} \tag{1.27}$$

As a result, it follows that.

$$\begin{aligned} &\varphi_j''' + j\varepsilon_j^{-1} \varphi_j \varphi_j'' = 0, \quad \psi_j'' + j\varepsilon_j \varphi_j \psi_j' = 0, \\ &\varphi_j(0) = 0, \quad \varphi_j'(\infty) = \frac{2}{j\varepsilon_j}, \quad \psi_j(\infty) = 0, \quad j = 1,2, \\ &\varphi_1'(0) = 2\theta_1 \frac{\varepsilon_2}{\varepsilon_1} \varphi_2'(0), \quad \varphi_2''(0) = -0,5\theta_2 \left(\frac{\varepsilon_1}{\varepsilon_2}\right)^2 \varphi_1''(0), \\ &\psi_1'(0) = \frac{\chi}{\varepsilon_0} \psi_2'(0), \quad \psi_1(0) + \psi_2(0) = 1, \\ &\theta_1 = \frac{u_{20}}{u_{10}}, \quad \theta_2 = \left(\frac{\mu_1}{\mu_2}\right) \left(\frac{v_1}{v_2}\right)^{-0,5} \left(\frac{u_{10}}{u_{20}}\right)^{1,5}, \quad \varepsilon_0 = \left(\frac{D_2\, u_{20}}{D_1\, u_{10}}\right)^{0,5}. \end{aligned} \tag{1.28}$$

The solution of (1.28) allows the determination of the interphase mass transfer rate between two phases with a moving phase boundary:

$$Sh_j = -\sqrt{Pe_j}\,\psi_j'(0), \quad Pe_j = \frac{u_{j0} L}{D_j}, \quad j = 1,2. \tag{1.29}$$

Problem (1.28) has been solved numerically [35]. In the case of a gas–liquid system an asymptotic solution using the perturbation method (see Page 414 and the next) has been developed [11, 36]. This asymptotic solution is a series of the orders of the small parameters θ_1 and θ_2. For $\psi_j'(0)$ (j = 1, 2) the following expressions are obtained as a first approximation regarding the small parameters θ_1 and θ_2:

$$\begin{aligned}\psi_1'(0) &= -\frac{2}{\varepsilon_1\varphi_{10}}\frac{1}{1+a} - \frac{2\theta_1}{\alpha\varphi_{10}^2\varepsilon_1}\frac{1}{(1+a)^2} - 8\theta_2\alpha\frac{\varepsilon_2}{\varepsilon_1}\frac{\bar{\varphi}_2}{\varphi_{10}}\frac{a}{(1+a)^2},\\ \psi_2'(0) &= -\frac{2}{\sqrt{\pi}}\frac{a}{1+a} - \theta_1\frac{2}{\sqrt{\pi}\alpha\varphi_{10}}\frac{a}{(1+a)^2} - 8\theta_2\frac{\alpha\varepsilon_2\bar{\varphi}_2}{\sqrt{\pi}}\frac{a^2}{(1+a)^2},\end{aligned} \tag{1.30}$$

where.

$$\varphi_{10} \approx \frac{3}{\sqrt[3]{Sc}}, \quad a = \frac{\chi\sqrt{\pi}}{\varepsilon_0\varepsilon_1\varphi_{10}}, \quad \alpha = 0,33205, \quad \bar{\varphi}_2 = \frac{1}{8}\sqrt{\frac{\pi}{Sc_2}}. \tag{1.31}$$

In the cases when the interphase mass transfer is limited by the mass transfer in the gas phase $\chi/\varepsilon_0 \to 0$, $a \to 0$, the Sherwood number can be expressed as.

$$Sh_1 = \sqrt{Pe_1}\left(\frac{2}{\varepsilon_1\varphi_{10}} + \frac{2\theta_1}{\varepsilon_1\alpha\varphi_{10}^2}\right). \tag{1.32}$$

When the interphase mass transfer is limited by the mass transfer in the liquid phase, the Sherwood number can be determined in a similar way, namely,

$$Sh_2 = \sqrt{Pe_2}\left(\frac{2}{\sqrt{\pi}} + 8\theta_2\frac{\alpha\varepsilon_2\bar{\varphi}_2}{\sqrt{\pi}}\right). \tag{1.33}$$

Similar results obtained for the hydrodynamics and the mass transfer in co-current flows for gas–liquid systems [35, 37–39] are in a good agreement with the experimental data.

2 Mass Transfer in Countercurrent Flows

Chemical technologies based on countercurrent flows in gas–liquid systems are widely encountered in practice. The analyses of such flows [9] reveal that there is a possibility to obtain asymptotic solutions for gas–liquid systems which are in agreement with the experimental data, obtained from thermoanemometrical measurements in the boundary layers. The correctness of the asymptotic method proposed in [9] was confirmed by numerical experiments concerning the exact solution of the problem [40]. The theoretical analysis of the countercurrent flow [41], for example, shows that it is a nonclassical problem of mathematical physics insufficiently discussed in the literature. The parabolic boundary value problem with changing direction of time [42, 43] is a typical problem of this type. It was shown in [41] that this nonclassical problem can be described as consisting of several classical subproblems.

2.1 Velocity Distribution

The mathematical description of the countercurrent flow in the boundary approximation is.

$$
\begin{aligned}
& u_j \frac{\partial u_j}{\partial x} + v_j \frac{\partial u_j}{\partial y} = v_j \frac{\partial^2 u_j}{\partial y^2}, \quad \frac{\partial u_j}{\partial x} + \frac{\partial v_j}{\partial y} = 0, \quad j = 1,2; \\
& x = 0, \quad y \geq 0, \quad u_1 = u_1^\infty; \qquad x = L, \quad y \leq 0, \quad u_2 = -u_2^\infty; \\
& y \to \infty, \quad 0 \leq x \leq L, \quad u_1 = u_1^\infty; \qquad y \to -\infty, \quad 0 \leq x \leq l, \quad u_2 = -u_2^\infty; \\
& y = 0, \quad 0 < x < L, \quad u_1 = u_2, \quad \mu_1 \frac{\partial u_1}{\partial y} = \mu_2 \frac{\partial u_2}{\partial y}, \quad v_1 = v_2 = 0.
\end{aligned}
\tag{2.1}
$$

Problem (2.1) can be represented in a dimensionless form using two different coordinate systems for the two phases. The flow in each phase is oriented along the longitudinal coordinate, which yields the following dimensionless variables and parameters:

$$
\begin{aligned}
& x = LX_1 = L - LX_2, \qquad y = \delta_1 Y_1 = -\delta_2 Y_2, \\
& u_1 = u_1^\infty U_1, \quad v_1 = u_1^\infty \frac{\delta_1}{L} V_1, \quad u_2 = -u_2^\infty U_2, \quad v_2 = -u_2^\infty \frac{\delta_2}{L} V_2, \\
& \delta_j = \sqrt{\frac{v_j L}{u_j^\infty}}, \quad j = 1,2, \quad \theta_1 = \frac{u_2^\infty}{u_1^\infty}, \quad \theta_2 = \left(\frac{\rho_1 \mu_1}{\rho_2 \mu_2}\right)^{\frac{1}{2}} \left(\frac{u_1^\infty}{u_2^\infty}\right)^{\frac{3}{2}}.
\end{aligned}
\tag{2.2}
$$

In the new coordinate systems, the model of countercurrent flows takes the form.

$$
\begin{aligned}
& U_j \frac{\partial U_j}{\partial X_j} + V_j \frac{\partial U_j}{\partial Y_j} = \frac{\partial^2 U_j}{\partial Y_j^2}, \quad \frac{\partial U_j}{\partial X_j} + \frac{\partial V_j}{\partial Y_j} = 0; \\
& X_j = 0, \quad U_j = 1; \quad Y_j \to \infty, \quad U_j = 1; \\
& Y_1 = Y_2 = 0, \quad U_1 = -\theta_1 U_2, \quad \theta_2 \frac{\partial U_1}{\partial Y_1} = \frac{\partial U_2}{\partial Y_2}, \quad V_j = 0; \quad j = 1,2.
\end{aligned}
\tag{2.3}
$$

Problem (2.3) cannot be solved directly, because the velocities $U_i (i = 1, 2)$ change their directions within the ranges $0 \leq X_i \leq 1$, $0 \leq Y_i < \infty$, $(i = 1, 2)$. This nonclassical problem of mathematical physics can be converted to a classical one by introduction of the following similarity variables:

$$
U_j = f_j', \quad V_j = \frac{1}{2\sqrt{X_j}} (\eta_j f_j' - f_j), \quad f_j = f_j(\eta_j), \quad \eta_j = \frac{Y_j}{\sqrt{X_j}}. \tag{2.4}
$$

The substitution of (2.4) into (2.3) yields.

$$2f_j''' + f_j f_j'' = 0, \quad f_j(0) = 0, \quad f_j(\infty) = 1, \quad j = 1,2,$$

$$f_1'(0) = -\theta_1 f_2'(0), \quad \theta_2 \sqrt{\frac{X_2}{X_1}} f_1''(0) = f_2''(0), \quad X_1 + X_2 = 1. \tag{2.5}$$

It is obvious from (2.5) that problem (2.3) has no solution in similarity variables.

However, problem (2.5) can be solved after the introduction of new parameter $\bar{\theta}_2$ for each $X_1 \in (0, 1)$:

$$\bar{\theta}_2 = \theta_2 \sqrt{\frac{1 - X_1}{X_1}}. \tag{2.6}$$

Hence, the problem has a local similarity solution. In this way problem (2.5) is substituted by several separate problems for each $X_1 \in (0, 1)$.

The solutions of these separate problems can be obtained after the introduction of the function.

$$F(a,b) = \int_6^7 (f_1' - 1)^2 d\eta_1 + \int_6^7 (f_2' - 1)^2 d\eta_2, \quad a = f_1'(0), \quad b = f_1''(0). \tag{2.7}$$

The solution of (2.5) for each $X_1 \in (0,1)$ is obtained after searching for the minimum of the function *F*(*a*,*b*). At each step of the minimization procedure, the boundary problem has to be solved:

$$2f_j''' + f_j f_j'' = 0, \quad f_j(0) = 0, \quad j = 1,2, \quad f_1'(0) = a, \quad f_2'(0) = -\frac{a}{\theta_1}, \quad f_1''(0) = b,$$
$$f_2''(0) = \bar{\theta}_2 b. \tag{2.8}$$

Problem (2.8) was solved numerically for countercurrent gas and liquid flows for the following parameters values: $\theta_1 = 0.1$ and $\theta_2 = 0.152$. In accordance with the requirement for a minimum of *F*(*a*,*b*) in (2.7), the boundary conditions *a*, *b* and *F*(*a*,*b*) were determined [40].

The energy dissipated in the laminar boundary layer [6, 7] is described for both phases by the equations.

$$e_j = \mu_j \int_0^l \int_0^{(-1)^{j+1}} \left(\frac{\partial u_j}{\partial y}\right)^2 dxdy, \quad E_j = -\int_0^1 \int_0^\infty \left(\frac{\partial U_j}{\partial Y_j}\right)^2 dY_j dX_j,$$
$$E_j = \frac{e_j \sqrt{u_j^\infty l / v_i}}{v_i \rho_i u_j^{\infty 2}}, \quad j = 1,2. \tag{2.9}$$

For gas–liquid countercurrent flows, the introduction of similarity variables leads to.

$$E_j = \int_0^1 \frac{1}{\sqrt{X_j}} \left[\int_0^\infty \left(f_j''\right)^2 d\eta_j \right] dX_j, \quad j = 1, 2. \tag{2.10}$$

In the case of co-current flows, the function $f_j''*$ does not depend on X_j and for energy dissipation one can obtain.

$$E_j* = 2 \int_0^\infty \left(f_i''*\right)^2 d\eta_i, \quad i = 1, 2. \tag{2.11}$$

Here f_j* (i = 1, 2) is the solution of equations (2.8) with boundary conditions for co-current flows:

$$\theta_1* = -0.1, \quad \theta_2* = \theta_2 = 0.152, \quad f_1' * (0) = 0.0908, \quad f_1'' * (0) = 0.32765. \tag{2.12}$$

The comparison of the energy dissipated in the laminar boundary layer [6, 7] for the case of gas–liquid countercurrent and co-current flows is shown in Table 1. These results reveal that the energy dissipation for the gas phase in co-current flows is lower than that in countercurrent flows, whereas in the liquid phase there are no significant changes.

2.2 Concentration Distribution

The mathematical model of mass transfer in gas–liquid systems with countercurrent flow in a laminar boundary layer with a flat phase boundary takes the following form:

$$\begin{aligned}
&u_j \frac{\partial c_j}{\partial x} + v_j \frac{\partial c_j}{\partial y} = D_j \frac{\partial^2 c_j}{\partial y^2}, \quad j = 1, 2;\\
&x = 0, \quad y \geq 0, \quad c_1 = c_1^\infty; \quad x = L, \quad y \leq 0, \quad c_2 = c_2^\infty;\\
&y \to \infty, \quad 0 \leq x \leq L, \quad c_1 = c_1^\infty; \quad y \to -\infty, \quad 0 \leq x \leq L, \quad c_2 = c_2^\infty;\\
&y = 0, \quad 0 < x < L, \quad c_1 = \chi c_2, \quad D_1 \frac{\partial c_1}{\partial y} = D_2 \frac{\partial c_2}{\partial y}.
\end{aligned} \tag{2.13}$$

Table 1 Comparison between countercurrent and co-current flows

$\theta_3 = 0$		$\theta_3 \to \infty$		$\theta_3 = 1$			
J_1	J_1*	J_2	J_2*	J_1	J_1*	J_2	J_2*
0.554	0.720	4.380	4.822	0.432	0.626	0.432	0.626
A_1	A_1*	A_2	A_2*	A_1	A_1*	A_2	A_2*
1.06	1.57	739	750	0.82	1.37	72.8	97.3
E_1	E_1*	E_2	E_2*	E_1	E_1*	E_2	E_2*
0.525	0.458	0.00593	0.00643	0.525	0.458	0.01328	0.00643

Here u_j and v_j (j = 1, 2) are the velocity components in the gas and in the liquid phase determined by the solution of (2.8).

The solution of (2.13) was performed [44] by introducing the following similarity variables:

$$\eta_j = (-1)^{j+1} y \sqrt{\frac{u_j^\infty}{\nu_j L X_j}}, \quad X_1 = \frac{x}{L}, \quad X_2 = \frac{L-x}{L}, \quad X_1 + X_2 = 1,$$
$$u_j = (-1)^{j+1} u_j^\infty f_j', \quad v_j = (-1)^{j+1} \frac{1}{2} \sqrt{\frac{\nu_j u_j^\infty}{L X_j}} \left(\eta_j f_j' - f_j\right), \quad f_j = f_j(\eta_j),$$
$$c_j = c_j^\infty - \chi^{1-j}(c_1^\infty - \chi c_2^\infty)\psi_j, \quad \psi_j = \psi_j(\eta_j), \quad j = 1, 2. \tag{2.14}$$

The substitution of equations (2.14) into equations (2.13) leads to.

$$2f_j''' + f_j f_j'' = 0, \quad 2\psi_j'' + Sc_j f_j \psi_j' = 0, \quad j = 1, 2,$$
$$f_1(0) = 0, \quad f_1'(0) = a, \quad f_1''(0) = b, \quad f_2(0) = 0, \quad f_2'(0) = -\frac{a}{\theta_1}, \quad f_{22}''(0) = \bar{\theta}_2 b,$$
$$\psi_1(0) + \psi_2(0) = 1, \quad \bar{\theta}_3 \psi_1'(0) = \psi_2'(0), \quad \psi_j(\infty) = 0, \quad j = 1, 2, \tag{2.15}$$

where.

$$Sc_j = \frac{\nu_j}{D_j} \quad (j = 1, 2), \quad \theta_1 = \frac{u_2^\infty}{u_1^\infty}, \quad \theta_2 = \left(\frac{\rho_1 \mu_1}{\rho_2 \mu_2}\right)^{\frac{1}{2}} \left(\frac{u_1^\infty}{u_2^\infty}\right)^{\frac{3}{2}},$$
$$\bar{\theta}_2 = \theta_2 \sqrt{\frac{X_2}{X_1}}, \quad \theta_3 = \chi \frac{D_1}{D_2} \sqrt{\frac{u_1^\infty \nu_2}{u_2^\infty \nu_1}}, \quad \bar{\theta}_3 = \theta_3 \sqrt{\frac{X_2}{X_1}}. \tag{2.16}$$

The boundary conditions a and b are determined by (2.8).

Obviously, it is evident from (2.15) that it is possible to develop a similarity solution for different values of $X_1 = 1 - X_2$.

The solution of (2.15) was carried out [44] under new boundary conditions for ψ_j(j = 1, 2):

$$\psi_1(0) = \alpha, \quad \psi_1'(0) = \beta, \quad \psi_2(0) = 1 - \alpha, \quad \psi_2'(0) = \bar{\theta}_3 \beta. \tag{2.17}$$

Here α and β are determined for different values of $X_1 = 1 - X_2$, so the conditions $\psi_j(\infty) = 0$, (j = 1, 2) have to be satisfied.

The Sherwood number may be obtained analogously to (1.25) as.

$$Sh_j = -\sqrt{Re_j} \int_0^1 \frac{\psi_1'(0)}{\sqrt{X_j}} dX_j, \quad Re_j = \frac{u_j^\infty L}{\nu_j}, \quad j = 1, 2. \tag{2.18}$$

The dimensionless diffusive flux has the form.

$$J_j = -\int_0^1 \frac{\psi_j'(0)}{\sqrt{X_j}} dX_j, \quad j = 1, 2. \tag{2.19}$$

2.3 Comparison Between Co-current and Countercurrent Flows

The comparison between the mass transfer rates in the countercurrent and the co-current flow modes will be performed by solving equation (2.15) with parameters corresponding to the case of co-current flow, namely,

$$\begin{aligned} &\theta_1 = -0.1, \quad \theta_2 = 0.152, \quad f_1'(0) = 0.0908, \quad f_1''(0) = 0.37265, \\ &\bar{\theta}_3 = \theta_3, \quad J_j^* = -2\psi_j'(0), \quad j = 1, 2. \end{aligned} \tag{2.20}$$

The results obtained for J_j^*(j = 1, 2) are summarized in Table 1. These results definitely indicate that with the co-current flow the mass transfer rate is higher than that exhibited under the countercurrent flow conditions.

The numerical results allow us to determine the ratio (A) between the mass transfer rate (Sh) and the corresponding energy dissipation (E) with both flow modes, namely,

$$A_i = \frac{Sh_i}{E_i}, \quad A_i^* = \frac{Sh_i^*}{E_i^*}, \quad i = 1, 2. \tag{2.21}$$

The data summarized in Table 1 indicate the better performance of the co-current flow mode with a higher mass transfer rate per unit energy dissipation required to perform the process. This theoretical analysis shows that the linear mass transfer theory permits us to predict the mass transfer resistance distribution in two-phase systems. In the diffusion boundary layer approximation, the mass transfer is governed by the parameter χ/ε_0 (see 1.28). The process is limited by the mass transfer in the first phase if $\chi/\chi\varepsilon_0 \geq 10^2.\varepsilon_0 \geq 10^?$; if $\chi/\chi\varepsilon_0.\varepsilon_0 \leq 10^{\ 2}$ the process is limited by the mass transfer in the second phase. The diffusion resistances are of the same order of magnitude if $\chi/\chi\varepsilon_0 \sim 1.\varepsilon_0 \sim 1$. In the film theory mass transfer approximation, the theory reveals that χ is the only governing parameter (see 1.26).

The results obtained so far represent the linear mass transfer theory in the boundary layer approximation with a flat phase boundary. Such problems (the linear mass transfer theory) can be solved in similar manners (in the boundary layer approximation) with conditions imposed by different forms of the interphase surface (wavy, spherical, cylindrical, and other shapes) in processes involving film flows, droplets, bubbles, jets, etc.

3 Nonlinear Mass Transfer

The theoretical analysis of linear mass transfer shows that such a process exists when the equation of convection–diffusion (1.4) is linear, i.e., the velocity (u, v) and the diffusivity (D) are independent of the concentration (c) of the transferred substance. These conditions are valid in systems where the mass transfer does not affect the hydrodynamics and the mass flux is related linearly to the concentration gradient. The linear mass transfer theory, built with these assumptions, has two main outcomes: (1) *the mass transfer coefficient is independent of the concentration and* (2) *the mass transfer rate is unaffected by the direction of interphase mass transfer.*

Any deviations in experiments apart from these two principal theoretical outcomes indicate that nonlinear effects are taking place. The latter are mainly due to secondary flows caused by high mass transfer rates or concentration effects on transport coefficients such as viscosity and diffusivity. The secondary flows can be caused by a concentration gradient (nonlinear mass transfer), a surface tension gradient (Marangoni effect), a density gradient (natural convection), or a pressure gradient (Stefan flow). Under such conditions, the effect of concentration on both the velocity field and the convection–diffusion equation becomes nonlinear. The secondary flows may affect the mass transfer rate through changes in the velocity field, i.e., the result is a changed ratio of the convective and the diffusive transfer in the convection–diffusion equation. The effect may increase many times when the system becomes unstable as a result of secondary flows and attains a new hierarchical state of self-organization (recall, the systems are dissipative structures).

One of the most interesting nonlinear effects arises from the conditions which are imposed by high concentration gradients. Such high gradients induce secondary flows at the phase boundaries. This effect is discussed in detail in this book for a variety of systems and is termed the "*nonlinear mass transfer effect*".

The development of modern processing (e.g., power production, chemical and oil processing, food engineering) calls for systems with intensive mass transfer. Most of these processes are carried out in two-phase systems and it is possible to use large concentration gradients of the transferred substance to develop process intensifications. Under such conditions, the large mass fluxes through the phase boundaries induce secondary flows at the vicinity of the interphase surfaces. The latter affect the flow conditions and change significantly both the mechanisms and the kinetics of mass transfer processes.

The analysis of the mechanism and kinetics of the interphase mass transfer in two-phase systems in many cases is possible if the velocity distribution is determined at the beginning. Then, after its substitution in the convection–diffusion equation, the rate of interface mass transfer can be determined. However, this procedure cannot be applied to systems with large concentration gradients because the flow at the interphase surface is induced by the intensive mass transfer. The velocity of this flow is perpendicular to the interphase surface and is directed towards the mass transfer direction.

The physical reason of this movement is the mechanical impulse transferred from one phase to the other through the particles responsible for the mass transfer. In the linear mass transfer theory, this impulse is considered to be insignificant, but when large mass fluxes cross the interphase surface, it has to be taken into account. Since the transferred impulse is proportional to the diffusive flux of particles involved in the mass transfer, the velocity field at the vicinity of the interphase boundary depends on the concentration field and the corresponding relationships have to be defined. The particular forms of such relationships are determined by a set of coupled equations of mass transfer and momentum transfer under boundary conditions relating the fluxes of mass and momentum at the interphase boundary.

3.1 Influence of Intensive Interphase Mass Transfer on the Hydrodynamics

The induced flow velocity at the *interphase* surface is determined by the hydrodynamic effects caused by the intensive mass transfer. This effect concerns mainly the boundary conditions relevant to the coupled equations of the hydrodynamics and the mass transfer. These equations cannot be solved independently, in contrast with the cases of low mass transfer rates at the phase boundaries. The relationship between the velocity of the induced flow at the phase boundary and the rate of the interphase mass transfer will be exemplified by an isothermal process of transfer of a dissolved substance from phase 1 into phase 2. Let us assume that each phase is a two-component mixture (a solution of substance m in the corresponding solvent) and that the two solvents are immiscible. The diffusive flux $\boldsymbol{j}_{\mathrm{m}}^{(\mathrm{i})}$ of substance m at each point of the space inside phase i is.

$$\boldsymbol{j}_{\mathrm{m}}^{(\mathrm{i})} = M_{\mathrm{m}} c_{\mathrm{m}}^{(\mathrm{i})} \left(\boldsymbol{v}_{\mathrm{m}}^{(\mathrm{i})} - \boldsymbol{v}^{(\mathrm{i})} \right), \quad \mathrm{i} = 1, 2. \tag{1}$$

Here $c_{\mathrm{m}}^{(\mathrm{i})}$ is the molar concentration of the transferred substance in phase i, M_{m} is the molecular mass of this substance, $\boldsymbol{v}_{\mathrm{m}}^{(\mathrm{i})}$ is the average statistical velocity of movement of particles of substance *m* in an arbitrary, but fixed coordinate system, and $\boldsymbol{v}^{(\mathrm{i})}$ is the velocity of the mass center of the whole liquid mixture in the same coordinate system. The velocity $\mathbf{v}^{(\mathrm{i})}$ is defined by the set of equations of hydrodynamics (in the case of laminar flow, the set of Navier–Stokes equations). Besides, by definition, this velocity is connected to the velocities of movement of the mixture components through the relationship.

$$\rho^{(\mathrm{i})} \boldsymbol{v}^{(\mathrm{i})} = M_{\mathrm{m}} c_{\mathrm{m}}^{(\mathrm{i})} \boldsymbol{v}_{\mathrm{m}} + M_0 c_0^{(\mathrm{i})} \boldsymbol{v}_0^{(\mathrm{i})}. \tag{3.2}$$

The variables with subscripts 0 refer to the corresponding solvent, and $\rho^{(\mathrm{i})} = M_{\mathrm{m}} c_{\mathrm{m}}^{(\mathrm{i})} + M_0 c_0^{(\mathrm{i})}$ is the common density of the solution in phase i (in the general case this density is a function of the space coordinates and time).

Let us represent each of the velocities in equation (3.2) as the sum of the velocity of movement at the interphase surface $d\boldsymbol{r}_{\mathrm{s}}/dt$ ($\boldsymbol{r}_{\mathrm{s}}$ is the radius vector of an arbitrary point at the phase boundary) and the velocity of movement in regard to this surface $\boldsymbol{v}_{\mathrm{rk}}^{(\mathrm{i})}(\mathrm{k}=\mathrm{m},0)$. Owing to the fact that the two solvents are immiscible, the normal components of velocities $\boldsymbol{v}_{\mathrm{r}0}^{(1)}$ and $\boldsymbol{v}_{\mathrm{r}0}^{(2)}$ at the interphase boundary must be equal to zero. Hence, when Eq. (3.2) is being projected in the direction of the normal vector ($\boldsymbol{n}$) to the interphase boundary at each point of this boundary, we have

$$\rho^{(\mathrm{i})}\left(\boldsymbol{v}_{\mathrm{r}}^{(\mathrm{i})},\boldsymbol{n}\right)=M_{\mathrm{m}}c_{\mathrm{m}}^{(\mathrm{i})}\left(\boldsymbol{v}_{\mathrm{rm}}^{(\mathrm{i})},\boldsymbol{n}\right). \tag{3.3}$$

Similarly, the projection of Eq. (3.1) on the direction of the normal vector $\boldsymbol{n}$ (taking into account relationship 3.3) yields

$$\left(\boldsymbol{v}_{\mathrm{s}}^{(\mathrm{i})},\boldsymbol{n}\right)=\left(\frac{d\boldsymbol{r}_{\mathrm{s}}}{dt},\boldsymbol{n}\right)+\frac{\left(\boldsymbol{j}_{\mathrm{ms}}^{(\mathrm{i})},\boldsymbol{n}\right)}{\rho_{\mathrm{s}}^{(\mathrm{i})}-M_{\mathrm{m}}c_{\mathrm{ms}}^{(\mathrm{i})}}. \tag{3.4}$$

This equation is correct at each point of the interphase surface. The subscript *s* denotes the variable related to the phase boundary. If the form of the surface is defined by the equations

$$y=f_1(x,t),\quad z=f_2(r,t),\quad r=f_3(\theta,t), \tag{3.5}$$

then for the first term on the right side of (3.4), the following expressions are valid:

$$\left(\frac{d\boldsymbol{r}_s}{dt},\boldsymbol{n}\right)=\begin{cases}1+\left(\frac{\partial f_1}{\partial x}\right)^{-\frac{1}{2}}\frac{\partial f_1}{\partial t} & \text{in Cartesian coordinate system,}\\ 1+\left(\frac{\partial f_2}{\partial r}\right)^{-\frac{1}{2}}\frac{\partial f_2}{\partial t} & \text{in cylindrical coordinate system,}\\ 1+\left(\frac{\partial f_3}{f_3\,\partial \theta}\right)^{-\frac{1}{2}}\frac{\partial f_3}{\partial t} & \text{in spherical coordinate system.}\end{cases} \tag{3.6}$$

From (3.4) it follows that in the case of high interphase mass transfer rates the solution of the hydrodynamic problem is not independent of the solution of the convection–diffusion problem because the velocity distribution is a function of the concentration distribution and the convection–diffusion equation is nonlinear (in linear mass transfer theory the hydrodynamic problem is independent of the diffusion problem).

The published literature refers to many systems of practical interest [71, 108] where intensive mass transfers lead to significant effects on the hydrodynamic conditions. Good examples of such systems are vapor condensation on a cooled wall [45, 46], evaporation of liquids from surfaces of droplets or bubbles [47–51], crystallization and dissolution of salts [52–56], and heat and mass transfer under conditions of intensive injection (or suction) of gases through a porous walls [57–61]. It is important to mention that the nonlinear mass transfer effects discussed further are mainly due to large concentration gradients in the liquid phase or large partial pressure gradients in the gas phase. In such cases the mass fluxes through the phase boundaries are determined by the mass transfer rates. In this context, such systems defined by large concentration gradients differ from Stefan systems [62, 63], where the flows are mainly induced by gradients of the general pressure.

3.2 Boundary Conditions of the Nonlinear Mass Transfer Problem

The mathematical formulation of mass transfer problems taking into account the mass transfer effect on the hydrodynamics was reported first in [64–66]. Equation (3.4) relates in general the velocity of the induced flow to the mass flux across the phase boundary, but in each particular case an exact specific relationship has to be defined.

Generally, with a two-phase system, in Cartesian coordinates [3] the phase boundary can be defined by the function $y = h(x)$ and differentiation of (3.4) can be performed under the assumption that the interphase surface does not vary in time, i.e., waves and any disturbances on the surfaces of growing droplets or bubbles are neglected. The mass flux of the transferred substance at each point of the phase discussed can be expressed by means of the average statistical velocity of this substance (molecules, atoms, ions) $\boldsymbol{v}$ and the mass center velocity of the mixture (phase) particles $\boldsymbol{v}_1$:

$$\boldsymbol{j} = Mc(\boldsymbol{v} - \boldsymbol{v}_1). \tag{3.7}$$

Velocity $\boldsymbol{v}_1$ should satisfy the hydrodynamic equations and should be related to the velocities of the mixture (phase) components through the equation

$$\rho\boldsymbol{v}_1 = M_0c_0\boldsymbol{v}_0 + Mc\boldsymbol{v}. \tag{3.8}$$

Here ρ is the phase (mixture) density, and subscript 0 denotes the phase (mixture) parameters in the absence of a transferred substance. In this way, the density of the phase discussed can be expressed as

$$\rho = M_0c_0 + Mc = \rho_0 + Mc. \tag{3.9}$$

The projection of the vector equation (3.8) on the normal to the interphase surface (vector $\boldsymbol{n}$) is

$$\rho^*\left(\boldsymbol{v}_1^*, \boldsymbol{n}\right) = Mc^*(\boldsymbol{v}^*, \boldsymbol{n}). \tag{3.10}$$

The asterisks denote the value of the function at the phase boundary. To obtain (3.10), the condition for complete mutual insolubility of both phases is used:

$$(\boldsymbol{v}_0, \boldsymbol{n}) = 0. \tag{3.11}$$

Equation (3.11) expresses the availability of a normal velocity component of the liquid or the gas $(\overset{*}{\boldsymbol{v}_1}, \boldsymbol{n})$ at the interphase surface, determined by the diffusion rate $(\boldsymbol{v}^*, \boldsymbol{n})$. The velocity at the interphase surface only has a tangential component.

The induced flow at the interphase surface creates a convective mass flux. This implies that the mass flux of the transferred substance through the interphase surface has both convective and diffusive components:

$$I = -MD\left(\frac{\partial c}{\partial n}\right)_{y=h} + Mc^*\left(\boldsymbol{v}_1^*, \boldsymbol{n}\right). \tag{3.12}$$

Here $\partial/\partial n$ is the derivative in the direction of the normal vector of the interphase surface.

The diffusion component may be expressed by means of the projection of the vector equation (3.7) on the normal vector of the interphase surface:

$$(\boldsymbol{j}^*, \boldsymbol{n}) = -MD\left(\frac{\partial c}{\partial n}\right)_{y=h} = Mc^*(\boldsymbol{v}, \boldsymbol{n}) - Mc^*\left(\boldsymbol{v}_1^*, \boldsymbol{n}\right). \tag{3.13}$$

From Eqs. (3.10–3.13) one can obtain

$$I = \rho^*\left(\boldsymbol{v}_1^*, \boldsymbol{n}\right) = -\frac{\rho^*}{\rho^* - Mc^*} MD\left(\frac{\partial c}{\partial n}\right)_{y=h}, \tag{3.14}$$

where

$$\rho^* = M_0 c_0^* + Mc^* = \rho_0^* + Mc^*. \tag{3.15}$$

For small concentrations of the transferred substance we have

$$\rho_0^* \approx \rho_0. \tag{3.16}$$

Expression (3.14) may be presented in the form

$$I = \rho^* \frac{v^* - h' u^*}{\sqrt{1 + h'^2}} = \frac{MD\rho^*}{\rho_0^*} \frac{h'\left(\frac{\partial c}{\partial x}\right)_{y=h} - \left(\frac{\partial c}{\partial y}\right)_{y=h}}{\sqrt{1 + h'^2}}. \tag{3.17}$$

Here u^* and v^* are the components of velocity $\boldsymbol{v}_1$ along the x- and y-axes, respectively.

Equation (3.17) gives the relation between the gas (or liquid) velocity at the interphase surface and the concentration gradient of the transferred substance. It will used further as a boundary condition for the Navier–Stokes equations. In the approximations of the linear mass transfer theory, relationship (3.17) represents a condition at an impermeable surface $y = h\ (x)$:

$$\frac{v^* - h' u^*}{\sqrt{1 + h'^2}} = 0. \tag{3.18}$$

The processes of nonlinear mass or heat transfer in multicomponent systems involving gas (liquid)–solid interphase surfaces will be exemplified by the case of streamlining of a semi-infinite flat plate in the boundary layer approximation. In this case, the nonlinear effect is encountered through the velocity of the induced flow v^* at the interface $y = 0$. This velocity is defined by (3.17) with $u^* = 0$ at $h \equiv 0$, namely,

$$v^* = -\frac{MD}{\rho_0^*}\left(\frac{\partial c}{\partial y}\right)_{y=0}. \tag{3.19}$$

3.3 Nonlinear Mass Transfer in the Boundary Layer

The kinetics of nonlinear mass transfer in the boundary layer approximations [67–69] will be discussed through solutions of momentum and convection–diffusion equations with boundary conditions imposing mass transfer effects on the hydrodynamics. In Cartesian coordinates, where

$y = 0$ corresponds to the gas (liquid)–solid interphase surface, the mathematical description of the nonlinear mass transfer is represented by

$$u\frac{\partial u}{\partial x}+v\frac{\partial u}{\partial y}=\nu\frac{\partial^2 u}{\partial y^2},\quad \frac{\partial u}{\partial x}+\frac{\partial v}{\partial y}=0,\quad u\frac{\partial c}{\partial x}+v\frac{\partial c}{\partial y}=D\frac{\partial^2 c}{\partial y^2};$$
$$x=0,\quad u=u_0,\quad c=c_0;\quad y=0,\quad u=0,\quad v=-\frac{MD}{\rho_0^*}\frac{\partial c}{\partial y},\quad c=c^*;$$
$$y\to\infty,\quad u=u_0,\quad c=c_0. \tag{3.20}$$

Here a potential flow with velocity u_0 streamlines the plate. The concentration of the transferred substance is assumed to be c_0. As a result of the rapid establishment of thermodynamic equilibrium, the concentration c^* is always constant on the solid surface. The normal component of the velocity at the interphase surface is determined by (3.19) as a consequence of intensive interface mass transfer.

The mass transfer rate for a plate of length L can be determined from the average mass flux as

$$J = Mk(c^* - c_0) = \frac{1}{L}\int_0^L I dx. \tag{3.21}$$

Here k is the mass transfer coefficient and I can be expressed from (3.17) as follows:

$$I = -\frac{MD\rho^*}{\rho_0^*}\left(\frac{\partial c}{\partial y}\right)_{y=0}. \tag{3.22}$$

To solve problem (3.20), it is necessary to introduce the similarity variables

$$u = 0.5u_0\varepsilon\varphi',\quad v = 0.5\left(\frac{u_0\nu}{x}\right)^{0.5}(\eta\varphi' - \varphi),\quad c = c_0 + (c^* - c_0)\psi,$$
$$y = \eta\left(\frac{u_0}{4Dx}\right)^{-0.5}, \tag{3.23}$$

where

$$\varepsilon = Sc^{0.5},\quad Sc = \frac{\nu}{D},\quad \varphi = \varphi(\eta),\quad \psi = \psi(\eta). \tag{3.24}$$

As a result of these substitutions, problem (3.20) has the following form:

$$\begin{aligned}
&\varphi''' + \varepsilon^{-1}\varphi\varphi'' = 0, \quad \psi'' + \varepsilon\varphi\psi' = 0, \qquad \theta = \frac{M(c^* - c_0)}{\varepsilon\rho_0^*},\\
&\varphi(0) = \theta\psi'(0), \quad \varphi'(0) = 0, \quad \varphi'(\infty) = 2\varepsilon^{-1}, \quad \psi(0) = 1, \quad \psi(\infty) = 0,\\
&\psi(0) = 1, \quad \psi(\infty) = 0.
\end{aligned} \tag{3.25}$$

Here θ is a small parameter that reflects the effect of the nonlinear mass transfer. In the linear theory of the diffusion boundary layer we have $\theta = 0$.

Considering the new variables and Eq. (3.21), one can obtain

$$Sh = \frac{kL}{D} = -\frac{\rho^*}{\rho_0^*} Pe^{0.5}\psi'(0), \quad Pe = \frac{u_0 L}{D}. \tag{3.26}$$

From (3.26) it is clear that the mass transfer kinetics is determined by the dimensionless diffusive flux $\psi'(0)$, which is a solution of (3.25). The solution was achieved using the perturbation method (see Sect. 7.1) by representing φ and ψ as power series of the small parameter θ [70]:

$$\varphi = \varphi_0 + \theta\varphi_1 + \theta^2\varphi_2 + \cdots, \qquad \psi = \psi_0 + \theta\psi_1 + \theta^2\psi_2 + \cdots. \tag{3.27}$$

If (3.27) is substituted into (3.25), then a series of boundary problems (solved in [67]) is the outcome. For the functions in (3.27) we may write that

$$\begin{aligned}
&\varphi_0(\eta) = f(z), \quad z = \frac{2}{\varepsilon}\eta, \quad \psi_0(\eta) = 1 - \frac{1}{\varphi_0}\int_0^z E(\varepsilon,p)dp, \quad E(\varepsilon,p) = \exp\left[-\frac{\varepsilon^2}{2}\int_0^p f(s)ds\right],\\
&\varphi_1(\eta) = -\frac{2}{\varepsilon\varphi_0}\varphi(z), \quad \psi_1(\eta) = \frac{\varepsilon\varphi_3}{\varphi_0^3}\int_0^z E(\varepsilon,p)dp - \frac{\varepsilon}{\varphi_0^2}\int_0^z\left[\int_0^p \varphi(s)ds\right]E(\varepsilon,p)dp,\\
&\varphi_2(\eta) = -\frac{2\varphi_3}{\varphi_0^3}\varphi(z) - \frac{4}{\varepsilon^2\varphi_0^2}\bar{\varphi}(z),\\
&\psi_2(\eta) = \left(-\frac{2\varepsilon^2\varphi_3^2}{\varphi_0^5} + \frac{\varepsilon^2\varphi_{33}}{2\varphi_0^4} + \frac{2\bar{\varphi}_{33}}{\varphi_0^4}\right)\int_0^z E(\varepsilon,p)dp + \frac{2\varepsilon^2\varphi_3}{\varphi_0^4}\int_0^z\left[\int_0^p \varphi(s)ds\right]E(\varepsilon,p)dp\\
&\qquad - \frac{\varepsilon^2}{2\varphi_0^3}\int_0^z\left[\int_0^p \varphi(s)ds\right]^2 E(\varepsilon,p)dp - \frac{2}{\varphi_0^3}\int_0^z\left[\int_0^p \bar{\varphi}(s)ds\right]E(\varepsilon,p)dp.
\end{aligned} \tag{3.28}$$

In (3.28) the functions f, φ, and $\bar{\phi}$ are solutions of the boundary problems:

$$\begin{aligned}
&2f''' + ff'' = 0, \quad 2\varphi''' + f\varphi'' + f''\varphi = 0, \quad 2\bar{\varphi}''' + f\bar{\varphi}'' + f''\bar{\varphi} = \varphi\varphi'';\\
&f(0) = 0, \quad f'(0) = 0, \quad f'(\infty) = 1, \quad f''(0) = 0,33205,\\
&\varphi(0) = 1, \quad \varphi'(0) = 0, \quad \varphi'(\infty) = 0, \quad \bar{\varphi}(0) = 0, \quad \bar{\varphi}'(0) = 0, \quad \bar{\varphi}'(\infty) = 0.
\end{aligned} \tag{3.29}$$

In (3.28) the parameters $\varphi_0, \varphi_3, \varphi_{33}, \bar{\varphi}_{33}$ are functions of the Schmidt number:

$$
\begin{aligned}
\varphi_0 &= \int\limits_0^\infty E(\varepsilon,p)dp \approx \begin{cases} 3.01\, Sc^{-0.35}\text{—for gases,} \\ 3.12\, Sc^{-0.34}\text{—for liquids,} \end{cases} \\
\varphi_3 &= \int\limits_0^\infty \left[\int\limits_0^p \varphi(s)ds\right] E(\varepsilon,p)dp \approx \begin{cases} 6.56\, Sc^{-0.80}\text{—for gases,} \\ 5.08\, Sc^{-0.67}\text{—for liquids,} \end{cases} \\
\varphi_{33} &= \int\limits_0^\infty \left[\int\limits_0^p \varphi(s)ds\right]^2 E(\varepsilon,p)dp \approx \begin{cases} 24.0\, Sc^{-1.3}\text{—for gases,} \\ 12.2\, Sc^{-1.0}\text{—for liquids,} \end{cases} \\
\bar{\varphi}_{33} &= \int\limits_0^\infty \left[\int\limits_0^p \bar{\varphi}(s)ds\right] E(\varepsilon,p)dp \approx \begin{cases} 0.326\, Sc^{-1.63}\text{—for gases,} \\ 0.035\, Sc^{-1.1}\text{—for liquids.} \end{cases}
\end{aligned}
\tag{3.30}
$$

The dimensionless diffusive flux in the Sherwood number (3.26) is obtained directly from (3.28):

$$
\psi'(0) = -\frac{2}{\varepsilon\varphi_0} + \theta\frac{2\varphi_3}{\varphi_0^3} + \theta^2\left(-\frac{4\varepsilon\varphi_3^2}{\varphi_0^5} + \frac{\varepsilon\varphi_{33}}{\varphi_0^4} + \frac{4\bar{\varphi}_{33}}{\varepsilon\varphi_0^4}\right). \tag{3.31}
$$

Equation (3.31) shows that the precision of this basic result from the asymptotic theory of the diffusion boundary layer significantly depends on θ and ε. If it is necessary to develop a theoretical result with an error less than 10%, then the second-order approximation of the small parameter θ should be smaller than one tenth of its zeroth-order approximation, i.e.,

$$
\left|\theta^2\left(-\frac{4\varepsilon\varphi_3^2}{\varphi_0^5} + \frac{\varepsilon\varphi_{33}}{\varphi_0^4} + \frac{4\bar{\varphi}_{33}}{\varepsilon\varphi_0^4}\right)\right| < (0.1)\left|-\frac{2}{\varepsilon\varphi_0}\right|. \tag{3.32}
$$

From (3.32) it follows that

$$
\begin{aligned}
&\varepsilon = 1, \quad \theta < 0,41; \quad \varepsilon = 2, \quad \theta < 0,23; \quad \varepsilon = 10, \quad \theta < 0,056; \\
&\varepsilon = 20, \quad \theta < 0,025;
\end{aligned}
\tag{3.33}
$$

To check the precision of the asymptotic theory of nonlinear mass transfer in a diffusion boundary layer, the finite problem (3.25) was solved through a numerical method [40].

In Table 2 results of the asymptotic theory $\psi'(0)$ are compared with results of the numerical experiments $\psi'_N(0)$. The data missing from the table correspond to cases when singular disturbances in the numerical solution of the problem increase. To this end, from (3.33) it is obvious that these cases go beyond the limits of the accepted precision of the asymptotic theory.

The results developed show that the direction of intensive mass transfer significantly affects the mass transfer kinetics and this cannot be predicted in the approximations of the linear theory

Table 2 Comparison of the results of the asymptotic theory $\psi'(0)$ with the results of the numerical experiment $\psi'_N(0)$

θ	$\varepsilon = 1$		$\varepsilon = 2$		$\varepsilon = 10$		$\varepsilon = 20$	
	$-\psi'_N(0)$	$-\psi'(0)$	$-\psi'_N(0)$	$-\psi'(0)$	$-\psi'_N(0)$	$-\psi'(0)$	$-\psi'_N(0)$	$-\psi'(0)$
0.00	0.664	0.664	0.535	0.535	0.314	0.305	0.250	0.246
+0.03	0.650	0.650	0.515	0.516	0.270	0.265	0.190	0.199
−0.03	0.679	0.679	0.553	0.555	0.384	0.365	0.406	0.363
+0.05	0.641	0.641	0.503	0.504	0.248	0.250	0.166	0.205
−0.05	0.689	0.689	0.572	0.570	0.459	0.415	–	0.479
+0.10	0.620	0.620	0.475	0.478	0.207	0.250	–	0.500
−0.10	0.716	0.716	0.616	0.611	–	0.581	–	0.903
+0.20	0.581	0.584	0.429	0.442	0.160	0.418	–	1.229
−0.20	0.779	0.776	0.736	0.707	–	1.080	–	2.325
+0.30	0.548	0.555	0.393	0.425	–	0.808	–	2.868
−0.30	0.855	0.843	0.936	0.822	–	1.800	–	4.512

($\theta = 0$). When the mass transfer is directed from the liquid bulk towards the phase boundary ($\theta < 0$), then the increase in the concentration gradient in the diffusion boundary layer ($c^* \; c_0$) leads to an increase in the diffusive mass transfer. In the cases when the mass transfer is directed from the phase boundary towards the bulk ($\theta > 0$), the increase in the concentration gradient leads to a decrease in the diffusive mass transfer.

The nonlinear effects in the mass transfer kinetics under conditions of intensive mass transfer occur in a thin layer near the interphase surface [67], having a thickness approximately three times smaller than that of the diffusion boundary layer. At the boundary of this "layer of nonlinear mass transfer" the local diffusive flux depends on the concentration gradient and the value of the parameter θ correspondingly. Inside the "nonlinear mass transfer layer" for $\theta < 0$ ($\theta > 0$) the flux increases (decreases) with increase in the absolute value of θ, and outside this layer this dependence is the opposite [71].

3.4 Two-Phase Systems

The interphase mass transfer in gas–liquid and liquid–liquid systems is associated primarily with absorption or extraction processes at industrial scales. The process intensification through the generation of large concentration gradients in the gas and the liquid leads to the manifestation of nonlinear effects in the kinetics of the mass transfer in the gas and liquid phases. In this way, the interphase mass transfer in gas–liquid and liquid–liquid systems becomes nonlinear.

Industrial gas absorption is most frequently performed in packed-bed columns. The sizes of the packing particles used being small, the interphase transfer of the absorbed material is effected through the thin layers bordering the phase boundary between the gas and the liquid. The main change in the absorbed material concentration takes place in these layers, which allows the theoretical analysis of the kinetics of nonlinear interphase mass transfer to be performed by making use of the approximation of the diffusion boundary layer.

The kinetics of the nonlinear interphase mass transfer in the cases of a flat phase interface and co-current movement of the gas and the liquid [3, 27] will be discussed. If the gas and the liquid are

designated as the first and the second phase respectively, Eqs. (3.20) take the form

$$\begin{aligned} &u_j \frac{\partial u_j}{\partial x} + v_j \frac{\partial u_j}{\partial y} = \nu_j \frac{\partial^2 u_j}{\partial y^2}, \quad \frac{\partial u_j}{\partial x} + \frac{\partial v_j}{\partial y} = 0, \\ &u_j \frac{\partial c_j}{\partial x} + v_j \frac{\partial c_j}{\partial y} = D_j \frac{\partial^2 c_j}{\partial y^2}, \quad j = 1, 2, \end{aligned} \tag{3.34}$$

with boundary conditions accounting for the continuity of the distribution of the velocities and the flows of momentum and mass at the interphase surface:

$$\begin{aligned} &x = 0, \quad u_j = u_{j0}, \quad c_j = c_{j0}, \quad j = 1, 2; \quad y = 0, \quad u_1 = u_2, \quad \mu_1 \frac{\partial u_1}{\partial y} = \mu_2 \frac{\partial u_2}{\partial y}, \\ &v_j = -\frac{M D_j}{\rho_{j0}^*} \frac{\partial c_j}{\partial y}, \quad j = 1, 2, \quad c_1 = \chi c_2, \quad \frac{D_1 \rho_1^*}{\rho_{10}} \frac{\partial c_1}{\partial y} = \frac{D_2 \rho_2^*}{\rho_{20}^*} \frac{\partial c_2}{\partial y}; \\ &y \to \infty, \quad u_1 = u_{10}, \quad c_1 = c_{10}; \quad y \to -\infty, \quad u_2 = u_{20}, \quad c_2 = c_{20}. \end{aligned} \tag{3.35}$$

The interphase mass transfer rate for a surface of length L is determined by averaging the local mass fluxes,

$$J = MK_1(c_{10} - \chi c_{20}) = -\frac{1}{L} \int_0^L I_1 dx = MK_2\left(\frac{c_{10}}{\chi} - c_{20}\right) = -\frac{1}{L} \int_0^L I_2 dx, \tag{3.36}$$

where K_j (j = 1,2) are the interphase mass transfer coefficients, and the local mass fluxes are obtained from (3.22):

$$I_j = -\frac{M D_j \rho_j^*}{\rho_{j0}^*} \left(\frac{\partial c_j}{\partial y}\right)_{y=0}, \quad j = 1, 2. \tag{3.37}$$

From (3.36) and (3.37) the Sherwood number is obtained as

$$Sh_j = \frac{K_j L}{D_j} = \frac{\rho_j^*}{\rho_{j0}^*} \frac{\chi^{j-1}}{c_{10} - \chi c_{20}} \int_0^L \left(\frac{\partial c_j}{\partial y}\right)_{y=0} dx, \quad j = 1, 2, \tag{3.38}$$

Equations (3.34) and (3.35) can be solved by introducing similarity variables:

$$
\begin{aligned}
& u_j = 0.5 j u_{j0} \varepsilon_j \varphi_j', \quad v_j = (-1)^{j-1} 0.5 j \left(\frac{u_{j0} v_j}{x}\right)^{0.5} \left(\xi_j \varphi_j' - \varphi_j\right), \\
& c_j = c_{j0} - (-\chi)^{1-j} (c_{10} - \chi c_{20}) \psi_j, \\
& \varphi_j = \varphi_j(\xi_j), \quad \psi_j = \psi_j(\xi_j)^{0.5}, \quad \xi_j = (-1)^{j-1} y \left(\frac{u_{j0}}{4 D_j x}\right)^{0,5}, \\
& \varepsilon_j = Sc_j^{0.5}, \quad Sc_j = \frac{v_j}{D_j}, \quad j = 1, 2.
\end{aligned}
\tag{3.39}
$$

Thus, as a result, we obtain

$$
\begin{aligned}
& \varphi_j''' + j \varepsilon_j^{-1} \varphi_j \varphi_j'' = 0, \quad \psi_j'' + j \varepsilon_j \varphi_j \psi_j' = 0, \\
& \varphi_j(0) = (-1)^j \theta_{j+2} \psi_j'(0), \quad \varphi_j'(\infty) = \frac{2}{j \varepsilon_j}, \quad \psi_j(\infty) = 0, \quad j = 1, 2; \\
& \varphi_1'(0) = 2\theta_1 \frac{\varepsilon_2}{\varepsilon_1} \varphi_2'(0), \quad \varphi_2''(0) = -0.5 \theta_2 \left(\frac{\varepsilon_1}{\varepsilon_2}\right)^2 \varphi_1''(0), \\
& \psi_2'(0) = \frac{\chi}{\varepsilon_0} \psi_1'(0), \quad \psi_1(0) + \psi_2(0) = 1,
\end{aligned}
\tag{3.40}
$$

where

$$
\begin{aligned}
& \theta_1 = \frac{u_{20}}{u_{10}}, \quad \theta_2 = \left(\frac{\mu_1}{\mu_2}\right) \left(\frac{v_1}{v_2}\right)^{-0.5} \left(\frac{u_{10}}{u_{20}}\right)^{1.5}, \\
& \theta_3 = \frac{M(c_{10} - \chi c_{20})}{\varepsilon_1 \rho_{10}^*}, \quad \theta_4 = \frac{M(c_{10} - \chi c_{20})}{2 \varepsilon_2 \rho_{20}^* \chi}, \quad \varepsilon_0 = \frac{\rho_{10}^* \rho_2^*}{\rho_{20}^* \rho_1^*} \left(\frac{D_2 u_{20}}{D_1 u_{10}}\right)^{0.5}.
\end{aligned}
\tag{3.41}
$$

It follows from (3.40) that the concentration of the absorbed material on the interphase surface ($y = 0$) is constant. This allows a set of new boundary conditions to be used:

$$
\psi_1(0) = A, \quad \psi_2(0) = 1 - A,
\tag{3.42}
$$

where A is determined from the conditions of the mass flow continuity on the phase interface. Thus, (3.42) permits the solution of (3.40) as two independent problems.

The parameters θ_1 and θ_2 account for the kinematic and dynamic interactions between the phases, whereas θ_3 and θ_4 account for the rate of the nonlinear effects in the gas and the liquid phases. For cases of practical interest, $\theta_k < 1$ (k = 1,...,4) is valid and the problem can be solved by making use of the perturbation method [36], expressing the unknown functions by an expansion of the following type:

$$
F = F^{(0)} + \theta_1 F^{(1)} + \theta_2 F^{(2)} + \theta_3 F^{(3)} + \theta_4 F^{(4)} + \cdots,
\tag{3.43}
$$

where F is a vector function

$$F = F(\varphi_1, \varphi_2, \psi_1, \psi_2, A). \tag{3.44}$$

The zeroth-order approximation is obtained from (3.40) when $\theta_k = 0$, k = 1,...,4 is substituted.

The first-order approximations are obtainable from the equations

$$\begin{gathered}\varphi_j^{(k)''''} + j\varepsilon_j^{-1}\left(\varphi_j^{(k)''}\varphi_j^{(0)} + \varphi_j^{(0)''}\varphi_j^{(k)}\right) = 0,\\ \psi_j^{(k)''} + j\varepsilon_j\left(\varphi_j^{(k)}\psi_j^{(0)'} + \varphi_j^{(0)}\psi_j^{(k)'}\right) = 0, \quad k = 1,\ldots,4, \quad j = 1,2,\end{gathered} \tag{3.45}$$

with boundary conditions

$$\begin{aligned}&\varphi_j^{(k)}(0) = 0,\ k = 1,2,\ j = 1,2;\quad \varphi_1^{(3)}(0) = -\psi_1^{(0)'}(0),\ \varphi_2^{(4)}(0) = -\psi_2^{(2)'}(0),\\ &\qquad \varphi_1^{(4)}(0) = 0,\ \varphi_2^{(3)}(0) = 0;\\ &\varphi_1^{(k)'}(0) = 0,\quad k = 2,3,4;\quad \varphi_1^{(1)'}(0) = 2\frac{\varepsilon_2}{\varepsilon_1}\varphi_2^{(0)'}(0);\quad \varphi_j^{(k)'}(0) = 0,\\ &\qquad k = 1,\ldots,4,\quad j = 1,2;\\ &\psi_j^{(k)'}(0) = A^{(k)},\qquad \psi_j^{(k)}(\infty) = 0,\quad k = 1,\ldots,4,\quad j = 1,2;\\ &\varphi_2^{(k)''}(0) = 0,\quad k = 1,3,4;\quad \varphi_2^{(2)''}(0) = -\frac{1}{2}\left(\frac{\varepsilon_1}{\varepsilon_2}\right)^2\varphi_1^{(0)''}(0);\quad \psi_2^{(k)}(0) = -A^{(k)},\quad k = 1,\ldots,4.\end{aligned} \tag{3.46}$$

The values for $A^{(k)}$ (k = 1,...,4) are calculated from the equation

$$\psi_2^{(k)\prime}(0) = \frac{\chi}{\varepsilon_0}\psi_1^{(k)\prime}(0), \quad k = 1,\ldots,4. \tag{3.47}$$

The solutions of problems of type (3.40) have been reported in a number of publications [11, 72–79]. Using these solutions, we can write

$$\varphi_1^{(0)}(\xi_1) = f(z),\ z = \frac{2}{\varepsilon_1}\xi_1,\ \psi_1^{(0)}(\xi_1) = A^{(0)}\left(1 - \frac{1}{\varphi_{10}}\right)\int_0^z E(\varepsilon_1, p)dp,$$

$$E(\varepsilon_1, p) = \exp\left[-\frac{\varepsilon_1^2}{2}\int_0^p f(s)ds\right],$$

$$\varphi_2^{(0)}(\xi_2) = \varepsilon_2^{-1}\xi_2,\quad \psi_2^{(0)}(\xi_2) = \left(1 - A^{(0)}\right) erfc\,\xi_2,\quad A^{(0)} = \frac{1}{1+a},\quad a = \frac{\chi\sqrt{\pi}}{\varepsilon_0\varepsilon_1\varphi_{10}},$$

$$\varphi_1^{(1)}(\xi_1) = \frac{1}{\alpha}f'(z),\quad \varphi_2^{(1)}(\xi_2) \equiv 0,\quad \alpha = f''(0),$$

$$\psi_1^{(1)}(\xi_1) = A^{(1)} + \frac{A^{(0)}}{\alpha\varphi_{10}}[1 - E(\varepsilon_1, z)] - \left(\frac{A^{(1)}}{\varphi_{10}} + \frac{A^{(0)}}{\alpha\varphi_{10}^2}\right)\int_0^z E(\varepsilon_1, p)\,dp,$$

$$
\begin{aligned}
&\psi_1^{(1)}(\xi_2) = -A^{(1)} erfc\,\xi_2, \quad A^{(1)} = -\frac{1}{\alpha\varphi_{10}}\frac{\alpha_0}{(1+a_0)^2}, \quad \varphi_1^{(2)}(\xi_1) \equiv 0,\\
&\varphi_2^{(2)}(\xi_2) = \alpha\sqrt{\pi}\int\limits_0^{\varepsilon_2/\mathrm{E}_2} erfcp\,dp,\\
&\psi_1^{(2)}(\xi_1) = A^{(2)}\left[1 - \frac{1}{\varphi_{10}}\int\limits_0^z E(\varepsilon_1, p)dp\right],\\
&\psi_2^{(2)}(\xi_2) = -A^{(2)} + \left[A^{(2)} - 4\alpha\varepsilon_2\left(1 - A^{(0)}\right)\bar{\varphi}_2\right] erfc\xi_2 + 4\alpha\varepsilon_2\left(1 - A^{(0)}\right)Q(\varepsilon_2, \xi_2),\\
&Q(\varepsilon_2, \xi_2) = \int\limits_0^{\xi_2}\left[\exp(-q^2)\int\limits_0^q\left(\int\limits_0^{p/\varepsilon_2} erfcs\,ds\right)dp\right]dq\\
&\approx \frac{1}{8}\sqrt{\frac{\pi}{Sc_2}} erf\,\xi_2 - \frac{1}{4\sqrt{Sc_2}}\xi_2\exp\left(-\xi_2^2\right), A^{(2)} = 4\alpha\varepsilon_2\bar{\varphi}_2\frac{a}{(1+a)^2},\\
&\bar{\varphi}_2 = Q(\varepsilon_2, \infty) = \frac{1}{8}\sqrt{\frac{\pi}{Sc_2}}, \quad \varphi_1^{(3)}(\xi_1) = \frac{2A^{(0)}}{\varepsilon_1\varphi_{10}}\varphi(z),\\
&\varphi_2^{(3)}(\xi_2) \equiv 0 \;\; \psi_1^{(3)}(\xi_1)\\
&= A^{(3)} - \left(\frac{A^{(3)}}{\varphi_{10}} + \frac{\varepsilon_1 A^{(0)}\varphi_{13}}{\varphi_{10}^3}\right)\int\limits_0^z E(\varepsilon_1, p)\,dp\\
&+ \frac{\varepsilon_1 A^{(0)}}{\varphi_{10}^2}\int\limits_0^z\left[\int\limits_0^p \varphi(s)ds\right]E(\varepsilon_1, p)dp,\\
&\psi_2^{(3)}(\xi_2) = -A^{(3)} erfc\xi_2, \quad A^{(3)} = -\frac{\varepsilon_1\varphi_{13}}{\varphi_{10}^2}\frac{a_0}{(1+a_0)^2},\\
&\varphi_1^{(4)}(\xi_1) \equiv 0, \quad \varphi_2^{(4)}(\xi_2) = -\frac{2}{\sqrt{\pi}}\left(1 - A^{(0)}\right),\\
&\psi_1^{(4)}(\xi_1) = A^{(4)}\left(1 - \frac{1}{\varphi_{10}}\int\limits_0^z E(\varepsilon_1, p)dp\right),\\
&A^{(4)} = -\frac{4\varepsilon_2}{\pi}\frac{a^2}{(1+a)^3}, \psi_2^{(4)}(\xi_2)\\
&= -A^{(4)} - \frac{4\varepsilon_2}{\pi}\left(1 - A^{(0)}\right)^2 + \frac{4\varepsilon_2}{\pi}\left(1 - A^{(0)}\right)^2\exp\left(-\xi_2^2\right)\\
&+ \left[A^{(4)} + \frac{4\varepsilon_2}{\pi}\left(1 - A^{(0)}\right)^2\right] erf\,\xi_2,
\end{aligned}
\tag{3.48}
$$

where f and φ are solutions of (1.29), and φ_{10} and φ_{13} are expressed as

$$\varphi_{10} = \int_0^\infty E(\varepsilon_1, p)dp \approx 3.01\, Sc_1^{-0.35},$$
$$\varphi_{13} = \int_0^\infty \left[\int_0^p \varphi(s)ds\right] E(\varepsilon_1, p)dp \approx 6.56\, Sc_1^{-0.8}, \tag{3.49}$$

i.e., their values can be obtained from φ_0 and φ_3 in (3.30) if $\varepsilon = \varepsilon_1$ ($Sc = Sc_1$) is substituted.

The nonlinear interphase mass transfer rate (the Sherwood number) is obtainable from (3.38),

$$Sh_{\mathrm{j}} = -\frac{\rho_{\mathrm{j}}^*}{\rho_{\mathrm{j0}}^*}\sqrt{Pe_{\mathrm{j}}}\psi_{\mathrm{j}}'(0), \quad Pe_{\mathrm{j}} = \frac{u_{\mathrm{j0}}L}{D_{\mathrm{j}}}, \quad \mathrm{j} = 1, 2, \tag{3.50}$$

where $\psi_1'(0)$ and $\psi_2'(0)$ can be determined from (3.48):

$$\psi_1'(0) = -\frac{2}{\varepsilon_1\varphi_{10}}\frac{1}{1+a} - \frac{2\theta_1}{\alpha\varphi_{10}^2\varepsilon_1}\frac{1}{(1+a)^2} - 8\theta_2\alpha\frac{\varepsilon_2}{\varepsilon_1}\frac{\bar{\varphi}_2}{\bar{\varphi}_{10}}\frac{a}{(1+a)^2}$$
$$- 2\theta_3\frac{\bar{\varphi}_{13}}{\bar{\varphi}_{10}^3}\frac{1}{(1+a)^2} + 8\theta_4\frac{\varepsilon_2}{\pi\varphi_{10}\varepsilon_1}\frac{a^2}{(1+a)^3}, \tag{3.51}$$

$$\psi_2'(0) = -\frac{2}{\sqrt{\pi}}\frac{a}{1+a} - \theta_1\frac{2}{\sqrt{\pi}\alpha\varphi_{10}}\frac{a}{(1+a)^2} - 8\theta_2\frac{\alpha\varepsilon_2\bar{\varphi}_2}{\sqrt{\pi}}\frac{a^2}{(1+a)^2}$$
$$- 2\theta_3\frac{\varepsilon_1\varphi_{13}}{\sqrt{\pi}0_{10}^2}\frac{a}{(1+u)^3} + 8\theta_4\frac{\varepsilon_2}{\pi\sqrt{\pi}}\frac{a^3}{(1+u)^3}.$$

In the cases where the rate of the interface mass transfer is limited by the diffusion resistance in the gas phase, from the last condition in (3.40) it follows that $\chi/\chi\varepsilon_0{\cdot}\varepsilon_0 \to 0$, i.e., $a \to 0$. Thus, the Sherwood number can be expressed in the form

$$Sh_1 = \frac{\rho_1^*}{\rho_{10}^*}Pe_1^{0.5}\left(\frac{2}{\varepsilon_1\varphi_{10}} + \frac{2\theta_1}{\varepsilon_1\alpha\varphi_{10}^2} + 2\theta_3\frac{\varphi_{13}}{\varphi_{10}^3}\right). \tag{3.52}$$

When the process is limited by the resistance in the liquid phase, $\chi/\chi\varepsilon_0{\cdot}\varepsilon_0 \to \infty$, $a \to \infty$, i.e.,

$$Sh_2 = \frac{\rho_2^*}{\rho_{20}^*}Pe_2^{0.5}\left(\frac{2}{\sqrt{\pi}} + 8\theta_2\frac{\alpha\varepsilon_2\bar{\varphi}_2}{\sqrt{\pi}} - 8\theta_4\frac{\varepsilon_2}{\pi\sqrt{\pi}}\right). \tag{3.53}$$

The comparison of the nonlinear effects in both the gas and the liquid [79] shows that the ratio of the parameters θ_3 and θ_4 takes the form

$$\frac{\theta_3}{\theta_4} = \frac{2\varepsilon_2 \rho_{20}^* \chi}{\varepsilon_1 \rho_{10}^*} \gg 1 \tag{3.54}$$

and is always greater than unity. The minimum value of this ratio occurs in cases of gases of high solubility, where θ_3 is greater than θ_4 by more than two orders of magnitude, i.e., for numerical calculation it is always possible to assume $\theta_4 = 0$.

A numerical solution of equations (3.45) and (3.46) was performed as a check of the asymptotic theory [79, 80]. The analysis of the results demonstrates that the nonlinear effects are most significant in cases where the nonlinear interphase mass transfer is limited by the mass transfer in the gas phase ($\chi/\varepsilon_0 = 0$). When the diffusion resistances are commensurable ($\chi/\varepsilon \approx 1$), the nonlinear effects are considerably smaller and their appearance in the liquid phase is a result of the hydrodynamic influence of the gas phase. However, these effects are totally absent when the process is limited by the mass transfer in the liquid phase. The influence of the direction of the interphase mass transfer on the kinetics of the mass transfer in gas–liquid systems is similar to that observed in systems with a gas (liquid)–solid interphase surface, i.e., the diffusive transfer in the case of absorption is greater than that in the case of desorption.

The results of the asymptotic theory (3.52) show that in the cases of absorption and desorption the deviation of the nonlinear mass transfer theory from the linear theory ($\theta_3 = 0$) is symmetric, whereas the numerical results show a nonsymmetric deviation. This "contradiction" with the asymptotic theory is possible to explain by the absence of the quadratic terms (proportional to θ_3^2).

It is evident that the asymptotic theory has to be made more precise and to include all the quadratic terms. In the case of a nonlinear interphase mass transfer limited by mass transfer in the gas phase, Eqs. (3.40) take the form

$$\begin{gathered}
\varphi_1''' + \varepsilon^{-1}\varphi_1\varphi_1'' = 0, \quad \varphi_2''' + 2\varepsilon_2^{-1}\varphi_2\varphi_2'' = 0, \quad \psi_1'' + \varepsilon_1\varphi_1\psi_1' = 0; \\
\varphi_1(0) = -\theta_3\psi_1'(0), \quad \varphi_2(0) = 0, \quad \varphi_1'(\infty) = \frac{2}{\varepsilon_1}, \quad \varphi_2'(\infty) = \frac{1}{\varepsilon_2}, \quad \varphi_1'(0) = 2\theta_1\frac{\varepsilon_2}{\varepsilon_1}\varphi_2''(0), \\
\varphi_2''(0) = -0.5\theta_2\left(\frac{\varepsilon_1}{\varepsilon_2}\right)^2\varphi_1''(0), \quad \psi_1(0) = 1, \quad \psi_1(\infty) = 0.
\end{gathered} \tag{3.55}$$

To solve problem (3.55), expansion (3.43) is used, where the terms $\theta_1^2F^{(11)} + \theta_3^2F^{(33)} + \theta_1\theta_3F^{(13)}$ should be added and $a_0 = 0$ should be substituted into relationships (3.48).

Approximations proportional to θ_1^2 were obtained in [11, 74]:

$$\begin{aligned}
&\varphi_1^{(11)}(\xi_1) = F(z), \quad \varphi_2^{(11)}(\xi_2) \equiv 0, \quad \psi_1^{(11)}(\xi_1) \\
&= \left(\frac{\varepsilon_1^4\varphi_{11}}{8\alpha^2\varphi_{10}^3} - \frac{\varepsilon_1^2\varphi_{12}}{2\varphi_{10}^2} - \frac{\varepsilon_1}{2\alpha^2\varphi_{10}^3}\right)\int_0^z E(\varepsilon_1,p)\,dp + \frac{\varepsilon_1^2}{2\varphi_{10}}\int_0^z\left[\int_0^p F(s)ds\right]E(\varepsilon_1,p)dp \\
&\quad + \frac{\varepsilon_1}{2\alpha^2\varphi_{10}^2}[1 - E(\varepsilon_1,z)] - \frac{\varepsilon_1^4}{8\alpha^2\varphi_{10}}\int_0^z f^2(p)E(\varepsilon_1,p)dp,
\end{aligned} \tag{3.56}$$

where the function F is the solution of the problem

$$2F''' + fF'' + f''F = -\frac{1}{\alpha^2}f'f''', \quad F(0) = F'(0) = F'(\infty) = 0 \tag{3.57}$$

and was tabulated in [34]. φ_{11} and φ_{12} were obtained in [11]:

$$\begin{aligned} \varphi_{11} &= \int_0^\infty f^2(p)E(\varepsilon_1,p)dp \approx 3.01\, Sc_1^{-1.608}, \\ \varphi_{12} &= \int_0^\infty \left[\int_0^p F(s)\, ds\right] E(\varepsilon_1,p)\, dp \approx 3.05\, Sc_1^{-1.285}. \end{aligned} \tag{3.58}$$

Approximations proportional to θ_3^2 were obtained in [81]:

$$\begin{aligned} \varphi_1^{(33)}(\xi_1) &= \frac{2\varphi_3}{\varphi_{10}^3}\varphi(z) - \frac{4}{\varepsilon_1^2\varphi_{10}^2}\bar{\varphi}(z), \qquad \varphi_2^{(33)}(\xi_2) \equiv 0, \\ \psi_1^{(33)}(\xi_1) &= \left(-\frac{\varepsilon_1^2\varphi_{13}^2}{\varphi_{10}^5} + \frac{\varepsilon_1^2\varphi_{133}}{2\varphi_{10}^4} + \frac{2\bar{\varphi}_{133}}{\varphi_{10}^4}\right)\int_0^z E(\varepsilon_{1,},p)\, dp \\ &\quad + \frac{\varepsilon_1^2\varphi_{13}}{\varphi_{10}^4}\int_o^z\left[\int_0^p \varphi(s)ds\right] E(\varepsilon_1,p)dp - \frac{\varepsilon_1^2}{2\varphi_{10}^3}\int_0^z\left[\int_0^p f(s)ds\right]^2 E(\varepsilon_1,p)dp \\ &\quad - \frac{2}{\varphi_{10}^3}\int_0^z\left[\int_0^p \bar{\varphi}(s)ds\right] E(\varepsilon_1,p)dp, \end{aligned} \tag{3.59}$$

where $\bar{\varphi}$ is the solution of (1.29). Thus, φ_{133} and $\bar{\varphi}_{133}$ take the forms

$$\begin{aligned} \varphi_{133} &= \int_0^\infty \left[\int_0^p \varphi(s)ds\right]^2 E(\varepsilon_1,p)dp \approx 24 Sc_1^{-1.3}, \\ \bar{\varphi}_{133} &= \int_0^\infty \left[\int_0^p \bar{\varphi}(s)ds\right] E(\varepsilon_1,p)dp \approx 0.326\, Sc_1^{-1.63}, \end{aligned} \tag{3.60}$$

i.e., they may be obtained from φ_{33} and $\bar{\varphi}_{33}$ in (3.30) via the substitution $\varepsilon = \varepsilon_1(Sc = Sc_1)$. From (3.28) and (3.59) it is obvious that $\psi_1^{(33)}(\xi_1) \equiv \psi_2(\eta)$, if $\varepsilon_1 = \varepsilon$. Approximations, proportional to $\theta_1\theta_3$, were obtained in [81]:

$$\varphi_1^{(13)} = \frac{1}{\alpha\varphi_{10}^2}\varphi(z) - \frac{2}{\varepsilon_1\alpha\varphi_{10}}\bar{\bar{\varphi}}(z), \qquad \varphi_2^{(13)}(\xi_2) \equiv 0,$$

$$\begin{aligned}\psi_1^{(13)}(\xi_1) = {} & \left(-\frac{\varepsilon_1\varphi_{13}}{2\alpha\varphi_{10}^4} + \frac{\varepsilon_1\varphi_{113}}{\alpha\varphi_{10}^3} + \frac{\varepsilon_1\bar{\varphi}_{113}}{\alpha\varphi_{10}^3} - \frac{2\varepsilon_1\varphi_{13}}{\alpha\varphi_{10}^4}\right)\int_0^z E(\varepsilon_1,p)\,dp \\ & + \left(\frac{\varepsilon_1}{\alpha\varphi_{10}^3} + \frac{\varepsilon_1^2}{2\alpha\varphi_{10}^3}\right)\int_0^z \left[\int_0^p \varphi(s)ds\right] E(\varepsilon_1,p)dp - \frac{\varepsilon_1}{\alpha\varphi_{10}^2} \\ & \times \int_0^z \left[\int_0^p \bar{\bar{\varphi}}(s)ds\right] E(\varepsilon_1,p)dp + \frac{\varepsilon_1}{\alpha\varphi_{10}^2}E(\varepsilon_1,z)\int_0^z \varphi(p)dp \\ & - \frac{\varepsilon_1}{\alpha\varphi_{10}^2}\int_0^z \varphi(p)E(\varepsilon_1,p)dp + \frac{\varepsilon_1\varphi_{13}}{\alpha\varphi_{10}^3}[1 - E(\varepsilon_1,z)],\end{aligned} \tag{3.61}$$

where $\bar{\bar{\varphi}}$ is the solution of the problem

$$2\bar{\bar{\varphi}}''' + f\bar{\bar{\varphi}}'' + f''\bar{\bar{\varphi}} = f'\varphi'' + f'''\varphi, \qquad \bar{\bar{\varphi}}(0) = \bar{\bar{\varphi}}'(0) = \bar{\bar{\varphi}}(\infty) = 0. \tag{3.62}$$

ϕ_{113} and $\bar{\phi}_{113}$ were obtained in [103]:

$$\begin{aligned}\varphi_{113} &= \int_0^\infty \left[\int_0^p \bar{\bar{\varphi}}(s)ds\right] E(\varepsilon_1,p)dp \approx Sc_1^{-1.3}, \\ \bar{\varphi}_{113} &= \int_0^\infty \varphi(p)E(\varepsilon_1,p)dp \approx 4.18Sc_1^{-0.46}.\end{aligned} \tag{3.63}$$

The expressions derived allow us to determine the rate of the nonlinear interphase mass transfer in a gas–liquid system when the process is limited by the mass transfer in the gas phase.

From (3.50), we find

$$Sh_1 = \frac{K_1L}{D_1} = \frac{\rho_1^*}{\rho_{10}^*}\sqrt{Pe_1}\psi_1'(0), \tag{3.64}$$

where $\psi_1'(0)$ is calculated taking all the quadratic approximations into account:

$$\begin{aligned}-\psi_1'(0) = {} & \frac{2}{\varepsilon_1\varphi_{10}} + \theta_1\frac{2}{\varepsilon_1\alpha\varphi_{10}^2} + \theta_3\frac{2\varphi_{13}}{\varphi_{10}^3} + \theta_1\theta_3\left(\frac{\varepsilon_1\varphi_{13}}{\alpha\varphi_{10}^4} - \frac{2\varphi_{113}}{\alpha\varphi_{10}^3} - \frac{2\bar{\varphi}_{113}}{\alpha\varphi_{10}^3} + \frac{4\varphi_{13}}{\alpha\varphi_{10}^4}\right) \\ & + \theta_1^2\left(-\frac{\varepsilon_1^3\varphi_{11}}{4\alpha^2\varphi_{10}^2} + \frac{\varepsilon_1\varphi_{12}}{\varphi_{10}^2} + \frac{2}{\varepsilon_1\alpha^2\varphi_{10}^3}\right) + \theta_3^2\left(\frac{2\varepsilon_1\varphi_{13}^2}{\varphi_{10}^5} - \frac{\varepsilon_1\varphi_{133}}{\varphi_{10}^4} - \frac{4\bar{\varphi}_{133}}{\varepsilon_1\varphi_{10}^4}\right).\end{aligned} \tag{3.65}$$

Expression (3.65) is the main result from the asymptotic theory of nonlinear interphase mass transfer in gas–liquid systems and is in good agreement with the results from the numerical solution of problem (3.40) obtained in [80].

3.5 Nonlinear Mass Transfer and the Marangoni Effect

Intensification of the mass transfer in industrial gas–liquid systems is obtained quite often by the creation of large concentration gradients. This can be achieved in a number of cases as a result of a chemical reaction of the transferred substance in the liquid phase. The thermal effect of the chemical reactions creates temperature gradients. The temperature and concentration gradients can considerably affect the mass transfer kinetics in gas–liquid systems. Hence, the experimentally obtained mass transfer coefficients differ significantly from those predicted by the linear mass transfer theory.

As shown in a number of papers [8, 82–93], the temperature and concentration gradients on the gas–liquid or liquid–liquid interphase surface can create a surface tension gradient. As a result of this, a secondary flow is induced. The velocity of the induced flow is directed tangentially to the interphase surface. It leads to a change in the velocity distribution in the boundary layer and therefore to a change in the mass transfer kinetics. These effects are thought to be of the Marangoni type and could provide an explanation for all experimental deviations from the prediction of the linear mass transfer theory.

Obviously, the Marangoni effect can exist together with the effect of the large concentration gradients. These two effects can manifest themselves separately as well as in combination. That is why their influence on the mass transfer kinetics has to be assessed.

Co-current gas and liquid flows in the laminar boundary layer along the flat interphase surface will be considered. One of the gas components is absorbed by the liquid and reacts with a component in the liquid phase. The chemical reaction rate is of first order. The thermal effect of the chemical reaction creates a temperature gradient, i.e., the mass transfer together with a heat transfer can be observed. Under these conditions, the mathematical model takes the following form:

$$\begin{gathered}
u_j \frac{\partial u_j}{\partial x} + v_j \frac{\partial u_j}{\partial y} = \nu_j \frac{\partial^2 u_j}{\partial y^2}, \quad \frac{\partial u_j}{\partial x} + \frac{\partial v_j}{\partial y} = 0, \quad u_j \frac{\partial c_j}{\partial x} + v_j \frac{\partial c_j}{\partial y} = D_j \frac{\partial^2 c_j}{\partial y^2} - (j-1) k c_j, \\
u_j \frac{\partial t_j}{\partial x} + v_j \frac{\partial t_j}{\partial y} = a_j \frac{\partial^2 t_j}{\partial y^2} + (j-1) \frac{q}{\rho_j c_{pj}} k c_j, \quad j = 1, 2,
\end{gathered} \tag{3.66}$$

where the indexes 1 and 2 refer to the gas and the liquid, respectively. The influence of the temperature on the chemical reaction rate is not included in (3.66) because it does not have a considerable effect in the comparative analysis of these two effects.

The boundary conditions of (3.66) determine the potential two-phase flows far from the phase boundary. Thermodynamic equilibrium and continuity of the velocity, stress tensor, and mass and heat fluxes can be detected on the phase boundary. It was shown in [94] that in gas–liquid systems the effect of nonlinear mass transfer is confined to the gas phase. Taking into account these considerations, the boundary conditions assume the following form:

$$\begin{aligned}&x=0,\quad u_j=u_{j0},\quad c_1=c_{10},\quad c_2=0,\quad t_j=t_0,\quad j=1,2;\\&y\to\infty,\quad u_1=u_{10},\quad c_1=c_{10},\quad t_1=t_0;\qquad y\to-\infty,\quad u_2=u_{20},\quad c_2=0,\quad t_2=t_0;\\&y=0,\quad u_1=u_2,\quad \mu_1\frac{\partial u_1}{\partial y}=\mu_2\frac{\partial u_2}{\partial y}-\frac{\partial\sigma}{\partial x},\quad v_1=-\frac{MD_1}{\rho_{10}^*}\frac{\partial c_1}{\partial y},\quad v_2=0,\quad \rho_1^*=\rho_{10}^*+Mc_1^*,\\&c_1=\chi c_2,\quad D_1\frac{\rho_1^*}{\rho_{10}^*}\frac{\partial c_1}{\partial y}=D_2\frac{\partial c_2}{\partial y},\quad t_1=t_2,\quad \lambda_1\frac{\partial t_1}{\partial y}+\rho_1c_{p1}v_1t_1=\lambda_2\frac{\partial t_2}{\partial y}.\end{aligned}\tag{3.67}$$

At high enough values of c_{10} a large concentration gradient directed normally to the interphase surface $(\partial c_1/\partial y)_{y=0}$ can be observed, which induces a secondary flow with rate v_1. The tangential concentration and temperature gradients along the phase boundary create a surface tension gradient,

$$\frac{\partial\sigma}{\partial x}=\frac{\partial\sigma}{\partial c_2}\frac{\partial c_2}{\partial x}+\frac{\partial\sigma}{\partial t_2}\frac{\partial t_2}{\partial x},\tag{3.68}$$

which induces a tangential secondary flow and its velocity is proportional to $\partial\sigma/\partial\sigma$. Later, the use of a substance which is not surface-active, i.e., $\partial\sigma/\partial c_2\approx 0$, will be examined.

The mass transfer rate (J_c) and the heat transfer rate (J_t) can be determined from the local mass (I_c) and heat (I_t) fluxes after taking the average of these fluxes along a length (L) of the interphase surface:

$$\begin{aligned}&J_c=k_cc_0=\frac{1}{L}\int_0^L I_c dx,\quad I_c=\frac{MD_1\rho_1^*}{\rho_{10}^*}\left(\frac{\partial c_1}{\partial y}\right)_{y=0},\\&J_t=k_tt_0=\frac{1}{L}\int_0^L I_t dx,\quad I_t=-\lambda_1\left(\frac{\partial t_1}{\partial y}\right)_{y=0}+\rho_1c_{p1}(v_1t_1)_{y=0},\end{aligned}\tag{3.69}$$

where c_1 and t_1 are determined upon solving problems (3.66–3.68). To do this, the following dimensionless (generalized) variables (see Sect. 4.1) are introduced:

$$\begin{aligned}&x=LX,\quad y=(-1)^{j+1}\delta_jY_j,\quad \delta_j=\sqrt{\frac{v_jL}{u_{j0}}},\quad u_j=u_{j0}U_j(X,Y_j),\quad v_j=(-1)^{j+1}u_{j0}\frac{\delta_j}{L}V_j(X,Y_j),\\&c_j=(-\chi)^{1-j}c_0C_j(X,Y_j),\quad t_j=t_0+(-1)^{j+1}t_0T_j(X,Y_i),\quad j=1,2.\end{aligned}\tag{3.70}$$

The introduction of (3.70) into (3.66) and (3.67) leads to the following equations:

$$U_j \frac{\partial U_j}{\partial X} + V_j \frac{\partial U_j}{\partial Y_j} = \frac{\partial^2 U_j}{\partial Y_j^2}, \quad \frac{\partial U_j}{\partial X} + \frac{\partial V_j}{\partial Y_j} = 0, \quad U_j \frac{\partial C_j}{\partial X} + V_j \frac{\partial C_j}{\partial Y_j} = \frac{1}{Sc_j} \frac{\partial^2 C_j}{\partial Y_j^2} - (j-1) Da C_j,$$
$$U_j \frac{\partial T_j}{\partial X} + V_j \frac{\partial T_j}{\partial Y_j} = \frac{1}{Pr_j} \frac{\partial^2 T_j}{\partial Y_j^2} + (j-1) Q Da C_j, \quad j = 1,2;$$
$$X = 0, \quad U_j = 1, \quad C_1 = 1, \quad C_2 = 0, \quad T_j = 0, \quad j = 1,2;$$
$$Y_1 \to \infty, \quad U_1 = 1, \quad C_1 = 1, \quad T_1 = 0; \qquad Y_2 \to \infty, \quad U_2 = 1, \quad C_2 = 0, \quad T_2 = 0;$$
$$Y_1 = Y_2 = 0, \quad U_1 = \theta_1 U_2, \quad \theta_2 \frac{\partial U_1}{\partial Y_1} = -\frac{\partial U_2}{\partial Y_2} + \theta_4 \frac{\partial T_2}{\partial X},$$
$$V_1 = -\theta_3 \frac{\partial C_1}{\partial Y_1}, \quad V_2 = 0, \quad C_1 + C_2 = 0, \quad T_1 + T_2 = 0, \quad \theta_5 \frac{\partial C_1}{\partial Y_1} = \frac{\partial C_2}{\partial Y_2}, \quad \theta_6 \frac{\partial T_1}{\partial Y_1} = \frac{\partial T_2}{\partial Y_2}, \tag{3.71}$$

where

$$Da = \frac{kl}{u_{20}}, \quad Q = \frac{q c_0}{\chi \rho_2 c_{p2} t_0}, \quad Sc_j = \frac{\nu_j}{D_j}, \quad Pr_j = \frac{\nu_j}{a_j}, \quad j = 1,2;$$
$$\theta_1 = \frac{u_{20}}{u_{10}}, \quad \theta_2 = \frac{\mu_1}{\mu_2} \sqrt{\frac{\nu_2}{\nu_1}} \left(\frac{u_{20}}{u_{10}} \right)^{\frac{3}{2}}, \quad \theta_3 = \frac{M c_0}{\rho_{10}^* Sc_1}, \tag{3.72}$$
$$\theta_4 = \frac{\partial \sigma}{\partial t_2} \frac{t_0}{u_{20} \mu_2} \sqrt{\frac{\nu_2}{u_{20} l}}, \quad \theta_5 = \chi \frac{D_1}{D_2} \frac{\rho_1^*}{\rho_{10}^*} \sqrt{\frac{u_{10} \nu_2}{u_{20} \nu_1}}, \quad \theta_6 = \frac{\lambda_1}{\lambda_2} \sqrt{\frac{u_{10} \nu_2}{u_{20} \nu_1}}.$$

From (3.69) and (3.70) the expressions for the Sherwood and Nusselt numbers are directly obtained:

$$Sh = \frac{k_c L}{D_1} = M \sqrt{Re_1} \int_0^L \left(1 + \theta_3 Sc_1 C_1^*\right) \left(\frac{\partial C_1}{\partial Y_1} \right)_{Y_1 = 0} dX,$$
$$Nu = \frac{k_t L}{\lambda_1} = -Re_1 \left[\int_0^L \left(\frac{\partial T_1}{\partial Y_1} \right)_{Y_1=0} dX + \theta_3 Pr_1 \int_0^L \left(1 + T_1^*\right) \left(\frac{\partial C_1}{\partial Y_1} \right)_{Y_1=0} dX \right],$$
$$C_1^* = C_1(X,0), \quad T_1^* = T_1(X,0), \quad Re_1 = \frac{u_{10} L}{\nu_1}. \tag{3.73}$$

The solution of (3.71) allows the determination of

$$J_1 = \int_0^L \left(\frac{\partial C_1}{\partial Y_1} \right)_{Y_1=0} dX, \quad J_2 = \int_0^L C_1(X,0) \left(\frac{\partial C_1}{\partial Y_1} \right)_{Y_1=0} dX,$$
$$J_3 = \int_0^L \left(\frac{\partial T_1}{\partial Y_1} \right)_{Y_1=0} dX, \quad J_4 = \int_0^L T_1(X,0) \left(\frac{\partial C_1}{\partial Y_1} \right)_{Y_1=0} dX. \tag{3.74}$$

The introduction of (3.74) into (3.73) allows us to determine the Sherwood and Nusselt numbers:

$$Sh = M\sqrt{Re_1}(J_1 + \theta_3 Sc_1 J_2), \quad Nu = -\sqrt{Re_1}[J_3 + \theta_3 Pr_1(J_1 + J_4)]. \tag{3.75}$$

Problem (3.71) can be solved conveniently using an iterative algorithm, where six problems are solved consecutively, until there is convergence with respect to the integral J_1 in (3.74):

$$\begin{aligned}
&U_1^{(k)}\frac{\partial U_1^{(k)}}{\partial X} + V_1^{(k)}\frac{\partial U_1^{(k)}}{\partial Y_1} = \frac{\partial^2 U_1^{(k)}}{\partial Y_1^2}, \quad \frac{\partial U_1^{(k)}}{\partial X} + \frac{\partial V_1^{(k)}}{\partial Y_1} = 0;\\
&X = 0, \quad U_1^{(k)} = 1; \qquad Y_1 = 0, \quad U_1^{(k)} = \theta_1 U_2^{(k-1)}, \quad V_1^{(k)} = -\theta_3\frac{\partial C_1^{(k-1)}}{\partial Y_1};\\
&\theta_1 = 0.1, \quad Y_{1\infty} = 6, \quad (\text{at the first iteration } \theta_1 = \theta_3 = 0).
\end{aligned} \tag{3.76}$$

$$\begin{aligned}
&U_2^{(k)}\frac{\partial U_2^{(k)}}{\partial X} + V_2^{(k)}\frac{\partial U_2^{(k)}}{\partial Y_2} = \frac{\partial^2 U_2^{(k)}}{\partial Y_2^2}, \quad \frac{\partial U_2^{(k)}}{\partial X} + \frac{\partial V_2^{(k)}}{\partial Y_2} = 0; \quad X = 0, \quad U_2^{(k)} = 1;\\
&Y_2 = 0, \quad \frac{\partial U_2^{(k)}}{\partial Y_2} = -\theta_2\left(\frac{\partial U_1^{(k)}}{\partial Y_1}\right)_{Y_1=0} + \theta_4\left(\frac{\partial T_2^{(k-1)}}{\partial X}\right)_{Y_2=0}, \quad V_2^{(k)} = 0;\\
&Y_2 \to \infty \ (Y_2 \geq Y_{2\infty}), \quad U_2^{(k)} = 1; \qquad 0 \leq X \leq 1, \quad 0 \leq Y_2 \leq Y_{2\infty};\\
&\theta_2 = 0.145, \quad Y_{2\infty} = 6, \quad (\text{at the first iteration } \theta_4 = 0).
\end{aligned} \tag{3.77}$$

$$\begin{aligned}
&U_1^{(k)}\frac{\partial C_1^{(k)}}{\partial X} + V_1\frac{\partial C_1^{(k)}}{\partial Y_1} = \frac{1}{Sc_1}\frac{\partial^2 C_1^{(k)}}{\partial Y_1^2};\\
&X = 0, \quad C_1^{(k)} = 1; \qquad Y_1 = 0, \quad C_1^{(k)} = -C_2^{(k-1)}(X, 0);\\
&Y_1 \to \infty\left(Y_1 \geq \overline{Y_1}\right), \quad C_1^{(k)} = 1; \qquad 0 \leq X \leq 1, \quad 0 \leq Y_1 \leq \overline{Y_1};\\
&Sc_1 = 0.735, \quad \overline{Y_1} = 7, \quad \left[\text{at the first iteration } C_2^{(k)}(X, 0) = 0\right].
\end{aligned} \tag{3.78}$$

$$\begin{aligned}
&U_2^{(k)}\frac{\partial C_2^{(k)}}{\partial X} + V_2^{(k)}\frac{\partial C_2^{(k)}}{\partial Y_2} = \frac{1}{Sc_2}\frac{\partial^2 C_2^{(k)}}{\partial Y_2^2} - DaC_2^{(k)};\\
&X = 0, \quad C_2^{(k)} = 0; \quad Y_2 = 0, \quad \frac{\partial C_2^{(k)}}{\partial Y_2} = \theta_5\left(\frac{\partial C_1^{(k)}}{\partial Y_1}\right)_{Y_1=0};\\
&Y_2 \to \infty \ \left(Y_2 \geq \overline{Y_2}\right), \quad C_2^{(k)} = 0; \quad 0 \leq X \leq 1, \quad 0 \leq Y_2 \leq \overline{Y_2};\\
&Sc_2 = 564, \quad \theta_5 = 18.3, \quad \bar{Y}_2 = 0.26, \quad Da = 10.
\end{aligned} \tag{3.79}$$

$$
\begin{aligned}
&U_1^{(k)}\frac{\partial T_1^{(k)}}{\partial X}+V_1^{k}\frac{\partial T_1^{(k)}}{\partial Y_1}=\frac{1}{Pr_1}\frac{\partial^2 T_1^{(k)}}{\partial Y_1^2};\\
&X=0,\quad T_1^{(k)}=0;\qquad Y_1=0,\quad T_1^{(k)}=-T_2^{(k-1)}(X,0);\\
&Y_1\to\infty\ \left(Y_1\ge\overline{\overline{Y_1}}\right),\quad T_1^{(k)}=0;\quad 0\le X\le 1,\quad 0\le Y_1\le\overline{\overline{Y_1}};\\
&Pr_1=0.666,\quad \overline{\overline{Y_1}}=7.4\quad \left[\text{at the first iteration } T_2^{(k)}(X,0)=0\right].
\end{aligned}
\tag{3.80}
$$

$$
\begin{aligned}
&U_2^{(k)}\frac{\partial T_2^{(k)}}{\partial X}+V_2^{(k)}\frac{\partial T_2^{(k)}}{\partial Y_2}=\frac{1}{Pr_2}\frac{\partial^2 T_2^{(k)}}{\partial Y_2^2}+QDaC_2^{(k)};\\
&X=0,\quad T_2^{(k)}=0;\qquad Y_2=0,\quad \frac{\partial T_2^{(k)}}{\partial Y_2}=\theta_6\left(\frac{\partial T_1^{(k)}}{\partial Y_1}\right)_{Y_1=0};\\
&Y_2\to\infty\left(Y_2\ge\overline{\overline{Y_2}}\right),\quad T_2^{(k)}=0;\quad 0\le X\le 1,\quad 0\le Y_2\le\overline{\overline{Y_2}};\\
&Pr_2=6.54,\quad \theta_6=0.034,\quad \overline{\overline{Y_2}}=2.4,\quad QDa=8.6.
\end{aligned}
\tag{3.81}
$$

The values of the parameters in (3.76–3.81) have been calculated for the process of absorption of NH_3 in water or water solutions of strong acids.

The results obtained by solving these problems are shown in Tables 3, 4 and 5.

The comparative analysis of the nonlinear mass transfer effect and the Marangoni effect in gas–liquid and liquid–liquid systems shows (Tables 3, 4) that the Marangoni effect does not affect the heat and mass transfer kinetics, because in real systems the parameter θ_4 is very small.

However, in cases where the velocity of the second phase is very low, the occurrence of the Marangoni effect is expected because of its velocity dependence from $u_{20}^{-3/2}$. To evaluate the case described above, systems with the velocity in the volume of the second phase equal to zero ($u_{20}=0$) have been investigated. The numerical results (Table 5) show that under these conditions the Marangoni effect is negligible.

The results obtained show that the Marangoni effect is negligible in two-phase systems with a movable phase boundary and the absence of surface-active agents. The deviations from the linear mass transfer theory has to be explained by the nonlinear mass transfer effect in conditions of large concentration gradients.

In this chapter the terms "mathematical description," "mathematical model," and "theory" are used. The difference between these terms is as follows:

- A *mathematical description* is a set of differential equations, where full correspondence between mathematical operators and physical effects exists.
- A *mathematical model* is a mathematical description, where all parameters are known or obtained (using experimental data).
- The solution of the model equations is *theory*.

The use of different *methods for solution* of the equations (analytical, numerical, asymptotic etc.) leads to *analytical*, *numerical*, and *asymptotic theories*.

Table 3 Influence of the nonlinear mass transfer effect and the Marangoni effect on the heat and mass transfer kinetics in gas–liquid systems ($\theta_1 = 0.1$, $\theta_2 = 0.145$)

No.	θ_3	θ_4	J_1	J_2	J_3	J_4
1	0	0	0.5671	0.09721	0.01855	−0.01337
2	0.2	0	0.6129	0.01155	0.02143	−0.01554
3	−0.2	0	0.5274	0.08542	0.01623	−0.01162
4	0	10^{-4}	0.5671	0.09721	0.01855	−0.01338
5	0	10^{-3}	0.5671	0.09721	0.01855	−0.01337
6	0	10^{-2}	0.5670	0.09718	0.01857	−0.01339
7	0	10^{-1}	0.5658	0.09696	0.01879	−0.01364
8	0	1	0.5658	0.09696	0.01879	−0.01364
9	0	5	0.5660	0.09696	0.01854	0.01345

Table 4 Influence of the nonlinear mass transfer effect and the Marangoni effect on the heat and mass transfer kinetics in liquid–liquid systems ($\theta_1 = 0.9$, $\theta_2 = 3$, $u_2\ (X,Y_2) = 1$)

No.	θ_{31}	θ_{32}	θ_4	J_1	J_2	J_3	J_4
1	0	0	0	21.1000	4.8778	0.3320	−0.0524
2	4×10^{-4}	-8×10^{-4}	0	22.5419	5.7854	0.4288	−0.0628
3	0	0	2×10^{-4}	21.1000	4.8778	0.3320	−0.0524
4	0	0	1×10^{-3}	21.0999	4.8778	0.3320	−0.0524
5	0	0	1×10^{-2}	21.0990	4.8774	0.3320	−0.0524
6	0	0	1×10^{-1}	21.0899	4.8736	0.3319	−0.0524
7	0	0	5	20.5698	4.6527	0.3291	−0.0513

Table 5 Influence of the nonlinear mass transfer effect and the Marangoni effect on the heat and mass transfer kinetics in liquid–liquid systems, when the second liquid is immobile ($\theta_1 = 1$, $\theta_2 = 1$, $u_2\ (X,Y_2) = 10^{-4}$)

No.	θ_{31}	θ_{32}	θ_4	J_1	J_2	J_3	J_4
1	0	0	0	16.9333	3.3960	0.3041	−0.0460
2	4×10^{-4}	-8×10^{-4}	0	18.3164	4.0715	0.3967	−0.0551
3	0	0	2×10^{-4}	16.9333	3.3960	0.3041	−0.0460
4	0	0	1×10^{-3}	16.9331	3.3959	0.3042	−0.0460
5	0	0	1×10^{-2}	16.9314	3.3952	0.3041	−0.0596
6	0	0	1×10^{-1}	16.9145	3.3885	0.3040	−0.0592
7	0	0	1	16.7421	3.3201	0.3026	−0.0456
8	0	0	5	15.8955	2.9669	0.2968	−0.0437

4 Examples

The nonlinear mass transfer effect of large concentration gradients is investigated [71] in the cases of:

- Heat transfer [95, 96] and mass transfer in countercurrent flows [98].
- Mass transfer in channels [99, 100] and liquid films [101, 102].

- Multicomponent mass transfer in gas (liquid)–solid [97], gas–liquid and liquid–liquid [103, 104], and gas–liquid film [78] systems.
- Large concentrations [36, 105, 106].
- Electrochemical systems [71, 108].

Heat transfer and multicomponent mass transfer will be considered as processes accompanying nonlinear mass transfer.

4.1 Heat Transfer in the Conditions of Nonlinear Mass Transfer

The intensive mass transfer leads to a change of the velocity distribution in the boundary layer. As a result, its influence on the heat and mass transfer should be analogous because the induced secondary flow at the interphase surface leads to convective heat and mass transfer. The theoretical analysis of the heat transfer kinetics that is accompanied by a nonlinear mass transfer was carried out [95, 96] for cases where the thermodiffusion and the diffusion thermal conductivity are not considered.

The influence of the nonlinear mass transfer on the heat transfer may be determined if problem (3.20) is completed with the equation of convection–conduction in the boundary layer approximations:

$$u\frac{\partial t}{\partial x}+v\frac{\partial t}{\partial y}=a\frac{\partial^2 t}{\partial y^2}x=0,\quad t=t_0;\quad y=0,\quad t=t*y\to\infty,\quad t=t_0,\tag{4.1}$$

where t_0 and t^* are the temperatures in the volume and on the solid surface, and a is the thermal diffusivity.

In (4.1) it is necessary to introduce similarity variables (3.23), and the result obtained is

$$T''+\bar{\varepsilon}_t\varphi T'=0,\qquad T(0)=1,\qquad T(\infty)=0,\tag{4.2}$$

where

$$T=T(\eta)=\frac{t-t_0}{t^*-t_0};\qquad \bar{\varepsilon}_t=\varepsilon\alpha_t;\quad \alpha_t=\frac{D}{a}=Le^{-1}.\tag{4.3}$$

In (4.2) it is necessary to substitute φ (η) from (3.27) and (3.28), which allows us to seek T (η) in the form of a series of the orders of the small parameter θ:

$$T=T_0+\theta T_1+\theta^2T_2+\cdots.\tag{4.4}$$

In this way, a series of finite problems for the separate approximations [95] are obtained and their solutions have the form

$$T_0(\eta) = 1 - \frac{1}{\varphi_{0t}} \int_0^x E(\varepsilon_t, p)dp, \quad z = \frac{2}{\varepsilon}\eta,$$

$$T_1(\eta) = \frac{\bar{\varepsilon}_t \varphi_{3t}}{\varphi_0 \varphi_{0t}^2} \int_0^z E(\varepsilon_t, p)dp - \frac{\bar{\varepsilon}_t}{\varphi_0 \varphi_{0t}} \int_0^z \left[\int_0^p \varphi(s)ds \right] E(\varepsilon_t, p)dp,$$

$$T_2(\eta) = -\left[\frac{\bar{\varepsilon}_t \varepsilon \varphi_{3t}}{\varphi_0^2 \varphi_{0t}^2} \left(\frac{\varphi_3}{\varphi_0} + \frac{\bar{\varepsilon}_t \varphi_{3t}}{\varepsilon \varphi_{0t}} \right) - \frac{\bar{\varepsilon}_t^2 \varphi_{33t}}{2\varphi_0^2 \varphi_{0t}^2} - \frac{\bar{\varepsilon}_t \bar{\varphi}_{33t}}{2\varepsilon \varphi_0^2 \varphi_{t0}^2} \right] \int_0^z E(\varepsilon_t, p)dp$$

$$+ \frac{\bar{\varepsilon}_t \varepsilon}{\varphi_0^2 \varphi_{0t}} \left(\frac{\varphi_3}{\varphi_0} + \frac{\bar{\varepsilon}_t \varphi_{3t}}{\varepsilon \varphi_{0t}} \right) \int_0^z \left[\int_0^p \varphi(s)ds \right] E(\varepsilon_t, p)dp$$

$$- \frac{\bar{\varepsilon}_t^2}{2\varphi_0^2 \varphi_{0t}} \int_0^z \left[\int_0^p \varphi(s)ds \right] E(\varepsilon_t, p)dp - \frac{\bar{\varepsilon}_t}{2\varepsilon \varphi_0^2 \varphi_{t0}} \int_0^z \left[\int_0^p \bar{\varphi}(s)ds \right] E(\varepsilon_t, p)dp, \tag{4.5}$$

where

$$\varepsilon_t = (\varepsilon \bar{\varepsilon}_t)^{0,5}, \quad E(\varepsilon_t, p) = \exp\left[-\frac{\varepsilon_t^2}{2} \int_0^p f(s)ds \right], \quad \varphi_{0t} = \int_0^\infty E(\varepsilon_t, p)dp \approx 3,01\varepsilon_t^{-0,7},$$

$$\varphi_{3t} = \int_0^\infty \left[\int_0^p \varphi(s)ds \right] E(\varepsilon_t, p)dp \approx 6,56\varepsilon_t^{-1,6},$$

$$\varphi_{33t} = \int_0^\infty \left[\int_0^p \varphi(s)ds \right]^2 E(\varepsilon_t, p)dp \approx 24\varepsilon_t^{-2,6},$$

$$\bar{\varphi}_{33t} = \int_0^\infty \left[\int_0^p \bar{\varphi}(s)ds \right] E(\varepsilon_t, p)dp \approx 0,326\varepsilon_t^{-3,26}. \tag{4.6}$$

Expressions (4.5) and (4.6) are correct for gases only. In [94] it was shown that for big values of the Lewis number for liquids ($a_t \approx 10^{-2}$) the nonlinear mass transfer does not influence the heat transfer, because the thickness of the diffusion boundary layer is much lower than the thickness of the temperature boundary layer. The nonlinear mass transfer influences the hydrodynamics only in the thin diffusion boundary layer and this influence is not essential for the heat transfer in the thick temperature boundary layer. This result may be obtained from (4.2), where the big hydrodynamic effects of the intensive mass transfer ($\varphi(0) \neq 0$) for high values of the Lewis number may be ignored.

The heat transfer rate in the gas at the boundary with a solid surface with length L is determined by the average value of the heat flux:

$$J_{\mathrm{t}} = k_{\mathrm{t}}(t^{*} - t_{0}) = \frac{1}{L}\int_{0}^{L} I_{\mathrm{t}} dx, \tag{4.7}$$

where I_t has a convective component as a result of the induced flow,

$$I_{\mathrm{t}} = -\lambda\left(\frac{\partial t}{\partial y}\right)_{y=0} + \rho_{0}^{*} c_{\mathrm{p}} (vt)_{y=0}. \tag{4.8}$$

Using (3.23), (4.3), (4.7), and (4.8), we can obtain an expression for the Nusselt number:

$$Nu = \frac{k_{\mathrm{t}} L}{\lambda} = -Pe^{0.5}\left[T'(0) + \theta\bar{\varepsilon}_{\mathrm{t}}\frac{t^{*}}{t^{*} - t_{0}}\psi'(0)\right], \tag{4.9}$$

where Pe is determined from (3.26). In (3.73) $\psi'(0)$ is determined from (3.31), and $T'(0)$ is obtained from (4.4) and (4.5):

$$T'(0) = -\frac{2}{\varepsilon\varphi_{0\mathrm{t}}} + \theta\frac{2\alpha_{\mathrm{t}}\varphi_{3\mathrm{t}}}{\varphi_{0}\varphi_{0\mathrm{t}}^{2}} - \theta^{2}\left[\frac{2\bar{\varepsilon}_{\mathrm{t}}\varphi_{3\mathrm{t}}}{\varphi_{0}^{2}\varphi_{0\mathrm{t}}^{2}}\left(\frac{\varphi_{3}}{\varphi_{0}} + \frac{\alpha_{\mathrm{t}}\varphi_{3\mathrm{t}}}{\varphi_{0\mathrm{t}}}\right) - \frac{\alpha_{\mathrm{t}}\bar{\varepsilon}_{\mathrm{t}}\varphi_{33\mathrm{t}}}{\varphi_{0}^{2}\varphi_{0\mathrm{t}}^{2}} - \frac{\alpha_{\mathrm{t}}\bar{\varphi}_{33\mathrm{t}}}{\varepsilon\varphi_{0}^{2}\varphi_{0\mathrm{t}}^{2}}\right]. \tag{4.10}$$

The check of the accuracy of the asymptotic theory for the joint heat and mass transfer in the boundary layer under conditions of intensive mass transfer was carried out in the solution of problem (4.2) through a numerical method [94].

The results of the asymptotic theory $T'(0)$ are in good agreement with the results of the numerical experiment $T_{\mathrm{N}}'(0)$ (Table 6). It is obvious that under conditions of intensive mass transfer ($\theta \neq 0$) the nonlinear mass transfer and heat transfer are not independent of each other, as follows from the linear theory ($\theta = 0$) for small concentration gradients. In cases where the nonlinear mass transfer is directed towards a rigid wall ($\theta < 0$), the increase of the concentration gradient leads to an increase of $T'(0)$. In contrast, the increase of θ leads to a decrease of $T'(0)$ when the intensive interphase mass transfer is from the solid wall towards the gas phase ($\theta > 0$).

4.2 Multicomponent Mass Transfer

The hydrodynamic nature of the nonlinear effect in the mass transfer kinetics under conditions of intensive mass transfer is a reason to assume that an analogous effect may occur under conditions of multicomponent mass transfer for all

Table 6 Comparison of the asymptotic theory prediction [$T'(0)$] with results from numerical experiments [$T'_N(0)$] for $\varepsilon = 1$, $\alpha_t = 2$, and $\bar{\varepsilon}_t = 2$

θ	$-T'_N(0)$	$-T'(0)$
0.0	0.864	0.847
+0.1	0.762	0.765
−0.1	0.943	0.945
+0.2	0.690	0.700
−0.2	1.063	1.059
+0.3	0.633	0.652
−0.3	1.212	1.190

components if there is a mass transfer of one of them in conditions of a very large concentration gradient.

The theory of diffusion in multicomponent systems [58, 63] shows that the independent diffusion approximation is valid in two cases: when the concentrations of the components are low and when their diffusivities do not differ considerably. This together with the mass transfer at a high concentration gradient provides a sufficient basis to discuss in an analogous way the mass transfer of the other components, for which the concentration gradients c_i (i = 1,...,n) are low. With this aim, the following should be added to Eq. (3.20):

$$u\frac{\partial c_i}{\partial x} + v\frac{\partial c_i}{\partial y} = D_i\frac{\partial^2 c}{\partial y^2};$$

$$x = 0,\quad c_i = c_{0i};\quad y = 0,\quad c_i = c_i^*;\quad y \to \infty,\quad c_i = c_{0i};\quad i = 1,\ldots,n. \tag{4.11}$$

Using the dimensionless variables (3.23), we can complete equation (3.25) as follows:

$$\psi_i'' + \bar{\varepsilon}_i\varphi\psi_i' = 0,\quad \psi_i(0) = 1,\quad \psi_i(\infty) = 0,\quad i = 1,\ldots,n, \tag{4.12}$$

where

$$\bar{\varepsilon}_i = \varepsilon\alpha_i,\quad \alpha_i = \left(\frac{\varepsilon_i}{\varepsilon}\right)^2,\quad \varepsilon_i = Sc_i^{0.5},\quad Sc_i = \frac{v}{D_i},\quad \psi_i = \psi_i(\eta) = \frac{c_i - c_{0i}}{c_i^* - c_{0i}},$$
$$i = 1,\ldots,n. \tag{4.13}$$

In (4.12) the function φ (η) is determined (3.27) in the form of a series of the orders of the small parameter θ, which allows us to seek the solution of (3.93) analogously:

$$\psi_i = \psi_i^{(0)} + \theta\psi_i^{(1)} + \theta^2\psi_i^{(2)} + \cdots,\qquad i = 1,\ldots,n. \tag{4.14}$$

The separate problems for the determination of the unknown functions in (4.14) are solved through the method of disturbances, where for φ_0, φ_1 and φ_2 (3.28) is used. In this way, the following expressions are obtained [97]:

$$\psi_i^{(0)}(\eta) = 1 - \frac{1}{\varphi_{0i}} \int_0^z E(\varepsilon_i, p) dp,$$

$$\psi_i^{(1)}(\eta) = \frac{\bar{\varepsilon}_i \varphi_{3i}}{\varphi_0 \varphi_{0i}^2} \int_0^z E(\varepsilon_i, p) dp - \frac{\bar{\varepsilon}_i}{\varphi_0 \varphi_{0i}} \int_0^z \left[\int_0^p \varphi(s) ds \right] E(\varepsilon_i, p) dp,$$

$$\psi_i^{(2)}(\eta) = -\left[\frac{\bar{\varepsilon}_i \varepsilon \varphi_{3i}}{\varphi_0^2 \varphi_{0i}^2} \left(\frac{\varphi_3}{\varphi_0} + \frac{\bar{\varepsilon}_i \varphi_{3i}}{\varepsilon \varphi_{0i}} \right) - \frac{\bar{\varepsilon}_i^2 \varphi_{33i}}{2\varphi_0^2 \varphi_{0i}^2} - \frac{\bar{\varepsilon}_i \bar{\varphi}_{33i}}{2\varepsilon \varphi_0^2 \varphi_{0i}^2} \right] \int_0^z E(\varepsilon_i, p) dp$$
$$+ \frac{\bar{\varepsilon}_i \varepsilon}{\varphi_0^2 \varphi_{0i}} \left(\frac{\varphi_3}{\varphi_0} + \frac{\bar{\varepsilon}_i \varphi_{3i}}{\varepsilon \varphi_{0i}} \right) \int_0^z \left[\int_0^p \varphi(s) ds \right] E(\varepsilon_t, p) dp$$
$$- \frac{\bar{\varepsilon}_i^2}{2\varphi_0^2 \varphi_{0i}} \int_0^z \left[\int_0^p \varphi(s) ds \right] E(\varepsilon_i, p) dp$$
$$- \frac{\bar{\varepsilon}_i}{2\varepsilon \varphi_0^2 \varphi_{0i}} \int_0^z \left[\int_0^p \bar{\varphi}(s) ds \right] E(\varepsilon_i, p) dp, \quad i = 1, \ldots, n, \tag{4.15}$$

where

$$E(\varepsilon_i, p) = \exp\left[-\frac{\varepsilon_i^2}{2} \int_0^p f(s) ds \right], \quad \varphi_{0i} = \int_0^\infty E(\varepsilon_i, p) dp \approx \begin{cases} 3,01\, Sc_i^{-0,35} \text{—for gases,} \\ 3,12\, Sc_i^{-0,34} \text{—for liquids,} \end{cases}$$
$$\varphi_{3i} = \int_0^\infty \left[\int_0^p \varphi(s) ds \right] E(\varepsilon_i, p) dp \approx \begin{cases} 6,56\, Sc_i^{-0,8} \text{—for gases,} \\ 5,08\, Sc_i^{-0,67} \text{—for liquids,} \end{cases}$$
$$\varphi_{33i} = \int_0^\infty \left[\int_0^p \varphi(s) ds \right]^2 E(\varepsilon_i, p) dp \approx \begin{cases} 24\, Sc_i^{-1,3} \text{—for gases,} \\ 12,2\, Sc_i^{-1} \text{—for liquids,} \end{cases}$$
$$\bar{\varphi}_{33i} = \int_0^\infty \left[\int_0^p \bar{\varphi}(s) ds \right] E(\varepsilon_i, p) dp \approx \begin{cases} 0,326\, Sc_i^{-1,63} \text{—for gases,} \\ 0,035\, Sc_i^{-1,1} \text{—for liquids,} \end{cases} \quad i = 1, \ldots, n. \tag{4.16}$$

The multicomponent mass transfer rate in gases (liquids) at a boundary with a solid surface with length L is determined by the average value of the mass flux,

$$J_i = M_i k_i \left(c_i^* - c_{0i} \right) = \frac{1}{L} \int_0^L I_i dx_i, \quad i = 1, \ldots, n, \tag{4.17}$$

which has a convective component as well as a result of the induced flow:

$$I_i = -M_i D_i \left(\frac{\partial c_i}{\partial y}\right)_{y=0} + M_i(c_i v) = -M_i D_i \left[\left(\frac{\partial c_i}{\partial y}\right)_{y=0} + \frac{M\alpha_i}{\rho_0^*}\left(c_i \frac{\partial c}{\partial y}\right)_{y=0}\right], \quad (4.18)$$
$$i = 1, \ldots, n.$$

The expression for the Sherwood number is derived from (3.23), (4.12), (4.17), and (4.18):

$$Sh_i = \frac{k_i L}{D_i} = Pe^{0.5}\left[\psi_i'(0) + \theta\bar{\varepsilon}_i \frac{c_i^*}{c_i^* - c_{0i}}\psi_i'(0)\right], \quad i = 1, \ldots, n, \quad (4.19)$$

where $\psi'(0)$ is determined from (3.31) and $\psi_i'(0)$ is determined from (4.15),

$$\psi_i'(0) = -\frac{2}{\varepsilon\varphi_{0i}} + \theta\frac{2\bar{\varepsilon}_i\varphi_{3i}}{\varepsilon\varphi_0\varphi_{0i}^2} - \theta^2\left[\frac{2\bar{\varepsilon}_i\varphi_{3i}}{\varphi_0^2\varphi_{0i}^2}\left(\frac{\varphi_3}{\varphi_0} + \frac{\bar{\varepsilon}_i\varphi_{3i}}{\varepsilon\varphi_{0i}}\right) - \frac{\bar{\varepsilon}_i^2\varphi_{33i}}{\varepsilon\varphi_0^2\varphi_{0i}^2} - \frac{\bar{\varepsilon}_i\bar{\varphi}_{33i}}{\varepsilon^2\varphi_0^2\varphi_{0i}^2}\right],$$
$$i = 1, \ldots, n.$$

(4.20)

The accuracy of the basic result (4.20) of the asymptotic theory for multicomponent mass transfer under conditions of intensive mass transfer depends on the parameters θ, ε and ε_i

(i =1,…,n). The limits of validation of this theory may be determined through an expression of type (3.32). For example, for $\varepsilon = 20$ and $\bar{\varepsilon}_i = 10$, the admissible values for θ are less than 0.033.

The evaluation of the accuracy of the asymptotic theory of multicomponent mass transfer (4.20) (under conditions of interphase mass transfer for one of the components) has been made using numerical solution of problem (4.12). The comparison of the results of the numerical experiment $\psi_{iN}'(0)$ [97] with the asymptotic theory data $\psi_i'(0)$ is shown in Tables 7 and 8. It can clearly be seen that intensive interphase mass transfer of one of the components from the volume towards the solid surface ($\theta < 0$) increases the diffusive mass transfer for all of the components. In the cases where the direction of intensive interface mass transfer is from the solid surface towards the volume ($\theta > 0$), the multicomponent mass transfer decreases. These effects do not depend on the change in the direction of the interphase mass transfer for components with low concentration gradients.

Table 7 Comparison of the asymptotic theory $[\psi_i'(0)]$ with the numerical experiments $\left[\psi_{iN}'(0)\right]$, $i = 1, \ldots, n$, for gases for $\varepsilon = 1$, $\alpha_i = 2$, and $\bar{\varepsilon}_i = 2$

θ	$-\psi_i'(0)$	$-\psi_{iN}'(0)$
0.0	0.845	0.847
+0.1	0.762	0.765
−0.1	0.943	0.945
+0.2	0.689	0.700
−0.2	1.060	1.061
+0.3	0.633	0.652
−0.3	1.212	1.190

Table 8 Comparison of the asymptotic theory $[\psi_i'(0)]$ with the numerical experiments $\left[\psi_{iN}'(0)\right]$, $i = 1,\ldots,n$, for liquids for $\varepsilon = 20$, $\alpha_i = 0.5$, and $\bar{\varepsilon}_i = 10$

θ	$-\psi_i'(0)$	$-\psi_{iN}'(0)$
0.00	0.198	0.194
+0.03	0.167	0.169
-0.03	0.275	0.250
+0.05	0.154	0.170
+0.10	0.132	0.234

4.3 Concentration Effects

The theoretical analysis of the nonlinear mass transfer and the hydrodynamic stability in the system

with intensive the mass transfer which has been done so far considers the "pure" effect of large concentration gradients. Under these conditions, however, the concentrations themselves are high and their influence is manifested by the concentration dependencies of density, viscosity, and diffusivity. To determine these effects, the use of the basic model of nonlinear mass transfer is needed.

The theoretical analysis of the dependence of the velocity of induced flow from the concentration gradient will be developed in the case of laminar gas flow flowing over a solid surface.

Consider a binary gas mixture, where gas 2 has partial density ρ_2 and flows over the surface with an average mass velocity of $\boldsymbol{w}_2$, whereas gas 1 is injected through the solid surface with an average mass velocity of $\boldsymbol{w}_1$ and has partial density ρ_1, i.e.,

$$\rho = \sum_{i=1}^{2} \rho_i, \qquad \boldsymbol{w} = \frac{\sum_{i=1}^{2} \rho_i \boldsymbol{w}_i}{\rho}. \tag{4.21}$$

The diffusion velocity $\boldsymbol{W}_i$ is the deviation of the velocity $\boldsymbol{w}_i$ from the gas mixture velocity $\boldsymbol{w}$:

$$\boldsymbol{W}_i = \boldsymbol{w}_i - \boldsymbol{w}, \quad i = 1, 2. \tag{4.22}$$

From (4.22) it directly follows that

$$\sum_{i=1}^{2} \rho_i \boldsymbol{w}_i = \boldsymbol{w} \sum_{i=1}^{2} \rho_i - \sum_{i=1}^{2} \rho_i \boldsymbol{W}_i = \boldsymbol{w}\rho + \sum_{i=1}^{2} \rho_i \boldsymbol{W}_i. \tag{4.23}$$

Hence, from (4.21) and (4.22) we obtain

$$\sum_{i=1}^{2} \rho_i \boldsymbol{W}_i = 0. \tag{4.24}$$

The law of conservation of mass holds for each component:

$$\operatorname{div}(\rho_i \boldsymbol{w}_i) = \operatorname{div}\left[\rho_i (\boldsymbol{w}_i + \boldsymbol{W})_i\right] = 0, \quad i = 1, 2, \tag{4.25}$$

and if i is a summation index for the mixture, we find

$$\operatorname{div}(\rho \boldsymbol{w}) = 0. \tag{4.26}$$

Consider isothermal diffusion of gas 1 into gas 2. The mass flux of gas 1 as a result of the diffusion is defined from the mass concentration gradient:

$$\boldsymbol{J}_1 = c_1 \boldsymbol{W}_1 = -D_{12}\, \boldsymbol{grad}\, c_1, \quad c_1 = \frac{\rho_1}{\rho}, \tag{4.27}$$

where D_{12} is the diffusivity and c_i is the mass concentration of gas 1 (weight factors).

From (4.22), (4.25), and (4.27) we obtain

$$\operatorname{div} \rho_1 \boldsymbol{w} = \operatorname{div}(\rho D_{12}\, \boldsymbol{grad}\, c_1). \tag{4.28}$$

Using the formula

$$\operatorname{div}(\boldsymbol{ab}) = \boldsymbol{b}\, \boldsymbol{grad}\, a + a \operatorname{div} \boldsymbol{b}, \tag{4.29}$$

we obtain the following expression directly:

$$\operatorname{div}(c_1 \rho \boldsymbol{w}) = \rho \boldsymbol{w}.\boldsymbol{grad}\, c_1 + c_1 \operatorname{div} \rho \boldsymbol{w}. \tag{4.30}$$

Taking into account (4.25), we obtain from (4.30) that

$$\rho \boldsymbol{w}.\boldsymbol{grad}\, c_1 = div(\rho D_{12}\, \boldsymbol{grad}\, c_1). \tag{4.31}$$

Having denoted by u and v the components of the velocity $\boldsymbol{w}$ and applying the boundary layer approximation $\left(\frac{\partial^2}{\partial x^2} \ll \frac{\partial^2}{\partial y^2}\right)$ from (4.31), we can obtain the following:

$$\rho\left(u\frac{\partial c_1}{\partial x} + v\frac{\partial c_1}{\partial y}\right) = \frac{\partial}{\partial y}\left(\rho D_{12}\frac{\partial c_1}{\partial y}\right), \tag{4.32}$$

where u and v satisfy the equations of the gas mixture motion.

Only one limitation is applied while obtaining the above results—the processes are assumed to be isothermal. Further, for definiteness one will be considered to be a potential gas flow with a constant velocity u_0 over a semi-infinite flat plate. Thus, the equations of motion of the gas mixture assume the following form:

$$\begin{gathered} \rho\left(u\frac{\partial u}{\partial x} + v\frac{\partial u}{\partial y}\right) = \frac{\partial}{\partial y}\left(\mu\frac{\partial u}{\partial y}\right), \quad \operatorname{div} \rho \boldsymbol{w} = 0; \\ x = 0, \quad u = u_0; \quad y = 0, \quad u = 0, \quad v = v_n; \quad y \to \infty, \quad u = u_0, \end{gathered} \tag{4.33}$$

where v_n is the induced flow velocity (induced by the intensive interphase mass transfer). This velocity is a normal component of the velocity $\boldsymbol{w}$ to the solid surface. If it is assumed that the second component of the gas does not go through the solid surface ($\boldsymbol{w}_2 = 0$), from (4.21) we can directly obtain

$$y = 0, \quad \boldsymbol{w} = -\boldsymbol{W}_2. \tag{4.34}$$

From (4.24) and (4.27) it is found that

$$-c_1\boldsymbol{W}_1 = c_2\boldsymbol{W}_2 = (1 - c_1)\boldsymbol{W}_2 = D_{12}\,\boldsymbol{grad}\,c_1 \tag{4.35}$$

and from (4.34) and (4.35) we obtain the following form:

$$\boldsymbol{w} = -\left(\frac{D_{12}}{1 - c_1}\boldsymbol{6}.c_1\right)_{y=0} = -\left(\frac{D_{12}\rho}{\rho_2}\boldsymbol{grad}\,c_1\right)_{y=0}. \tag{4.36}$$

If we express the concentration of gas 1 in (4.36) in moles per unit volume, and if for D_{12}, ρ and ρ_2 on the solid surface (y = 0) the denotations D, ρ^* and ρ_0^* are used, we have

$$c = \frac{\rho_1}{M}, \quad D = D_{12}, \quad \rho^* = \rho, \quad \rho_0^* = \rho_2 \quad (y = 0). \tag{4.37}$$

Thus, from (4.36) we can obtain the normal component of the velocity on the phase boundary:

$$v_{\mathrm{n}} = -\frac{MD\rho^*}{\rho_0^*}\left[\frac{\partial}{\partial y}\left(\frac{c}{\rho}\right)\right]_{y=0}. \tag{4.38}$$

The results obtained so far, i.e., (4.32), (4.33), (4.37), and (4.38), allow us to formulate in general the mass transfer in the diffusion boundary layer in the case of gas or liquid flow over a semi-infinite flat plate in the boundary layer approximations:

$$\begin{gathered}
\rho\left(u\frac{\partial u}{\partial x} + v\frac{\partial u}{\partial y}\right) = \frac{\partial}{\partial y}\left(\mu\frac{\partial u}{\partial y}\right), \qquad u\frac{\partial \rho}{\partial x} + v\frac{\partial \rho}{\partial y} + \rho\left(\frac{\partial u}{\partial x} + \frac{\partial u}{\partial y}\right) = 0, \\
\rho\left(u\frac{\partial c}{\partial x} + v\frac{\partial c}{\partial y}\right) = \frac{\partial}{\partial y}\left(\rho D\frac{\partial c}{\partial y}\right); \\
x = 0, \quad u = u_0, \quad c = c_0; \quad y \to \infty, \quad u = u_0, \quad c = c_0; \\
y = 0, \quad u = 0, \quad v = -\frac{MD}{\rho_0^*}\left(\frac{\partial c}{\partial y}\right)_{y=0} - \frac{D(\rho^* - \rho_0^*)}{\rho^*\rho_0^*}\left(\frac{\partial \rho}{\partial y}\right)_{y=0}, \qquad c = c^*,
\end{gathered} \tag{4.39}$$

where

$$\rho^* = \rho_0^* + Mc^*. \tag{4.40}$$

An equation for ρ should be added to (4.39). At constant pressure and temperature, ρ cannot be obtained from the equation of state. According to (4.21), ρ depends only on the concentration of the components:

$$\rho = \rho_2 + Mc. \tag{4.41}$$

Comparison of (4.39–4.41) with models (3.20) shows that the results of the asymptotic theory of nonlinear mass transfer in systems with intensive interphase mass transfer were obtained in the following approximations:

$$\rho = \text{const.}, \quad \mu = \text{const.}, \quad D = \text{const.} \tag{4.42}$$

The theoretical analysis of the influence of the diffused substance concentration on the density, viscosity, and diffusivity shows that in a number of cases these effects can be considered as small [105] (first-order approximation):

$$\rho = \rho_0(1 + \bar{\rho}C), \quad \mu = \mu_0(1 + \bar{\mu}C), \quad D = D_0(1 + \bar{D}C), \quad C = \frac{c - c_0}{c^* - c_0}, \tag{4.43}$$

where $\bar{\rho}$, $\bar{\mu}$ and $\bar{D}$ are small parameters defined from the experimental data for the dependence of ρ, μ and D in c.

The introduction of (4.43) into (4.39) leads to a complete mathematical description of the hydrodynamics and the mass transfer in systems with intensive interface mass transfer. These are practically simplified Oberbeck–Boussinesq equations, where the gravitational effect is negligible in the boundary layer approximations in the case of a horizontal flat plate if the following condition is valid:

$$\frac{g}{u_0^2}\sqrt{\frac{\nu_0 l}{u_0}} < 10^{-2}, \quad \nu_0 = \frac{\mu_0}{\rho_0}, \tag{4.44}$$

i.e., the second Navier–Stokes equation has the usual form in the boundary layer approximations:

$$\frac{\partial p}{\partial y} = 0. \tag{4.45}$$

The concentration effects were investigated within the concentration range $0 - c_{\max}$, where there is a significant effect of nonlinear mass transfer ($\theta = 0.3$). The concentration difference Δc can be determined by

$$\Delta c = \frac{M(c^* - c_0)}{\rho_0} \tag{4.46}$$

and can be used to normalize a scale of concentrations:

$$C = \frac{c}{\Delta c}, \quad C^* = \frac{c^*}{\Delta c}, \quad C_{\max} = \frac{c_{\max}}{\Delta c}. \tag{4.47}$$

From Table 9 it is evident that for the diffusion of ammonia into air, the influence of the ammonia concentration on the density ($\bar{\rho}C_{\max}$) and viscosity ($\bar{\mu}C_{\max}$) is about 15–16%. For a more precise analysis, these effects must be taken into account. For gas mixtures the effect of concentration on the diffusivity is negligible.

For liquid mixtures the effect of the concentration on the density (Table 9) is a few percent (under 5%), which is valid for a great number of completely mixing

Table 9 Maximum concentration effect on the density, viscosity, and diffusivity

System	Ammonia/ air	Acetic acid/ water	Acetic acid/ toluene	Acetone/ water	Water/ acetone
c_{max} (kmol/ k molm^3 m^3)	0.0134	3.80	3.40	3.68	10.60
θ	0.3	0.3	0.3	0.3	0.3
Δc (kmol/ k molm^3 m^3)	0.0160	3.92	3.52	3.81	10.9
C_{max}	0.837	0.969	0.967	0.967	0.968
$\bar{\rho}$	−0.149	0.0134	0.0420	−0.518	0.0461
$\bar{\rho}C_{max}$	−0.125	0.0130	0.0420	−0.050	0.0450
$\bar{\mu}$	−0.190	0.0208	0.263	−0.0854	0.746
$\bar{\mu}C_{max}$	−0.159	0.0200	0.254	−0.0830	0.722
$\bar{D}$	0	0	0	−0.336	−0.843
$\bar{D}C_{max}$	0	0	0	−0.325	−0.816

pairs of liquids. There is a similar effect of the viscosity, but with many exceptions (acetic acid/toluene, water/acetone).

The effect of the concentration on the diffusivity for a liquid diffused in another one is often significant [107]. For acetone/water (Table 9), the dependence deviates from linearity. In cases of the parameters $\bar{\rho}$, $\bar{\mu}$ and $\bar{D}$ exceeding 0.3, the linear approximation (4.43) is not accurate enough, since neglecting the second-order approximations leads to error of more then 10%.

The analysis of the approximations of nonlinear mass transfer theory [106] which was developed for $\bar{\rho} = \bar{\mu} = \bar{D} = 0$ shows that the results are valid in the cases where these parameters are small enough (e.g., less than 0.05). It is valid for different systems: gas (liquid)–solid, gas–liquid, and liquid–liquid. At these conditions the hydrodynamics and the mass transfer depend on the concentration gradient (θ). When the parameters $\bar{\rho}$, $\bar{\mu}$ and $\bar{D}$ are within the interval 0.1–0.3, the concentration affects the mass transfer and this effect can be considered by introducing the linear approximation (4.43). For systems in which these parameters assume values greater than 0.3, the nonlinear terms should be added to (4.43).

4.4 Influence of High Concentration on the Mass Transfer Rate

The mass transfer rate can be expressed by the mass transfer coefficient. It will be defined from the average diffusive flux through an interphase surface with specific length L:

$$J = Mk(c^* - c_0) = \frac{M}{L}\frac{\rho^*}{\rho_0^*}\int_0^L D\left(\frac{\partial c}{\partial y}\right)_{y=0} dx. \tag{4.48}$$

The thickness of the diffusion boundary layer in gases and liquids is of a different order of magnitude. That is why different numerical algorithms [94] are used.

The thickness of the laminar and diffusion boundary layers in gases is of the same order of magnitude, so one characteristic scale can be applied:

$$\delta_0 = \sqrt{\frac{D_0 L}{u_0}}. \tag{4.49}$$

Problem (4.39) can be expressed in terms of the following dimensionless variables:

$$x = LX, \quad y = \delta_0 Y, \quad u = u_0 U, \quad v = v_0 \frac{\delta_0}{L} V, \quad c = c_0 + (c^* - c_0)C. \tag{4.50}$$

The introduction of (4.50) into (4.39) leads to the following equations:

$$\begin{aligned}
&\varphi\left(U\frac{\partial U}{\partial X} + V\frac{\partial U}{\partial Y}\right) = Sc\frac{\partial}{\partial Y}\left(\psi\frac{\partial U}{\partial Y}\right), \quad \frac{\partial}{\partial X}(\varphi U) + \frac{\partial}{\partial Y}(\varphi V) = 0,\\
&\varphi\left(U\frac{\partial C}{\partial X} + V\frac{\partial C}{\partial Y}\right) = \frac{\partial}{\partial Y}\left(\varphi\omega\frac{\partial C}{\partial Y}\right); \quad X = 0, \quad U = 1, \quad C = 0;\\
&Y = 0, \quad U = 0, \quad V = -\theta\frac{\partial}{\partial Y}\left(\frac{\frac{c_0}{\Delta c} + C}{\varphi}\right),\\
&C = 1; \quad Y \to \infty, \quad U = 1, \quad C = 0,
\end{aligned} \tag{4.51}$$

where

$$\begin{aligned}
&\theta_0 = \frac{M\Delta c_0}{\rho_0^*}\varphi(1)\omega(1), \quad \rho^* = \rho_0\varphi(1), \quad \rho_0^* = \rho_0\varphi(1) - Mc^*, \Delta c_0 = c^* - c_0,\\
&Sc = \frac{\mu_0}{\rho_0 D_0}, \varphi = \varphi(C) = \frac{\rho}{\rho_0}, \quad \psi = \psi(C) = \frac{\mu}{\mu_0},\\
&\omega = \omega(C) = \frac{D}{D_0}, \quad \varphi(0) = 1, \psi(0)
\end{aligned} \tag{4.52}$$

The solution of problem (4.51) can be obtained after introducing the similarity variables:

$$\begin{aligned}
&\varphi U = \Phi', \quad \varphi V = \frac{1}{2\sqrt{X}}(\Phi'\eta - \Phi), \quad C = F,\\
&\Phi = \Phi(\eta), \quad F = F(\eta), \quad \eta = \frac{Y}{\sqrt{X}}, \quad \Phi' = \frac{d\Phi}{d\eta}.
\end{aligned} \tag{4.53}$$

Hence, directly from (4.51), we obtain the following:

$$
\begin{aligned}
&2Sc\,\varphi^2\psi\Phi''' + \varphi^2\Phi\Phi'' - \varphi\varphi'\Phi\Phi'F' + 2Sc\,\varphi(\varphi\psi' - \varphi'\psi)\Phi''F' \\
&\quad - 2Sc\,\varphi'(\varphi\psi' - 2\varphi'\psi)\Phi'F'^2 = 0, \\
&2\varphi\omega F'' + 2(\varphi'\omega + \varphi'\psi)F'^2 + \Phi F' = 0, \quad \Phi(0) = -\theta F'(0), \quad \Phi'(0) = 0, \quad \Phi'(\infty) = 1, \\
&F(0) = 1, \quad F(\infty) = 0, \quad \theta = 2\theta_0 \frac{\Delta c_0 \varphi(1) - c^* \varphi'(1)}{\Delta c_0 \varphi(1)}.
\end{aligned}
\tag{4.54}
$$

The functions ϕ, ψ and ω in (4.54) are set outwardly by spline approximations of the experimental dependencies of ρ, μ and D on c. For a wide range of gas mixtures these functions can be obtained with enough accuracy through the linear approximation:

$$
\varphi = 1 + \bar{\rho}C, \quad \psi = 1 + \bar{\mu}C, \quad \omega = 1 + \bar{D}C. \tag{4.55}
$$

The introduction of (4.55) into (4.54) leads to the following equations:

$$
\begin{aligned}
&2Sc(1+\bar{\rho}F)^2(1+\bar{\mu}F)\Phi''' + (1+\bar{\rho}F)^2\Phi\Phi'' - \bar{\rho}(1+\bar{\rho}F)\Phi\Phi'F' + \\
&\quad + 2Sc(1+\bar{\rho}F)[\bar{\mu}(1+\bar{\rho}F) - \bar{\rho}(1+\bar{\mu}F)]\Phi''F' - 2Sc\bar{\rho}[\bar{\mu}(1+\bar{\rho}F) - 2\bar{\rho}(1+\bar{\mu}F)]\Phi'F'^2 = 0, \\
&2(1+\bar{\rho}F)(1+\bar{D}F)F'' + 2[\bar{\rho}(1+\bar{D}F) + \bar{D}(1+\bar{\rho}F)]F'^2 + \Phi F' = 0, \\
&\theta = 2\theta_0 \frac{1 - \frac{c_0}{\Delta c_0}\bar{\rho}}{1+\bar{\rho}}.
\end{aligned}
\tag{4.56}
$$

The parameters $\bar{\rho}$ and $\bar{\mu}$ in (4.56) are small, whereas $\bar{D} = 0$. Omitting the square terms regarding the small parameters $\bar{\rho}$ and $\bar{\mu}$ leads to

$$
\begin{aligned}
&2Sc(1 + 2\bar{\rho}F + \bar{\mu}F)\Phi''' + (1 + 2\bar{\rho}F)\Phi\Phi'' - \bar{\rho}\Phi\Phi'F' + 2Sc(\bar{\mu} - \bar{\rho})\Phi''F' = 0, \\
&2(1 + \bar{\rho}F)F'' + 2\bar{\rho}F'^2 + \Phi F' = 0, \\
&\Phi(0) = -\theta F'(0), \quad \Phi'(0) = 0, \quad \Phi'(\infty) = 1, \quad F(0) = 1, \quad F(\infty) = 0.
\end{aligned}
\tag{4.57}
$$

Problem (4.57) can be solved conveniently using the following algorithm:

1. Determination of the zeroth approximations of Φ and F by solving the boundary problem:

$$
\begin{aligned}
&2\Phi'''^{(0)} + \Phi^{(0)}\Phi''^{(0)} = 0, \quad \Phi^{(0)}(0) = 0, \quad \Phi'^{(0)}(0) = 0, \quad \Phi''^{(0)}(0) = 0.33206, \\
&\left[\Phi'^{(0)}(\infty) = 1\right], \quad 2F''^{(0)} + \Phi^{(0)}F'^{(0)} = 0, \quad F^{(0)}(0) = 1, \\
&F'^{(0)}(0) = 0.33205, \quad \left[F^{(0)}(\infty) = 0\right].
\end{aligned}
\tag{4.58}
$$

2. Determination of Φ at the kth iteration:

$$\begin{aligned}&2Sc\left(1+2\bar{\rho}F^{(k-1)}+\bar{\mu}F^{(k-1)}\right)\Phi'''^{(k)}+\left(1+2\bar{\rho}F^{(k-1)}\right)\Phi^{(k)}\Phi''^{(k)}-\\&-\bar{\rho}\Phi^{(k-1)}\Phi'^{(k-1)}F'^{(k-1)}+2Sc(\bar{\mu}-\bar{\rho})\Phi''^{(k-1)}F'^{(k-1)}=0,\\&\Phi^{(k)}(0)=-\theta F'^{(k-1)}(0),\quad \Phi'^{(k)}(0)=0,\quad \Phi'^{(k)}(\infty)=1.\end{aligned}\tag{4.59}$$

The value of $\Phi'^{(k)}(0)$ is varied till the condition $\Phi'^{(k)}(6)=1$ is reached with accuracy 10^{-3}.

3. Determination of F at the kth iteration:

$$\begin{aligned}&2\left(1+\bar{\rho}F^{(k-1)}\right)F''^{(k)}+2(\bar{\rho}F'^{(k-1)})^2+\Phi^{(k)}F'^{(k)}=0,\quad F^{(k)}(0)=1,\\&F^{(k)}(\infty)=0.\end{aligned}\tag{4.60}$$

The value of $F'^{(k)}(0)$ is varied till $F^{(k)}(\infty)=0$ with accuracy 10^{-3}.

4. The calculation procedure (from step 2 of the algorithm) is repeated until a result confirming

$$\left|\Phi''^{(k)}(0)-\Phi''^{(k-1)}(0)\right|\le 10^{-3},\quad \left|F'^{(k)}(0)-F'^{(k-1)}(0)\right|\le 10^{-3}\tag{4.61}$$

is obtained.

The integration of (4.58–4.60) is performed numerically with step $h = 10^{-2}$ in the interval $0\le\eta\le 6$.

The results for $\Phi''(0)$ and $F'(0)$ in the case of $Sc = 1$ are shown in Table 10 for different values of θ, $\bar{\rho}$, $\bar{\mu}$. They are obtained with three to four iterations.

The mass transfer rate in gases can be determined from the data in Table 10. To do this, (4.54) and (4.57) are introduced into (4.52):

$$Sh=\frac{kL}{D_0}=-2\frac{\rho*}{\rho_0}Pe^{1/2}F'(0),\quad Pe=\frac{u_0L}{D_0}.\tag{4.62}$$

The results given in Table 10 show that the dependence of $\Phi''(0)$ and $F'(0)$ on θ, $\bar{\rho}$ and $\bar{\mu}$ is monotonous. The change in viscosity $\bar{\mu}$ has no effect of practical importance on the mass transfer rate $[F'(0)]$, whereas the effect of the density $\bar{\rho}$ is six to seven times greater than that of the nonlinear mass transfer (θ).

The thicknesses of the laminar and diffusion boundary layers in liquids are of different orders of magnitude, so two specific scales should be applied:

$$\delta_1=\sqrt{\frac{\mu_0L}{\rho_0u_0}},\ \delta_2=\sqrt{\frac{D_0L}{u_0}},\quad \frac{\delta_1}{\delta_2}=\varepsilon=Sc^{1/2}.\tag{4.63}$$

Table 10 Comparative data for the momentum transfer $[\Phi''(0)]$ and the mass transfer $[F'(0)]$ at high concentrations [effect due to density $(\bar{\rho} \neq 0)$, viscosity $(\bar{\mu} \neq 0)$ and large concentration gradients $(\theta \neq 0)$ in gases] for $Sc = 1$

θ	$\bar{\rho}$	$\bar{\mu}$	$\Phi''(0)$	$-F'(0)$
0	0	0	0.332	0.332
0.3	0	0	0.301	0.299
−0.3	0	0	0.373	0.372
0.3	0.15	0	0.356	0.187
0	0.15	0	0.379	0.198
−0.3	−0.15	0	0.329	0.531
0.3	0	0.2	0.264	0.292
0	0	0.2	0.290	0.322
−0.3	0	−0.2	0.447	0.386
0.3	0.15	0.2	0.320	0.187
0	0.15	0.2	0.340	0.198
−0.3	0.15	0.2	0.362	0.211
0	−0.15	0	0.280	0.446
0	0	−0.2	0.394	0.343
0	−0.15	−0.2	0.347	0.469
−0.3	−0.15	−0.2	0.417	0.558

Considering these two scales, we should introduce the following dimensionless variables:

$$x = LX, \quad y = \delta_1 Y_1 = \delta_2 Y_2, \quad u = u_0 U_1(X, Y_1) = u_0 U_2(X, Y_2),$$
$$v = u_0 \frac{\delta_1}{L} V_1(X, Y_1) = u_0 \frac{\delta_2}{L} V_2(X, Y_2), \quad c = c_0 + \Delta c_0 C_1(X, Y_1) = c_0 + \Delta c_0 C_2(X, Y_2), \tag{4.64}$$

where

$$\begin{aligned} &Y_2 = \varepsilon Y_1, \quad U_2(X, Y_2) = U_1\left(X, \varepsilon^{-1} Y_2\right), \; U_1(X, Y_1) = U_2(X, \varepsilon Y_1), \\ &V_2(X, Y_2) = \varepsilon V_1\left(X, \varepsilon^{-1} Y_2\right), \; V_1(X, Y_1) = \varepsilon^{-1} V_2(X, \varepsilon Y_1), \\ &C_2(X, Y_2) = C_1\left(X, \varepsilon^{-1} Y_2\right), \; C_1(X, Y_1) = C_2(X, \varepsilon Y_1). \end{aligned} \tag{4.65}$$

In the new variables, the problem has the following form:

$$\begin{aligned} &\varphi 1\left(U_1 \frac{\partial U_1}{\partial X} + V_1 \frac{\partial V_1}{\partial Y_1}\right) = \frac{\partial}{\partial Y_1}\left(\psi_1 \frac{\partial U_1}{\partial Y_1}\right), \qquad \frac{\partial}{\partial X}(\varphi_1 U_1) + \frac{\partial}{\partial Y_1}(\varphi_1 V_1) = 0, \\ &\varphi_2\left(U_2 \frac{\partial C_2}{\partial X} + V_2 \frac{\partial C_2}{\partial Y_2}\right) = \frac{\partial}{\partial Y_2}\left(\varphi_2 \omega_2 \frac{\partial C_2}{\partial Y_2}\right); \qquad X_1 = 0, \quad U_1 = U_2 = 1, \quad C_1 = C_2 = 0; \\ &Y_1 = Y_2 = 0, \quad U_1 = U_2 = 0, \quad C_1 = C_2 = 0, \quad V_2 = -\theta_0 \frac{\partial}{\partial Y_2}\left(\frac{\frac{C_0}{\Delta c} + C_2}{\varphi_2}\right); \\ &Y_1 = Y_2 \to \infty, \quad U_1 = U_2 = 1, \quad C_1 = C_2 = 0. \end{aligned} \tag{4.66}$$

This boundary problem can be expressed by the following similarity variables:

$$\varphi_1 U_1 = \Phi_1'(\eta_1), \quad \varphi_2 U_2 = \Phi_2'(\eta_2), \quad \eta_1 = \frac{Y_1}{\sqrt{X}}, \quad \eta_2 = \frac{Y_2}{\sqrt{X}},$$
$$\varphi_1 V_1 = \frac{1}{2\sqrt{X}}\left(\Phi_1'\eta_1 - \Phi_1\right), \quad \varphi_2 V_2 = \frac{1}{2\sqrt{X}}\left(\Phi_2'\eta_2 - \Phi_2\right), \tag{4.67}$$
$$C_1 = F_1(\eta_1), \quad C_2 = F_2(\eta_2), \quad \eta_2 = \varepsilon\eta_1.$$

For ϕ, ψandω linear approximations can be used:

$$\varphi_i = 1 + \bar{\rho}F_i, \quad \psi_i = 1 + \bar{\mu}F_i, \quad \omega_i = 1 + \bar{D}F_i, \quad i = 1, 2. \tag{4.68}$$

In the new variables (4.66) gets the following form

$$2(1 + 2\bar{\rho}F_1 + \bar{\mu}F_1)\Phi_1''' + (1 + 2\bar{\rho}F_1)\Phi_1\Phi_1'' - \bar{\rho}\Phi_1\Phi_1'F_1' + 2(\bar{\mu} - \bar{\rho})\Phi_1''F_1' = 0,$$
$$2(1 + \bar{\rho}F_2 + \bar{D}F_2)F_2'' + 2(\bar{\rho} + \bar{D})F_2'^2 + \Phi_2 F_2' = 0,$$
$$\Phi_2(0) = -\theta F_2'(0), \quad \Phi_1'(0) = 0, \quad \Phi_1'(\infty) = 1, \quad F_2(0) = 1, \quad F_2(\infty) = 0, \tag{4.69}$$

where

$$F_1(\eta_1) = F_2(\eta_2) = F_2(\varepsilon\eta_1), \quad F_1'(\eta_1) = \varepsilon F_2'(\varepsilon\eta_1),$$
$$\Phi_2(\eta_2) = \varepsilon\Phi_1(\eta_1) = \varepsilon\Phi_1\left(\varepsilon^{-1}\eta_2\right), \quad \Phi_2(0) = \varepsilon\Phi_1(0) = -\theta F_2'(0). \tag{4.70}$$

Problem (4.69) can be directly solved using the following algorithm:

1. Determination of the zeroth approximations of $\Phi_1(\eta_1)$ by integration of the equation

$$\Phi_1'''^{(0)} + \Phi_1^{(0)}\Phi_1''^{(0)} = 0, \quad \Phi_1^{(0)}(0) = 0, \quad \Phi_1'^{(0)}(0) = 0, \quad \Phi_1'^{(0)}(0) = 1, \tag{4.71}$$

   with step $h_1 = 0.06/\varepsilon$ in the interval $0 \le \eta_1 \le 6. \Phi_1'^{(0)}$is varied until the condition $\Phi_1'^{(0)}(6) \ge 0.999$ is satisfied.
2. Determination of the zeroth approximations of $\Phi_2(\eta_2)$:

$$\Phi_2^{(0)}(\eta_2) = \varepsilon\Phi_1^{(0)}(\eta_1), \quad \eta_2 = \varepsilon\eta_1, \quad 0 \le \eta_1 \le 6. \tag{4.72}$$

3. Determination of the zeroth approximations of $F_2(\eta_2)$ by integration of the equation

$$F_2''^{(0)} + \Phi_2^{(0)}F_2'^{(0)} = 0, \quad F_2^{(0)}(0) = 1, \quad F_2^{(0)}(\infty) = 0, \tag{4.73}$$

   with step $h_2 = 0.06$ in the interval $0 \le \eta_2 \le 60$. To do this $F_2'^{(0)}(0)$ is varied until the condition $F_2^{(0)}(60) \le 0.001$ is satisfied.
4. Determination of the zeroth approximations of $F_1(\eta_1)$and $F_1'(\eta_1)$:

$$F_1^{(0)}(\eta_1) = F_2^{(0)}(\eta_2) = F_2^{(0)}(\varepsilon\eta_1), \quad F_1'^{(0)}(\eta_1) = \varepsilon F_2'^{(0)}(\eta_2) = \varepsilon F_2'^{(0)}(\varepsilon\eta_1). \tag{4.74}$$

5. Determination of $\Phi_1(\eta_1)$ at the kth iteration:

$$2\left(1+2\bar{\rho}F_1^{(k-1)}+\bar{\mu}F_1^{(k-1)}\right)\Phi_1^{\prime\prime\prime(k)}+\left(1+2\bar{\rho}F_1^{(k-1)}\right)\Phi_1^{(k)}\Phi_1^{\prime\prime(k)}-$$
$$-\bar{\rho}\Phi_1^{(k-1)}\Phi_1^{\prime(k-1)}F_1^{\prime(k-1)}+2(\bar{\mu}-\bar{\rho})\Phi_1^{\prime\prime(k-1)}F_1^{\prime(k-1)}=0, \qquad (4.75)$$
$$\Phi_1^{(k)}(0)=-\frac{\theta}{\varepsilon}F_2^{\prime(k-1)}(0),\quad \Phi_1^{\prime(k)}(0)=0,\quad \Phi_1^{\prime(k)}(\infty)=1.$$

The value of $\Phi_1^{''(k)}(0)$ is varied till the condition $\Phi_1^{'(k)}(6)\geq 0.999$ is reached.

6. Determination of $\Phi_2(\eta_2)$ at the kth iteration:

$$\Phi_2^{(k)}(\eta_2)=\varepsilon\Phi_1^{(k)}(\eta_1)=\varepsilon\Phi_1^{(k)}\left(\varepsilon^{-1}\eta_2\right),\quad 0\leq\eta_2\leq 60. \qquad (4.76)$$

7. Determination of $F_2(\eta_2)$ at the kth iteration with step h_2 in the interval $0\leq\eta_2\leq 60$:

$$2\left(1+\bar{\rho}F_2^{(k-1)}+\bar{D}F_2^{(k-1)}\right)F_2^{\prime\prime(k)}+2(\bar{\rho}+\bar{D})\left(F_2^{\prime(k-1)}\right)^2+\Phi_2^{(k)}F_2^{\prime(k)}=0,$$
$$F_2^{(k)}(0)=1,\quad F_2^{(k)}(\infty)=0. \qquad (4.77)$$

The value of $F_2^{'(k)}(0)$ is varied till the condition $F_1^{(k)}(60)\leq 0.001$ is satisfied.

8. Determination of $F_1(\eta_1)$and$F_1^{'}(\eta_1)$ at the kth iteration:

$$F_1^{(k)}(\eta_1)=F_2^{(k)}(\eta_2)=F_2^{(k)}(\varepsilon\eta_1),\quad F_1^{\prime(k)}(\eta_1)=\varepsilon F_2^{\prime(k)}(\eta_2)=\varepsilon F_2^{\prime(k)}(\varepsilon\eta_1),$$
$$0\leq\eta_1\leq 6. \qquad (4.78)$$

9. The calculation procedure (from step 5 of the algorithm on) is repeated until convergence is reached:

$$\left|\Phi_1^{\prime\prime(k)}(0)-\Phi_1^{\prime\prime(k-1)}(0)\right|\leq 10^{-3},\quad \left|F_2^{\prime(k)}(0)-F_2^{\prime(k-1)}(0)\right|\leq 10^{-3}, \qquad (4.79)$$

The results obtained for $\Phi_1^{''}(0)$and $F_2^{'}(0)$ at $\varepsilon=10$ and for different values of θ, $\bar{\rho}$, $\bar{\mu}$ and $\bar{D}$ are shown in Table 11. They are obtained with three to four iterations.

The mass transfer rate in liquids can be determined from the data in Table 11. To do this, (4.64) and (4.67) are introduced into (4.48):

$$Sh=-2\frac{\rho^*}{\rho_0}(1+\bar{D})Pe^{1/2}F_2^{\prime}(0). \qquad (4.80)$$

The results given in Table 11 show that the influence of density $\bar{\rho}$ and viscosity $\bar{\mu}$ on the hydrodynamics $[\Phi_1^{''}(0)]$ is similar to that observed in the case of gases, whereas their influence on the mass transfer rate $[F_2^{'}(0)]$ is practically insignificant. The change in diffusivity $\bar{D}$ does not affect $\Phi_1^{'}(0)$ as much as $F_2^{'}(0)$.

The theoretical analysis of the influence of high concentration gradients of the transferred substance on the hydrodynamics $[\Phi^{''}(0)]$ and mass transfer $[F^{'}(0)]$ through the concentration dependencies of density $(\bar{\rho})$, viscosity $(\bar{\mu})$, and

Table 11 Comparative data for the momentum transfer [$\Phi''(0)$] and the mass transfer [$F'(0)$] at high concentrations [effect due to density ($\bar{\rho} \neq 0$), viscosity ($\bar{\mu} \neq 0$), and large concentration gradients ($\theta \neq 0$)] in liquids for $Sc = 100$

θ	$\bar{\rho}$	$\bar{\mu}$	$\bar{D}$	$\Phi''(0)$	$-F'(0)$
0	0	0	0	0.332	0.332
0.03	0	0	0	0.330	0.176
−0.03	0	0	0	0.334	0.206
0	0.15	0	0	0.397	0.194
0	−0.15	0	0	0.201	0.181
0	0	0.2	0	0.272	0.186
0	0	−0.2	0	0.418	0.194
0	0	0	0.30	0.332	0.192
0	0	0	0.30	0.332	0.186
0.03	0.15	0.2	0.30	0.272	0.177
−0.03	0.15	0.2	0.30	0.275	0.200
0.03	−0.15	−0.2	−0.30	0.243	0.164
−0.03	−0.15	−0.2	−0.30	0.247	0.206
0.3	0	0	0	0.318	0.135
−0.1	0	0	0	0.342	0.268

diffusivity ($\bar{D}$) shows that the change of the density with the concentration affects the hydrodynamics in gases and liquids but does not influence the mass transfer in gases. The change in the viscosity with the concentration affects the hydrodynamics in gases and liquids and the mass transfer. The change in the diffusivity with the concentration does not influence the hydrodynamics and the mass transfer.

These results show that the predictions of the nonlinear theory of mass transfer at constant values of density, viscosity, and diffusivity [94] are of acceptable accuracy for gases and liquids if the density of the transferred substance is not sufficiently different from the density of the gas mixture. That is why the models of mass transfer in systems with intensive interphase mass transfer could be considerably simplified.

4.5 Nonlinear Mass Transfer in Countercurrent Flows

In this case the mathematical description can be obtained [98] using (2.15) and (3.40), i.e.,

$$\begin{aligned}
&2f_i''' + f_i f_i'' = 0, \quad 2\varphi_i'' + Sc_i f_i \varphi_i' = 0, \quad i = 1, 2, \\
&f_i(0) = -\theta^{(i)} \varphi_i'(0), \quad f_1'(0) = -\theta_1 f_2'(0), \quad \bar{\theta}_2 f_1''(0) = f_2''(0), \qquad (4.81)\\
&\varphi_1(0) + \varphi_2(0) = 1, \quad \bar{\theta}_3 \varphi_1'(0) = \varphi_2'(0), \quad \varphi_i(\infty) = 0, \quad i = 1, 2,
\end{aligned}$$

where

$$Sc_{\mathrm{i}} = \frac{\upsilon_{\mathrm{i}}}{D_{\mathrm{i}}}, \quad \theta_1 = \frac{u_2^{\infty}}{u_1^{\infty}}, \quad \theta_2 = \left(\frac{\rho_1\mu_1}{\rho_2\mu_2}\right)^{1/2}\left(\frac{u_1^{\infty}}{u_2^{\infty}}\right)^{3/2}, \quad \theta^{(\mathrm{i})} = \frac{2(c_1^{\infty} - \chi c_2^{\infty})\chi^{1-\mathrm{i}}}{\rho_{0\mathrm{i}}^{*} Sc_{\mathrm{i}}}, \quad \mathrm{i} = 1, 2,$$
$$\bar{\theta}_2 = \theta_2\sqrt{\frac{X_2}{X_1}}, \quad \theta_3 = \chi\frac{D_1}{D_2}\sqrt{\frac{u_1^{\infty}\upsilon_2}{u_2^{\infty}\upsilon_1}}, \quad \bar{\theta}_3 = \theta_3\sqrt{\frac{X_2}{X_1}}. \tag{4.82}$$

In gas–liquid systems it was shown (Sect. 3.3) that the nonlinear effect in the liquid phase may be neglected in comparison with that in the gas phase $\left(\theta^{(2)} = 0\right)$, i.e., it manifests itself when the mass transfer is limited by the mass transfer in the gas phase (θ_3 = 0). At these conditions it directly follows that $\varphi_2(\eta_2) \equiv 0$, i.e.,

$$\begin{aligned}
&2f_1''' + f_1 f_1'' = 0, \quad 2\varphi_1'' + Sc_1 f_1 \varphi_1' = 0, \quad 2f_2''' + f_2 f_2'' = 0,\\
&f_{\mathrm{i}}(0) = \theta\varphi_{\mathrm{i}}'(0), \quad f_2(0) = 0, \quad f_1'(0) = -\theta_1 f_2'(0), \quad \bar{\theta}_2 f_1''(0) = f_2''(0),\\
&\varphi_1(0) = 1, \quad \varphi_1(\infty) = 0, \quad f_{\mathrm{i}}'(\infty) = 1, \quad \mathrm{i} = 1, 2.
\end{aligned} \tag{4.83}$$

The numerical results are presented in [98].

References

1. Schlichting H, Gerstein K (2000) Boundary layer theory, 8th edn. Springer, Berlin
2. Nernst W (1904) Z Phys Chem 47:52
3. Boyadjiev Chr, Beshkov V (1988) Mass transfer in following liquid films. Mir, Moscow (in Russian)
4. Boyadjiev Chr, Beshkov V (1984) Mass transfer in liquid film flows. Bulgarian Academy of Sciences, Sofia
5. Langmuir I (1912) Phys Rev 34:321
6. Lewis WK, Whitman WG (1924) Ind Eng Chem 16:1215
7. Higbie R (1935) Trans Am Inst Chem Eng 31:365
8. Sternling CV, Scriven LE (1959) AIChE J 5:514
9. Boyadjiev Chr, Pl Mitev, Beshkov V (1976) Int J Multiphase Flow 3:61
10. Boyadjiev Chr, Levich VG, Krylov VS (1968) Int Chem Eng 8:393
11. Boyadjiev Chr (1971) Int Chem Eng 11:459
12. Prandtl L (1910) Z Phys 2:1072
13. Taylor GI (1916) British Advisory Communications for Aeronautics, R. and M. No 272
14. Kishinevsky MKh, Pamfilov AV (1949) J Appl Chem (Russia) 22:1173
15. Kishinevsky MKh (1951) J Appl Chem (Russia) 24:542
16. Danckwerts PV (1951) Ind Eng Chem 43:1960
17. Toor HL, Marchello JM (1958) AIChE J 4:97
18. Ruckenstein E (1958) Chem Eng Sci 7:265
19. Ruckenstein E (1963) Chem Eng Sci 18:233
20. Ruckenstein E (1967) Chem Eng Sci 22:474
21. Kishinevsky MKh (1965) Int Heat Mass Transf 8:1181
22. Reichardt H (1951) Z Angew Math Mech 31:208
23. Elrod HG (1957) J Aeronaut Sci 24:468

24. Wasan DT, Tien CL, Wilke CR (1963) AIChEJ 9:567
25. Dilman VV (1967) Theor Fundam Chem Technol (Russia) 1:438
26. Deissler RG (1955) NACA report no 1210
27. Levich VG (1962) Physicochemical hydrodynamics. Prentice Hall, New York
28. Son JS, Hanratty TJ (1967) AIChEJ 13:689
29. Van Driest ER (1956) J Aeronaut Sci 23:1007
30. Loytsiansky LG (1960) Appl Math Mech (Russia) 24:156
31. Reichardt H (1957) National Advisory Communications Aeronautics. Technical note 1408
32. Boyadjiev Chr, Toshev E (1989) Hung J Ind Chem 17:457
33. Pohlhausen E (1921) Z Angew Math Mech 1:115
34. Boyadjiev Chr, Piperova M (1971) Int Chem Eng 11:479
35. Boyadjiev Chr, Mitev Pl (1977) Chem Eng J 14:225
36. Boyadjiev Chr, Vulchanov N (1988) Int J Heat Mass Transf 31:795
37. Boyadjiev Chr, Mitev Pl, Sapundjiev Tsv (1976) Int J Multiphase Flow 3:51
38. Mitev Pl, Boyadjiev Chr (1976) Int J Multiphase Flow 3:57
39. Mitev Pl, Boyadjiev Chr (1978) Lett Heat Mass Transf 5:349
40. Boyadjiev Chr, Doichinova M (2000) Int J Heat Mass Transf 43:2701
41. Boyadjiev Chr, Vabishchevich P (1992) J Theor Appl Mech (Bulgaria) 23:114
42. Tersenov SA (1985) Parabolic equations with changing direction of time. Science, Novosibirsk (in Russian)
43. Larkin IA, Novikov VA, Ianenko NN (1983) Nonlinear equations of changed type. Science, Novosibirsk (in Russian)
44. Doichinova M, Boyadjiev Chr (2000) Int J Heat Mass Transf 43:2707
45. Krylov VS, Bogoslovsky VE, Mihnev NN (1976) J Appl Chem (Russia) 49:1769
46. Bird RB, Stewart WE, Lightfoot EN (2007) Transport phenomena, 2nd edn. Willey, New York
47. Brounstain BI, Fishbain GA (1977) Hydrodynamics. mass and heat transfer in disperse systems, Himia, Leningrad (in Russian)
48. Chang WS (1973) Int J Heat Mass Transf 16:811
49. Duda JL, Vrentas JS (1971) Int J Heat Mass Transf 14:395
50. Parlange JY (1973) Acta Mech 18:157
51. Ranz WE, Dickinson PF (1965) Ind Eng Chem Fundam 4:345
52. Golovin AM, Rubinina NM, Hohrin VM (1971) Theor Fundam Chem Technol (Russia) 5:651
53. Emanuel AS, Olander DR (1964) Int J Heat Mass Transf 7:539
54. Nienow AW, Unahabhokha R, Mullin JW (1968) Chem Eng Sci 24:1655
55. Olander DR (1962) Int J Heat Mass Transf 5:765
56. Unahabhokha R, Nienow AW, Mullin JW (1972) Chem Eng Sci 26:357
57. Uan SU (1963) In: Czja-Czjao L (ed) Cooling by means of liquid films—turbulence flows and heat transfer. Inostrannaja Literatura, Moscow (in Russian)
58. Olander DR (1962) J Heat Transf Trans ASME Ser C 84:185
59. Ross SM (1974) J Fluid Mech 63:157
60. Sparrow EW, Gregg JL (1960) J Heat Transf Trans ASME Ser C 82:294
61. Yuan SW, Finkelstein AB (1956) J Heat Transf Trans ASME Ser C 78:719
62. Hirshfelder J, Kertis E, Berd R (1961) Molecular theory of gases and liquids. Inostrannaja Literatura, Moscow (in Russian)
63. Franc-Kamenetskii VA (1969) Diffusion and heat transfer in chemical kinetics. Plenum, New York
64. Krylov VS, Davidov AD (1972) In: Proceedings of advanced methods in electrochemical machining. Shtiinca, Kishinev, p 13 (in Russian)
65. Krylov VS, Malienko VN (1972) Thermodynamics of irreversible processes and its applications. In: Proceedings of the 1st USSR conference, Chernovcy, p 69 (in Russian)
66. Krylov VS, Malienko VN (1973) Electrochemistry (Russia) 9:3
67. Boyadjiev Chr, Vulchanov N (1987) C R Acad Bulg Sci 40(11):35

68. Boyadjiev Chr, Vulchanov N (1990) Int J Heat Mass Transf 33:2039
69. Vulchanov N, Boyadjiev Chr (1988) Theor Appl Mech 19(4):74
70. Boyadjiev Chr (1991) J Eng Phys (Russia) 60:845
71. ChrB Boyadjiev, Babak VN (2000) Non-linear mass transfer and hydrodynamic stability. Elsevier, Amsterdam
72. Boyadjiev Chr (1972) Theor Fundam Chem Technol (Russia) 6:118
73. Boyadjiev Chr, Velchev L (1971) Theor Fundam Chem Technol (Russia) 5:912
74. Boyadjiev Chr (1971) Int Chem Eng 11:464
75. Boyadjiev Chr (1971) Int Chem Eng 11:470
76. Boyadjiev Chr, Elenkov D (1971) Int Chem Eng 11:474
77. Krylov VS, Boyadjiev Chr, Levich VG (1967) C R USSR Acad Sci 175:156
78. Boyadjiev Chr (1992) Russ J Eng Thermophys 2:289
79. Vulchanov N, Boyadjiev Chr (1988) Int J Heat Mass Transf 31:801
80. Vulchanov N, Boyadjiev Chr (1990) Int J Heat Mass Transf 33:2045
81. Boyadjiev Chr (1998) Hung J Ind Chem 26:181
82. Linde H, Schwarz E, Groger K (1967) Chem Eng Sci 22:823
83. Ruckenstein E, Berbente C (1964) Chem Eng Sci 19:329
84. Thomas WJ, McNicholl EK (1969) Trans Inst Chem Eng.47(10):325
85. Dilman VV, Kulov NN, Lothov VA, Kaminski VA, Najdenov VI (1998) Theor Fundam Chem Technol (Russia) 32:377
86. Porter KE, Cantwell ADC, McDermott C (1971) AIChE J 17:536
87. Hennenberg M, Bisch PM, Vignes-Adler M, Sanfeld A (1979) In: Sorensen TS (ed) Lecture notes in physics, vol 105. Springer, Berlin, p 229
88. Linde H, Schwartz P, Wilke H (1979) In: Sorensen TS (ed) Lecture notes in physics, vol 105. Springer, Berlin, p 75
89. Sanfeld A, Steinchen A, Hennenberg M, Bisch PM, Van Lamswerde-Galle D, Dall-Vedove W (1979) In: Sorensen TS (ed) Lecture notes in physics, vol 105. Springer, Berlin, p 168
90. Savistowski H (1981) Ber Bunsenges Phys Chem 85:905
91. Sorensen TS, Hennenberg M (1979) In: Sorensen TS (ed) Lecture notes in physics, vol 105. Springer, Berlin, p 276
92. Scriven LE, Sterling CV (1960) Nature 127(4733):186
93. Velarde J, Gastillo L (1981) In: Zierep J, Oertel H (eds) Transport and reactive phenomena leading to interfacial instability, convective transport and instability phenomena, Braun, Karlsruhe
94. Boyadjiev Chr, Halatchev I (1998) Int J Heat Mass Transf 41:939
95. Boyadjiev Chr (1990) Hung J Ind Chem 18:1
96. Boyadjiev Chr, Vulchanov N (1988) C R Acad Bulg Sci 41(10):49
97. Boyadjiev Chr (1990) Hung J Ind Chem 18:7
98. Doichinova M, Boyadjiev Chr (2001) Int J Heat Mass Transf 44:2121
99. Krylov VS (1980) Adv Chem (Russia) 49:118
100. Toshev E, Boyadjiev Chr (1994) Hung J Ind Chem 22:81
101. Boyadjiev Chr (1990) J Eng Phys (Russia) 59:92
102. Boyadjiev Chr, Toshev E (1990) J Eng Phys (Russia) 59:277
103. Boyadjiev Chr (1998) Hung J Ind Chem 26:245
104. Sapundjiev Tsv, Boyadjiev Chr (1993) Russ J Eng Thermophys (Russia) 3:185
105. Boyadjiev Chr (1984) Int J Heat Mass Transf 27:1277
106. Boyadjiev Chr (1993) Bulg Chem Commun 26:33
107. Sherwood TK, Pigford RL, Wilke CR (1975) Mass transfer. McGraw-Hill, New York
108. Krylov VS, Boyadjiev Chr (1996) Non-linear mass transfer. Institute of Thermophysics, Novosibirsk (in Russian)

Part II
Theoretical Analysis of Models

The models need to be analyzed theoretically prior to starting the simulations. The analysis helps us to understand the process mechanism, define the minimum number of model parameters, and solve the problem under consideration in the context of scale-up, model adequacy, etc.

Qualitative Analysis

The qualitative analysis addresses a generalized analysis of models [1] to obtain a measure of the contributions of the elementary processes such as diffusion and convection that contribute to the modeled process. An order of magnitude analysis allows us to estimate the order of the terms in the model representing physical effects. As a result of the order of magnitude analysis, some terms remain, whereas others are neglected, which finally results in reduced models, i.e., reduced with respect to the starting mathematical descriptions where both significant and insignificant physical effects are represented by terms (operators). Commonly, terms representing particular physical effects contributing to the process whose orders of magnitudes are less than 0.01 (i.e., the effect of this process is not possible to measure experimentally) are insignificant, whereas those remaining in the submodel have orders of magnitude of unity.

1 Generalized Analysis

The generalized analysis uses generalized variables [2], where the scales used to perform the adimensionalization of the model should be chosen in a manner ensuring that the dimensionless terms (generalized variables) have orders of magnitude of unity, $O(1)$. All those with order of magnitude 0.01 are omitted as insignificant. Hence, all dimensionless terms remaining have order of magnitude $O(1)$ and the submodel is well balanced.

1.1 Generalized Variables

The selection of the scales used for the generalized variables needs knowledge of the physics of the process and to some extent is an art. After selection of the scales,

C. Boyadjiev, *Theoretical Chemical Engineering*,

DOI: 10.1007/978-3-642-10778-8_4,

adimensionalization of the model equations has to be carried out. The next step is to perform inspection of the dimensionless variables with respect to their order of magnitude. As mentioned already, only those with $O(1)$ remain in the dimensionless equations, i.e., the maximum orders of magnitude of the different terms (mathematical operators) in the model equations must be equal to 1. In the context of these general remarks, let us see how these initial steps have to be performed.

Consider we have x and $y = (x)$ in domains $x_1 \le x_2 \le x_3$ and $y_1 \le y_2 \le y_3$ with $y_1 = y(x_1)$ and $y_2 = y(x_2)$. In this case the generalized variables are

$$X = \frac{x - x_1}{x_2 - x_1}, \quad Y = \frac{y - y_1}{y_2 - y_1}. \tag{1.1}$$

The adimensionalization uses as scales the domain widths $(x_2 - x_1)$ and $(y_2 - y_1)$, respectively, i.e., the maximum variations of the dimensional variables x and y, respectively. This change of variables yields unified domains for both X and Y, i.e., $0 \le X \le 1$ and $0 \le Y \le 1$. If $x_1 = 0$ and $y_1 = 0$, the characteristic scales are the maximum values x_2 and y_2. In cases where the minimum and maximum values of the variables are unknown, the scales are the average values.

1.2 Mass Transfer with a Chemical Reaction

Let us consider a transient mass transfer with a bulk chemical reaction [1] within a domain defined by $0 \le x \le L$, $0 \le y \le \delta$ and where the fluid flows along the x-axis. The mathematical description of the process is

$$\frac{\partial c}{\partial t} + u_x \frac{\partial c}{\partial x} + u_y \frac{\partial c}{\partial y} = D\left(\frac{\partial^2 c}{\partial x^2} + \frac{\partial^2 c}{\partial y^2}\right) \pm kc, \quad \frac{\partial u_x}{\partial x} + \frac{\partial u_y}{\partial y} = 0. \tag{1.2}$$

The initial and boundary conditions are

$$t = 0, c = 0; \quad x = 0, c = 0; \quad y = 0, c = 0; \quad x = L, c = c^*; \quad y = \delta, c = c^*. \tag{1.3}$$

The use of scales of the variables involved in (1.2) allows us to transform them into dimensionless ones.

$$T = \frac{t}{\tau}, \quad X = \frac{x}{L}, \quad Y = \frac{y}{\delta}, \quad U_x = \frac{U_x}{U_0}, \quad U_y = \frac{U_y}{V_0}, \quad C = \frac{c}{c^*}, \tag{1.4}$$

where τ is the process timescale and U_0 and V_0 are velocity scales along the x and y directions, respectively. The length scales are L along the x-axis and δ along the y-axis. In other words, this initial step of adimensionalization (1.4) simply means a change of variables from x to X and from y to Y. This change of variables also transforms the initial domain from $x_1 \le x_2 \le x_3$ into $0 \le X \le 1$ and from $y_1 \le y_2 \le y_3$ into $0 \le Y \le 1$.

With the new variables X and Y defined by (1.4), Eq. 1.2 can be expressed as

$$\frac{\partial U_x}{\partial X} + \frac{V_0 L}{U_0 \delta}\frac{\partial U_y}{\partial Y} = 0. \tag{1.5}$$

If all characteristic scales in (1.4) are known and $\frac{V_0 L}{U_0 \delta} \ll 1$, from (1.5) it follows that

$$\frac{\partial U_x}{\partial X} = 0, \quad U_y \equiv 0, \tag{1.6}$$

i.e., the flow is stratified and the scale V_0 is unnecessary. In the opposite case $\frac{V_0 L}{U_0 \delta} \gg 1$ and the stratified flow is directed in the Y direction. In all other cases the characteristic velocity V_0 must be obtained from the condition

$$\frac{V_0 L}{U_0 \delta} \sim 1, \tag{1.7}$$

i.e.,

$$V_0 = U_0 \frac{\delta}{L}, \quad \frac{\partial U_x}{\partial X} + \frac{\partial U_y}{\partial Y} = 0. \tag{1.8}$$

The results obtained show that *the characteristic scales must be known constants (determinant scales) or must be obtained as a combination from known characteristic scales (determined scales)*. Introducing (1.4) and (1.8) into (1.2) and (1.3) leads to a model in generalized variables

$$\begin{gathered} Sth\frac{\partial C}{\partial T} + U_x\frac{\partial C}{\partial X} + U_y\frac{\partial C}{\partial Y} = Fo\left(\alpha^2\frac{\partial^2 C}{\partial X^2} + \frac{\partial^2 C}{\partial Y^2}\right) \pm DaC; \\ T = 0, C = 0; \quad X = 0, C = 0; \quad Y = 0, C = 0; \quad X = 1, C = 1; \quad Y = 1, C = 1; \\ Sth = \frac{L}{U_0 \tau}, \quad Fo = \frac{DL}{U_0 \delta^2}, \quad \alpha = \frac{\delta}{L}, \quad Da = \frac{KL}{U_0}. \end{gathered} \tag{1.9}$$

Here *Sth*, *Fo*, and *Da* are the Sthruchal, Fourier, and Damkohler numbers, respectively. When there is a large difference in the length scales, i.e., $0 = a^2 < 10^{-2}$, the longitudinal mass transfer (along the X-axis) is negligible with respect to that in the Y direction.

Each term in the first equation in (1.9) represents a physical effect and the order of magnitude of this effect is equal to the parameter value (dimensionless prefactor) in this term. If there is no prefactor, like in the case with the first term in (1.5), which is actually equal to 1, then the order of magnitude of the entire term is $O(1)$. If the model equation contains dimensionless parameters greater than 1, this equation must be divided by the parameter with maximum value. As a result, all effects in the complicated process are of three types: *main effects* (the terms do not contain dimensionless parameters), *small effects* (the parameter values are in the range $[10^{-2}, 10^{-1}]$), and *negligible effects* (the parameter values are less than 10^{-2}).

1.3 Nonstationary Processes

In the cases of mass transfer in thin layers ($0 = a^2 \leq 10^{-2}$), the model equation has the form

$$
\begin{aligned}
&Sth\frac{\partial C}{\partial T} + U_x\frac{\partial C}{\partial X} + U_y\frac{\partial C}{\partial Y} = Fo\frac{\partial^2 C}{\partial Y^2} \pm DaC; \\
&T = 0, C = 0; \quad X = 0, C = 0; \quad Y = 0, C = 0; \quad Y = 1, C = 1.
\end{aligned} \tag{1.10}
$$

For nonstationary processes the timescale can be expressed as the ratio of the longitudinal length scale L and the convection velocity scale U_0, i.e., $\tau = L/U_0$. With a timescale defined in this manner, we have $Sth = 1$, which reduces the number of parameters (prefactors) in (1.10).

With short-term processes, the real time is too short and $Sth \gg 1$. Then, all the terms in (1.10) have to be divided by Sth and Eq. 1.10 has to be represented by the approximation $0 = Sth^{-1} < 10^{-2}$, which yields

$$
\begin{aligned}
&\frac{\partial C}{\partial T} = \frac{Fo}{Sth}\frac{\partial^2 C}{\partial Y^2} \pm \frac{Da}{Sth}C; \\
&T = 0, C = 0; \quad X = 0, C = 0; \quad Y = 0, C = 0; \quad Y = 1, C = 1.
\end{aligned} \tag{1.11}
$$

The form of (1.11) reveals that terms representing the fluid convection are neglected. In fact with short-term processes, the convection does not affect the mass transfer process.

With long-term processes ($Sth \ll 1$) Eq. 1.10 has to be represented by the approximation $0 = Sth < 10^{-2}$, which yields

$$
\begin{aligned}
&U_x\frac{\partial C}{\partial X} + U_y\frac{\partial C}{\partial Y} = Fo\frac{\partial^2 C}{\partial Y^2} \pm DaC; \\
&X = 0, C = 0; \quad Y = 0, C = 0; \quad Y = 1, C = 1.
\end{aligned} \tag{1.12}
$$

The form of (1.12) reveals that the transient processes practically approach steady-state conditions when $\tau \gg L/U_0$.

1.4 Steady-State Processes

With slow diffusion or a very large velocity scale, the flow can be approximated by (with $0 = Fo < 10^{-2}$)

$$U_x \frac{\partial C}{\partial X} + U_y \frac{\partial C}{\partial Y} = \pm DaC; \quad X = 0, C = 0; \quad Y = 0, C = 0. \tag{1.13}$$

When $Fo \sim 1$, the diffusion boundary layer thickness (see Sect. 1.3 in Chap. 1) can be defined as

$$\delta = \sqrt{\frac{DL}{U_0}}, \tag{1.14}$$

where δ is the order of magnitude. The diffusion boundary layer thickness for transient cases (see 1.11) follows from the condition $(Fo/Sth) \sim 1$, which yields

$$\delta = \sqrt{D\tau}. \tag{1.15}$$

1.5 Effect of the Chemical Reaction Rate

The effect of the chemical reaction rate is negligible if $Da \ll 1$ and (1.12), which requires the approximation $0 = Da < 10^{-2}$, is used, namely,

$$U_x \frac{\partial C}{\partial X} + U_y \frac{\partial C}{\partial Y} = Fo \frac{\partial^2 C}{\partial Y^2}; \quad X = 0, C = 0; \quad Y = 0, C = 0; \quad Y = 1, C = 1. \tag{1.16}$$

When fast chemical reactions take place ($Da \gg 1$), the terms in (1.12) have to be divided by Da and the approximation $0 = Da^{-1} < 10^{-2}$ has to be applied. The results is

$$0 = \frac{Fo}{Da} \frac{\partial^2 C}{\partial Y^2} \pm C; \quad Y = 0, C = 0; \quad Y = 1, C = 1. \tag{1.17}$$

Hence, the flow does not affect the mass transfer. From (1.17) it is possible to define the diffusional length scale δ (the order of magnitude of the diffusion boundary layer), namely,

$$Fo \sim Da \Rightarrow \delta = \sqrt{\frac{D}{K}}. \tag{1.18}$$

Introducing this characteristic scale into (1.17) leads to

$$0 = \frac{\partial^2 C}{\partial Y^2} \pm C; \quad Y = 0, C = 0; \quad Y = 1, C = 1. \tag{1.19}$$

The cases $Fo \gg Da$ and $Fo \ll Da$ correspond to processes dominated by molecular diffusion and a chemical reaction at the interphase surface.

From (1.16) it follows that the model is valid when $Fo \sim 1$. The limits $Fo \gg 1$ and $Fo \ll 1$ correspond to diffusion in an immobile medium and complete mixing ($C = 0$), respectively.

The examples of qualitative analysis exemplified above by means of model (1.9) do not represent all possible versions since we have four parameters (pre-factors) and each of them may be of order approximately 1 or much less than 1 (or much greater than 1). The main lesson is that the scales have to be defined preliminarily either by the macroscopic process variables and the geometry of the domain or as combinations of the former ones and transport coefficients.

2 Mechanism of Gas–Liquid Chemical Reactions

Chemical processes in gas–liquid systems [3, 4] are widely encountered in chemical processing, power generation, food processing, material synthesis, separations by absorption and for heat potential augmentation in heat pumps. Generally, all these processes span various situations where one or several gaseous components are absorbed by liquids accompanied by chemical reactions in the liquid phase. The reaction can occur either between the gaseous components or some of them and a liquid-phase component. Generally, these processes cover a wide range of situations where one or several components in the gas phase are absorbed by a liquid, accompanied by chemical reactions between either the gaseous components themselves or some of them and the liquid phase.

When chemical reactions take place in the liquid phase, they affect significantly the overall mass transfer through the gas–liquid interface. The reaction rate defines the mass transfer across the interface surface and the entire mass transfer mechanism could be changed. The increase in the reaction rate, for instance, yields augmentation of mass transfer across the interface. As a consequence of the increased reaction rate, a high concentration gradient occurs in the vicinity of the interface, which macroscopically results in a secondary flow [5].

The effects of the reaction kinetics in the liquid phase on both the mechanism and the rate of the mass transfer across the interphase surface will be analyzed next. The analysis will address both the linear and the nonlinear reaction kinetics laws as well as reversible and irreversible chemical transformations. In all cases homogenous catalytic reactions are considered.

2.1 Irreversible Chemical Reactions

Gas absorption accompanied by a chemical reaction in the liquid phase is a widely encountered process for separation of gas mixtures performed either in bubble or in packed-bed columns. With a packed-bed column we have to take into consideration the size of the packing elements. The trickling liquid spreads as thin films over the surfaces of the packing elements, thus forming the gas–liquid interface. Since the diffusion boundary layer theory [6] is widely used for analyses of chemical reactions in liquids, we will use this theoretical tool for the further analysis.

Let us consider gas absorption in a liquid with an irreversible chemical reaction in the liquid phase. The evaluation of the effect of the chemical reaction rate on the overall mass transfer mechanism will be performed by convection–diffusion equations with a volumetric term of a first-order chemical reaction:

$$\tilde{u}\frac{\partial\tilde{c}}{\partial x}+\frac{\partial\tilde{c}}{\partial y}=\tilde{D}\frac{\partial^2\tilde{c}}{\partial y^2},\qquad u\frac{\partial c}{\partial x}+v\frac{\partial c}{\partial y}=D\frac{\partial^2 c}{\partial y^2}-kc. \tag{2.1}$$

If a potential co-current flow of both phases is assumed and the interphase surface is flat, the boundary conditions of Eq. 2.1 are

$$\begin{aligned}&x=0,\tilde{c}=\tilde{c}_0,c=0;\quad y=0,\tilde{c}=\chi c,\tilde{D}\frac{\partial\tilde{c}}{\partial y}=D\frac{\partial c}{\partial y};\\&y\to\infty,\tilde{c}=\tilde{c}_0;\quad y\to-\infty,c=0.\end{aligned}\tag{2.2}$$

Here constant concentrations $\tilde{c}_0$ of the absorbed substance at the flow inlet as well as in the bulk of the gas flow are assumed. The thermodynamic equilibrium and the mass flow continuity are satisfied at the interphase surface $y = 0$.

The analysis of Eqs. 2.1 and 2.2 requires the following dimensionless (generalized) variables to be introduced:

$$\begin{aligned}&X=\frac{x}{L},Y=\frac{y}{\delta},\tilde{Y}=\frac{y}{\tilde{\delta}},\tilde{U}=\frac{\tilde{u}}{\tilde{u}_0},\tilde{V}=\frac{\tilde{v}}{\tilde{\varepsilon}\tilde{u}_0},\tilde{C}=\frac{\tilde{c}}{\tilde{c}_0},\\&U=\frac{u}{u_0},V=\frac{\tilde{v}}{\varepsilon u_0},C=\frac{c}{c_0},\varepsilon=\frac{\delta}{L},\tilde{\varepsilon}=\frac{\tilde{\delta}}{L},c_0=\frac{\tilde{c}_0}{\chi},\end{aligned}\tag{2.3}$$

where L is the length scale at the gas–liquid interface, δ and $\tilde{\delta}$ are the thicknesses of the diffusion boundary layers in the liquid and the gas, u_0 and $\tilde{u}_0$ are the velocities of the potential flows in the bulk of the gas and the liquid respectively, and χ is Henry's constant. The scales used in (2.3) allow the magnitude of all the dimensionless functions and their derivatives to be of the order of unity. In this way, from (2.1) to (2.3) we obtain

$$\begin{aligned}&\tilde{U}\frac{\partial\tilde{C}}{\partial X}+\tilde{V}\frac{\partial\tilde{C}}{\partial\tilde{Y}}=\tilde{Fo}\frac{\partial^2\tilde{C}}{\partial\tilde{Y}^2},\quad U\frac{\partial C}{\partial X}+V\frac{\partial^2 C}{\partial Y^2}=Fo\frac{\partial^2 C}{\partial Y^2}-KC,\\&X=0,\tilde{C}=1,C=0;\quad Y=\tilde{Y}=0,\tilde{C}=C,\frac{\chi}{\varepsilon_0}\frac{\partial\tilde{C}}{\partial\tilde{Y}}=\frac{\partial C}{\partial Y};\\&\tilde{Y}\to\infty,\tilde{C}=1;\quad Y\to-\infty,C=0,\end{aligned}\tag{2.4}$$

where

$$Fo=\frac{DL}{u_0\delta^2},\quad \tilde{Fo}=\frac{\tilde{D}L}{\tilde{u}_0\tilde{\delta}^2},\quad \varepsilon_0=\frac{D\tilde{\delta}}{\tilde{D}\delta},\quad K=\frac{kL}{u_0}. \tag{2.5}$$

In Eq. 2.4 all the functions and their derivatives are of the order of magnitude of unity and the contribution of the terms, i.e., the effect of the physical and chemical phenomena, is determined by the order of the dimensionless parameters.

The mass transfer in the gas phase is a result of the balance between the convective and the diffusive transport; thus, in the first equation in (2.4) the left and the right sides must have equal orders of magnitude. Taking into account that the order of magnitude of the left side is unity, the order of magnitude of the Fourier number must be of unity too:

$$\tilde{F}o \sim 1. \tag{2.6}$$

Equations (2.5) and (2.6) allow us to determine the order of magnitude of the diffusion boundary layer thickness in the gas phase, namely,

$$\tilde{\delta} = \sqrt{\frac{\tilde{D}L}{\tilde{u}_0}}. \tag{2.7}$$

The second equation in (2.4) shows that for slow chemical reactions, where $K < 10^{-2}$, the term KC may be omitted. Thus, in the convection–diffusion equation in the liquid phase $K = 0$ may be assumed and (2.4) becomes a mathematical description of the physical absorption (in accordance with the approximation of the boundary layer). From the above considerations it follows that at

$$\frac{kL}{u_2} < 10^{-2} \tag{2.8}$$

the chemical reaction has no practical effect on the mass transfer rate in the liquid phase. In these cases ($K = 0$), it follows from (2.4) that

$$Fo \sim 1. \tag{2.9}$$

Thus, the order of magnitude of the diffusion boundary layer thickness in the liquid phase becomes

$$\delta = \sqrt{\frac{DL}{u_0}}. \tag{2.10}$$

The chemical reaction rate affects the mass transfer rate when $K > 10^{-2}$. In the range:

$$10^{-2} < K < 1 \tag{2.11}$$

the effect of the chemical reaction rate in the liquid is always less than that of the convective transport if both the convective and the diffusive transport are of equal orders of magnitude. In cases when condition (2.11) is satisfied, the orders of magnitude of the Fourier number and the boundary layer thickness can be evaluated from (2.9) to (2.10). Thus, the order of magnitude of the boundary layer thickness in the liquid can be determined from (2.10) when

$$\frac{kL}{u_0} < 1. \tag{2.12}$$

The effect of the chemical reaction rate on the mass transfer, when $K > 1$, may be analyzed if the second equation in (2.4) is expressed in the form

$$\frac{1}{K}\left(U\frac{\partial C}{\partial X} + V\frac{\partial C}{\partial Y}\right) = \frac{Fo}{K}\frac{\partial^2 C}{\partial Y^2} - C. \tag{2.13}$$

From (2.13) it follows that the increase in the chemical reaction rate leads to a decrease in the convective transport, whereas the other two effects (the diffusive transport and the chemical reaction) should have equal orders of magnitude if $K \gg 1$, i.e.,

$$K^{-1} \ll 1, \quad \frac{Fo}{K} \sim 1. \tag{2.14}$$

From (2.5) to (2.14) it follows also that in the cases when

$$\frac{kL}{u_0} \gg 1, \tag{2.15}$$

the order of magnitude of the boundary layer thickness may be determined from (2.14), i.e.,

$$\delta = \sqrt{\frac{D}{k}}. \tag{2.16}$$

The analysis of (2.13) shows that at high rates of the chemical reaction ($K > 10^2$), the convective transport may be neglected with respect to the effect of the chemical reaction. In this case the hydrodynamics does not affect the mass transport if the condition

$$\frac{kL}{u_0} > 10^2 \tag{2.17}$$

is satisfied. When $K^{-1} = 0$ is substituted in (2.13), the boundary layer thickness can be determined from (2.16).

It is evident from (2.4) that the parameter $\frac{\chi}{\varepsilon_0}$ determines the distribution of the diffusion resistance in both the liquid phase and the gas phase. In a situation where

$$\frac{\chi}{\varepsilon_0} > 10^2, \tag{2.18}$$

the interphase mass transfer is limited by the transport in the liquid, whereas if

$$\frac{\chi}{\varepsilon_0} < 10^{-2}, \tag{2.19}$$

the mass transfer in the gas phase is the limiting effect. In the range

$$10^{-2} < \frac{\chi}{\varepsilon_0} < 10^2 \tag{2.20}$$

both diffusion resistances are comparable.

The terms in (2.5) indicate that the parameter ε_0 depends on δ, i.e., on the chemical reaction rate:

$$K < 1, \varepsilon_0 = \sqrt{\frac{Du_0}{\tilde{D}\tilde{u}_0}}; \quad K > 1, \varepsilon_0 = \sqrt{\frac{Du_0 K}{\tilde{D}\tilde{u}_0}}. \tag{2.21}$$

Hence, reasonable cases occur when

$$\frac{D}{\tilde{D}} \sim 10^{-4}, \quad \frac{u_0}{\tilde{u}_0} \sim 10^{-1}. \tag{2.22}$$

Thus, from (2.21) it follows that $\varepsilon_0 \sim 10^{-1}$ when $K < 1$.

The distribution of the diffusion resistances strongly depends on the physical solubility of the gas, i.e., on Henry's constant (χ). For low-solubility gases (N_2, O_2, CH_4) the value of χ is between 20 and 60. For gases with medium solubility (CO_2, C_2H_2, Cl_2, H_2S, Br_2, SO_2) Henry's constant is in the range 0.02–2. For highly soluble gases (HCl, NH_3) the values of χ are in the range 1×10^{-3} to 2×10^{-3}.

The analysis of the mutual effects of the physical solubility of the gas (χ) and the chemical reaction rate (K) on the distribution of the diffusion resistance between a gas and a liquid shows that for highly soluble gases $\frac{\chi}{\varepsilon_0} \sim 1$ if $K < 1$. This contradicts the experimental data [3] showing the interphase mass transfer during absorption of HCl and NH_3 is limited by the mass transfer in the gas phase. Thus, the process in the gas may limit the interphase mass transfer only in cases when the reaction rate in the liquid phase is high ($K > 1$). From (2.19), (2.21), and (2.22) it follows that in the case of highly soluble gases the interphase mass transfer is limited by the process in the gas if $K > 10^3$.

When $K > 1$ and $10^{-3} < \chi < 10^2$, the ratio $\frac{\chi}{\varepsilon_0}$ depends on K, i.e., the increase of the chemical reaction rate may lead to a change of the limiting stage of the process. It is evident from (2.15) to (2.19) that at

$$K > 1, \quad \chi\sqrt{\frac{\tilde{D}\tilde{u}_0}{Du_0 K}} < 10^{-2} \tag{2.23}$$

the interphase mass transfer is limited by the process in the gas phase and its rate is independent of the chemical reaction rate.

Figure 1 shows the variations of $\frac{\chi}{\varepsilon_0}$ with variations of K for several cases of practical interest (2.22) and various values of χ. The data plotted allow us to determine 12 areas with different interphase mass transfer mechanisms:

1. The chemical reaction rate does not affect the mass transfer and the limiting stage is the process in the liquid.
2. Absorption of low-solubility gas is limited by the mass transfer in the liquid and it is strongly dependent on the chemical reaction rate.

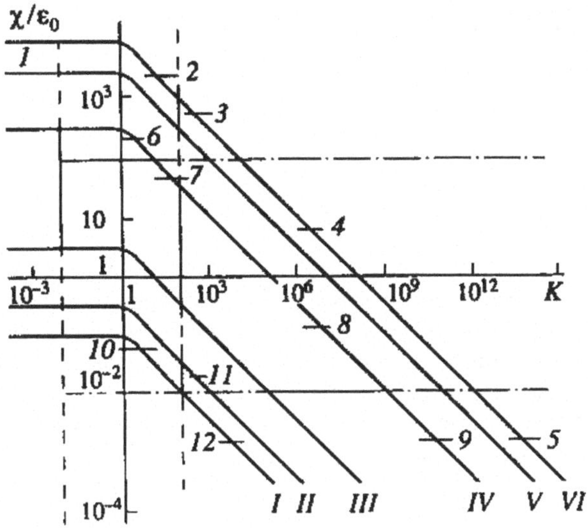

Fig. 1 Distribution of the diffusion resistance in the gas and liquid phases for various chemical reaction rates (in the liquid phase)

3. The case is the same as that in point 2, but the mass transfer rate is independent of the liquid hydrodynamics.
4. The case is the same as that in point 3, but the diffusion resistances in both phases are comparable.
5. Absorption of low-solubility gas is limited by the mass transfer in the gas phase and it is independent of the chemical reaction rate.
6. Absorption of gases of medium solubility is limited by the mass transfer in the liquid phase and it is affected by the chemical reaction rate.
7. The case is the same as that in point 6, but the diffusion resistances in both phases are comparable.
8. The case is the same as that in point 7, but the mass transfer is independent of the liquid hydrodynamics.
9. Absorption of gases of medium solubility is limited by the mass transfer in the gas phase and is independent of the chemical reaction rate.
10. There is absorption of highly soluble gas with comparable diffusion resistances in both phases and a significant effect of the chemical reaction rate.
11. The case is the same as that in point 10, but the liquid hydrodynamics does not affect the mass transfer.
12. Absorption of highly soluble gases is limited by the mass transfer in the gas phase and there are no effects of the chemical reaction rate in the liquid phase.

It is clear that the chemical reaction rate affects the mass transfer rate. The lower boundary of the range of variations is the rate of the physical absorption limited by the mass transfer in the liquid phase. The upper boundary is the rate of the physical absorption limited by the mass transfer in the gas phase. The lower boundary of the range in the case of highly soluble gases is a special situation.

Table 1 The values of the parameters in model (2.4)

$K < 10^{-2}$	$Fo = 1, K = 0$
$K < 1$	$Fo = 1$
$K > 1$	$\frac{Fo}{K} = 1$
$K > 10^2$	$\frac{Fo}{K} = 1, \quad K^{-1} = 0$
$\frac{\chi}{\varepsilon_0} < 10^{-2}$	$\frac{\chi}{\varepsilon_0} = 0, \quad C \equiv 0$
$\frac{\chi}{\varepsilon_0} > 10^{-2}$	$\frac{\chi}{\varepsilon_0} = 0, \quad \tilde{C} \equiv 1$

The situations listed above have mathematical descriptions following from (2.4) with suitable substitutions of the parameters and functions listed in Table 1.

The effect of the chemical reaction rate on the interphase mass transfer mechanism becomes more complicated if irreversible chemical reactions of arbitrary order occur. This case is discussed in [7] with the following chemical reaction:

$$a_1 A_1 + a_2 A_2 = a_3 A_3, \tag{2.24}$$

where component A_1 of the gas mixture reacts with component A_2 of the liquid absorbent and the yield is substance A_3.

The reaction rate is determined by

$$r_i = -k_i c_1^{\mathrm{m}} c_2^{\mathrm{n}}, \quad \frac{k}{a_i} = k_0 > 0, \quad i = 1, 2, \tag{2.25}$$

where k_i are the rate constants and c_i are the concentrations of substances A_i ($i = 1, 2$) in the liquid.

In fact substances A_2 and A_3 are nonvolatile and for the determination of their concentrations only equations for the liquid phase are required. Thus, the mathematical description of the absorption process may be obtained form (2.1), where $\tilde{c}$ is the concentration of substance A_1 in the gas, whereas the desired equation for c must be replaced with two equations for c_1 and c_2 in the liquid:

$$\begin{aligned}
&u\frac{\partial c_i}{\partial x} + v\frac{\partial c_i}{\partial y} = D_i\frac{\partial^2 c_i}{\partial y^2} - k_i c_1^{\mathrm{m}} c_2^{\mathrm{n}}, \quad i = 1, 2;\\
&x = 0,\ \tilde{c} = \tilde{c}_0,\ c_2 = c_{02};\\
&y = 0,\ \tilde{c} = \chi c_1,\ \tilde{D}\frac{\partial \tilde{c}}{\partial y} = D_1\frac{\partial c_1}{\partial y},\ \frac{\partial c_2}{\partial y} = 0;\\
&y \to \infty,\ \tilde{c} = \tilde{c}_0; \quad y \to -\infty,\ c_1 = 0,\ c_2 = c_{02},
\end{aligned} \tag{2.26}$$

where the boundary conditions follow from (2.2) taking into account the nonvolatility of substance A_2.

The theoretical analysis of the effect of the irreversible reaction on the interphase mass transfer mechanism needs the dimensionless variables defined by (2.3) to taken into consideration with the assumption that there are two diffusing substances in the liquid phase:

$$Y_i = \frac{y}{\delta_i},\ V_i = \frac{v}{\varepsilon_i u_0},\ \varepsilon_i = \frac{\delta_i}{L},\ C_i = \frac{c_i}{c_{01}},\quad i = 1, 2, \tag{2.27}$$

where

$$c_{01} = \frac{\tilde{c}_0}{\chi}. \tag{2.28}$$

Hence, Eq. 2.4 take the form

$$\begin{aligned}
&\tilde{U}\frac{\partial \tilde{C}}{\partial X} + \tilde{V}\frac{\partial \tilde{C}}{\partial \tilde{Y}} = \tilde{F}o\frac{\partial^2 \tilde{C}}{\partial \tilde{Y}^2},\ U\frac{\partial C_i}{\partial X} + V_i\frac{\partial C_i}{\partial Y_i} = Fo_i\frac{\partial^2 C_i}{\partial Y_i^2} - K_i C_1^{\mathrm{m}} C_2^{\mathrm{n}},\quad i = 1, 2;\\
&X = 0,\ \tilde{C} = 1,\ C_1 = 0,\ C_2 = 1;\\
&\tilde{Y} = Y_1 = Y_2 = 0,\ \tilde{C} = C_1,\ \frac{\partial C_2}{\partial Y_2} = 0,\ \frac{\chi}{\varepsilon_{01}}\frac{\partial \tilde{C}}{\partial \tilde{Y}} = \frac{\partial C_1}{\partial Y_1};\\
&\tilde{Y} \to \infty,\ \tilde{C} = 1;\quad Y_1 \to -\infty,\ C_1 = 0;\quad Y_2 \to -\infty,\ C_2 = 1,
\end{aligned} \tag{2.29}$$

where

$$Fo_i = \frac{D_i L}{u_0 \delta_i^2},\ K_i = \frac{k_i L}{u_0}\frac{\tilde{c}_0^{\mathrm{m}-2+i} c_{02}^{\mathrm{n}+1-i}}{\chi^{\mathrm{m}-2+i}},\ \varepsilon_{01} = \frac{D_1 \tilde{\delta}}{\tilde{D}\delta_1},\quad i = 1, 2 \tag{2.30}$$

and $\tilde{\delta}$ may be obtained from (2.7).

The dimensionless rate constants of the chemical reaction are always interrelated, namely,

$$\frac{K_1}{K_2} = \frac{c_{02}/a_2}{\tilde{c}_0/\chi a_1}, \tag{2.31}$$

where the ratios $\frac{c_{02}}{a_2}$ and $\frac{\tilde{c}_0}{\chi a_1}$ may be considered as the maximum values of the concentrations of substances A_1 and A_2 in the liquid. It is clear that $\frac{K_1}{K_2}$ may vary in an unlimited range.

Equations (2.29) show that in the case of slow chemical reactions, we have $K_1 < 10^{-2}$, which allows the term $K_1 C_1^{\mathrm{m}} C_2^{\mathrm{n}}$ on left-hand side of the equation to be neglected. In this way it is possible to substitute $K_1 = 0$ in (2.29), but the result is that the equation for C_2 becomes redundant because the interphase mass transfer of substance A_1 between the gas and the liquid no longer depends on the mass transfer of substance A_2 in the liquid phase. Hence, it follows that at

$$\frac{k_1 L}{u_0}\frac{\tilde{c}_0^{\mathrm{m}-1} c_{02}^{\mathrm{n}}}{\chi^{\mathrm{m}-1}} < 10^{-2} \tag{2.32}$$

the change in the chemical reaction rate, the change in the initial concentrations of the reagents, as well as the change in the physical solubility of the absorbing gas have no effects on the mass transfer process in the liquid phase.

From (2.32) it follows that it is possible to substitute $K_1 = 0$ in (2.29), which yields

$$Fo_1 \sim 1. \tag{2.33}$$

This allows us to establish the order of magnitude of the diffusion boundary layer thickness:

$$\delta_1 = \sqrt{\frac{D_1 L}{u_0}}. \tag{2.34}$$

It is evident that the chemical reaction rate and the initial concentrations of the reagents as well the physical solubility of the absorbing substance affect the mass transfer rate in the liquid phase if $K_1 > 10^{-2}$. Within the range,

$$10^{-2} < K_1 < 1, \tag{2.35}$$

the effect of the chemical reaction in the liquid is always less than the effect of the convective transport. Taking into account that the convective and the diffusive transport have to be of equal orders of magnitude, the value of δ_1 should be evaluated from (2.34) when

$$K_1 < 1. \tag{2.36}$$

It is clear from (2.31) that within the range defined by (2.35) the order of K_2 may vary in an unlimited range. If $K_2 < 10^{-2}$, then $K_2 = 0$ is assumed in (2.29). Thus, for C_2 the result is

$$C_2 \equiv 1. \tag{2.37}$$

If (2.29) is substituted in (2.37), the result is a mathematical description of an absorption process with a chemical reaction of pseudo-order m. For $m = 1$ the resulting equation is of pseudo-first order [8].

When $10^{-2} < K < 1$, it is possible to obtain in a similar way the following relations:

$$Fo_2 \sim 1, \quad \delta_2 = \sqrt{\frac{D_2 L}{u_0}}. \tag{2.38}$$

The effects of the parameters K_1 and K_2 on the interphase mass transfer mechanism are considerable when $K_1 > 1$ and $K_2 > 1$. For the purposes of the theoretical analysis of the absorption process under these conditions, the equation for C_i $(i = 1, 2)$ should be represented in the form:

$$\frac{1}{K_i}\left(U\frac{\partial C_i}{\partial X} + V_i\frac{\partial C_i}{\partial Y_i}\right) = \frac{Fo_i}{K_i}\frac{\partial^2 C_i}{?_i^2} - C_1^m C_2^n, \quad i = 1, 2. \tag{2.39}$$

It follows from (2.39) that in the range defined by (2.35) the increase in $K_2 > 1$ leads to a decrease in the convective transport of substance A_2, while at the same time the effects of both the diffusive transport and the chemical reaction must be of equal orders of magnitude:

$$\frac{Fo_2}{K_2} \sim 1. \tag{2.40}$$

Thus, when $K_2 > 1$, the thickness of the diffusion boundary layer for substance A_2 has an order of magnitude defined by

$$\delta_2 = \sqrt{\frac{D_2}{K_2} \frac{\chi^{m/2}}{\tilde{c}_0^{m/2} c_{02}^{(n-1)/2}}}. \tag{2.41}$$

When $K_2 > 10^2$, the hydrodynamics does not affect the interphase mass transfer of substance A_2 and one may substitute $K_2^{-1} = 0$ in (2.29).

The increase in the parameter $K_1 > 1$ leads to significant changes in the mass transfer mechanism for substance A_1 in the liquid as well as in the overall gas–liquid interphase mass transfer mechanism. It is evident from (2.39) that this is accompanied by a decreasing effect of the convective transport. In this case it is possible to establish the thickness of the diffusion boundary layer with respect to substance A_1 in the liquid phase:

$$\delta_1 = \sqrt{\frac{D_1}{K_1} \frac{\chi^{(m-1)/2}}{c_0^{(m-1)/2} c_{02}^{n/2}}}, \tag{2.42}$$

i.e.,

$$Fo_1 \sim K_1. \tag{2.43}$$

There are no hydrodynamic effects on the mass transfer when $K_1 > 10^2$. This allows us to substitute $K_1^{-1} = 0$ in (2.29). Under all conditions when $K_1 > 1$, the parameter K_2 may take values within an unrestricted range. Despite this, all the conclusions about the effect of K_2 on the transport of substance A_2 in the liquid already have drawn for $K_1 < 1$ are valid in the case when $K_1 > 1$ too.

The distribution of the diffusion resistances in both phases is determined like in the case of first-order chemical reactions. It is clear from (2.29) that the transport processes in the liquid (in the gas) limit the mass transfer if $\frac{\chi}{\varepsilon_{01}} > 10^2$ $\left(\frac{\chi}{\varepsilon_{01}} < 10^{-2}\right)$. The diffusion resistances are commensurable if $10^{-2} < \frac{\chi}{\varepsilon_{01}} < 10^2$.

If $K_1 > 1$, the parameter $\frac{\chi}{\varepsilon_{01}}$ depends on the chemical reaction rate and the initial concentrations of the reagents. Thus, it is possible to derive conditions under which the process is only limited by the mass transport in the gas phase,

$$\chi\sqrt{\frac{\tilde{D}\tilde{u}_0}{D_1 u_0 K_1}} = \frac{\chi^{(m+1)/2}}{c_0^{(m-1)/2} c_{02}^{n/2}} \sqrt{\frac{\tilde{D}\tilde{u}_0}{D_1 K_1 L}} < 10^{-2}, \tag{2.44}$$

and the mass transfer rate is independent of the reaction rate.

Table 2 The values of the parameters in model (2.29)

$K_1 < 10^{-2}$	$K_1 = 0$	$K_2 < 10^{-2}$	$K_2 = 0 \quad C_2 \equiv 0$
	$Fo_1 = 0$	$K_2 < 1$	$C_2 \equiv 0$
		$K_2 > 1$	$C_2 \equiv 0$
		$K_2 > 10^2$	$K_2^{-1} = 0 \quad C_2 \equiv 0$
$K_1 < 1$	$Fo_1 = 1$	$K_2 < 10^{-2}$	$K_2 = 0 \quad C_2 \equiv 1$
		$K_2 < 1$	$Fo_2 = 1$
		$K_2 > 1$	$Fo_2/K_2 = 1$
		$K_2 > 10^2$	$K_2^{-1} = 0 \quad Fo_2/K_2 = 1$
$K_1 > 1$	$\frac{Fo_1}{K_1} = 1$	$K_2 < 10^{-2}$	$K_2 = 0 \quad C_2 \equiv 1$
		$K_2 < 1$	$Fo_2 = 1$
		$K_2 > 1$	$Fo_2/K_2 = 1$
		$K_2 > 10^2$	$K_2^{-1} = 0 \quad Fo_2/K_2 = 1$
$K_1 > 10^2$	$K_1^{-1} = 0$	$K_2 < 10^{-2}$	$K_2 = 0 \quad C_2 \equiv 1$
	$\frac{Fo_1}{K_1} = 1$	$K_2 < 1$	$Fo_2 = 1$
		$K_2 > 1$	$Fo_2/K_2 = 1$
		$K_2 > 10^2$	$K_2^{-1} = 0 \quad Fo_2/K_2 = 1$
$10^{-2} < K_1 < 10^2$	$\frac{\chi}{\varepsilon_{01}} < 10^{-2}$	$\frac{\chi}{\varepsilon_{01}} = 0$	$C_1 \equiv 1$
	$\frac{\chi}{\varepsilon_{01}} > 10^2$	$\frac{\varepsilon_{01}}{\chi} = 0$	$\tilde{C}_1 \equiv 1$

The effect of a chemical reaction rate of arbitrary order on the interphase mass transfer mechanism discussed here depends on the physical solubility of the gas. The relationship may be obtained (see Fig. 1) by replacing ε_0 and K with ε_{01} and K_1, i.e., the zones in the figure remain the same.

The existence of mass transport of substance A_2 in the liquid phase is the reason for various possible interphase mass transfer mechanisms and the respective mathematical models in comparison with the pseudo-first-order chemical reaction. All of these mathematical descriptions of the process may be obtained from (2.29) if suitable parameters and functions are used (see Table 2). In Table 2, it is noted by convention that $C_2 \equiv 0$ because when $K_1 < 10^{-2}$ the equation for C_2 in (2.29) is unnecessary.

The effect of the chemical reactions on the interphase mass transfer mechanism becomes more complicated with increase of the number of reactions considered.

2.2 Homogenous Catalytic Reactions

Homogenous catalytic processes in gas–liquid systems are widely used for the synthesis of a new substance [3], as well as for enhancement of the separation of gas mixtures by absorption [4]. In these cases there are at least two chemical reactions in the liquid phase. Consider a simple example of a homogenous catalytic process:

$$A_1 + A_3 = A_4, \quad A_2 + A_4 = A_5 + A_3, \tag{2.45}$$

where A_3 is the catalyst for the interaction between A_1 and A_2 in the liquid phase. The reaction rates may be expressed as

$$\begin{aligned} r_1 &= -\frac{\partial c_1}{\partial t} = k_1 c_1 c_3, \quad r_2 = -\frac{\partial c_2}{\partial t} = k_2 c_2 c_4, \\ r_3 &= -\frac{\partial c_3}{\partial t} = r_1 - r_2, \quad r_4 = -\frac{\partial c_4}{\partial t} = -r_1 + r_2, \end{aligned} \tag{2.46}$$

where c_i are the concentrations of substances A_i $(i = 1, \ldots, 4)$.

The equations describing the process are

$$\begin{aligned} &\tilde{u}\frac{\partial \tilde{c}_i}{\partial x} + \tilde{v}\frac{\partial \tilde{c}_i}{\partial y} = \tilde{D}_i \frac{\partial^2 c_i}{\partial y^2}; \quad u\frac{\partial c_i}{\partial x} + v\frac{\partial c_i}{\partial y} = D_i \frac{\partial^2 c_i}{\partial y^2} - r_i; \quad u\frac{\partial c_j}{\partial x} + v\frac{\partial c_j}{\partial y} = D_j \frac{\partial^2 c_j}{\partial y^2} - r_j; \\ &x = 0,\ \tilde{c}_i = \tilde{c}_{0i},\ c_i = 0,\ c_j = c_{0j},\ (c_{04} = 0); \\ &y = 0,\ \tilde{c}_i = \chi_i c_i,\ \tilde{D}_i \frac{\partial \tilde{c}_i}{\partial y} = D_i \frac{\partial c_i}{\partial y},\ \frac{\partial c_j}{\partial y} = 0; \\ &y \to \infty,\ \tilde{c}_{\mathrm{i}} = \tilde{c}_{0i}; \quad y = -\infty,\ c_i = 0,\ c_j = c_{0j},\ (c_{04} = 0); \\ &i = 1, 2; \quad j = 3, 4, \end{aligned} \tag{2.47}$$

where the symbols are the same as those used in (2.1) and (2.2).

The analysis of (2.47) requires dimensionless variables such as those in (2.3):

$$\begin{aligned} &X = \frac{x}{L}, \tilde{Y}_i = \frac{y}{\tilde{\delta}_i}, Y_i = \frac{y}{\delta_i}, \tilde{U} = \frac{\tilde{u}}{\tilde{u}_0}, V_i = \frac{\tilde{v}}{\tilde{\varepsilon}_i \tilde{u}_0}, \tilde{C}_i \frac{\tilde{c}_i}{\tilde{c}_{0i}}, \\ &U = \frac{u}{u_0}, V_i = \frac{v}{\varepsilon_i u_0}, V_j = \frac{v}{\varepsilon_j u_0}, C_i = \frac{c_i}{c_{0i}}, C_j = \frac{c_j}{c_{03}}, c_{0\mathrm{i}} = \frac{\tilde{c}_{0i}}{\chi_i}, \\ &\tilde{\varepsilon}_i = \frac{\tilde{\delta}_i}{L}, \varepsilon_i = \frac{\delta_i}{L}, \varepsilon_j = \frac{\delta_j}{L}, \quad i = 1, 2, \quad j = 3, 4. \end{aligned} \tag{2.48}$$

In this way six coupled boundary problems are obtained:

$$\begin{aligned} &\tilde{U}\frac{\partial \tilde{C}_i}{\partial X} + V_i \frac{\partial \tilde{C}}{\partial \tilde{Y}_i} = \tilde{F}o_i \frac{\partial^2 C_i}{\partial \tilde{Y}_i^2}; \\ &X = 0,\ \tilde{C}_i = 1; \quad \tilde{Y}_i = 0; \quad \tilde{C}_i = C_i; \quad \tilde{Y}_i \to \infty,\ \tilde{C}_i = 1; \quad i = 1, 2. \end{aligned} \tag{2.49}$$

$$U\frac{\partial C_i}{\partial X}+V_i\frac{\partial C_i}{\partial Y_i}=Fo_i\frac{\partial^2 C_i}{\partial Y_i^2}-R_i;$$

$$X=0,\ C_i=0,\quad Y_i=0,\frac{dC_i}{dY_i}=\frac{\chi_i}{\varepsilon_{0\mathrm{i}}}\left(\frac{\partial \tilde{C}_i}{\partial \tilde{Y}_i}\right)_{\tilde{Y}_i=0};\quad Y_i\to-\infty,\ C_i=0;\quad i=1,2. \tag{2.50}$$

$$U\frac{\partial C_j}{\partial X}+V_j\frac{\partial C_j}{\partial Y_j}=Fo_j\frac{\partial^2 C_j}{\partial Y_j^2}-R_j;\quad X=0,\ C_3=1,\ C_4=0;\quad Y_j=0,\frac{\partial C_j}{\partial Y_j}=0;$$
$$Y_j\to-\infty,\quad C_3=1,\ C_4=0,\quad j=3,4. \tag{2.51}$$

$$\tilde{F}o_i=\frac{\tilde{D}_iL}{\tilde{u}\delta_i^2},\ Fo_i=\frac{D_iL}{u\delta_i^2},\ Fo_j=\frac{D_jL}{u\delta_j^2},\ K_i=\frac{k_iL}{u_0},\ \varepsilon_{0i}=\frac{D_i\tilde{\delta}_i}{\tilde{D}_i\delta_i},\quad i=1,2,\quad j=3,4,$$
$$R_1=K_1c_{03}C_1C_3,\ R_2=K_2c_{03}C_2C_4,\ R_3=K_1\frac{c_{01}}{\chi_1}C_1C_3-K_2\frac{\tilde{c}_{02}}{\chi_2}C_2C_4,$$
$$R_4=-K_1\frac{\tilde{c}_{01}}{\chi_1}C_1C_3+K_2\frac{\tilde{c}_{02}}{\chi_2}C_2C_4. \tag{2.52}$$

The boundary problems (2.49)–(2.51) allow us to determine the effect of the chemical reaction rate (and the physical solubility of substances A_1 and A_2 in the liquid too) on the interphase mass transfer mechanism by the method described above.

In situations where the following conditions are satisfied,

$$K_1c_{03}>10^2,\quad K_2c_{03}>10^2,\quad K_1\frac{\tilde{c}_{01}}{\chi_1}\sim K_2\frac{\tilde{c}_{02}}{\chi_2}>10^2, \tag{2.53}$$

Eqs. 2.50 and 2.51 take the form

$$\frac{\partial^2 C_i}{\partial Y_i^2}=C_iC_{i+2};\quad Y_i=0,\ \frac{\partial C_i}{\partial Y_i}=\frac{\chi_i}{\varepsilon_{0i}}\left(\frac{\partial C_i}{\partial \tilde{Y}_i}\right)_{\tilde{Y}_i=0};\quad Y_i\to-\infty,\ C_i=0;\quad i=1,2. \tag{2.54}$$

$$\frac{\partial^2 C_3}{\partial Y_3^2}=C_1C_3-aC_2C_4;\quad \frac{\partial^2 C_4}{\partial Y_4^2}=-C_1C_3+aC_2C_4;$$
$$Y_j=0,\ \frac{\partial C_j}{\partial Y_j}=0;\quad Y_j\to-\infty,\ C_3=1,\ C_4=0, \tag{2.55}$$

where

$$a=\frac{K_2\tilde{c}_{02}\chi_1}{K_1\tilde{c}_{01}\chi_2},\quad \delta_i\sqrt{\frac{D_iL}{u_0K_ic_{03}}},\quad i=1,2;\quad \delta_j=\sqrt{\frac{D_jL\chi_1}{u_0K_1\tilde{c}_{01}}},\quad j=3,4. \tag{2.56}$$

From (2.55) it follows that if $D_3 \approx D_4$, one may write $C_4 \approx 1 - C_3$.

When reactions (2.45) are fast, $\delta_i = 0$, $\varepsilon_{0i} \to \infty$ ($i = 1, 2$), the homogenous reaction becomes a heterogeneous one. In these cases reactions (2.45) may occur at the liquid surface and the mass transfer in the gas phase limits the entire process.

2.3 Reversible Chemical Reactions

Reversible gas absorption is applied in cases when the regeneration of absorbents is needed. In practical situations, the absorption is reversible owing to the reversibility of the chemical reaction. The rates of both the forward and the backward reactions affect the interphase mass transfer mechanism as in the case of irreversible processes. These effects become more complicated owing to simultaneous reversibilities of both the physical and the chemical parts of the absorption process.

An example of reversible absorption was described in [9] with a simple reversible chemical reaction:

$$a_1A_1 + a_2A_2 \Leftrightarrow a_3A_3. \tag{2.57}$$

Here, the assumption that absorbed substance A_1 from the gas reacts with substance A_2 in the liquid is used. Far from the equilibrium, the rate of interaction between substances A_1 and A_2 is

$$r_i = -\frac{\partial c_i}{\partial t} = k_i c_1^{\mathrm{m}} c_2^{\mathrm{n}} - k_{3i} c_3^{\mathrm{p}}, \tag{2.58}$$

$$\frac{k_1}{a_1} = \frac{k_2}{a_2} = -\frac{k_3}{a_3} = k_0 > 0. \tag{2.59}$$

Here c_i are the concentrations of substances A_i ($i = 1, 2, 3$).

The introduction of the chemical equilibrium constant

$$R = \frac{k_{3i}}{k_i}, \quad i = 1, 2, 3 \tag{2.60}$$

from (2.58) leads to

$$r_i = k_i\left(c_1^{\mathrm{m}} c_2^{\mathrm{n}} - R c_3^{\mathrm{p}}\right), \quad i = 1, 2, 3, \tag{2.61}$$

i.e., the equilibrium constants and the forward reaction rate constant may be used to express the overall process rate.

The mathematical description of absorption with a reversible chemical reaction in the liquid phase is

$$\tilde{u}\frac{\partial\tilde{c}}{\partial x}+\tilde{v}\frac{\partial\tilde{c}}{\partial y}=\tilde{D}\frac{\partial^2\tilde{c}}{\partial y^2};\quad u\frac{\partial c_i}{\partial x}+v\frac{\partial c_i}{\partial y}=D_i\frac{\partial^2 c_i}{\partial y^2}-k_i\left(c_1^m c_2^n-Rc_3^p\right),\quad i=1,2,3;$$
$$x=0,\ \tilde{c}=\tilde{c}_0,\ c_1=0,\ c_2=c_{02},\ c_3=0;\quad y=0,\ \tilde{c}=\chi c_1,\ \tilde{D}\frac{\partial\tilde{c}}{\partial y}=D_1\frac{\partial c_1}{},$$
$$\frac{\partial c_2}{\partial y}=0,\ \frac{\partial c_3}{\partial y}=0;\quad y\to\infty,\ \tilde{c}=\tilde{c}_0;\quad y\to-\infty,\ c_1=0,\ c_2=c_{02},\ c_3=0.$$
(2.62)

The analysis of the effect of the reversible reaction on the interphase mass transfer mechanism needs the following dimensionless variables:

$$X=\frac{x}{L},\ \tilde{Y}=\frac{y}{\tilde{\delta}},\ \tilde{C}=\frac{\tilde{c}}{\tilde{c}_0},\ Y_i=\frac{y}{\delta_i},\ C_i=\frac{c_i}{c_{0i}},\ \left(c_{01}=\frac{\tilde{c}_0}{\chi}\right),$$
$$\tilde{U}=\frac{\tilde{u}}{\tilde{u}_0},\ \tilde{V}=\frac{\tilde{v}}{\tilde{\varepsilon}\tilde{u}_0},\ U=\frac{u}{u_0},\ V_i=\frac{v}{\varepsilon_i u_0},\ \tilde{\varepsilon}=\frac{\tilde{\delta}}{L},\ \varepsilon_i=\frac{\delta_i}{L},\ r_0=\frac{k_0\tilde{c}_0^m c_{02}^n}{\chi^m},$$
$$\tilde{Fo}=\frac{\tilde{D}L}{\tilde{u}_0\tilde{\delta}},\ Fo_i=\frac{D_i L}{u_0\delta_i^2},\ \alpha=\frac{r_0 L}{u_0},\ \alpha_1=\frac{a_1\alpha}{c_{01}},\ \alpha_2=\frac{a_2\alpha}{c_{02}},\ \alpha_3=-\frac{a_3\alpha}{c_{03}},\qquad(2.63)$$
$$c_{03}=\frac{c_{01}^{m/p}c_{02}^{n/p}}{\chi^{m/p}},\quad \varepsilon_{01}=\frac{D_1\tilde{\delta}}{\tilde{D}\delta_1},\quad i=1,2,3.$$

In this manner, Eq. 2.62 take the form

$$\tilde{U}\frac{\partial\tilde{C}}{dX}+\tilde{V}\frac{\partial\tilde{C}}{d\tilde{Y}}=\tilde{Fo}\frac{\partial^2\tilde{C}}{d\tilde{Y}^2};\quad U\frac{\partial C_i}{dX}+V_i\frac{\partial C_i}{dY_i}=Fo_i\frac{\partial^2 C_i}{dY_i^2}-\alpha_i\left(C_1^m C_2^n-RC_3^p\right);$$
$$X=0,\ \tilde{C}=1,\ C_1=0,\ C_2=1,\ C_3=0;$$
$$\tilde{Y}=Y_i=0,\ \tilde{C}=C_1,\ \frac{\partial C_1}{1}=\frac{\chi}{\varepsilon_{01}}\frac{\partial\tilde{C}}{\partial\tilde{Y}},\ \frac{\partial C_2}{\partial Y_2}=\frac{\partial C_3}{\partial Y_3}=0;$$
$$\tilde{Y}\to\infty,\ \tilde{C}=1;_i\to-\infty,\ C_1=0,\ C_2=1,\ C_3=0.$$
(2.64)

From (2.64) it follows that the process depends on the parameters K, α_i, Fo_i, and $\frac{\chi}{\varepsilon_{01}}$. Their values determine the interphase mass transfer mechanism.

If $R>10^2$, the chemical reactions are practically irreversible. If we substitute $R=0$ in (2.64), the problem of determination of C_3 results. Thus, for $R>10^2$ the equations for C_i ($i=1, 2, 3$) become

$$U\frac{\partial C_i}{\partial X}+V_i\frac{i}{\partial Y_i}=Fo_i-R\alpha_i\left(\frac{1}{R}C_1^m C_2^n-C_3^p\right),\quad i=1,2,3,\qquad(2.65)$$

Here it is possible to substitute $R^{-1}=0$. From the boundary problem for C_3 it follows that $C_3\equiv 0$ and consequently $C_2\equiv 1$. For C_1 the resulting equations describe the case of physical absorption of substance A_1.

Reversible absorption occurs if $10^{-2} < R < 10^{2}$ and the mass transfer mechanism depends strongly on the rate of the forward reaction, i.e., on the parameters α_i $(i = 1, 2, 3)$.

If $\alpha_1 < 10^{-2}$, it is possible to substitute $\alpha_1 = 0$ in (2.64), which corresponds to the case of physical absorption (at various values of α_2 and α_3). If $\alpha_2 < 10^{-2}$, it follows from (2.64) that $\alpha_2 = 0$ and $C_2 \equiv 1$, i.e., this is the case of a forward reaction of pseudo mth order. When $\alpha_3 < 10^{-2}$, it follows that $C_3 \equiv 0$ and the case corresponds to the situation of an irreversible reaction.

The correlation between the effects of both the diffusive and the convective transport as well as the diffusion boundary layer thickness depends on the values of the parameters α_i and R.

At $\alpha_i > 10^{-2}$ and $10^{-2} < \mathrm{R} < 1$ $(i = 1,2,3)$ it is possible to substitute $\alpha_i^{-1} = 0 (i = 1, 2, 3)$ in (2.64). This means that the convective transport may be neglected and the thickness of the diffusion boundary layer may be evaluated from the condition

$$Fo_i \sim \alpha_i, \quad i = 1, 2, 3, \tag{2.66}$$

i.e.,

$$\delta_i = \sqrt{\frac{D_i L}{u_0 \alpha_i}}, \quad i = 1, 2, 3. \tag{2.67}$$

At $\alpha_i R > 10^2$ and $1 < R < 10^2$ one may substitute $(\alpha_i R)^{-1} = 0$ $(i = 1, 2, 3)$ in (2.64) and neglect the left-hand sides of the equations for C_i $(i = 1, 2, 3)$. In this case the result for the boundary layer thickness is

$$Fo_i \sim \alpha_i R, \quad i = 1, 2, 3, \tag{2.68}$$

i.e.,

$$\delta_i = \sqrt{\frac{D_i L}{u_0 \alpha_i R}}, \quad i = 1, 2, 3. \tag{2.69}$$

In the case of slow reactions $\alpha_i < 1$, $\alpha_i R < 1$, $i = 1, 2, 3$, the thickness of the diffusion boundary layer may be obtained from a relationship similar to (2.10), namely,

$$\delta_i = \sqrt{\frac{D_i L}{u_0}}, \quad i = 1, 2, 3. \tag{2.70}$$

The distribution of the diffusion resistances is determined by the conditions (2.18–2.20), where ε must be replaced by ε_{01} from (2.63). The determination of $\tilde{\delta}$and δ_1 must be performed from (2.7), (2.67), (2.69), and (2.70).

The analysis of the boundary problem (2.64) does not consider processes in which

$$C_1^m C_2^n - R C_3^p \approx 0, \quad \alpha_1 \left(C_1^m C_2^n - R C_3^p \right) \neq 0. \tag{2.71}$$

This corresponds to a situation when in the bulk of the liquid there is a chemical equilibrium and the rates of both the forward and the backward reactions are significant, but the difference between them significantly affects the mass transfer. In this situation the solution must take into account both the physical and the chemical equilibrium.

2.4 Relationships Between the Chemical Equilibrium and the Physical Equilibrium During Absorption

The absorption process in systems such as NH_3–H_2O and SO_2–H_2O is characterized by the fast establishment of the chemical equilibrium in the liquid phase. The chemical reaction significantly increases the gas–liquid mass transfer rate. The mathematical description uses a model of physical absorption, where Henry's constant may be assumed as a conditional value relating the physical solubility of the gas and the chemical equilibrium in the liquid [41]. Under these assumptions, the liquid absorbs substance A_1 from the gas. After that, substance A_1 reacts with the liquid component A_2 and the reaction is followed by dissociation of the reaction product. Generally, the reversible reaction may be expressed [9] as

$$A_1 + a_2 A_2 \Leftrightarrow A_3 + a_4 A_4, \qquad (2.72)$$

where A_1 represents the reagent molecule in the gas (NH_3, SO_2, HCl, etc.), A_2 is the molecule of the reagent in the liquid (H_2O), A_3 is the ionic form of molecule A_1, and A_4 is a hydroxyl or hydrogen ion.

The kinetic equations are

$$r_i = -\frac{\partial c_i}{\partial t} = k_i \left(c_1^{\mathrm{m}} c_2^{\mathrm{n}} - R c_3^{\mathrm{p}} c_4^{\mathrm{q}} \right), \quad i = 1, \ldots, 4, \qquad (2.73)$$

where

$$k_1 = \frac{k_2}{a_2} = -k_3 = -\frac{k_4}{a_4} = k_0 > 0 \qquad (2.74)$$

and R is the chemical equilibrium constant.

Substituting r_i $(i = 1, \ldots, 4)$ from (2.73) in (2.62) yields the mathematical description of an absorption process with a reversible chemical reaction (2.72). It should take into account that the boundary conditions for c_3 and c_4 are equal.

In the situations where the rate of establishment of the chemical equilibrium is high and the corresponding rates of both reactions (forward and backward) are also significant, Eq. 2.73 indicate that

$$c_1^{\mathrm{m}} c_2^{\mathrm{n}} - R c_3^{\mathrm{p}} c_4^{\mathrm{q}} \approx 0, \quad k_1 \left(c_1^{\mathrm{m}} c_2^{\mathrm{n}} - R c_3^{\mathrm{p}} c_4^{\mathrm{q}} \right) \neq 0, \quad i = 1, \ldots, 4. \qquad (2.75)$$

The conditions of the physical and chemical equilibrium are

$$c_1 = \frac{\tilde{c}}{\chi}, \quad c_1^{m} c_2^{n} - R c_3^{p} c_4^{q} = 0 \tag{2.76}$$

and as a result

$$c_1 + c_3 = \frac{\tilde{c}}{\chi}\left(1 + \frac{\tilde{c}^{m/p-1} c_2^{n/p}}{\chi^{m/p-1} R^{1/p} c_4^{q/p}}\right). \tag{2.77}$$

The concentration of substance A_1 in the liquid (c) may be expressed by the concentrations of its molecular and ionic forms,

$$c = c_1 + c_3, \tag{2.78}$$

whereas their diffusivities may be assumed to be practically equal:

$$D_1 = D_3 = D. \tag{2.79}$$

From (2.57) to (2.62) it can be seen that in the case of the reversible chemical reaction (2.72)

$$a_1 = a_3 = 1, \quad k_1 = -k_3. \tag{2.80}$$

This allows a summation in (2.62) for the equations for c_1 and c_3:

$$u\frac{\partial c}{\partial x} + v\frac{\partial c}{\partial y} = D\frac{\partial^2 c}{\partial y^2} \tag{2.81}$$

with boundary conditions

$$\begin{aligned} &x = 0,\ c = 0; \quad y \to \infty,\ c = 0; \\ &y = 0,\ c = \frac{\tilde{c}}{\chi}\left(1 + \frac{\tilde{c}^{m/p-1} c_2^{n/p}}{\chi^{m/p-1} R^{1/p} c_4^{q/p}}\right), \ D\frac{\partial c}{\partial y} = \tilde{D}\frac{\partial \tilde{c}}{\partial y}. \end{aligned} \tag{2.82}$$

The boundary conditions follow from the conditions imposed by the physical and chemical equilibriums:

$$c_1 = \tilde{c}/\chi, {}_1^{m} c_2^{n} - R c_3^{p} c_4^{q} = 0. \tag{2.83}$$

The boundary conditions (2.82) permit us to obtain [41] an apparent Henry's constant:

$$H = \chi\left(1 + \frac{\tilde{c}^{m/p-1} c_2^{n/p}}{\chi^{m/p-1} R^{1/p} c_4^{q/p}}\right)^{-1}, \tag{2.84}$$

which takes into account both the physical equilibrium and the chemical equilibrium.

In most of the cases of practical interest substance A_2 is water, so c_2 has a large value that does not change during the process. This means that the reaction is of pseudo-first order:

$$c_2 \equiv c_{02} = \text{const.}, \quad \mathrm{m} = \mathrm{n} = \mathrm{p} = \mathrm{q} = 1. \tag{2.85}$$

Moreover, it follows from (2.72) to (2.76) that

$$c_1 = c - c_3, \quad c_3 = c_4/a_4. \tag{2.86}$$

Hence, the mathematical description of the process may be expressed in the following form:

$$\begin{aligned}
&\tilde{u}\frac{\partial \tilde{c}}{\partial x} + \tilde{v}\frac{\partial \tilde{c}}{\partial y} = D\frac{\partial^2 \tilde{c}}{\partial y^2}, \quad u\frac{\partial c}{\partial x} + v\frac{\partial c}{\partial y} = D\frac{\partial^2 c}{\partial y^2},\\
&u\frac{\partial c_4}{\partial x} + v\frac{\partial c_4}{\partial y} = D_4\frac{\partial^2 c_4}{\partial y^2} - k_4\left(c_{02}c - \frac{1}{a_4}c_{02}c_4 - R\frac{1}{a_4}c_4^2\right);\\
&x = 0,\ \tilde{c} = \tilde{c}_0,\ c = 0,\ c_4 = 0; \quad y = 0; \quad \tilde{c} = HC,\ \tilde{D}\frac{\partial \tilde{c}}{\partial y} = D\frac{\partial c}{\partial y},\ \frac{\partial c_4}{\partial y} = 0;\\
&y \to \infty,\ \tilde{c} = \tilde{c}_0; \quad y \to -\infty,\ c = 0,\ c_4 = 0,
\end{aligned} \tag{2.87}$$

where

$$H = \chi\left(1 + \frac{c_{02}}{Rc_4}\right)^{-1}. \tag{2.88}$$

In the cases when there is no chemical reaction in the liquid phase ($R \to \infty$), it follows from (2.84) that $H = \chi$. When the chemical reaction is irreversible ($R = 0$) and the rate constant is high, relationship (2.84) gives $H = 0$, i.e., the mass transfer is limited by the mass transfer in the gas phase.

In the situations when $c_4 = c_{04} = \text{const.}$ (there is absorption of SO_2 or NH_3 by means of buffer solutions and c_4 is the concentration of the hydrogen or hydroxyl ions), the equation for c_4 in (2.87) is not needed. A model of physical absorption may describe the process with a Henry's constant given by (2.88).

The evaluations of the boundary layer thickness as well as the distribution of the diffusion resistances in both phases are similar to those discussed above.

3 Comparative Qualitative Analysis for Process Mechanism Identification

The existence of experimental data on the process kinetics permits is to decipher the process mechanism. The solution of this problem is made difficult by the possibility of the existence of different physical effects and their combination in the process mechanism (e.g., different nonlinear effects). The comparative qualitative analysis permits us to reject the negligible effects. For example, in many cases there is a big difference between the experimental data and the predictions of the linear mass transfer theory as a result of nonlinear effects. These effects are caused by secondary flows, where the velocity is a function of the concentration. This influence of the mass transfer on the hydrodynamics is result of a big concentration gradient, a surface tension gradient (Marangoni effect), a pressure gradient (Stefan flow), or a vertical density gradient (natural convection).

3.1 Comparison of the Nonlinear Effects

The theoretical analysis of nonlinear mass transfer was developed in [10]. The main idea follows from the nonlinearity of the convection–diffusion equation:

$$\rho(c)\boldsymbol{W}(c)\boldsymbol{grad}\, c = \mathrm{div}[\rho(c)D(c)\boldsymbol{grad}\, c] + kc^{\mathrm{n}}. \tag{3.1}$$

The velocity $\boldsymbol{W}$ is governed by the hydrodynamic equations. However, the principal nonlinear phenomenon is due to the concentration effects on the velocity $W(c)$, density $\rho(c)$, viscosity $\mu(c)$, diffusivity $D(c)$, and the chemical reaction rate kc^{n} (for $\mathrm{n} \neq 1$).

It was shown [10, 11] that there are a number of cases with nonlinear mass transfer behavior. The well-known linear mass transfer theory can be successfully applied in these cases. However, in the case of two-phase interphase mass transfer with a flat interface, the above equation permits a nonlinear mass transfer model to be derived by means of the boundary layer approximation:

$$\begin{gathered}
\rho_i\left(u_j\frac{\partial u_j}{\partial x}+v_j\frac{\partial u_j}{\partial y}\right)=\mu_j\frac{\partial^2 u_j}{\partial y^2}+A_i,\ \frac{\partial u_j}{\partial x}+\frac{\partial u_j}{\partial y}=0,\ u_j\frac{\partial c_j}{\partial x}+v_j\frac{\partial c_j}{\partial y}=D\frac{\partial^2 c_j}{\partial y^2}+B_j,\ j=1,2,\\
x=0,\ u_j=u_{j0},\ c_j=c_{j0};\ y=0,\ u_1=u_2,\ \mu_1\frac{\partial u_1}{\partial y}=\mu_2\frac{\partial u_2}{\partial y},\ v_j=0,\\
c_1=\chi c_2,\ D_1\frac{\partial c_1}{\partial y}=D_2\frac{\partial c_2}{\partial y};\ \ y=(-1)^{j+1}\infty,\ u_j=u_{j0},\ c_j=c_{j0};\ \ j=1,2,
\end{gathered}\tag{3.2}$$

where the index *1* is used to denote the gas or the liquid phase, whereas the index *2* designates the liquid or the solid phase. The terms A_j and B_j ($j = 1, 2$) are the contributions of some additional physical effects.

There are a number of processes where u_j, v_j, μ_j, ρ_j, D_j, A_j, and B_j are independent of the concentration c_j ($j = 1, 2$). These situations are the basis of the linear mass transfer theory.

In systems with high concentrations, and exhibiting large concentration gradients, the deviations from the linear Fick's diffusion law are significant. Under such conditions, the higher concentrations can affect the diffusivity, viscosity, and density:

$$D_j = D_j(c_j),\ v_j = \mu_j(c_j),\ \rho_j = \rho_j(c_j),\quad j = 1, 2. \tag{3.3}$$

The concentration effects introduce a nonlinearity in the convection–diffusion equations, discussed in detail in [12, 13].

The other nonlinear effect, due to the nonuniform concentration distributions

$$A_j = g(\rho_j - \rho_{0j}),\ \rho_j = \rho_j(c_j), \tag{3.4}$$

leads to a natural convection [14, 15].

The next reason why the mass transfer process may be intensified is the existence of a chemical reaction with rate B_j in the bulk of the phases:

$$B_j = B_j(c_j),\ j = 1, 2. \tag{3.5}$$

The studies reported in [10, 16] show that in gas–liquid systems with a chemical reaction $B_1 = 0$, whereas $B_2 = kc^{\mathrm{n}}$. Moreover, the chemical reaction rate can significantly affect the interface mass transfer mechanism between the phases.

The thermal effect of the chemical reactions can lead to temperature nonuniformity at the interface and to subsequent surface tension gradients. This calls for new boundary conditions taking into account the equality of the tangential components of the stress tensor at the interface:

$$y = 0, \quad \mu_1 \frac{\partial u_1}{\partial y} = \mu_2 \frac{\partial u_2}{\partial y} - \frac{\partial \sigma}{\partial x}. \tag{3.6}$$

The investigation of this effect (Marangoni effect) [17, 18] shows that it is negligible when there are no surfactants in the system.

One of the most interesting nonlinear effects arises from the conditions imposed by the high concentration gradients. The latter induce secondary flows at the interface. This effect is discussed in detail in [10] for a large number of systems as examples and it has been termed the "nonlinear mass transfer effect."

Under the conditions imposed by high concentration gradients, secondary flows are induced. They cause convective components of the mass transfer flux in addition to the main mass flux. In this case the mass transfer rate is

$$J = \frac{MD\rho^*}{L\rho_0} \int\limits_0^L \left(\frac{\partial c}{\partial y}\right)_{y=0} dx, \tag{3.7}$$

where the secondary flow affects both the diffusive mass transfer $D\left(\frac{\partial c}{\partial y}\right)_{y=0}$ and the convective mass transfer $\frac{M\rho^*}{\rho_0}$. In gas–liquid and liquid–liquid systems [19, 20] the nonlinearity is the effect of the induced secondary flow on the diffusive transfer. In liquid–solid systems the induced flow affects mainly the convective transfer. These effects are clearly demonstrated in the electrochemical systems [10, 16] owing to the high molecular mass of the metals.

All of the nonlinear effects influence the velocity fields, which leads to changes in the hydrodynamic stability of the system. The loss of stability could cause an increase of the amplitudes of the random disturbances until a new stable state or a stable periodic process is reached [10]. The latter is a self-organizing dissipative structure with a mass transfer rate growing sharply, which is not the case for conventional systems. The problem is discussed in detail in [14, 15] in the case of nonstationary absorption of pure gases in an immobile liquid layer with a flat interface.

Nonlinear effects in the mass transfer kinetics, induced by the secondary flows, lead to qualitative changes of the mass transfer rate, since they are related to physical mechanisms inducing secondary flows. In this sense, the most interesting secondary flows are the Stefan flow [21] (occurring as volume change at the interface), the flow induced by large concentration gradients [10, 22], and the flow resulting from surface tension gradients [17, 18]. These secondary flows require the introduction of new boundary conditions at the interface (between gas–solid, gas–liquid, liquid–liquid, and liquid–solid phases). A comparative analysis of the occurrence of these secondary flows will be presented further.

Stefan flow occurs in the cases of heterogeneous reactions at the interface between two phases as a result of the disappearance (or the generation) of substances at this interface. The "disappearance" (or "generation") of substances might be a consequence of surface reactions such as adsorption (or desorption processes), a liquid–vapor phase transition (boiling or condensation), or interphase mass transfer.

Some of the above-mentioned heterogeneous reactions lead to changes of the volume (and the pressure too) of the phase at the interface, which lead to the occurrence of a pressure gradient and as a result a hydrodynamic flow, called Stefan flow. These reactions at the interphase surface are chemical reactions when a difference between the number of molecules of the reagents and reaction products exists, and a liquid–vapor phase transition in the cases of boiling and condensation.

Let us consider a heterogeneous chemical reaction [22] expressed by the stoichiometric equation

$$\sum_{i=1}^{N} \upsilon_i A_i = 0, \tag{3.8}$$

where A_i and υ_i $(i = 1, \ldots, N)$ correspond to the substances participating in the reaction and their stoichiometric coefficients. For the initial substances (reagents) $\upsilon_i > 0$, whereas for the reaction products $\upsilon_i < 0$. The rate of the heterogeneous reaction j_i (mol/m^{2}s) is defined for the separate substances $(i = 1, \ldots, N)$, where N is their total number. For the reagents $j_i > 0$, and for the reaction products $j_i < 0$.

The reagents (reaction products) are supplied to (taken from) the surface reaction by a diffusion and a convection:

$$\boldsymbol{j} = -D\, \textbf{\textit{grad}}\, c + \boldsymbol{v} c, \tag{3.9}$$

where $\boldsymbol{j}$ is the vector of the mass transfer rate, D (m^2/s) is the diffusivity, ***grad*** is the vector of the gradient, c (mol/m^3) is the molar concentration, and $\boldsymbol{v}$ is the velocity vector. For the separate substances the molar flux has the following form:

$$j_i = -D_i \, \boldsymbol{grad} \, c_i + \boldsymbol{v} c_i, \quad i = 1, \ldots N. \tag{3.10}$$

The projection of the vectors in the vector Eq. 3.10 over the normal vector of the interphase surface $\boldsymbol{n}$ can be expressed as

$$j_i = (\boldsymbol{j}_i \cdot \boldsymbol{n}), \quad \frac{\partial c_i}{\partial n} = (\boldsymbol{grad} \, c_i \cdot \boldsymbol{n}), \quad v = (\boldsymbol{v} \cdot \boldsymbol{n}), \tag{3.11}$$

where j_i (mol/m^{2}s) are molar fluxes (which have to be equal to the rates of the reactions of the separate substances), $\frac{\partial c_i}{\partial n}$ is the normal derivate at the interface, and v is the rate induced as a result of the heterogeneous reaction at certain conditions and is termed "the velocity of the Stefan flow." It is positive when $\boldsymbol{v}$ is oriented towards the phase boundary and negative in the opposite case.

The introduction of (3.11) into (3.10) leads to

$$j_i = -D_i \frac{\partial c_i}{\partial n} + v c_i, \quad i = 1, \ldots, \mathrm{N}, \tag{3.12}$$

where j_i ($i = 1, \ldots, N$) should satisfy the condition for the stoichiometry of the flows:

$$\frac{j_1}{\upsilon_1} = \frac{j_2}{\upsilon_2} = \cdots = \frac{j_N}{\upsilon_N}. \tag{3.13}$$

From (3.13) it can be seen that the stoichiometric coefficients υ_i (mol/m^{2}s) represent the number of moles of the substances ($i = 1, \ldots, N$) which participate in the heterogeneous reaction (per unit area, per unit time). They can be represented as

$$\upsilon_i = \frac{v_i}{w_i}, \quad i = 1, \ldots, \mathrm{N}, \tag{3.14}$$

where v_i (m^3/m^{2}s) is the volume reaction rate of the substances in the gas (vapor) phase and w_i (m^3/mol) is their molar volume.

The gas (vapor)–liquid (solid) systems where the Stefan flow occurs in the gas (vapor) phase will be considered below, because it is practically not physically applicable in liquid and solid phases.

The summation of the stoichiometric coefficients leads to

$$\upsilon = \sum_{i=1}^{N} \upsilon_i, \tag{3.15}$$

where $\upsilon > 0$ ($\upsilon < 0$) means the increase (decrease) of the number of moles (the volume) of the reaction mixture as a result of the heterogeneous reaction.

From (3.13) it follows directly that

$$j_i = \frac{\upsilon_i}{\upsilon_1} j_1, \quad i = 1, \ldots, \mathrm{N}. \tag{3.16}$$

The summation of (3.16) yields

$$\sum_{i=1}^{N} j_i = \gamma j_1, \quad \gamma = \frac{\upsilon}{\upsilon_1}, \tag{3.17}$$

where we consider substance A_1 as limiting, i.e., the rate of its reaction limits the rate of the heterogeneous reaction.

In the case of gases and vapors, we can express the concentration through the partial pressure:

$$c_i = \frac{P_i}{RT}, \quad i = 1, \ldots, \mathrm{N}, \tag{3.18}$$

where R is the universal gas constant and T is the temperature. Then, from (3.12) it directly follows that

$$j_i = \frac{D_i}{RT} \frac{\partial P_i}{\partial n} + \frac{vP_i}{RT}, \quad i = 1, \ldots, \mathrm{N}. \tag{3.19}$$

The summation of (3.19) yields

$$\sum_{i=1}^{\mathrm{N}} j_i = -\frac{1}{RT} \sum_{i=1}^{N} D_i \frac{\partial P_i}{\partial n} + \frac{vP}{RT}, \tag{3.20}$$

where $P = \sum_{i=1}^{\mathrm{N}} P_i$ is the total pressure of the mixture.

The velocity of the Stefan flow is obtained directly from (3.17) to (3.20):

$$v = \frac{RT}{P} \gamma j_1 + \frac{1}{P} \sum_{i=1}^{\mathrm{N}} D_i \frac{\partial P_i}{\partial n}. \tag{3.21}$$

In the case of two-component mixtures and $D_1 = D_2 = D$,

$$v = \frac{RT}{P} \gamma j_1 + \frac{D}{P} \frac{\partial P}{\partial n}. \tag{3.22}$$

It can be seen from (3.22) that the velocity of the Stefan flow is determined by the relative change in the volume of the reaction mixture γ as a result of changes in the volume velocity v_i or in the case of a phase transition (the change of the molar volume w_i). The velocity decreases as a result of hydraulic resistance $\left(\frac{\partial P}{\partial n} < 0\right)$. In the absence of a hydraulic resistance $P = \text{const.}$ and the velocity of the Stefan flow takes the form

$$v = \frac{RT}{P} \gamma j_1. \tag{3.23}$$

In the case of a reduction of the reaction mixture volume (as a result of a heterogeneous reaction), the Stefan flow is oriented towards the reaction interface

($\gamma > 0$, $v > 0$). In the opposite case ($\gamma < 0$, $v < 0$), it is oriented away from the reaction interface.

In the case of heterogeneous chemical reactions without phase transitions

$$w_1 = w_2 = \cdots = w_N, \quad \gamma = \frac{\sum_{i=1}^{N} v_i}{v_1}, \tag{3.24}$$

i.e., $v > 0$, when the total volume rate of the chemical reaction of the substances in the mixture is positive and the volume increases. In the opposite case $v < 0$.

In the cases when the heterogeneous reaction has a liquid–vapor phase transition at the interface (boiling, condensation), the molar rates and (as a result) their volume rates are equal:

$$v_1 = v_2, \quad \gamma = \frac{\frac{1}{w_1} - \frac{1}{w_2}}{\frac{1}{w_1}} = \frac{w_2 - w_1}{w_2}, \tag{3.25}$$

i.e., in the case of condensation (boiling) $w_1 > w_2$, $\gamma < 0$ ($w_2 > w_1$, $\gamma > 0$) and the Stefan flow is oriented towards (away from) the interface.

In cases when the heterogeneous reaction involves adsorption (desorption) $v_1 = v_2$, $\gamma = 0$, i.e., the conditions for the Stefan flow do not exist. In the analogous situation of absorption (desorption), the product transfer occurs into the second phase, i.e., $\upsilon_1 = \upsilon_2$, $v = 0$.

The result obtained (3.25) shows that the Stefan flow at the interface arises when the heterogeneous reaction leads to changes of the total volume of the reaction mixture. Obviously, this could happen only at the phase boundary of the gas (or the vapor) phase and it is practically impossible at the boundary of the liquid (or the solid) phase.

A significant nonlinear mass transfer effect occurs in systems with intense interphase mass transfer, where large concentration gradients induce secondary flows whose velocities are oriented normally to the interface.

For simplicity of explanation, a two-component fluid will be considered [10, 16]. Component A is a substance dissolved in component B (solvent). The density of the solution ρ (kg/m^3) can be represented through the mass concentrations of component A (Mc) and solvent B (M_0c_0):

$$\rho = M_0c_0 + Mc = \rho_0 + Mc, \tag{3.26}$$

where M and M_0 are the molar masses (kg/mol) of components A and B and c and c_0 are their molar concentrations (mol/m^3).

Any elementary volume of the solution has velocity $\boldsymbol{V}$, which can be expressed through the velocities of the substances $A(\boldsymbol{v})$ and $B(\boldsymbol{v}_0)$. Thus, the velocity of the mass flow transferred by any elementary volume is the sum of the mass flows of components A and B:

$$\rho \boldsymbol{V} = \rho_0 \boldsymbol{v}_0 + Mc\boldsymbol{v}. \tag{3.27}$$

Equation (3.28) can be projected on the normal vector $\boldsymbol{n}$ of the interphase surface:

$$\rho^*(\boldsymbol{V}\cdot\boldsymbol{n}) = \rho_0^*(\boldsymbol{v}_0\cdot\boldsymbol{n}) + Mc^*(\boldsymbol{v}\cdot\boldsymbol{n}), \tag{3.28}$$

where the asterisks denote the values at the interface. From (3.28) the velocity of the secondary flow v (m/s), induced by the diffusion (large concentration gradient), can be determined as

$$v = (\boldsymbol{V}\cdot\boldsymbol{n}). \tag{3.29}$$

At the boundary between two immiscible phases the mass flux is zero, i.e.,

$$(\boldsymbol{v}_0\cdot\boldsymbol{n}) = 0. \tag{3.30}$$

The molar flux of the dissolved substance (at the interface) N (mol/m^{2}s) can be expressed through the rate (mol/m^{2}s) of the diffusive and convective transfer terms:

$$N = c^*(\boldsymbol{v}\cdot\boldsymbol{n})^* = -D\left(\frac{\partial c}{\partial n}\right)^* + vc^*. \tag{3.31}$$

The introduction of (3.29–3.31) into (3.28) yields

$$v = -\frac{MD}{\rho_0^*}\left(\frac{\partial c}{\partial n}\right)^*, \quad \rho_0^* = M_0c_0^*, \tag{3.32}$$

where c_0^* is the molar concentration of component B at the interface. For a flat phase boundary $y = 0$ can be obtained directly [10, 16]:

$$v = -\frac{MD}{\rho_0^*}\left(\frac{\partial c}{\partial y}\right)_{y=0}. \tag{3.33}$$

Some approximations, assumed in [22], allow the flux N to be expressed only by its diffusion component:

$$N = -D\left(\frac{\partial c}{\partial y}\right)_{y=0}. \tag{3.34}$$

Thus, the velocity of the secondary flow is

$$v = \frac{MD}{\rho_0^*}\left(\frac{\partial c}{\partial y}\right)_{y=0}, \quad \rho^* = \rho_0^* + Mc^*. \tag{3.35}$$

Obviously, the results obtained (3.33, 3.35) coincide at $c^* = 0$ (e.g., at a desorption of gases).

It can be seen from (3.33) that in systems with intense interphase mass transfer the normal component of the velocity is not zero (as in the systems with linear mass transfer). Moreover, it depends on the concentration of the transferred substance, i.e., the convection–diffusion equation is nonlinear. The latter requires the boundary condition at $y = 0$ ($v = 0$) to be replaced by (3.33).

The result (3.33) shows that the local mass flux at the phase boundary has diffusive and convective components:

$$i = -MD\left(\frac{\partial c}{\partial y}\right)_{y=0} + Mvc^{*} = MD\frac{\rho^{*}}{\rho_0^{*}}\left(\frac{\partial c}{\partial y}\right)_{y=0}. \tag{3.36}$$

The mass transfer rate can be directly determined by averaging the mass flux i (kg/m^{2}s) over the interface.

The comparison between the Stefan flow velocity (3.23) and the velocity of the secondary flow induced by large concentration gradients (3.35) indicates that Stefan flow arises in the gas (or the vapor) phase as a result of changes in the phase volume (pressure gradient). Such changes occur in some heterogeneous reactions accompanied by changes of the reaction mixture volumes and phase transitions (boiling and condensation).

The secondary flow induced by large concentration gradients (partial pressure gradients in gas phases) is a result of the intense interphase mass transfer that can be observed in gas and liquid phases (Stefan flow in a liquid phase is physically impossible).

The Marangoni effect is a result of a secondary flow whose velocity is oriented tangentially to the interface and it is induced by a surface tension gradient. The latter is a result of concentration (or temperature) gradients at the interface. Here, the Marangoni effect will be concerned only with gas–liquid systems.

The influence of the secondary flows on the mass transfer rate is a result of their velocity component oriented normally to the interface. It creates an intense convective transfer which contributes to the total transfer process (the diffusive transfer exists in its absence). Upon provoking the Marangoni effect, the induced flow is tangential and the normal component appears from the flow continuity equation:

$$\frac{\partial u}{\partial x} + \frac{\partial v}{\partial y} = 0. \tag{3.37}$$

The flows in the boundary layer are characterized by two characteristic scales of velocity (u_0, v_0) and two linear scales (δ, L), which are related to the dimensionless variables of the flow:

$$u = u_0 U, \quad v = v_0 V, \quad x = LX, \quad y = \delta Y, \quad \delta = \sqrt{\frac{\mu L}{u_0 \rho}}. \tag{3.38}$$

The introduction of (3.38) into (3.37) leads to a dimensionless equation:

$$\frac{\partial U}{\partial X}+\frac{v_0 L}{u_0 \delta}\frac{\partial V}{\partial Y}=0. \tag{3.39}$$

Here, the flow continuity is expressed as a ratio of characteristic scales:

$$\frac{v_0 L}{u_0 \delta}=1, \quad v_0=\frac{\delta}{L}u_0=u_0\sqrt{\frac{\upsilon}{u_0 L}}. \tag{3.40}$$

Let us suppose that the Marangoni effect is a result of the temperature gradient at the interface. In this case the characteristic velocity of the Marangoni effect can be determined from the equation [37]

$$\mu\left(\frac{\partial u}{\partial y}\right)_{y=0}=\frac{\partial \sigma}{\partial x}=\frac{\partial \sigma}{\partial t}\frac{\partial t}{\partial x}. \tag{3.41}$$

If (3.38) is introduced into (3.41) (and the temperature scale is Δt), we can obtain the condition for the existence of the Marangoni effect and its characteristic velocity:

$$u_0=\frac{\delta}{L}\frac{\Delta t}{\mu}\frac{\partial \sigma}{\partial t}. \tag{3.42}$$

The introduction of (3.42) into (3.40) allows us to find the characteristic velocity of the secondary flow responsible for the increase of the mass transfer rate:

$$v_0=\frac{\Delta t}{\rho u_0 L}\frac{\partial \sigma}{\partial t}. \tag{3.43}$$

For example, in the case of absorption of CO_2 in H_2O and a temperature changes due to a chemical reaction, the order of magnitude of the velocity v_0 can be evaluated as

$$v_0 \sim 10^{-8}\,\mathrm{m/s}. \tag{3.44}$$

In the cases of nonlinear mass transfer, the characteristic scales have to be introduced into (3.33) as

$$v=v_0 V, \quad c=\Delta c C, \quad y=\delta_c Y, \quad \delta_c=\sqrt{\frac{\delta l}{u_0}}. \tag{3.45}$$

From (3.33) to (3.45) it follows directly that the condition for the existence of the nonlinear mass transfer effect and its characteristic velocity is

$$v_0=\frac{M\Delta c}{\rho_0^*}\sqrt{\frac{u_0 D}{L}}. \tag{3.46}$$

The order of the magnitude of the velocity v_0 in (3.46) under similar conditions imposed by (3.44) can determined directly:

$$v_0 \sim 10^{-6}\mathrm{m/s}. \tag{3.47}$$

The results obtained (3.44 and 3.47) show that in systems with an intense interphase mass transfer the nonlinear effects are a result of the concentration gradients oriented normally to the interface and are not the result of the temperature gradients at the phase boundary. The difference between the effect of nonlinear mass transfer and the Marangoni effect is based mainly on the following three reasons:

1. The normal component of the velocity v_0 is always smaller than the tangential velocity component (see 3.40).
2. The small temperature gradients Δt, due to the small heat effect of the absorption.
3. The absence of surface-active substances.

The nonlinear mass transfer effect and the Marangoni effect can affect the mass transfer rate not only by additional convective flows, but also by a loss of stability. In such cases the accidental perturbations lead to self-organized dissipative structures with very intense mass transfer [10]. The stability of such systems depends mainly on v_0, which is the reason why the process is limited again by the nonlinear mass transfer.

Natural convection arises in fluids as a result of the vertical gradient of the density:

$$\rho = \rho_0 + \rho_1(c, \theta), \tag{3.48}$$

where ρ_1 is a function of the concentration and temperature. If the coordinate x-axis is vertical and directed upward, natural convection exists when $\partial\rho_1/\partial x > 0$. The mathematical description of the process can be obtained from (1.2.24) if the Newtonian force is replaced by the buoyant force:

$$\begin{aligned}
&\frac{\partial u_x}{\partial t} + u_x\frac{\partial u_x}{\partial x} + u_y\frac{\partial u_x}{\partial y} + u_z\frac{\partial u_x}{\partial z} = g\frac{\rho_1}{\rho} - \frac{1}{\rho}\frac{\partial p}{\partial x} + \nu\left(\frac{\partial^2 u_x}{\partial x^2} + \frac{\partial^2 u_x}{\partial y^2} + \frac{\partial^2 u_x}{\partial z^2}\right),\\
&\frac{\partial u_y}{\partial t} + u_x\frac{\partial u_y}{\partial x} + u_y\frac{\partial u_y}{\partial y} + u_z\frac{\partial u_y}{\partial z} = -\frac{1}{\rho}\frac{\partial p}{\partial y} + \nu\left(\frac{\partial^2 u_y}{\partial x^2} + \frac{\partial^2 u_y}{\partial y^2} + \frac{\partial^2 u_y}{\partial z^2}\right),\\
&\frac{\partial u_z}{\partial t} + u_x\frac{\partial u_z}{\partial x} + u_y\frac{\partial u_z}{\partial y} + u_z\frac{\partial u_z}{\partial z} = -\frac{1}{\rho}\frac{\partial p}{\partial z} + \nu\left(\frac{\partial^2 u_z}{\partial x^2} + \frac{\partial^2 u_z}{\partial y^2} + \frac{\partial^2 u_z}{\partial z^2}\right).
\end{aligned} \tag{3.49}$$

From (3.49) it can be seen, that a solution $u_x = u_y = u_z = 0$ exists if the boundary conditions (the velocities on the border of the system) are equal to zero too. In this case it is possible for the system to be stable and

$$\frac{\partial p}{\partial x} = g\rho_1. \tag{3.50}$$

For large values of the gradient $\partial \rho_1/\partial x > 0$ the system is unstable and small disturbances lead to natural convection in a form of Benard cells [23]. All secondary flows at the border of the system provoke natural convection too.

3.2 Nonstationary Absorption Mechanism

The theoretical analysis of the Marangoni effect in the cases of gas absorption shows [11] that the characteristic velocity (3.43) of the secondary flow increases if the characteristic velocity of the main flow decreases, i.e., the maximum Marangoni effect will exist in an immovable liquid. The experimental investigation [24] of the mass transfer between two immovable phases in the case of absorption of pure gases in an immovable liquid shows that under these conditions three processes can take place: natural convection, nonlinear mass transfer, and the Marangoni effect.

Let us consider a vertical tube with radius r_0 in which an immovable liquid (H_2O) contacts an immovable gas (CO_2, SO_2, NH_3). The gas is absorbed in the liquid, and the process is accompanied by a thermal effect. As a result, several effects in the liquid may occur, having the form of secondary flows owing to the large concentration gradients (nonlinear mass transfer), the difference in density (natural convection), and the surface tension gradient (Marangoni effect).

The mathematical description of this process will be done in cylindrical coordinates [25–28]. The influence of the density gradient, the concentration gradient, and the surface tension gradient will be considered. In this way, the problem has the form

$$\begin{gathered}
\rho\left(\frac{\partial v_z}{\partial t} + v_z\frac{\partial v_z}{\partial z} + v_r\frac{\partial v_z}{\partial r}\right) = -\frac{\partial p}{\partial z} + \mu\left(\frac{\partial^2 v_z}{\partial z^2} + \frac{1}{r}\frac{\partial v_z}{\partial r} + \frac{\partial^2 v_z}{\partial r^2}\right) + g(\rho - \rho_0), \\
\rho\left(\frac{\partial v_r}{\partial t} + v_z\frac{\partial v_r}{\partial z} + v_r\frac{\partial v_r}{\partial r}\right) = -\frac{\partial p}{\partial r} + \mu\left(\frac{\partial^2 v_r}{\partial z^2} + \frac{1}{r}\frac{\partial v_r}{\partial r} + \frac{\partial^2 v_r}{\partial r^2}\right), \\
\frac{\partial \rho}{\partial t} + \frac{\partial(\rho v_z)}{\partial z} + \frac{\rho v_r}{r} + \frac{\partial(\rho v_r)}{\partial r} = 0, \qquad \rho = \rho_0\left[1 + \frac{c}{\rho_0} - \beta(\theta - \theta_0)\right], \\
\frac{\partial c}{\partial t} + v_z\frac{\partial c}{\partial z} + v_r\frac{\partial c}{\partial r} = D\left(\frac{\partial^2 c}{\partial z^2} + \frac{1}{r}\frac{\partial c}{\partial r} + \frac{\partial^2 c}{\partial r^2}\right), \\
\rho c_p\left(\frac{\partial \theta}{\partial t} + v_z\frac{\partial \theta}{\partial z} + v_r\frac{\partial \theta}{\partial r}\right) = \lambda\left(\frac{\partial^2 \theta}{\partial z^2} + \frac{1}{r}\frac{\partial \theta}{\partial r} + \frac{\partial^2 \theta}{\partial r^2}\right),
\end{gathered} \tag{3.51}$$

where ρ_0 is the density, μ is the viscosity, D is the diffusivity, λ is the conductivity, and θ_0 is the initial liquid temperature. Equations (3.51) permit us to obtain the Oberbeck–Boussinesq approximations ($c \equiv 0, 0 = \beta < 10^{-2}$).

The boundary conditions account for the induced secondary flows at the liquid surface as a result of the significant gradients of the concentration (c) and the surface tension (σ). Moreover, the change of the temperature is a result of the heat effect of absorption (q), under the assumption that the chemical reaction (gas–liquid) is fast and only exists at the interface ($z = 0$). Thus, the initial and the boundary conditions are

$$t = 0,\ v_z = v_r = c = 0,\ \theta = \theta_0; \quad z \to \infty,\ v_z = v_r = c = 0,\ \theta = \theta_0;$$

$$z = 0,\ v_z = -\frac{D}{\rho_0}\frac{\partial c}{\partial z},\ \mu\frac{\partial v_r}{\partial z} = \frac{\partial \sigma}{\partial r} = \frac{\partial \sigma}{\partial \theta}\frac{\partial \theta}{\partial r},\ c = c^*,\ \lambda\frac{\partial \theta}{\partial z} = qD\frac{\partial c}{\partial z}; \tag{3.52}$$

$$r = 0,\ v_r = 0,\ \frac{\partial v_z}{\partial r} = \frac{\partial c}{\partial r} = \frac{\partial \theta}{\partial r} = 0; \quad r = r_0,\ v_z = v_r = 0,\ \frac{\partial c}{\partial r} = \frac{\partial \theta}{\partial r} = 0,$$

where c^* is the gas solubility at the initial temperature θ_0.

The process considered has some characteristic scales that are known:

$$t_0 \sim 10^2\text{s}, \quad r_0 \sim 10^{-2}\text{m}, \quad c^* \sim 10^2\text{kg/m}^3, \quad \theta_0 \sim 10°\text{C}. \tag{3.53}$$

The characteristic scales of the velocity (v_0) and the depth where the main changes of the velocity (l), concentration (δ), and the temperature (h) take place must be determined from the analysis of the process mechanism. Thus, the dimensionless variables are

$$t = t_0 T, \quad z = lZ_1 = \delta Z_2 = hZ_3, \quad r = r_0 R, \quad p = \rho_0 v_0^2 P,$$

$$v_z(t,z,r) = v_0 V_z(T,Z_1,R) = v_0 \tilde{V}_z(T,Z_2,R) = v_0 \tilde{\tilde{V}}_z(T,Z_3,R),$$

$$v_r(t,z,r) = \frac{v_0 r_0}{l} V_r(T,Z_1,R) = \frac{v_0 r_0}{\delta}\tilde{V}_r(T,Z_2,R) = \frac{v_0 r_0}{h}\tilde{\tilde{V}}_r(T,Z_3,R), \tag{3.54}$$

$$c(t,z,r) = c^* C(T,Z_1,R) = c^*\tilde{C}(T,Z_2,R) = c^*\tilde{\tilde{C}}(T,Z_3,R),$$

$$\theta(t,z,r) = \theta_0\Theta(T,Z_1,R) = \theta_0\tilde{\Theta}(T,Z_2,R) = \theta_0\tilde{\tilde{\Theta}}(T,Z_3,R).$$

The dimensionless (generalized) variables (3.54) allow the hydrodynamic, the diffusion, and the heat transfer processes to be described in terms of the scales. Moreover, they permit us to evaluate the limiting states in the mechanism of this complicated process.

The mechanisms of the nonstationary diffusion with a nonlinear mass transfer, natural convection, and the Marangoni effect may be elucidated by substitution of (3.54) in (3.51) and (3.52). The resulting dimensionless parameters of the different terms (differential operators) permit us to evaluate the roles of each elementary process taking part in the complex process. From this point of view, the dimensionless parameters are of an order of magnitude of unity, lower than unity and negligible compared with unity. The later allows an evaluation of the weight of each term (elementary process): a significant role, a low role, and an insignificant role. From these positions the problem may be rewritten as follows:

$$
\left[1+\frac{c^*}{\rho_0}C-\beta\theta_0(\Theta-1)\right]\left[\frac{v_0\rho_0}{gt_0c^*}\frac{\partial V_z}{\partial T}+\frac{v_0^2\rho_0}{glc^*}\left(V_z\frac{\partial V_z}{\partial Z_1}+V_r\frac{\partial V_z}{}\right)\right]
$$

$$
=-\frac{v_0^2\rho_0}{glc^*}\frac{\partial P}{\partial Z_1}+\frac{0}{gl^2c^*}\left[\frac{\partial^2 V_z}{\partial Z_1^2}+\frac{l^2}{r_0^2}\left(\frac{1}{R}\frac{\partial V_z}{\partial R}+\frac{\partial^2 V_z}{\partial R^2}\right)\right]+C-\frac{\rho_0\beta\theta_0}{c^*}(\Theta-1),
$$

$$
\left[1+\frac{c^*}{\rho_0}C-\beta\theta_0(\Theta-1)\right]\left[\frac{v_0\rho_0}{gt_0c^*}\frac{\partial V_r}{\partial T}+\frac{v_0^2\rho_0}{glc^*}\left(V_z\frac{\partial V_r}{\partial Z_1}+V_r\frac{\partial V_r}{}\right)\right]
$$

$$
=-\frac{l^2}{r_0^2}\frac{v_0^2\rho_0}{glc^*}\frac{}{\partial R}+\frac{\mu v_0}{gl^2c^*}\left[\frac{\partial^2 V_r}{\partial Z_1^2}+\frac{l^2}{r_0^2}\left(\frac{1}{R}\frac{\partial V_r}{\partial R}+\frac{\partial^2 V_r}{\partial R^2}\right)\right],
$$

$$
\frac{lc^*}{\rho_0v_0t_0}\frac{\partial C}{\partial T}+\frac{\beta\theta_0 l}{t_0v_0}\frac{\partial\Theta}{\partial T}+\left(\frac{\partial V_z}{\partial Z_1}+\frac{V_r}{R}+\frac{\partial V_r}{\partial R}\right)\left[1+\frac{c^*}{\rho_0}C-\beta\theta_0(\Theta-1)\right]+
$$

$$
+V_z\left(\frac{c^*}{\rho_0}\frac{\partial C}{\partial Z_1}-\beta\theta_0\frac{\partial\Theta}{\partial Z_1}\right)-V_r\left(\frac{c^*}{\rho_0}\frac{}{\partial R}-\beta\theta_0\frac{\partial\Theta}{\partial R}\right)=0,
$$

$$
\frac{\partial\tilde{C}}{\partial T}+\frac{v_0t_0}{\delta}\left(\tilde{V}_z\frac{\partial\tilde{C}}{\partial Z_2}+\tilde{V}_r\frac{\partial\tilde{C}}{\partial R}\right)=\frac{Dt_0}{\delta^2}\left[\frac{\partial^2\tilde{C}}{\partial Z_2^2}+\frac{\delta^2}{r_0^2}\left(\frac{1}{R}\frac{\partial\tilde{C}}{\partial R}+\frac{\partial^2\tilde{C}}{\partial R^2}\right)\right],
$$

$$
\left[1+\frac{c^*}{\rho_0}\tilde{\tilde{C}}-\beta\theta_0\left(\tilde{\tilde{\Theta}}-1\right)\right]\left[\frac{\partial\tilde{\tilde{\Theta}}}{\partial T}+\frac{v_0t_0}{h}\left(\tilde{\tilde{V}}_z\frac{\partial\tilde{\tilde{\Theta}}}{\partial Z_3}+\tilde{\tilde{V}}_r\frac{\partial\tilde{\tilde{\Theta}}}{\partial R}\right)\right]
$$

$$
=\frac{at_0}{h^2}\left[\frac{\partial^2\tilde{\tilde{\Theta}}}{_3^2}+\frac{h^2}{r_0^2}\left(\frac{1}{R}\frac{\partial\tilde{\tilde{\Theta}}}{\partial R}+\frac{\partial^2\tilde{\tilde{\Theta}}}{\partial R^2}\right)\right];
$$

$$
T=0,_{z}=V_r=\tilde{C}=0,\quad\tilde{\tilde{\Theta}}=1;\quad Z_1=Z_2=Z_3\to\infty,\ V_z=V_r=\tilde{C}=0,\ \tilde{\tilde{\Theta}}=1;
$$

$$
Z_1=Z_2=Z_3=0,\quad V_z=-\frac{Dc^*}{v_0\rho_0\delta}\frac{\partial\tilde{C}}{\partial Z_2},\ \frac{\mathrm{r}}{\partial Z_1}=\frac{\partial\sigma}{\partial\theta}\frac{\theta_0}{\mu v_0}\frac{l^2}{r_0^2}\frac{\partial\tilde{\tilde{\Theta}}}{\partial R},
$$

$$
\tilde{C}=1,\ \frac{\partial\tilde{\tilde{\Theta}}}{\partial Z_3}=\frac{qDhc^*}{\delta\lambda\theta_0}\left(1+\frac{c^*}{\rho_0}\right)\frac{\partial\tilde{C}}{_2};
$$

$$
R=0,\quad V_r=0,\quad\frac{\partial V_z}{\partial R}=\frac{\partial\tilde{C}}{\partial R}=\frac{\partial\tilde{\tilde{\Theta}}}{\partial R}=0;\quad R=1,\ V_z=V_r=0,\frac{\partial\tilde{C}}{\partial R}=\frac{\partial\tilde{\tilde{\Theta}}}{\partial R}=0, \tag{3.55}
$$

where $a=\frac{\lambda}{\rho_0c_p}\sim 10^{-7}\mathrm{m}^2/\mathrm{s}$ and $q=3\times 10^2$ kcal/kg.

Under the assumption that the flow is a result of natural convection, induced by diffusion and heat transfer, the characteristic linear scales may be determined from the conditions

$$\frac{\mu v_0}{gl^2c^*}=1,\quad \frac{Dt_0}{\delta^2}=1,\quad \frac{at_0}{h^2}=1. \tag{3.56}$$

The characteristic velocity scale may be determined from the condition

$$\frac{Dc^*}{v_0\rho_0\delta}=1 \tag{3.57}$$

if the nonlinear mass transfer effect is important.

From (3.56) to (3.57) it may be evaluated that the unknown characteristic scales are

$$v_0=\frac{c^*}{\rho_0}\sqrt{\frac{D}{t_0}}\sim 10^{-7}\text{m/s.}\quad l=\sqrt{\frac{\mu}{\rho_0 g}\sqrt{\frac{D}{t_0}}}\sim 10^{-7}\text{m,}$$
$$\delta=\sqrt{Dt_0}\sim 10^{-4}\text{m},\quad h=\sqrt{at_0}\sim 10^{-3}\text{m}. \tag{3.58}$$

The latter allows determination of the order of magnitude of all the dimensionless parameters in (3.55) and the estimation of the role of each elementary process in the complex one as well:

$$\varepsilon=\frac{c^*}{\rho_0}\sim 10^{-1},\quad \frac{v_0\rho_0}{gt_0c^*}\sim 10^{-9},\quad \frac{v_0^2\rho_0}{glc^*}\sim 10^{-7},\quad \frac{l^2}{r_0^2}\sim 10^{-10},$$
$$\frac{lc^*}{\rho_0 v_0 t_0}\sim 10^{-3},\quad \frac{v_0t_0}{\delta}=\varepsilon\sim 10^{-1},\quad \frac{\delta^2}{r_0^2}\sim 10^{-4},\quad \frac{h^2}{r_0^2}\sim 10^{-2},$$
$$\frac{v_0t_0}{h}\sim 10^{-2},\quad \frac{\partial\sigma}{\partial\theta}\frac{\theta_0}{\mu v_0}\frac{l^2}{r_0^2}\sim 10^{-3},\quad \alpha_1=\frac{qDh\rho_0}{\delta\lambda\theta_0}\sim 1, \tag{3.59}$$
$$\beta\theta_0\sim 10^{-3},\quad \frac{\partial\sigma}{\partial\theta}\sim 10^{-5}\text{kg/m}^{2\circ}\text{C},\quad \frac{\rho_0\beta\theta_0}{c^*}\sim 10^{-2},\quad \frac{\beta\theta_0 l}{t_0v_0}\sim 10^{-5}.$$

From (3.56) to (3.59) evidently the parameter of the Marangoni effect $\left(\frac{\partial\sigma}{\partial\theta}\frac{\theta_0}{\mu v_0}\frac{l^2}{r_0^2}\sim 10^{-3}\right)$ has an order of 10^{-2}, whereas the effects of the natural convection $\left(\frac{\mu v_0}{gl^2c^*}\right)$ and the nonlinear mass transfer $\left(\frac{Dc^*}{v_0\rho_0\delta}\right)$ are of the order of unity.

If the characteristic velocity is limited by natural convection $\frac{lc^*}{\rho_0 r_0 t_0}\sim 1$, e.g., $v_0=\frac{\upsilon c^*}{\rho_0 g t_0}\sim 10^{-12}$m/s, $l=\frac{\upsilon}{gt_0}\sim 10^{-9}$m and the parameter of the nonlinear mass transfer is obtained as $\frac{Dc^*}{v_0\rho_0\delta}\sim 10^6$, then the characteristic velocity is limited by the nonlinear mass transfer.

Taking into account (3.59), we may express problem (3.55) in a zeroth-order approximation for a parameter of order lower than 10^{-2}:

$$\frac{\partial^2 V_z}{\partial Z_1^2}+C=0,\quad \frac{\partial^2 V_r}{\partial Z_1^2}=0,\quad (1+\varepsilon C)\left(\frac{\partial V_z}{\partial Z_1}+\frac{V_r}{R}+\frac{\partial V_r}{\partial R}\right)+\varepsilon\left(V_z\frac{\partial C}{\partial Z_1}+V_r\frac{\partial C}{\partial R}\right)=0,$$

$$\frac{\partial \tilde{C}}{\partial T}+\varepsilon\left(\tilde{V}_z\frac{\partial \tilde{C}}{\partial Z_2}+\tilde{V}_r\frac{\partial \tilde{C}}{\partial R}\right)=\frac{\partial^2 \tilde{C}}{\partial Z_2^2},\quad \left(1+\varepsilon\tilde{C}\right)\frac{\partial \tilde{\tilde{\Theta}}}{\partial T}=\frac{\partial^2 \tilde{\tilde{\Theta}}}{\partial Z_3^2};\quad T=0,\quad \tilde{C}=0,\quad \tilde{\tilde{\Theta}}=1;$$

$$Z_1=Z_2=Z_3=0,\quad V_z=-\frac{\partial \tilde{C}}{\partial Z_2},\frac{\partial V_r}{\partial Z_1}=0,\quad \tilde{C}=1,\quad \frac{\partial \tilde{\tilde{\Theta}}}{\partial Z_3}=\varepsilon\alpha_1(1+\varepsilon)\frac{\partial \tilde{C}}{\partial Z_2};$$

$$Z_1=Z_2=Z\to\infty,\quad V_z=V_r=\tilde{C}=0,\quad \tilde{\tilde{\Theta}}=1;\quad R=0,\quad V_r=0,\quad \frac{\partial \tilde{C}}{\partial R}=0. \tag{3.60}$$

From (3.60) it is clear that both the natural convection and the nonlinear mass transfer are independent of the heat transfer.

In (3.60) ε is a small parameter and the solutions may be obtained by a perturbation method:

$$\begin{aligned} V_z &= V_z^{(0)}+\varepsilon V_z^{(1)}+\ldots,\quad V_r = V_r^{(0)}+\varepsilon V_r^{(1)}+\ldots,\\ \tilde{C} &= \tilde{C}^{(0)}+\varepsilon\tilde{C}^{(1)}+\ldots,\quad \tilde{\tilde{\Theta}} = \tilde{\tilde{\Theta}}^{(0)}+\varepsilon\tilde{\tilde{\Theta}}^{(1)}+\ldots. \end{aligned} \tag{3.61}$$

The zeroth-order approximations ($\varepsilon = 0$) may be obtained directly from (3.60):

$$\frac{\partial^2 V_z^{(0)}}{\partial Z_1^2}+C^{(0)}=0;\quad Z_1=0,\quad V_z^{(0)}=-\left(\frac{\partial \tilde{C}^{(0)}}{\partial Z_2}\right)_{Z_2=0};\quad Z_1\to\infty,\quad V_z^{(0)}=0. \tag{3.62}$$

$$\frac{\partial V_r^{(0)}}{\partial R}+\frac{V_r^{(0)}}{R}=-\frac{\partial V_z^{(0)}}{\partial Z_1};\quad R=0,\quad V_r^{(0)}=0. \tag{3.63}$$

$$\frac{\partial \tilde{C}^{(0)}}{\partial T}=\frac{\partial^2 \tilde{C}^{(0)}}{\partial Z_2^2};\quad Z_2=0,\quad \tilde{C}^{(0)}=1;\quad Z_2\to\infty,\quad \tilde{C}^{(0)}=0;\quad T=0,\quad \tilde{C}^{(0)}=0. \tag{3.64}$$

$$\frac{\partial \tilde{\tilde{\Theta}}^{(0)}}{\partial T}=\frac{\partial^2 \tilde{\tilde{\Theta}}^{(0)}}{\partial Z_3^2};\quad Z_3=0,\quad \frac{\partial \tilde{\tilde{\Theta}}^{(0)}}{\partial Z_3}=0;\quad Z_3\to\infty,\quad \tilde{\tilde{\Theta}}^{(0)}=1;\quad T=0,\quad \tilde{\tilde{\Theta}}=1. \tag{3.65}$$

Initially, the solution must start with (3.64) and the solution is

$$\tilde{C}^{(0)}=erfc\frac{Z_2}{2\sqrt{T}}. \tag{3.66}$$

From (3.54) it is clear that

$$Z_2 = \alpha Z_1, \quad \alpha = \frac{l}{\delta} \sim 10^{-3}, \quad C = \tilde{C}, \tag{3.67}$$

i.e., $C^{(0)}(Z_1, T) = \tilde{C}^{(0)}(\alpha Z_1, T)$. For small values of α it follows that

$$C^{(0)} = erfc \frac{\alpha Z_1}{2\sqrt{T}} \approx 1 - \frac{\alpha Z_1}{\sqrt{\pi T}} \approx 1 \quad (\alpha \sim 10^{-3}). \tag{3.68}$$

Thus, (3.62) takes the form

$$\frac{\partial^2 V_z^{(0)}}{\partial Z_1^2} = -1; \quad Z_1 = 0, \quad V_z^{(0)} = \frac{1}{\sqrt{\pi T}}; \quad Z_1 = 1, \quad V_z^{(0)} = 0. \tag{3.69}$$

In (3.69) the boundary condition at infinity is replaced with a condition for velocity attenuation at $z \geq l$. The solution of problem (3.69) is

$$V_z^{(0)} = -\frac{1}{2} Z_1^2 + \left(\frac{1}{2} - \frac{1}{\sqrt{\pi T}} \right) Z_1 + \frac{1}{\sqrt{\pi T}}. \tag{3.70}$$

Taking into account (3.70), we may express problem (3.63) as

$$\frac{\partial V_r^{(0)}}{\partial R} + \frac{V_r^{(0)}}{R} = Z_1 + \frac{1}{\sqrt{\pi T}} - \frac{1}{2}; \quad R = 0, \quad V_r^{(0)} = 0 \tag{3.71}$$

and for $V_r^{(0)}$ the result is

$$V_r^{(0)} = \left[\frac{1}{2} Z_1 + \frac{1}{\sqrt{\pi T}} - \frac{1}{2} \right] R. \tag{3.72}$$

The zeroth-order approximation for temperature may be obtained from (3.65),

$$\begin{gathered} \frac{\partial \tilde{\tilde{\Theta}}^{(0)}}{\partial T} = \frac{\partial^2 \tilde{\tilde{\Theta}}^{(0)}}{\partial Z_3^2}; \quad T = 0, \quad \tilde{\tilde{\Theta}}^{(0)} = 1; \\ Z_3 = 0, \quad \frac{\partial \tilde{\tilde{\Theta}}^{(0)}}{\partial Z_3} = 0; \quad Z_3 \to \infty, \quad \frac{\partial \tilde{\tilde{\Theta}}^{(0)}}{\partial Z_3} = 0, \end{gathered} \tag{3.73}$$

and the solution is

$$\tilde{\tilde{\Theta}}^{(0)} \equiv 1. \tag{3.74}$$

The first approximation of the concentration $\tilde{C}^{(1)}$ follows from (3.60), taking into account (3.66) and (3.70):

$$\frac{\partial \tilde{C}^{(1)}}{\partial T} = \frac{\partial^2 \tilde{C}^{(1)}}{\partial Z_2^2} - \tilde{V}_z^{(0)} \frac{\partial \tilde{C}^{(0)}}{\partial Z_2};$$
$$T = 0, \quad \tilde{C}^{(1)} = 0; \quad Z_2 = 0, \quad \tilde{C}^{(1)} = 0; \quad Z_2 \to \infty, \quad \tilde{C}^{(1)} = 0, \tag{3.75}$$

where

$$\tilde{V}_z^{(0)}(Z_2, T) = V_z^{(0)}(Z_1, T), \quad Z_1 = \frac{1}{\alpha} Z_2. \tag{3.76}$$

From (3.69), (3.70) and (3.76) it is clear that $V_z^{(0)}$ differs from zero in the interval $0 \leq Z_1 < 1$, $0 \leq Z_2 < \alpha$, i.e., in (3.75) the volume source (mass production rate per unit volume) is a surface source ($Z_2 = 0$) because

$$\tilde{V}_z^{(0)} \frac{\partial \tilde{C}^{(0)}}{\partial Z_2} \neq 0 \text{ for } 0 \leq Z_2 < \alpha \sim 10^{-3}. \tag{3.77}$$

The volumetric convective mass transfer $\tilde{V}_z^{(0)} \frac{\partial \tilde{C}^{(0)}}{\partial Z_2}$ near the interface ($Z_2 < 10^{-3}$) may be replaced with a surface convective mass transfer ($Z_2 = 0$):

$$\left[\tilde{V}_z^{(0)} \frac{\partial \tilde{C}^{(0)}}{\partial Z_2}\right]_{Z_2 \prec 10^{-3}} = -\left(\tilde{V}^{(0)} \cdot \tilde{C}^{(0)}\right)_{Z_2 = 0} = -\frac{1}{\sqrt{\pi T}}. \tag{3.78}$$

Thus, problem (3.75) takes the form

$$\frac{\partial \tilde{C}^{(1)}}{\partial T} = \frac{\partial^2 \tilde{C}^{(1)}}{\partial Z_2^2};$$
$$T = 0, \quad \tilde{C}^{(1)} = 0; \quad Z_2 = 0, \quad \frac{\partial \tilde{C}^{(1)}}{\partial Z_2} = -\frac{1}{\sqrt{\pi T}}; \quad Z_2 \to \infty, \quad \tilde{C}^{(1)} = 0. \tag{3.79}$$

It is evident that the expressions for $\tilde{C}^{(1)}$ and $V_z^{(1)}$ are identical to those for $\tilde{C}^{(0)}$ and $V_z^{(0)}$:

$$\tilde{C}^{(1)} \equiv \tilde{C}^{(0)} = erfc \frac{Z_2}{2\sqrt{T}}, \quad V_z^{(1)} \equiv V_z^{(0)} = -\frac{Z_1^2}{2} + \left(-\frac{1}{\sqrt{\pi T}} + \frac{1}{2}\right) Z_1 + \frac{1}{\sqrt{\pi T}}. \tag{3.80}$$

The average absorption rate J for a time interval t_0 may be expressed by means of the mass transfer coefficient k. It may be determined from the average mass flux I:

$$J = kc^* = \frac{1}{t_0} \int_0^{t_0} I dt, \quad I = -\frac{D\rho^*}{\rho_0} \left(\frac{\partial c}{\partial z}\right)_{z=0}, \quad \rho^* = \rho_0 + c^*. \tag{3.81}$$

Thus, Eq. 3.81 may be used to obtain the Sherwood number for nonstationary diffusion:

$$Sh = \frac{kt_0}{D} = -(1+\varepsilon)\sqrt{\frac{t_0}{D}} \int_0^1 \left(\frac{\partial \tilde{C}}{\partial Z_2}\right)_{Z_2=0} dT = 2\sqrt{\frac{t_0}{\pi D}}(1+\varepsilon)^2. \qquad (3.82)$$

The amount of gas absorbed (Q, kg/m^2) for time interval t_0 (s) is

$$Q = \int_0^{t_0} I dt = 2c^* \sqrt{\frac{Dt_0}{\pi}}(1+\varepsilon)^2. \qquad (3.83)$$

The results of the qualitative analysis show that in the case of mass transfer between two immovable phases (absorption of pure gases, when the mass transfer is limited by the diffusion resistance in the liquid phase), the effects of the nonlinear mass transfer and the natural convection are comparable, whereas the Marangoni effect is negligible in comparison. These two effects influence the mass transfer kinetics and induce a secondary flow in the liquid. It would be expected for them to affect the flow hydrodynamic stability too.

In cases where the liquid surface is heated [29] with sufficient intensity, the Marangoni effect occurs within a thin layer of the liquid phase (2×10^{-3} to 3×10^{-3} m).

3.3 Nonstationary Evaporation Kinetics

The studies described in the previous sections addressed the absorption (desorption) of low-solubility gases when the liquid mass transfer resistance limits the mass transfer rates. Obviously, it is interesting to focus the investigations on a situation where the mass transfer is limited by the gas phase. An adequate example is the case of nonstationary evaporation of a stagnant liquid layer in a stagnant gas phase above it. Detailed experimental investigations of such systems were reported in [30].

The nonstationary evaporation of a liquid with a moderate partial pressure (water, methanol, ethanol, and isopropyl alcohol) at 20 °C in an inert gas (nitrogen, argon, and helium) is investigated. The process occurs in a thermostatic condition, corresponding to the experiments reported in [30]. Under such conditions, the nonstationary mass transfer of the liquid vapors in the gas phase limits the process rate.

The mechanism of the nonstationary evaporation may be considered as a nonstationary diffusion complicated by additional effects of a variable temperature at the liquid surface (as a result of the thermal effect of the evaporation phenomenon) and a convection (secondary Stefan flow) as well as a natural convection. The effects of these phenomena on the evaporation rate will be analyzed subsequently.

The investigations [30] on the evaporation rates of liquids (H_2O, CH_3OH, C_2H_5OH, i-C_3H_7OH) show a time-dependent average liquid temperature. In fact, the process depends only on the surface temperature (θ^*, °C).

The temperature distribution in a layer of an evaporating liquid is described by

$$\frac{\partial\theta}{\partial t} = a\frac{\partial^2\theta}{\partial z^2}; \quad t = 0, \quad \theta = \theta_0; \quad z = 0, \quad \lambda\frac{\partial\theta}{\partial z} = qJ; \quad z = h, \quad \theta = \theta_0, \quad (3.84)$$

where the coordinate z-axis is oriented normally to the liquid–gas interface ($z = 0$), θ and θ_0 (°C) are the temperatures of the liquid and its initial value, t (s) is time, λ (kcal/ms °C) is the thermal conductivity of the liquid, a (m^2/s) is the temperature diffusivity, q (kcal/kg) is the latent heat of the evaporation, J (kg/m^{2}s) is the evaporation rate, and h (m) is the thickness of the evaporating liquid.

The evaporation rate J in (3.84) may be determined from experimental data concerning the amount of evaporated liquid Q (kg/m^2) at time t (s). Such data are available in [30]. After a sufficiently large initial time interval (where J has its greatest value), the relationship may be expressed as

$$Q_e = A_e\sqrt{t}, \quad (3.85)$$

where A_e (kg/m^{2}s^{½}) may be determined on the basis of the experimental data reported in [30] for systems such as H_2O–N_2, H_2O–He, H_2O–Ar, CH_3OH–Ar, C_2H_5OH–Ar, and i-C_3H_7OH–Ar (see Table 3). This allows us to define (by means of 3.85) the values of J_e as

Table 3 Characteristic parameters of gas–liquid systems (20 °C)

Parameters	Systems					
	H_2O–N_2	H_2O–He	H_2O–Ar	CH_3OH–Ar	C_2H_5OH–Ar	i-C_3H_7OH–Ar
q (kcal/kg)	584.3	584.3	584.3	280.0	217.9	179.1
$a \times 10^7$ (m^2/s)	1.43	1.43	1.43	1.05	0.888	0.752
$\lambda \times 10^5$ (kcal/ms °C)	1.448	1.448	1.448	4.875	4.015	3.657
$D \times 10^5$ (m^2/s)	2.41	2.57	8.86	0.98	1.0	0.846
$\theta_0 - \theta^*$ (°C)	0.3	0.3	0.2	3.4	0.6	0.4
$A_e \times 10^4$ (kg/m^{2}s$^{1/2}$)	2.051	2.190	1.633	19.47	3.963	3.287
$A_D \times 10^4$ (kg/m^{2}s$^{1/2}$)	0.936	0.967	1.79	5.09	3.89	1.51
$A \times 10^4$ (kg/m^{2}s$^{1/2}$)	0.954	0.961	1.94	5.92	4.28	1.57
c^* (kg/m^3)	0.0142	0.0143	0.0161	0.142	0.0995	0.0939
c_0^* (kg/m^3)	1.13	1.66	0.162	1.48	1.57	1.63
$\vartheta \times 10^5$ (m^2/s)	1.441	1.360	12.12	1.360	1.360	1.360
ρ_0 (kg/m^3)	1.16	1.66	0.166	1.66	1.66	1.66
$Q_{max} \times 10^2$ (kg/m^2)	0.3660	0.3684	0.4129	3.644	2.558	2.414
α	−0.555	0.778	−1.216	−0.246	0.133	0.335

$$J_e = \frac{dQ_e}{dt} = \frac{A_e}{2\sqrt{t}}. \tag{3.86}$$

The substitution of (3.86) into (3.86) permits us to define [31] the temperature distribution within the evaporating liquid layer with thickness h:

$$\theta = \theta_0 - \frac{qA_e\sqrt{\pi a}}{2\lambda}\sum_{n=0}^{\infty}(-1)^n\left[erfc\frac{2nh+z}{2\sqrt{at}} - erfc\frac{2(n+1)h-z}{2\sqrt{at}}\right]. \tag{3.87}$$

Equation (3.87) allows us to determine the temperature variations at the liquid top surface ($z = 0$) as a result of the evaporation process:

$$\theta_0 - \theta^* = \frac{qA_e\sqrt{\pi a}}{2\lambda}\left[1 + 2\sum_{n=1}^{\infty}(-1)^n erfc\frac{nh}{\sqrt{at}}\right], \tag{3.88}$$

where the liquid thickness was assumed to be 3×10^{-3} m. It follows from (3.88) that the maximum temperature at the interface may be reached at the limiting situations of $t \to 0$ or $h \to 0$:

$$\theta_0 - \theta^* = \frac{qA_e\sqrt{\pi a}}{2\lambda}. \tag{3.89}$$

The result (3.89) can also be obtained in the case when the liquid layer is not under thermostatic conditions. In such a situation the last boundary condition in (3.84) becomes

$$z \to \infty, \theta = \theta_0\left(z > h_a = \sqrt{at}\right), \tag{3.90}$$

where h_a is the thickness of the thermal boundary layer. Thus, the temperature distribution is

$$\theta = \theta_0 - \frac{qA_e\sqrt{\pi a}}{2\lambda}erfc\frac{z}{2\sqrt{at}}. \tag{3.91}$$

Equation (3.91) shows that the results (3.89) follow directly at $z = 0$.

The results concerning the interface temperature $\theta - \theta_0$ are listed in Table 3. It is clear that the temperature at the liquid surface practically remains unchanged as a result of the evaporation process. Some deviations from that "rule" are demonstrated by the CH_3OH–Ar system, but they change the partial pressure of the vapors above the liquid in the range of 10%, which should be neglected (the data in [30] have the same experimental error).

The experimental relationship $\theta\left(\sqrt{t}\right)$ (obtained in [30]) shows that the asymptotic value $Q = Q_{max}$ is reached at large values of $\sqrt{t}$. It allows the determination of the exact vapor concentration at the interface $c^* = \frac{Q_{max}}{h}\left(kg/m^3\right)$. The results are shown in Table 3. For example, the results show that the CH_3OH surface temperature is 15–16 °C.

The experimental data reported in [30] allow the value of A_e in (3.85) to be determined and the results are summarized in Table 3. If the evaporation rate is limited by the nonstationary diffusion, the concentration distribution is [32–35]

$$c = c^* erfc \frac{z}{2\sqrt{Dt}}, \tag{3.92}$$

where c and c^* (kg/m^3) are the concentrations of the vapors in the gas phase and at the interface respectively, and D (m^2/s) is the diffusivity. Solution (3.92) permits us to obtain the rate of the diffusion:

$$J_D = -D\left(\frac{\partial c}{\partial z}\right)_{z=0} = c^*\sqrt{\frac{D}{\pi t}} = \frac{A_D}{2\sqrt{t}}, \tag{3.93}$$

where

$$A_D = 2c^*\sqrt{\frac{D}{\pi}}. \tag{3.94}$$

The values of A_D are collected in Table 3. The comparison between the values of A_D and A_e indicates significant differences that may be attributed to the occurrence of a Stefan flow [11]. Some strange behaviors are demonstrated by the C_2H_5OH–Ar ($A_e \approx A_D$) and H_2O–He ($A_e < A_D$) systems.

The difference between the evaporation rate and the rate of the nonstationary diffusion indicates that a convective contribution exists. The evaporation of a liquid in an inert gas is a result of a liquid–vapor phase transition, so there is a volumetric effect of a heterogeneous reaction at the interface [11] that creates the Stefan flow. If the process occurs in thermostatic conditions and it is limited by both the diffusive and the convective transports in the gas phase, the evaporation rate can be expressed as

$$J = -D\left(\frac{\partial c}{\partial z}\right)_{z=0} + c^* v_s, \tag{3.95}$$

where v_s (m/s) is the velocity of the Stefan flow.

The mass transfer rate of the inert gas (in the gas phase) in the case of a gas–vapor binary system may be expressed in a similar manner:

$$J_0 = -D\left(\frac{\partial c_0}{\partial z}\right)_{z=0} + c_0^* v_s, \tag{3.96}$$

where c_0 and c_0^* (kg/m^3) are the concentrations of the inert gas in the bulk of the gas phase and at the interface. If the evaporating liquid is saturated by the inert gas (like in the experiments described in [30]), it follows that

$$J_0 = 0, \quad c_0 + c = \rho = \rho_0 + \alpha c, \quad \frac{\partial c_0}{\partial z} = -(1-\alpha)\frac{\partial c}{\partial z}, \quad \alpha = \frac{\rho_V - \rho_0}{\rho_V}, \tag{3.97}$$

where ρ_V (kg/m^3) is the density of the vapor of the liquid, ρ_0 (kg/m^3) is the density of the inert gas, and ρ (kg/m^3) is the density of the gas phase. In this way, we obtain from Eqs. (3.95) and (3.96) that

$$v_s = -\frac{D(1-\alpha)}{c_0^*}\left(\frac{\partial c}{\partial z}\right)_{z=0}, \quad J = -D\frac{\rho_0}{c_0^*}\left(\frac{\partial c}{\partial z}\right)_{z=0}. \tag{3.98}$$

The comparison between the velocity of the Stefan flow (3.98) and the velocity of the secondary flow induced by the large concentration gradients [10] shows that they are different when there is evaporation of a liquid in inert gases under isothermal conditions (a thermostated system) since $\rho_0 \neq \rho^*$. Here

$$\rho^* = \rho_0 + \alpha c^*. \tag{3.99}$$

The convective mass transfer upon nonstationary evaporation from a stagnant liquid into a stagnant gas above it (within a large initial time interval) could be attributed to the Stefan flow and the natural convection. Let us consider a gaseous layer above a stagnant liquid. The momentum equations of the gas phase and the convection–diffusion equations of the liquid vapors (under the assumption of a one-dimensional approximation) are

$$\frac{\partial v}{\partial t} + v\frac{\partial v}{\partial z} = \vartheta\frac{\partial^2 v}{\partial z^2} - \frac{1}{\rho_0}\frac{\partial p}{\partial z} - \frac{g\alpha c}{\rho_0}, \quad \frac{\partial c}{\partial t} + v\frac{\partial c}{\partial z} = D\frac{\partial^2 c}{\partial z^2}; \quad t=0, \quad v=c=0;$$
$$z=0, \quad v = -\frac{D(1-\alpha)}{c_0^*}\left(\frac{\partial c}{\partial z}\right)_{Z=0}, \quad c=c^*; \quad z\to\infty, \quad v=c=0, \tag{3.100}$$

where ϑ (m^2/s) is the kinematic viscosity .

The z coordinate is oriented vertically upwards and the liquid interface is $z=0$. In the cases when the Stefan flow does not exist (see 3.98), its velocity is zero,

$$v(0,t) = 0, \tag{3.101}$$

which leads to a stable solution of (3.100):

$$\bar{v} \equiv 0, \quad \bar{c} = a_1 z + a_2, \quad \frac{\partial \bar{p}}{\partial z} = g\alpha c, \tag{3.102}$$

i.e., the gas phase is stagnant, the concentration distribution is linear, and the pressure gradient depends on the concentration distribution along the gaseous layer depth [36]. This is a stable state of the system, but small disturbances could lead to a new stable state, where the motion of the gas phase is a result of the natural convection.

It is possible to introduce into (3.100) the dimensionless (generalized) variables

$$t = t_0 T, \quad z = \delta Z, \quad v = u_0 V, \quad p = \rho_0 u_0^2 P, \quad c = c^* C, \tag{3.103}$$

where t_0 (s) is the characteristic timescale of the process. The length δ denotes the depth of the gaseous layer above the liquid, where the principal changes of both the concentration and the velocity distributions occur. The value of u_0 is the characteristic velocity scale. The results is

$$\frac{\partial V}{\partial T}+\frac{u_0 t_0}{\delta}V\frac{\partial V}{\partial Z}=-\frac{u_0 t_0}{\delta}\frac{\partial P}{\partial Z}+\frac{\vartheta t_0}{\delta^2}\frac{\partial^2 V}{\partial Z^2}-\frac{g\alpha t_0 c^*}{\rho_0 u_0}C,\quad \frac{\partial C}{\partial T}+\frac{u_0 t_0}{\delta}V\frac{\partial C}{\partial Z}=\frac{D t_0}{\delta^2}\frac{\partial^2 C}{\partial Z^2};$$
$$T=0,\quad V=C=0;\quad Z=0,\quad V=-\frac{Dc^*(1-\alpha)}{c_0^*\delta u_0}\left(\frac{\partial C}{\partial Z}\right)_{z=0},\quad C=1;$$
$$Z\to\infty,\quad V=C=0. \tag{3.104}$$

The existence of the Stefan flow leads to the occurrence of flow inside the gas phase whose characteristic velocity is defined by condition (3.104):

$$\frac{Dc^*(1-\alpha)}{c_0^*\delta u_0}\sim 1,\quad u_0=\frac{Dc^*(1-\alpha)}{c_0^*\delta}. \tag{3.105}$$

If the evaporation rate is limited by the nonstationary diffusion, the parameters of both the nonstationary and the diffusion terms of the diffusion equation in (3.104) should have equal orders of magnitude:

$$\frac{Dt_0}{\delta^2}\sim 1,\quad \delta=\sqrt{Dt_0}\sim 10^{-2}\text{m if } t_0\sim 10^2\text{s}. \tag{3.106}$$

In this way the characteristic velocity of (3.105) is

$$u_0=\frac{c^*}{c_0^*}\sqrt{\frac{D}{t_0}}\sim 10^{-5}\text{m/s}. \tag{3.107}$$

From the first equation in (3.104) and (3.106) it follows that

$$\frac{\vartheta t_0}{\delta^2}=Sc\sim 1,\quad Sc=\frac{\vartheta}{D}, \tag{3.108}$$

If we suggest that the flow is limited by the natural convection, the first equation in (3.104) becomes

$$\frac{\rho_0 u_0}{g\alpha t_0 c^*}\frac{\partial V}{\partial T}+\frac{\rho_0 u_0^2}{g\alpha\delta c^*}V\frac{\partial V}{\partial Z}=-\frac{\rho_0 u_0^2}{g\alpha\delta c^*}\frac{\partial P}{\partial Z}+\frac{\vartheta\rho_0 u_0}{\delta^2 g\alpha c^*}\frac{\partial^2 V}{\partial Z^2}-C. \tag{3.109}$$

In this particular case the parameters of the last two terms in (3.109) should have equal orders of magnitude:

$$\frac{\vartheta\rho_0 u_0}{\delta^2 g\alpha c^*}\sim 1,\quad \delta=\sqrt{\frac{\nu\rho_0}{g\alpha c_0^*}\sqrt{\frac{D}{t_0}}}\sim 10^{-5}\text{m}. \tag{3.110}$$

Conditions (3.109) and (3.110) indicate that the effects of the Stefan flow and the natural convection occur in different zones of the gaseous layer above the liquid surface. This fact permits us to separate these effects if the velocity, pressure, and concentration in the first equation in (3.100) may be expressed in the form

$$v = v + \bar{v}, \quad p = \bar{p}, \quad c = \bar{c}. \tag{3.111}$$

Here $\bar{v}$, $\bar{p}$ and $\bar{c}$are determined by (3.102), whereas (3.100) (with neglected last two terms of the first equation) determines the values of v, p, and c. In this way, the form of (3.104) becomes

$$\frac{\partial V}{\partial T} + \beta V \frac{\partial V}{\partial Z} = Sc \frac{\partial^2 V}{\partial Z^2}, \quad \frac{\partial C}{\partial T} + \beta V \frac{\partial C}{\partial Z} = \frac{\partial^2 C}{\partial Z^2};$$
$$T = 0, \quad V = C = 0; \quad Z = 0, \quad V = -\frac{\partial C}{\partial Z}, \quad C = 1; \quad Z \to \infty, \quad V = C = 0, \tag{3.112}$$

where β follows directly from (3.104) to (3.106–3.108):

$$\beta = (1 - \alpha)\frac{c^*}{c_0^*} \sim 10^{-1}. \tag{3.113}$$

Obviously Eq. 3.112 are valid within a broad initial time interval t_0 when the thickness of the diffusion boundary layer $\delta = \sqrt{Dt_0}$ is less than of the depth of the gas phase l (in the cases studied in [30], $l = 0.257$ m).

The solution of (3.112) may be obtained as a series of the powers of a small parameter β:

$$V = V_0 + \beta V_1, \quad C = C_0 + \beta C_1. \tag{3.114}$$

Thus, the zeroth-order approximation is

$$\frac{\partial V_0}{\partial T} = Sc \frac{\partial^2 V_0}{\partial Z^2}, \quad \frac{\partial C_0}{\partial T} = \frac{\partial^2 C_0}{\partial Z^2};$$
$$T = 0, \quad V_0 = C_0 = 0; \quad Z = 0, \quad V_0 = -\frac{\partial C_0}{\partial Z}, \quad C_0 = 1; \quad Z \to \infty, \tag{3.115}$$
$$V_0 = C_0 = 0.$$

The solution for C_0 is

$$C_0 = erfc \frac{Z}{2\sqrt{T}}. \tag{3.116}$$

The problem for V_0 is

$$\frac{\partial V_0}{\partial T} = Sc \frac{\partial^2 V_0}{\partial Z^2}; \quad T = 0, \quad V_0 = 0; \quad Z = 0, \quad V_0 = \varphi(T) = \frac{1}{\sqrt{\pi T}};$$
$$Z \to \infty, \quad V_0 = 0. \tag{3.117}$$

The solution of (3.117) may be obtained by Green's function [37]:

$$V_0 = \varphi(T)\exp\left(-\frac{Z^2}{4ScT}\right) - \int_0^T \frac{\varphi(\tau) + 2\tau\varphi'(\tau)}{\sqrt{\tau}} \left[\int_0^{\frac{\tau Z}{\sqrt{4ScT\tau(T-\tau)}}} \exp(-u^2)du \right] d\tau, \tag{3.118}$$

This permits us to determine directly the value of V_0 at $\varphi = (\pi T)^{-1/2}$:

$$V_0 = \frac{\exp\left(-\frac{Z^2}{4ScT}\right)}{\sqrt{\pi T}}. \tag{3.119}$$

The problem formulation for C_1 follows from (3.112):

$$\frac{\partial C_1}{\partial T} - \frac{\partial^2 C_1}{\partial Z^2} = V_0 \frac{\partial C_0}{\partial Z} = -\frac{\exp\left[-\frac{Z^2}{4T}\left(1 + \frac{1}{Sc}\right)\right]}{\pi T}; \tag{3.120}$$
$$T = 0, \quad C_1 = 0; \quad Z = 0, \quad C_1 = 0; \quad Z \to \infty, \quad C_1 = 0.$$

The solution of (3.120) is obtained through Green's functions [37]:

$$C_1 = \frac{\exp\left(-\frac{Z^2}{4T}\right)}{2\pi\sqrt{\pi}} \int_0^T \frac{1}{\tau\sqrt{T-\tau}} \int_0^\infty \left\{ \exp\left[-\frac{\xi(T - Z\tau)^2}{4T\tau(T-\tau)}\right] - \exp\left[-\frac{(\xi T + Z\tau)^2}{4T\tau(T-\tau)}\right] \right\} \times \exp\left(-\frac{\xi^2}{4Sc\tau}\right) d\xi d\tau. \tag{3.121}$$

From (3.121) it follows that

$$\left(\frac{\partial C_1}{\partial Z}\right)_{Z=0} = -\frac{2}{\pi}\sqrt{\frac{Sc}{\pi T}} \operatorname{arctg}\left(Sc^{-1/2}\right). \tag{3.122}$$

The amount of evaporated liquid Q (kg/m^2) from (3.98), (3.114), (3.116), and (3.122) is

$$Q = \int_0^{t_0} Jdt = -c^* \frac{\rho^*}{c_0^*}\sqrt{Dt_0} \int_0^1 \left[\left(\frac{\partial C_0}{\partial Z}\right)_{Z=0} + \frac{c^*(1-\alpha)}{c_0^*} \left(\frac{\partial C_1}{\partial Z}\right)_{Z=0} \right] dT = A\sqrt{t_0}, \tag{3.123}$$

where

$$A = 2c^* \frac{\rho_0}{c_0^*} \left(\sqrt{\frac{D}{\pi}} + \frac{2c^*(1-\alpha)}{\pi c_0^*} \sqrt{\frac{\vartheta}{\pi}} \operatorname{arctg}\left(Sc^{-1/2}\right) \right). \tag{3.124}$$

Expression (3.124) permits us to calculate the parameter A for various systems, summarized in Table 3. The values are close to those of A_D, but in the dominating situations are quite different from the values of A_e. This indicates that the existence of an additional convective transport, which could be induced by a loss of stability of the system. Thus, the small disturbances grow to the establishment of stable amplitudes and the dissipative structures formed have a greater rate of the transport processes.

4 Example

4.1 Sulfuric Acid Alkylation Process in a Film Flow Reactor

Film flow reactors are usually designed as a bundle of tubes with liquid flowing down on their inner surfaces. The flow is oriented upward and the absorption occurs in a countercurrent mode. A cooling agent cools the bundle of tubes enclosed in a cylindrical shell.

Film flow reactors are employed for gas–liquid reactions in two principal cases.

1. When the gas absorbed by the liquid reacts with the reagents of the absorbent.
2. When the liquid absorbs two components of a gas mixture, which react thereafter in the bulk of the absorbent.

In the second case the reaction is usually a homogenous catalytic reaction, where the liquid plays the role of the catalyst.

In film flow reactors the conditions allow intense heat exchange (cooling). In this way these reactors are suitable for carrying out gas–liquid reactions accompanied by high thermal effects. An example for such a type of reaction is the alkylation of isobutane with butene with concentrated sulfuric acid as a catalyst [38].

The process is performed in the following manner. The gas mixture of isobutane (A_1) and butene (A_2) flows downward in a co-current mode with the liquid film inside a cylindrical tube. Both gases are absorbed in the liquid, where homogeneous catalytic reactions of alkylation and oligomerization take place.

$$\begin{aligned} A_1 + A_2 &= A_3, \\ A_2 + A_2 &= A_4. \end{aligned} \tag{4.1}$$

The reaction products are isooctane (A_3) and octene (A_4). The first reaction in (4.1) gives the desired products, whereas in the second reaction a by-product is generated.

Reactions (4.1) are exothermic with large thermal effects. The cooling is effected by water flowing on the outer surface of the tube. In this way a constant temperature along the tube length is maintained.

The mathematical description of the mass transfer process [38, 39] may be obtained by means of (2.1) and (2.2):

$$\tilde{u}\frac{\partial \tilde{c}_i}{\partial x} + \tilde{v}\frac{\partial \tilde{c}_i}{\partial y} = \tilde{D}_i \frac{\partial^2 \tilde{c}_i}{\partial y^2}; \quad u\frac{\partial c_i}{\partial x} = D_i \frac{\partial^2 c_i}{\partial y^2} - r_i; \quad y = 0, \quad c_i = 0;$$
$$y = h_0, \quad \tilde{c}_i = \chi_i c_i, \quad \tilde{D}_i \frac{\partial \tilde{c}_i}{\partial y} = D_i \frac{\partial c_i}{\partial y}; \quad y \to \infty, \quad \tilde{c}_i = \tilde{c}_{01}, \quad i = 1,2, \tag{4.2}$$

where

$$r_1 = -\frac{\partial c_1}{\partial t} = k_1 c_1 c_2, \quad r_2 = -\frac{\partial c_2}{\partial t} = k_1 c_1 c_2 + k_2 c_2^2. \tag{4.3}$$

The solution of the boundary problem requires dimensionless variables and parameters as follows:

$$X = \frac{x}{L}, \quad \tilde{Y}_i = \frac{y - h_0}{\tilde{\delta}}, \quad Y_i = \frac{h_0 - y}{\delta_i}, \quad \tilde{U} = \frac{\tilde{u}}{\tilde{u}_0}, \quad \tilde{V}_i = \frac{\tilde{v}}{\tilde{\varepsilon}_i \tilde{u}_0}, \quad U = \frac{u}{u_0},$$
$$\tilde{C}_i = \frac{\tilde{c}_i}{\tilde{c}_{0i}}, \quad C_i = \frac{c_i}{c_{0i}}, \quad c_{0i} = \frac{\tilde{c}_{0i}}{\chi_i}, \quad \tilde{F}o_i = \frac{\tilde{D}_i L}{\tilde{u}_0 \tilde{\delta}_i^2}, \quad Fo_i = \frac{D_i L}{u_0 \delta_i^2}, \quad K_i = \frac{k_i L}{u_0},$$
$$R_i = \frac{r_i L}{u_0 c_{0i}}, \quad \tilde{\varepsilon}_i = \frac{\tilde{\delta}_i}{L}, \quad \varepsilon_i = \frac{\delta_i}{L}, \quad \varepsilon_{0i} = \frac{D_i \tilde{\delta}_i}{\tilde{D}_i \delta_i}, \quad i = 1,2, \tag{4.4}$$

where $\tilde{\delta}_i, \delta_i (i = 1,2)$ are the orders of magnitude of the diffusion boundary layer thicknesses in the gas and the liquid. In this way (4.2) takes the form

$$\tilde{U}\frac{\partial \tilde{C}_i}{\partial X} + \tilde{V}_i \frac{\partial \tilde{C}_i}{\partial \tilde{Y}_i} = \tilde{F}o_i \frac{\partial^2 \tilde{C}_i}{\partial \tilde{Y}_i^2}; \quad U\frac{\partial C_i}{\partial X} = Fo_i \frac{\partial^2 \tilde{C}_i}{\partial \tilde{Y}_i^2} - R_i; \quad X = 0, \quad \tilde{C}_i = 1,$$
$$C_i = 0;$$
$$\tilde{Y}_i = Y_i = 0, \quad \tilde{C}_i = C_i, \quad \frac{\partial \tilde{C}_i}{\partial \tilde{Y}_i} = \frac{\varepsilon_{0i}}{\chi_i}\frac{\partial C_i}{\partial Y_i};$$
$$\tilde{Y}_i \to \infty, \quad \tilde{C}_i = 1; \quad Y_i \to \infty; \quad C_i = 1; \quad i = 1,2, \tag{4.5}$$

where

$$R_1 = K_1 c_{02} C_1 C_2, \quad R_2 = K_2 c_{01} C_1 C_2 + K_2 c_{02} C_2^2. \tag{4.6}$$

It is possible to express the mass transfer rate as

$$J_i = \tilde{\beta}_i \tilde{c}_{0i} - \beta_i c_{0i}, \quad i = 1,2. \tag{4.7}$$

This rate is determined by means of the average value of the diffusive flux. For a liquid film of length L the results are

$$J_i = \frac{1}{L}\int_0^L \tilde{D}_i\left(\frac{\partial \tilde{c}_i}{\partial y}\right)_{y=h} dx = \frac{1}{L}\int_0^L D_i\left(\frac{\partial c_i}{\partial y}\right)_{y=h} dx, \quad i = 1,2. \tag{4.8}$$

or for the Sherwood number,

$$\widetilde{Sh}_i = \frac{\tilde{\beta}_i L}{\tilde{D}_i} = \tilde{\varepsilon}_i^{-1}\int_0^1 \left(\frac{\partial \tilde{C}_i}{\partial \tilde{Y}_i}\right)_{\tilde{Y}=0} dX, \quad Sh_i = \frac{\beta_i L}{D_i} = \varepsilon_i^{-1}\int_0^1 \left(\frac{\partial C_i}{\partial Y_i}\right)_{Y=0} dX, \tag{4.9}$$

$$i = 1,2.$$

Relationships (4.9) permit us to determine the kinetics of the sulfuric acid alkylation process on the basis of the solution of (4.5).

The second reaction in (4.1) is undesirable. Because of that, in practical situations the condition $c_{02} \ll c_{01}$ is satisfied. From experimental data [40] for the microkinetics of the reaction (4.1) and on the basis of the findings of the experiments reported in [38, 39] ($u_0 = 0.224$ m/s, $L = 2$ m) one obtains that

$$K_1 = 0,4 \times 10^7, \quad K_2 = 1,6 \times 10^7. \tag{4.10}$$

From (4.5), (4.6), and (4.10) it is evident that the liquid hydrodynamics does not affect the mass transfer.

In practical situations, the condition $K_1 c_{01} > K_2 c_{02}$ is satisfied and the orders of magnitude of the diffusion boundary layers thicknesses can be determined:

$$\widetilde{Fo}_i \approx 1, \quad Fo_i = K_1 \frac{c_{01} c_{02}}{c_{0i}}, \quad \tilde{\delta}_i = \sqrt{\frac{\tilde{D}_i L}{\tilde{u}_0}}, \quad \delta_i = \sqrt{\frac{D_i L c_{0i}}{u_0 K_1 c_{01} c_{02}}}, \quad i = 1,2. \tag{4.11}$$

The experimental conditions reported in [38, 39] ($\tilde{u}_0 = 0.23\mathrm{m/s}$) and the data published in [40] permit us to establish that

$$\frac{\varepsilon_{01}}{\chi_1} = 0,53, \quad \frac{\varepsilon_{02}}{\chi_2} = 0,36, \tag{4.12}$$

i.e., the diffusion resistances are located in both phases.

The results obtained (1.10–1.12) allow us to solve problem (4.5) subsequently for the gas and the liquid phases:

$$\tilde{U}\frac{\partial \tilde{C}_i}{\partial X} + \tilde{V}_i\frac{\partial \tilde{C}_i}{\partial \tilde{Y}_i} = \tilde{F}o_i\frac{\partial^2 \tilde{C}_i}{\partial \tilde{Y}_i^2}; \quad X = 0, \quad \tilde{C}_i = 1;$$

$$\tilde{Y}_i = 0, \quad \tilde{C}_i = \tilde{C}_i^*; \quad \tilde{Y}_i \to \infty, \quad \tilde{C}_i = 1; \quad i = 1,2. \tag{4.13}$$

$$\frac{\partial^2 C_1}{\partial Y_2^2} = \frac{\delta_2}{\delta_1} C_1 C_2; \quad \frac{\partial^2 C_2}{\partial Y_2^2} = C_1 C_2 + \frac{k_2 c_{20}}{k_1 c_{10}} C_2^2;$$

$$Y_2 = 0, \quad C_i = \tilde{C}_i^*; \quad Y_2 \to \infty, \quad C_i = 0; \quad i = 1, 2, \tag{4.14}$$

where Y_1 is replaced by Y_2. The unknown constants $\tilde{C}_i^*(i = 1, 2)$ must be determined in a way allowing satisfaction by $\tilde{C}_i$ and $C_i(i = 1, 2)$ of the following boundary condition:

$$\left(\frac{\partial \tilde{C}_i}{\partial \tilde{Y}_i}\right)_{\tilde{Y}_i=0} = \frac{\varepsilon_{01}}{\chi_i} \left(\frac{\partial C_i}{\partial Y_i}\right)_{Y_i=0}, \quad i = 1, 2. \tag{4.15}$$

Problem (4.13) was solved in [39]. In this case (4.15) takes the form

$$\left(\frac{\partial C_i}{\partial Y_2}\right)_{Y_2=0} = -A_i(1 - \tilde{C}_i^*)\chi_i \left(\frac{1}{\varphi_{01}} + \frac{\theta_1}{\alpha \varphi_{01}^2}\right), \quad i = 1, 2, \tag{4.16}$$

where

$$A_i = \sqrt{\frac{\tilde{D}_i \tilde{u}_0 \chi_1 D_2}{D_i^2 u_0 K_1 \tilde{c}_{01}}}, \quad i = 1, 2, \quad \theta_1 = \frac{u_0}{\tilde{u}_0}, \quad \alpha = 0,332 \tag{4.17}$$

and φ_{01} is a function of the Schmidt number of the gas phase [5].

Problem (4.14) was solved numerically [38, 39] by an iterative procedure. The values of $\tilde{C}_i^*(i = 1, 2)$ were varied to satisfy condition (4.4.16). After the determination of $\tilde{C}_i^*(i = 1, 2)$, the Sherwood number was established by means of (4.9):

$$Sh_1 = \tilde{S}h_{0i}(1 - \tilde{C}_i^*), \quad i = 1, 2, \tag{4.18}$$

where

$$\tilde{S}h_{0i} = \frac{\chi_i}{D_i} \sqrt{\tilde{D}_i L \tilde{u}_0} \left(\varphi_{0i}^{-1} + \frac{\theta}{\alpha} \varphi_{0i}^{-2}\right), \quad i = 1, 2, \tag{4.19}$$

is the Sherwood number in the case of a mass transfer limited by the transport in the gas. The term $\left(1 - \tilde{C}_i^*\right), i = 1, 2$, indicates that the diffusion resistance in the liquid plays an important role.

The process discussed here depends strongly on the effective utilization of the butene. This indicates that the values of Sh_2 and β_2 are enough to establish the macrokinetics. Because of the fact that butene reacts vigorously during its dissolution in sulfuric acid, the evaluation of Henry's constant is practically impossible. It may, however, be done through a comparison of Sh_2 with experimental data.

The analysis of the experimental data for the Sherwood number (Sh_{exp}) published in [38] shows that the turbulence in the gas phase must be taken into account (a linear relationship between the coefficient of a turbulent diffusion and the gas velocity). Thus, $\tilde{D}_2$ in (4.2) may be replaced by $\tilde{D}_2^t$:

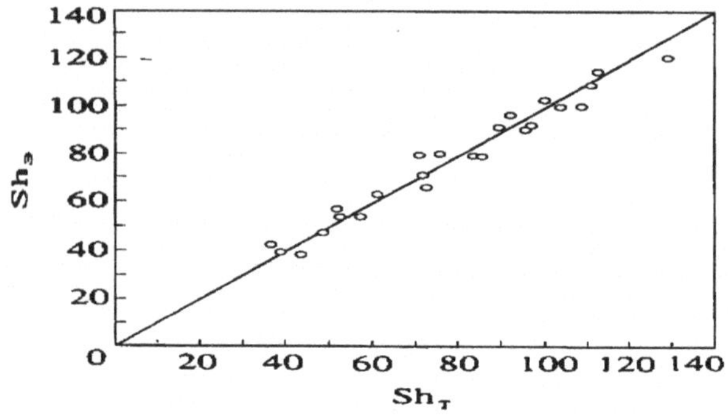

Fig. 2 Comparison of the theoretical and experimental values of the Sherwood number

$$\tilde{D}_2^t = d\tilde{D}_2 u_0, \tag{4.20}$$

where d is a constant.

This demonstrates the possibility to obtain a theoretical relationship for the Sherwood number by means of (4.19):

$$Sh_{\mathrm{T}} = \frac{\chi_2 \sqrt{d\tilde{u}_0}\sqrt{\tilde{D}_2 L}}{D_2}\left(\varphi_{02}^{-1} + \frac{\theta}{\alpha}\varphi_{02}^{-2}\right)\left(1 - \tilde{C}_2^*\right). \tag{4.21}$$

The comparison of the experimental data and Sh_{T} permits us to establish that

$$\chi_2\sqrt{d} = 14.5 \tag{4.22}$$

and to substitute it in (4.22). A parity plot of the theoretical and experimental values of the Sherwood number is shown in Fig. 2.

References

1. Boyadjiev Chr (1993) Fundamentals of modeling and simulation in chemical engineering and technology. Bulgarian Academy of Sciences, Sofia (in Bulgarian)
2. Guhman AA (1973) Introduction to similarity theory. Vishaya Shkola, Moskow (in Russian)
3. Danckwerts FRE (1970) Gas-liquid reactions. McGraw-Hill, New York
4. Ramm VM (1966) Gas absorption. Khimia, Moscow (in Russian)
5. Chr Boyadjiev, Beschkov V (1988) Mass transfer in liquid film flows. Mir, Moscow (in Russian)
6. Chr Boyadjiev (1983) Chem Eng Sci 38:641
7. Boyadjiev Chr (1990) Chem Technol (Russia) 5:53
8. Boyadjiev Chr (1990) Chem Technol (Russia) 5:48
9. Boyadjiev Chr (1985) Hung J Ind Chem 13:163
10. ChrB Boyadjiev, Babak VN (2000) Non-linear mass transfer and hydrodynamic stability. Elsevier, Amsterdam
11. Chr Boyadjiev (2002) Hung J Ind Chem 30:13
12. Chr Boyadjiev (1996) Hung J Ind Chem 24:35
13. Chr Boyadjiev, Halatchev I (1998) Int J Heat Mass Transf 41:939
14. Chr Boyadjiev (2000) Int J Heat Mass Transf 43:2749

15. Chr Boyadjiev (2000) Int J Heat Mass Transf 43:2759
16. Krylov VS, ChrB Boyadjiev (1996) Non-linear mass transfer. Institute of Thermophysics, Novosibirsk (in Russian)
17. Chr Boyadjiev, Halatchev I (1998) Int J Heat Mass Transf 41:197
18. Chr Boyadjiev, Doichinova M (1999) Hung J Ind Chem 27:215
19. Chr Boyadjiev (1998) Hung J Ind Chem 26:181
20. Ts Sapundzhiev, ChrB Boyadjiev (1993) Russ J Eng Thermophys 3:185
21. Frank-Kamenetskii DA (1967) Diffusion and heat transfer in chemical kinetics. Science, Moscow (in Russian)
22. Bird RB, Stewart WE, Lightfoot EN (1965) Transport phenomena. Wiley, New York
23. Benard H (1901) Ann Chim Phys 23(7):62
24. Dilman VV, Kulov NN, Lothov VA, Kaminski VA, Najdenov VI (1998) Theor Fundam Chem Technol 32(4):377 (in Russian)
25. Chr Boyadjiev (2000) Int J Heat Mass Transf 43:2749
26. Chr Boyadjiev (2000) Int J Heat Mass Transf 43:2759
27. Chr Boyadjiev (2001) Int J Heat Mass Transf 44:1119
28. Chr Boyadjiev (2001) Int J Heat Mass Transf 44:2505
29. Pearson JKA (1958) J Fluid Mech 4:489
30. Dilman VV, Lothov VA, Kulov NN, Najdenov VI (2000) Theor Found Chem Eng 34:227
31. Carslaw HS, Jaeger JC (1959) Conduction of heat in solids, 2nd edn edn, Oxford University Press, London
32. Chr Boyadjiev (2000) Int J Heat Mass Transf 43:2749
33. Chr Boyadjiev (2000) Int J Heat Mass Transf 43:2759
34. Chr Boyadjiev (2001) Int J Heat Mass Transf 43:1119
35. Boyadjiev B, Chr Boyadjiev (2001) Int J Heat Mass Transf 43:2505
36. Gershuni GZ, Zhuhovicki EM (1972) Convection stability of a non-compressible liquid. Nauka, Moscow (in Russian)
37. Chr Boyadjiev, Beschkov V (1984) Mass transfer in liquid film flows. Bulgarian Academy of Sciences, Sofia
38. Parvez IM (1988) PhD thesis. Higher Institute of Chemical Technology, Sofia (in Bulgarian)
39. Parvez IM, Chr Karagiozov, Chr Boyadjiev (1991) Chem Eng Sci 46:1589
40. Lee L, Harriott P (1977) Ind Eng Chem Process Des Dev 16:282
41. Chr Boyadjiev (1987) Hung J Ind Chem 15.479

Quantitative Analysis

The quantitative analysis of the models is the last step before using the models for modeling and simulation of the real processes. The quantitative results obtained from small-scale laboratory models must be "remade" for large-scale real process simulation. This "scale-up" from the models to the real processes is a very important stage in the modeling and simulation.

The quantitative results obtained from the models are random numbers because the experimental data used for parameter identification are random numbers too. A statistical analysis of the significance of the parameters and the adequacy of the model is needed. *The model becomes a model after the statistical proof of its adequacy has been done.*

1 Scale-Up

As mentioned earlier, the quantitative description of industrial processes is the main problem in chemical engineering. The solutions of such problems need experimental data obtained from laboratory models and permit us to study the process mechanisms or to realize their optimal design, control, or improvement. In physical modeling, these experimental data are used directly through scaling coefficients. In mathematical modeling, however, the experimental data are needed to determine the model parameters, whereas the quantitative description is a result of a mathematical (numerical) experiment. The quantitative description of an industrial process through a quantitative description of the laboratory model is the essence of the chemical engineering scale-up [1].

The design of industrial apparatuses, especially those with high capacities, faces two principal scale-up problems:

1. The occurrence of incompatible similarity criteria.
2. The scale-up effect.

C. Boyadjiev, *Theoretical Chemical Engineering*,
DOI: 10.1007/978-3-642-10778-8_5,

The theory of similarity [2] shows that the increase of the linear scale in the transition from the laboratory model to the industrial prototype, in most cases, leads to an incompatibility of the similarity criteria. This occurs because the criteria values cannot be identical (equal) for both the model and the prototype. The reason is that in such situations there are big differences between the geometrical (linear) scales. This inconvenience can be avoided (in mathematical modeling) by the direct introduction into the model of the linear scale of the prototype, i.e., by nonemployment of scaling.

However, the scale-up theory shows that in the mathematical modeling the problem of the scale effect arises if the size of the modeled device increases. The most frequently occurring scale-up effect is the decrease of the efficiency of the industrial process with respect to the laboratory model. The reason is that a physical effect occurs only in industrial devices. This indicates that during the determination of the model parameters, on the basis of the experimental data from the laboratory model, the scale effect is not taken into account. Therefore, the mathematical model is not able to predict the scale effect.

The solution of the scale effect requires answers to several questions about the similarity and the scale-up, the physical essence of the scale effect, and the possibilities of the scale-up theory and hydrodynamic modeling to diminish and to predict the scale effect [1].

1.1 Similarity and Scale-Up

It was demonstrated earlier that the introduction of proper (characteristic) scales of the process leads to a mathematical description in terms of generalized (dimensionless) variables (a set of generalized equations). As a result, the mathematical description contains dimensionless parameters. For concrete values of the system parameters, the set represents the mathematical description of a *generalized individual case* [2]. Thus, it concerns many processes that are similar owing to their identical mathematical descriptions and equal values of the dimensionless parameters. For a particular set of parameters, the generalized descriptions become a description of a particular process through generalized variables. Therefore, each process of the generalized individual case may be used as a model for the other ones.

The theory of similarity is widely used for *single-phase systems*. For example, the fluid flow along a semi-infinite plate has the following mathematical description [3]:

$$u_x \frac{\partial u_x}{\partial x} + u_y \frac{\partial u_x}{\partial y} = \nu \frac{\partial^2 u_x}{\partial y^2}, \quad \frac{\partial u_x}{\partial x} + \frac{\partial u_y}{\partial y} = 0, \tag{1.1}$$

$$x = 0, u_x = u_0; \quad y = 0, u_x = u_y = 0; \quad y \to \infty, u_x = u_0.$$

The introduction of the characteristic scales

$$x = LX, \quad y = LY, \quad u_x = u_0 U_x, \quad u_y = u_0 U_y, \tag{1.2}$$

into (1.1) leads to

$$\begin{gathered} U_x \frac{\partial U_x}{\partial X} + U_y \frac{\partial U_x}{\partial Y} = \frac{1}{Re} \frac{\partial^2 U_x}{\partial Y^2}, \quad \frac{\partial U_x}{\partial X} + \frac{\partial U_y}{\partial Y} = 0, \\ X = 0,\ U_x = 1; \quad Y = 0,\ U_x = U_y = 0; \quad Y \to \infty,\ U_x = 1, \end{gathered} \tag{1.3}$$

where $Re = \frac{u_0 L}{\nu}$. Here, L is a predetermined length along the plate.

Thus, for $Re = Re_0$ all the flows described by the generalized individual case have identical dimensionless velocity profiles obtained by solving (1.3):

$$U_x = U_x(X, Y, Re_0), \quad U_y = U_y(X, Y, Re_0). \tag{1.4}$$

The various velocity profiles of the generalized individual case follow from (1.4) through the introduction of the scale coefficients:

$$u_x = u_0 U_x\left(\frac{x}{L}, \frac{y}{L}, Re_0\right), \quad u_y = u_0 U_y\left(\frac{x}{L}, \frac{y}{L}, Re_0\right). \tag{1.5}$$

Solutions (1.5) represent the velocity profiles of the different flows of a generalized individual case with $Re_0 = \text{const.}$, i.e., they are similar flows.

Let us consider two flows described by (1.5) with characteristic linear dimensions L_1 and L_2 ($L_1 > L_2$). Besides, the first flow is a model of the second one. If the conditions of similarity are satisfied, it is necessary that the following condition to held:

$$\frac{u_{01} L_1}{\nu} = \frac{u_{02} L_2}{\nu} = Re_0. \tag{1.6}$$

Thus, the characteristic velocities are

$$u_{01} = \frac{Re_0 \nu}{L_1}, \quad u_{02} = \frac{\mathrm{Re}_0 \nu}{L_2}. \tag{1.7}$$

The above relationships indicate that if the velocity profile of the model flow (with characteristic parameters $u_0 = u_{01}$ and $L = L_1$) is known, then we have

$$u_{x1} = u_{01} U_x\left(\frac{x}{L_1}, \frac{y}{L_1}, Re_0\right), \quad u_{y1} = u_{01} U_y\left(\frac{x}{L_1}, \frac{y}{L_1}, Re_0\right) \tag{1.8}$$

The velocity distribution of the prototype follows directly from (9.1.8) in the form

$$u_{x2} = u_{02} U_x\left(\frac{x}{L_2}, \frac{y}{L_2}, Re_0\right), \quad u_{y2} = u_{02} U_y\left(\frac{x}{L_2}, \frac{y}{L_2}, Re_0\right). \tag{1.9}$$

The results obtained above show that the scale-up is done correctly if the characteristic scale of velocity u_0 decreases when the linear scale L is increased, so the following condition is satisfied:

$$u_{01}L_1 = u_{02}L_2 = \nu Re_0 = \text{const.} \tag{1.10}$$

As a second example, let us consider gravity-driven film flow down to an inclined plate. The force balance yields an equation defining the velocity profile:

$$u_x \frac{\partial u_x}{\partial x} + u_y \frac{\partial u_x}{\partial y} = \nu \frac{\partial^2 u_x}{\partial y^2} + g.\sin\alpha, \tag{1.11}$$

where α is the angle of plate inclination with respect to the horizon. Under the new conditions, problem (1.3) takes the form

$$\begin{gathered} U_x \frac{\partial U_x}{\partial X} + U_y \frac{\partial U_x}{\partial Y} = \frac{1}{Re}\frac{\partial^2 U_x}{\partial Y^2} + \frac{\sin\alpha}{Fr}, \quad \frac{\partial U_x}{\partial X} + \frac{\partial U_y}{\partial Y} = 0, \\ X = 0,\ U_x = 1; \quad Y = 0,\ U_x = U_y = 0; \quad Y \to \infty,\ U_x = 1, \end{gathered} \tag{1.12}$$

where $Fr = \frac{u_0^2}{gL}$ is the Froude number.

The comparison between Re and Fr indicates that these similarity criteria are incompatible with the scaling rules, because two conditions must be satisfied simultaneously:

$$u_0 L = \nu Re = \text{const.}, \quad \frac{u_0^2}{L} = gFr = \text{const.} \tag{1.13}$$

The first condition requires the increase of the linear scale L to be compensated by the reduction of the characteristic velocity u_0. On the other hand, the second condition needs just the opposite behavior of the characteristic scales.

The above example shows that if single-phase flows have a characteristic velocity (incorporated in Re) and depend on the gravity (incorporated in Fr), the latter considers the fact that the increase of L cannot be compensated by a significant change of the fluid kinematic viscosity to satisfy conditions (1.13). Therefore, the similarity theory cannot be employed for a scale-up procedure of all the flows depending on Re and Fr only owing to the incompatibility of both dimensionless numbers under the imposed conditions (1.13).

Industrial processes in gas–liquid and liquid–liquid two-phase systems are carried out in dispersions with a continuous phase and dispersed phase. This indicates that both phases have different characteristic linear scales. The dispersion phase moves through the continuous one in the form of bubbles, drops, liquid films, jets, etc. This specific form of movement determines a characteristic length scale d usually employed as a constant linear dimension in the scaling. The continuous phase occupies almost the entire apparatus volume and consequently its characteristic length scale depends on the linear dimension L of the vessel (column diameter or height).

In *two-phase systems* [3], the flow similarity is determined by

- The momentum equations of both phases

$$u_x^{(i)} \frac{\partial u_x^{(i)}}{\partial x} + u_y^{(i)} \frac{\partial u_x^{(i)}}{\partial y} = \nu^{(i)} \frac{\partial^2 u_x^{(i)}}{\partial y^2}, \qquad i = 1, 2; \tag{1.14a}$$

- The boundary conditions at the interphase surface $y = 0$

$$y = 0, \quad u_x^{(1)} = u_x^{(2)}, \quad \mu^{(1)} \frac{\partial u_x^{(1)}}{\partial y} = \mu^{(2)} \frac{\partial u_x^{(2)}}{\partial y}. \tag{1.14b}$$

assuming a continuity of both the velocity and the stress tensor.

In a way similar to (1.2), the characteristic scales

$$x = L^{(i)} X^{(i)},\ y = L^{(i)} Y^{(i)},\ u_x^{(i)} = u_0^{(i)} U_x^{(i)},\ u_y^{(i)} = u_0^{(i)} U_y^{(i)} \tag{1.15}$$

transform (1.14) into dimensionless forms:

$$U_x^{(i)} \frac{\partial U_x^{(i)}}{\partial X} + U_y^{(i)} \frac{\partial U_x^{(i)}}{\partial Y} = \frac{1}{Re^{(i)}} \frac{\partial^2 U_x^{(i)}}{\partial Y^2}, \qquad i = 1, 2; \tag{1.16a}$$

$$Y^{(1)} = Y^{(2)} = 0,\ U_x^{(1)} = \vartheta_1 U_x^{(2)}, \quad \vartheta_2 \frac{\partial U_x^{(1)}}{\partial Y} = \frac{\partial U_x^{(2)}}{\partial Y}, \tag{1.16b}$$

where

$$Re^{(i)} = \frac{u_0^{(i)} L^{(i)}}{\nu^{(i)}}, \qquad i = 1, 2, \tag{1.17a}$$

$$\vartheta_1 = \frac{u_0^{(2)}}{u_0^{(1)}}, \tag{1.17b}$$

$$\vartheta_2 = \frac{\mu_1 u_0^{(1)} L^{(2)}}{\mu_2 u_0^{(2)} L^{(1)}}. \tag{1.17c}$$

Let us consider the first phase is the dispersion phase ($L^{(1)} = d =$ const.) and the second phase is the continuous one with a characteristic scale $L^{(2)} = L$. The scale-up procedure concerns the increase of $L^{(2)}$ from $L_1^{(2)}$ to $L_2^{(2)}$. The condition $Re_0^{(2)} =$ const. requires that

$$u_{01}^{(2)} L_1 = u_{02}^{(2)} L_2 = \nu_2 Re_0^{(2)} = \text{const.} \tag{1.18}$$

Thus, the increase of $L^{(2)}$ needs a reduction (decrease) of $u_0^{(2)}$.

Expression (1.17c) indicates that the similarity condition $\vartheta_2 = \text{const}$ yields

$$\frac{L_1}{u_{01}^{(2)}} = \frac{L_2}{u_{02}^{(2)}} = \vartheta_2 \frac{\mu_2 d}{\mu_1 u_0^{(1)}} = \text{const.} \tag{1.19}$$

Condition (1.19) states that the increase of $L = L^{(2)}$ must be compensated by an increase of $u_0^{(2)}$ and vice versa.

Conditions (1.18) and (1.19) form two incompatible similarity criteria, $Re^{(2)}$ and ϑ_2, that must be satisfied with the variations of the linear dimension $L = L^{(2)}$. Therefore, the data obtained from the model with characteristic scale $L_1^{(2)}$ cannot be employed for a quantitative description of the prototype with characteristic dimension $L_2^{(2)}$.

The results of this example show that the theory of similarity cannot be employed for a scale-up of two-phase systems with a significant hydrodynamic interphase interaction ($\vartheta_2 \sim 1$). This restriction is removed in the case of weak interphase interactions ($\vartheta_2 < < 1$ or $\vartheta_2 > > 1$, when the flows do not depend on ϑ_2).

The *mass transfer complicated by a chemical reaction is a typical case of when incompatibility of criteria occurs after the scale-up procedure.* The basic equation represents a mass balance between convective transfer, diffusive transfer, and chemical reaction rate (as a volume source):

$$u_x \frac{\partial c}{\partial x} + u_y \frac{\partial c}{\partial y} = D \frac{\partial^2 c}{\partial y^2} + kc\,. \tag{1.20}$$

The use of the characteristic scales (1.2) yields

$$U_x \frac{\partial C}{\partial X} + U_y \frac{\partial C}{\partial Y} = \frac{1}{Pe} \frac{\partial^2 C}{\partial Y^2} + Da\, C\,, \tag{1.21}$$

where two dimensionless number occur,

$$Pe = \frac{u_0 L}{D}, \tag{1.22a}$$

where *Pe* is the Péclet number, and

$$Da = \frac{kL}{u_0}, \tag{1.22b}$$

where *Da* is the Damkohler number. Obviously, both dimensionless numbers are incompatible in the scale-up procedure owing to the requirement for the following conditions to be satisfied simultaneously:

$$u_{01} L_1 = u_{02} L_2 = Pe_0 D = \text{const.}, \tag{1.23a}$$

$$\frac{L_1}{u_{01}} = \frac{L_2}{u_{02}} = \frac{Da_0}{k} = \text{const.} \tag{1.23b}$$

The three examples discussed above span a large number of processes with incompatible similarity criteria during the scale-up procedure. This hinders the employment of the physical modeling approach for the scale-up problems. The mathematical modeling allows us to overcome the problem. Three basic assumptions form its skeleton:

1. *The first assumption is that the mathematical model is not related to the linear scale of the process.* Thus, the numerical experiments are addressed to the prototype (in an industrial scale) as an object of modeling. Despite this, the scaling sets up problems related to the experimental data needed to determine the model parameters.
2. *The hierarchical approach is the second basic assumption allowing the mathematicalmodeling to solve the scale-up problem.* The existence of incompatible similarity criteria in the mathematical description needs experiments to be carried out on the industrial prototype. The scale-up problems could be avoided by the use of the hierarchical modeling approach. It allows a separate determination of the model parameters that are incompatible similarity criteria in the scale-up. For example, *Pe* and *Da* in (9.1.21) (i.e., *D* and *k*) could be determined experimentally with different laboratory models. The hierarchical approach is inapplicable to the physical modeling because the hierarchical models do not satisfy the similarity conditions.
3. *The employment of both the scale-up theory and the hydrodynamic modeling* is the third and the most important assumption that allows the scale-up problems to be solved. The reason is that there are significant discrepancies between the efficiencies of the process performed in laboratory and that performed on industrial scales.

1.2 Scale Effect

The increase of the dimensions of apparatuses from the laboratory model to the industrial prototype leads to a decrease of the process efficiency. The reduction of the process efficiency usually has hydrodynamic origins. Practically, there are several reasons for that: the velocity nonuniformity in the cross-sectional area of larger devices, the increase of the turbulence scale, etc. Usually these reasons manifest themselves through an increased axial mixing in the apparatuses.

The scale-up coefficient is the ratio of the height of the transfer units in the industrial prototype to that of the laboratory model. It can vary over a very large range. For example, in liquid–liquid extraction columns with plates [1], it varies between 5 and 10, whereas the use of valve plates and perforated bubble plates for gas–liquid systems shrinks the range from 2 to 3.

The scale effect can have other manifestations as well. For example, in chemical reactors the appearance of new undesirable by-products usually has a scale-up origin.

The nonuniformity of both the velocity and the phase holdups in the cross sections of industrial devices is the hydrodynamic origin of the scale effect. For example, the increase of the radial nonuniformity of the velocity (as a result of the increase of the dimensions the apparatus) leads to an augmentation of the axial mixing, an increase of the turbulence scale, stagnant (dead) zone formation, bypass currents, etc.

The flows in industrial apparatuses are generally turbulent and two types of turbulence pulsations characterize them [3]:

1. *Large-scale pulsations*. They play the principal role in the flow formation and introduce most of the kinetic energy. The scale of the large-scale pulsations depends on the characteristic length of the region where the flow passes. This length is a basic (important) scale of the turbulent motion. The large-scale pulsation transmits energy to the low-scale pulsations without energy dissipation.
2. *Low-scale pulsations*. They are not important for the general flow structure. They contribute a small amount of the flow kinetic energy and their scale could be different. The minimum scale depends on the Reynolds flow number, i.e., the pulsations decrease with the increase of *Re*. The flow energy dissipation (the transformation of kinetic energy into heat) occurs at the level of the low-scale pulsations. Usually the dissipated energy is a small amount of the flow kinetic energy.

Very often, the scale-up changes the main scale of the turbulent flow. For example, in a columnar device with plates, the increase of the column diameter is followed by an increase of the plate spacing to satisfy the condition of geometrical similarity. However, this changes the scale of the large-scale pulsations of the flow between the plates.

The radial flow nonuniformity is also a manifestation of the scale effect occurring parallel to the changes of the turbulence scale. It can be predicted theoretically and expressed by the axial mixing [1].

The physical modeling cannot detect the occurrences of stagnation zones and recirculation currents as well as the bypass flows. The same effects cannot be predicted theoretically by the mathematical model too. Therefore, they must be minimized by the method of hydrodynamic modeling and their consequences must be expressed through the axial mixing.

The analysis of the manifestations of the scale-up effect given above indicates that the nonuniformities can be arranged in four groups:

1. The first type is *the nonuniformities at the boundaries*. They represent the nonuniform flow distributions (in the apparatus cross section) at both the inlet and the outlet of the device. The main reasons for such nonuniformities are the designs of the distributors, the diffusers, etc.

2. The second type represents *the internal nonuniformities* due to specific properties of the two-phase flows and the design of the contactors, where the interaction between the phases takes place. Some examples are the liquid distribution over random packing, liquid flow bypass along the column walls, the axial flow mixing, the wave formation on bubbling plates, and the recirculation flows in bubble columns and fluidized beds.
3. The third type is *nonuniformities due to incorrect assembly* of the contacting devices. Usually, they can be observed at nonhorizontal contacting plates, nonuniformly arranged packing, or catalysts in columns, etc.
4. The fourth type is *nonuniformities due to a number of design imperfections* of the contacting devices. The bypass flows or the stagnation zones are results of such types of nonuniformities.

The types of nonuniformity described above can exist in various combinations in the contacting devices and they are often specific for a particular type of contactor. For example, the first three types of nonuniformities occur in countercurrent packed columns. On the other hand, the boundary effects are not significant in contactors with mechanical flow stirring (by impellers). In this case the turbulence pulsation scale determines the scale effect, the axial mixing intensity, and the dimensions of the circulation loops.

The types of nonuniformities considered cannot be modeled on small-diameter apparatuses, because they are typical for large-scale (large-diameter) industrial devices. Attempts to predict these effects have met with no success owing to the effect of the contactor design on the flow structure that cannot be predicted theoretically with existing engineering tools. The only reasonable approach in such situations is:

- To take measures through the contactor design to minimize the nonuniformities.
- To incorporate the residual scale effect in the mathematical model for determining the industrial apparatus efficiency.

The approach can be performed by the *approximate scale-up theory* [1].

1.3 Diffusion Model

The use of the diffusion model to solve the scale-up problems is made very conditionally [1], and so the terms need to be formulated exactly.

The equalization of the concentration in a volume is possible by mixing (diffusion mechanism) and by stirring (convection mechanism). In the first case, the effect is represented in the model as a diffusion term (second derivative of the concentration), whereas in the second case the effect is presented as a convection term (first derivative of the concentration). In the theory presented [1] different stirring effects are substituted for the diffusion equivalents. This exchange of physical effects and their mathematical descriptions is possible on the basis of an experimentally obtained parameter, but its value is a constant within a very small interval of the experimental conditions.

The diffusion model is widely used for modeling of column apparatuses. In the case of a single-phase flow, the model expresses the convective mass transfer (through the velocity w) and the diffusive transfer (through the axial diffusivity $D_{||}$):

$$D_{||}\frac{d^2c}{dz^2} - w\frac{dc}{dz} = 0; \quad z = 0,\ w(c_0 - c) = -D_{||}\frac{dc}{dz}; \quad z = L,\ \frac{dc}{dz} = 0. \tag{1.24}$$

Here, w does not depend on the apparatus radius r (i.e., it is the superficial velocity). In this case $D_{||}$ is the *axial diffusion coefficient.*

The presence of a radial nonuniformity of the velocity $w(r)$ leads to a radial diffusion with a *radial diffusivity* $D_{\perp}$:

$$\begin{gathered} D_{\perp}\frac{1}{r}\frac{\partial}{\partial r}\left(r\frac{\partial c}{\partial r}\right) + D_{||}\frac{\partial^2 c}{\partial z^2} - w(r)\frac{\partial c}{\partial z} = 0; \quad z = 0,\ w(c_0 - c) = -D_{||}\frac{\partial c}{\partial z}; \\ z = L,\ \frac{\partial c}{\partial z} = 0; \quad r = 0,\ \frac{\partial c}{\partial r} = 0; \quad r = r_0,\ \frac{\partial c}{\partial r} = 0. \end{gathered} \tag{1.25}$$

The boundary conditions of (1.25) assume that the column wall ($r = r_0$) is impermeable for the diffusing substance; the concentration profile is symmetric with respect to the column axis at $r = 0$ [if the velocity profile $w(r)$ is symmetric too].

A common characteristic feature of the single-parameter model (1.24) and the two-parameter model (1.25) is that if the diffusive transfer is determined by the molecular and turbulent diffusion, then $D_{||} = D_{\perp}$. Addition of the mass transfer as a result of the large-scale eddies (pulsations) leads to different diffusivity values ($D_{||} \neq D_{\perp}$).

The *diffusion model with a chemical reaction* is the core of the plug-flow model of chemical reactors. It follows from the above-mentioned models through the introduction of a volumetric mass source term, which is equal to the chemical reaction rate $R(c)$. The latter depends on the concentration of the transported substance:

$$D_{||}\frac{d^2c}{dz^2} - w\frac{dc}{dz} \pm R(c) = 0; \quad z = 0,\ w(c_0 - c) = -D_{||}\frac{dc}{dz}; \quad z = L,\ \frac{dc}{dz} = 0. \tag{1.26}$$

If a radial nonuniformity $w(r)$ exists, the source term $R(c)$ contributes in a way similar to that employed in the building of model (1.25).

The case of a mass transfer between two countercurrently moving phases can be described by (1.25), denoting the concentration in the phases by $c = c_i$ ($i = 1, 2$) and taking into account the phases holdup ε_i ($i = 1, 2$), where $\varepsilon_1 + \varepsilon_2 = 1$. In that case the source term $R(c)$ expresses the mass transfer rate between the phases:

$$R = k(c_1 - mc_2), \tag{1.27}$$

where k is the volumetric mass transfer coefficient and m is Henry's constant (or a coefficient of distribution between the liquid phases). If a radial nonuniformity exists, the two-phase diffusion model with an interphase mass transfer has the form

$$\begin{aligned}
&\varepsilon_i D_{i\perp}\frac{1}{r}\frac{\partial}{\partial r}\left(r\frac{\partial c_i}{\partial r}\right)+\varepsilon_i D_{i\|}\frac{\partial^2 c_i}{\partial z^2}-\varepsilon_i w_i(r)\frac{\partial c_i}{\partial z}-(-1)^{i-1}k(c_1-mc_2)=0, \quad i=1,2;\\
&z=0,\ w_1(r)(c_{10}-c_1)=-D_{1\|}\frac{\partial c_1}{\partial z},\quad \frac{\partial c_2}{\partial z}=0;\quad z=L,\\
&\quad w_2(r)(c_{20}-c_2)=-D_{2\|}\frac{\partial c_2}{\partial z},\quad \frac{\partial c_1}{\partial z}=0;\\
&r=0,\quad \frac{\partial c_i}{\partial r}=0,\quad i=1,2;\quad r=r_0,\quad \frac{\partial c_i}{\partial r}=0,\quad i=1,2.
\end{aligned} \tag{1.28}$$

1.4 Scale-Up Theory

The formulation of the mathematical models described in the previous sections encounters substantial difficulties owing to the scale effect especially in the case of high-capacity contacting devices. The reasons are the various types of nonuniformities that can be avoided through the design, but that cannot be eliminated completely. This means that the determination of the model parameters must be done on the basis of experimental data obtained from the prototype (industrial scale) under industrial conditions (technical experiment). Obviously, this is too expensive and too inaccurate, and sometimes it is an impossible way to solve the problem. The scale-up theory has tools that overcome the situation. It can explain the scale effects and can find suitable design solutions. Moreover, it creates a method of modeling (hydrodynamic) that does not use data from industrial-scale technological experiments for model formulation and evaluation of the contactor efficiency.

The scale-up theory shows the hydrodynamic nature of the scale effect and the ways for its elimination through hydrodynamic modeling [1]. According to this approach, the mathematical modeling gives the "ideal" industrial apparatus, whereas the hydrodynamic modeling provides a real possibility to come closer to the "ideal" one. Two principal problems have to be solved for that purpose:

1. *The reduction of the scale effect*, i.e., the efficiency of the industrial device must be close to that achieved with the laboratory model.
2. *Information for the flow structure in the industrial apparatus.* This information is needed owing to the impossibility to achieve equal hydrodynamic conditions in both the laboratory and the industrial device. The latter permits us to determine the residual scale effect and to evaluate the industrial apparatus efficiency under these conditions.

The approach is applicable to different types of models (of processes and apparatuses) in chemical engineering [1]. Its universality lies in the fact that the principal aim is not to achieve similarity, but to obtain the identity of the specific flow patterns in both the model and the prototype. The hydrodynamics conditions imposed concern mainly equal mean velocities in both the model and the prototype, as well as a uniform velocity distribution in the prototype.

The method of the scale-up theory will be considered through an example of mass transfer in columnar contactors with a countercurrent flow mode.

Let us consider, for simplicity of explanation, the longitudinal and the radial mixing in one of the phases having a nonuniform velocity profile in the radial direction [1]. The diffusion model (1.25) assumes that the concentration distribution depends on the balance of the convective mass transfer $w(r)\frac{\partial c}{\partial z}$, the axial diffusion mass transfer $D_{\|}\frac{\partial^2 c}{\partial z^2}$, and the radial diffusion mass transfer $D_{\perp}\frac{1}{r}\frac{\partial}{\partial r}\left(r\frac{\partial c}{\partial r}\right)$.

The axial diffusion represents a pseudo-diffusion mass transfer combining the simultaneous effects of both the turbulent diffusion and the convective macroscopic flows. The macroscopic flow length scale L is defined by the inequality $L \leq d_a \leq H$, where da is the column diameter and H is the height of the contacting zone. Thus, $D_{\|}$ can be expressed as follows :

$$D_{\|} = D_{\mathrm{T}} + D_{\mathrm{AS}}, \tag{1.29}$$

where D_{T} is the turbulent diffusivity and D_{AS} expresses the axial stirring effect.

The radial nonuniformity of the velocity profile $w(r)$ leads to a mass transfer that can be expressed as a superposition of both the turbulent diffusion and the convective stirring:

$$D_{\perp} = D_{\mathrm{T}} + D_{\mathrm{RS}}, \tag{1.30}$$

where D_{RS} expresses the radial stirring effect.

1.5 Axial Mixing

A model incorporating an effective diffusive transport represented by the effective axial diffusion can replace the diffusion model (1.25). Let us consider as an example non-steady-state diffusions in both the axial (along z) and the radial (along r) directions in a device with an arbitrary cross section in a coordinate system which moves with the mean flow velocity $\bar{w}$:

$$\frac{\partial c}{\partial t} + (w - \bar{w})\frac{\partial c}{\partial z} = D_{\|}\frac{\partial^2 c}{\partial z^2} + D_{\perp}\frac{\partial^2 c}{\partial r^2}\,. \tag{1.31}$$

The contactor walls at $r = 0$ and at $r = d$ are impermeable, so the diffusive fluxes are zero:

$$r = 0,\ \frac{\partial c}{\partial r} = 0; \quad r = d,\ \frac{\partial c}{\partial r} = 0\,. \tag{1.32}$$

The averaging of (1.31) with respect to r over the range from *0* to d yields

$$\frac{\partial \bar{c}}{\partial t} = \frac{\partial}{\partial z}\left[D_{\mathrm{AM}}(z,t)\frac{\partial \bar{c}}{\partial z}\right], \tag{1.33}$$

where

$$\bar{c} = \frac{1}{d}\int_0^{\mathrm{d}} c\, dr. \tag{1.34}$$

Thus, $D_{\mathrm{AM}}(t)$ can be presented as a sum of two terms [1]:

$$D_{\mathrm{AM}}(z,t) = D_{||} + D_{\mathrm{RNU}}(z,t), \tag{1.35}$$

where $D_{\mathrm{RNU}}(z,t)$ is a coefficient which transforms the effect of the radial nonuniformity of the velocity into an additional axial mixing effect. Thus, the coefficient $D_{\mathrm{AM}}(z,t)$ takes the form:

$$D_{\mathrm{AM}}(z,t) = D_{\mathrm{T}} + D_{\mathrm{AS}} + D_{\mathrm{RNU}}(z,t), \tag{1.36}$$

Expression (1.36) unifies the contributions of the turbulent diffusion, the axial stirring, and the radial nonuniformity, i.e., $D_{\mathrm{AM}}(z,t)$ contains both diffusive and convective components. It combines the diffusion mixing with the axial stirring in the so-called *axial mixing coefficient.*

The analysis of (1.33) indicates [1] that $D_{\mathrm{RNU}}(z,t)$ decreases as the radial diffusion increases:

$$D_{\mathrm{RNU}} = \frac{f_0 \Delta w^2 d_{\mathrm{a}}^2}{D_{|}}, \tag{1.37}$$

where the velocity profile nonuniformity $\Delta w = w_1 - w_0$ is expressed as the difference between the maximum mean velocity w_1 and the minimum mean velocity w_0. The coefficient f_0 depends on the character of the nonuniformity and d_{a} is the apparatus diameter. Therefore, the axial mixing coefficient can be expressed as

$$D_{\mathrm{AM}} = D_{||} + \frac{f_0 \Delta w^2 d_{\mathrm{a}}^2}{D_{\perp}}. \tag{1.38}$$

The axial diffusivity can be expressed (in a way similar to that applied to the molecular and the turbulent diffusion) by a specific linear scale $L_{||}$ (i.e., the mean free path of molecules or the scale of turbulence pulsation) and a effective velocity w_{eff} (i.e., the mean velocity of molecules or the velocity of the turbulence pulsation) [1]:

$$D_{||} = \kappa_{||} L_{||} w_{\mathrm{eff}}. \tag{1.39}$$

For example for packed-bed or catalyzed-bed columns $\kappa = 0.5$, $L_{||}$ is the packing (catalyzer particles) size, and w_{eff} is the average velocity. The parameters in (1.39) can be determined also for a number of columnar contactors such as perforated plate columns and rotary disk columns.

The radial diffusivity is expressed in a similar way as

$$D_{\perp} = \kappa_{\perp} L_{\perp} w_{\text{eff}}. \tag{1.40}$$

Therefore, the axial mixing coefficient is

$$D_{\text{AM}} = \kappa_{||} L_{||} w_{\text{eff}} + \frac{f_0 \Delta w^2 d_{\text{a}}^2}{\kappa_{\perp} L_{\perp} w_{\text{eff}}}. \tag{1.41}$$

The form of (1.41) indicates that $D_{\text{AM}} = f(w_{\text{eff}})$. It reaches a minimum at a certain value of w_{eff} that defines the optimal effective velocity $\left(w_{\text{eff}}^{\text{opt}}\right)$ and the minimum value of D_{AM}:

$$w_{\text{eff}}^{\text{opt}} = \Delta w d_{\text{a}} \sqrt{\frac{f_0}{\kappa_{||} L_{||} \kappa_{\perp} L_{\perp}}}. \tag{1.42}$$

1.6 Evaluation of the Scale Effect

In the case of a laboratory device, d_{a} is small parameter and (1.41) can be expressed as

$$D_{\text{AM}}^{\text{lab}} = \kappa_{||} L_{||}^{\text{lab}} w_{\text{eff}}. \tag{1.43}$$

This approach allows us to evaluate the scale effect ΔD:

$$\Delta D = D_{\text{AM}} - D_{\text{AM}}^{\text{lab}} = \kappa_{||} \left(L_{||} - L_{||}^{\text{lab}} \right) w_{\text{eff}} + \frac{f_0 \Delta w^2 d_{\text{a}}^2}{\kappa_{\perp} L_{\perp} w_{\text{eff}}}. \tag{1.44}$$

Expression (1.44) indicates that the scale effect has two main origins [1]:

1. The radial flow nonuniformity Δw, if the specific scales of the contactor (packing size or plate parameter) in both the model and the prototype are equal, i.e., $L_{||} = L_{||}^{\text{lab}}$.
2. The increase of the specific linear dimension of the prototype $\left(L_{||} > L_{||}^{\text{lab}}\right)$, if the radial diffusion ($D_{\perp}$) is so intense that the effect of the radial nonuniformity of the velocity can be neglected (i.e., $\Delta w \approx 0$). In this case ΔD is determined only by the first term in (1.44).

The above comments show [1] that the increase of the packed column diameter without change of the packing size $d_{\text{p}} \left(L_{||} = L_{||}^{\text{lab}} = d_{\text{p}} \right)$ leads to

$$\Delta D = \frac{f_0 \Delta w^2 d_a^2}{\kappa_\perp L_\perp w_{\rm eff}}. \qquad (1.45)$$

Thus, the scale effect grows with the increase of both the column diameter and the velocity nonuniformity, and diminishes with the increase of the effective velocity.

For example, the increase of the dimensions of a rotary extractor [1], characterized by a strong radial mixing, reduces the scale effect to

$$\Delta D = \kappa_{||}\left(L_{||} - L_{||}^{\rm lab}\right) w_{\rm eff}, \qquad (1.46)$$

because the distance between disks $L = L_{||}$ increases with increase of the column diameter $d_{\rm p}$ to satisfy the conditions of the geometrical similarity. However, in many cases the scale effect is determined by both terms in (1.44).

1.7 Hydrodynamic Modeling

The hydrodynamic nature of the scale effect allows us to employ hydrodynamic approaches to solve two principal problems of the hydrodynamic modeling:

1. The reduction of the scale effect through rational design of the contactors
2. The evaluation of the influence of the residual scale effect on the efficiency of industrial devices

The hydrodynamic modeling takes into account the specificity of the particular device under consideration. However, it follows a slender common scheme independent of the contactor type [1]:

- The laboratory-scale experiments must determine the mass transfer efficiency, the hydrodynamic characteristics, and the axial mixing coefficient under the operating conditions.
- The laboratory-scale experiments must define the hydrodynamic characteristics (phase holdup, response curve, residence time distribution, etc.) by means of model liquids.
- The experiments on a hydrodynamics stand (a test rig) must allow us develop the design to obtain uniform velocity distributions of the phase in the cross section of an industrial-scale contactor with a small height.
- The experiments on a hydrodynamics stand (a test rig) must define the hydrodynamic characteristics (phase holdup, response curve, residence time distribution, etc.) by means of model liquids.

The approach does not need technological testing procedures for the industrial devices. Their efficiencies can be determined numerically, taking into account the deviation of the hydrodynamic characteristics obtained with the laboratory model and the hydrodynamic stand.

The scale-up theory and the hydrodynamic modeling allow us not only to solve the design problems of high-capacity contactors, but also to determine the way to enhance their efficiencies.

The discussed approach above allows the formulation of the conditions that must be satisfied to increase the efficiencies of industrial-scale devices [1]:

1. Formation of uniform velocity profiles (Δw) at the apparatus inlets using distributors, diffusers, structured or nonstructured packed beds, etc.
2. The increase of the radial mixing ($D_{\perp}$) by means of additional pulsations, vibrations, stirrers, plates with directed motion of the phases, etc.
3. Operating conditions concerning the optimal velocity $\left(w_{\text{eff}}^{\text{opt}}\right)$ range to minimize the axial mixing D_{AM}.
4. Regime intensification (high velocities w_{eff}) if conditions for invariability of the turbulence scale ($L_{||}$) can be created.
5. Reduction of the radial mixing by decreasing of the turbulence scale of the recirculation currents and circulation flows ($L_{||}$) through the apparatus design.
6. Reduction of the bypass effects by formation of short contact sections in packed-bed columns.
7. Reduction of the axial mixing (and of the effect of wave formation) on perforated bubbling plates.
8. Elimination of the liquid bypass on the bubbling plates through the contactor design.
9. Reduction of the scale of recirculation loops in bubble columns and fluidized beds by dishes or grids.
10. If the elimination of the scale effect is impossible, an alternative solution is to employ a number of smaller devices.

The above recommendations were formulated by Rosen [1] for a number of scale-up problems concerning fixed-bed reactors, packed columns, moving-bed adsorption columns, bubble plate absorbers, bubble columns, fluidized beds, etc.

The efficiency of the method of hydrodynamic modeling depends on the type of contacting device considered. However, it leads to very useful results:

1. For example, the height of the transfer units of a packed distillation column [1] with a diameter of 0.3 m has been reduced 10 times as a result of the uniform liquid distribution inside the packing.
2. In the case of a packed-bed absorber of 4.5-m diameter, the good distribution of the phases increases the mass transfer coefficient by 36%.
3. The shutter rotary plates in a sieve-plate extraction column reduce 5 times the height of the transfer unit. The reduction from 0.030 to 0.015 m of the deviation from the horizontal surface of a plate with a diameter of 3.2 m increases the degree of absorption by 15%. There are reports [4] of hydrodynamic stands for testing of plates with a diameter of 20 m.

2 Average Concentration Model and Scale-Up

The approximate scale-up theory shows the hydrodynamic nature of the scale effect and the ways for its elimination through hydrodynamic modeling [1], where the scale effect is related to the axial mixing coefficient. The defects of this approach are the expression of the stirring effects as mixing (diffusion-type) effects and the use of the concentration (1.34), averaged with respect to r, because it is not possible to measure this average concentration experimentally. These shortcomings can be eliminated by use of a diffusion-type model, where the concentration is averaged for the column apparatus's cross-sectional area, and the effect of the radial nonuniformity of the velocity is represented as a convective mass transfer. On the basis of a simple model, we will present the influence of the radial nonuniformity of the velocity on the efficiency of chemical processes, the scale-up, and the possibility of modeling the scale effect.

2.1 Diffusion-Type Model

Let us consider gas motion in a column with radius r_0 through a layer of catalyzer particles. One of the gas components reacts on the catalytic interface. If the volume concentration of the active sites at the catalytic interface is very large, a volume chemical reaction of first order is possible.

The volume chemical reaction and radial nonuniformity of the velocity lead to convective and diffusive mass transfer, i.e., a convection–diffusion equation with volume reaction [5, 6] can be used for the mathematical description of the process:

$$u\frac{\partial c}{\partial z} = D\left(\frac{\partial^2 c}{\partial z^2} + \frac{1}{r}\frac{\partial c}{\partial r} + \frac{\partial^2 c}{\partial r}\right) - kc, \qquad (2.1)$$

where $u(r)$ and $c(r, z)$ are the velocity and concentration distributions in the column.

The radial component of the gas velocity is equal to zero if the catalytic particle distribution in the column is uniform. The volume reaction rate $v = kc$ (mol m^{-3} s^{-1}) is obtained using the surface catalytic reaction rate v_s (mol m^{-2} s^{-1}) and specific catalytic particle surface a (m^2 m^{-3}), i.e., $v = av_s$.

The boundary conditions are the inlet concentration c_0 and mass balance of the active gas component:

$$z = 0,\ c = c_0,\ \bar{u}c_0 = uc_0 - D\frac{\partial c}{\partial z}; \quad r = 0,\ \frac{\partial c}{\partial r} = 0; \quad r = r_0,\ \frac{\partial c}{\partial r} = 0, \qquad (2.2)$$

where $\bar{u}$ is the average velocity at the cross-sectional area of the column. In (2.2) it is supposed that a symmetric radial velocity distribution will lead to a symmetric concentration distribution.

Different expressions for the velocity distribution in the column apparatuses permit us to analyze [7] the influence of the radial nonuniformities of the velocity distribution on the process efficiency:

$$u_{\mathrm{n}}(r) = \bar{u}\left(\frac{\mathrm{n}+1}{\mathrm{n}} - \frac{2\,r^2}{\mathrm{n}\,r_0^2}\right), \quad \mathrm{n} = 1, 2, \ldots, \infty; \quad u_i(r) = \bar{u}\left(1 + a_i\frac{r^2}{r_0^2} + b_i\frac{r^4}{r_0^4}\right),$$
$$i = 1, 2, \tag{2.3}$$

where $n = 1$ (Poiseuille flow), $\mathrm{n} = 2$, $\mathrm{n} \to \infty$ (plug flow), $a_1 = 2$, $b_1 = -3$, $a_2 = -2$, $b_2 = 3$. The velocity distributions in (2.3) have identical average velocities (for the cross-sectional area), equal to the average velocity of the Poiseuille flow $\bar{u}$, i.e., to the plug flow velocity.

As the mass transfer efficiency of the column, we will use the amount of reacted substance q, i.e., the difference between the inlet and the outlet average convective mass flux:

$$q = \bar{u}c_0 - \frac{2}{r_0^2}\int_0^{r_0} ruc(r, l)dr, \quad \bar{u} = \frac{2}{r_0^2}\int_0^{r_0} ru(r)dr, \tag{2.4}$$

where l is the column height (catalytic zone height).

2.2 *Influence of the Radial Nonuniformity of the Velocity Distribution on the Process Efficiency*

The solution of problem (2.1) and (2.2) permits us to obtain the mass transfer efficiency q in the column under the influence of the radial nonuniformity of the velocity distribution. For this purpose dimensionless variables must be used:

$$r = r_0R, \quad z = lZ, \quad u(r) = \bar{u}U(R), \quad c(r, z) = c_0C(R, Z). \tag{2.5}$$

Introducing (2.5) into (2.1) and (2.2), the dimensionless problem has the form

$$U\frac{\partial C}{\partial Z} = Fo\left(\beta\frac{\partial^2 C}{\partial Z^2} + \frac{1}{R}\frac{\partial C}{\partial R} + \frac{\partial^2 C}{\partial R^2}\right) - DaC,$$

$$Z = 0, \quad C = 1, \quad 1 = U - \frac{1}{Pe}\frac{\partial C}{\partial Z}; \quad R = 0, \quad \frac{\partial C}{\partial R} = 0; \quad R = 1, \quad \frac{\partial C}{\partial R} = 0, \tag{2.6}$$

where Fo and Da are similar to the Fourier and Damkohler numbers:

$$Fo = \frac{Dl}{\bar{u}r_0^2}, \quad Da = \frac{kl}{\bar{u}}, \quad \beta = \left(\frac{r_0}{l}\right)^2. \tag{2.7}$$

The parameter β is very small and the solution of (2.6) can be obtained in the approximation $0 = \beta < 10^{-2}$:

$$U\frac{\partial C}{\partial Z} = Fo\left(\frac{1}{R}\frac{\partial C}{\partial R} + \frac{\partial^2 C}{\partial R^2}\right) - DaC;$$
$$Z = 0,\ C = 1; \quad R = 0,\ \frac{\partial C}{\partial R} = 0; \quad R = 1,\ \frac{\partial C}{\partial R} = 0. \tag{2.8}$$

The numerical solution of problem (2.8) permits us to obtain the relative mass transfer efficiency of the column (the conversion degree):

$$Q = \frac{q}{\bar{u}c_0} = 1 - 2\int\limits_0^1 RU(R)C(R,1)dR. \tag{2.9}$$

From (2.3) it is possible to obtain different dimensionless velocity distributions:

$$U_0(R) = 1, \quad U_1(R) = 2 - 2R^2, \quad U_2(R) = 1 + 2R^2 - 3R^4,$$
$$U_3(R) = 1 - 2R^2 + 3R^4, \quad U_4(R) = \frac{3}{2} - R^2. \tag{2.10}$$

The differences between the maximum and minimum velocity values $\Delta U_j = U_j^{max} - U_j^{min} (j = 1,\ldots,4)$ are the *radial nonuniformity parametersof thevelocity distribution* ($\Delta U_1 = 2, \Delta U_2 = \Delta U_3 = \frac{4}{3}$, $\Delta U_4 = 1$).

Figure 1 shows velocity distributions U_j, $j = 0, 1, \ldots, 4$.

The concentration distributions after the solution of (2.8) for $Da = 2$ and $Fo = 0.1$ are shown in Fig. 2 and 3.

Numerical simulation, based on the mathematical model (2.8), using different velocity profiles (2.10), gives the effect of the radial nonuniformity of the velocity on the process efficiency (see Table 1, where $Da = 2$). It can be seen in Table 1 [$Da = 2$, $Fo = 0.1$("laboratory" column) and $Fo = 0.01$ ("industrial" column)] that the process efficiency decreases if the column radius and the velocity distribution radial nonuniformity parameter increase.

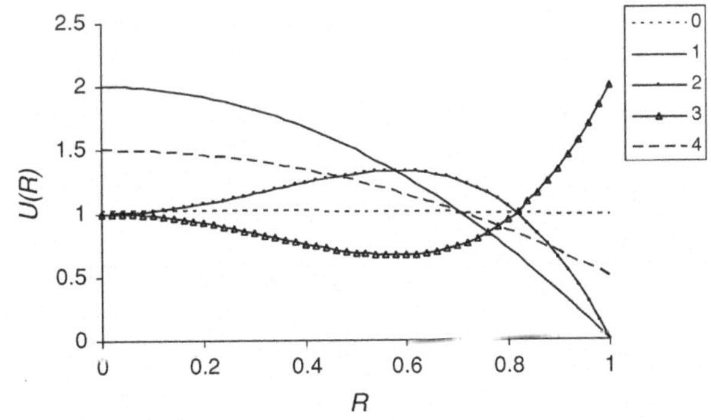

Fig. 1 Velocity distributions: *0*U_0, *1*U_1, *2*U_2, *3*U_3, *4*U_4

Let us consider the effect of the radial nonuniformity of the velocity ΔU_j, $j = 1,\ldots,4$, on the *relative process efficiency decrease* (in comparison with the plug flow):

$$\Delta Q_j = \frac{Q_0 - Q_j}{Q_j} \times 100\%, \quad j = 1,\ 2,\ 3. \tag{2.11}$$

The results obtained show (see Table 2) the influence of the increase of the column radius and the radial nonuniformity of the velocity distribution.

Let us consider the column heights H_j ($j = 1,\ldots,4$) for column efficiency $Q_0 = 0.8643$, $Fo = 0.1$, and $Da = 2$, i.e., the necessary column heights H_j ($j = 1,\ldots,4$)

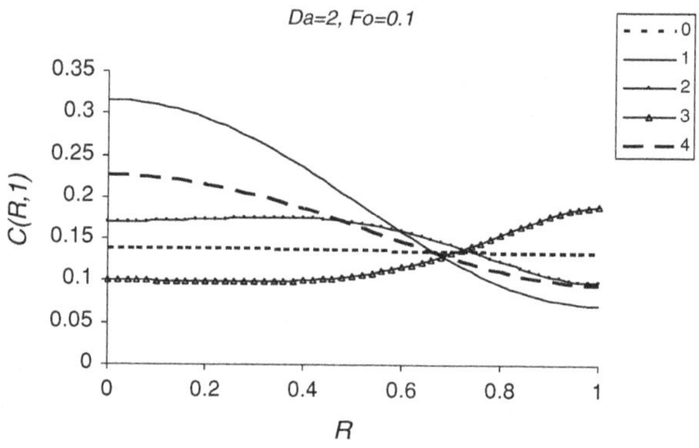

Fig. 2 Concentration distributions using the velocity profiles: $0U_0$, $1U_1$, $2U_2$, $3U_3$, $4U_4$

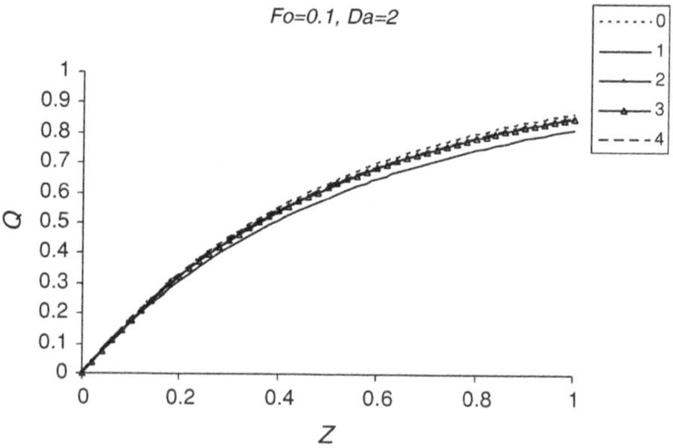

Fig. 3 Process efficiency for different radial nonuniformities of the velocity: $0U_0$, $1U_1$, $2U_2$, $3U_3$, $4U_4$

Table 1 Process efficiency Q at $Z = 1$ and column height $H = Z$ at $Q_0 = 0.8643$

Fo	U_0	U_1	U_2	U_3
0.1 ("laboratory")	$Q_0 = 0.8643$	$Q_1 = 0.8143$	$Q_2 = 0.8516$	$Q_3 = 0.8513$
	$H_0 = 1$	$H_1 = 1.2$	$H_2 = 1.05$	$H_3 = 1.05$
0.01 ("industrial")	$Q_0 = 0.8645$	$Q_1 = 0.7870$	$Q_2 = 0.8349$	$Q_3 = 0.8371$
	$H_0 = 1$	$H_1 = 1.34$	$H_2 = 1.12$	$H_3 = 1.12$

Table 2 Effect of the radial nonuniformity of the velocity on the relative process efficiency and column height

Fo	U_1	U_2	U_3
0.1 ("laboratory")	$\Delta Q_1 = 6\%$	$\Delta Q_2 = 1.4\%$	$\Delta Q_3 = 1.5\%$
	$\Delta H_1 = 20\%$	$\Delta H_2 = 5\%$	$\Delta H_3 = 5\%$
0.01 ("industrial")	$\Delta Q_1 = 9.8\%$	$\Delta Q_2 = 3.5\%$	$\Delta Q_3 = 3.3\%$
	$\Delta H_1 = 34\%$	$\Delta H_2 = 12\%$	$\Delta H_3 = 12\%$

to realize plug flow column efficiency. The result obtained shows an increase of the column heights (see Table 1) as a result of the radial nonuniformity of the velocity. The *relative column height increases* ΔH_j, j = 1,...,4, can be obtained from

$$\Delta H_j = \frac{H_j - H_0}{H_0} \times 100\%, \quad j = 1, 2, 3. \tag{2.12}$$

The numerical results (Table 2) show the necessity of an essential augmentation of the column height to compensate for the effect of the radial nonuniformity of the velocity distribution.

The comparison of the results in Tables 1 and 2 shows that the effects of $\Delta U_2$and $\Delta U_3$are similar, i.e., the effects of the radial nonuniformity of the velocity distribution are caused by the *nonuniformity of the velocity* $\Delta U_j = U_j^{max} - U_j^{min}$ (j = 1,...,3), but not by the *velocity distribution* U_j, (j = 1,...,3).

2.3 Scale Effect

Let us consider a "laboratory" column ($Da = 2$, $Fo = 0.1$, $r_0 = 0.2$ m) and an "industrial" column ($Da = 2$, $Fo = 0.01$, $r_0 = 0.5$ m). The *relative scale effect* $\Delta Q^{scale-up}$ can be obtained using Table 1:

$$\Delta Q_j^{scale-up} = \frac{Q_j^{laboratory} - Q_j^{industrial}}{Q_j^{industrial}} \times 100\%, \quad j = 1, 2, 3. \tag{2.13}$$

The results obtained are shown in Table 3.

The comparison between these two columns shows that the scale-up leads to a decrease of the column efficiency (for a constant column height). If we consider columns with constant process efficiency, it leads to a column height increase as result of the radius increase. The scale effect on the column height (at $Q = Q_0 = 0.8643$) can be obtained using Table 1:

$$\Delta H_j^{scale-up} = \frac{H_j^{industrial} - H_j^{laboratory}}{H_j^{laboratory}} \times 100\%, \quad j = 1, \ldots, 4 \tag{2.14}$$

Table 3 Comparison of the scale effect for different velocity profiles

	U_1	U_2	U_3
$\Delta Q^{\text{scale-up}}$ (%)	3.5	1.9	1.7
$\Delta H^{\text{scale-up}}$ (%)	11.6	6.6	6.6

and the results are shown in Table 3.

The results in Tables 1, 2 and 3 demonstrate the influence of the radial nonuniformity of the velocity on the column process efficiency and scale-up. Very complicated hydrodynamic behavior in industrial column apparatuses is an obstacle to the use of Eqs. 2.1 and 2.2 for the scale effect modeling and average concentration models must be used.

2.4 Average Concentration Model

At the page 84 it was shown that the velocity and concentration in diffusion-type models can be replaced by the average velocity and concentration (3.9) for the cross-sectional area of the column:

$$\bar{u} = \frac{2}{r_0^2} \int_0^{r_0} r u(r) dr, \quad \bar{c}(z) = \frac{2}{r_0^2} \int_0^{r_0} r c(r, z) dr, \tag{2.15}$$

i.e., the velocity and concentration distributions may be represented with the help of the average functions:

$$u(r) = \bar{u}\tilde{u}(r), \quad c(r, z) = \bar{c}(z)\tilde{c}(r, z), \tag{2.16}$$

where

$$\frac{2}{r_0^2} \int_0^{r_0} r\tilde{u}(r) dr = 1, \quad \frac{2}{r_0^2} \int_0^{r_0} r\tilde{c}(r, z) dr = 1. \tag{2.17}$$

The average concentration model may be obtained if put (2.16) into (2.1) and (2.2), multiply by r, and integrate over r within the interval [0, r_0]:

$$\alpha(r_0, z)\bar{u}\frac{d\bar{c}}{dz} + \frac{d\alpha}{dz}\overline{uc} = D\frac{d^2\bar{c}}{dz^2} - k\bar{c}; z = 0, \quad \bar{c} = c_0, \quad \frac{d\bar{c}}{dz} = 0, \tag{2.18}$$

where the *scale effect function* $\alpha(r_0, z)$ is the result of the radial nonuniformity of the velocity and the concentration,

$$\alpha(r_0, z) = \frac{2}{r_0^2} \int_0^{r_0} r\tilde{u}(r)\tilde{c}(r, z) dr, \quad \alpha(r_0, 0) = 1. \tag{2.19}$$

The function α can be obtained [8] using the solution of problem (2.1) and (2.2) in dimensionless variables (2.5). If we put (2.16) and (2.5) into (2.5), the dimensionless form of α is the following:

$$A(Z) = \alpha(r_0, lz) = \frac{2}{\bar{C}(Z)} \int_0^1 RU(R)C(R,Z)dR, \tag{2.20}$$

where the dimensionless concentration $C(R,Z)$ is solution of (2.8) and the average dimensionless concentration $\bar{C}(Z)$ is obtained from

$$\bar{C}(Z) = 2 \int_0^1 RC(R,Z)dR. \tag{2.21}$$

The values of the average concentration $\bar{C}$ and the scale effect function $A(Z)$ were obtained in the case of $Fo = 0.1$, $Da = 2$ ($r_0 = 0.2$ m) and $Fo = 0.01$, $Da = 2$ ($r_0 = 0.5$ m) (see Fig. 4).

From Fig. 4 it can be seen that the maximum scale effect exists in the case of a Poiseuille flow ($U = U_1$), whereas for plug flow ($U = U_0$) the scale effect is absent ($\tilde{u} = \tilde{c} = 1$, i.e., $A \equiv 1$). Because of this we will only consider the modeling of the Poiseuille flow scale effect.

The scale effect function can be represented using the linear or quadratic approximation:

$$A(Z) = 1 + aZ, \quad A(Z) = 1 + a_1 Z + a_2 Z^2, \tag{2.22}$$

where the approximation parameters are shown in Table 4.

Figure 5 shows a comparison between the function $A(Z)$ and its polynomial (linear and quadratic) approximations.

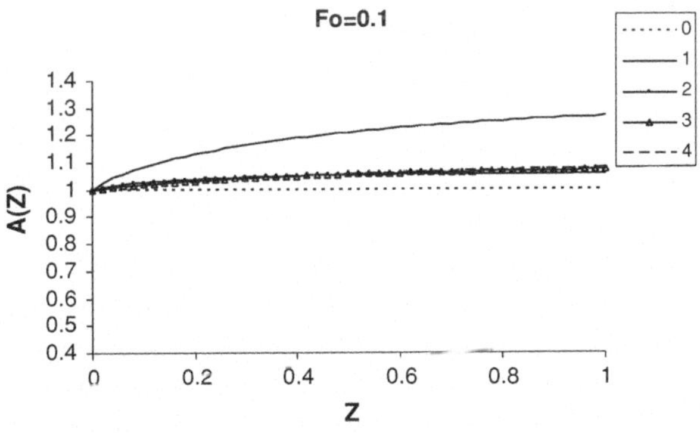

Fig. 4 Scale effect function. $0U_0$, $1U_1$, $2U_2$, $3U_3$, $4\ U_4$

Table 4 Values of the approximation parameters

$Fo = 0.1$	$a = 0.4347$
	$a_1 = -0.3732$
	$a_2 = 0.6221$
$Fo = 0.01$	$a = 0.6778$
	$a_1 = -0.5362$
	$a_2 = 0.9538$

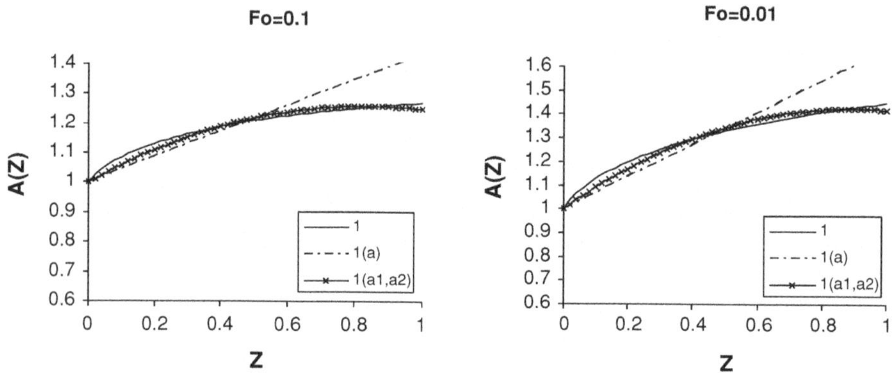

Fig. 5 Comparison between function *A*(*Z*) and its polynomial (linear and quadratic) approximations for different values of Fourier number. $1U_1$

2.5 Scale Effect Modeling

The connection between the scale effect and the radial nonuniformity of the velocity shows that the model established which gives the radial nonuniformity will allow the modeling of the scale effect, i.e., the influence of radius r_0 on α.

The influence of the column radius r_0 on the function $A(z) = \alpha(r_0, lz)$ is a result of the influence of *Fo* on the solution *C*(*R*,*Z*) (see 2.8).

Consider the dimensionless form of (2.18) using dimensionless variables:

$$r = r_0 R, \quad z = lZ, \quad u(r) = \bar{u}U(R), \quad c(r,z) = c_0 C(R,Z), \\ \bar{c}(z) = \bar{c}(lZ) = c_0 \bar{C}(Z), \tag{2.23}$$

i.e.,

$$A(Z)\frac{\partial \bar{C}}{\partial Z} = \frac{1}{Pe}\frac{\partial^2 \bar{C}}{\partial Z^2} - \left(Da + \frac{\partial A}{\partial Z}\right)\bar{C}; \quad Z = 0, \quad \bar{C} = 1, \quad \frac{d\bar{C}}{dZ} = 0, \tag{2.24}$$

where $Pe = \bar{u}l/D$, i.e., $Pe^{-1} = \beta Fo$.

Comparison of models (2.8) and (2.24) shows that the average concentration model is equivalent to the longitudinal diffusion model with a volume reaction, where the chemical reaction rate is corrected by the effect of the radial nonuniformity of the velocity and concentration distributions.

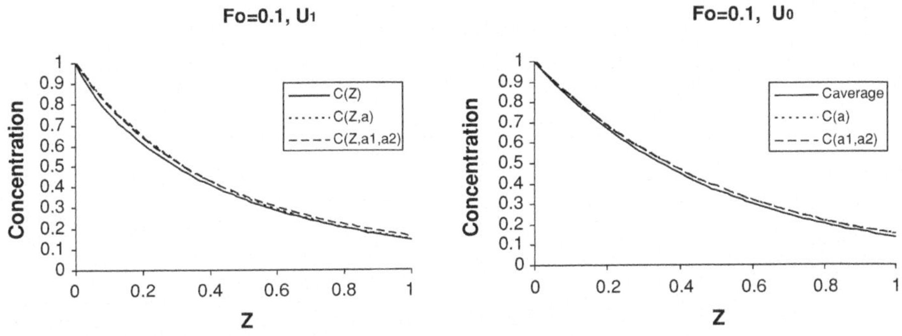

Fig. 6 Comparison of concentration distribution functions calculated for different approximations [*C(a)*, linear; *C(a1,a2)*, quadratic] of the scale effect function *A*(*Z*) and different values of Fourier number

The determination of $A(z)$ in polynomial form (2.22) permits us to obtain the average concentration $\bar{C}$ in the column using (2.21). Figure 6 shows a comparison of the values for $\bar{C}$ obtained using (2.21).

2.6 Scale-Up Parameter Identification

The results in Fig. 6 demonstrate the possibility to represent the scale effect by one parameter (a).

Let us consider the case of a linear approximation for $A(Z)$. As a result, (1.24) has the form

$$(1 + aZ)\frac{\partial \bar{C}}{\partial Z} = \frac{1}{Pe}\frac{\partial^2 \bar{C}}{\partial Z^2} - (Da + a)\bar{C}; \quad Z = 0, \quad \bar{C} = 1, \quad \frac{d\bar{C}}{dZ} = 0. \tag{2.25}$$

The identification of the scale effect parameter a is possible using experimental data for the average concentration. Here we will use "experimental data" $\bar{C}^{\exp}(Z_i)$, i = 1,…,N, obtained from the exact solution of model (2.8), and random numbers δ_{j} (j = 1,…,M), obtained using a generator for random numbers:

$$\bar{C}_{\mathrm{ij}}^{\exp}(Z_i) = \bar{C}(Z_i)[0.95 + \delta_{\mathrm{j}}], \quad 0 \le \delta_{\mathrm{j}} \le 1, \quad i = 1, \ldots, \mathrm{N}, \ \mathrm{j} = 1, \ldots, \mathrm{M} \quad (\mathrm{N} = 10). \tag{2.26}$$

Obviously the maximum relative error of the "experimental data" is ± 5%.

The parameter identification of model (2.25) will be made by minimization of the least-squares function Φ :

$$\Phi(a) = \sum_{i=1}^{\mathrm{N}} \sum_{\mathrm{j}=1}^{\mathrm{M}} [\bar{C}(Z_i) - \bar{C}_{\mathrm{ij}}^{\exp}(Z_i)]^2. \tag{2.27}$$

Table 5 Values of a

	$Z = 0.2$	$Z = 0.4$	$Z = 0.6$	$Z = 0.8$	$Z = 1$
$Fo = 0.1$	$a = -0.573$	$a = 1.331$	$a = 6.333$	$a = 14.656$	$a = 25.175$
$Fo = 0.01$	$a = -0.537$	$a = 1.475$	$a = 5.988$	$a = 13.317$	$a = 22.296$

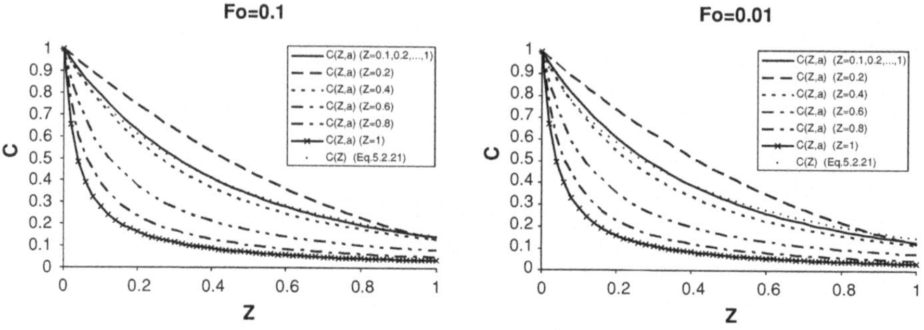

Fig. 7 Comparison of concentration distribution functions calculated for different values of a and different values of Fourier number

Model (2.25) is characterized by a small parameter ($Pe^{-1} < 10^{-3}$) at the highest derivative, i.e., the inverse identification problem will be incorrect (ill-posed) and for its solution we must use a regularization method [9].

The next calculations are made for the case

$$Fo = 10^{-1} \left(Fo = 10^{-2}\right), \quad Da = 2, \quad \mathrm{N} = 10, \quad Z_{\mathrm{j}} = 0.1\mathrm{j}, \quad \mathrm{j} = 1, \ldots, 10, \tag{2.28}$$

and the values of the parameter a which minimize the least-squares function Φ (2.27) are 0.6357 ($Fo = 0.1$) and 0.6773 ($Fo = 0.01$).

The next case is parameter identification using ten "experimental data" values only at one point ($Z = 0.2, 0.4, 0.6, 0.8, 1$) (see Table 5).

In Fig. 7 the concentration distribution functions, calculated using (2.25), are shown, where the parameter a was obtained using "experimental data" at ten points ($Z = 0.1, 0.2, \ldots, 1$) and at separate points ($Z = 0.2, Z = 0.4, \ldots, Z = 1$).

The results obtained (Fig. 7) show that the scale effect is related to one parameter which can be calculated using experimental data for the average concentration at some different points of the column height. It is possible to use some different values for the average concentration at one point at the middle column height.

3 Statistical Analysis

The use of models for quantitative descriptions of real processes as well as for subsequent engineering designs sets up a demand concerning the accuracy of the information obtained through simulations. The accuracy of the information depends on both the exactness of the mathematical model chosen and the determination of the model parameters. On the other hand, the exactness of the model parameters depends on the accuracy of the experimental data and the calculation method. The errors of the experimental data have a random nature and the consequently the calculated model parameters and objective functions are random values too.

The stochastic nature of the errors during the experimental determination of the objective function leads to subsequent errors of the model parameters and the calculated values of the objective function during simulations. This sets up two basic questions with respect to the model:

1. *Significance of the model parameters*. Are they different from zero owing to the accumulated errors of the experimental data only?
2. *Is the model adequate or not*? This question concerns the adequacy of the objective function because of the simulation. What are the error dispersions of both the calculated and the experimentally determined function? Are they equal or not?

The answers to both questions can be found out through a statistical analysis of the models [3] developed next.

3.1 Basic Terms

The statistical determination of a particular quantity X looks for its true μ_X. The experimentally measured value X differs from μ_X and the error $X - \mu_X$ is a result of the measurements and is practically unknown *since* μ_X*is unknown.* The errors should be classified as *coarse errors*, *systematic errors,* and *random errors*. The former two types are relevant to the methods employed for the measurements and the information registration. They are *external* with respect to the process investigated. The random errors follow from the nature of the process and very often are the result of uncontrollable and unpredictable variations of both the process parameters and the operation of the measuring device.

The experimental data employed for the purposes of the modeling should not contain coarse and systematic errors, so the further explanation will only focus on the random errors.

The observation of the values of a particular variable X is a *stochastic event*. If N observations were made and X was observed m times, the frequency of the event is $p = \frac{m}{N}$. As the number of the observations N increases, the frequency of the event tends to a limited value of its probability P $(p \to P)$. This allows us to define a sum of events as a *cumulativeobserved event* as well as a *product of events* representing a simultaneous observation of several events. The *law of its random distribution* gives the relationship between the random values of X and the probability of its experimental determination. Each *random variable* is characterized by its *distribution function* $F(x)$ (sometimes called a *cumulative* distribution function), which represents the probability P that the measured X is lower than x:

$$F(x) = P(X \leq x), -\infty < x < \infty, \tag{3.1}$$

where the probability $P(X \leq x)$ means the probability associated with event$\{X \leq x\}$. In other words, when the experiments are done, the measured value of X (the random variable) should not take on a value larger than the number x, i.e., $-\infty < X \leq x$. Obviously, the distribution function varies within the interval [0, 1] and $0 \leq F(X) \leq 1$ for all x. Moreover, $F(X)$ is not decreasing, which means if $x_1 < x_2$, then $F(x_1) \leq F(x_2)$. Since X takes only finite values, then $\lim_{x \to \infty} F(X) = 1$ and $\lim_{x \to -\infty} F(X) = 0$.

A random (stochastic) variable is *a discrete* one if it can take at most a countable number of values. *Countable* means that the set can be put *one to one* with a set of positive integers. An example of an *uncountable set* is all real numbers between 0 and 1. Therefore, if the random variable takes only a finite number of values $x_1, x_2, \ldots, x_n$, it is a discrete one. The probability that the discrete random value X takes on the value x_i is $p(x_i) = P(X = x_i)$ for i = 1,2,. . ., which means that it must have $\sum_{i=1}^{\infty} p(x_i) = 1$. The probability $p(x)$ allows us to compute all the probability statement of X, i.e., the so-called *the probability mass function* for the discrete random variable X. If $I = [a, b]$, where a and b are real numbers such that $a \leq b$, then

$$P(X \in I) = \sum_{a \leq x_i \leq b} p(x_i), \tag{3.2}$$

where the symbol $\in$ means "contained in." The summation concerns the addition of $p(x_i)$, such that $a \leq x_i \leq b$. The distribution function $F(X)$ for the discrete random variable X is

$$F(X) = \sum_{x_i \leq x} p(x_i), \quad -\infty < x < \infty. \tag{3.3}$$

If a random variable X can takes each value within a given interval. i.e., an *uncountable* infinite number of different values (all also negative real numbers), it can be considered as a *continuous* random variable X. Furthermore, this statement means that a random variable X is said to be *continuous* if there exists a nonnegative function *f*(*x*) such that for any set of real numbers B

$$P(X \in B) = \int_B f(x)dx \text{ and } \int_{-\infty}^{\infty} f(x) = 1. \tag{3.4}$$

All probability statements about X can be calculated from $f(x)$, termed the *probability density function* for the *continuous random variable X* .

If X takes on values within the interval $[x_1, x_2)$, such that $x_1 \leq X < x_2$, that are predetermined by the experiment designed, the probability is

$$P(x_1 \leq X < x_2) = P(X < x_1) = F(x_2) - F(x_1) = \Delta F. \qquad (3.5)$$

Taking into account that X is a continuous random variable, the *probability density function* is

$$f(x) = \frac{dF}{dx} = \lim_{\Delta x \to 0} \frac{\Delta F}{\Delta x}, \quad \Delta x = x_2 - x_1. \qquad (3.6)$$

Thus, by means of (3.2) and (3.3) it is possible to find that

$$P(x_1 \leq X < x_2) = \int_{x_1}^{x_2} f(x)\,dx. \qquad (3.7)$$

Since $F(x)$ is not a decreasing function of X [taking into account that $F(-\infty) = 0$ and $F(\infty) = 1$] (3.4), we can say that

$$f(x) \geq 0, \quad \int_{-\infty}^{\infty} f(x)\,dx = 1. \qquad (3.8)$$

The distribution function $F(x)$ for a continuous random variable X is

$$F(x) = P(X \in [-\infty, x]) = \int_{-\infty}^{x} f(x)dx \text{ for all } -\infty < x < \infty. \qquad (3.9)$$

Thus, $f(x) = F'(x)$ [the derivative of $F(x)$]. Moreover, if the interval is defined as $I = [a, b]$, where a and b are real numbers, such that $a < b$, then we have

$$P(X \in I) = \int_{a}^{b} f(x)dx = F(b) - F(a), \qquad (3.10)$$

since $f(x) = F'(x)$, which is an example of the application of *the fundamental theorem of calculus*.

If X and Y are discrete random variables, then

$$p(x, y) = P(X = x, \quad Y = y) \qquad (3.11)$$

for all x,y, where $p(x, y)$ is the *joint probability mass function* of X and Y. Both variables are *independent* if

$$p(x, y) = p_x(x)p_y(y) \qquad (3.12)$$

for all x,y, where $p_x(x)$ and $p_y(y)$ are the (marginal) probability mass functions of X and Y, defined such that

$$p_x(x) = \sum_{\text{ally}} p(x,y), \; p_y(y) = \sum_{\text{allx}} p(x,y). \tag{3.13}$$

The random variables are *jointly continuous* if there exists a nonnegative function $f(x)$, termed the *joint probability density function* of X and Y, such that for all sets of real numbers A and B

$$P(X \in A, Y \in B) = \int_B \int_A f(x,y)\, dxdy. \tag{3.14}$$

The random variables X and Y are independent if $f(x,y) = f_x(x) \,.\, f_y(y)$ for all x, y, where $f_x(x)$ and $f_y(y)$ are the (marginal) probability density functions of X and Y, such that

$$f_x(x) = \int_{-\infty}^{\infty} f(x,y)dy, f_y(y) = \int_{-\infty}^{\infty} f(x,y)dx. \tag{3.15}$$

Generally, from an intuitive point of view, the random variables X and Y, discrete or continuous, are *independent* if the value of one of them is known; this does not inform us about the distribution of the other. Thus, if X and Y are not independent, they are dependent random variables.

The probability density function cannot be determined easily from the experimental data, but it allows us to introduce parameters characterizing the random variable, such as the *mathematical expectation* and the *dispersion*.

If a random variable X takes different values $X_1, \ldots, X_n$, the random value of $m_x = M[X]$ can be defined in a such manner that at $n \rightarrow \infty$, $m_x$will approach its true value, i.e., $\lim\limits_{n\rightarrow\infty} m_x = \mu_x$. The *mean* or *expected value* of the random variable X_i $(i = 1, \ldots, n)$ is defined as

$$m_x = M[X] = \lim_{n\rightarrow\infty} \frac{1}{n} \sum_{i=1}^{n} X_i, \tag{3.16}$$

which follows from a feature of a converging series of numbers $X_1, \ldots, X_n$ having a convergence point m_x. This allows us to create a converging series:

$$X_1, \frac{X_1 + X_2}{2}, \ldots, \frac{X_1 + X_2 + \ldots X_n}{n}, \ldots \tag{3.17}$$

which also approaches μ_x when $n \rightarrow \infty$.

Taking into account the definitions of the probability density function, we can define the *meanvalue* [mathematical *expectation*, sometimes denoted here as $M(X_i)$] as

$$\mu_x = \begin{cases} \sum_{j=1}^{\infty} x_j p_{x_i}(x_i) & \text{if } X_i \text{ is discrete;} \\ \int_{-\infty}^{\infty} x f_{x_i}(x)dx & \text{if } X_i \text{ is continuous.} \end{cases} \tag{3.18}$$

The mean is a measure of the central tendency, similar in sense to gravitational acceleration.

If a_i is a real number, the important properties of the mean are

$$M(aX) = aM(x);$$

$$M\left(\sum_{i=1}^{n} a_i X_i\right) = \sum_{i=1}^{n} a_i M(X_i) \text{ even } X_i \text{ are dependent.} \tag{3.19}$$

The *median* $x_{0.5}$ of a random variable X_i is an alternative to the measure of the central tendency and is defined as the smallest value of x such that $F_x(X) \geq 0.5$. If X_i is a *continuous*, then $F_x(X) = 0.5$. Sometimes the median can better represent the central tendency than the mean, e.g., in the case of very small or very large values. The latter means that extreme values (very small or very large) greatly affect the *mean*; such is *not the case with the median.*

The *variance* of a random variable Y_i, frequently denoted by σ_i^2 or $Var(Y_i)$, is a measure of the dispersion of the random variable about its mean. The definition considers σ_i^2 as a mathematical dispersion of the random variable about its mean. The definition also considers σ_i^2 as a mathematical expectation of the square of the differences $\overset{0}{Y_i} = (Y_i - \mu_y)$, such that

$$\sigma_y^2 = M\left[\overset{0}{Y_i}\right] = \lim_{n \to \infty} \frac{1}{n} \sum_{i=1}^{n} (Y_i - \mu_y) \text{or } \sigma_y^2 = M\left[(Y_i - \mu_y)^2\right] = M(Y_i^2) - \mu_y^2. \tag{3.20}$$

The larger the variance, the further the random variable is from the mean. The definition of the variance allows us to define the following properties:

$$Var(Y) \geq 0, \quad Var(aY) = a^2 Var(Y), \quad Var\left(\sum_{i=1}^{n} a_i Y_i\right) = \sum_{i=1}^{n} Var(Y_i), \tag{3.21}$$

if the random values of Y_i are independent. The standard deviation of the random variable Y_i is defined as $\sigma_y = \sqrt{\sigma_i^2}$.

If we have two random variables X_i andX_j, where i = 1,. . ., n; j = 1,. . ., n, the question is how to measure their dependence (linear *dependence*). The *covariance*, denoted by C_{ij} or $Cov(X_i, Y_j)$, is defined by

$$C_{ij} = M\left[(X_i - \mu_i)(X_j - \mu_j)\right] = M(X_i X_j) - \mu_i \mu_j. \tag{3.22}$$

The covariances are symmetric, such that $C_{ij} = C_{ji}$. Moreover, if $i = j$, then $C_{ij} = C_{ii} = \sigma_i^2$. If $C_{ij} = 0$, the random variables are *uncorrelated* (or independent—see the third property of the variance). The opposite statement is not generally true, but if X_i and Y_j are *jointly normally distributed* (Gauss distribution—see below) with $C_{ij} = 0$, they are also independent.

If $C_{ij} > 0$, then X_i and Y_j are *positively correlated.* If the random variables are positively correlated, then $X_i > \mu_i$ and $X_j > \mu_j$ tend to occur together (and $X_i < \mu_i$ and $X_j < \mu_j$also tend to do so). This means at $C_{ij} > 0$ if the random variable X_i is large, the other random variable Y_j is also likely to be large.

The *negatively correlated* random variables exhibit $C_{ij} < 0$. In this case $X_i > \mu_i$ and $X_j < \mu_j$ tend to occur together (and $X_i < \mu_i$ and $X_j > \mu_j$ tend to occur together). Therefore, for negatively correlated random variables if one is large, the other is likely to be small.

The main difficulty in applying the covariance C_{ij} is that this is *not a dimensionless* measure of the dependence between X_i and Y_j, which causes difficulties in the interpretations. For example, if X_i and Y_j are in units of Newton's, C_{ij} in units of Newton's squared. This problem is easily eliminated by the use of the *correlation* ρ_{ij} defined as

$$\rho_{ij} = \frac{C_{ij}}{\sqrt{\sigma_i^2 \sigma_j^2}}, \quad i = 1,2,\ldots,n; \quad j = 1,2,\ldots,n \tag{3.23}$$

as a measure of the linear dependence of the random variables X_i and Y_j. The denominator of ρ_{ij} is *always positive*, which means it has the same sign as C_{ij}.

The correlation ρ_{ij} varies within the interval $-1 \le \rho_{ij} \le 1$ for all i and j. Thus, if $\rho_{ij} \to +1$, then X_i and Y_j are positively correlated, whereas at $\rho_{ij} \to -1$, they are highly negatively correlated.

The mathematical expectation and the dispersion characterize the random variable for a particular distribution law. There are different distribution laws. The most important, in fact the most frequently utilized, is the *Gaussian law*, or a *normal* density function. It is valid when the random variable depends on a set of independent or weakly dependent variables with arbitrary distribution laws, but without dominating one of them (the central limit theorem) [10].

The normal density function is

$$f(x) = \frac{1}{\sigma\sqrt{2\pi}} \exp\left[-\frac{1}{2\sigma^2}(x - m_x)^2\right]. \tag{3.24}$$

It reaches a maximum $\frac{1}{\sigma\sqrt{2\pi}}$ at $x = m_x$.

The probability that a particular random variable with a normal distribution is associated with the interval $(-\infty, x)$ follows from (3.7) and (3.24):

$$F(X) = \int\limits_{-\infty}^{X} f(x)\,dx = \frac{1}{\sigma\sqrt{2\pi}} \int\limits_{-\infty}^{X} \exp\left[-\frac{1}{2\sigma^2}(x - m_x)^2\right] dx$$

$$= \frac{1}{\sqrt{2\pi}} \int\limits_{-\infty}^{\frac{X-m}{\sigma}} \exp\left(-\frac{t^2}{2}\right) dt = \Phi\left(\frac{x - m_x}{\sigma}\right), \qquad (3.25a)$$

where Φ is the Laplacian function

$$\Phi(z) = \frac{1}{\sqrt{2\pi}} \int\limits_{-\infty}^{Z} \exp\left(-\frac{t^2}{2}\right) dt. \qquad (3.25b)$$

Similarly, if the random variable is associated with the interval $[X_1, X_2]$, the normal density distribution gives

$$P(x_1 \le X < x_2) = \Phi\left(\frac{x_1 - m_x}{\sigma}\right) - \Phi\left(\frac{x_2 - m_x}{\sigma}\right).. \qquad (3.26)$$

If we substitute $x_1 = m_x - 3\sigma$ and $x_2 = m_x + 3\sigma$, the value of P is about 0.998. Therefore, this means that practically almost all values of the random variable should be associated with the interval $m_x \pm 3\sigma$ (the rule of 3σ). If this is not satisfied, the observation is assumed as one containing a strong error.

There are many important features of the Gaussian density function and some of them are:

1. It is a model of many random physical phenomena, so it is possible to be justified theoretically.
2. The linear combination of Gaussian random variables is also a Gaussian random variable.
3. The process described by the Gaussian random variable as a model can be specified completely from a statistical point of view by the first and the second moments only if it is not true for the other processes.
4. In the analysis of complex systems the Gaussian law provides a suitable model for both linear and nonlinear processes.

Both the Gaussian density function $f(x)$ and the Gaussian distribution functions are shown in Fig. 8. The most important points for these curves are:

1. The only maximum $\frac{1}{\sigma\sqrt{2\pi}}$ coincides with the mean value (see 3.24).
2. The density function is symmetric with respect to the mean.
3. The width of the density function is directly proportional to the *standard deviation* σ. For example, the weight of 0.607 of the maximum values corresponds to 2σ, whereas at 3σ almost all the band random variations of the variable are spanned.

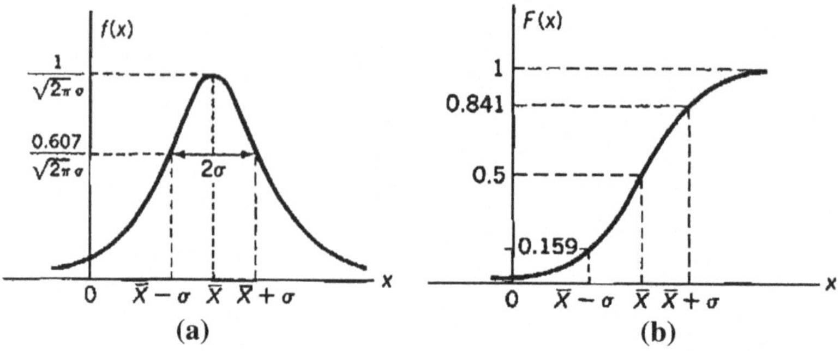

Fig. 8 Gaussian random variable: **a** density function and **b** distribution function

4. The maximum of the Gaussian density function is proportional to inverse value of the standard deviation σ (see 3.24).
5. The density function has an area of unity and can be successfully applied to represent the impulse delta function. The delta function δ follows from (6.1.25) at $\sigma \to 0$:

$$\delta\,(x - m_x) = \lim_{\sigma \to 0} \frac{1}{\sqrt{2\pi\sigma}} \exp\left[\frac{-(x - m_x)^2}{2\sigma^2}\right]. \tag{3.27}$$

This form of the delta function has the advantage that it is infinitely differentiable.

A function closely related to the Gaussian density function is the Laplacian function defined above. It is usually tabulated in many textbooks for only positive values of x. This requires an additional relationship to be introduced:

$$\Phi\,(-x) = 1 - \Phi\,(x). \tag{3.28}$$

A more convenient expression related to $\Phi\,(x)$ is the Q function:

$$Q\,(x) = \frac{1}{\sqrt{2\pi}} \int_{x}^{\infty} \exp\left(-\frac{u^2}{2}\right) du, \tag{3.29a}$$

$$Q(-x) = 1 - Q(x). \tag{3.29b}$$

Thus,

$$Q\,(x) = 1 - \Phi\,(x), \tag{3.30}$$

or if we make a comparison with the definition of the Laplacian function, (3.25b) becomes

$$F(x) = 1 - Q\left(\frac{x - m_x}{\sigma}\right). \tag{3.31}$$

Some alternative types of notation are often used for both functions defined above. Usually,

$$erf\,(x) = \frac{2}{\sqrt{\pi}} \int_0^x \exp(-u^2)\,du \tag{3.32}$$

is called the *error function*, whereas

$$erfc\,(x) = 1 - erf(x) \tag{3.33}$$

is termed the *complementary error function*. Sometimes the definition of the error function (see, e.g., the comment in [10]) is given by

$$erf\,(x) = 1 - 2\Phi\left(\sqrt{2}x\right). \tag{3.34}$$

Despite the fact the both $\Phi(x)$ and $Q(x)$ are widely tabulated, some advantages in calculations are provided by the utilization of $Q(x)$ if some deficiency of values occurs (the values needed are not in the table). A simple calculation procedure is explained in detail in [10].

The *Gaussian density function* has for many reasons a famous position among the *probability density functions*. However, there are many other density functions arising from practical situations and they can be derived from the Gaussian law.

When a current flows through a circuit and the voltage is a random variable, the power dissipation W in the resistor is proportional to the square of the current I:

$$W = RI^2. \tag{3.35}$$

A similar situation exists when a fluid is flowing through a tube. The pressure drop ΔP is proportional to a square of the superficial velocity V (or the volumetric flow rate Q_V), whereas the power dissipated is proportional to the product (ΔPQ_V). The latter gives $N_p \equiv \Delta PQ \equiv Q^3$, or $N_p \sim V^2$ if the superficial velocity is employed.

Despite the variety of practical situations leading to power laws, the further explanation will be developed on the basis of (3.35). We have $I(t_1)$ and assume that the density function $f_i(i)$ is Gaussian. We want to find the probability density function $f_W(W)$, so we have [10]

$$f_w(w) = \frac{1}{2\sqrt{Rw}}\left[f_i\left(\sqrt{\frac{w}{R}}\right) + f_i\left(-\sqrt{\frac{w}{R}}\right)\right], \quad w \geq 0; \quad f_w(w) = 0, \quad w < 0. \tag{3.36a}$$

If I is Gaussian with a zero mean assumed, then

$$f_i(i) = \frac{1}{\sqrt{2\pi}\sigma_I} \exp\left(-\frac{i^2}{2\sigma_I^2}\right), \tag{3.36b}$$

where σ_I is the variance of I. Therefore, the variance σ_I physically signifies the root mean square value of the current. Taking into account the symmetry of the density function $f_i(i) = f_i(-i)$, we can easily to recognize that the two terms on the right-hand side of (3.36) are identical, so

$$f_w(w) = \frac{1}{\sigma_I\sqrt{2\pi Rw}}\exp\left(-\frac{w}{2R\sigma_I^2}\right), \quad w \geq 0; \quad f_w(w) = 0, \quad w < 0. \tag{3.37}$$

The density function is a decreasing function of w and the calculation of the mean power dissipated yields

$$\bar{W} = M\left[RI^2\right] = R\sigma_I^2 \tag{3.38a}$$

and its variance is

$$\sigma_w^2 = \bar{W} - (\bar{W})^2 = M\left[R^2I^4\right] - (\bar{W})^2 = 3R^2\sigma_I^4 - \left(R\sigma_I^2\right)^2 = 2R^2\sigma_I^4. \tag{3.38b}$$

The classical thermodynamic problems considering the probability density function of the velocity of the molecules in a perfect gas lead to the *Maxwell distribution.* The principal assumption considers that each velocity component is Gaussian with zero mean, whereas the variance is $\sigma^2 = \frac{kT}{m_0}$, where k is the Boltzmann constant, T is the absolute temperature, and m_0 the mass of the molecule. The total velocity $V = \sqrt{V_x^2 + V_y^2 + V_z^2}$ has a *Maxwell distribution* with a probability density function

$$f_V(v) = \sqrt{\frac{2}{\pi}\frac{v^2}{\sigma^3}}\exp\left(-\frac{v^2}{2\sigma^2}\right), \quad v \geq 0; \quad f_V(v) = 0, \quad v < 0. \tag{3.39}$$

The mean value of a random variable exhibiting a Maxwell distribution gives the average velocity of the molecules:

$$\bar{V} = \sqrt{\frac{8}{\pi}}\sigma. \tag{3.40}$$

The mean squares and the variance are

$$\bar{V}^2 = 3\sigma^2, \quad \sigma_V^2 = \overline{V^2} - (\bar{V})^2 = \left(3 - \frac{8}{\pi}\right)\sigma^2 = 0.435\sigma^2, \quad \sigma^2 = \frac{kT}{m_0}. \tag{3.41}$$

Thus, the mean kinetic energy, proportional to $\overline{V^2}$, and its expectation are

$$e = \frac{1}{2}m_0V^2, \quad M[e] = \frac{1}{2}m_0\overline{V^2} = 3m_0\sigma^2 = \frac{3}{2}kT, \tag{3.42}$$

which is the Maxwell classical result.

If a random variable is defined as the sum of the squares of independent Gaussian random variables with zero mean and variance 1

$$X^2 = Y_1^2 + Y_2^2 + \ldots + Y_n^2, \tag{3.43}$$

we have a random variable X^2 with a *χ^2distribution.* The probability density function with *ndegrees of freedom* is

$$f(x^2) = \frac{(x^2)^{\frac{n}{2}-1}}{2^{\frac{n}{2}}\left(\frac{n}{2}-1\right)!}\exp\left(-\frac{x^2}{2}\right), \quad x^2 \geq 0; \quad f(x^2) = 0, \quad x^2 < 0. \tag{3.44}$$

With a suitable normalization of the random variable to obtain unit variance, *the power distribution* presented above is in fact a χ^2 distribution with $n = 1$. Moreover, the Maxwell distribution of the square of the velocity (V^2) is χ^2 with $n = 3$.

The mean and the variance of the χ^2 random variable are

$$\overline{X^2} = n, \left(\sigma_{x^2}\right)^2 = 2n. \tag{3.45}$$

These simple results are due to the initial assumption of unit variance of components.

χ^2 arises in many signal detection problems where one is trying to determine if just noise or a signal is detected. If the observed random variable is a signal, the mean values of the samples are not zero, whereas the noise has a zero mean. The random variable defined by (3.43) has a *noncentralχ^2* distribution.

Sometimes the random variables are defined as logarithms of other random variables:

$$Y = \ln X \text{ or } X = e^Y. \tag{3.46}$$

The assumption of a Gaussian Y with a mean $\bar{Y}$ and a variance σ_Y^2 leads to the *log-normal* density function:

$$f_X(x) = \frac{1}{x\sqrt{2\pi}\sigma_Y}\exp\left[-\frac{(\ln x - \bar{Y})^2}{2\sigma_Y^2}\right], \quad x \geq 0; \quad f_X(x) = 0, \quad x < 0. \tag{3.47}$$

It cannot be expressed in terms of elementary functions. If calculations are required, numerical integrations are often necessary.

The mean and the variance of a log-normal variable are

$$\bar{X} = \exp\left(\bar{Y} + \frac{1}{2}\sigma_Y^2\right), \quad \sigma_X^2 = \left[\exp\left(\sigma_y^2\right) - 1\right]\exp 2\left(\bar{Y} + \frac{1}{2}\sigma_Y^2\right). \tag{3.48}$$

If p is the probability an event will happen in any simple trial (usually called a *success*) and $q = 1 - p$ is the probability of an event not happening (usually called *failure*), the probability the event will happen X times in N trials, i.e., X*successes* and N – X*failures* will happen, is

$$p(X) = \binom{N}{X} p^X q^{N-X} = \frac{N!}{X!(N-X)!} p^X q^{N-X}, \tag{3.49}$$

where $X = 0, 1, 2, \ldots, N$; $N! = N(N-1)(N-2)\ldots1$; and $0! = 1$.

The discrete probability function (3.49) is often called the *binomial distribution* since $X = 0,1,2,\ldots, N$ corresponds to the terms of the *binomial formula* (*binomial expansion*):

$$(q+p)^N = q^N + \binom{N}{1}q^{N-1}p + \binom{N}{2}q^{N-1}p^2 + \ldots + p^N, \tag{3.50}$$

where $1, \binom{N}{1}, \binom{N}{2}, \ldots$ are the *binomial coefficients*.

Distribution (3.49) is also called the Bernoulli distribution after James Bernoulli.

If N is large and neither p nor q is too close to zero, the binomial distribution can be closely approximated by a normal distribution with a standardized variable expressed as

$$z = \frac{X - Np}{\sqrt{Npq}}. \tag{3.51}$$

In fact if both Np and $\sqrt{N}q$are greater than 5, a good approximation is achieved [11].

In some applications (e.g., the test of hypotheses), it is important to know the sampling distribution of the difference of the means $(\bar{X}_1 - \bar{X}_2)$ of two samples. In a similar way, the distribution of the *difference of variances* $\left(s_1^2 - s_1^2\right)$ may be considered. The latter is rather complicated, so it is easier to consider the distribution of the ratio $\left(s_1^2/s_2^2\right)$. The main idea is that *small or large* ratios would indicate *large* differences, whereas a ratio close to 1 would indicate small differences. The sampling distribution in such a case is called an *F distribution.*

Let us take samples of size N_1 and N_2 drawn from two *normal* (or nearly normal) populations with variances σ_1^2 and σ_1^2, respectively. The *F-statistic* is defined as

$$F = \frac{\left(\frac{\hat{s}_1^2}{\sigma_1^2}\right)}{\left(\frac{\hat{s}_2^2}{\sigma_2^2}\right)} = \frac{\frac{N_1 s_1}{(N_1-1)\sigma_1^2}}{\frac{N_2 s_2}{(N_2-1)\sigma_2^2}}, \tag{3.52}$$

where

$$\hat{s}_1^2 = \frac{N_1 s_1^2}{N_1 - 1}, \quad \hat{s}_2^2 = \frac{N_2 s_2^2}{N_2 - 1} \tag{3.53}$$

are the modified variances. To elucidate the problem, we note that the sample variance s^2 is a *biased* estimate of the population balance σ^2. With the help of the modified variance $\hat{s}^2$, we find an *unbiased* estimate of σ^2.

The sampling distribution F is called Fisher's F *distribution* (briefly F *distribution*) with degrees of freedom $\nu_1 = N_1 - 1$ and $\nu_2 = N_2 - 1$. The distribution is given by

$$Y = \frac{CF^{\frac{\nu_1}{2}-1}}{(\nu_1 F + \nu_2)^{\frac{\nu_1+\nu_2}{2}}}. \tag{3.54}$$

The constant C depends on ν_1 and ν_2.

In addition to the density functions related to the Gaussian ones, many *other functions arise in engineering practice*. Some of them are presented below to simplify the further discussions.

The *uniform distribution* very often arises in a physical situation where there are no preferable values for the random variable. The uniform probability density function can generally be expressed as

$$f(x) = \frac{1}{x_2 - x_1}, \quad x_1 < x \leq x_2; \quad f(x) = 0, \text{ otherwise}. \tag{3.55}$$

The mean and the dispersion are

$$\bar{X} = \frac{x_1 + x_2}{2}, \quad \sigma_x^2 = \frac{1}{12}(x_2 - x_1)^2. \tag{3.56}$$

The probability distribution function follows from $f(x)$ through integration:

$$F_x(x) = 0, \quad x \leq x_1; \quad F_x(x) = \frac{x - x_1}{x_2 - x_1}, \quad x_1 < x \leq x_2; \quad F_x(x) = 1, \quad x > x_2. \tag{3.57}$$

One important application of the uniform distribution is to describe the dispersion of the errors of the models (see later).

3.2 Statistical Treatment of Experimental Data

The parameters of the random variable defined above can be derived from an infinite number of experimental data. Obviously, this is unrealistic because often experiments provide a limited number of observations, i.e., these limited experimental data will give not the exact values of the parameter, but some *estimates* of them.

Let Y be random variable and a a parameter of its random distribution. The value $\tilde{a}$ is an estimate of the real parameter a if it satisfies the following conditions:

- *Consistency*, meaning that $\tilde{a}$ approaches a as the number of observations is increased.
- It is *unbiased*, i.e., $M[\tilde{a}] = a$ (it has no systematic errors).

- *Efficiency*, i.e., $\sigma^2(\tilde{a}) \rightarrow$ min(the dispersion of $\tilde{a}$ is minimal with respect to the other estimates). If the sampling distributions of two statistics have the same mean (or expectation), then the statistic with the smaller variance is called an *efficient estimator* of the mean, whereas the other statistic is an *inefficient estimator*. The corresponding values of the statistics are termed *efficient estimates*, respectively.

The estimates can be determined from a sample drawn (with size n) from the population of the random variable Y - $y_1, y_2, \ldots, y_n$ corresponding to n observations .

3.3 Estimates of the Expectation and the Dispersion

As already defined, the estimates are parameters defined from samples with a finite number of data derived from experiments:

$$\tilde{m}_y = \frac{1}{n}\sum_{u=1}^{n} y_u, \quad \tilde{\sigma}^2 = \frac{1}{n}\sum_{u=1}^{n}\left(y_u - \tilde{m}_y\right)^2. \tag{3.58}$$

Here, the estimate $\tilde{\sigma}_y^2$ is a *biased* one, but an *unbiased* one can be determined as

$$s_y^2 = \frac{n}{n-1}\tilde{\sigma}^2 = \frac{1}{n-1}\sum_{u=1}^{n}\left(y_u - \tilde{m}_y\right)^2 = \frac{1}{v}\sum_{u=1}^{n}\left(y_n - \tilde{m}_y\right)^2, \tag{3.59}$$

where $v = n - 1$ is the degree of freedom.

The estimates defined above are called *point estimates*. Each one them is given by a single number of the parameter. In other cases, an estimate of the population parameter given by an interval defined by two numbers with which the parameter may be associated is termed an *interval estimate* of the parameter.

Therefore, considering an interval of length of 2δ, where the difference $|a - \tilde{a}|$ should lie, i.e., $|a - \tilde{a}| < \delta$, with probability β (of about 0.90–0.99), we have

$$P(|a - \tilde{a}| < \delta) = \beta. \tag{3.60}$$

The latter means the probability of an observation taking a value outside the interval $(\tilde{a} - \delta,\ \tilde{a} + \delta)$ is very low. In other words, the *level of the significance* α is defined as $\alpha = 1 - \beta$, where β is the *confidence limit*, or *fiducial limit*. The percentage confidence is often called the *confidence level*. The numbers of the confidence limits define the *confidence intervals* for estimating the parameter a.

The definition of the confidence intervals faces some difficulties since the law of distribution of $\tilde{a}$ depends on the *unknown value* of a (i.e., the parameters of the probability distribution of $\tilde{a}$ depend on the parameters of a). This difficulty can be avoided if a particular function of $y_1, \ldots, y_n$ is defined. The definition of such a function should be managed in a way that allows its distribution to be *independent of the value of a*, but it should depend only on the sample size and on the distribution law of Y. If Y has a *normal law of distribution*, the value

$$t = \sqrt{\mathrm{n}}\frac{m_{\mathrm{y}} - \tilde{m}_{\mathrm{y}}}{s_{\mathrm{y}}} \tag{3.61}$$

is the *Student distribution* law [3, 10].

The Student distribution depends on the degree of freedom (i.e., on the sample size), which is defined by the number of the observations N, reduced by the *number of the linear relationships* between them. Here, the *estimate* of the mathematical expectation $\tilde{m}_{\mathrm{y}}$ (see 3.58) is one linear relationship between $y_1, \ldots, y_{\mathrm{n}}$. Thus, the degree of freedom is $v = n - 1$. The *Student distribution* depends on the *degree of freedom* and when $v \to \infty$ (very often $v > 30$), it coincides with the Gaussian law.

If a degree of freedom v and a confidence level β (or a level of significance $\alpha = 1 - \beta$) are preliminarily defined, it is possible to determine the confidence limits of t:

$$-t\left(v, \frac{\alpha}{2}\right) \leq t \leq t\left(v, \frac{\alpha}{2}\right). \tag{3.62}$$

The latter means that the limit β defines the interval where the true value of t belongs. The values of β can be determined from tabulated data of t [10].

The introduction of the variable t (3.61) and the interval (3.62) allows us to determine the confidence interval of the mathematical expectation:

$$\tilde{m}_{\mathrm{y}} - t\left(v, \frac{\alpha}{2}\right)\frac{s_{\mathrm{y}}}{\sqrt{\mathrm{n}}} \leq \tilde{m}_{\mathrm{y}} \leq \tilde{m}_{\mathrm{y}} + t\left(v, \frac{\alpha}{2}\right)\frac{s_{\mathrm{y}}}{\sqrt{\mathrm{n}}}. \tag{3.63}$$

The length of the confidence interval is 2δ, where δ is defined as

$$\delta = t\left(v, \frac{\alpha}{2}\right)\frac{s}{\sqrt{\mathrm{n}}}. \tag{3.64}$$

Obviously, the value of δ decreases as the number of the experimental observations (the sample size) increases.

The confidence interval of the dispersion can be defined in a similar manner. For that purpose, we need the χ^2 distribution:

$$\chi^2 = \sum_{\mathrm{u}=1}^{\mathrm{n}}\left(\frac{y_{\mathrm{u}} - \tilde{m}_{\mathrm{y}}}{\sigma}\right)^2. \tag{3.65}$$

As discussed earlier (see 3.43) the value of χ^2 represents the distribution of the random variable $y_1^2, \ldots, y_{\mathrm{n}}^2$ (where $y_1, \ldots, y_{\mathrm{n}}$ have a *Gaussian distribution*) depending on the degree of freedom v and the level of significance α, so

$$\chi^2\left(v, \frac{\alpha}{2}\right) \geq 0. \tag{3.66}$$

The χ^2 distribution is *noncentral*, which means it is an asymmetric one within the interval $(0, \infty)$. The increase of the value of v diminishes the asymmetry. The confidence limits of the confidence interval where χ^2 belongs are

$$-\chi^2\left(v, \frac{\alpha}{2}\right) \leq \chi^2 \leq \chi^2\left(v, 1 - \frac{\alpha}{2}\right). \tag{3.67}$$

The forms of (3.59) and (3.65) allow us to define

$$s_y^2 = \frac{\sigma^2}{v}\chi^2. \tag{3.68}$$

Therefore, the final form of the confidence interval of the dispersion is

$$\frac{v\, s_y^2}{\chi^2\left(v, 1 - \frac{\alpha}{2}\right)} \leq \sigma^2 \leq \frac{v\, s_y^2}{\chi^2\left(v, \frac{\alpha}{2}\right)}. \tag{3.69}$$

3.4 Tests of Hypotheses

The building of the models is related to various hypotheses and their tests [3, 10]. First, the hypothesis concerning the structure of the model should be considered, which involves a test of the model adequacy. Other important hypothesis concerns the factors affecting the process, i.e., a check of *model parameter significance.*

Despite the variety of hypotheses, their tests concern some principal consequences of them, very often represented by random variables. Generally, we suggest that a particular statistical distribution is associated with the hypothesis, i.e., we suppose that this statistical hypothesis is true. However, if the results of the random sample differ from those expected under the hypothesis chosen, then we need to decide if the observed differences are *significant* or not. This test requires a sequence of rules, or in other words *criteria of* hypothesis *significance.*

The procedures that allow us to decide whether the observed sample differ from the results expected, or in other words whether we should *accept* or *reject* the hypothesis, are often called *tests ofhypotheses*, *tests of significance*, *rules of decision*, or simply *decision rules.*

If the hypothesis that *should be accepted* was *rejected* owing to the test, a *type I error* occurred. Otherwise, if we accept a hypothesis that should be rejected, this means a *type II error* occurred. Therefore, the decision rules should be designed in a manner that minimizes the possibility of wrong decisions. Very often, this is not a simple procedure, since if we reduce some types of errors in a particular sample, we obviously increase the other types of errors. The only general rule to minimize both types of errors is to *increase the sample size* if this is possible.

In the test of a given hypothesis, the *maximum probability* of minimizing the risk of a wrong decision, e.g., *type I*, is called the *level of significance* or the *significance level* of the test performed. This probability, denoted by α, is generally defined before the sample is drawn. This means that the result obtained does not affect the choice.

It is customary to use significance levels of 0.05 and 0.01. This common practice does not prevent the use of other values of α. The value of $\alpha = 0.05$ (5%) is often chosen at the beginning of test. This means that there are five chances in 100 to reject the correct hypothesis or, in other words, we are about 95%*confident* that the decision is right. Otherwise, if the hypothesis is rejected under the prolific assumption of a 0.05 significance level, this means that the hypothesis has 5% probability of being wrong.

The main idea of the test supposes that the parameter of the random variable distribution Y takes a particular value a_0. This is the so-called *null hypothesis*:

$$H_0 : a = a_0. \tag{3.70}$$

Any hypotheses differing from a given hypothesis are called *alternative hypotheses*. For example, if one hypothesis is $p = 0.5$, alternatives could be $p = 0.9$, $p \neq 0.5$, or $p > 0.5$. The alternative hypothesis to the null hypothesis is often denoted by H_1:

$$H_1 : a \neq a_0. \tag{3.71}$$

The test of H_0 looks for a random variable $\xi\ (a)$ depending on the parameter a. First, the distribution density of $\xi\ (a)$ must be determined *assuming* that the hypothesis is *true*. The next step is to determine the *confidence interval* of $\xi\ (a)$. Under the assumption of a *confidence limit* β and a level of significance $\alpha = 1 - \beta$, the confidence interval is

$$\xi\left(v, \frac{\alpha}{2}\right) \leq \xi\ (a_0) \leq \xi\left(v, 1 - \frac{\alpha}{2}\right). \tag{3.72}$$

The next step is to define the *estimate* $\tilde{a}$ of the parameter a. If the hypothesis is *true*, $\xi\ (\tilde{a})$ satisfies condition (3.72), i.e., $\xi\ (\tilde{a})$ belongs to the confidence interval.

In many cases of the statistical analysis of the models, a *test of a null hypothesis of the variances* is required. Obviously, the null hypothesis concerns

$$H_0 : \sigma_1^2 = \sigma_2^2. \tag{3.73}$$

The procedure needs the application of the distribution of the F statistic (3.52):

$$F = \frac{\left(\frac{s_1^2}{s_2^2}\right)}{\left(\frac{\sigma_1^2}{\sigma_2^2}\right)}. \tag{3.74}$$

If the null hypothesis is true, (3.74) becomes

$$F = \frac{s_1^2}{s_2^2}. \tag{3.75}$$

The confidence interval at a level of significance 2α is

$$F'(\alpha, \nu_1, \nu_2) \leq \frac{s_1^2}{s_2^2} \leq F(\alpha, \nu_1, \nu_2). \tag{3.76}$$

If we utilize the fact that for the F statistic

$$F'(\alpha, \nu_1, \nu_2) = \frac{1}{F(\alpha, \nu_1, \nu_2)}, \tag{3.77}$$

it follows immediately that

$$\frac{1}{F(\alpha, \nu_1, \nu_2)} \leq \frac{s_1^2}{s_2^2} \leq F(\alpha, \nu_1, \nu_2). \tag{3.78}$$

The tabulated values of F (see, e.g., [12, 13]) indicate that $F(\alpha, \nu_1, \nu_2) \geq 1$, i.e., $1/F \leq 1$. This allows us to apply a one-sided criterion in the case of $s_1^2 > s_2^2$. This means that we have

$$\frac{s_1^2}{s_2^2} \leq F(\alpha, \nu_1, \nu_2), \quad (\nu_1 = n_1 - 1, \quad \nu_2 = n_2 - 1). \tag{3.79}$$

If condition (3.79) is satisfied, the null hypothesis is true; otherwise, it must be rejected. Condition (3.79) means that the dispersions are equal and $F \geq 1$ is the maximum deviation of the ratio $\left(s_1^2/s_2^2\right)$ from 1 owing to random errors.

In some tests concerning the statistical analysis of the models, the problem is to verify the *homogeneity of the dispersion*. This requires applying the null hypothesis:

$$H_0 : \sigma_1^2 = \sigma_2^2 = \ldots = \sigma_k^2. \tag{3.80}$$

For that purpose the random variable

$$G = \frac{(s_{\max})^2}{s_1^2 + s_2^2 + \ldots + s_k^2} \tag{3.81}$$

must be created. Here, s_k^2 is the maximum value of the estimates $s_1^2, \ldots, s_k^2$.

The calculated value of G must be compared with the tabulated values of $G(\alpha, k, \nu)$ [12] at a particular significance level under the condition imposed by the number of the samples drawn from the population k and sizes n ($\nu = n - 1$). If the comparison between the calculated and the tabulated values gives

$$G \leq G(\alpha, k, \nu), \tag{3.82}$$

the null hypothesis is true; otherwise, it should be rejected.

If *coarse errors* occur in the sample, this could lead to wrong results despite the correct statistical tests having been performed. The tests assume that all the errors have a *random nature*, so the preliminary detection of the coarse errors and their rejections is an important step of model analysis. The deviation of the coarse error $y*$ from the estimate of the mathematical expectation $\tilde{m}_y$ and the ratio

$$\tau = \frac{|y* - \tilde{m}_y|}{s_y} \tag{3.83}$$

allow us to detect coarse errors among the sample drawn.

The values of $\tau\ (1 - \alpha,\ n)$ at given significance level α and a sample size n are available in a tabulated form [12]. If $\tau\ \leq \tau\ (1 - \alpha,\ n)$, the deviation $y*$ is not a result of a coarse error; otherwise, the observation must be rejected.

3.5 Dispersion Analysis

In many cases, the analysis of the dispersion enables us to detect some important facts about it [3, 9]. For example, if we consider a particular model, it is important to minimize the number of factors determining the objective function. Usually, the analysis looks for factors whose magnitudes are comparable with the order of magnitude of the experimental error. They should be rejected as unimportant for the model.

The dispersion analysis concerning the effect of a *factor* A on the *objective function* y needs experimental data y_{ij} at different levels of the factor $A_i(\mathrm{i} = 1,\ldots,\mathrm{k})$. Each level of the factor requires n experiments $(\mathrm{j} = 1,\ldots,\mathrm{n})$. If $\mu_i(\mathrm{i} = 1,\ldots,\mathrm{k})$ denotes the mathematical expectation of the experimental data at each ith level of A, the test needed is the *null hypothesis*, i.e.,

$$H_0 : \mu_1 = \mu_2 = \ldots = \mu_{\mathrm{k}}. \tag{3.84}$$

Obviously, if the null hypothesis is true, the factor A does not affect the objective function and it should be omitted from the model.

The mathematical expectations μ_i require us to determine the arithmetic means of the experiments performed (i.e., the mean of the samples drawn from the experiments):

$$\bar{y}_1 = \frac{1}{\mathrm{n}}\sum_{\mathrm{j}=1}^{\mathrm{n}} y_{i\mathrm{j}}\ , \quad \bar{y} = \frac{1}{\mathrm{k}}\sum_{i=1}^{\mathrm{k}} \bar{y}_i = \frac{1}{\mathrm{nk}}\sum_{i=1}^{\mathrm{k}}\sum_{\mathrm{j}=1}^{\mathrm{n}} y_{i\mathrm{j}}. \tag{3.85}$$

Let us consider the dispersion $S^2$of the random variable Y following from (3.59) at $\tilde{m}_y = \bar{y}$:

$$S^2 = \frac{Q}{v}, \tag{3.86}$$

where

$$Q = \sum_{i=1}^{\mathrm{k}}\sum_{\mathrm{j}=1}^{\mathrm{n}} \left(y_{\mathrm{ij}} - \bar{y}\right)^2, \quad v = \mathrm{kn} - 1. \tag{3.87}$$

The degrees of freedom $v = \mathrm{kn} - 1$ result from the number of experiments kn*reduced* by one linear relationship (3.85) between $y_{\mathrm{ij}}(\mathrm{i} = 1,\ldots\mathrm{k},\ \mathrm{j} = 1,\ldots,\mathrm{n})$ in the expression for $\bar{y}$.

The sum Q may be expressed as

$$Q = \sum_{i=1}^{k} \sum_{j=1}^{n} \left(y_{ij} - \bar{y}_i + \bar{y}_i - \bar{y}\right)^2. \tag{3.88}$$

This allows us to obtain directly that

$$\begin{aligned} Q = Q_1 + Q_\varepsilon, \quad Q_1 &= \sum_{i=1}^{k} \sum_{j=1}^{n} \left(\bar{y}_1 - \bar{y}\right)^2 = n \sum_{i=1}^{k} \left(\bar{y}_i - \bar{y}\right)^2, \\ Q_\varepsilon &= \sum_{i=1}^{k} \sum_{j=1}^{n} \left(y_{ij} - \bar{y}_i\right)^2, \end{aligned} \tag{3.89}$$

where the following relationship is valid:

$$\sum_{i=1}^{k} \sum_{j=1}^{n} \left(y_{ij} - \bar{y}_i\right)\left(\bar{y}_i - \bar{y}\right) = \sum_{i=1}^{k} B_i\left(\bar{y}_i - \bar{y}\right) = 0, \tag{3.90}$$

since from (3.85) we have

$$B_i = \sum_{j=1}^{n} \left(y_{ij} - \bar{y}_i\right) = \sum_{j=1}^{n} y_{ij} - n\bar{y}_i = 0. \tag{3.91}$$

The degree of freedom v_1 results from Q_1, where the random variables $\bar{y}_i$ (i = 1,. . ., k) participate. One linear relationship exists between them in the form (3.89) (i.e., the relationship between $\bar{y}$ and $\bar{y}_i$, i = 1,. . ., k, so we have $v_1 = k - 1$). In a similar way it is possible to determine $v_\varepsilon = nk - k = k(n - 1)$ since the number of experiments in Q_ε is kn and the linear relationships between them are k through $\bar{y}_i$ (i = 1,. . ., k).

The above manipulations of the formulae permit us to determine the dispersion estimates:

$$S_1^2 = \frac{Q_1}{v_1}, \quad S_\varepsilon^2 = \frac{Q_\varepsilon}{v_\varepsilon}, \quad S^2 = \frac{Q}{v}. \tag{3.92}$$

The form of (3.92) indicates that both the error of the experiments and that of the factor A effect contribute to the dispersion S^2. On the other hand, in S_1^2 and S_ε^2 these effects are separated.

Let us suppose that factor A does not affect the random variable y if the dispersion σ_1^2 relevant to its effect equals the dispersion of the experimental errors, i.e., $\sigma_1^2 = \sigma_\varepsilon^2$. In this case, a more convenient way is to utilize the random variable (3.75), expressed through the F statistic:

$$F = \frac{S_1^2}{S_\varepsilon^2}, \tag{3.93}$$

where S_1^2 and S_ε^2 have χ^2 distributions since Q_1 and Q_ε are sums of random variable with *normal density distributions*. The ratio of two χ^2-distributed random variables has an F distribution and if F from (3.93) satisfies the condition

$$F \leq F(\alpha, v_1, v_2), \tag{3.94}$$

the factor A does not affect the random variable, since the null hypothesis H_0 : $\sigma_1^2 = \sigma_2^2$ is true.

Otherwise, the effect of the factor A is *significant*.

3.6 Significance of Parameter Estimates and Model Adequacy

The main problem arising during the statistical analysis of models is the test of the *significances ofthe parameter estimates and model adequacy*. The problem solution needs testing for two statistical hypotheses:

1. The first hypothesis is $H : b = 0$, where b is the estimate of the parameter under consideration.
2. The second step is to test the hypothesis that $H : \sigma^2 = \sigma_\varepsilon^2$, where σ^2 is the dispersion of the model and σ_ε^2 is the dispersion of the experimental data.

The test concerning the significance of the parameter estimates and the model adequacy can be performed through an example of a regression model (see, e.g., 2.5.5). The regression models permit us to perform the test in a more general form. The approach will be described in detail in the cases of particular models discussed further.

The test of the significance of the regression model (2.5.5) requires the hypothesis of the existence of null coefficients to be proved [12]. At the beginning, several experimental values of the objective function y_{u_0}, $u_0 = 1,\ldots, n_0 (n_0 = 5 \div 10)$, obtained under the same conditions, are required. The latter means the factors $x_{u_0} = (x_{1u_0}, \ldots, x_{mu_0})$ must be kept constant. These experiments should be designed either as part of the experiments required determining the parameters $b_i(i = 1, \ldots, k)$ or as special experiments concerning only the statistical analysis of the model.

Let us assume that the coefficients $b_i(i = 1, \ldots, k)$ of the model (2.6.6) are *independent*. The *insignificance* of the estimate b_{i_0} can be determined by utilizing the Student density distribution (3.61), which in this particular case is

$$t_{i_0} = \frac{b_{i_0} - \beta_{i_0}}{s(b_{i_0})}. \tag{3.95}$$

It is necessary to put $\beta_{i_0} = 0$ in (3.95) and for the dispersion estimate $s(b_{i_0})$ of the coefficients b_{i_0} to be expressed through the dispersion of error of the experimental data:

$$s(b_{i_0}) = \sqrt{c_{i_0 i_0}}\, s_{\varepsilon}, \tag{3.96}$$

where c_{ij} are diagonal elements of the matrix **C**, i.e., inverse matrix of **A** (see 3.6.9):

$$C = \|c_{ij}\| = A^{-1}, \quad i = 1,\ldots,k; \quad j = 1,\ldots,k. \tag{3.97}$$

As a result,

$$t_{i_0} = \frac{b_{i_0}}{s_{\varepsilon}\sqrt{c_{i_0 i_0}}}, \quad i = 1,\ldots,k, \tag{3.98}$$

where s_{ε}^2 is the biased estimate of the dispersion of the random variable y taking values of y_{u_0}, $u_0 = 1,\ldots,n_0$,

$$s_{\varepsilon} = \sqrt{\frac{1}{n_0 - 1}\sum_{u_0=1}^{n_0}\left(y_{u_0} - \tilde{m}_y\right)^2}, \tag{3.99}$$

determined through the *estimate* of the mathematical expectation of y,

$$\tilde{m}_y = \frac{1}{n_0}\sum_{u_0=1}^{n_0} y_{u_0}. \tag{3.100}$$

At a preliminarily chosen level of significance α, it is possible to determine the tabulated value [9] of the *Student distribution* $t\left(\frac{\alpha}{2}, v_{\varepsilon}\right)$, where the degree of freedom is $v_{\varepsilon} = n - 1$. The

$$\|t_{i_0}\| \leq t\left(\frac{\alpha}{2}, v_{\varepsilon}\right) \tag{3.101}$$

condition concerning the coefficient b_{i_0}*confirms* the hypothesis $b_{i_0} = 0$, i.e., it proves the *insignificance* of the coefficient.

The employment of models for process simulation is the correct approach if their *adequacy* is proved. The term *adequacy* means the calculated and the experimental values of the objective function are *congruent*, i.e., the variances of the errors of the calculated and experimental values of the objective function are equal. The adequacy of the model depends both on the chance of building a suitable mathematical structure and the subsequent correct calculation of the model parameters. Obviously, as already mentioned for the parameter significance, the adequacy depends on the *right choice* of the *function* $f_i(i = 1,\ldots,k)$ utilized in (2.5.5).

The methods testing the adequacy do not depend on the model structure and the form of its expression. They are defined by data from several measurements of the objective function, at fixed values of the factors, performed for few typical regimes of the process. In many cases, however, it is necessary to compare the effect of the *nonadequacy* on the calculated values of the objective function. Thus, that effect of the nonadequacy needs to be compared with the effect of the random errors on the objective function. The latter means that a hypothesis test concerning the equality of the dispersion of both the model and the experimental error is required.

Let us consider, as the previous example, that the test of adequacy is performed based on n additional experiments performed at fixed values of the factors. The model parameters are determined preliminarily through treatment of the data of N experiments. Therefore, the residual sum of squares [12, 13] can be formed:

$$Q_{\mathrm{res}} = \sum_{\mathrm{u}=1}^{\mathrm{N}} (y_{\mathrm{u}} - \hat{y}_{\mathrm{u}})^2, \tag{3.102}$$

where $\hat{y}_{\mathrm{u}}$ values of y are calculated through model (2.5.6). It is clear that Q_{res} is a random variable representing the error of the model. The estimation of the dispersion of Q_{res} is

$$s_{\mathrm{res}} = \frac{Q_{\mathrm{res}}}{v_{\mathrm{res}}} = \frac{Q_{\mathrm{res}}}{\mathrm{N}-\mathrm{k}}. \tag{3.103}$$

The degree of freedom $v_{\mathrm{res}} = \mathrm{N} - \mathrm{k}$ depends on both the number of experiments N and the number of linear relationships between y_u in (3.6.9). Obviously, the number of linear relationships equals the number of the model coefficients k.

The random error dispersion s_ε^2 from the experiments should be calculated through (3.99) and (3.100), which allows us to form the dispersion ratio

$$F = \frac{s_{\mathrm{res}}^2}{s_\varepsilon^2}. \tag{3.104}$$

Very often, $s_{\mathrm{res}}^2 > s_\varepsilon^2$ since s_{res}^2 accumulate both the model nonadequacy and the experimental errors.

The adequacy condition means

$$F \le F(\alpha, v_{\mathrm{res}}, v_\varepsilon), \tag{3.105}$$

where $F(\alpha, v_{\mathrm{res}}, v_\varepsilon)$ is a tabulated value of the F distribution [12, 13]

A second important case is the situation where there are multiple observations under the conditions of several regimes. This case does not require additional experiments for the evaluation of the dispersion estimate s_ε^2.

If p is the number of experiments $(\mathrm{p} = 1, \ldots, \mathrm{n})$ and q is the number of observations during these p experiments $(\mathrm{q} = 1, \ldots, \mathrm{V_p})$, the total number of experimental data is

$$\mathrm{N} = \sum_{\mathrm{p}=1}^{\mathrm{n}} \mathrm{V_p} = \mathrm{n} + \sum_{\mathrm{p}=1}^{\mathrm{n}} (\mathrm{V_p} - 1). \tag{3.106}$$

In this case [12, 13]

$$Q_{\mathrm{res}} = Q_{\mathrm{L}} + Q_\varepsilon. \tag{3.107}$$

Here, Q_L is relevant to the model nonadequacy (a systematic error), whereas Q_ε is the experimental error (random error).Thus,

$$Q_\varepsilon = \sum_{p=1}^{n}\sum_{q=1}^{V_p}\left(y_{pq} - \bar{y}_p\right)^2, \quad \bar{y}_p = \frac{1}{V_p}\sum_{q=1}^{V_p} y_{pq},$$
$$Q_L = Q_{res} - Q_\varepsilon = \sum_{p=1}^{n} V_p\left(\bar{y}_p - \hat{y}_p\right)^2. \tag{3.108}$$

The degrees of freedom are [13]

$$\nu_\varepsilon = \sum_{p=1}^{n}\left(V_p - 1\right) = N - n, \quad \nu_L = \nu_{res} - \nu_\varepsilon = N - k - (N - n) = n - k. \tag{3.109}$$

The *adequacy test* considers both the estimates of the dispersions and the dispersion ratio:

$$s_L^2 = \frac{Q_L}{\nu_L}, \quad s_\varepsilon^2 = \frac{Q_\varepsilon}{\nu_\varepsilon}, \quad F = \frac{s_L^2}{s_\varepsilon^2}. \tag{3.110}$$

Comparison with the tabulated data of the F distribution $F(\alpha, \nu_L, \nu_\varepsilon)$ should lead to the *adequacycondition:*

$$F \leq F(\alpha, \nu_L, \nu_\varepsilon). \tag{3.111}$$

The existence of multiple observations through several experiments (regimes) permits us to test the dispersion homogeneity. The latter means that the distribution of the dispersion of the experimental error should be uniform over the area where the factors vary. The procedure concerns forming the criterion G (see 3.81). The condition $G \leq G(\alpha, k, \nu)$ is a necessary condition of the model adequacy.

3.7 Model Suitability

In some situations, there is no possibility to repeat the experiments several times to define s_ε^2 as was mentioned earlier. Very often this situation follows from passive experiments concerning the determination of the parameters b_i or when the experiments are very expensive or a long time is needed for a single experiment (e.g., a biotechnology experiment). In this case the model suitability, from the point of view of process simulation, can be evaluated on the basis of the coefficient of multiple correlations R. The latter requires the sum

$$Q = \sum_{u=1}^{N}(y_u - \bar{y})^2, \quad \bar{y} = \frac{1}{N}\sum_{u=1}^{N} y_u \tag{3.112}$$

and its decomposition [13]

$$Q = Q_R + Q_{res}, \quad Q_R = \sum_{u=1}^{N} (\hat{y}_u - \bar{y})^2, \quad Q_{res} = \sum_{u=1}^{N} (y_u - \hat{y}_u)^2. \tag{3.113}$$

Obviously, the degrees of freedom of the sums Q, Q_R, and Q_{res} are

$$\nu = N - 1, \quad \nu_R = k - 1, \quad \nu_{res} = N - k. \tag{3.114}$$

The coefficient of multiple correlation [2, 12, 13] is

$$R = \sqrt{\frac{Q_R}{Q}} = \sqrt{1 - \frac{Q_{res}}{Q}}. \tag{3.115}$$

The condition for model *suitability* requires R to be very close to 1. Sometimes $R = 1$ but the model is not adequate [12, 13]. The reliability of the estimation of the model adequacy increases as the degree of freedom ν_{res} increases. This can be estimated through the ratio of dispersions:

$$F = \frac{\left(\frac{Q - Q_{res}}{k-1}\right)}{\left(\frac{Q_{res}}{N-k}\right)} = \frac{R^2(N-k)}{(1-R^2)(k-1)}. \tag{3.116}$$

The tabulated values of the F distribution allow two conclusions to be drawn:

1. If $F > F(\alpha, \nu_R, \nu_{res})$, the coefficient of multiple correlation is *significant* and the model is *suitable* for simulation.
2. If $F \leq F(\alpha, \nu_R, \nu_{res})$, the coefficient of multiple correlation is *insignificant* and the model is *unsuitable* for simulation.

The tests concerning the significance of the estimates of the model coefficients and the adequacy of the criteria models are similar to those of the regression models after taking logarithms.

3.8 Adequacy of the Theoretical Models and Model Theories

The parameters of the theoretical models and model theories are considered as exactly defined, i.e., they are not random variables. Therefore, there is no need to perform procedures for their determination as well as tests of their significances. The analysis of the significance of the dimensionless parameters is a problem that was especially discussed (See page 187 and the next). Thus, the problem concerning the adequacy of the theoretical models will be discussed only. The results are valid also for the model theories and vice versa.

Let the objective function of a theoretical model (model theory) is expressed in a general form:

$$\hat{y} = \phi(x_1, \ldots, x_m ; \beta_1, \ldots, \beta_k), \tag{3.117}$$

where the parameters $\beta_1, \ldots, \beta_k$ participate with their true values (they are not random variables) and very often ϕ is nonlinear with respect the parameters $\beta_1, \ldots, \beta_k$.

The existence of experimentally determined values of the objective function y_u derived under different regimes $x_{1u}, \ldots, x_{mu}$, $u = 1, \ldots, N$ allows us to calculate the residual sum of the squares:

$$Q_{res} = \sum_{u=1}^{N} (y_u - \hat{y}_u)^2, \tag{3.118}$$

where $\hat{y}_u$ are calculated through the model (3.117), i.e.,

$$\hat{y}_u = \phi(x_{1u}, \ldots, x_{mu}; \beta_1, \ldots, \beta_k), \quad u = 1, \ldots, N. \tag{3.119}$$

In this case $\hat{y}_u(u = 1, \ldots, N)$ are *not random variables*, since they are calculated through the exact values of x and β. Therefore, the degree of freedom of Q_{res} and ν_{res} is N, which allows us to determine the dispersion estimate:

$$s^2_{res} = \frac{Q_{res}}{N}. \tag{3.120}$$

The dispersion of the random experimental error s^2_ε can be determined in a way similar to that applied to the regression models (3.99). The determination the dispersion ratio

$$F = \frac{s^2_{res}}{s^2_\varepsilon} \tag{3.121}$$

permits us to apply the condition of model adequacy:

$$F \leq F(\alpha, N, n - 1), \tag{3.122}$$

where $F(\alpha, N, n - 1)$ is the tabulated value [12] of the F distribution at significance level α.

The analysis of the model adequacy would be very successful if multiple measurements from several experiments were available. Such a situation would allow us to determine the uniform distribution of the dispersion of the experimental error over the area where the factors vary. Under such conditions, the adequacy can be determined through a method similar to that applied to the regression models, but $\nu_{res} = N\,(\nu_L = n)$. This requires applying randomized experimental plans like those ones discussed further (See page 268 and the next).

The nonlinear relationship between $\hat{y}$ and $\beta_1, \ldots, \beta_k$ does not permit us to apply the coefficient of multiple correlations for estimation of the model suitability when there are no additional observations. In other words, the multiple observations performed under different regimes are obligatory for the analysis of the adequacy of the theoretical models.

4 Example

4.1 Statistical Analysis of Diffusion Type Models

The parameters of the *diffusion-type models* are determined using experimental data and as a result they are random numbers, but the test of significance can be performed, like in the case of the theoretical models.

The statistical analysis of the model adequacy faces difficulties mainly because of the nonlinearity of the objective function (with respect to the parameters). This leads to methods applicable (as discussed earlier) to the theoretical models. However, there is a small difference since $\hat{y}_u(n = 1, \ldots, N)$ are random variables because of the errors of the determination of $b_1, \ldots, b_k$. The latter sets up the problem of the degree of freedom v_{res}. The problem can be solved with a certain approximation like in the case of the linear regressions, i.e., $v_{res} = N - k$. The idea is that close to the area of the parameters determined the objective function can be expressed through a Taylor series with respect to the parameters. Thus, (3.120) can be expressed as

$$s_{res}^2 = \frac{Q_{res}}{N - k}. \tag{4.1}$$

The further analysis is like that discussed for the theoretical models.

The statistical analysis of the model adequacy permits us to *confirm* or to *reject* a particular mathematical structure employed for a process description. In the case where the mathematical structure follows directly from the process mechanism, the proof of the model adequacy can be utilized to test whether the *suggested mechanism* is adequate or not. Thus, the modeling can be applied for the analysis of the process mechanism by means of experimental data.

The methods of statistical model analysis presented are valid under three conditions:

1. The objective function is a *random variable* with a *Gaussian* density distribution.
2. The dispersion of the experimental error (determined through multiple measurements) is the same, i.e., *does not vary* because of the regime.
3. The errors occurring during the measurements of the factors are *negligible* with respect to the errors introduced when the measurement of the objective function take places.

There are methods for solving the problems when these conditions are not satisfied.

References

1. Rosen AM (ed) (1980) Scale-up in chemical technology. Khimia, Moscow (in Russian)
2. Gukhman AA (1973) Introduction to the similarity theory. Vishaia Shkola, Moscow (in Russian)
3. Boyadjiev Chr (1993) Fundamentals of modeling and simulation in chemical engineering and technology. Bulgarian Academy of Sciences, Sofia (in Bulgarian)
4. Weiler D, Bonnett F, Leavitt F (1971) Chem Eng Prog 69(9):86
5. Chr Boyadjiev (2006) Int J Heat Mass Transf 49:796
6. Boyadjiev Chr (2006) Int J Heat Mass Transf 49:2053
7. Panayotova K, Doichinova M, Boyadjiev Chr (2009) Int J Heat Mass Transf 52:543
8. Panayotova K, Doichinova M, Boyadjiev Chr (2009) Int J Heat Mass Transf 52:2358
9. Boyadjiev Chr, Dimitrova E (2005) Comput Chem Eng 29:941
10. Cooper GR, McGillem CD (1986) Probabilistic methods of signal and system analysis, 2nd edn. CBS, Tokyo
11. Clark AB, Disney RL (1970) Probability and random processes for engineers and scientists. Wiley, New York
12. Vuchkov I, Stoyanov S (1986) Mathematical modeling and optimization of technological objects. Technika, Sofia (in Bulgarian)
13. Draper NR, Smith H (1998) Applied regression analysis, 3rd edn. Wiley series on probability and statistics. Wiley-Interscience, New York

Stability Analysis

The theoretical analysis of the non-linear mass transfer shows (See page 140 and the next) that the systems with intensive mass transfer are characterized by a number of nonlinear effects. They change significantly both the kinetics and the mass transfer mechanism as a result of the mass transfer effects on the hydrodynamics of the system. The change may have a significantly greater effect if the system loses its stability and reaches a new stable state (a self-organization of a dissipative system). The mathematical description of these systems may be done on the basis of the stability theory.

1 Stability Theory

Various problems concern the behavior of systems (mechanical, chemical, physical, and economic) when they are far from their equilibrium state. The behavior depends on the system stability, i.e., it is related to the ability of a system to undergo a sharp change for a smooth change of the external conditions.

The system stability is also a feature of its mathematical description. This needs a short description of the theory of the mathematical stability required for further development of the hydrodynamic stability theory. However, in all situations the stability will be considered as a specific feature of a particular process.

1.1 Evolution Equations

Let us assume that the features of the systems may be determined by the quantities x_i $(i = 1,\ldots,n)$. This permits us to consider the state of the system as a point in n-dimensional space with coordinates x_i $(i = 1,\ldots,n)$ (a phase space) [1–3]. The rate of change of the system features in time is a vector in the n-dimensional

C. Boyadjiev, *Theoretical Chemical Engineering*,
DOI: 10.1007/978-3-642-10778-8_6, © Springer-Verlag Berlin Heidelberg 2010

space. The projections on the coordinate axis $\frac{dx_i}{dt}$ $(i = 1, \cdots, n)$ satisfy the *evolution law* of the system:

$$\frac{dx_i}{dt} = X_i(x_1, \cdots, x_n, t), \quad x_i(0) = x_{i0}, \quad i = 1, \cdots, n. \tag{1.1}$$

The evolution Eq. 1.1 for systems with laws independent of time are termed *autonomous equations*:

$$\frac{dx_i}{dt} = X_i(x_1, \cdots, x_n), \quad x_i(0) = x_{i0}, \quad i = 1, \cdots, n. \tag{1.2}$$

The components of the *phase velocity* X_i ($i = 1,\ldots,n$) are the coordinates of the vector field of the same phase velocity and determine the velocity of the system in the *phase space*. The points $x_i(t)$, $i = 1, \cdots, n$, represent a curve (a *phase trajectory*) in the scalar phase space (field).

Let us consider the autonomous equation for simplicity of explanation:

$$\frac{dx}{dt} = X(x), \quad x(0) = x_0. \tag{1.3}$$

Let us assume that

$$X(x) = 0, \tag{1.4}$$

i.e.,

$$X(a) = 0. \tag{1.5}$$

It follows from (1.5) that the point $x = a$ may be considered as a *stationary point* (the system velocity is zero). If $a = x_0$, it is clear that

$$x(t) \equiv a \tag{1.6}$$

is a solution of (1.3) for $a = x_0$, where a is a *singular point*.

Let us consider the linear version of Eq. 1.3 and its solution for exactness:

$$\frac{dx}{dt} = \lambda x, \quad x(0) = x_0, \quad x = x_0 \exp(\lambda t). \tag{1.7}$$

It is clear from (1.7) that $x = 0$ is a singular point, i.e., $x_0 = 0$ and the solution of (1.7) has the following features (see Fig. 1):

$$\lambda < 0, \quad \lim_{t\to\infty} x(t) = 0, \quad \forall x_0; \tag{1.8}$$

$$\lambda = 0, \quad x = x_0, \quad \forall x_0; \tag{1.9}$$

$$\lambda > 0, \quad x = x_0 \, \text{and} \, x = 0;$$

$$\lambda > 0, \quad \lim_{t\to\infty} x(t) \to \infty, \quad \forall x_0 > 0;$$

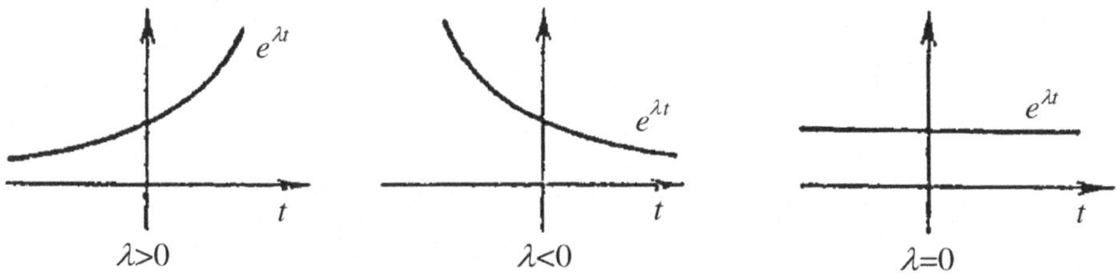

Fig. 1 Solution of Eq. 1.7

$$\lambda > 0, \quad \lim_{t\to\infty} x(t) \to -\infty, \quad \forall x_0 < 0. \tag{1.10}$$

The multiformity of the solution at $\lambda > 0$ is not a result of its nonuniqueness, but t is due to the solution instability with respect of the small perturbation of the initial condition (x_0).

Inequality (1.8) leads to the following conclusions:

1. The solution (the process) is unstable at $\lambda > 0$ and the small deviations of the initial state $x_0 \neq 0$ lead to deviations of the solution $x = 0$.
2. At $\lambda \leq 0$ the solution is unstable for each x_0.
3. At $\lambda \leq 0$ the solution approaches the singular point $x = 0$, i.e., the stationary point becomes a focus of attraction of the solution (an *attractor*).

Linear equation (1.7) together with the conditions for the solution stability are attractive ones because they provide the basis of the kinetics models of many important processes (evolution of organisms, nuclear processes, chemical reactions, etc.). These features in the area of real numbers (R) become more interesting in the complex area (C), where Eq. 1.7 has the form

$$\frac{dz}{dt} = \lambda z, \quad z \in C, \quad \lambda \in C, \quad t \in R,$$

$$z(0) = z_0, \quad z(t) = z_0 \exp(\lambda t). \tag{1.11}$$

It follows from (1.11) that if λ is a real number, the same is valid for z.

In the cases when λ is an imaginary number ($\lambda = i\omega$, $i^2 = -1$), the solution of (1.11) is a complex number because the Euler formula gives

$$z = z_0 \exp(\lambda t) = z_0 \exp(i\omega t) = z_0(\cos \omega t + i \sin t) = \zeta + i\eta. \tag{1.12}$$

Thus, the solution (1.12) is a circle in the plane of the complex numbers (Fig. 2) and the phase points moves along that circular trajectory clockwise ($\omega < 0$) or in the opposite direction ($\omega > 0$).

When λ is complex number,

$$\lambda = \alpha + i\omega t, \tag{1.13}$$

it follows directly from (5.11) that

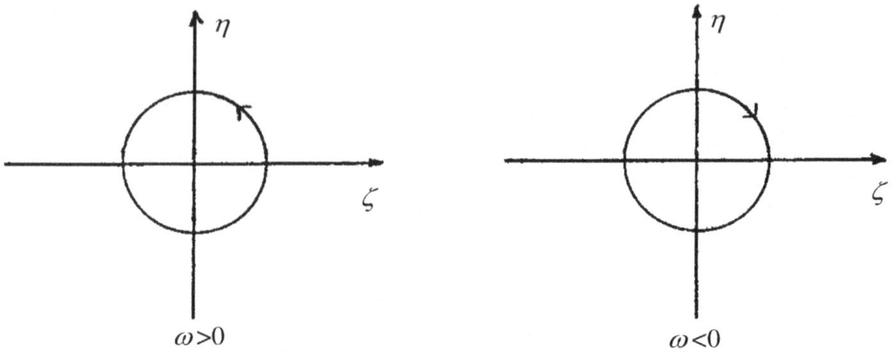

Fig. 2 The solution of Eq. 1.12 in the plane of complex numbers

$$z = z_0 \exp(\lambda t) = z_0 \exp(\alpha t) \exp(i\omega t) = z_0 \exp(\alpha t)(\cos \omega t + i \sin \omega t) = x + iy, \tag{1.14}$$

i.e., the solution is a complex number. However, this periodic solution has variable amplitude $z_0 \exp(\alpha t)$ depending on α. At $\alpha > 0$ the solution is unstable $\left[\lim_{t\to\infty} z(t) = \pm\infty\right]$.

At $\alpha \leq 0$ the solution (1.14) is stable (see Fig. 3). The solution (1.14) shows that at $z = 0$ is the unique singular point, termed *a focus*. The focuses may be stable or unstable depending on whether they are or are not attractors for the solution (see Fig. 4). At $\alpha < 0$ the focuses are stable, whereas at $\alpha > 0$ they are unstable.

At $\alpha \neq 0$ and $\omega \neq 0$ the phase curves are circles (Fig. 2) and the singular point is their center.

The use of complex variables provides a number of advantages for the mathematical analysis of the process stability. However, with real processes the real

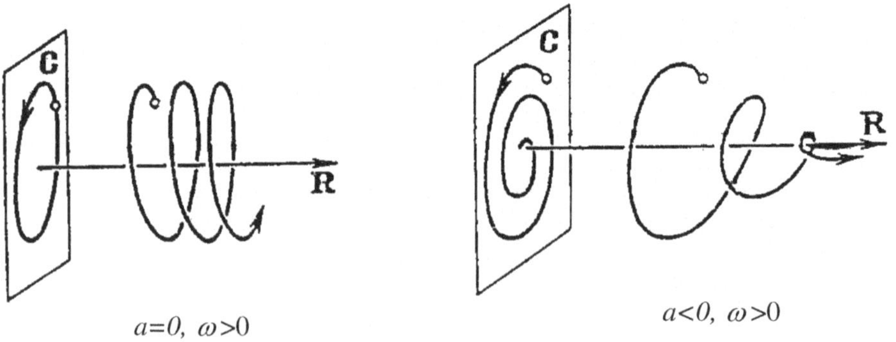

Fig. 3 Complex solution (1.14). *C* denotes the complex plane (*x*,*y*) and *R* corresponds to the real axis *t*

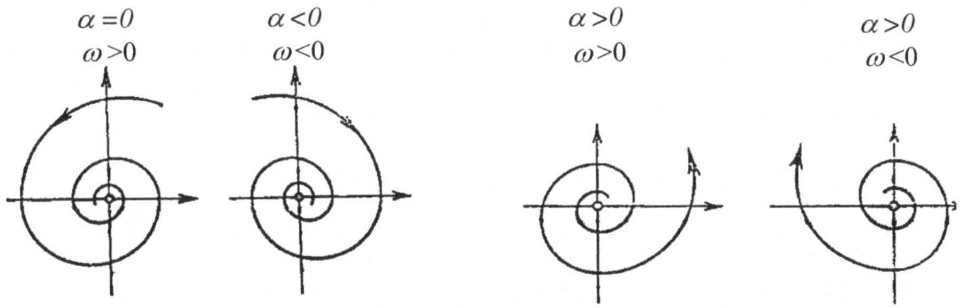

Fig. 4 Stable ($\alpha < 0$) and unstable ($\alpha > 0$) focuses

parts make sense, i.e., the physical solutions are identical to the real parts of the mathematical solutions. Thus, for real processes it follows from (5.14) that

$$z = r\,\exp(\alpha t)\cos\omega t, \quad r = z_0. \tag{1.15}$$

The solution obtained is a periodic solution and it may be stable ($\alpha < 0$) or unstable ($\alpha > 0$), as shown in Fig. 5.

The simulation of stable processes may be carried out with stable models. The unstable models are applicable for investigations on the transitions from one stable state to the next one. They have been applied for the simulation of the transition from stable to unstable states (processes such as explosions) and provide the basis of the theory of catastrophes.

1.2 Bifurcation Theory

The bifurcation theory [4] is wide applied for investigations of jump reactions of systems as responses to smooth changes of the external conditions. For real

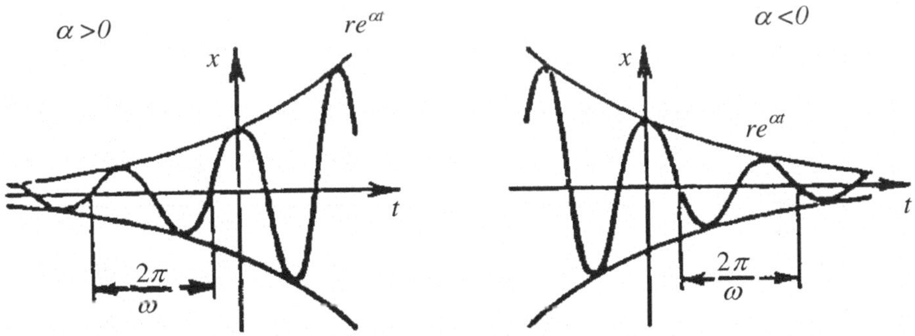

Fig. 5 Stable ($\alpha < 0$) and unstable ($\alpha > 0$) periodic solutions

systems it has been developed recently as a theory of catastrophes. Here, the bifurcation theory will only be considered in two-dimensional phase space.

For clarity of explanation, let us consider a real evolutionary process occurring in the phase plane (x,y). Its corresponding model is

$$\frac{dx}{dt} = X(x, y, \mu), \quad \frac{dy}{dt} = Y(x, y, \mu), \quad x(0) = x_0, \quad y(0) = y_0. \tag{1.16}$$

The system evolution in time is represented by the phase trajectory (the trajectory of the phase point) of the process:

$$F(x, y, \mu) = 0, \tag{1.17}$$

where $x(t)$ and $y(t)$ in (1.16) are determined from the solution of (1.16). Depending on the form of the relationships for X and Y in (1.16), the parameter μ, and the initial conditions x_0 and y_0, various phase trajectories are possible.

The variations of the parameter μ lead to several interesting cases of the solution of (1.16), as shown in Fig. 6. The case shown in Fig. 6a corresponds to a periodic process which attenuates with time and approaches a focus (a stationary state point). If another value of μ is chosen, the process might be unstable and periodic (Fig. 6b). The stable periodic processes (*limit cycles*) have closed trajectories in the phase space (Fig. 6c). The change of the initial state (y_0) of the stable processes leads to attenuating processes approaching a stable periodic state (Fig. 6d).

Figure 6 may be developed further for more complicated cases (see Fig. 7). It is possible for two limit cycles (periodic processes and solutions) to exist, where one of them (the internal one) is stable if the initial conditions are in the entire internal area of the large cycle. The internal cycle attracts all the solutions, whereas the external cycle is the unstable one (Fig. 7a).

The variations of the parameter μ may lead to a junction of both cycles (Fig. 7b). The junction of both an unstable cycle and a stable cycle (as those in Fig. 7a) may lead to an abnormal limit cycle (Fig. 7b). In this case the solutions go from the initial conditions in the internal area, approach the cycle, and then owing

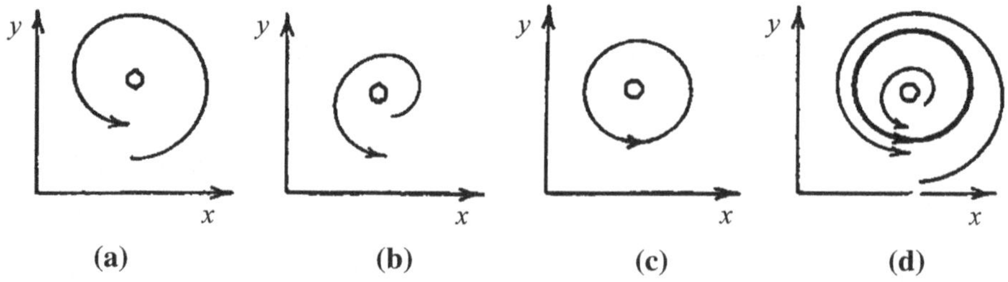

Fig. 6 Phase trajectories: **a** periodic process which attenuates with time and approaches a focus (a stationary state point); **b** unstable and periodic process; **c** stable periodic processes (*limit cycles*); **d** stable periodic state

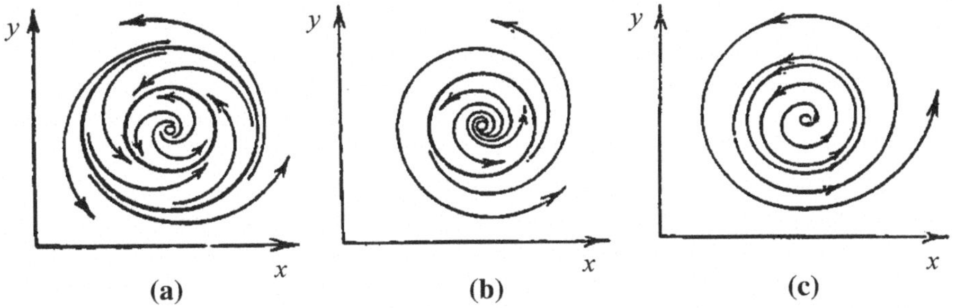

Fig. 7 Limit cycles: **a** two limit cycles (periodic processes and solutions); **b** junction of both an unstable cycle and a stable cycle; **c** limit cycle disappears and the process becomes unstable

to small perturbations may go out of the cycle, so the process becomes an unstable one.

The further changes of μ may lead to a situation where the limit cycle disappears and the process becomes unstable (Fig. 7c).

The results obtained here show that the bifurcation theory considers qualitatively the changes of the movement of a phase point as a result of a continuous variation of the model parameters. Parallel to the existence of stable points (focuses), stable cycles exist. They describe stationary periodic oscillations of the systems (*self-oscillations*). They differ from the free oscillations (e.g., of a pendulum), where the system does not interact with the environment as well as from the forced oscillations provoked by external periodic impacts.

The focuses and the limit cycles attracting the solution (the phase point) are termed *attractors*.

The phase trajectory (1.17) depends on μ because from (1.16) it follows that

$$x = x(t, \mu), \quad y = y(t, \mu). \tag{1.18}$$

Let us assume that $\bar{x}$ and $\bar{y}$ are the coordinates of a singular point moving with a liquid having a dynamic viscosity μ:

$$\bar{x} = x(\infty, \mu) < \infty, \quad \bar{y} = y(\infty, \mu) < \infty, \tag{1.19}$$

The different values of μ determine different singular points (1.19) forming a continuous curve in the phase space:

$$F(\bar{x}, \bar{y}, \mu) = F[p(\mu)] = 0. \tag{1.20}$$

Here $p(\mu)$ is a continuous function and is the geometric locus of the singular points for various values of the parameter μ.

Let us assume that the point $[\mu, p(\mu)]$ attracts the solution for $\mu > \mu_0$. In this case the point $[\mu_0, p(\mu_0)]$ is pitchfork point (a *bifurcation point*) of the flux $F(x,y,\mu)$ in the vector field determined by (1.16). This means that at $t \to \infty$ the trajectory of the flux approaches $p(\mu)$ for $\mu < \mu_0$. At $\mu > \mu_0$ the singular point $p(\mu)$ is unstable. Further, such bifurcations leading to stable regimes for $\mu > \mu_0$ will be considered.

Let us suppose that there are several curves $p_i(\mu)$, $i = 1, 2, \ldots$, where $F(p_i(\mu)) = 0$, $i = 1, 2, \ldots$. At $\mu = \mu_0$ it is possible to find a common point of the curves $p_1(\mu_0) = p_2(\mu_0) = \ldots$. Moreover, it is possible for some of these curves to be stable ones at $\mu > \mu_0$, so they are a locus of singular points. Thus, different types of bifurcations are possible. The further discussion will consider a bifurcation leading to a developed cycle from a focus that is important for hydrodynamic stability.

Figure 8 shows bifurcations of cycle transitions from focuses; the space map of $F(x,y,\mu)$ is used. The case in Fig. 8a corresponds to a supercritical bifurcation (stable closed trajectories), whereas the case in Fig. 8b corresponds to a *subcritical bifurcation* (unstable and closed trajectories).

In Fig. 8a the points (x,y,μ) are singular at $x = 0$, $y = 0$, $\mu \leq 0$ [i.e., $F(0,0,\mu) = 0$] and become stable at $\mu \leq 0$. The points $(0,0,\mu)$ at $(\mu > 0)$ are unstable singular points. The trajectories $F(x,y,\mu_0) = 0$ at $\mu_0 > 0$ are closed and stable. Moreover, it is clear from Fig. 8a that owing to the shape of the surface $F(x,y,\mu) = 0$ there are closed unstable trajectories $F(x,y,\mu_0) = 0$ at $\mu_0 < 0$.

Further, Fig. 8a shows the mechanism of a transition from a stable point (focus) towards a stable orbit (cycle). This type of bifurcation is shown in Fig. 9. The stages of the transition are (1) a stable point, (2) the occurrence of a closed trajectory, and (3) an increase of the amplitude of the closed trajectory. This order leads to the existence of a stable three-dimensional torus.

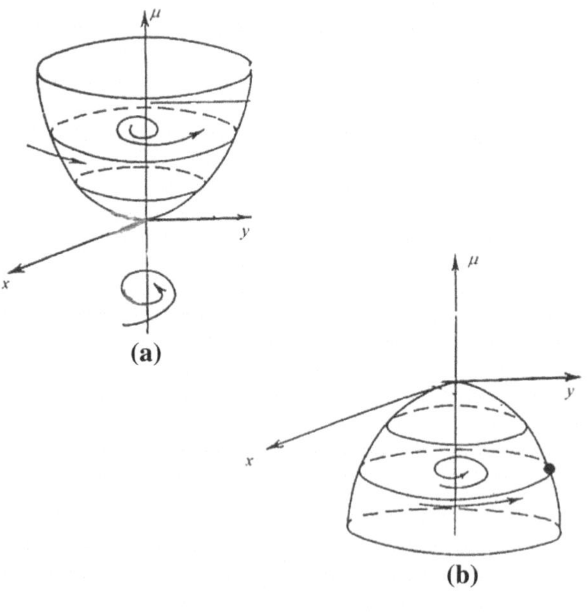

Fig. 8 Bifurcation of cycle transitions from focuses: **a** supercritical bifurcation (stable closed trajectories); **b** *subcritical bifurcation* (unstable and closed trajectories)

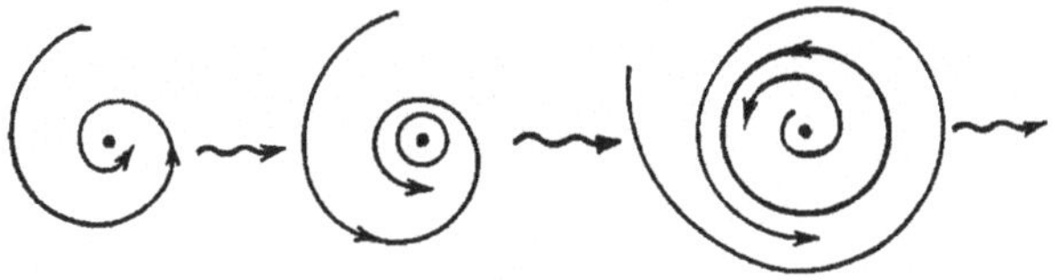

Fig. 9 Mechanism of a transition from a focus towards a cycle

1.3 Eigenvalue Problems

The analysis of the processes and the systems made in the previous section concerns the stability of the solutions as functions of the model parameters. This requires a solution of differential equations with parameters. When the boundary conditions contain function values at two different points, this leads to eigenvalue problems. The solution of such a problem will be demonstrated by an example of a homogeneous equation in the real numbers area:

$$y' + [f(x) - \lambda g(x)]\, y = 0, \tag{1.21}$$

with boundary condition

$$y(b) = \alpha y(a), \quad \alpha \neq 0, \tag{1.22}$$

where λ is a parameter $y(a) = C$ is an arbitrary number.

The solution of (1.21) is well known:

$$y = C \exp\left[- \int_{a}^{x} (f - \lambda\, g)\, dx \right]. \tag{1.23}$$

The substitution of (1.23) into (1.22) (made to estimate the constant C) shows that condition (1.22) is satisfied only in the case of $\lambda = \lambda_0$:

$$\lambda_0 = \frac{\ln \lambda + \int_a^b f dx}{\int_a^b g dx}, \tag{1.24}$$

which is well known as an eigenvalue. The substitution of (1.24) into (1.23) leads to an eigenfunction. Thus, for example, at $f \equiv 0$ and $g \equiv 1$ it follows directly that

$$\lambda_0 = \frac{\ln \alpha}{b - a}, \quad y = C \exp\left[- \frac{\ln \alpha (x - a)}{b - a} \right]. \tag{1.25}$$

It is well demonstrated in the differential equation theory [5] that if $\int_a^b g\, dx \neq 0$, there is an infinite set of eigenvalues:

$$\lambda_k = \lambda_0 + \frac{2k\pi i}{\int\limits_a^b g dx}, \quad k = 0, \pm 1, \pm 2, \cdots \tag{1.26}$$

2 Hydrodynamic Stability

Most industrial-scale processes depend on the stability of the fluid flow. The equations describing such fluid (gas or liquid) flows are typical evolutionary equations and relate to the change of both the velocity and the pressure with time. This permits us to use the approaches developed already for stability analysis of evolutionary equations for these hydrodynamic equations [6].

2.1 Fundamental Equations

Let us consider a fluid (liquid or gas) flow with kinematic viscosity ν under the action of external forces $\boldsymbol{F}(\boldsymbol{x},t)$ or due to the movement of the boundary $S(t)$ of a closed volume $V(t)$. The velocity field $\boldsymbol{U}(\boldsymbol{x},t)$ is determined by the Navier–Stokes equations expressed for the velocity $\boldsymbol{U}$ and the pressure π in a volume V:

$$\frac{\partial \boldsymbol{U}}{\partial t} + (\boldsymbol{grad} \cdot \boldsymbol{U})\boldsymbol{U} - \nu \nabla^2 \boldsymbol{U} + \boldsymbol{grad}\, \pi - \boldsymbol{F}(\boldsymbol{x}, t) = 0, \quad div\, \boldsymbol{U} = 0, \tag{2.1}$$

where $\boldsymbol{x}$ is the coordinate vector. The boundary conditions for (2.1) are

$$\boldsymbol{U}(\boldsymbol{x}, t) = \boldsymbol{U}_s(\boldsymbol{x}, t), \quad \boldsymbol{x} \in S(t), \quad t \geq 0, \tag{2.2}$$

The corresponding initial conditions are:

$$\boldsymbol{U}(\boldsymbol{x}, 0) = \boldsymbol{U}_0(\boldsymbol{x}), \quad \boldsymbol{x} \in V(0). \tag{2.3}$$

The solution of (2.1)–(2.3) is the function

$$\boldsymbol{U} = \boldsymbol{U}(\boldsymbol{x}, t; \nu, \boldsymbol{U}_0). \tag{2.4}$$

In (2.4) there is no term for π because the pressure is determined directly by the velocity function solution $\boldsymbol{U}$ and it may be omitted [7] in (2.1).

The hydrodynamic stability will be considered as the flow stability (solution stability) under the variation of the parameters $\boldsymbol{U}_0$ and ν. Two solutions at a given value of ν and different initial conditions will be considered:

$$\boldsymbol{U} = \boldsymbol{U}(\boldsymbol{x}, t; \boldsymbol{U}_0), \quad \boldsymbol{U}^a = U(\boldsymbol{x}, t; \boldsymbol{U}_0 + \boldsymbol{u}_0), \tag{2.5}$$

where $\boldsymbol{U}$ is the velocity of the main flow, $\boldsymbol{U}^a$ is the velocity of the disturbed flow, and $\boldsymbol{u}_0 = \boldsymbol{u}_0(\boldsymbol{x})$ is a perturbation. The difference between the velocity of the main flow and that of the disturbed flow is a function representing the perturbation evolution:

$$\boldsymbol{u}(x,t) = \boldsymbol{U}^a - \boldsymbol{U}. \tag{2.6}$$

It may be determined directly from (2.1) to (2.6) that

$$\frac{\partial \boldsymbol{u}}{\partial t} + (\boldsymbol{grad} \cdot \boldsymbol{U})\boldsymbol{u} + (\boldsymbol{grad} \cdot \boldsymbol{u})\boldsymbol{U} + (\boldsymbol{grad} \cdot \boldsymbol{u})\boldsymbol{u} - \nu\nabla^2\boldsymbol{u} + \boldsymbol{grad}\, P = 0,$$

$$div\, \boldsymbol{u} = 0, \quad \boldsymbol{u}|_s = 0, \quad \boldsymbol{u}|_{t=0} = \boldsymbol{u}_0(x), \quad P = \pi^{\mathrm{a}} - \pi. \tag{2.7}$$

Equations (2.7) are the vector forms of the evolutionary hydrodynamics equations. If the flow is stable, one of the solutions $\boldsymbol{u}^{\mathrm{a}}$ confluences with $\boldsymbol{U}$ at $t \to \infty$, i.e.,

$$\boldsymbol{u}(x,t) \equiv 0 \quad \text{at} \quad t \to \infty. \tag{2.8}$$

The problem concerning the stability of $\boldsymbol{U}$ at $t \to \infty$ with respect to the perturbation of the initial conditions $\boldsymbol{u}_0 \neq 0$ leads to the problem of the stability of the zeroth-order solution of the system (2.7). This problem may be solved in the case of arbitrary perturbations (nonlinear stability) or small perturbations (linear stability). The linear theory does not predict the perturbation amplitudes. The problem may be solved by the nonlinear theory, which usually employs power conditions.

2.2 Power Theory

The mean kinetic energy of the perturbation is

$$E(t) = \frac{1}{2}\left\langle |\boldsymbol{u}|^2 \right\rangle, \tag{2.9}$$

where the symbol $\langle\rangle$ means a suitable (usually integrating) averaging procedure. In accordance with the power theory [6], the condition for the stability of the zeroth-order solution $\boldsymbol{u}(\boldsymbol{x},t)$ with respect to the perturbations of the initial conditions $\boldsymbol{u}_0(\boldsymbol{x})$ is

$$\lim_{t\to\infty} \frac{E(t)}{E(0)} = 0. \tag{2.10}$$

Here $E(0)$ is the initial energy of the perturbation, i.e., $\boldsymbol{u} \to 0$ at $t \to 0$. Condition (2.10) is a criterion for an asymptotic stability if

$$E(0) < \delta, \quad \delta > 0. \tag{2.11}$$

For every $\delta > 0$ there is a set of initial perturbations $\boldsymbol{u}_0$ that attract the solution $\boldsymbol{u}_0 \equiv 0$, i.e., δ is the radius of attraction of the conditionally stable solution $\boldsymbol{u}_0 \equiv 0$. If $\delta \to \infty$, the zeroth-order solution is absolutely (globally) stable. If the solution is asymptotically stable and $\frac{dE}{dt} \leq 0$ (at $t > 0$), the zeroth-order solution is a monotonous stable solution.

There is a second formulation of the stability conditions in accordance with the power theory, i.e.,

$$\langle |\boldsymbol{U}(\boldsymbol{x},t;\boldsymbol{U}_0) - \boldsymbol{U}(\boldsymbol{x},t;\boldsymbol{U}_0 + \boldsymbol{u}_0)|\rangle \to 0 \, \text{at}\, t \to \infty, \quad \text{when} \left\langle |\boldsymbol{u}|^2 \right\rangle < 2\delta. \tag{2.12}$$

The solution $\boldsymbol{U} = \boldsymbol{U}(\boldsymbol{x},t;\nu,\boldsymbol{U}_0)$ of the system (2.1)–(2.3) at fixed external conditions ($\boldsymbol{u}_0$) gives a one-parameter family of solutions with a variable parameter ν (the Reynolds number in the dimensionless form of the equations). The solution of the problem for the absolute stability requires values of ν and δ which allow $\boldsymbol{u}(\boldsymbol{x},t;\nu,\boldsymbol{u}_0)$ to be a stable solution of (1.33), i.e., $\boldsymbol{U} = \boldsymbol{U}(\boldsymbol{x},t;\nu,\boldsymbol{U}_0)$ is the stable solution of (2.1)–(2.3).

For every value of δ, condition (2.12) may be disturbed for various values of $\nu = \nu_c$, which depend on δ and will be termed critical conditions. The stability limit may be obtained similarly to the relationship $F(\nu_c,\delta) = 0$, i.e., $\delta(\nu_c)$ and $\nu_c(\delta)$. This allows us to express (2.11) in the form

$$E(0) < \delta(\nu_c). \tag{2.13}$$

In this way, the power theory formulates various critical values of the viscosity ν_c. They are shown in Fig. 10. For clarity of presentation, a coordinate axis ν^{-1} which is proportional to the Reynolds number at fixed external flow conditions is introduced. Thus, the following four zones are defined in the figure:

1. $\nu > \nu_E$ is the area of the monotonic and global stability.
2. $\nu_G < \nu < \nu_E$ is the area of the global stability because it is possible to achieve $\frac{dE}{dt} > 0$ at $t > 0$ for given perturbations.
3. $\nu_L < \nu < \nu_G$ is the area of the conditional stability with a radius of attraction $\delta(\nu)$.
4. $\nu < \nu_L$ and $\nu < \nu_c(\delta)$ determine the area of the instability.

Fig. 10 Zones of stability

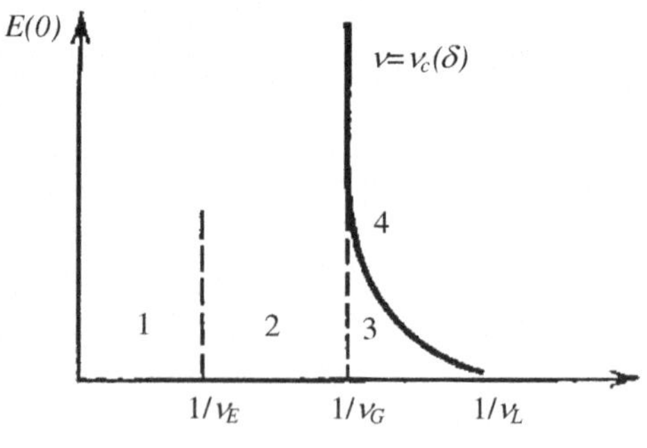

There are also absolutely stable flows, i.e., $\nu_c = 0$. This is possible under special conditions such as:

- A flow in a tube when the perturbations do not vary along its longitudinal axis.
- Some flows with infinitely small perturbations.
- Flows without velocity gradients under arbitrary perturbations.

In many cases the stable solutions of (2.1)–(2.3) are independent of the initial conditions, i.e.,

$$\boldsymbol{U}(\boldsymbol{x}, t; \nu, \boldsymbol{U}_0 + \boldsymbol{u}_0) \to \boldsymbol{U}(\boldsymbol{x}, t; \nu) \quad \text{for} \quad \left\langle |\boldsymbol{u}|^2 \right\rangle < 2\delta. \tag{2.14}$$

The boundary flow $\boldsymbol{U}(\boldsymbol{x},t;\nu)$ depends mainly on the boundary conditions (2.12). When $\nu > \nu_c$, the flow is called a basic flow.

The analysis of the flow stability by means of the power theory shows that the flow is stable if the perturbation energy is transmitted to the main flow. On the other hand, if the energy of the main flow is transmitted to the perturbation, the flow becomes unstable. Thus, the first critical viscosity ν_E shows the existence (or not) of perturbations whose energy at the initial stage grows with time. At $\nu > \nu_E$ there is unique stable flow and all other flows are approaching it.

In the situations where $\nu < \nu_E$ there are perturbations with energy growing with time, but at $\nu > \nu_G$ that energy is approaching zero. Thus, ν_G is the limit of the global stability.

If $\nu \leq \nu_G$, there is more than one stable solution. However, questions about the number of these solutions, their dependence on the parameters, and the stability conditions could be set. These problems are solved particularly by the bifurcation theory. The latter is a nonlinear theory of periodic motions with small but finite amplitudes. By its application one can follow the behavior of the stable solutions, which are branched out from the main flow in the case of loss of stability owing to the perturbations with infinitely small amplitudes. Under these conditions the perturbations grow, i.e., their amplitudes increase with time, approaching a finite value. The decrease of ν leads to a decrease of the stability of these bifurcations and causes the onset of new bifurcations. This demonstrates that the bifurcation theory starts with the linear stability analysis with respect to infinitely small perturbations.

2.3 Linear Theory

The linear theory [6] of hydrodynamic stability considers the main flow $\boldsymbol{U}(\boldsymbol{x},\nu)$ and its nonstationary perturbation $\boldsymbol{u}(\boldsymbol{x},t;\nu)$ satisfying the set of equations (1.33). If one assumes that the perturbations are small, it is possible to write

$$\boldsymbol{u} = \varepsilon \boldsymbol{v}, \quad P = \varepsilon p, \quad \varepsilon \ll 1. \tag{2.15}$$

The substitution of (2.15) into (2.7) leads to

$$\frac{\partial \boldsymbol{v}}{\partial t} + L[\boldsymbol{U}, v] + \varepsilon \cdot (\boldsymbol{grad} \cdot \boldsymbol{v})\boldsymbol{v} + \boldsymbol{grad}\, p = 0, \quad div\, \boldsymbol{v} = 0, \quad \boldsymbol{v}|_S = 0, \quad \boldsymbol{v}|_{t=0} = \boldsymbol{v}_0, \tag{2.16}$$

where L is a linear differential operator that represents all the differential operators (gradients, Laplacians, etc.).

At $\varepsilon = 0$ Eq. 2.16 become an autonomous linear set of equations with solutions in an exponential form,

$$\boldsymbol{v}(\boldsymbol{x}, t) = \exp(-\sigma t)\xi(\boldsymbol{x}), \tag{2.17}$$

upon setting the condition that there are numbers σ for which the spectral problem (an eigenvalue problem) with respect to ξ is

$$\sigma\, \xi + \mathrm{L}[\boldsymbol{U}, v]\xi + \boldsymbol{grad}\, p = 0, \quad div\, \xi = 0, \quad \xi|_{\mathrm{S}} = 0. \tag{2.18}$$

The problem has a nontrivial solution $\xi \neq 0$. The values of σ are the eigenvalues of Eq. 2.18 and ξ are the corresponding eigenfunctions (for every σ).

In the general situation σ may be a complex number and the eigenvalues may be expressed as

$$\sigma = \sigma_n = \sigma_{nr} + i\sigma_{ni}, \quad n = 1, 2, \cdots, \tag{2.19}$$

i.e.,

$$\boldsymbol{v}(\boldsymbol{x}, t) = \prod_{n=1}^{\infty} \exp(-\sigma_{nr} t) \exp(-i\sigma_{ni} y)\xi(\boldsymbol{x}). \tag{2.20}$$

The solution of (2.18) leads to the determination of σ_{nr}, σ_{ni}, and $\xi(\boldsymbol{x})$, $n = 1, 2 \cdots$ It follows from (2.20) that

1. The flow is stable if $\sigma_{nr} > 0, \quad n = 1, 2, \cdots$
2. The flow is unstable if $\sigma_{nr} < 0, \quad n = n_0$.
3. The flow is neutrally stable when $\sigma_{n_0 r} = 0$, $\sigma_{nr} > 0$, $n = 1, 2, \cdots, n_0 - 1, n_0 + 1, \cdots$
4. The flow is stationary stable when $\sigma_{nr} = 0, \quad \sigma_{ni} \neq 0, \quad \sigma_{nr} > 0, n = 1, 2, \cdots$
5. When $\sigma_{n_0 r} = 0, \quad \sigma_{n_0 i} \neq 0, \quad \sigma_{nr} > 0, \; n = 1, 2, \cdots n_0 - 1, n_0 + 1$, the flow is periodically stable.

The eigenvalues may be ordered as follows:

$$\sigma_{1r} < \sigma_{2r} < \cdots \tag{2.21}$$

In this case σ_{1r} is the principal eigenvalue (the perturbation with greatest amplitude), σ_{2r} is the second eigenvalue, etc.

The eigenvalues depend on the viscosity:

$$\sigma_{nr} = \sigma_{nr}(\nu), \quad \sigma_{ni} = \sigma_{ni}(\nu), \quad n = 1, 2, \cdots \tag{2.22}$$

This allows us to define the first critical viscosity ν_{L}. The value of ν_{L} is the greatest value of ν (the minimum critical Reynolds number) that allows σ_{1r} $(\nu_{\mathrm{L}}) = 0$ to be satisfied. Thus, ν_{L} coincides with the critical viscosity of the nonlinear (power) theory for a conditional stability at $\delta \ll 1$.

The hydrodynamic stability of a periodic main flow may be investigated in a similar way:

$$\boldsymbol{U}(\boldsymbol{x}, t, \nu) = \boldsymbol{U}(\boldsymbol{x}, t + T, \nu). \tag{2.23}$$

Then

$$\boldsymbol{v}(\boldsymbol{x}, t) = \exp(-\gamma t)\xi(\boldsymbol{x}, t), \tag{2.24}$$

where

$$\gamma = \gamma_r + i\gamma_i \tag{2.25}$$

are Floke's powers. When $\gamma_r > 0$, the periodic flow is stable, i.e., the stable periodic flow is superposed by a secondary stable periodic perturbation.

Comparison of the linear and the nonlinear (power) theories shows that

$$\nu_{\mathrm{L}} \le \nu_{\mathrm{E}} \tag{2.26}$$

and several main conclusions follow:

1. If the flow is unstable in accordance with the linear theory ($\nu < \nu_{\mathrm{L}}$), it is unstable according to the nonlinear theory too.
2. If the flow is stable in accordance with the linear theory ($\nu > \nu_{\mathrm{L}}$), it may be unstable according to the nonlinear theory [E(0) $> \delta$] or may conditionally stable [E(0) $< \delta$], where δ is the radius of attraction.
3. The linear theory does not predict the value of δ and the condition ($\nu > \nu_{\mathrm{L}}$) does not guarantee global stability.
4. The linear theory guarantees only the instability at $\sigma_{1r} < 0$ ($\nu < \nu_{\mathrm{L}}$), which explains the cases of disagreement between the prediction of the linear theory and the experimental results (where the perturbation amplitudes are not infinitely small).

2.4 Stability, Bifurcations, and Turbulence

The invariant form of the flow corresponds to its stable periodic solution (neutral stability) that occurs as a solution of (2.18) at $\sigma_{1r} = 0$. Then

$$\sigma_1 = \sigma_1(\nu_L) = \pm i\,\omega_0. \tag{2.27}$$

If $\omega_0 = 0$ at $\varepsilon_0 = 0$, problem (2.16) becomes a linearized problem with a unique stationary solution with amplitude

$$\varepsilon^2 = E(t) = E(0). \tag{2.28}$$

This is a boundary solution (at $\varepsilon = 0$) for the one-parameter family of stationary branched-out solutions of the nonlinear problem (2.20).

If $\omega_0 \neq 0$, the linearized (at $\varepsilon = 0$) problem (2.16) has periodic (with time t) complex conjugate solutions:

$$\boldsymbol{v}(\boldsymbol{x}, t, v) = \exp(\pm i\omega_0 t)\, \boldsymbol{\xi}(\boldsymbol{x}, t). \tag{2.29}$$

In this case there is a unique family of one-parameter periodic solutions of (2.16) with a parameter ε that branches out from the solution for $\boldsymbol{U}$. Different methods are available for the determination of ε, but the most convenient approach is to express it as the energy of the stationary branched out solution (the average energy of the cycle for one period):

$$\varepsilon^2 = \frac{1}{T}\int_0^T E(t)dt. \tag{2.30}$$

The set of Eq. 2.16 does not have solutions for every ε. Because of that, the problem is focused on a family of solutions with a parameter ε, ensuring that for every small value of ε there is a corresponding value of $v(\varepsilon)$. Moreover, the values of $v(\varepsilon)$ allow a periodic solution of (2.16). Thus, $v(\varepsilon)$ is a bifurcation curve.

If $v(\varepsilon) > v_{\mathrm{L}}$, it is possible to obtain a periodic flow with invariant form $\boldsymbol{U}(\boldsymbol{x}, v) + \boldsymbol{u}(\boldsymbol{x}, t, \varepsilon)$ with amplitude ε (sufficiently small) which permits the perturbation energy to be constant within the time. The branched-out solutions at $v(\varepsilon) > v_{\mathrm{L}}$ (the linear theory guarantees stability in that range) are subcritical, whereas at $v(\varepsilon) < v_{\mathrm{L}}$ the bifurcations are supercritical.

The stability of the secondary stable periodic flows has been investigated too [5, 6]. Let the branched-out solution be considered:

$$\boldsymbol{U}(\boldsymbol{x}; v(\varepsilon)) + \boldsymbol{u}(\boldsymbol{x}, s; \varepsilon), \tag{2.31}$$

where $\boldsymbol{u}$ is periodic with respect to s with a period of 2π. A condition for the stability of $\boldsymbol{U} + \boldsymbol{u}$ with respect to the small perturbations $\mathbf{q} = \mathbf{q}(\mathbf{x}, t)$ is required. The problem may be solved by an equation similar to (2.16). After linearization, (2.16) takes the form:

$$\frac{\partial \boldsymbol{q}}{\partial t} + L\boldsymbol{q} + \boldsymbol{grad}\, p = 0, \quad div\, \boldsymbol{q} = 0, \quad \boldsymbol{q}|_s = 0, \tag{2.32}$$

where L is a linear differential operator representing all linear operators,

$$\mathrm{L} = \mathrm{L}[\boldsymbol{U} + \boldsymbol{u}; \nu]. \tag{2.33}$$

Here $\boldsymbol{U}(\boldsymbol{x}; \nu(\varepsilon))$, $\boldsymbol{u}(\boldsymbol{x}, s; \varepsilon)$, and $\nu(\varepsilon)$ are calculated at a fixed value of ε. It was proved in [5, 6] that for small ε the subcritical solutions (bifurcations) are unstable, whereas the supercritical solutions are stable. In these cases $\nu = \nu(\varepsilon^2)$.

The laminar flow of a fluid may turn into a turbulent flow if the flow parameter changes (e.g., the viscosity). However, in real situations there is a continuous transition with the reduction of the viscosity. The continuous decrease of the fluid viscosity leads to a continuous transition from an organized (laminar regime) flow through a self-organized flow (dissipative structures) towards a complex nonorganized flow (turbulent regime). This continuous transition is a series of supercritical bifurcations. The following sequence of physical phenomena takes place:

1. At $\nu > \nu_\mathrm{L}$ the basic stationary flow exists.
2. At $\nu = \nu_\mathrm{L}$ the flow loses its stability and a secondary (more complex) periodic flow occurs. This secondary flow is stable at $\nu < \nu_\mathrm{L}$ and its amplitude approaches zero when $(\nu_\mathrm{L} - \nu) \to 0$.
3. At $\nu = \nu_2 < \nu_\mathrm{L}$ (the second critical point) the secondary flow becomes unstable and a next solution pitchfork occurs. The next tertiary flow is more complex and is stable until $\nu < \nu_2$ (when the next bifurcation will start).

The transition from one stable flow regime to the next one through a series of supercritical bifurcations is a continuous process, because the amplitudes of the subsequent flows approach zero when $(\nu - \nu_\mathrm{L}) \to 0$ and $(\nu - \nu_2) \to 0$. Thus, there are "sharp" changes of the flow regime.

There is the possibility of a transition to a turbulent flow through a series of subcritical bifurcations. In this case $\nu < \nu_\mathrm{L}$ and the branched-out solution is unstable and there is no area of attraction. An arbitrary initial perturbation "goes away" from the basic flow, branches out into a secondary unstable flow, and approaches a flow (or a family of flows) with large amplitude.

The transition from a laminar flow to a turbulent flow regime as a continuous process of supercritical bifurcations is a basic idea of Landau and Hopf [4, 6, 11, 12] for the onset of the turbulence.

2.5 *Stability of Parallel Flows*

The theory of hydrodynamic stability uses the stability analysis of various flow types. Here, several results of the nonlinear theory will be discussed to investigate parallel fluid flows (Poiseuille and Couette flows).

The stability of flows in pipes far away from the inlet region depends on the Reynolds number:

$$Re = \frac{U_m(b - a)}{\nu}, \tag{2.34}$$

where U_m is the maximum velocity, and a and b are the radii of the coaxial tubes forming the annulus when the fluid flows. At $a = 0$ there is Hagen–Poiseuille flow in the pipe. In a parallel-plate channel, $a = 0$ and b is the channel width (parallel Poiseuille flow). In these cases there are several critical values of the Reynolds number:

1. The flow is globally and monotonously stable at $Re < Re_{\mathrm{E}} \approx 85 \div 100$.
2. The flow is globally stable at $Re_{\mathrm{E}} < Re < Re_{\mathrm{G}}$, where $Re_{\mathrm{G}} \approx 2,000 \div 2,300$.
3. The flow is conditionally stable at $Re_{\mathrm{G}} < Re < Re_{\mathrm{L}} > 11,000$ (40, 000, 10^6, etc.).
4. The flow is unstable at $Re > Re_{\mathrm{L}}$.

When $Re_{\mathrm{E}} > Re_{\mathrm{G}}$, there are various stable and periodic flows having equal coefficients of wall drag resistance. The state is termed *a stable turbulence*.

The next example of a parallel flow is the Couette flow between two rotating cylinders. At given Reynolds numbers a laminar flow exists. The increase of the Reynolds number is followed by the occurrence of Taylor vortexes (a stationary bifurcation), which is in fact a stable periodic flow. The further increase of Re leads to the next stable periodic flow—wavy Taylor vortexes (nonstationary bifurcations).

A large class of problems analyzed by the hydrodynamic stability theory are related to the Oberbeck–Boussinesq equations. They followed from the Navier–Stokes equations by means of the introduction of additional terms taking into consideration the natural convection as a result of the density differences. These differences may occur owing to concentration or temperature gradients and require additional equations considering the diffusion and the heat transfer [6, 8]. Among these problems, Benard cell convection has been the subject of extensive investigation [6, 8]. In fact the phenomenon is a series of supercritical bifurcations (a consequence of transitions into secondary, tertiary flows, etc.).

Under controllable conditions the thermoconvective instability may pass into a thermocapillary instability [9] and approach the Marangoni effect.

3 Orr–Sommerfeld Equation

The linear analysis of hydrodynamic stability may be applied in the cases of parallel fluid flows (Poiseuille or Couette flows) [6, 7, 10] as well as to almost parallel flows such as flows in jets or in laminar boundary layers [6, 10, 11]. All these situations lead to the solution of the Orr–Sommerfeld equation. It may be derived from (2.18) with the substitutions

$$\boldsymbol{v} = (u, v), \quad \boldsymbol{x} = (x, y), \quad \varepsilon = 0, \quad \boldsymbol{U} = U(y), \quad u = \frac{\partial \Psi}{\partial y}, \quad v = -\frac{\partial \Psi}{\partial x},$$

$$\Psi(x, y, t) = \varphi(y) \exp[i(\alpha x - \beta t)], \tag{3.1}$$

where φ, α, and β/α are the perturbation amplitude, the wave number, and the phase velocity respectively. It is assumed in (3.2) that the amplitude and the frequency are complex, whereas λ is the wavelength of the perturbation:

$$\varphi = \varphi_r + i\varphi_i, \quad \alpha = \frac{2\pi}{\lambda}, \quad \beta = \beta_r + i\beta_i, \tag{3.2}$$

where β_r is the critical frequency and β_i is the increment factor. Thus, the current function of the perturbation may be expressed as

$$\Psi(x, y, t) = \varphi(y)\exp(\beta_i t)\exp[i(\alpha x - \beta_r t)], \tag{3.3}$$

where φ may be determined from the Orr–Sommerfeld equation:

$$(u - c)(\varphi'' - \alpha^2\phi) - U''\varphi = \frac{i}{\alpha Re}\varphi^{IV} - 2\alpha^2\varphi'' + \alpha^4\varphi. \tag{3.4}$$

The boundary conditions of (3.4) have various forms [14] for parallel flows and almost parallel flows. In the former cases $\varphi(\pm 1) = \varphi'(\pm 1) = 0$, whereas in the latter ones $\varphi(0) = \varphi'(0) = \varphi(\infty) = \varphi'(\infty) = 0$.

3.1 Parallel Flows

In the case of parallel flow, the parameters of the Orr–Sommerfeld equations are

$$c = \frac{\beta}{\alpha} = c_r + ic_i, \quad Re = \frac{U_m\delta}{\nu}. \tag{3.5}$$

Equation (3.4) is a typical eigenvalue problem with respect to α, Re, c_r, and c_i. At a given value of Re (or α) there is a need for variations of α (or Re) to find c_r satisfying the condition $c_i = 0$. The $\alpha(Re)$ neutral curve plot obtained (see Fig. 11) allows determination of the critical Reynolds numbers.

The analysis of the hydrodynamic stability of parallel flow shows that the stability is a feature of the velocity profile shape $U(y)$ changing by Re. At $Re = Re_{cr}$ this profile becomes unstable.

Fig. 11 The neutral curve

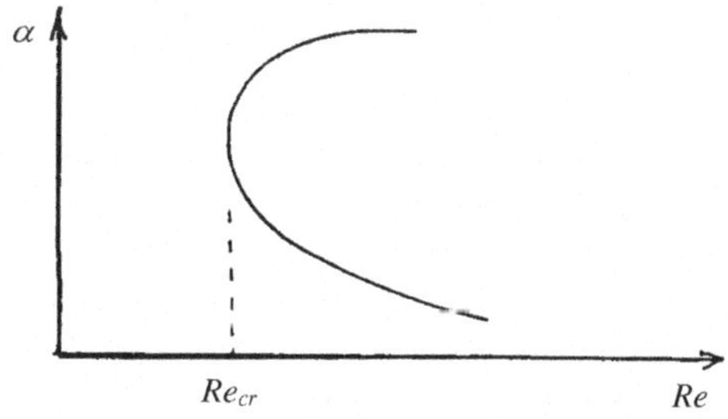

3.2 Almost Parallel Flows

The almost parallel flows in jets and laminar boundary layers are characterized by the fact that the velocity profile $U(y)$ changes along the boundary layer length. At a certain distance it becomes unstable, i.e., at a certain length the perturbation amplitude starts to grow. Finally, this leads to turbulence developing further along the flow direction at lengths greater than its critical value. If the boundary layer is laminar the value of δ in (3.5) is a‘ function of the distance along the flow. Thus, there are coupled critical values of the length and the Reynolds number.

The relationship between the velocity of the main flow and the velocity components $U = [u(x, y), v(x, y)]$ leads to additional terms [16] in the Orr–Sommerfeld equation:

$$\left(u - \frac{\beta}{\alpha}\right)(\varphi'' - \alpha\varphi) - \frac{\partial^2 u}{\partial y^2}\varphi = -\frac{iv}{\alpha}\left(\varphi'' - 2\alpha^2\varphi'' + \alpha^4\varphi\right) + \frac{i}{\alpha}\left[v\varphi''' + \left(\frac{\partial^2 u}{\partial x \partial y} - \alpha^2 v\right)\varphi'\right], \tag{3.6}$$

where u and v satisfy the Prandtl equations

$$u\frac{\partial u}{\partial x} + v\frac{\partial u}{\partial y} = v\frac{\partial^2 u}{\partial y^2}, \quad \frac{\partial u}{\partial x} + \frac{\partial v}{\partial y} = 0;$$

$$x = 0, \quad u = U_\infty; \quad y = 0, \quad u = v = 0; \quad y \to \infty, \quad u = U_\infty. \tag{3.7}$$

The results obtained are the basis of further theoretical development of the stability analysis in the case of systems with intensive mass transfer.

3.3 Linear Stability and Nonlinear Mass Transfer

Several problems has been discussed concerning the influence of large concentration gradients on the velocity fields in the laminar boundary layer (See page 145 and the next). The results obtained for the mass transfer rate coincide qualitatively with the experimental data, which deviate from the linear mass transfer theory predictions. However, sometimes the theory of non-linear mass transfer misses the experimental results quantitatively. This could be explained with the lost flow stability which leads to the significant mass transfer rate arising. The induced secondary flow on the phases interface is the cause of the above phenomenon, which depends on the mass transfer direction, produces suction (injection) from (to) the boundary layer and leads to increase (decreasing) the hydrodynamic stability in the laminar boundary layer [7]. This effect will be discussed for different systems with intensive interphase mass transfer.

Theoretical studies of the influence of the suction (injection) from (into) the boundary layer on the hydrodynamics and hydrodynamic stability have been carried out in cases where the normal component of the velocity on the phase boundary is constant along the boundary layer. In cases of nonlinear mass transfer the rate of the suction (injection) effects on the local diffusive flux (see page 145 and the next) and changes from ∞ to 0 when x changes from 0 to ∞. The letter leads to a significant change in the flow stability.

The hydrodynamic stability in gas (liquid)–solid systems will be demonstrated in the case of nonlinear mass transfer in a stream flow along a semi-infinite plate [13, 14, 18]. In this case the mathematical model takes the following form:

$$u\frac{\partial u}{\partial x}+v\frac{\partial u}{\partial y}=\nu\frac{\partial^2 u}{\partial y^2},\quad \frac{\partial u}{\partial x}+\frac{\partial v}{\partial y}=0,\quad u\frac{\partial c}{\partial x}+v\frac{\partial c}{\partial y}=D\frac{\partial^2 c}{\partial y^2};$$

$$x=0,\quad u=u_0,\quad c=c_0;\quad y=0,\quad u=0,\quad v=-\frac{MD}{\rho_0^*}\frac{\partial c}{\partial y},\quad c=c^*; \tag{3.8}$$

$$y\to\infty,\quad u=u_0,\quad c=c_0.$$

The solution of problem (3.8) can be obtained if the following similarity variables are used:

$$u=0.5u_0\varepsilon\Phi',\quad v=0.5\left(\frac{u_0\nu}{x}\right)^{0.5}(\eta\Phi'-\Phi),\quad c=c_0+(c^*-c_0)\Psi,\quad \Phi=\Phi(\eta),$$
$$\Psi=\Psi(\eta),$$

$$\eta=y\left(\frac{u_0}{4Dx}\right)^{0.5},\quad \varepsilon=Sc^{0.5},\quad Sc=\frac{\nu}{D}, \tag{3.9}$$

where Sc is the Schmidt number.

Substitution of (3.9) into (3.9) leads to a system of ordinary differential equations:

$$\Phi'''+\varepsilon^{-1}\Phi\Phi''=0,\quad \Psi''+\varepsilon\Phi\Psi'=0;\quad \Phi(0)=\theta\Psi'(0),\quad \Phi'(0)=0,$$
$$\Phi'(\infty)=2\varepsilon^{-1},$$

$$\Psi(0)=1,\quad \Psi(\infty)=0, \tag{3.10}$$

where θ is a small parameter which characterizes the nonlinearity of the mass transfer and depends on the intensity of the interphase mass transfer.

Problem (3.10) has been solved [14, 18, 20] numerically and asymptotically as well. The results obtained by asymptotic theory [18] were confirmed through direct numerical experiments [18, 20] and show that the secondary flow with rate $\Phi(0)=\theta\Psi'(0)$ does not change the character of the flow in the boundary layer but only the shape of the velocity profile $\Phi(\eta)$ [14]. This can also be proven by the following theoretical evaluations. The induction of secondary flows on the interface surface has the effect of injection into (suction from) the boundary layer, depending on the direction of the interphase mass transfer. This effect affects the potentiality of the flow at $y \to \infty$ and is not in contradiction with the boundary layer approximations used [7]:

$$v_0 < u_0 Re_L^{-1/2}, \quad Re_L = \frac{u_0 L}{\nu}, \tag{3.11}$$

where v_0 is the mean rate of injection (suction) through a solid surface of a length L,

$$v_0 = \frac{1}{L}\int_0^L v dx, \quad v = \frac{\mathrm{MD}}{\rho_0^*}\left(\frac{\partial c}{\partial y}\right)_{y=0}. \tag{3.12}$$

The introduction (3.9) into (3.12) leads to the following expression:

$$v_0 = -\theta\, u_0 Re_L^{-1/2}\Psi'(0). \tag{3.13}$$

Comparison of (3.12) with Eq. 3.13 shows that (3.13) is valid if

$$|\theta\Psi'(0)| < 1. \tag{3.14}$$

Taking into account that $|\Psi'(0)| < 1$, it is obvious that at $|\theta| < 1$ condition (3.14) is always valid.

Analytical and numerical solutions of problem (3.10) for different values of ε and θ allow the initial values of Φ and its derivatives to be found,

$$\Phi(0) = a, \quad \Phi'(0) = 0, \quad \Phi''(0) = b, \tag{3.15}$$

and these values [18] are shown in Table 1.

The linear analysis was made easier by considering (3.10) as a Cauchy problem.

Table 1 also shows the initial conditions a and b as the effect of the mass transfer on the velocity profiles in the boundary layer. They depend considerably on the magnitude and the direction of the rate of the induced flow, i.e., on the direction and the rate of the intensive interphase mass transfer.

At high values of θ in the case of liquids ($\varepsilon \gg 1$), the numerical solution cannot converge, owing to an increasing singular perturbation (or stiffness) of the solution in the boundary layer.

It can be seen from Table 1 that $\theta > 0(\theta < 0)$ corresponds to injection into (suction from) the boundary layer and according to the theory of hydrodynamic stability [7] a decrease (increase) of the hydrodynamic stability of the flow in the boundary layer should be expected.

The influence of intensive interphase mass transfer on the hydrodynamic stability of the flows in a laminar boundary layer was investigated by applying the linear stability theory [7, 10]. This theory will be applied also for a almost parallel flow in a boundary layer, such as was done in [6, 21] taking into account two linear scales:

Table 1 Initial values of Φ, its derivatives, and parameter k in (3.33)

ε	θ	a	b	k
1	−0.30	0.2546	1.710	1.232
	−0.20	0.1557	1.557	1.414
	−0.10	0.07162	1.432	1.576
	0.0	0.0	1.329	1.718
	0.10	−0.06196	1.239	1.849
	0.20	−0.1162	1.162	1.968
	0.30	−0.1643	1.095	2.076
10	−0.05	0.02295	0.01359	1.673
	0.0	0.0	0.01328	1.718
	0.05	−0.01237	0.01309	1.745
	0.10	−0.02074	0.01298	1.763
	0.20	−0.03196	0.01281	1.786
20	−0.05	0.02395	0.003389	1.668
	−0.03	0.01219	0.003375	1.697
	0.0	0.0	0.003321	1.718
	0.03	−0.00570	0.003321	1.734

$$x \text{ and } \delta = \sqrt{\frac{\nu x}{u_0}}. \tag{3.16}$$

The relation of these two scales for $x = L$ is connected with the Reynolds number:

$$Re_{\mathrm{L}} = \frac{u_0 L}{\nu} = \left(\frac{L}{\delta}\right)^2 \gg 1. \tag{3.17}$$

The approximations of the boundary layer (3.8) are zeroth-order approximations regarding the small parameter $(\delta/L)^2$, i.e., the following relations are applicable

$$\frac{\partial^2 v}{\partial x \partial y} \approx \frac{\partial^2 u}{\partial x^2} \ll \frac{\partial^2 u}{\partial y^2}, \quad \frac{\partial^2 v}{\partial x^2} \ll \frac{\partial^2 v}{\partial y^2}, \tag{3.18}$$

and will be used in the subsequent analysis.

The linear stability analysis considers a nonstationary flow (U, V, P, C), obtained as a combination of a basic stationary flow (u, v, c) and two-dimensional periodic disturbances (u_1, v_1, p_1, c_1) with small amplitudes ($\omega \ll 1$):

$$U(x,y,t) = u(x,y) + \omega u_1(x,y,t), \quad V(x,y,t) = v(x,y) + \omega v_1(x,y,t),$$
$$P(x,y,t) = \omega p_1(x,y,t), \quad C(x,y,t) = c(x,y) + \omega c_1(x,y,t). \tag{3.19}$$

The nonstationary flow thus obtained satisfies the full system of Navier–Stokes equations:

$$\frac{\partial U}{\partial t}+U\frac{\partial U}{\partial x}+V\frac{\partial U}{\partial y}=-\frac{1}{\rho}\frac{\partial P}{\partial x}+\nu\left(\frac{\partial^2 U}{\partial x^2}+\frac{\partial^2 U}{\partial y^2}\right),$$

$$\frac{\partial V}{\partial t}+U\frac{\partial V}{\partial x}+V\frac{\partial V}{\partial y}=-\frac{1}{\rho}\frac{\partial P}{\partial y}+\nu\left(\frac{\partial^2 V}{\partial x^2}+\frac{\partial^2 V}{\partial y^2}\right),\quad \frac{\partial U}{\partial x}+\frac{\partial V}{\partial y}=0,$$

$$\frac{\partial C}{\partial t}+U\frac{\partial C}{\partial x}+V\frac{\partial C}{\partial y}=D\left(\frac{\partial^2 C}{\partial x^2}+\frac{\partial^2 C}{\partial y^2}\right);$$

$$x=0,\quad U=u_0,\quad V=0,\quad P=p_0;\quad y=0,\quad U=0,\quad V=-\theta A_0\frac{\partial C}{\partial y};$$

$$y\to\infty,\quad U=u_0,\quad V=0,\quad P=p_0, \tag{3.20}$$

where

$$A_0=\frac{\varepsilon D}{c^*-c_0}. \tag{3.21}$$

After linearization of (3.20), i.e., in the zeroth approximation of the small parameters ω^2 and $\theta\omega$, the substitution of (3.18) and (3.19) into (3.20) leads to the following problem:

$$\frac{\partial u_1}{\partial t}+u\frac{\partial u_1}{\partial x}+v\frac{\partial u_1}{\partial y}+u_1\frac{\partial u}{\partial x}+v_1\frac{\partial u}{\partial y}=-\frac{1}{\rho}\frac{\partial p_1}{\partial x}+\nu\left(\frac{\partial^2 u_1}{\partial x^2}+\frac{\partial^2 u_1}{\partial y^2}\right),$$

$$\frac{\partial v_1}{\partial t}+u\frac{\partial v_1}{\partial x}+v\frac{\partial v_1}{\partial y}+u_1\frac{\partial v}{\partial x}+v_1\frac{\partial v}{\partial y}=-\frac{1}{\rho}\frac{\partial p_1}{\partial y}+\nu\left(\frac{\partial^2 v_1}{\partial x^2}+\frac{\partial^2 v_1}{\partial y^2}\right),$$

$$\frac{\partial u_1}{\partial x}+\frac{\partial v_1}{\partial y}=0;\quad x=0,\quad u_1=0,\quad v_1=0,\quad p_1=p_0;$$

$$y=0,\quad u_1=0,\quad v_1=0,\quad p_1=p_0;\quad y\to\infty,\quad u_1=0,\quad v_1=0. \tag{3.22}$$

Equations (3.22) skip the equation for c_1 since in the linear approximation ($\theta\omega=0$) the disturbances in the velocity do not depend on the disturbances in the concentration.

The differentiation for y and x of the first two equations provides the opportunity to exclude the pressure p_1. The stability of the basic flow will be examined considering periodic disturbances of the form

$$u_1=F'(y)\exp i(\alpha x-\beta t),\quad v_1=-i\alpha F(y)\exp i(\alpha x-\beta t), \tag{3.23}$$

where $F(y)$ is the amplitude of a one-dimensional disturbance (regarding y); α and β/α are its wave number and phase velocity, respectively:

$$\alpha = \frac{2\pi}{\lambda}, \quad \beta = \beta_r + i\beta_i. \tag{3.24}$$

In expression (3.24) λ is the wavelength, β_r is the circle frequency, and β_i is the increment. Obviously, the condition for stability of the flow is

$$\beta_i < 0. \tag{3.25}$$

In the case of $\beta_i > 0$ the basic flow is unstable (the amplitude grows with time).

Introducing (3.23) into Eq. 3.22 leads to Orr–Sommerfeld-type equations [11, 22, 23] for the amplitude of the disturbances:

$$\begin{gathered}\left(u - \frac{\beta}{\alpha}\right)\left(F'' - \alpha^2 F\right) - \frac{\partial^2 u}{\partial y^2} F = -\frac{i\nu}{\alpha}\left(F^{IV} - 2\alpha^2 F'' + \alpha^4 F\right) \\ + \frac{i}{\alpha}\left[\nu F''' + \left(\frac{\partial^2 u}{\partial x \partial y} - \alpha^2 v\right) F'\right], \\ y = 0, \quad F = 0, \quad F' = 0, \quad y \to \infty, \quad F = 0, \quad F' = 0.\end{gathered} \tag{3.26}$$

In (3.26) $F = F(y)$, whereas **u** and **v** depend on y and νx; hence, the dependence on x is insignificant. This gives us an opportunity to consider x as a parameter [11]. There are four constants in Eq. 3.26, where ν and α are known beforehand, whereas the eigenvalues β_r and β_i of the eigenfunction $F(y)$ are the sought ones. Obviously, the eigenvalues β_r and β_i thus determined depend on x, and at some x_{cr}

$$\beta_i(x_{cr}) = 0, \tag{3.27}$$

i.e., the velocity profile $u(x, y)$ becomes unstable.

The assumption that the variable x is a parameter in Eq. 3.26 allows a new variable to be introduced:

$$\xi = \frac{y}{\delta} = y\left(\frac{u_0}{\nu x}\right)^{0.5} = \frac{2}{\varepsilon}\eta. \tag{3.28}$$

Hence, all functions in Eq. 3.26 can be expressed by the new variable ξ (3.28):

$$\begin{gathered}u = u_0 f'(\xi), \quad v = 0.5\left(\frac{u_0 \nu}{x}\right)^{0.5}(\xi f' - f), \quad F(y) = \varphi(\xi), \quad F^{(j)} = \delta^{-j}\varphi^{(j)}, \\ j = 1, \cdots, 4.\end{gathered} \tag{3.29}$$

It can be seen from (3.10), (3.11), and (3.29) that f can be determined from

$$2f''' + ff'' = 0, \quad f(0) = a, \quad f'(0) = 0, \quad f''(0) = \frac{\varepsilon^2}{4} b. \tag{3.30}$$

The introduction of (3.29) into Eq. 3.26 leads to the following Orr–Sommerfeld type of equation:

$$(f' - C)(\varphi' - A^2\varphi) - f'''\varphi$$
$$= -\frac{i}{ARe}\left\{(\varphi^{IV} - 2A^2\varphi'' + A^4\varphi) - \frac{1}{2}(\xi f' - f)\varphi''' + \left[\frac{1}{2}(\xi f''' + f'') + \frac{A^2}{2}(\xi f' - f)\right]\varphi'\right\},$$

$$\varphi(0) = 0, \quad \varphi'(0) = 0, \quad \varphi(\infty) = 0, \quad \varphi'(\infty) = 0, \tag{3.31}$$

where

$$A = \alpha\delta, \quad C = \frac{\beta}{\alpha u_0} = C_r + iC_i, \quad Re = 1.72\frac{u_0\delta}{\nu}. \tag{3.32}$$

The linear analysis of the hydrodynamic stability of a laminar boundary layer under the condition of intensive interphase mass transfer is finally reduced to determining C_r and $\phi(\xi)$ at $C_i = 0$ when Re and A are given. The minimum Reynolds number, i.e., the critical Reynolds number Re_{cr} at which the flow becomes unstable, can be obtained from the dependence $C_r(Re)$.

Problem (3.31) is an eigenvalue problem about C when Re and A are given. The imaginary part of the eigenvalue C determines whether or not the basic flow is stabile relative to the infinitesimal disturbances. Since this is a linear eigenvalue problem, in this theory it can be solved for $C = C(Re, A)$. The solutions of this problem are usually presented in two ways: (1) for specific values of the parameters A and Re, the corresponding values of C are tabulated and (2) the locus plane where $C_i = 0$ (the "neutral stability curve") is plotted on (Re, A). The critical Reynolds number is the minimum Reynolds number at which an infinitesimal disturbance will grow. The growing with time disturbances are applied when Re and A are given real values, whereas the parameter C is the complex eigenvalue searched for.

To solve problem (3.31) numerically in an infinite interval, the boundary conditions $(\varphi(\infty) = 0, \quad \varphi'(\infty) = 0)$ are assumed valid at finite distance $\xi = \xi_\infty \gg 1$ far from the plate. The boundary conditions there will be replaced with two differential equations. To obtain these equations, the solution [7] of (3.30) is used at high values of ξ:

$$f(\xi) = \xi - k + 0.231\int_\infty^\xi d\xi \int_\infty^\xi \exp\left[-\frac{1}{4}(\xi - k)^2\right]d\xi. \tag{3.33}$$

Comparison of the numerical solutions of (3.30) and (3.33) shows that an accuracy of 10^{-4} to 10^{-6} is achieved when ξ is greater than 6, and we can assume

$$f' = 1, f'' = f''' = 0, \quad \xi f' - f = k, \quad \xi f''' = 0. \tag{3.34}$$

Thus, introducing (3.34) into (3.31), we obtain the following expression, which is valid in case of $\xi \geq 6$:

$$(1-C)(\varphi''-A^2\varphi)=-\frac{\mathrm{i}}{ARe}\left[\left(\varphi^{IV}-2A^2\varphi''+A^4\varphi\right)-\frac{1}{2}k\varphi'''+\frac{A^2}{2}k\varphi'\right]. \tag{3.35}$$

The solution of (3.35) depends on four constants [14, 20], two of them being equal to zero, because two of the solutions of the characteristic Eq. 3.35 are positive, i.e., conditions $\varphi(\infty)=\varphi'(\infty)=0$ are satisfied:

$$\varphi=C_1\exp(-A\xi)+C_2\exp(-\gamma\xi), \tag{3.36}$$

where the constants C_1 and C_2 are determined using boundary conditions. The exclusion of these constants from (3.36) leads to the following relations for $\xi\geq\xi_\infty=6$,

$$(\varphi''-A^2\varphi)'-\gamma(\varphi''-A^2\varphi)=0,\quad \xi=\xi_\infty,$$

$$(\varphi''-\gamma^2\varphi)'+A(\varphi''-\gamma^2\varphi)=0, \tag{3.37}$$

and for γ the following is obtained:

$$\gamma=\frac{k}{4}-\frac{\sqrt{k^2+16A[A+iRe(1-C)]}}{4}. \tag{3.38}$$

The numerical solution of (3.30) for different values of θ shows that k depends on θ (Table 1). In the case of $\theta=0$, comparison of $Re_{\mathrm{cr}}\approx 500$ obtained in [26, 27] in the approximations of parallel flows and $Re_{\mathrm{cr}}=501$ obtained by us in the case of almost parallel flows shows that Re_{cr} depends slightly on k. Analogous results have been obtained at $\theta\neq 0$.

The matrix form of Eq. 3.31 is as follows:

$$\begin{bmatrix} b_1\\ b_2\\ b_3\\ b_4\end{bmatrix}'+\begin{bmatrix}0 & -1 & 0 & 0\\ 0 & 0 & -1 & 0\\ 0 & 0 & 0 & -1\\ a_1 & a_2 & a_3 & a_4\end{bmatrix}\begin{bmatrix} b_1\\ b_2\\ b_3\\ b_4\end{bmatrix}=0, \tag{3.39}$$

and a_j ($j=1,\ldots,4$) are obtained directly from Eq. 3.31:

$$a_1=\left[iA^3Re(f'-B)-iARef'''+A^4\right],$$

$$a_2=\frac{1}{2}(\xi f'''-f'')+\frac{A^2}{2}(\xi f'-f),\quad a_3=-\left[iARe(f'-B)+2A^2\right],$$

$$a_4=-\frac{1}{2}(\xi f'-f), \tag{3.40}$$

where $b_j(j=1,\cdots,4)$ are $b_1=\varphi,\quad b_2=\varphi',\quad b_3=\varphi'',\quad b_4=\varphi'''$.

The boundary conditions are transformed in

$$\begin{bmatrix} 1 & 0 & 0 & 0 \\ 0 & 1 & 0 & 0 \end{bmatrix} \begin{bmatrix} b_1 \\ b_2 \\ b_3 \\ b_4 \end{bmatrix} = 0, \quad \xi = 0; \tag{3.41}$$

and

$$\begin{bmatrix} -\gamma A^2 & -A^2 & \gamma & 1 \\ -A\gamma^2 & -\gamma^2 & A & 1 \end{bmatrix} \begin{bmatrix} b_1 \\ b_2 \\ b_3 \\ b_4 \end{bmatrix} = 0, \quad \xi = \xi_\infty = 6, \tag{3.42}$$

respectively.

Using the substitutions

$$b_j = \varphi^{(j)}(\xi), \quad (j = 1, \ldots, 4), \quad \boldsymbol{B} = (b_1, b_2, b_3, b_4)^{\mathrm{T}},$$

we can rewrite the eigenvalue problem (3.41, 3.42) in the form

$$\boldsymbol{B}'(\xi) + \boldsymbol{A}(\xi; C)\boldsymbol{B}(\xi) = 0, \quad \xi \in [0, \xi_\infty];$$

$$\boldsymbol{\Psi}_0^{\mathrm{T}}\boldsymbol{B} = 0, \quad \xi = 0; \quad \boldsymbol{\Psi}_1^{\mathrm{T}}\boldsymbol{B} = 0, \quad \xi = \xi_\infty, \tag{3.43}$$

where $\boldsymbol{A}(\xi;\ \mathrm{C})$ is 4×4 matrix of the continuous components of $\xi \in [0, \infty]$ and depends on C; $\boldsymbol{\Psi}_1^{\mathrm{T}}$ and $\boldsymbol{\Psi}_0^{\mathrm{T}}$ are scalar matrixes of order 4×2 ($\boldsymbol{\Psi}^{\mathrm{T}}$ denotes the transposed matrix of $\boldsymbol{\Psi}$).

To solve the eigenvalue problem (3.43) the method proposed by Abramov [15–17] is used. Let $\mathbf{B}(\xi;\ C)$ be an arbitrary solution of the system (5.103) satisfying the boundary condition at $\xi = \xi_\infty$. Then, as shown in [15], the solution $\boldsymbol{\Psi}(\xi; C)$ of the initial value problem

$$\boldsymbol{\Psi}' - (A^T + \boldsymbol{\Psi}(\boldsymbol{\Psi}^T\boldsymbol{\Psi})^{-1}\boldsymbol{\Psi}^T A^T)\boldsymbol{\Psi} = 0, \quad \xi \in [0, \infty], \quad \boldsymbol{\Psi} = \boldsymbol{\Psi}_1, \quad \xi = \xi_\infty, \tag{3.44}$$

satisfies

$$\boldsymbol{\Psi}(\xi; C)\boldsymbol{B}(\xi; C) = 0 \text{ for any } \xi \in [0, \infty], \tag{3.45}$$

i.e., it can have the boundary conditions at $\xi = \xi_\infty$ transferred to any $\xi \in [0, \infty]$.

Hence, integrating (3.44) up to $\xi = 0$, we obtain the required eigenvalue relation in the form

$$\det\begin{pmatrix} \boldsymbol{\Psi}_0^T \\ \boldsymbol{\Psi}_{1,0}^T(C) \end{pmatrix} = 0, \tag{3.46}$$

where $\boldsymbol{\Psi}_{1,0}(C)$ denotes the solution of (3.44) at $\xi = 0$.

The proposed method is reliable and $\Psi\Psi^T = \text{const}$ along the integration path. The basic procedure is to iterate C until the solution C^* of the characteristic

Eq. 3.46 is obtained with a given accuracy. The same procedure has to be repeated at greater ξ_∞ with a view to achieving convergence of the successive approximations C^*. When convergence has been established with the prescribed accuracy, the last computed C^* is taken as an eigenvalue of the original problem (3.31). The numerical experiments show that an accuracy of 10^{-4} to 10^{-6} is achieved when ξ_∞ is greater than 5–6.

The neutral curves for A as a function of Re and for C as a function of Re are shown in Figs. 12, 13, 14 and 15. They were obtained for gases ($\varepsilon = 1$) and for liquids ($\varepsilon = 10$).

The critical Reynolds numbers Re_{cr} corresponding to the wave velocities C_r and wave numbers A have been obtained. $C_{r\ min}$ and A_{min} were obtained from these results too. Let $C_{r\ min}$ and A_{min} denote the minimum values of the wave velocities and wave numbers at which the flow is stable for any value of the Reynolds number. They are shown in Table 2 for the dependence on the magnitude and on the direction of the concentration gradient under the conditions of an intensive interphase mass transfer.

It can be seen from Figs. 12, 13, 14 and 15 and from Table 2 that the intensive interphase mass transfer directed towards the phase boundary ($\theta < 0$) (the effect of "suction") stabilizes the flow, i.e., the increase of the concentration difference $|c_0 - c^*|$ leads to an increase of Re_{cr} and to a decrease of $C_{r\ min}$ and A_{min}. In the case of intensive interphase mass transfer directed from the phase boundary towards the volume ($\theta > 0$) (the effect of "injection") a destabilization of the flow is observed, i.e., the increase of the concentration difference $|c_0 - c^*|$ leads to a decrease of Re_{cr} and to an increase of $C_{r\ min}$ and A_{min}.

The high concentration gradients have a significantly stabilizing effect at $\theta < 0$; the destabilizing effect occurs in the case of a change in the direction of the mass transfer ($\theta > 0$).

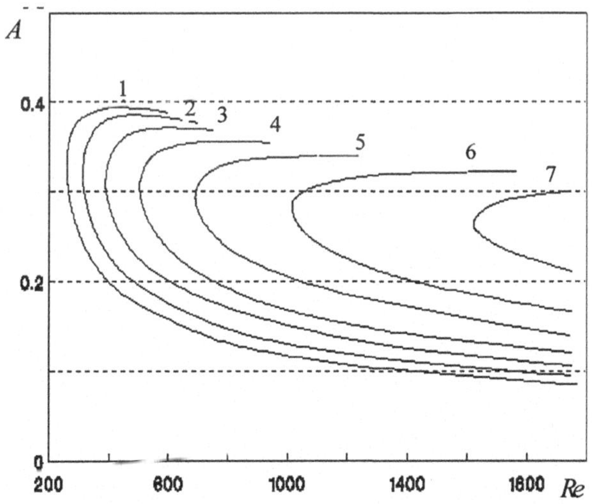

Fig. 12 The neutral curve for the wave number A as a function of the Reynolds number Re in the case of $\varepsilon = 1$: *1* $\theta = 0.3$, *2* $\theta = 0.2$, *3* $\theta = 0.1$, *4* $\theta = 0.0$, *5* $\theta = -0.1$, *6* $\theta = -0.2$, *7* $\theta = -0.3$

In the above discussions the fact that the diffusive fluxes through the interface surface at ($\theta < 0$) increase with the rise of the concentration difference $|c_0 - c^*|$, while at ($\theta > 0$) they decrease with the rise of $|c^* - c_0|$.

The results obtained can be used for clarification of the mechanism and the kinetics of a number of practically interesting processes. For instance, in liquid–solid systems the anode dissolution of metals in the electrolyte flow under the condition of intensive interphase mass transfer can increase substantially before flow turbulence for comparatively small values of the Reynolds number, whereas the electrode deposition of metals out of concentrated solutions can be implemented under laminar conditions at high values of the Reynolds number. Intensive

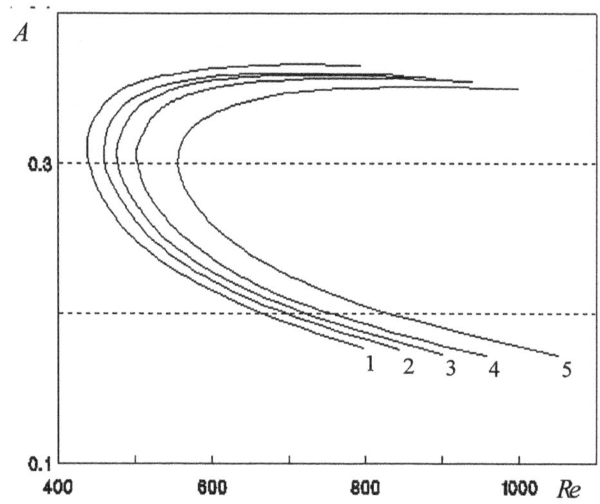

Fig. 13 The neutral curve for the wave number A as a function of the Reynolds number Re in the case of $\varepsilon = 10$: *1* $\theta = 0.2$, *2* $\theta = 0.1$, *3* $\theta = 0.05$, *4* $\theta = 0.0$, *5* $\theta = 0.05$

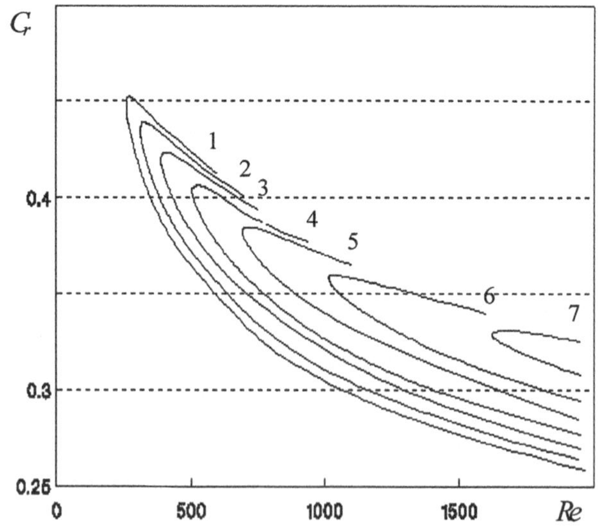

Fig. 14 The neutral curve for the wave velocity C_r as a function of the Reynolds number Re in the case of $\varepsilon = 1$: *1* $\theta = 0.3$, *2* $\theta = 0.2$, *3* $\theta = 0.1$, *4* $\theta = 0.0$, *5* $\theta = -0.1$, *6* $\theta = -0.2$, *7* $\theta = -0.3$

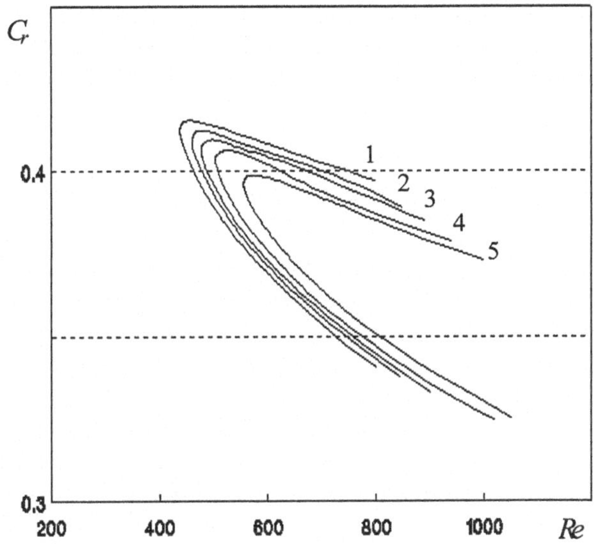

Fig. 15 The neutral curve for the wave velocity C_r as a function of the Reynolds number Re in the case of $\varepsilon = 10$: *1* $\theta = 0.2$, *2* $\theta = 0.1$, *3* $\theta = 0.05$, *4* $\theta = 0.0$, *5* $\theta = 0.05$

interphase mass transfer is of interest for the process of ablation (e.g., launching a spacecraft in denser atmospheric layers). Intensive evaporation of a substance from a solid surface leads to an increase of the interphase heat transfer coefficients, i.e., to a decrease of the "undesired" heat flux towards the rounded fuselage nose of spacecraft (missiles). It is evident that from the results obtained under these conditions that turbulence of the gas at considerably small Reynolds numbers is possible, which will also affect the rate of the interphase heat transfer.

Table 2 Values of the critical Reynolds number Re_{cr}, corresponding to the wave velocities C_r, wave numbers A, and $C_{r\,min}$ and A_{min} obtained

ε	θ	Re_{cr}	A	C_r	A_{min}	$C_{r\,min}$
1	−0.30	1619	0.259	0.3281	0.301	0.3310
	−0.20	1014	0.285	0.3587	0.322	0.3599
	−0.10	689	0.290	0.3816	0.340	0.3848
	0.0	501	0.305	0.4035	0.359	0.4067
	0.10	386	0.309	0.4196	0.373	0.4243
	0.20	310	0.320	0.4351	0.387	0.4396
	0.30	258	0.331	0.4488	0.398	0.4526
10	−0.05	555	0.300	0.3960	0.351	0.3990
	0.0	501	0.305	0.4035	0.359	0.4067
	0.05	476	0.305	0.4062	0.360	0.4097
	0.10	459	0.305	0.4085	0.361	0.4124
	0.20	437	0.310	0.4123	0.367	0.4155
20	−0.05	558	0.305	0.3959	0.351	0.3978
	−0.03	528	0.305	0.4010	0.354	0.4037
	0.0	501	0.305	0.4035	0.359	0.4067
	0.03	488	0.305	0.4064	0.362	0.4099

The observed influence of intensive interphase mass transfer on the hydrodynamic stability in gas (liquid)–solid systems is much more interesting for systems with a movable phase boundary (gas–liquid, liquid–liquid).

4 Self-Organizing Dissipative Structures

In Sect. 2.4 it was shown, that in the cases of supercritical bifurcations ($Re > Re_{cr}$) the amplitudes of the small disturbances (perturbations) increase and reach a stable amplitude. In these mass transfer in the flow is very intensive. These flow regimes are very useful because the energy dissipation is less than that in turbulent flow regimes.

The theoretical analysis of the self-organizing dissipative structures in the approximations of the linear stability theory is embarrassed by the absence of a condition for amplitude determination. It is possible to be use an additional physical hypothesis and an equivalent mathematical condition or experimental data.

Theoretical analysis of systems with intensive interphase mass transfer as a result of large concentration gradients shows that the large mass flux induces secondary flow at the interphase surface [69]. The velocity of this flow is directed normally to this interface. In the cases of interphase mass transfer between a gas (liquid) and a solid interface, the following was obtained [69]:

$$v = -\frac{D}{\rho_0^*}\left(\frac{d\tilde{c}}{dn}\right)_s, \tag{4.1}$$

where D is the diffusivity, ρ_0* is the density of gas (liquid) at the interface, and $(d\tilde{c}/dn)_s$ is the normal derivative of the concentration at the interface.

The solution of this problem in the approximation of the boundary layer uses the Prandtl equations and the convection–diffusion equation

$$\tilde{u}\frac{\partial \tilde{u}}{\partial x} + \tilde{v}\frac{\partial \tilde{u}}{\partial y} = \upsilon\frac{\partial^2 \tilde{u}}{\partial y^2}, \quad \frac{\partial \tilde{u}}{\partial x} + \frac{\partial \tilde{v}}{\partial y} = 0, \quad \tilde{u}\frac{\partial \tilde{c}}{\partial x} + \tilde{v}\frac{\partial \tilde{c}}{\partial y} = D\frac{\partial^2 \tilde{c}}{\partial y^2} \tag{4.2}$$

with boundary condition

$$x = 0, \quad \tilde{u} = \bar{u}, \quad \tilde{c} = \tilde{c}_0; \quad y = 0, \quad \tilde{u} = 0, \quad \tilde{v} = -\frac{D}{\rho_0^*}\frac{\partial \tilde{c}}{\partial y}, \quad \tilde{c} = \tilde{c}^*;$$

$$y \to \infty, \quad \tilde{u} = \bar{u}, \quad \tilde{c} = \tilde{c}_0. \tag{4.3}$$

To solve problem (4.2, 4.3), it is necessary to introduce the similarity variables

$$\tilde{u} = \bar{u}F', \quad \tilde{v} = \left(\frac{\bar{u}\upsilon}{4x}\right)^{0.5}(\eta F' - F), \quad \tilde{c} = \tilde{c}_0 + (\tilde{c}^* - \tilde{c}_0)\psi,$$

$$y = \eta \left(\frac{\bar{u}}{\upsilon x} \right)^{-0.5}, \quad F = F(\eta), \quad \psi = \psi(\eta). \tag{4.4}$$

The introduction of (4.4) into (4.2, 4.3) leads to

$$2F'' + FF'' = 0, \quad \psi'' + \frac{Sc}{2} F\psi' = 0, \quad F(0) = \theta\psi'(0), \quad F'(0) = 0,$$
$$F'(\infty) = \frac{2}{Sc},$$

$$\psi(0) = 1, \quad \psi(\infty) = 0, \quad Sc = \frac{\upsilon}{D}, \tag{4.5}$$

where θ is a nonlinear mass transfer parameter:

$$\theta = \frac{\tilde{c}^* - \tilde{c}_0}{\rho_0^* \sqrt{Sc}}. \tag{4.6}$$

Problem 4.5) was solved for different systems [69] and the solutions obtained permit us to use the next boundary conditions for hydrodynamic problems:

$$F(0) = a, \quad F'(0) = b, \quad F''(0) = c, \tag{4.7}$$

where a, b, and c are obtained (see Table 3) for the linear mass transfer case ($a = b = \theta = 0$), gas–solid ($Sc = 1$, $b = 0$) and liquid–solid ($Sc = 400$, $b = 0$) systems [70], and gas–liquid (in the gas phase, $Sc = 1$) and liquid–liquid ($Sc = 400$) systems [29–31, 36].

The secondary flows obtained as a result of the large mass flux are suction from ($\theta < 0$) or injection into ($\theta > 0$) the boundary layer and change the hydrodynamic stability of the flow [69].

The linear stability analysis [7–9] shows that critical Reynolds number $Re_{cr} = 1.72\sqrt{\frac{\bar{u}x}{\upsilon}}$ increases ($\theta < 0$, stabilization) or decreases ($\theta > 0$, destabilization) for different mass transfer directions (see Table 4).

The comparative analysis of the nonlinear mass transfer effect (as a result of the large concentration gradient) and the Marangoni effect (as a result of the surface

Table 3 Boundary condition values in (4.7)

Sc	θ	a	b	c
	0	0	0	0.33205
1	−0.2	0.1557	0	0.3892
1	−0.2	−0.1162	0	0.2905
400	−0.03	0.01219	0	0.3389
400	0.03	−0.00570	0	0.3321
1	−0.2	0.1703	0.1083	0.3800
1	0.2	−0.1283	0.1059	0.2710

Table 4 Critical Reynolds number

θ	Re_{cr}
−0.3	1,014
−0.2	1,619
0	501
0.2	310
0.3	258

tension gradient) shows that the Marangoni effect is negligible in systems with moving phase boundaries [28, 43].

The theoretical deviation of the nonlinear mass transfer rate from the prediction of the linear mass transfer theory is practically 10–30%, but in many cases the deviation of the experimental data is larger. This deviation may be explained by the loss of stability of the system, i.e., small disturbances increase to form a new stable state [6, 8]. As a result, the amplitude of the disturbances of this self-organized dissipative structure is a constant. This problem was solved in the cases of gas absorption [58, 65, 66] and liquid evaporation [67, 68], where the amplitudes of the disturbances were obtained using experimental data for the mass transfer rate.

The main problem of the theoretical analysis of self-organized structures is to obtain the amplitude of the disturbances [10]. There are practically two approaches—to use experimental data and to use an additional theoretical condition. In [71] a method was proposed for stability analysis of the nonlinear mass transfer processes using a balance between kinetic energies of the main flow and the disturbance.

4.1 Nonlinear Mass Transfer in the Boundary Layer

Let us consider the velocity distribution in the laminar boundary layer flow $(\tilde{u}, \tilde{v})$. The existence of disturbances in the system leads to their interaction with the main flow $(\tilde{u}, \tilde{v})$. If the main flow is unstable, a small disturbance can accept the energy from the main flow $(\tilde{u}, \tilde{v})$ and increases to form a stable state (dissipative structure). As a result, the main flow $(\tilde{u}, \tilde{v})$ loses energy and reduces to new main flow (u, v, p). This flow and disturbance are a new flow (U, V, P). The new flow (U, V, P) is nonstationary because of the nonstationary character of the disturbance:

$$U(x,y,t) = u(x,y,t) + u'(x,y,t), \quad V(x,y,t) = v(x,y,t) + v'(x,y,t),$$
$$P(x,y,t) = p(x,y,t) + p'(x,y,t). \tag{4.8}$$

It is obvious that the kinetic energies E of the main flow $(\tilde{u}, \tilde{v})$ and the nonstationary flow (U, V, P) are equal. In the cases when the main flow is unstable, the disturbances increase on account of the energy of the main flow [6]. As a result, the nonstationary flow is a superposition of the "new main flow" (u, v, p) with kinetic energy E_0 and flow of disturbances with constant amplitude (u', v', p') and kinetic energy E_1, i.e.,

$$E = E_0 + E_1. \tag{4.9}$$

Condition (4.9) permits us to obtain the amplitude of the disturbances.

The new flow (4.8) satisfies the full system of Navier–Stokes equations:

$$\frac{\partial U}{\partial t} + U\frac{\partial U}{\partial x} + V\frac{\partial U}{\partial y} = -\frac{1}{\rho}\frac{\partial P}{\partial x} + \upsilon\left(\frac{\partial^2 U}{\partial x^2} + \frac{\partial^2 U}{\partial y^2}\right),$$

$$\frac{\partial V}{\partial t} + U\frac{\partial V}{\partial x} + V\frac{\partial V}{\partial y} = -\frac{1}{\rho}\frac{\partial P}{\partial y} + \upsilon\left(\frac{\partial^2 V}{\partial x^2} + \frac{\partial^2 V}{\partial y^2}\right), \quad \frac{\partial U}{\partial x} + \frac{\partial V}{\partial y} = 0. \tag{4.10}$$

The introduction of (4.8) into (4.10) permits us to eliminate the pressure, using the differentiation and subsequent subtraction of the first two equations in (4.10), and the following result is obtained:

$$\frac{\partial \psi}{\partial t} + (u + u')\frac{\partial \psi}{\partial x} + (v + v')\frac{\partial \psi}{\partial y} = \upsilon\left(\frac{\partial^2 \psi}{\partial x^2} + \frac{\partial^2 \psi}{\partial y^2}\right), \quad \frac{\partial u}{\partial x} + \frac{\partial v}{\partial y} = 0,$$

$$\frac{\partial \psi'}{\partial t} + (u + u')\frac{\partial \psi'}{\partial x} + (v + v')\frac{\partial \psi'}{\partial y} = \upsilon\left(\frac{\partial^2 \psi'}{\partial x^2} + \frac{\partial^2 \psi'}{\partial y^2}\right), \quad \frac{\partial u'}{\partial x} + \frac{\partial v'}{\partial y} = 0, \tag{4.11}$$

where

$$\psi = \frac{\partial u}{\partial y} - \frac{\partial v}{\partial x}, \quad \psi' = \frac{\partial u'}{\partial y} - \frac{\partial v'}{\partial x}. \tag{4.12}$$

Thus, (4.11) stresses the segregation of the effects:

- The first equation provides an account of the influence of the disturbance on the new main flow.
- The second equation provides an account of the influence of the new main flow on the disturbance.

However, the influences in both equations have nonlinear character.

The particular solutions of (4.11) in the form *of "normal" disturbances* [8], i.e., periodic disturbances whose amplitude depends exponentially on the time, will be discussed further:

$$u(x, y, t) = \exp(-\omega t)u_0(x, y), \quad v(x, y, t) = -\exp(-\omega t)\int \frac{\partial u_0}{\partial x}dy,$$

$$u'(x, y, t) = \exp(-\omega t)[v_0(x, y) + u_1(x, y)\sin nx + v_1(x, y)\cos nx],$$

$$v'(x,y,t) = -\exp(-\omega t)\int\left[\frac{\partial v_0}{\partial x} + \left(\frac{\partial u_1}{\partial x} - nv_1\right)\sin nx + \left(\frac{\partial v_1}{\partial x} + nu_1\right)\cos nx\right]dy, \tag{4.13}$$

where $n = 2\pi/\lambda$ is the wave number of the disturbance and λ is the wavelength.

The substitution of (4.13) into (4.11) allows the determination of a stable dissipative structure as a partial solution at $\omega = 0$. As a result, two equations for u_0, v_0, u_1, v_1, corresponding to (4.11) were obtained, where $\cos^2 nx = 1 - \sin^2 nx$. From these equations it is possible to obtain a *set of equations*, if we put the aperiodic parts and all parts containing $\sin nx$, $\cos nx$, $\sin^2 nx$, and $\sin nx \cos nx$ to be equal to zero.

From the aperiodic parts we can obtain directly two equations to determine u_0 and v_0:

$$(u_0 + v_0)\left(\frac{\partial^2 u_0}{\partial x \partial y} + \int \frac{\partial^3 u_0}{\partial x^3}dy\right) - \left(\int \frac{\partial u_0}{\partial x}dy + \int \frac{\partial v_0}{\partial x}dy\right)\left(\frac{\partial^2 u_0}{\partial x^2} + \frac{\partial^2 u_0}{\partial y^2}\right)$$
$$= \upsilon\left(\int \frac{\partial^4 u_0}{\partial x^4}dy + 2\frac{\partial^3 u_0}{\partial x^2 \partial y} + \frac{\partial^3 u_0}{\partial y^3}\right), \tag{4.14a}$$

$$(u_0 + v_0)\left(\frac{\partial^2 v_0}{\partial x \partial y} + \int \frac{\partial^3 v_0}{\partial x^3}dy\right) - \int \frac{\partial u_0}{\partial x}dy\left(\frac{\partial^2 v_0}{\partial x^2} + \frac{\partial^2 v_0}{\partial y^2}\right)$$
$$= \upsilon\left(\int \frac{\partial^4 v_0}{\partial x^4}dy + 2\frac{\partial^3 v_0}{\partial x^2 \partial y} + \frac{\partial^3 v_0}{\partial y^3}\right)$$
$$- v_1\left[\frac{\partial^2 v_1}{\partial x \partial y} + n\frac{\partial u_1}{\partial y} + \int\left(\frac{\partial^3 v_1}{\partial x^3} + 3n\frac{\partial^2 u_1}{\partial x^2} - 3n^2\frac{\partial v_1}{\partial x} - n^3 u_1\right)dy\right]$$
$$+ \left(\frac{\partial^2 v_1}{\partial x^2} + \frac{\partial^2 v_1}{\partial y^2} + 2n\frac{\partial u_1}{\partial x} - n^2 v_1\right)\int\left(\frac{\partial v_1}{\partial x} + nu_1\right)dy. \tag{4.14b}$$

The periodic part in the first equation in (4.11) contains terms with $\sin nx$ and $\cos nx$. After summing up separately the parts containing $\sin x$ and $\cos x$ and equalization of these sums through division by zero, we obtain

$$u_1\left(\frac{\partial^2 u_0}{\partial x \partial y} + \int \frac{\partial^3 u_0}{\partial x^3}dy\right) + \left(\frac{\partial^2 u_0}{\partial x^2} + \frac{\partial^2 u_0}{\partial y^2}\right)\int\left(nv_1 - \frac{\partial u_1}{\partial x}\right)dy = 0, \tag{4.15a}$$

$$v_1\left(\frac{\partial^2 u_0}{\partial x \partial y} + \int \frac{\partial^3 u_0}{\partial x^3}dy\right) - \left(\frac{\partial^2 u_0}{\partial x^2} + \frac{\partial^2 u_0}{\partial y^2}\right)\int\left(nu_1 + \frac{\partial v_1}{\partial x}\right)dy = 0. \tag{4.15b}$$

By analogy, from the terms with $\sin nx$ and $\cos nx$ in the second equation in (4.11), we obtain two equations for the determination of u_1 and v_1:

$$u_0\left[\frac{\partial^2 u_1}{\partial x\partial y}-n\frac{\partial v_1}{\partial y}+\int\left(\frac{\partial^3 u_1}{\partial x^3}-3n\frac{\partial^2 v_1}{\partial x^2}-3n^2\frac{\partial u_1}{\partial x}+n^3 v_1\right)dy\right]$$
$$-\int\frac{\partial u_0}{\partial x}dy\left(\frac{\partial^2 u_1}{\partial y^2}+\frac{\partial^2 u_1}{\partial x^2}-2n\frac{\partial v_1}{\partial x}-n^2 u_1\right)=\upsilon\left[2\frac{\partial^3 u_1}{\partial x^2\partial y}-4n\frac{\partial^2 v_1}{\partial x\partial y}-2n^2\frac{\partial u_1}{\partial y}\right.$$
$$\left.+\int\left(\frac{\partial^4 u_1}{\partial x^4}-4n\frac{\partial^3 v_1}{\partial x^3}-6n^2\frac{\partial^2 u_1}{\partial x^2}+4n^3\frac{\partial v_1}{\partial x}+n^4 u_1\right)dy+\frac{\partial^3 u_1}{\partial y^3}\right]$$
$$-u_1\left(\frac{\partial^2 v_0}{\partial x\partial y}+\int\frac{\partial^3 v_0}{\partial x^3}dy\right)-v_0\left(\frac{\partial^2 u_1}{\partial x\partial y}-n\frac{\partial v_1}{\partial y}\right)-v_0\int\left(\frac{\partial^3 u_1}{\partial x^3}-3n\frac{\partial^2 v_1}{\partial x^2}-3n^2\frac{\partial u_1}{\partial x}+n^3 v_1\right)dy$$
$$+\left(\frac{\partial^2 u_1}{\partial x^2}+\frac{\partial^2 u_1}{\partial y^2}-2n\frac{\partial v_1}{\partial x}-n^2 u_1\right)\int\frac{\partial v_0}{\partial x}dy+\left(\frac{\partial^2 v_0}{\partial x^2}+\frac{\partial^2 v_0}{\partial y^2}\right)\int\left(\frac{\partial u_1}{\partial x}-nv_1\right)dy, \tag{4.16a}$$

$$u_0\left[\frac{\partial^2 v_1}{\partial x\partial y}+n\frac{\partial u_1}{\partial y}+\int\left(\frac{\partial^3 v_1}{\partial x^3}-3n\frac{\partial^2 u_1}{\partial x^2}-3n^2\frac{\partial v_1}{\partial x}-n^3 u_1\right)dy\right]$$
$$-\int\frac{\partial u_0}{\partial x}dy\left(\frac{\partial^2 v_1}{\partial x^2}+\frac{\partial^2 v_1}{\partial y^2}+2n\frac{\partial u_1}{\partial x}-n^2 v_1\right)=\upsilon\left[2\frac{\partial^3 v_1}{\partial x^2\partial y}+4n\frac{\partial^2 u_1}{\partial x\partial y}-2n^2\frac{\partial v_1}{\partial y}\right.$$
$$\left.+\int\left(\frac{\partial^4 v_1}{\partial x^4}+4n\frac{\partial^3 u_1}{\partial x^3}-6n^2\frac{\partial^2 v_1}{\partial x^2}-4n^3\frac{\partial u_1}{\partial x}+n^4 v_1\right)dy+\frac{\partial^3 v_1}{\partial y^3}\right]$$
$$-v_1\left(\frac{\partial^2 v_0}{\partial x\partial y}+\int\frac{\partial^3 v_0}{\partial x^3}dy\right)-v_0\left(\frac{\partial^2 v_1}{\partial x\partial y}+n\frac{\partial u_1}{\partial y}\right)$$
$$-v_0\int\left(\frac{\partial^3 v_1}{\partial x^3}+3n\frac{\partial^2 u_1}{\partial x^2}-3n^2\frac{\partial v_1}{\partial x}-n^3 u_1\right)dy$$
$$+\left(\frac{\partial^2 v_1}{\partial x^2}+\frac{\partial^2 v_1}{\partial y^2}+2n\frac{\partial u_1}{\partial x}-n^2 v_1\right)\int\frac{\partial v_0}{\partial x}dy+\left(\frac{\partial^2 v_0}{\partial x^2}+\frac{\partial^2 v_0}{\partial y^2}\right)\int\left(\frac{\partial v_1}{\partial x}+nu_1\right)dy. \tag{4.16b}$$

From the terms with $\sin^2 nx$ and *sin nx cos nx* in the second equation in (4.11), we obtain

$$\begin{aligned}&u_1\left[\frac{\partial^2 u_1}{\partial x\partial y}-n\frac{\partial v_1}{\partial y}+\int\left(\frac{\partial^3 u_1}{\partial x^3}-3n\frac{\partial^2 v_1}{\partial x^2}-3n^2\frac{\partial u_1}{\partial x}+n^3 v_1\right)dy\right]\\&-\left(\frac{\partial^2 u_1}{\partial x^2}+\frac{\partial^2 u_1}{\partial y^2}-2n\frac{\partial v_1}{\partial x}-n^2 u_1\right)\int\left(\frac{\partial u_1}{\partial x}-nv_1\right)dy\\&-v_1\left[\frac{\partial^2 v_1}{\partial x\partial y}+n\frac{\partial u_1}{\partial y}+\int\left(\frac{\partial^3 v_1}{\partial x^3}+3n\frac{\partial^2 u_1}{\partial x^2}-3n^2\frac{\partial v_1}{\partial x}-n^3 u_1\right)dy\right]\\&+\left(\frac{\partial^2 v_1}{\partial x^2}+\frac{\partial^2 v_1}{\partial y^2}+2n\frac{\partial u_1}{\partial x}-n^2 v_1\right)\int\left(\frac{\partial v_1}{\partial x}+nu_1\right)dy=0,\end{aligned} \tag{4.17a}$$

$$
\begin{aligned}
& u_1\left[\frac{\partial^2 v_1}{\partial x \partial y}+n\frac{\partial u_1}{\partial y}+\int\left(\frac{\partial^3 v_1}{\partial x^3}+3n\frac{\partial^2 u_1}{\partial x^2}-3n^2\frac{\partial v_1}{\partial x}-n^3 u_1\right)dy\right] \\
& +v_1\left[\frac{\partial^2 u_1}{\partial x \partial y}-n\frac{\partial v_1}{\partial y}+\int\left(\frac{\partial^3 u_1}{\partial x^3}-3n\frac{\partial^2 v_1}{\partial x^2}-3n^2\frac{\partial u_1}{\partial x}+n^3 v_1\right)dy\right] \\
& -\left(\frac{\partial^2 v_1}{\partial x^2}+\frac{\partial^2 v_1}{\partial y^2}+2n\frac{\partial u_1}{\partial x}-n^2 v_1\right)\int\left(\frac{\partial u_1}{\partial x}-n v_1\right)dy \\
& -\left(\frac{\partial^2 u_1}{\partial x^2}+\frac{\partial^2 u_1}{\partial y^2}-2n\frac{\partial v_1}{\partial x}-n^2 u_1\right)\int\left(\frac{\partial v_1}{\partial x}+n u_1\right)dy=0.
\end{aligned}
\tag{4.17b}
$$

Here it must be noted that the right side of (4.16b) is different if we put $\sin^2 nx = 1 - \cos^2 nx$. On the other hand, from (4.16a) it can be seen that for these two cases the right sides are equal.

The *similarity variables* (4.4) can be also introduced into (4.14a)–(4.17b):

$$
u_0=\bar{u}f_0'(\eta),\quad v_0=\bar{u}\beta\varphi_0'(\eta),\quad u_1=\bar{u}\beta f'(\eta),\quad v_1=\bar{u}\beta\varphi'(\eta),\quad \beta=\frac{\bar{u}_1}{\bar{u}},
\tag{4.18}
$$

where $\bar{u}$ and $\bar{u}_1$ are characteristic velocities of the main flow and disturbances and β is the dimensionless amplitude of the disturbances.

The problem will be solved in an approximation of the laminar boundary layer theory, where γ is a small parameter:

$$
\gamma=\left(\frac{\delta}{x}\right)=Re^{-1/2},\quad \delta=\sqrt{\frac{\upsilon x}{\bar{u}}},\quad Re=\frac{\bar{u}x}{\upsilon},\quad A=\frac{n\delta}{\gamma}.
\tag{4.19}
$$

In this approximation $\gamma^2 = 0$ and the parameters of the problem are A and β. The parameter β will be considered as small and the problem will be solved in the zeroth approximation of β^2 ($\beta^2 = 0$).

The introduction of similarity variables (4.4) and the approximation ($\gamma^2 = 0$) into (4.14a) leads to the equations for f_0 and φ_0:

$$
2f_0'''+(f_0+\beta\varphi_0)f_0''=0,
\tag{4.20a}
$$

$$
2\varphi_0'''+(f_0+\beta\varphi_0)\varphi_0''=\beta[2A(f''\varphi'-f\varphi''')+\varphi'\varphi''+\varphi\varphi'''].
\tag{4.20b}
$$

The introduction of similarity variables (4.4) and the approximation ($\gamma^2 = 0$) into (4.15a) leads to the conditions

$$
2f_0'''+\frac{2f'}{f-2A\varphi}f_0''=0,\quad 2f_0'''+\frac{2\varphi'}{\varphi+2Af}f_0''=0.
\tag{4.21}
$$

From (4.21) to (4.20a) the following conditions are obtained:

$$
f_0+\beta\varphi_0=\frac{2f'}{f-2A\varphi}=\frac{2\varphi'}{\varphi+2Af}.
\tag{4.22}
$$

The introduction of (4.22) into (4.20a) leads to new equations for f_0 and φ_0:

$$2f_0''' + \frac{2f'}{f - 2A\varphi} f_0'' = 0,$$
$$2\varphi_0''' + \frac{2\varphi'}{\varphi + 2Af}\varphi_0'' = \beta[2A(f''\varphi' - f\varphi''') + \varphi'\varphi'' + \varphi\varphi'''].\tag{4.23}$$

The introduction of similarity variables (4.4) and the approximation ($\gamma^2 = 0$) into (4.16a) leads to the equations for f and φ:

$$\begin{aligned}&2f^{IV} + f_0'f'' + f_0f''' + 2Af_0'\varphi'' + \beta\left(f'\varphi_0'' + f''\varphi_0' + f'''\varphi_0 + f\varphi_0'''\right)\\&+ 2\beta A\left(\varphi_0'\varphi'' - \varphi'''\varphi_0\right)\\&= 0,\end{aligned}$$

$$\begin{aligned}&2\varphi^{IV} + f_0'\varphi'' + f_0\varphi''' - 2Af_0'f'' + \beta\left(\varphi'\varphi_0'' + \varphi''\varphi_0' + \varphi'''\varphi_0 + \varphi\varphi_0'''\right)\\&- 2\beta A\left(\varphi_0'f'' - f\varphi_0'''\right) = 0.\end{aligned}\tag{4.24}$$

In the approximations used ($\beta^2 = 0$ and $\gamma^2 = 0$) the equations in (4.17a) are eliminated.

The boundary conditions are obtained with the help of two conditions:

1. The boundary conditions for F and f_0 are equal.
2. The dependencies of the disturbances from the main flow are identical in the volume and at the interface.

From (4.5) to (4.7) it follows that

$$F(0) = a, \quad F'(0) = b, \quad F''(0) = c, \quad F'''(0) = -\frac{F''(0)F(0)}{2} = -\frac{ac}{2},$$

$$F^{IV}(0) = -\frac{bc}{2} + \frac{a^2c}{4}, \quad F^{V}(0) = -\frac{c^2}{2} + \frac{3abc}{4} - \frac{a^3c}{8},\tag{4.25}$$

where $F^{IV}(0)$ and $F^{V}(0)$ were obtained after double differentiation of (4.5).

From (4.20a) it follows that

$$\varphi_0 = -\frac{2f_0'''}{\beta f_0''} - \frac{f_0}{\beta}.\tag{4.26}$$

The double differentiation of (4.26) and the introduction of $\eta = 0$ into φ_0, $\varphi_0{}'$ и $\varphi_0{}''$ and use of (4.26) allows the determination of the boundary conditions in f_0 and φ_0:

$$f_0(0) = a, \quad f_0'(0) = b, \quad f_0'' = c; \quad \varphi_0(0) = 0, \quad \varphi_0'(0) = 0, \quad \varphi_0'' = 0.\tag{4.27}$$

From conditions (4.22), we directly obtain

$$2f' = (f_0 + \beta\varphi_0)(f - 2A\varphi), \quad 2\varphi' = (f_0 + \beta\varphi_0)(\varphi + 2Af).\tag{4.28}$$

If $\alpha_1 = f(0)$ and $\alpha_2 = \varphi(0)$, the boundary conditions for (4.24) after differentiation (4.28) and use of (4.27) can be determined from (4.28):

$$f(0) = \alpha_1, \quad f'(0) = \frac{a}{2}(\alpha_1 - 2A\alpha_2),$$
$$f''(0) = \frac{\alpha_1}{2}\left(b + \frac{a^2}{b} - 2A^2a^2\right) - A\alpha_2\left(a^2 + b\right),$$

$$f'''(0) = \frac{\alpha_1}{2}\left(c + \frac{3ab}{2} - 6abA^2 + \frac{1}{4}a^3 - 3a^3A^2\right) - \alpha_2 A\left(c + 3ab + \frac{3}{4}a^3 - a^3A^2\right); \tag{4.29a}$$

$$\varphi(0) = \alpha_2, \quad \varphi'(0) = \frac{a}{2}(\alpha_2 + 2A\alpha_1),$$
$$\varphi''(0) = A\alpha_1\left(a^2 + b\right) + \frac{\alpha_2}{2}\left(b + \frac{1}{2}a^2 - 2a^2A^2\right). \tag{4.29b}$$

The *parameter β might be determined* from the condition that the kinetic energy (E) of the main flow is distributed between the energy of the new main flow E_0 and the energy of the disturbance E_1 (see 4.9).

The kinetic energies are proportional to the squares of the velocities, which are integrated in the area of the boundary layer (s). In the case $\omega = 0$, one obtains

$$E \sim \iint_{(s)} \tilde{u}^2 dxdy, \quad E_0 = \iint_{(s)} u^2 dxdy, \quad E_1 = \iint_{(s)} u'^2 dxdy. \tag{4.30}$$

$$E \sim \int_0^{\lambda}\int_0^{\delta} \tilde{u}^2 dydx = \bar{u}^2\delta\lambda \int_0^6 F'^2 d\xi, \quad E_0 \sim \int_0^{\lambda}\int_0^{\delta} u_0^2 dydx = \bar{u}^2\delta\lambda \int_0^6 f_0'^2 d\xi,$$
$$E_1 \sim \int_0^{\lambda}\int_0^{\delta} (v_0 + u_1 \sin nx + v_2 \cos nx)^2 dydx = \bar{u}^2\delta\lambda\beta^2\left(\int_0^6 \varphi_0'^2 d\xi + \frac{1}{2}\int_0^6 f'^2 d\xi + \frac{1}{2}\int_0^6 \varphi'^2 d\xi\right). \tag{4.31}$$

The introduction of (4.31) into (4.9) allows the determination of β:

$$\beta = \sqrt{\frac{\int_0^6 F'^2 d\xi - \int_0^6 f_0'^2 d\xi}{\int_0^6 \varphi_0'^2 d\xi + \frac{1}{2}\int_0^6 f'^2 d\xi + \frac{1}{2}\int_0^6 \varphi'^2 d\xi}}. \tag{4.32}$$

Equations (4.23) and (4.24) with boundary conditions (4.27) and (4.29a) represent an *eigenvalue problem* where α_1, α_2, β, and A are eigenvalues. The amplitude of the disturbances (β) can be obtained using an iterative procedure if the normal velocity component (α_1, α_2) and the wavelength (A) of the disturbances are specified. The procedure starts from initial value β_0, followed by solution of the problem (4.23, 4.24, 4.27, 4.29a), and calculation of β from (4.32). The calculated value of β is an initial value for the next step of the iterative procedure.

Numerical results were obtained for mass transfer in a gas–solid system ($Sc = 1$, $b = 0$). As a result, three cases are possible:

1. A self-organized structure exists if the calculated value of β is in the interval $0 < \beta < 0.5$ (β must be a small parameter).
2. A self-organized structure does not exist if $\beta^2 \leq 0$.
3. A self-organized structure does not exist if the calculated value of β oscillates between two values ($\beta_1 < \beta < \beta_2$). For β_0 in the interval (β_1, β_2), calculated values of $\beta^2 < 0$ exist.

In the case of linear mass transfer ($\theta = 0$) the solution of the problem (4.23, 4.24, 4.27, 4.29a) shows that $\beta^2 < 0$. The same results were obtained in the cases of injection into the boundary layer as a result of the large mass flux ($\theta > 0$).

In the cases of suction from the boundary layer ($\theta < 0$) the nonlinear mass transfer effect stabilizes the flow. The dimensionless amplitude of the disturbances (β) depends on the wave number A (Table 5) and the intensity of the disturbance $\alpha_1 = \alpha_2 = \alpha$. For some wave numbers self-organized structures are not possible. The parameter β^2 is negative for large values of α.

The influence of α and A is shown in the Table 6 for different intensities of the nonlinear mass transfer effect ($\theta < 0$). The increase of the intensity of the disturbances at the interface (α) leads to an increase of the suction effect and as a result the disturbance amplitude decreases (stabilization of the flow).

The results obtained show that self-organized dissipative structures exist for $\theta < 0$, i.e., in the cases when the mass transfer is directed from the volume towards the interphase surface.

In gas–liquid systems $\theta < 0$ is equivalent to gas absorption and many experimental data show [69] that the mass transfer rate in gas absorption is larger than the desorption rate.

Theoretical analysis of the gas absorption between gas and liquid immobile layers shows [58, 65, 66] that absorption rate is larger than that predicted by the linear mass transfer theory as a result of self-organized dissipative structures. The desorption rate is equivalent to that predicted by the linear mass transfer theory.

Table 5 Amplitude values for different wave numbers

A ($\theta = -0.2$, $a = 0.1557$)	β ($\alpha = 1$, $b = 0$, $c = 0.3892$)
0.217	0.015888
0.218	–[a]
0.219	0.029576
0.220	0.033220
0.221	0.041680
0.222	0.049298
0.223	0.055280
0.224	0.055725
0.225	0.057790

[a] $\beta_1 < \beta < \beta_2$

Table 6 Amplitude values for different intensities of the disturbances and the nonlinear mass transfer effect

α	A	β		A	β		A	β	
		$\theta = -0.2$	$\theta = -0.3$		$\theta = -0.2$	$\theta = -0.3$		$\theta = -0.2$	$\theta = -0.3$
0.1	0.217	–[a]	0.5599	0.222	–[b]	0.7305	0.226	–[b]	0.8609
0.2		0.0824	0.2799		0.2481	0.3653		–[b]	0.4304
0.4		0.0397	0.13961		0.1232	0.1826		0.1594	0.2152
0.5		0.0318	0.1117		0.0986	0.1461		0.1275	0.1722
0.9		0.0176	0.0620		0.0548	0.0812		0.0736	0.0956
1.0		0.0159	0.0558		0.0493	0.0731		0.0662	0.0861
3.0		0.0053	0.0186		0.0164	0.0244		0.0221	0.0287
0.1	0.219	0.3208	0.6333	0.223	0.5108	0.7512	0.227	0.6901	–[b]
0.2		0.1533	0.3166		0.2764	0.3756		0.3450	–[b]
0.4		0.0739	0.1583		0.1382	0.1878		0.1652	–[b]
0.5		0.0591	0.1266		0.1106	0.1502		0.1321	0.1772
0.9		0.0328	0.0704		0.0614	0.0835		0.0768	–[b]
1.0		0.0296	0.0633		0.0553	0.0751		0.0691	–[b]
3.0		00098	0.0211		–[a]	0.0249		0.0230	–[b]
0.1	0.220	0.2027	0.6613	0.224	0.5108	0.7952	0.228	0.7153	0.9208
0.2		0.1876	0.3307		0.2764	0.3976		0.3576	0.4604
0.4		0.0831	0.1653		0.1382	0.1988		0.1772	0.2302
0.5		0.0665	0.1323		0.1106	0.1590		0.1418	0.1842
0.9		0.0369	0.0735		0.0614	0.0884		0.0796	0.1018
1.0		0.0332	0.0661		0.0553	0.0795		0.0716	0.0921
3.0		0.0111	0.0220		–[a]	0.0271		0.0239	0.0309
0.1	0.221	0.4419	0.6992	0.225	0.5779	0.8349	0.230	–[b]	–[b]
0.2		0.1981	0.3496		0.2889	0.4175		–[b]	–[b]
0.4		0.1042	0.1748		0.1445	0.2087		–[b]	–[b]
0.5		0.0834	0.1398		0.1155	0.1670		–[b]	–[b]
0.9		0.0463	0.0777		0.0642	0.0928		–[b]	–[b]
1.0		0.0417	0.0699		0.0578	0.0835		0.0688	–[b]
3.0		0.0139	0.0233		0.0193	0.0278		0.0229	0.0311

[a] $\beta^2 < 0$
[b] $\beta_1 < \beta < \beta_2$

The existence of self-organized dissipative structures in laminar boundary layer flow [72] for $\theta < 0$ creates the condition for the mass transfer rate increasing when the critical Reynolds number increases. In this case the self-organized dissipative structure is supercritical bifurcation and intensifies the mass transfer processes.

4.2 Gas Absorption

Experimental studies of systems with intensive mass transfer show in many cases serious deviations from the linear theory of mass transfer, which presumes independence of the velocity field from the fields of concentration and temperature.

These effects are usually considered as Marangoni-type effect [18, 20, 49] and are explained by the occurrence of tangential secondary flow on the phase boundary, caused by the surface tension gradient as a result of surface gradients of concentration or (and) temperature on the mass transfer surface [56, 57].

Theoretical studies of systems with intensive interphase mass transfer [69] as a result of large concentration gradients show [14, 23, 24, 31, 36] that these gradients induce normal secondary flows on the phase boundary. On this basis the nonlinear theory of mass transfer [33] was built and provides a satisfactory explanation for the deviations of the experimental results from the predictions of the linear theory of mass transfer.

The above-mentioned results illustrate the possibility for a simultaneous or an independent role of two mechanisms of heat mass transfer in systems with intensive mass transfer. This needs the definition of the conditions for the occurrence of the Marangoni effect and effect of nonlinear mass transfer, which would allow the creation of adequate models of chemical engineering processes under the condition of intensive interphase mass transfer between two phases.

The linear analysis of the hydrodynamic stability in systems with intensive interphase mass transfer was given in [16, 34, 35, 40, 69] The normal and the tangential components of the velocity on the phase boundary influence the hydrodynamic stability of the flows in the boundary layer, but the influence of the normally directed velocity component is significantly greater. This suggests that a considerable difference in the intensities of the Marangoni effect and the effect of nonlinear mass transfer is possible.

The comparative analysis of these two effects was made in the cases of mass transfer between two phases (gas–liquid and liquid–liquid) [28, 43], when a substance from the first phase goes into the second phase and the chemical reaction takes place. The large concentration gradients of the transferred substance create a normally directed secondary flow on the phase boundary. On the other hand, the thermal effect of the chemical reaction creates a gradient of the surface tension as a result of the temperature inhomogeneity at the phase interface.

The theoretical results show [28, 43] that the Marangoni effect is negligible compared with the effect of the nonlinear mass transfer, i.e., the kinetics of mass transfer and the hydrodynamic stability do not depend on the surface tension gradient, caused by the temperature inhomogeneity at the phase interface as a result of the thermal effect of the chemical reaction. It should be noted, however, that the parameter representing the Marangoni effect increases with decrease of the characteristic velocity in the second phase. The results obtained [28] show in the case of an immovable phase that under these conditions the Marangoni effect is also considerably smaller. Thus, the Marangoni effect may be expected in the *limited case of interphase mass transfer between two immovable phases (e.g., in the absorption of pure gases in an immovable liquid). Under these conditions three processes are likely to take place: natural convection, nonlinear mass transfer, and the Marangoni effect.*

Actually, these three effects may exhibit a dual influence on the *mechanism and kinetics of the heat and mass transfer in systems with intensive interphase mass transfer*. The first influence is relatively weak and it is a result of the secondary flows, which change the velocity field. However, this may lead to changes in the hydrodynamic stability and therefore to the creation of self-organizing dissipative structures, having the form of stable periodic flows, which have a very strong influence on the mechanism and the kinetics of mass transfer. These two types of influences will be discussed next.

Let us consider a vertical tube with radius r_0, in which an immovable liquid (H_2O) contacts an immovable gas (CO_2, SO_2, NH_3) [48]. The gas is absorbed in the liquid, and the process is accompanied by a thermal effect. As a result, several effects in the liquid may occur in the form of secondary flows owing to the large concentration gradients on the phase boundary (nonlinear mass transfer), a density gradient in the volume (natural convection), and a surface tension gradient (Marangoni effect) [48, 58].

The mathematical description of the process uses the *Oberbeck–Boussinesq-type equations* [6, 8, 33], where the influence of the density gradient [50, 51], the concentration gradient [39], and the surface tension gradient will be considered [52–54]. In this way, in cylindrical coordinates the problem assumes the form

$$\rho\left(\frac{\partial v_z}{\partial t}+v_z\frac{\partial v_z}{\partial z}+v_r\frac{\partial v_z}{\partial r}+\frac{v_\varphi}{r}\frac{\partial v_z}{\partial \varphi}\right)=-\frac{\partial p}{\partial z}+\mu\left(\frac{\partial^2 v_z}{\partial z^2}+\frac{1}{r}\frac{\partial v_z}{\partial r}+\frac{\partial^2 v_z}{\partial r^2}+\frac{1}{r^2}\frac{\partial^2 v_z}{\partial \varphi^2}\right)+g(\rho-\rho_0),$$

$$\rho\left(\frac{\partial v_r}{\partial t}+v_z\frac{\partial v_r}{\partial z}+v_r\frac{\partial v_r}{\partial r}+\frac{v_\varphi}{r}\frac{\partial v_r}{\partial \varphi}-\frac{v_\varphi^2}{r}\right)=-\frac{\partial p}{\partial r}+\mu\left(\frac{\partial^2 v_r}{\partial z^2}+\frac{1}{r}\frac{\partial v_r}{\partial r}+\frac{\partial^2 v_r}{\partial r^2}-\frac{v_r}{r^2}+\frac{1}{r^2}\frac{\partial^2 v_r}{\partial \varphi^2}-\frac{2}{r^2}\frac{\partial v_\varphi}{\partial \varphi}\right),$$

$$\rho\left(\frac{\partial v_\varphi}{\partial t}+v_z\frac{\partial v_\varphi}{\partial z}+v_r\frac{\partial v_\varphi}{\partial r}+\frac{v_\varphi}{r}\frac{\partial v_\varphi}{\partial \varphi}+\frac{v_r v_\varphi}{r}\right)=-\frac{1}{r}\frac{\partial p}{\partial \varphi}+\mu\left(\frac{\partial^2 v_\varphi}{\partial z^2}+\frac{1}{r}\frac{\partial v_\varphi}{\partial r}+\frac{\partial^2 v_\varphi}{\partial r^2}-\frac{v_\varphi}{r^2}+\frac{1}{r^2}\frac{\partial^2 v_\varphi}{\partial \varphi^2}+\frac{2}{r^2}\frac{\partial v_r}{\partial \varphi}\right),$$

$$\frac{\partial \rho}{\partial t}+\frac{\partial(\rho v_z)}{\partial z}+\frac{\partial(\rho v_r)}{\partial r}+\frac{\rho v_r}{r}+\frac{1}{r}\frac{\partial(\rho v_\varphi)}{\partial \varphi}=0,$$

$$\frac{\partial c}{\partial t}+v_z\frac{\partial c}{\partial z}+v_r\frac{\partial c}{\partial r}+\frac{v_\varphi}{r}\frac{\partial c}{\partial \varphi}=D\left(\frac{\partial^2 c}{\partial z^2}+\frac{1}{r}\frac{\partial c}{\partial r}+\frac{\partial^2 c}{\partial r^2}+\frac{1}{r^2}\frac{\partial^2 c}{\partial \varphi^2}\right),$$

$$\rho c_p\left(\frac{\partial \theta}{\partial t}+v_z\frac{\partial \theta}{\partial z}+v_r\frac{\partial \theta}{\partial r}+\frac{v_\varphi}{r}\frac{\partial \theta}{\partial r}\right)=\lambda\left(\frac{\partial^2 \theta}{\partial z^2}+\frac{1}{r}\frac{\partial \theta}{\partial r}+\frac{\partial^2 \theta}{\partial r^2}+\frac{1}{r^2}\frac{\partial^2 \theta}{\partial \varphi^2}\right),$$

$$\rho=\rho_0\left[1+\frac{c}{\rho_0}-\beta(\theta-\theta_0)\right], \tag{4.33}$$

with the corresponding initial and boundary conditions,

$$t = 0, \quad v_z = v_r = v_\varphi = c = 0, \quad \theta = \theta_0;$$

$$z = 0, \quad v_z = -\frac{D}{\rho_0}\frac{\partial c}{\partial z}, \quad \mu\left(\frac{\partial v_r}{\partial z} + \frac{\partial v_z}{\partial r}\right) = \frac{\partial \sigma}{\partial r} = \frac{\partial \sigma}{\partial \theta}\frac{\partial \theta}{\partial r},$$

$$\mu\left(\frac{\partial v_\varphi}{\partial z} + \frac{1}{r}\frac{\partial v_z}{\partial \varphi}\right) = \frac{1}{r}\frac{\partial \sigma}{\partial \varphi} = \frac{1}{r}\frac{\partial \sigma}{\partial \theta}\frac{\partial \theta}{\partial \varphi}, \quad c = c^*, \quad \lambda\frac{\partial \theta}{\partial z} = qD\frac{\partial c}{\partial z};$$

$$z \to \infty, \quad v_z = v_r = v_\varphi = c = 0, \quad \theta = \theta_0;$$

$$r = 0, \quad v_z, v_r, v_\varphi, c, \theta - \text{finite}; \quad r = r_0, \quad v_z = v_r = v_\varphi = 0, \quad \frac{\partial c}{\partial r} = \frac{\partial \theta}{\partial r} = 0. \tag{4.34}$$

Let us assume that for angle φ the processes are periodic ones with period 2π.

Equations (4.33) and (4.34) refer to the natural convection by means of the Archimedean force $g(\rho - \rho_0)$, the large concentration gradients through the connection between the velocity v_z and the concentration gradient $\frac{\partial c}{\partial z}$, and the surface tension (σ) gradient by means of its components on r and φ and their connection with the tangential components of the stress tensor at the surface $z = 0$.

The problem (4.33, 3.34) may be represented in a dimensionless form if the individual scales of the physical independent and dependent variables are used. These characteristic scales should be selected in such a way that the values of the dimensionless variables and parameters are not greater in order of magnitude than unity.

The characteristic scales may be set in advance and for the example discussed they are of the following order for the time, radial coordinate, concentration, and temperature:

$$t_0 \sim 10^2\text{s}, \quad r_0 \sim 10^{-2}\text{m}, \quad c^* \sim (1 - 100)\text{kg/m}^3 \text{ (for different gases)}, \quad \theta_0 \sim 10\,^\circ\text{C}. \tag{4.35}$$

The *characteristic scales* will be known in advance or will be unknown and will be determined as a result of the qualitative analysis of the model (4.33, 4.34). If the characteristic scale of the velocity along the z-axis is indicated by u_0, then the scales of the other velocity components are determined in a way so that the equation of continuity is satisfied in dimensionless variables, and for a characteristic scale of pressure the dynamic pressure $\rho_0 u_0^2$ is used.

The difference in the orders of magnitude of μ, D, and λ shows that the basic changes of the velocity, the concentration, and the temperature will be reached at different water depths in the tube. These characteristic depths for the velocity (l), concentration (δ), and temperature (h) will be determined by the qualitative analysis of (4.33).

Using the above-mentioned considerations, we obtain the following dimensionless variables:

$$t = t_0 T, \quad z = lZ_1 = \delta Z_2 = hZ_3, \quad r = r_0 R, \quad \varphi = 2\pi\Phi, \quad p = \rho_0 u_0^2 P,$$

$$v_z(t,z,r,\varphi) = u_0 V_z(T,Z_1,R,\Phi) = u_0 \tilde{V}_z(T,Z_2,R,\Phi) = u_0 \tilde{\tilde{V}}_z(T,Z_3,R,\Phi),$$

$$v_r(t,z,r,\varphi) = \frac{u_0 r_0}{l} V_r(T,Z_1,R,\Phi) = \frac{u_0 r_0}{l} \tilde{V}_r(T,Z_2,R,\Phi) = \frac{u_0 r_0}{l} \tilde{\tilde{V}}_r(T,Z_3,R,\Phi),$$

$$\begin{aligned} v_\varphi(t,z,r,\varphi) &= 2\pi \frac{u_0 r_0}{l} V_\varphi(T,Z_1,R,\Phi) = 2\pi \frac{u_0 r_0}{l} \tilde{V}_\varphi(T,Z_2,R,\Phi) \\ &= 2\pi \frac{u_0 r_0}{l} \tilde{\tilde{V}}_\varphi(T,Z_3,R,\Phi), \end{aligned}$$

$$c(t,z,r,\varphi) = c^* C(T,Z_1,R,\Phi) = c^* \tilde{C}(T,Z_2,R,\Phi) = c^* \tilde{\tilde{C}}(T,Z_3,R,\Phi),$$

$$\theta(t,z,r,\varphi) = \theta_0 \Theta(T,Z_1,R,\Phi) = \theta_0 \tilde{\Theta}(T,Z_2,R,\Phi) = \theta_0 \tilde{\tilde{\Theta}}(T,Z_3,R,\Phi). \quad (4.36)$$

The introduction of (4.36) into (4.33, 4.34) converts the problem to a dimensionless form, where the dimensional characteristic parameters (scales) are grouped in such a way that the dimensionless parameters obtained are on the order of unity, less than unity (10^{-1}), and many times less than unity (10^{-2} or much less):

$$\begin{aligned} &\left[1 + \frac{c^*}{\rho_0} C - \beta\theta_0(\Theta - 1)\right] \left[\frac{u_0 \rho_0}{g t_0 c^*} \frac{\partial V_z}{\partial T} + \frac{u_0^2 \rho_0}{g l c^*} \left(V_z \frac{\partial V_z}{\partial Z_1} + V_r \frac{\partial V_z}{\partial R} + \frac{V_\varphi}{R} \frac{\partial V_z}{\partial \Phi}\right)\right] \\ &= -\frac{u_0^2 \rho_0}{g l c^*} \frac{\partial P}{\partial Z_1} + \frac{\mu u_0}{g l^2 c^*} \left[\frac{\partial^2 V_z}{\partial Z_1^2} + \frac{l^2}{r_0^2} \left(\frac{1}{R} \frac{\partial V_z}{\partial R} + \frac{\partial^2 V_z}{\partial R^2} + \frac{1}{4\pi^2} \frac{1}{R^2} \frac{\partial^2 V_z}{\partial \Phi^2}\right)\right] + C - \frac{\rho_0 \beta \theta_0}{c^*} (\Theta - 1), \end{aligned} \quad (4.37)$$

$$\begin{aligned} &\left[1 + \frac{c^*}{\rho_0} C - \beta\theta_0(\Theta - 1)\right] \left[\frac{u_0 \rho_0}{g t_0 c^*} \frac{\partial V_r}{\partial T} + \frac{u_0^2 \rho_0}{g l c^*} \left(V_z \frac{\partial V_r}{\partial Z_1} + V_r \frac{\partial V_r}{\partial R} + \frac{v_\varphi}{R} \frac{\partial V_r}{\partial \Phi} - 4\pi^2 \frac{V_\varphi^2}{R}\right)\right] \\ &= -\frac{l^2}{r_0^2} \frac{u_0^2 \rho_0}{g l c^*} \frac{\partial P}{\partial R} + \frac{\mu u_0}{g l^2 c^*} \left[\frac{\partial^2 V_r}{\partial Z_1^2} + \frac{l^2}{r_0^2} \left(\frac{1}{R} \frac{\partial V_r}{\partial R} + \frac{\partial^2 V_r}{\partial R^2} - \frac{V_r}{R^2} + \frac{1}{4\pi^2} \frac{\partial^2 V_r}{\partial \Phi^2} - \frac{2}{R} \frac{\partial V_\varphi}{\partial \Phi}\right)\right], \end{aligned} \quad (4.38)$$

$$\begin{aligned} &\left[1 + \frac{c^*}{\rho_0} C - \beta\theta_0(\Theta - 1)\right] \left[\frac{u_0 \rho_0}{g t_0 c^*} \frac{\partial V_\varphi}{\partial T} + \frac{u_0^2 \rho_0}{g l c^*} \left(V_z \frac{\partial V_\varphi}{\partial Z_1} + V_r \frac{\partial V_\varphi}{\partial R} + \frac{V_\varphi}{R} \frac{\partial V_\varphi}{\partial \Phi} + \frac{V_r V_\varphi}{R}\right)\right] \\ &= -\frac{l^2}{r_0^2} \frac{u_0^2 \rho_0}{g l c^*} \frac{1}{R} \frac{\partial P}{\partial \Phi} + \frac{\mu u_0}{g l^2 c^*} \left[\frac{\partial^2 V_\varphi}{\partial Z_1^2} + \frac{l^2}{r_0^2} \left(\frac{1}{R} \frac{\partial V_\varphi}{\partial R} + \frac{\partial^2 V_\varphi}{\partial R^2} - \frac{V_\varphi}{R^2} + \frac{1}{4\pi^2} \frac{1}{R^2} \frac{\partial^2 V_\varphi}{\partial \Phi^2} + \frac{1}{2\pi^2} \frac{1}{R^2} \frac{\partial V_r}{\partial \Phi}\right)\right], \end{aligned} \quad (4.39)$$

$$\frac{lc^*}{\rho_0 u_0 t_0}\frac{\partial C}{\partial T} - \frac{\beta\theta_0 l}{t_0 u_0}\frac{\partial \Theta}{\partial T} + \left(\frac{\partial V_z}{\partial Z_1} + \frac{V_r}{R} + \frac{\partial V_r}{\partial R} + \frac{1}{R}\frac{\partial V_\varphi}{\partial \Phi}\right)\left[1 + \frac{c^*}{\rho_0}C - \beta\theta_0(\Theta - 1)\right]$$
$$+ V_z\left(\frac{c^*}{\rho_0}\frac{\partial C}{\partial Z_1} - \beta\theta_0\frac{\partial \Theta}{\partial Z_1}\right) + V_r\left(\frac{c^*}{\rho_0}\frac{\partial C}{\partial R} - \beta\theta_0\frac{\partial \Theta}{\partial R}\right) + \frac{V_\varphi}{R}\left(\frac{c^*}{\rho_0}\frac{\partial C}{\partial \Phi} - \beta\theta_0\frac{\partial \Theta}{\partial \Phi}\right) = 0, \tag{4.40}$$

$$\frac{\partial \tilde{C}}{\partial T} + \frac{u_0 t_0}{\delta}\left(\tilde{V}_z\frac{\partial \tilde{C}}{\partial Z_2} + \tilde{V}_r\frac{\partial \tilde{C}}{\partial R} + \frac{\tilde{V}_r}{R}\frac{\partial \tilde{C}}{\partial \Phi}\right) = \frac{Dt_0}{\delta^2}\left[\frac{\partial^2 \tilde{C}}{\partial Z_2^2} + \frac{\delta^2}{r_0^2}\left(\frac{1}{R}\frac{\partial \tilde{C}}{\partial R} + \frac{\partial^2 \tilde{C}}{\partial R^2} + \frac{1}{4\pi^2}\frac{1}{R^2}\frac{\partial^2 \tilde{C}}{\partial \Phi^2}\right)\right], \tag{4.41}$$

$$\left[1 + \frac{c^*}{\rho_0}\tilde{\tilde{C}} - \beta\theta_0\left(\tilde{\tilde{\Theta}} - 1\right)\right]\left[\frac{\partial \tilde{\tilde{\Theta}}}{\partial T} + \frac{u_0 t_0}{h}\left(\tilde{\tilde{V}}_z\frac{\partial \tilde{\tilde{\Theta}}}{\partial Z_3} + \tilde{\tilde{V}}_r\frac{\partial \tilde{\tilde{\Theta}}}{\partial R} + \frac{\tilde{\tilde{V}}_\varphi}{R}\frac{\partial \tilde{\tilde{\Theta}}}{\partial \Phi}\right)\right]$$
$$= \frac{at_0}{h^2}\left[\frac{\partial^2 \tilde{\tilde{\Theta}}}{\partial Z_3^2} + \frac{h^2}{r_0^2}\left(\frac{1}{R}\frac{\partial \tilde{\tilde{\Theta}}}{\partial R} + \frac{\partial^2 \tilde{\tilde{\Theta}}}{\partial R^2} + \frac{1}{4\pi^2}\frac{1}{R^2}\frac{\partial^2 \tilde{\tilde{\Theta}}}{\partial \Phi^2}\right)\right]; \tag{4.42}$$

$$T = 0, \quad V_z = V_r = V_\varphi = \tilde{C} = 0, \quad \tilde{\tilde{\Theta}} = 1; \tag{4.43}$$

$$Z_1 = Z_2 = Z_3 = 0, \quad V_z = -\frac{Dc^*}{u_0\rho_0\delta}\frac{\partial \tilde{C}}{\partial Z_2}, \quad \frac{\partial V_r}{\partial Z_1} + \frac{l^2}{r_0^2}\frac{\partial V_z}{\partial R} = \frac{\partial \sigma}{\partial \theta}\frac{\theta_0}{\mu u_0}\frac{l^2}{r_0^2}\frac{\partial \tilde{\tilde{\Theta}}}{\partial R},$$

$$\frac{\partial V_\varphi}{\partial Z_1} + \frac{l^2}{4\pi^2 r_0^2}\frac{1}{R}\frac{\partial V_z}{\partial \Phi} = \frac{\partial \sigma}{\partial \theta}\frac{\theta_0}{\mu u_0}\frac{l^2}{4\pi^2 r_0^2}\frac{1}{R}\frac{\partial \tilde{\tilde{\Theta}}}{\partial \Phi},$$

$$\tilde{C} = 1, \quad \frac{\partial \tilde{\tilde{\Theta}}}{\partial Z_3} = \frac{qDhc^*}{\delta\lambda\theta_0}\left(1 + \frac{c^*}{\rho_0}\right)\frac{\partial \tilde{C}}{\partial Z_2}; \tag{4.44}$$

$$Z_1 = Z_2 = Z_3 \to \infty, \quad V_z = V_r = V_\varphi = \tilde{C} = 0, \quad \tilde{\tilde{\Theta}} = 1; \tag{4.45}$$

$$R = 0, \quad V_z, V_r, V_\varphi, \tilde{P}, \tilde{C}, \tilde{\Theta} - \text{finite}; \tag{4.46}$$

$$R = 1, \quad V_z = V_r = V_\varphi = 0, \quad \frac{\partial \tilde{C}}{\partial R} = \frac{\partial \tilde{\tilde{\Theta}}}{\partial R} = 0. \tag{4.47}$$

The qualitative analysis of the model (4.37–4.47) begins with the determination of the unknown characteristic scales. The process under consideration is a result of the absorption of the gas and its thermal effect, i.e., the fields of concentration and temperature are determined by the diffusion and the heat transfer. It directly follows that the parameters in front of the Laplacians in (4.41) and (4.42) should be of the order of magnitude of unity:

$$\frac{Dt_0}{\delta^2} = 1, \quad \frac{at_0}{h^2} = 1, \tag{4.48}$$

which makes the determination of the characteristic linear scales of δ and h possible,

$$\delta = \sqrt{Dt_0} \sim 10^{-4}\mathrm{m}, \quad h = \sqrt{at_0} \sim 10^{-3}\mathrm{m}. \tag{4.49}$$

As a result of the diffusion and the heat transfer, conditions for a natural convection arise, the influence of which on the velocity field appears when the parameter in front of the Laplacian in (4.37)–(4.39) is of the order of magnitude of unity (viscous flow):

$$\frac{\mu u_0}{g l^2 c^*} = 1. \tag{4.50}$$

From (4.50) it is obvious that it is not necessary to determine the characteristic velocity of the flow, which depends on the limitation process. Natural convection cannot limit the velocity, because for diffusion and heat transfer in a stagnant liquid [6, 8, 32] there is a mechanical equilibrium ($v_z = v_r = v_\varphi \equiv 0$) and the natural convection is only a result of the loss of stability.

The large concentration gradients induce a secondary flow, the characteristic velocity of which may be determined if the parameter of the nonlinear mass transfer in (4.44) is of the order of magnitude of unity:

$$\frac{Dc^*}{u_0 \rho_0 \delta} = 1. \tag{4.51}$$

In this way the characteristic scales l and u_0 are obtained directly from (4.50) to (4.51):

$$u_0 = \frac{c^*}{\rho_0}\sqrt{\frac{D}{t_0}} \sim 10^{-7}\mathrm{m/s}, \quad l = \sqrt{\frac{\mu}{\rho_0 g}\sqrt{\frac{D}{t_0}}} \sim 10^{-7}\mathrm{m}. \tag{4.52}$$

The attempt to define the characteristic velocity from the Marangoni effect, i.e., from the condition

$$\frac{\partial \sigma}{\partial \theta}\frac{\theta_0}{\mu u_0}\frac{l^2}{r_0^2} = 1, \tag{4.53}$$

is not successful because there is no value of u_0 that can satisfy both (4.50) and (4.53). The natural convection and the Marangoni effect can arise simultaneously in cases where the characteristic radius is very small:

$$r_0 = \sqrt{\frac{\partial \sigma}{\partial \theta}\frac{\theta_0}{\rho_0 \varepsilon g}} \sim 10^{-3}\mathrm{m}. \tag{4.54}$$

The characteristic scales determined in this way in (4.49) and (4.52) allow the determination of the order of magnitude of the parameters in (4.37)–(4.45):

$$\varepsilon = \frac{c^*}{\rho_0} \sim 10^{-1}, \quad \frac{u_0\rho_0}{gt_0c^*} \sim 10^{-9}, \quad \frac{u_0^2\rho_0}{glc^*} \sim 10^{-7}, \quad \frac{l^2}{r_0^2} \sim 10^{-10},$$

$$\frac{lc^*}{\rho_0 u_0 t_0} \sim 10^{-3}, \quad \frac{u_0 t_0}{\delta} = \varepsilon \sim 10^{-1}, \quad \frac{\delta^2}{r_0^2} \sim 10^{-4}, \quad \frac{h^2}{r_0^2} \sim 10^{-2},$$

$$\frac{u_0 t_0}{h} \sim 10^{-2}, \quad \frac{\partial\sigma}{\partial\theta}\frac{\theta_0}{\mu\, u_0}\frac{l^2}{r_0^2} \sim 10^{-3}, \quad \alpha = \frac{qDh\rho_0}{\delta\lambda\theta_0} \sim 1,$$

$$\beta\theta_0 \sim 10^{-3}, \quad \frac{\partial\sigma}{\partial\theta} \sim 10^{-5}\mathrm{kg/m^2\,{}^\circ C}, \quad \frac{\rho_0\beta\theta_0}{c^*} \sim 10^{-2}, \quad \frac{\beta\theta_0 l\rho_0}{t_0\mu} \sim 10^{-5}. \quad (4.55)$$

The dimensionless parameters determined in this way in the model (4.37–4.47) are not of an order of magnitude greater than unity, which is a necessary condition for validity of the result from the qualitative analysis.

From (4.55) it is evident that the parameter of the Marangoni effect (4.50) is of order of magnitude 10^{-3} and does not influence the velocity, the concentration, and the temperature field. Similar results have been obtained [52–54] in the analysis of the simultaneous influence of the natural convection and the Marangoni effect in cases of a fixed thickness of the water layer, greater than 10^{-3} m. This result shows that under the condition of intensive mass transfer the natural convection and the nonlinear mass transfer lead to a flow whose characteristic velocity is two orders greater than the velocity at which the Marangoni effect may occur.

Another significant result of (4.55) is that $\beta\theta_0 \ll \frac{c^*}{\rho_0} < 1$, i.e., the temperature change does not influence the density ρ, and further we will assume $\beta = 0$. This result explains the original approximation in the *Oberbeck–Boussinesq equations* [6, 8, 33], where (see 4.33) $\beta = 0$ in the expression for ρ and $\beta \neq 0$ in the Archimedean force $g(\rho - \rho_0)$.

The different effects in the complex process take place when their corresponding parameters are greater than 10^{-2}, i.e., the problem (4.37–4.47) may be expressed in a zeroth-order approximation regarding the parameters of order lower than 10^{-2} (and smaller). In this way from (4.37)–(4.47) to (4.55) it follows that

$$\frac{\partial^2 V_z}{\partial Z_1^2} + C = 0, \quad \frac{\partial^2 V_r}{\partial Z_1^2} = 0, \quad \frac{\partial^2 V_\varphi}{\partial Z_1^2} = 0,$$

$$(1+\varepsilon C)\left(\frac{\partial V_z}{\partial Z_1} + \frac{V_r}{R} + \frac{\partial V_r}{\partial R} + \frac{1}{R}\frac{\partial V_\varphi}{\partial \Phi}\right) + \varepsilon\left(V_z\frac{\partial C}{\partial Z_1} + V_r\frac{\partial C}{\partial R} + \frac{1}{R}V_\varphi\frac{\partial C}{\partial \varphi}\right) = 0,$$

$$\frac{\partial \tilde{C}}{\partial T} + \varepsilon\left(\tilde{V}_z\frac{\partial \tilde{C}}{\partial Z_2} + \tilde{V}_r\frac{\partial \tilde{C}}{\partial R} + \frac{\tilde{V}_\varphi}{R}\frac{\partial \tilde{C}}{\partial \Phi}\right) = \frac{\partial^2 \tilde{C}}{\partial Z_2^2}, \quad \left(1+\varepsilon\tilde{\tilde{C}}\right)\frac{\partial \tilde{\tilde{\Theta}}}{\partial T} = \frac{\partial^2 \tilde{\tilde{\Theta}}}{\partial Z_3^2};$$

$$T = 0, \quad \tilde{C} = 0, \quad \tilde{\tilde{\Theta}} = 1;$$

$$Z_1 = Z_2 = Z_3 = 0, \quad V_z = -\frac{\partial \tilde{C}}{\partial Z_2}, \quad \frac{\partial V_r}{\partial Z_1} = 0, \quad \frac{\partial V_\varphi}{\partial Z_1} = 0, \quad \tilde{C} = 1,$$
$$\frac{\partial \tilde{\tilde{\Theta}}}{\partial Z_3} = \varepsilon \alpha_1 (1+\varepsilon) \frac{\partial \tilde{C}}{\partial Z_2};$$

$$Z_1 = Z_2 = Z_3 \to \infty, \quad V_z = V_r = V_\varphi = \tilde{C} = 0, \quad \tilde{\tilde{\Theta}} = 1; \quad R = 0, \quad V_r = 0. \tag{4.56}$$

The solution of (4.56) depends on two parameters (α, ε), where ε is a small parameter, and the solutions may be expressed in an expansion of ε in the form

$$V_z = V_z^{(0)} + \varepsilon V_z^{(1)} + \cdots, \quad V_r = V_r^{(0)} + \varepsilon V_r^{(1)} + \cdots, \quad V_\varphi = V_\varphi^{(0)} + \varepsilon V_\varphi^{(1)} + \cdots,$$

$$\tilde{C} = \tilde{C}^{(0)} + \varepsilon \tilde{C}^{(1)} + \cdots, \quad \tilde{\tilde{\Theta}} = \tilde{\tilde{\Theta}}^{(0)} + \varepsilon \tilde{\tilde{\Theta}}^{(1)} + \cdots \tag{4.57}$$

The introduction of (4.57) into (4.56) allows us to find the zeroth-order approximations $\varepsilon = 0$:

$$\frac{\partial^2 V_z^{(0)}}{\partial Z_1^2} + C^{(0)} = 0; \quad Z_1 = 0, \quad V_z^{(0)} = -\left(\frac{\partial \tilde{C}^{(0)}}{\partial Z_2}\right)_{Z_2=0}; \quad Z_1 \to \infty, \quad V_z^{(0)} = 0. \tag{4.58}$$

$$\frac{\partial^2 V_\varphi^{(0)}}{\partial Z_1^2} = 0; \quad Z_1 = 0, \quad \frac{\partial V_\varphi^{(0)}}{\partial Z_1} = 0; \quad Z_1 \to \infty, \quad V_\varphi^{(0)} = 0. \tag{4.59}$$

$$\frac{\partial V_r^{(0)}}{\partial R} + \frac{V_r^{(0)}}{R} = -\frac{\partial V_z^{(0)}}{\partial Z_1} - \frac{1}{R}\frac{\partial V_\varphi^{(0)}}{\partial \Phi}; \quad R = 0, \quad V_r^{(0)} = 0 \quad \text{(finite)}. \tag{4.60}$$

$$\frac{\partial \tilde{C}^{(0)}}{\partial T} = \frac{\partial^2 \tilde{C}^{(0)}}{\partial Z_2^2}; \quad T = 0, \quad \tilde{C}^{(0)} = 0; \quad Z_2 = 0,$$
$$\tilde{C}^{(0)} = 1; \quad Z_2 \to \infty, \quad \tilde{C}^{(0)} = 0. \tag{4.61}$$

$$\frac{\partial \tilde{\tilde{\Theta}}^{(0)}}{\partial T} = \frac{\partial^2 \tilde{\tilde{\Theta}}^{(0)}}{\partial Z_3^2}; \quad T = 0, \quad \tilde{\tilde{\Theta}}^{0} = 1; \quad Z_3 = 0, \quad \frac{\partial \tilde{\tilde{\Theta}}^{(0)}}{\partial Z_3} = 0; \quad Z_3 \to \infty,$$
$$\tilde{\tilde{\Theta}}^{(0)} = 1. \tag{4.62}$$

The solutions of (4.59), (4.61), and (4.62) are obtained directly:

$$V_\varphi^{(0)} = 0, \quad \tilde{C}^{(0)} = erfc \frac{Z_2}{2\sqrt{T}}, \quad \tilde{\tilde{\Theta}}^{(0)} \equiv 1. \tag{4.63}$$

From (4.36), (4.49), and (4.52) it is clear that

$$C^{(0)} = erfc\left(\frac{l}{\delta}\frac{Z_1}{2\sqrt{T}}\right) \approx 1, \quad \alpha_0 = \frac{l}{\delta} \sim 10^{-3}. \tag{4.64}$$

The introduction of (4.64) into (4.58) allows the determination of $V_z^{(0)}$, replacing the infinity condition with $V_z(1) = 0$:

$$V_z^{(0)} = -\frac{1}{2}Z_1^2 + \left(\frac{1}{2} - \frac{1}{\sqrt{\pi T}}\right)Z_1 + \frac{1}{\sqrt{\pi T}}. \tag{4.65}$$

Substitution of (4.63) and (4.65) into (4.60) leads to

$$V_r^{(0)} = \left[\frac{1}{2}Z_1 + \frac{1}{2\sqrt{\pi T}} - \frac{1}{4}\right]R. \tag{4.66}$$

The problem for determination of the first approximation of the concentration $\tilde{C}^{(1)}$ is of the type

$$\begin{aligned} &\frac{\partial \tilde{C}^{(1)}}{\partial T} = \frac{\partial^2 \tilde{C}^{(1)}}{\partial Z_2^2} - \tilde{V}_z^{(0)}\frac{\partial \tilde{C}^{(0)}}{\partial Z_2}; \quad T = 0, \quad \tilde{C}^{(1)} = 0; \quad Z_2 = 0, \quad \tilde{C}^{(1)} = 0; \\ &\quad Z_2 \to \infty, \quad \tilde{C}^{(1)} = 0, \end{aligned} \tag{4.67}$$

where

$$\tilde{V}_z^{(0)}(Z_2, T) = V_z^{(0)}(Z_1, T), \quad Z_1 = \frac{1}{\alpha_0}Z_2. \tag{4.68}$$

From (4.68) it directly follows that the volume source in (4.67) is

$$\tilde{V}_z^{(0)}\frac{\partial \tilde{C}^{(0)}}{\partial Z_2} \neq 0 \quad \text{for} \quad 0 \le Z_2 < \alpha_0 \sim 10^{-3}, \tag{4.69}$$

i.e., its influence on the mass transfer is practically confined to the interface ($Z_2 = 0$) and may be replaced by a surface flow as>

$$S = \int_0^{\alpha} \tilde{V}_z^0 \frac{\partial \tilde{C}^{(0)}}{\partial Z_2} dZ_2. \tag{4.70}$$

Thus, problem (4.67) takes the form

$$\begin{aligned} &\frac{\partial \tilde{C}^{(1)}}{\partial T} = \frac{\partial^2 \tilde{C}^{(1)}}{\partial Z_2^2}; \quad T = 0, \quad \tilde{C}^{(1)} = 0; \quad Z_2 = 0, \quad \frac{\partial \tilde{C}^{(1)}}{\partial Z_2} = -S; \quad Z_2 \to \infty, \\ &\quad \tilde{C}^{(1)} = 0; \end{aligned}$$

$$S = -\sqrt{\frac{T}{\pi}}\frac{\alpha_0 e^{-\frac{\alpha^2}{4T}} - \sqrt{\pi} erf\frac{\alpha}{2\sqrt{T}}}{\alpha_0^2} - \left(\sqrt{\frac{T}{\pi}} - \frac{2}{\pi}\right)\frac{e^{-\frac{\alpha^2}{4T}} - 1}{\alpha_0} - \frac{1}{\sqrt{\pi T}} erf\frac{\alpha_0}{2T}. \quad (4.71)$$

From (4.71) is clear that for small values of α_0 ($\alpha_0 \sim 10^{-3}$), $S \approx 0$; therefore,

$$\tilde{C}^{(1)} \equiv 0. \quad (4.72)$$

It is not difficult to show that

$$V_z^{(1)} \equiv 0, \quad V_r^{(1)} \equiv 0, \quad V_\varphi^{(1)} \equiv 0, \quad \tilde{\tilde{\Theta}}^{(1)} \equiv 0. \quad (4.73)$$

The average absorption rate J (per unit interface) for a time interval t_0 may be expressed by means of the mass transfer coefficient k. It may be determined from the average mass flux I:

$$J = kc^* = \frac{1}{\pi r_0^2 t_0}\int_0^{t_0} I\,dt, \quad I = \pi r_0^2 i, \quad i = -\frac{D\rho^*}{\rho_0}\left(\frac{\partial c}{\partial z}\right)_{z=0}, \quad \rho^* = \rho_0 + c^*,$$
$$\frac{\rho^*}{\rho_0} = 1 + \varepsilon. \quad (4.74)$$

Thus, Eq. 4.74 may be used to obtain the Sherwood number for nonstationary diffusion:

$$Sh = \frac{kl}{D} = -(1+\varepsilon)\sqrt{\frac{\nu}{gt_0}}\sqrt{\frac{1}{Dt_0}}\int_0^1 \left(\frac{\partial\tilde{C}}{\partial Z_2}\right)_{Z_2=0} dT, \quad (4.75)$$

e.g.,

$$Sh = 2(1+\varepsilon)\sqrt{\frac{\nu}{\pi g t_0}}\sqrt{\frac{1}{Dt_0}}. \quad (4.76)$$

The amount of the gas absorbed Q (kg/m^2) for the time interval t_0 (s) is

$$Q = \frac{1}{\pi r_0^2}\int_0^{t_0} Idt = 2(1+\varepsilon)\,c^*\sqrt{\frac{Dt_0}{\pi}}. \quad (4.77)$$

The results reported (4.63, 4.65, 4.66) show that the temperature has practically a constant value and does not influence the fields of velocity and concentration. These results differ from the solution of the Benard problem [32, 50, 51], where $V_z = V_r = V_\varphi = 0$, because the effect of the nonlinear mass transfer does not allow the existence of a mechanical equilibrium, where the liquid may remain stagnant.

The experimental results from the absorption of CO_2 in an immobile layer of water [55] show that the rate of the absorption is significantly greater than that which can be determined from (4.77). This fact indicates that the nonstationary process that is described by Eqs. 4.63, 4.65, and 4.66 (analogous to the Benard problem) is unstable regarding small periodic disturbances. Their increase may lead to new periodic flows with constant amplitude, which will evidently change the mechanism and the kinetics of mass and heat transfer.

The results obtained show that in cases of absorption of pure gases in a cylindrical liquid column, a second flow is induced as a result of a natural convection and a nonlinear mass transfer. Under these conditions the Marangoni effect is negligible and for the velocity, temperature, and concentration the following expressions in dimension form have been obtained:

$$v_z = \varepsilon\left[-\frac{g}{2\nu}z^2 + \left(\frac{1}{2} - \sqrt{\frac{t_0}{\pi t}}\right)\sqrt{\frac{g}{\nu}}\sqrt{\frac{D}{t_0}}z + \sqrt{\frac{D}{\pi t}}\right],$$

$$v_{\mathrm{r}} = \varepsilon\left[+\frac{g}{2\nu}z - \left(\frac{1}{4} - \frac{1}{2}\sqrt{\frac{t_0}{\pi t}}\right)\sqrt{\frac{g}{\nu}}\sqrt{\frac{D}{t_0}}\right]r,$$

$$v_\varphi \equiv 0, \quad p \equiv 0, \quad c^* = erfc\frac{z}{2\sqrt{Dt}}, \quad \theta \equiv \theta_0, \quad \varepsilon = \frac{c^*}{\rho_0}, \quad \nu = \frac{\mu}{\rho_0}. \tag{4.78}$$

These results differ significantly from the Benard problem [50, 51], where under certain conditions a mechanical equilibrium ($v_z = v_r = v_\varphi = 0$) is possible. The reason for this difference is the nonlinear mass transfer, i.e., the large mass flux induces a secondary flow on the phase boundary,

$$z = 0, \quad v_z = \sqrt{\frac{D}{\pi t}}, \tag{4.79}$$

and in this way violates the necessary condition for a mechanical equilibrium [8, 32].

The process described by expressions (4.78) as may be expected, analogous to the Benard problem, is unstable regarding small disturbances, which makes the usage of the *linear stability analysis* [64, 66] possible.

A process represented as a superposition of the basic process (4.78) and small disturbances in the velocity (v_z', v_r', v_φ'), pressure (p'), concentration (c'), and temperature (θ') will be considered:

$$v_z + v_z', \quad v_r + v_r', \quad v_\varphi + v_\varphi', \quad p + p', \quad c + c', \quad \theta + \theta'. \tag{4.80}$$

This new process should satisfy (as well as the basic one) the Oberbeck–Boussinesq equations (4.33, 4.34). Introducing (4.78) and (4.80) into Eqs. 4.33 and 4.34, we obtain a system of equations concerning v_z', v_r', v_φ', p', c', and θ'. We will analyze the process in a linearized form with regard to these small disturbances:

$$\left(1+\frac{c}{\rho_0}\right)\left(\frac{\partial v_z'}{\partial t}+v_z'\frac{\partial v_z}{\partial z}+v_z\frac{\partial v_z'}{\partial z}+v_r\frac{\partial v_z'}{\partial r}\right)$$
$$=-\frac{1}{\rho_0}\frac{\partial p'}{\partial z}+\nu\left(\frac{\partial^2 v_z'}{\partial z^2}+\frac{1}{r}\frac{\partial v_z'}{\partial r}+\frac{\partial^2 v_z'}{\partial r^2}+\frac{1}{r^2}\frac{\partial^2 v_z'}{\partial \varphi^2}\right)+\frac{g}{\rho_0}c',$$

$$\left(1+\frac{c}{\rho_0}\right)\left(\frac{\partial v_r'}{\partial t}+v_z'\frac{\partial v_r}{\partial z}+v_z\frac{\partial v_r'}{\partial z}+v_r'\frac{\partial v_r}{\partial r}+v_r\frac{\partial v_r'}{\partial r}\right)=-\frac{1}{\rho_0}\frac{\partial p'}{\partial r}$$
$$+\nu\left(\frac{\partial^2 v_r'}{\partial z^2}+\frac{1}{r}\frac{\partial v_r'}{\partial r}+\frac{\partial^2 v_r'}{\partial r^2}-\frac{v_r'}{r^2}+\frac{1}{r^2}\frac{\partial^2 v_r'}{\partial \varphi^2}-\frac{2}{r^2}\frac{\partial v_\varphi'}{\partial \varphi}\right),$$

$$\left(1+\frac{c}{\rho_0}\right)\left(\frac{\partial v_\varphi'}{\partial t}+v_z\frac{\partial v_\varphi'}{\partial z}+v_r\frac{\partial v_\varphi'}{\partial r}+\frac{1}{r}v_r v_\varphi'\right)=-\frac{1}{\rho_0 r}\frac{\partial p'}{\partial \varphi}$$
$$+\nu\left(\frac{\partial^2 v_\varphi'}{\partial z^2}+\frac{1}{r}\frac{\partial v_\varphi'}{\partial r}+\frac{\partial^2 v_\varphi'}{\partial r^2}-\frac{v_\varphi'}{r^2}+\frac{1}{r^2}\frac{\partial^2 v_\varphi'}{\partial \varphi^2}+\frac{2}{r^2}\frac{\partial v_r'}{\partial \varphi}\right); \quad (4.81)$$

$$\frac{\partial c'}{\partial t}+(\rho_0+c)\left(\frac{\partial v_z'}{\partial z}+\frac{\partial v_r'}{\partial r}+\frac{v_r'}{r}+\frac{1}{r}\frac{\partial v_\varphi'}{\partial \varphi}\right)+v_z'\frac{\partial c}{\partial z}+v_z\frac{\partial c'}{\partial z}+v_r\frac{\partial c'}{\partial r}=0; \quad (4.82)$$

$$\frac{\partial c'}{\partial t}+v_z'\frac{\partial c}{\partial z}+v_z\frac{\partial c'}{\partial z}+v_r\frac{\partial c'}{\partial r}=D\left(\frac{\partial^2 c'}{\partial z^2}+\frac{1}{r}\frac{\partial c'}{\partial r}+\frac{\partial^2 c'}{\partial r^2}+\frac{1}{r^2}\frac{\partial^2 c'}{\partial \varphi^2}\right),$$

$$\left(1+\frac{c}{\rho_0}\right)\left(\frac{\partial \theta'}{\partial t}+v_z\frac{\partial \theta'}{\partial z}+v_r\frac{\partial \theta'}{\partial r}\right)=a\left(\frac{\partial^2 \theta'}{\partial z^2}+\frac{1}{r}\frac{\partial \theta'}{\partial r}+\frac{\partial^2 \theta'}{\partial r^2}+\frac{1}{r^2}\frac{\partial^2 \theta'}{\partial \varphi^2}\right); \quad (4.83)$$

$$t=0, \quad v_z'=v_r'=v_\varphi'=c'=\theta'=0;$$

$$z=0, \quad v_z'=-\frac{D}{\rho_0}\frac{\partial c'}{\partial z}, \quad \mu\left(\frac{\partial v_r'}{\partial z}+\frac{\partial v_z'}{\partial r}\right)=\frac{\partial \sigma}{\partial \theta}\frac{\partial \theta'}{\partial r},$$

$$\mu\left(\frac{\partial v_\varphi'}{\partial z}+\frac{1}{r}\frac{\partial v_z'}{\partial \varphi}\right)=\frac{1}{r}\frac{\partial \sigma}{\partial \theta}\frac{\partial \theta'}{\partial \varphi}, \quad c'=0, \quad \lambda\frac{\partial \theta'}{\partial z}=qD\frac{\partial c'}{\partial z};$$

$$z\to\infty, \quad v_z'=v_r'=v_\varphi'=c'=\theta'=0; \quad r=0, \quad v_z', v_r', v_\varphi', c', p', \theta' \ \text{– finite};$$

$$r=r_0, \quad v_z'=v_r'=v_\varphi'=0, \quad \frac{\partial c'}{\partial r}=\frac{\partial \theta'}{\partial r}=0. \quad (4.84)$$

Equations (4.82) are obtained from the equation of continuity using the condition $\beta \ll 1$. Boundary conditions for the pressure are not used because it will be eliminated in Eq. 4.81. Boundary conditions regarding the coordinate φ are not included because periodic disturbances regarding φ will be discussed.

The set of Eqs. 4.81–4.84 have partial solutions ("normal" disturbances) which depend exponentially on time:

$$v'_z = \bar{v}_z(t,z,r,\varphi)\exp(-\omega t), \quad p' = \bar{p}(t,z,r,\varphi)\exp(-\omega t),$$

$$v'_z = \bar{v}_r(t,z,r,\varphi)\exp(-\omega t), \quad c' = \bar{c}(t,z,r,\varphi)\exp(-\omega t),$$

$$v'_\varphi = \bar{v}_\varphi(t,z,r,\varphi)\exp(-\omega t), \quad \theta' = \bar{\theta}(t,z,r,\varphi)\exp(-\omega t), \tag{4.85}$$

where the pre-exponential parts depend on time because the basic process (4.78) is nonstationary. The disturbances presented in this way decrease or increase with time, depending on the value of ω, and for $\omega = 0$ the disturbances are "neutral," i.e., a process which neither slows down nor intensifies with time. The mathematical description of this process is obtained from (4.81) to (4.84) after introducing (4.85) and $\omega = 0$:

$$\left(1+\frac{c}{\rho_0}\right)\left(\frac{\partial \bar{v}_z}{\partial t}+\bar{v}_z\frac{\partial v_z}{\partial z}+v_z\frac{\partial \bar{v}_z}{\partial z}+v_r\frac{\partial \bar{v}_z}{\partial r}\right)$$
$$= -\frac{1}{\rho_0}\frac{\partial \bar{p}}{\partial z}+\nu\left(\frac{\partial^2 \bar{v}_z}{\partial z^2}+\frac{1}{r}\frac{\partial \bar{v}_z}{\partial r}+\frac{\partial^2 \bar{v}_z}{\partial r^2}+\frac{1}{r^2}\frac{\partial^2 \bar{v}_z}{\partial \varphi^2}\right)+\frac{g}{\rho_0}\bar{c},$$

$$\left(1+\frac{c}{\rho_0}\right)\left(\frac{\partial \bar{v}_r}{\partial t}+\bar{v}_z\frac{\partial v_r}{\partial z}+v_z\frac{\partial \bar{v}_r}{\partial z}+\bar{v}_r\frac{\partial v_r}{\partial r}+v_r\frac{\partial \bar{v}_r}{\partial r}\right) = -\frac{1}{\rho_0}\frac{\partial \bar{p}}{\partial r}$$
$$+\nu\left(\frac{\partial^2 \bar{v}_r}{\partial z^2}+\frac{1}{r}\frac{\partial \bar{v}_r}{\partial r}+\frac{\partial^2 \bar{v}_r}{\partial r^2}-\frac{\bar{v}_r}{r^2}+\frac{1}{r^2}\frac{\partial^2 \bar{v}_r}{\partial \varphi^2}-\frac{2}{r^2}\frac{\partial \bar{v}_\varphi}{\partial \varphi}\right),$$

$$\left(1+\frac{c}{\rho_0}\right)\left(\frac{\partial \bar{v}_\varphi}{\partial t}+v_z\frac{\partial \bar{v}_\varphi}{\partial z}+v_r\frac{\partial \bar{v}_\varphi}{\partial r}+\frac{1}{r}v_r\bar{v}_\varphi\right) = -\frac{1}{\rho_0 r}\frac{\partial \bar{p}}{\partial \varphi}$$
$$+\nu\left(\frac{\partial^2 \bar{v}_\varphi}{\partial z^2}+\frac{1}{r}\frac{\partial \bar{v}_\varphi}{\partial r}+\frac{\partial^2 \bar{v}_\varphi}{\partial r^2}-\frac{\bar{v}_\varphi}{r^2}+\frac{1}{r^2}\frac{\partial^2 \bar{v}_\varphi}{\partial \varphi^2}+\frac{2}{r^2}\frac{\partial \bar{v}_r}{\partial \varphi}\right); \tag{4.86}$$

$$\frac{\partial \bar{c}}{\partial t}+(\rho_0+c)\left(\frac{\partial \bar{v}_z}{\partial z}+\frac{\partial \bar{v}_r}{\partial r}+\frac{\bar{v}_r}{r}+\frac{1}{r}\frac{\partial \bar{v}_\varphi}{\partial \varphi}\right)+\bar{v}_z\frac{\partial c}{\partial z}+v_z\frac{\partial \bar{c}}{\partial z}+v_r\frac{\partial \bar{c}}{\partial r} = 0; \tag{4.87}$$

$$\frac{\partial \bar{c}}{\partial t}+\bar{v}_z\frac{\partial c}{\partial z}+v_z\frac{\partial \bar{c}}{\partial z}+v_r\frac{\partial \bar{c}}{\partial r} = D\left(\frac{\partial^2 \bar{c}}{\partial z^2}+\frac{1}{r}\frac{\partial \bar{c}}{\partial r}+\frac{\partial^2 \bar{c}}{\partial r^2}+\frac{1}{r^2}\frac{\partial^2 \bar{c}}{\partial \varphi^2}\right),$$

$$\left(1+\frac{c}{\rho_0}\right)\left(\frac{\partial \bar{\theta}}{\partial t}+v_z\frac{\partial \bar{\theta}}{\partial z}+v_r\frac{\partial \bar{\theta}}{\partial r}\right) = a\left(\frac{\partial^2 \bar{\theta}}{\partial z^2}+\frac{1}{r}\frac{\partial \bar{\theta}}{\partial r}+\frac{\partial^2 \bar{\theta}}{\partial r^2}+\frac{1}{r^2}\frac{\partial^2 \bar{\theta}}{\partial \varphi^2}\right); \tag{4.88}$$

$$z = 0, \quad \bar{v}_z = -\frac{D}{\rho_0}\frac{\partial \bar{c}}{\partial z}, \quad \mu\left(\frac{\partial \bar{v}_r}{\partial z}+\frac{\partial \bar{v}_z}{\partial r}\right) = \frac{\partial \sigma}{\partial \theta}\frac{\partial \bar{\theta}}{\partial r},$$

$$\mu\left(\frac{\partial \bar{v}_\varphi}{\partial z}+\frac{1}{r}\frac{\partial \bar{v}_z}{\partial \varphi}\right) = \frac{1}{r}\frac{\partial \sigma}{\partial \theta}\frac{\partial \bar{\theta}}{\partial \varphi}, \quad \bar{c} = 0, \quad \lambda\frac{\partial \bar{\theta}}{\partial z} = qD\frac{\partial \bar{c}}{\partial z};$$

$$z \to \infty, \quad \bar{v}_z = \bar{v}_r = \bar{v}_\varphi = \bar{c} = \bar{\theta} = 0; \quad r = 0, \quad \bar{v}_z, \bar{v}_r, \bar{v}_\varphi, \bar{p}, \bar{c}, \bar{\theta} - \text{finite};$$

$$r = r_0, \quad \bar{v}_z = \bar{v}_r = \bar{v}_\varphi = 0, \quad \frac{\partial \bar{c}}{\partial r} = \frac{\partial \bar{\theta}}{\partial r} = 0. \tag{4.89}$$

Problem (4.86)–(4.89) obviously has partial solutions for which the velocity, the concentration, and the temperature harmonically depend on φ, i.e., the following range of neutral disturbances may be introduced into (4.86)–(4.89):

$$\bar{v}_z = \sum_{n=0}^{\infty} v_n(t, z, r) \cos(n\varphi), \quad \bar{v}_r = \bar{v}_\varphi = 0, \quad \bar{p} = \sum_{n=0}^{\infty} p_n(t, z, r) \cos(n\varphi),$$

$$\bar{c} = \sum_{n=0}^{\infty} c_n(t, z, r) \cos(n\varphi), \quad \bar{\theta} = \sum_{n=0}^{\infty} \theta_n(t, z, r) \cos(n\varphi). \tag{4.90}$$

Introducing (4.90) into (4.86)–(4.89), we obtain the following eigenvalue problem:

$$(1+\varepsilon)\left(\frac{\partial v_n}{\partial t} + v_n \frac{\partial v_z}{\partial z} + v_z \frac{\partial v_n}{\partial z} + v_r \frac{\partial v_n}{\partial r}\right)$$
$$= -\frac{1}{\rho_0}\frac{\partial p_n}{\partial z} + \nu\left(\frac{\partial^2 v_n}{\partial z^2} + \frac{1}{r}\frac{\partial v_n}{\partial r} + \frac{\partial^2 v_n}{\partial r^2} - \frac{n^2}{r^2} v_n\right) + \frac{g}{\rho_0} c_n,$$

$$(1+\varepsilon)\frac{\partial v_r}{\partial z} v_n = -\frac{1}{\rho_0}\frac{\partial p_n}{\partial r}, \quad \frac{\partial p_n}{\partial \varphi} = 0; \tag{4.91}$$

$$\frac{\partial c_n}{\partial t} + \rho_0(1+\varepsilon)\frac{\partial v_n}{\partial z} + v_n \frac{\partial c}{\partial z} + v_z \frac{\partial c_n}{\partial z} + v_r \frac{\partial c_n}{\partial r} = 0; \tag{4.92}$$

$$\frac{\partial c_n}{\partial t} + v_n \frac{\partial c}{\partial z} + v_z \frac{\partial c_n}{\partial z} + v_r \frac{\partial c_n}{\partial r} = D\left(\frac{\partial^2 c_n}{\partial z^2} + \frac{1}{r}\frac{\partial c_n}{\partial r} + \frac{\partial^2 c_n}{\partial r^2} - \frac{n^2}{r^2} c_n\right),$$

$$(1+\varepsilon)\left(\frac{\partial \theta_n}{\partial t} + v_z \frac{\partial \theta_n}{\partial z} + v_r \frac{\partial \theta_n}{\partial r}\right) = a\left(\frac{\partial^2 \theta_n}{\partial z^2} + \frac{1}{r}\frac{\partial \theta_n}{\partial r} + \frac{\partial^2 \theta_n}{\partial r^2} - \frac{n^2}{r^2}\theta_n\right); \tag{4.93}$$

$$z = 0, \quad v_n = -\frac{D}{\rho_0}\frac{\partial c_n}{\partial z}, \quad c_n = 0, \quad \lambda \frac{\partial \theta_n}{\partial z} = qD\frac{\partial c_n}{\partial z}; \quad z \to \infty,$$
$$v_n = c_n = \theta_n = 0;$$

$$r = 0, \quad v_n, c_n, \theta_n \text{ — finite}; \quad r = r_0, \quad v_n = 0, \quad \frac{\partial c_n}{\partial r} = \frac{\partial \theta_n}{\partial r} = 0; \tag{4.94}$$
$$n = 0, 1, 2, \ldots, \infty.$$

In (4.91)–(4.93) $c = c^*$ is accepted because the thickness of the velocity change layer is many times less than the thickness of the concentration change layer.

The pressure in (4.91) may be eliminated if the second equation is integrated regarding r and then differentiated in regard to z:

$$(1+\varepsilon)\int \frac{\partial v_r}{\partial z}\frac{\partial v_n}{\partial z}dr = -\frac{1}{\rho_0}\frac{\partial p_n}{\partial z}, \quad n = 0,1,2,\ldots,\infty. \tag{4.95}$$

The introduction of (4.95) into (4.91) and (4.92) into (4.93) leads to the final form of the equations for determination of the "neutral" velocity, concentration, and temperature disturbances:

$$(1+\varepsilon)\left(\frac{\partial v_n}{\partial t} + v_n\frac{\partial v_z}{\partial z} + v_z\frac{\partial v_n}{\partial z} + v_r\frac{\partial v_n}{\partial r}\right)$$
$$= (1+\varepsilon)\int \frac{\partial v_r}{\partial z}\frac{\partial v_n}{\partial z}dr + \nu\left(\frac{\partial^2 v_n}{\partial z^2} + \frac{1}{r}\frac{\partial v_n}{\partial r} + \frac{\partial^2 v_n}{\partial r^2} - \frac{n^2}{r^2}v_n\right) + \frac{g}{\rho_0}c_n;$$

$$\nu\left(\frac{\partial^2 v_n}{\partial z^2} + \frac{1}{r}\frac{\partial v_n}{\partial r} + \frac{\partial^2 v_n}{\partial r^2} - \frac{n^2}{r^2}v_n\right) + \frac{g}{\rho_0}c_n;$$

$$-(1+\varepsilon)\frac{\partial v_n}{\partial z} = \frac{D}{\rho_0}\left(\frac{\partial^2 c_n}{\partial z^2} + \frac{\partial^2 c_n}{\partial r^2} + \frac{1}{r}\frac{\partial c_n}{\partial r} - \frac{n^2}{r^2}c_n\right);$$

$$\frac{\partial^2 \theta_n}{\partial z^2} + \frac{\partial^2 \theta_n}{\partial r^2} + \frac{1}{r}\frac{\partial \theta_n}{\partial r} - \frac{n^2}{r^2}\theta_n = 0; \quad n = 0,1,2,\ldots,\infty \tag{4.96}$$

with boundary conditions

$$z = 0, \quad v_n = -\frac{D}{\rho_0}\frac{\partial c_n}{\partial z}, \quad c_n = 0, \quad \frac{\partial \theta_n}{\partial z} = \frac{qD}{\lambda}\frac{\partial c_n}{\partial z}; \quad z \to \infty,$$
$$v_n = c_n = \theta_n - 0;$$

$$r = 0, \quad v_n, c_n, \theta_n - \text{finite}; \quad r = r_0, \quad v_n = 0, \quad \frac{\partial c_n}{\partial r} = \frac{\partial \theta_n}{\partial r} = 0; \tag{4.97}$$
$$n = 0,1,2,\ldots,\infty.$$

Problem (4.96) will be solved by introducing the dimensionless variables and a partial separation of the variables:

$$t = t_0 T, \quad z = lZ, \quad r = r_0 R, \quad v_n = u_0[V_n(Z,T) - Bf_n(R)],$$

$$c_n = c^*[C_n(Z,T) + Zf_n(R)], \quad \theta_n = \frac{qDc^*}{\lambda}f_n(R), \quad B = \frac{D\varepsilon}{u_0 l}, \quad n = 0,1,2,\ldots,\infty, \tag{4.98}$$

where the dependence of the disturbances on the coordinates is supposed to be analogous to the basic process (4.78) for small values of z.

The introduction of (4.98) into (4.96) and (4.97) leads to

$$(1+\varepsilon)\left\{\frac{u_0}{\varepsilon t_0 g}\frac{\partial V_n}{\partial T}+\frac{u_0^2}{\varepsilon\,\mathrm{lg}}\left[(V_n-Bf_n)\frac{\partial V_z}{\partial Z}+V_z\frac{\partial V_n}{\partial Z}-BV_rf_n\right]\right\}$$
$$=(1+\varepsilon)\frac{u_0^2r_0^2}{\varepsilon l^3 g}\int\frac{\partial V_r}{\partial Z}\frac{\partial V_n}{\partial Z}dR+\frac{\nu u_0}{\varepsilon l^2 g}\frac{\partial^2 V_n}{\partial Z^2}-\frac{\nu u_0}{\varepsilon r_0^2 g}V_n\frac{n^2}{R^2}+C_n+Zf_n,$$

$$f_n''+\frac{1}{R}f_n'-\frac{n^2}{R^2}f_n=0,$$

$$-(1+\varepsilon)\frac{u_0 l}{\varepsilon D}\frac{\partial V_n}{\partial Z}=\frac{\partial^2 C_n}{\partial Z^2}-\frac{l^2}{r^2}\frac{n^2}{R^2}C_n;\quad n=0,1,2,\ldots,\infty; \tag{4.99}$$

with boundary conditions

$$Z=0,\quad V_n=-B\frac{\partial C_n}{\partial Z},\quad C_n=0;\quad R=0,\ f_n-\text{finite};\quad R=1,\ f_n'=0;$$
$$n=0,1,2,\ldots,\infty, \tag{4.100}$$

where

$$u_0=\varepsilon\sqrt{\frac{D}{t_0}}\sim 10^{-7}\mathrm{m/s},\quad l=\sqrt{\frac{\nu}{g}\sqrt{\frac{D}{t_0}}}\sim 10^{-7}\mathrm{m},\quad \frac{u_0}{\varepsilon g t_0}\sim 10^{-9},\quad \frac{u_0^2}{\varepsilon g t_0}\sim 10^{-7},$$
$$\frac{u_0^2 r_0^2}{\varepsilon g l^3}\sim 10^3,$$

$$\frac{\nu u_0}{\varepsilon g l^2}\sim 1,\quad \frac{\nu u_0}{\varepsilon g r_0^2}\sim 10^{-9},\quad \frac{u_0 l}{\varepsilon D}\sim 10^{-8},\quad \varepsilon\sim 10^{-1}. \tag{4.101}$$

The solution of the Euler equation in (4.99) is obtained through Green's functions [5], searching for the eigenvalues and the eigenfunctions for $n = 0, 1, 2,\ldots,\infty$:

$$f_0=\text{const.};\quad f_n=\frac{\xi^n+\xi^{-n}}{2n}R^n,\quad R<\xi;\quad f_n=\frac{\xi^n}{2n}(R^n+R^{-n}),\quad R>\xi;$$

$$f_n=\frac{\xi^{2n}}{2n}+\frac{1}{2n},\quad R=\xi,\quad 0<\xi<1,\quad n=1,2,\ldots,\infty, \tag{4.102}$$

where the eigenvalue ξ is a parameter that cannot be determined in the approximations of the linear stability theory.

Having in mind the order of the dimensionless variables in (4.101), one can directly obtain from (4.99)

$$\frac{\partial^2 C_n}{\partial Z^2}=0;\quad Z=0,\quad C_n=0;\quad C_n=\gamma_n Z;$$

$$\frac{\partial V_n}{\partial Z} = 0; \quad Z = 0, \quad V_n = -B\frac{\partial C_n}{\partial Z}; \quad V_n = -B\gamma_n; \quad n = 0, 1, 2, \ldots, \infty, \tag{4.103}$$

where the eigenvalue $\gamma_n < 0$ cannot be determined in the approximations of the linear stability analysis.

The solutions obtained (4.78, 4.102, 4.103) allow us to produce the final expressions for the velocity, the concentration, and the temperature:

$$v_z = \frac{c^*}{\rho_0}\left\{-\frac{g}{2\nu}z^2 + \left(\frac{1}{2} - \sqrt{\frac{t_0}{\pi t}}\right)\sqrt{\frac{g}{\nu}\sqrt{\frac{D}{t_0}}}z + \sqrt{\frac{D}{\pi t}} - \sqrt{\frac{gD}{\nu}\sqrt{Dt_0}}\left[\gamma + \sum_{n=1}^{\infty}(\gamma_n + f_n)\cos n\varphi\right]\right\},$$

$$c = c^*\left\{erfc\frac{z}{2\sqrt{Dt}} + z\sqrt{\frac{g}{\nu}\sqrt{\frac{t_0}{D}}}\left[\gamma + \sum_{n=1}^{\infty}(\gamma_n + f_n)\cos n\varphi\right]\right\},$$

$$\theta = \theta_0 + \frac{qc^*D}{\lambda}\left[f_0 + \sum_{n=1}^{\infty}f_n\left(\frac{r}{r_0}\right)\cos n\varphi\right], \quad \gamma = \gamma_0 + f_0. \tag{4.104}$$

From (4.104) is possible to determine the mass flow at a given moment:

$$i = -\frac{D\rho^*}{\rho_0}\left(\frac{\partial c}{\partial z}\right)_{z=0} = \frac{D\rho^* c^*}{\rho_0}\left\{\frac{1}{\sqrt{\pi Dt}} - \sqrt{\frac{g}{\nu}\sqrt{\frac{t_0}{D}}}\left[\gamma + \sum_{n=1}^{\infty}(\gamma_n + f_n)\cos n\varphi\right]\right\},$$

$$\rho^* = \rho_0 + c^*, \quad f_n = f_n\left(\frac{r}{r_0}\right), \quad n = 1, 2, \ldots, \infty. \tag{4.105}$$

The amount of absorbed substance that passes through the cross-sectional area is determined directly from (4.105), integrating over φ in the range (0, 2π), and having in mind that the integrals of the harmonic functions are equal to zero:

$$I = \pi r_0^2 c^*(1+\varepsilon)\left(\sqrt{\frac{D}{\pi t}} - \bar{\gamma}\right), \quad \bar{\gamma} = \gamma D\sqrt{\frac{g}{\nu}\sqrt{\frac{t_0}{D}}}. \tag{4.106}$$

From (4.106) the absorption rate (J), the Sherwood number (Sh), and the mass of the absorbed substance (Q) for a period of time t_0 through a unit surface are directly obtained:

$$J = kc^* = \frac{1}{\pi r_0^2 t_0}\int_0^{t_0} I dt = (1+\varepsilon)c^*\left(2\sqrt{\frac{D}{\pi t_0}} - \bar{\gamma}\right), \tag{4.107}$$

$$Sh = \frac{kl}{D} = (1+\varepsilon)\left(2\sqrt{\frac{\nu}{\pi g t_0}}\sqrt{\frac{1}{Dt_0}} - \gamma\right), \tag{4.108}$$

$$Q = \frac{1}{\pi r_0^2}\int_0^{t_0} I dt = (1+\varepsilon)c^*\left(2\sqrt{\frac{Dt_0}{\pi}} - \bar{\gamma} t_0\right), \quad \bar{\gamma} = \frac{\gamma D}{l}, \tag{4.109}$$

where k is the mass transfer coefficient of the nonstationary absorption.

In this way Eqs. 4.107–4.109 allow the determination of the absorption rate with an accuracy defined by the parameter $\bar{\gamma}$, the value of which cannot be determined in the approximations of the linear stability analysis. The parameter $\bar{\gamma}$ may be determined by introducing an additional physical condition, or from experimental data.

The study of nonstationary absorption of pure CO_2 in H_2O [55] provides experimental data for the dependence of Q on $\sqrt{t_0}$. They have been used for the *determination of* $\bar{\gamma}$ in (4.109) by means of the least-squares method. The value of $\bar{\gamma}$ was calculated as -1.787×10^{-6} m/s. In Fig. 16 Eq. 4.109 is shown for $\bar{\gamma} = -1.787 \times 10^{-6}$ m/s, and the dots are experimental data from [55].

An attempt to explain the discrepancy between the experimental data for nonstationary absorption of pure CO_2 in H_2O and the linear theory of mass transfer with the Marangoni effect was made in [55]. There it was correctly shown that $(\theta^* - \theta_0) \approx 0.02$ °C (θ^*—is the temperature of the phase boundary), but the assumption that the fluctuations of θ^* as a result of a surface tension gradient are enough to cause the loss of stability is unreasonable. The use of the experimental data for the determination of the characteristic velocity of the flow $u_0 = 1.12 \times 10^{-6}$ m/s shows that the velocity obtained is very close to the characteristic one in the cases when it is a result of a nonlinear mass transfer:

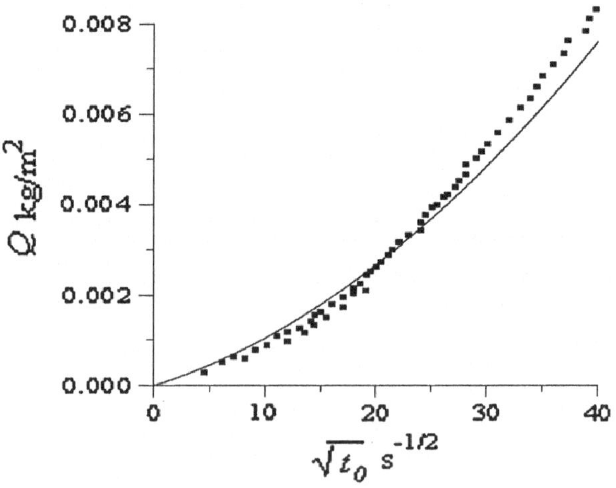

Fig. 16 Relation between the amount of absorbed gas (Q kg/m^2) and the time $\left(\sqrt{t_0}s^{-1/2}\right)s^{-1/2}$ ($D = 1.78 \times 10^{-9}$, $c^* = 1.6$ kg/m^3, $\bar{\gamma} = -1.787 \times 10^{-6}$ m/s, $\varepsilon = 1.6 \times 10^{-3}$). *Dots* experimental data from [55]

$$u_0 = \frac{c^*}{\rho_0}\sqrt{\frac{D}{t_0}} = 0.876 \times 10^{-6}\ \mathrm{m/s}, \quad t_0 = 10\,\mathrm{s}. \tag{4.110}$$

The solution of the Benard problem taking into account the surface tension gradient [8, 60–62] shows that the Marangoni effect may occur in layers with thickness not greater than several millimeters. Further, it is evident that in the case of nonstationary absorption at large concentration gradients the occurrence of the Marangoni effect cannot be expected.

The interphase mass transfer in a stagnant gas–liquid system has been investigated, as follows: in the case of an irreversible chemical reaction [59]; when an absorbed component is surface-active [60]; in the case of stagnant liquid evaporation [61], in the case of the presence of waves at the interface [62], etc.

Most of the experimental studies [55, 63, 64] indicate that under gas absorption the mass transfer rate is significantly greater than that predicted by the linear theory. On the other hand, under desorption such differences have not been observed. This fact has been explained in different ways [55]. It is possible to demonstrate that the *difference between the absorption and desorption rates* follows directly from the nonlinear theory of the mass transfer and the hydrodynamic stability [65].

Carbon dioxide desorption from a saturated stagnant water solution into a nitrogen gas phase has been investigated experimentally by several research groups [55, 63, 64]. In this case the CO_2 concentration in the gas phase changes from the equilibrium value at the interface to zero in the bulk of the gas. Thus, it is necessary to use the Oberbeck–Boussinesq equations (4.33) for both phases, where the temperature is practically constant. Thus, (4.33) gives the following equations for the gas ($i = 1$) and the liquid ($i = 2$) phase, respectively:

$$\rho_i\left(\frac{\partial v_z^{(i)}}{\partial t} + v_z^{(i)}\frac{\partial v_z^{(i)}}{\partial z} + v_r^{(i)}\frac{\partial v_z^{(i)}}{\partial r} + \frac{v_\varphi^{(i)}}{r}\frac{\partial v_z^{(i)}}{\partial \varphi}\right)$$
$$= -\frac{\partial p_i}{\partial z} + \mu_i\left(\frac{\partial^2 v_z^{(i)}}{\partial z^2} + \frac{1}{r}\frac{\partial v_z^{(i)}}{\partial r} + \frac{\partial^2 v_z^{(i)}}{\partial r^2} + \frac{1}{r^2}\frac{\partial^2 v_z^{(i)}}{\partial \varphi^2}\right) + g(\rho_i - \rho_{0i}),$$

$$\rho_i\left(\frac{\partial v_r^{(i)}}{\partial t} + v_z^{(i)}\frac{\partial v_r^{(i)}}{\partial z} + v_r\frac{\partial v_r^{(i)}}{\partial r} + \frac{v_\varphi}{r}\frac{\partial v_r^{(i)}}{\partial \varphi} - \frac{v_\varphi^{(i)2}}{r}\right)$$
$$= -\frac{\partial p_i}{\partial r} + \mu_i\left(\frac{\partial^2 v_r^{(i)}}{\partial z^2} + \frac{1}{r}\frac{\partial v_r^{(i)}}{\partial r} + \frac{\partial^2 v_r^{(i)}}{\partial r^2} - \frac{v_r^{(i)}}{r^2} + \frac{1}{r^2}\frac{\partial^2 v_r^{(i)}}{\partial \varphi^2} - \frac{2}{r^2}\frac{\partial v_\varphi^{(i)}}{\partial \varphi}\right),$$

$$\rho_i\left(\frac{\partial v_\varphi^{(i)}}{\partial t} + v_z^{(i)}\frac{\partial v_\varphi^{(i)}}{\partial z} + v_r^{(i)}\frac{\partial v_\varphi^{(i)}}{\partial r} + \frac{v_\varphi^{(i)}}{r}\frac{\partial v_\varphi^{(i)}}{\partial \varphi} + \frac{v_r^{(i)}v_\varphi^{(i)}}{r}\right)$$
$$= -\frac{1}{r}\frac{\partial p_i}{\partial \varphi} + +\mu_i\left(\frac{\partial^2 v_\varphi^{(i)}}{\partial z^2} + \frac{1}{r}\frac{\partial v_\varphi^{(i)}}{\partial r} + \frac{\partial^2 v_\varphi^{(i)}}{\partial r^2} - \frac{v_\varphi^{(i)}}{r^2} + \frac{1}{r^2}\frac{\partial^2 v_\varphi^{(i)}}{\partial \varphi^2} + \frac{2}{r^2}\frac{\partial v_r^{(i)}}{\partial \varphi}\right),$$

$$\frac{\partial \rho_i}{\partial t} + \frac{\partial\left(\rho_i v_z^{(i)}\right)}{\partial z} + \frac{\partial\left(\rho_i v_r^{(i)}\right)}{\partial r} + \frac{\rho_i v_r^{(i)}}{r} + \frac{1}{r}\frac{\partial\left(\rho_i v_\varphi^{(i)}\right)}{\partial \varphi} = 0,$$

$$\frac{\partial c_i}{\partial t} + v_z^{(i)}\frac{\partial c_i}{\partial z} + v_r^{(i)}\frac{\partial c_i}{\partial r} + \frac{v_\varphi^{(i)}}{r}\frac{\partial c_i}{\partial \varphi} = D_i\left(\frac{\partial^2 c_i}{\partial z^2} + \frac{1}{r}\frac{\partial c_i}{\partial r} + \frac{\partial^2 c_i}{\partial r^2} + \frac{1}{r^2}\frac{\partial^2 c_i}{\partial \varphi^2}\right),$$

$$\rho_i = \rho_{0i}\left(1 + \frac{a^{2-i}c_i}{\rho_{0i}}\right), \quad i = 1, 2, \tag{4.111}$$

where ρ_{01} is the water density and ρ_{02} is the density of nitrogen.

Under CO_2 diffusion in N_2 (20 °C) a is determined through the densities of both gases:

$$a = \frac{\rho_{CO_2} - \rho_{N_2}}{\rho_{CO_2}} = 0.367. \tag{4.112}$$

The boundary conditions of (4.111) follow from (4.34), taking into account the interaction between the phases during the desorption process:

$$t = 0, \quad v_z^{(i)} = v_r^{(i)} = v_\varphi^{(i)} = c_1 = 0, \quad c_2 = c_{20};$$

$$z = 0, \quad v_z^{(i)} = -\frac{D_i}{\rho_{0i}}\frac{\partial c_i}{\partial z}, \quad v_r^{(1)} = v_r^{(2)},\ v_\varphi^{(1)} = v_\varphi^{(2)},$$

$$\mu_1\left(\frac{\partial v_r^{(1)}}{\partial z} + \frac{\partial v_z^{(1)}}{\partial \varphi}\right) = \mu_2\left(\frac{\partial v_r^{(2)}}{\partial z} + \frac{\partial v_z^{(2)}}{\partial \varphi}\right),$$

$$\mu_1\left(\frac{\partial v_\varphi^{(1)}}{\partial z} + \frac{1}{r}\frac{\partial v_z^{(1)}}{\partial \varphi}\right) = \mu_2\left(\frac{\partial v_\varphi^{(2)}}{\partial z} + \frac{1}{r}\frac{\partial v_z^{(2)}}{\partial \varphi}\right),$$

$$c_1 = \chi c_2, \quad \frac{D_1 \rho_1^*}{\rho_{01}}\frac{\partial c_1}{\partial z} = \frac{D_2 \rho_2^*}{\rho_0^*}\frac{\partial c_2}{\partial z};$$

$$z \to \infty, \quad v_z^{(1)} = v_r^{(1)} = v_\varphi^{(1)} = c_1 = 0; \quad z \to -\infty, \quad v_z^{(2)} = v_r^{(2)} = v_\varphi^{(2)} = c_2 = c_0;$$

$$r = 0, \quad v_z^{(i)}, v_r^{(i)}, v_\varphi^{(i)}, c_i - \text{finite}; \quad r = r_0, \quad v_z^{(i)} = v_r^{(i)} = v_\varphi^{(i)} = 0, \quad \frac{\partial c_i}{\partial r} = 0;$$
$$i = 1, 2. \tag{4.113}$$

The solution of problems (4.111) and (4.113) requires *dimensionless variables* such as (4.36) for both phases:

$$t = t_0 T, \quad z = l_i Z_1^{(i)} = \delta_i Z_2^{(i)}, \quad r = r_0 R, \quad \varphi = 2\pi\Phi, \quad p_i = \rho_{0i} u_{0i}^2 P^{(i)},$$

$$
\begin{aligned}
v_z^{(i)}(t,z,r,\varphi) &= u_{0i} V_z^{(i)}\left(T, Z_1^{(i)}, R, \Phi\right) = u_{0i} \tilde{V}_z^{(i)}\left(T, Z_2^{(i)}, R, \Phi\right), \\
v_r^{(i)}(t,z,r,\varphi) &= \frac{u_{0i} r_0}{l_i} V_r^{(i)}\left(T, Z_1^{(i)}, R, \Phi\right) = \frac{u_{0i} r_0}{\delta_i} \tilde{V}_r^{(i)}\left(T, Z_2^{(i)}, R, \Phi\right), \\
v_\varphi^{(i)}(t,z,r,\varphi) &= 2\pi \frac{u_{0i} r_0}{l_i} V_\varphi^{(i)}\left(T, Z_1^{(i)}, R, \Phi\right) = 2\pi \frac{u_{0i} r_0}{\delta_i} \tilde{V}_\varphi^{(i)}\left(T, Z_2^{(i)}, R, \Phi\right), \\
c^{(i)}(t,z,r,\varphi) &= c_{0i}^* C^{(i)}\left(T, Z_1^{(i)}, R, \Phi\right) = c_{0i}^* \tilde{C}^{(i)}\left(T, Z_2^{(i)}, R, \Phi\right), \quad i = 1,2,
\end{aligned}
\tag{4.114}
$$

where the order of the following characteristic scales is known,

$$
t_0 \sim 10^2\text{s}, \quad r_0 \sim 10^{-2}\text{m}, \quad \chi = 1.06, \quad c_0 = 1.6\ \text{kg/m}^3, \quad c_{01}^* = \chi c_0, \quad c_{02}^* = c_0. \tag{4.115}
$$

The substitution of (4.114) into (4.112) and (4.113) converts the problem into a dimensionless form:

$$
\begin{aligned}
&\left[1 + \frac{\bar{c}_{0i}^*}{\rho_{0i}} C^{(i)}\right] \left[\frac{u_{0i}\rho_{0i}}{g t_0 \bar{c}_{0i}^*} \frac{\partial V_z^{(i)}}{\partial T} + \frac{u_{0i}^2 \rho_{0i}}{g l_i \bar{c}_{0i}^*} \left(V_z^{(i)} \frac{\partial V_z^{(i)}}{\partial Z_1^{(i)}} + V_r^{(i)} \frac{\partial V_z^{(i)}}{\partial R} + \frac{V_\varphi^{(i)}}{R} \frac{\partial V_z^{(i)}}{\partial \Phi}\right)\right] \\
&\quad = -\frac{u_{0i}^2 \rho_{0i}}{g l_i \bar{c}_{0i}^*} \frac{\partial P^{(i)}}{\partial Z_1^{(i)}} + \frac{\mu_i u_{0i}}{g l_i^2 \bar{c}_{0i}^*} \left[\frac{\partial^2 V_z^{(i)}}{\partial Z_1^{(i)2}} + \frac{l_i^2}{r_0^2} \left(\frac{1}{R} \frac{\partial V_z^{(i)}}{\partial R} + \frac{\partial^2 V_z^{(i)}}{\partial R^2} + \frac{1}{4\pi^2} \frac{1}{R^2} \frac{\partial^2 V_z^{(i)}}{\partial \Phi^2}\right)\right] + C^{(i)},
\end{aligned}
$$

$$
\begin{aligned}
&\left[1 + \frac{\bar{c}_{0i}^*}{\rho_{0i}} C^{(i)}\right] \left[\frac{u_{0i}\rho_{i0}}{g t_0 \bar{c}_{0i}^*} \frac{\partial V_r^{(i)}}{\partial T} + \frac{u_{0i}^2 \rho_{0i}}{g l_i \bar{c}_{0i}^*} \left(V_z^{(i)} \frac{\partial V_r^{(i)}}{\partial Z_1^{(i)}} + V_r^{(i)} \frac{\partial V_r^{(i)}}{\partial R} + \frac{V_\varphi^{(i)}}{R} \frac{\partial V_r^{(i)}}{\partial \Phi} - 4\pi^2 \frac{V_\varphi^{(i)2}}{R}\right)\right] \\
&\quad = -\frac{l_i^2}{r_0^2} \frac{u_{0i}^2 \rho_{0i}}{g l_i \bar{c}_{0i}^*} \frac{\partial P^{(i)}}{\partial R} + \frac{\mu_i u_{0i}}{g l_i^2 \bar{c}_{0i}^*} \left[\frac{\partial^2 V_r^{(i)}}{\partial Z_1^{(i)2}} + \frac{l_i^2}{r_0^2} \left(\frac{1}{R} \frac{\partial V_r^{(i)}}{\partial R} + \frac{\partial^2 V_r^{(i)}}{\partial R^2} - \frac{V_r^{(i)}}{R^2} + \frac{1}{4\pi^2} \frac{\partial^2 V_r^{(i)}}{\partial \Phi^2} - \frac{2}{R} \frac{\partial V_\varphi^{(i)}}{\partial \Phi}\right)\right],
\end{aligned}
$$

$$
\begin{aligned}
&\left[1 + \frac{\bar{c}_{0i}^*}{\rho_{0i}} C^{(i)}\right] \left[\frac{u_{0i}\rho_{0i}}{g t_0 \bar{c}_{0i}^*} \frac{\partial V_\varphi^{(i)}}{\partial T} + \frac{u_{0i}^2 \rho_{0i}}{g l_i \bar{c}_{0i}^*} \left(V_z^{(i)} \frac{\partial V_\varphi^{(i)}}{\partial Z_1^{(i)}} + V_r^{(i)} \frac{\partial V_\varphi^{(i)}}{\partial R} + \frac{V_\varphi^{(i)}}{R} \frac{\partial V_\varphi^{(i)}}{\partial \Phi} + \frac{V_r^{(i)} V_\varphi^{(i)}}{R}\right)\right] \\
&\quad = -\frac{l_i^2}{r_0^2} \frac{u_{0i}^2 \rho_{0i}}{g l_i \bar{c}_{0i}^*} \frac{1}{R} \frac{\partial P^{(i)}}{\partial \Phi} + \frac{\mu_i u_{0i}}{g l_i^2 \bar{c}_{0i}^*} \left[\frac{\partial^2 V_\varphi^{(i)}}{\partial Z_1^{(i)2}} + \frac{l_i^2}{r_0^2} \left(\frac{1}{R} \frac{\partial V_\varphi^{(i)}}{\partial R} + \frac{\partial^2 V_\varphi^{(i)}}{\partial R^2} - \frac{V_\varphi^{(i)}}{R^2} + \frac{1}{4\pi^2} \frac{1}{R^2} \frac{\partial^2 V_\varphi^{(i)}}{\partial \Phi^2} + \frac{1}{2\pi^2} \frac{1}{R^2} \frac{\partial V_r^{(i)}}{\partial \Phi}\right)\right],
\end{aligned}
$$

$$
\begin{aligned}
&\frac{l_i c_{0i}^*}{\rho_{0i} u_{0i} t_0} \frac{\partial C^{(i)}}{\partial T} - \left(\frac{\partial V_z^{(i)}}{\partial Z_1^{(i)}} + \frac{V_r^{(i)}}{R} + \frac{\partial V_r^{(i)}}{\partial R} + \frac{1}{R} \frac{\partial V_\varphi^{(i)}}{\partial \Phi}\right) \left[1 + \frac{\bar{c}_{0i}^*}{\rho_{0i}} C^{(i)}\right] \\
&\quad + V_z^{(i)} \frac{\bar{c}_{0i}^*}{\rho_{0i}} \frac{\partial C^{(i)}}{\partial Z_1^{(i)}} + V_r^{(i)} \frac{\bar{c}_{0i}^*}{\rho_{0i}} \frac{\partial C^{(i)}}{\partial R} + \frac{V_\varphi^{(i)}}{R} \frac{\bar{c}_{0i}^*}{\rho_{0i}} \frac{\partial C^{(i)}}{\partial \Phi} = 0,
\end{aligned}
$$

$$\frac{\partial \tilde{C}^{(i)}}{\partial T} + \frac{u_{0i}t_0}{\delta_i}\left(\tilde{V}_z^{(i)}\frac{\partial \tilde{C}^{(i)}}{\partial Z_2^{(i)}} + \tilde{V}_r^{(i)}\frac{\partial \tilde{C}^{(i)}}{\partial R} + \frac{\tilde{V}_r^{(i)}}{R}\frac{\partial \tilde{C}^{(i)}}{\partial \Phi}\right)$$
$$= \frac{D_i t_0}{\delta_i^2}\left[\frac{\partial^2 \tilde{C}^{(i)}}{\partial Z_2^{(i)2}} + \frac{\delta_i^2}{r_0^2}\left(\frac{1}{R}\frac{\partial \tilde{C}^{(i)}}{\partial R} + \frac{\partial^2 \tilde{C}^{(i)}}{\partial R^2} + \frac{1}{4\pi^2}\frac{1}{R^2}\frac{\partial^2 \tilde{C}^{(i)}}{\partial \Phi^2}\right)\right];$$

$$T = 0, \quad V_z^{(i)} = V_r^{(i)} = V_\varphi^{(i)} = \tilde{C}^{(1)} = 0, \quad \tilde{C}^{(2)} = 1;$$

$$Z_1^{(i)} = Z_2^{(i)} = 0, \quad V_z^{(i)} = -\frac{D_i c_{0i}^*}{u_{0i}\rho_{0i}\delta_i}\frac{\partial \tilde{C}^{(i)}}{\partial Z_2^{(i)}}, \quad C^{(1)} = C^{(2)},$$
$$\frac{\partial C^{(1)}}{\partial Z_1^{(1)}} = \frac{D_2\rho_2^*\rho_{01}\delta_1}{D_1\rho_1^*\rho_{02}\delta_2\chi}\frac{\partial \tilde{C}^{(2)}}{\partial Z_1^{(2)}},$$

$$\frac{\partial V_r^{(1)}}{\partial Z_1^{(1)}} + \frac{l_1^2}{r_0^2}\frac{\partial V_z^{(1)}}{\partial R} = \frac{\mu_2 u_{02} l_1^2}{\mu_1 u_{01} l_2^2}\left(\frac{\partial V_r^{(2)}}{\partial Z_1^{(2)}} + \frac{l_2^2}{r_0^2}\frac{\partial V_z^2}{\partial R}\right),$$

$$\frac{\partial V_\varphi^{(1)}}{\partial Z_1^{(1)}} + \frac{l_1^2}{4\pi^2 r_0^2}\frac{1}{R}\frac{\partial V_z^{(1)}}{\partial \Phi} = \frac{\mu_2 u_{02} l_1^2}{\mu_1 u_{01} l_2^2}\left(\frac{\partial V_\varphi^{(2)}}{\partial Z_1^{(2)}} + \frac{l_2^2}{r_0^2 4\pi^2}\frac{\partial V_z^{(2)}}{\partial R}\right);$$

$$Z_1^{(1)} \to \infty, \quad V_z^{(1)} = V_r^{(1)} = V_\varphi^{(1)} = C^{(1)} = 0; \quad Z_1^{(2)} \to -\infty,$$
$$V_z^{(2)} = V_r^{(2)} = V_\varphi^{(2)} = 0, \quad C^{(2)} = 1;$$

$$R = 0, \quad V_z^{(i)}, V_r^{(i)}, V_\varphi^{(i)}, c^{(i)} \text{ - finite}; \quad R = 1, \quad V_z^{(i)} = V_r^{(i)} = V_\varphi^{(i)} = 0,$$
$$\frac{\partial C^{(i)}}{\partial R} = 0; \quad i = 1, 2, \tag{4.116}$$

where $\bar{c}_{0i}^* = a^{2-i}c_{0i}^*, \quad i = 1, 2.$

The *qualitative analysis* of (4.116) may be performed in a way similar to that employed when considering the absorption. This allows us to evaluate the characteristic scales for both phases as

$$\frac{D_i t_0}{\delta_i^2} = 1, \quad \delta_i = \sqrt{D_i t_0}, \quad \delta_1 \sim 10^{-2}\text{m}, \quad \delta_2 \sim 10^{-4}\text{m}, \quad \frac{\mu_i u_{0i}}{g l_i^2 \bar{c}_{0i}^*} = 1,$$
$$\frac{D_i c_{0i}^*}{u_{0i}\rho_{0i}\delta_i} = 1,$$

$$u_{0i} = \frac{c_i^*}{\rho_{0i}}\sqrt{\frac{D_i}{t_0}}, \quad u_{0i} \sim 10^{-4}\text{m/s}, \quad u_{02} \sim 10^{-8}\text{m/s},$$

$$l_i = \sqrt{\frac{a^{2-i}\mu_i}{\rho_{0i} g}\sqrt{\frac{D_i}{t_0}}}, \quad l_1 \sim 10^{-5}\text{m}, \quad l_2 \sim 10^{-7}\text{m}. \tag{4.117}$$

The characteristic scales (4.117) permit the evaluation of the parameter orders in (4.116):

$$\varepsilon = \frac{c_i^*}{\rho_{0i}} \sim \bar{\varepsilon}_i = \frac{\bar{c}_i^*}{\rho_{0i}} \sim [1, 10^{-3}], \quad \frac{u_{0i}\rho_{0i}}{gt_0\bar{c}_i^*} \sim [10^{-7}, 10^{-9}],$$

$$\frac{u_{0i}^2}{gl_i\bar{\varepsilon}_i} \sim [10^{-3}, 10^{-7}], \quad \frac{l_i^2}{r_0^2} \sim [10^{-6}, 10^{-10}], \quad \frac{\delta_i^2}{r_0^2} \sim [1, 10^{-4}],$$

$$\frac{l_i}{t_0 u_{0i}\bar{\varepsilon}_i} \sim [10^{-3}, 10^{-4}], \quad \frac{u_{0i}t_0}{\delta_i} = \varepsilon_i \sim [1, 10^{-3}],$$

$$b = \frac{\mu_2 u_{02} l_1^2}{\mu_1 u_{01} l_2^2} \sim 1, \quad \frac{D_2\rho_2^*\rho_{01}\delta_1}{D_1\rho_1^*\rho_{02}\delta_2\chi} \sim 10^{-2}, \tag{4.118}$$

where the values in square brackets are for the gas phase and the liquid phase $[i = 1, i = 2]$, respectively.

The order of magnitude (10^{-2}) of the last parameter in (4.118) shows that the mass transfer under desorption of CO_2 from saturated water is limited by the mass transfer in the liquid phase, i.e., $C^{(1)} \equiv 0$. Further, only the equations for the liquid phase will be considered. For simplicity, the superscript 2 will be omitted. In this way, the set (4.116) gives

$$\frac{\partial^2 V_z}{\partial Z_1^2} + C = 0, \quad \frac{\partial^2 V_r}{\partial Z_1^2} = 0, \quad \frac{\partial^2 V_\varphi}{\partial Z_1^2} = 0, \quad \frac{\partial V_z}{\partial Z_1} + \frac{V_r}{R} + \frac{\partial V_r}{\partial R} + \frac{1}{R}\frac{\partial V_\varphi}{\partial \Phi} = 0,$$
$$\frac{\partial \tilde{C}}{\partial T} = \frac{\partial^2 \tilde{C}}{\partial Z_2^2};$$

$$T = 0, \quad \tilde{C} = 0; \quad Z_1 = Z_2 = 0, \quad V_z = -\frac{\partial \tilde{C}}{\partial Z_2}, \quad \frac{\partial V_r}{\partial Z_1} = 0, \quad \frac{\partial V_\varphi}{\partial Z_1} = 0, \quad \tilde{C} = 0;$$

$$Z_1 = Z_2 \to -\infty, \quad V_z = V_r = V_\varphi = 0, \quad \tilde{C} = 1; \quad R = 0, \quad V_r - \text{finite}. \tag{4.119}$$

It follows from (4.119) that the solution for $\tilde{C}$ is the following:

$$\tilde{C} = -erf\frac{Z_2}{2\sqrt{T}} \quad (Z_2 \le 0), \quad \left\langle \tilde{C} = erfc\frac{Z_2}{2\sqrt{T}} \quad (Z_2 \ge 0) \right\rangle, \quad Z_2 = \alpha_0 Z_1,$$
$$\alpha_0 = \frac{l}{\delta} \sim 10^{-3}. \tag{4.120}$$

Solution (4.120) gives

$$C = -erf\left(\alpha_0\frac{Z_1}{2\sqrt{T}}\right) \approx 0, \quad \left\langle C = erfc\left(\alpha\frac{Z_1}{2\sqrt{T}}\right) \approx 1 \right\rangle. \tag{4.121}$$

Here the results concerning absorption are shown in angled brackets.

The substitution of (4.121) into (4.119) shows that in the case of desorption of gas from a stagnant liquid there are no conditions allowing a natural convection. Thus, for the flow velocity components (induced by the mass transfer in the liquid phase), we obtain

$$\frac{\partial^2 V_z}{\partial Z_1^2} = 0; \quad \left\langle \frac{\partial^2 V_z}{\partial Z_1^2} = 1 \right\rangle : \quad Z_1 = 0, \quad V_z = -\left(\frac{\partial \tilde{C}}{\partial Z_2}\right)_{Z_2=0}; \quad Z_1 = -1,$$
$$V_z = 0.$$

$$\frac{\partial V_r}{\partial R} + \frac{V_r}{R} = -\frac{\partial V_z}{\partial Z_1} - \frac{1}{R}\frac{\partial V_\varphi}{\partial \Phi}; \quad R = 0, \quad V_r^{(0)} \text{ -- finite.}$$

$$\frac{\partial^2 V_\varphi}{\partial Z_1^2} = 0; \quad Z_1 = 0, \quad \frac{\partial V_\varphi}{\partial Z_1} = 0; \quad Z_1 = -1, \quad V_\varphi = 0. \tag{4.122}$$

In the above problems the boundary condition at $(-\infty)$ is substituted by the condition at (-1), i.e., at the border of the boundary layer. The solutions were obtained in [65]:

$$V_z = \frac{1}{\sqrt{\pi T}}(Z_1 + 1), \quad (Z_1 \leq 0),$$
$$\left\langle V_z = -\frac{1}{2}Z_1^2 + \left(\frac{1}{2} - \frac{1}{\sqrt{\pi T}}\right)Z_1 + \frac{1}{\sqrt{\pi T}} \quad (Z_1 \geq 0) \right\rangle,$$

$$V_r = -\frac{1}{2\sqrt{\pi T}}R, \quad \left\langle V_r = \left(\frac{1}{2}Z_1 + \frac{1}{2\sqrt{\pi T}} - \frac{1}{4}\right)R \right\rangle, \quad V_\varphi \equiv 0. \tag{4.123}$$

The result (4.123) indicates that a larger concentration gradient in the liquid (at the interface) induces the flow in a liquid bulk.

The velocity field (4.123) and the concentration distribution in the liquid (4.120) may be expressed as

$$v_z = \varepsilon_0 \sqrt{\frac{D}{\pi t}}\left(\frac{z}{l} + 1\right), \quad v_r = -\frac{\varepsilon_0}{2l}\sqrt{\frac{D}{\pi t}}r,$$

$$\varepsilon_0 = \frac{c_0}{\rho_0}, \quad v_\varphi \equiv 0, \quad c = -c_0 erf\frac{z}{2\sqrt{Dt}}, \quad l = \sqrt{\frac{\mu}{\rho_0 g}\sqrt{\frac{D}{t_0}}}. \tag{4.124}$$

The *mass transfer rate* may be obtained in a manner similar to that employed for (4.74). The relationships for the Sherwood number and the amount of desorbed substance are similar to those obtained for the absorption process, where in case of desorption $c^* = 0$ and $\rho^* = \rho_0$:

$$Sh = \frac{kl}{D} = 2\sqrt{\frac{\nu}{\pi g\, t_0}\sqrt{\frac{1}{D t_0}}}, \quad Q = 2c_0\sqrt{\frac{D\, t_0}{\pi}}\text{kg/m}^2, \tag{4.125}$$

The *linear stability* of the desorption process may be studied by means of small perturbations of the axial velocity, the pressure, and the concentration,

$$v_z + v_z', \quad p + p', \quad c + c', \tag{4.126}$$

in the complete set of equations of Oberbeck and Boussinesq (4.111, 4.113). The perturbations may be expressed through Fourier series of eigenfunctions, where ω and n are eigenvalues:

$$v_z' = \exp(\omega t) \sum_{n=0}^{\infty} v_n(t,z,r) \cos(n\varphi), \quad p' = \exp(\omega t) \sum_{n=0}^{\infty} p_n(t,z,r) \cos(n\varphi),$$

$$c' = \exp(\omega t) \sum_{n=0}^{\infty} c_n(t,z,r) \cos(n\varphi). \tag{4.127}$$

There are stable periodic solutions at $\omega = 0$. After elimination of the pressure, the eigenvalue problem takes the form

$$\frac{\partial v_n}{\partial t} + v_n \frac{\partial v_z}{\partial z} + v_z \frac{\partial v_n}{\partial z} + v_r \frac{\partial v_n}{\partial r} = \nu \left(\frac{\partial^2 v_n}{\partial z^2} + \frac{1}{r}\frac{\partial v_n}{\partial r} + \frac{\partial^2 v_n}{\partial r^2} - \frac{n^2}{r^2} v_n \right)$$
$$+ \frac{g}{\rho_0} c \left\langle + \int \frac{\partial v_r}{\partial z}\frac{\partial v_n}{\partial z} dr \right\rangle,$$

$$-\frac{\partial v_n}{\partial z} = \frac{D}{\rho_0} \left(\frac{\partial^2 c_n}{\partial z^2} + \frac{\partial^2 c_n}{\partial r^2} + \frac{1}{r}\frac{\partial c_n}{\partial r} - \frac{n^2}{r^2} c_n \right);$$

$$z = 0, \quad v_n = -\frac{D}{\rho_0}\frac{\partial c_n}{\partial z}, \quad c_n = 0; \quad z = -\infty, \quad v_n = -\frac{D}{\rho_0}\frac{\partial c_n}{\partial z};$$

$$r = 0, \quad v_n, c_n - \text{finite}; \quad r = r_0, \quad v_n = 0, \quad \frac{\partial c_n}{\partial r} = 0; \quad n = 0, 1, 2, \ldots, \infty. \tag{4.128}$$

Comparison of (4.128) and (4.96) shows that the difference between the absorption and the desorption processes is determined by the velocity distribution in the main flow (under desorption $\frac{\partial v_r}{\partial z} = 0$).

The solution of (4.128) may be presented in the form of (4.98):

$$v_n = u_0[V_n(Z,T) - Bf_n(R)], \quad c_n = c_0[C_n(Z,T) + Zf_n(R)], \quad n = 0, 1, 2, \ldots, \infty.$$

$$T = \frac{t}{t_0}, \quad Z = \frac{z}{l}, \quad R = \frac{r}{r_0}, \quad B = \frac{D\varepsilon_0}{u_0 l}, \quad \varepsilon_0 = \frac{c_0}{\rho_0}. \tag{4.129}$$

The introduction of the new variables into (4.128) leads to

$$\frac{l^2}{\nu t_0}\frac{\partial V_n}{\partial T}+\frac{u_0 l}{\nu}\left[(V_n - Bf_n)\frac{\partial V_z}{\partial Z}+V_z\frac{\partial V_n}{\partial Z}-BV_r f_n'\right]$$
$$=\frac{\partial^2 V_n}{\partial Z^2}-\frac{l^2}{r_0^2}\frac{n^2}{R^2}V_n+\frac{\varepsilon_0 l^2 g}{\nu u_0}(C_n+Zf_n)\quad\left\langle+\frac{u_0 l}{\nu}\frac{r_0^2}{l^2}\int\frac{\partial V_r}{\partial Z}\frac{\partial V_n}{\partial Z}dR\right\rangle,$$

$$\frac{u_0 l}{\varepsilon_0 D}\frac{\partial V_n}{\partial Z}=\frac{\partial^2 C_n}{\partial Z^2}-\frac{l^2}{r_0^2}\frac{n^2}{R^2}C_n;\quad f_n''+\frac{1}{R}f_n'-\frac{n^2}{R^2}f_n=0;$$

$$Z=0,\quad V_n=-B\frac{\partial C_n}{\partial Z},\quad C_n=0;\quad Z\to-\infty,\quad V_n=-B\frac{\partial C_n}{\partial Z};$$

$$R=0,\quad f_n - \text{finite};\quad R=1,\quad f_n'=0,\quad n=0,1,2,\ldots,\infty, \tag{4.130}$$

where (4.102) is the solution for $f_n(n = 0, 1, 2, \ldots, \infty)$.

The orders of magnitude of the dimensionless parameters in (4.130) are as follows:

$$\frac{l^2}{\nu t_0}\sim 10^{-10},\quad \frac{u_0 l}{\nu}\sim 10^{-9},\quad \frac{l^2}{r_0}\sim 10^{-10},\quad \frac{\varepsilon_0 l^2 g}{\nu u_0}\sim 10^{-2},\quad \frac{u_0 l}{\varepsilon_0 D}\sim 10^{-3},$$
$$\left\langle\frac{u_0 l}{\nu}\frac{r_0^2}{l^2}\sim 10\right\rangle. \tag{4.131}$$

The small parameters in (4.130) may be assumed to be zero and the resulting set for the determination of V_n and C_n $(n = 0, 1, 2,\ldots, \infty)$ is

$$\frac{\partial^2 V_n}{\partial Z^2}=0\quad\left\langle\frac{\partial V_n}{\partial Z}=0\right\rangle;\quad\frac{\partial^2 C_n}{\partial Z^2}=0;$$

$$Z=0,\quad V_n=-B\frac{\partial C_n}{\partial Z},\quad C_n=0;\quad Z\to-\infty,\quad V_n=-B\frac{\partial C_n}{\partial Z};\quad n=0,1,2,\ldots,\infty. \tag{4.132}$$

Problem (4.132) has a solution for $V_n(n = 0, 1, 2, \ldots, \infty)$ depending linearly on Z. This leads to a result similar to (4.103):

$$V_n=-B\gamma_n,\quad C_n=\gamma_n Z,\quad \gamma_n\leq 0,\quad Z\leq 0,\quad n=0,1,2,\ldots,\infty, \tag{4.133}$$

where the velocity and the concentration are determined with the accuracy of an unspecified constant that could not be obtained in the approximation of the linear stability theory.

The result obtained [65] shows that under desorption of CO_2 from stagnant saturated water the desorption rate may be expressed by a relationship similar to one obtained for the absorption. So, the process rate can be determined from (4.107) to (4.109). This result could be derived more precisely by taking into account the weak dependence of $\bar{\gamma}$ on t_0 $\left(\bar{\gamma} \sim \sqrt{4}t_0\right)$. Thus, from (4.109) to (4.106), it directly follows that

$$Sh = \frac{kl}{D} = 2\sqrt{\frac{\nu}{\pi g t_0}\sqrt{\frac{1}{Dt_0}}} - \gamma, \quad Q = c_0\left[2\sqrt{\frac{Dt_0}{\pi}} - \gamma\left(\frac{g}{\nu}\right)^{\frac{1}{2}} D^{\frac{3}{4}} t_0^{\frac{5}{4}}\right]. \qquad (4.134)$$

The result obtained (4.134) is valid in case of absorption as well, when $c_0 = c^*(\varepsilon \ll 1)$.

The eigenvalue γ is determined by the least-squares method applied to the experimental data obtained in [55, 63, 64]. In the case of absorption $\gamma = -4.204 \times 10^{-4}$, whereas in the case of desorption $\gamma = 3.032 \times 10^{-5}$. This result shows that the desorption process is stable, in contrast to the absorption process. In this case the mass transfer rate may be determined by (4.124). Figure 17 presents the relationship $Q = Q(\sqrt{t_0})$ in (4.134) for absorption, desorption, and according to the linear theory of mass transfer, i.e., $\gamma = 0$, compared with the experimental data [55, 63, 64].

The comparative analysis of both processes shows that under desorption of CO_2 from stagnant saturated water there are no conditions allowing a natural convection. As a result, the axial velocity component depends linearly on the axial coordinate (see 4.125), whereas the radial component is independent of the same coordinate. This result is opposite that obtained under absorption, where the relationship of the axial coordinate is a power of 2. Thus, the axial perturbations of the concentration attenuate and the respective axial perturbations of the velocity attenuate too. The radial perturbations are symmetric and do not affect the mass transfer rate. The concentration gradient at the interface induces a flow, but its velocity is small and has no effect on the mass transfer rate. This fact together with the absence of a natural convection in desorption make the induced flow stable with respect to the axial perturbations provoked by the perturbations of the concentration.

Owing to the stability of the desorption process and the absence of a nonlinear mass transfer effect, the process rate may be determined by the linear theory of nonstationary mass transfer (4.25). This result is confirmed by the experimental data [55, 63, 64] shown in Fig. 17.

The theoretical analysis of the mechanism and the kinetics of the transport processes in systems with intensive mass transfer shows that in the case of gas absorption at great concentration gradients and a chemical reaction in the liquid phase, the mass transfer rate is significantly higher than that predicted by the linear theory of mass transfer. In the absence of surface-active agents and the availability of a temperature field, caused by the thermal effect of the chemical reaction, the surface tension gradient is not enough for the occurrence of the Marangoni effect. In the case of nonstationary absorption of a gas in a stagnant liquid, a flow is induced as a result of a natural convection and a nonlinear mass transfer (a density gradient in the volume and a large mass flux through the phase boundary). This problem differs significantly from the Benard problem, as the large concentration gradient at the interphase induces a secondary flow, oriented normally to this

Fig. 17 Comparison of the absorption and desorption rates (4.134) of CO_2 in H_2O: *1* absorption ($\gamma = -4.204 \times 10^{-4}$), *2* desorption ($\gamma = 3.029 \times 10^{-5}$), *3* linear theory ($\gamma = 0$). Experimental data: *squares* absorption, *triangles* desorption

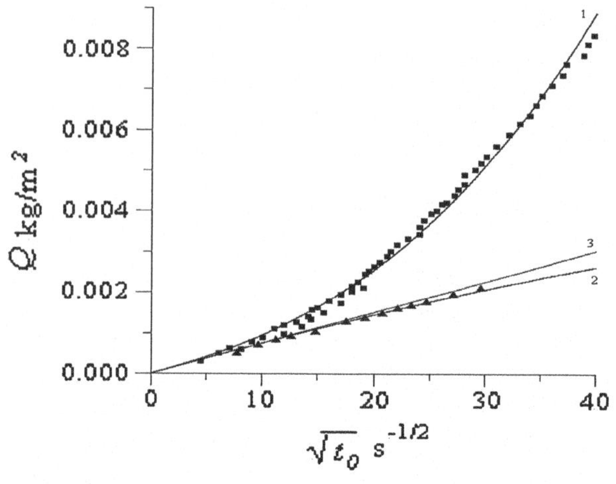

surface, and in this way does not allow the existence of a mechanical equilibrium (diffusion in an immobile liquid). In this way the basic process considered (simultaneous momentum, mass and heat transfer) is unstable regarding disturbances (that may not be small). As a result, the process becomes unstable and it is transformed into a periodic stable process, i.e., a self-organizing dissipative structure (velocity, concentration, and temperature field). The mass transfer rate is significantly greater than that predicted by the linear theory. In the case of desorption, the process is stable and the mass transfer kinetics is determined according to the linear theory. These results are confirmed by a large amount of experimental data.

The theoretical results obtained show [58, 65] that the mass transfer rate under nonstationary absorption of concentrated gas admixtures by a stagnant liquid layer may be determined (see 4.109) by the amount of substance absorbed per time t_0 (s):

$$Q = (1+\varepsilon)c^* \left[2\left(\frac{Dc_0}{\pi}\right)^{\frac{1}{2}} + \gamma\left(\frac{g}{\nu}\right)^{\frac{1}{2}} D^{\frac{3}{4}} t_0^{\frac{5}{4}} \right], \tag{4.135}$$

where for low-solubility gases $\varepsilon \ll 1$ and c^* is the concentration at the liquid surface. In the case of a desorption process (see 4.134) from a saturated solution (e.g., a saturated aqueous solution of CO_2 contacting with a gas phase containing N_2), $c^* = 0$ and the amount of desorbed substance is

$$Q = c_0 \left[2\left(\frac{Dc_0}{\pi}\right)^{\frac{1}{2}} + \gamma\left(\frac{g}{\nu}\right)^{\frac{1}{2}} D^{\frac{3}{4}} t_0^{\frac{5}{4}} \right], \tag{4.136}$$

where c_0 is the initial gas concentration in the solution.

In the relationships for Q, the first term depends on the rate of the nonstationary diffusion in the stagnant liquid. The second term occurs owing the loss of stability

Table 7 Systems considered in [55, 63, 64]

Process	System	T (°C)	D ($\times 10^9$ m²/s)	ν ($\times 10^6$ m²/s)	c (kg/m³)	γ ($\times 10^4$)	γ_{av} ($\times 10^4$)
Absorption	100% CO_2–H_2O	23	1.88	0.97	1.60	3.70	4.20
Absorption	30% $CO_2(N_2)$–H_2O	23	1.88	0.97	0.471	2.73	2.68
Absorption	7.1% $CO_2(N_2)$–H_2O	23	1.88	0.97	0.112	1.90	1.20
Absorption	100% Ar–H_2O	20	0.47	1.00	0.0599	1.63	1.79
Absorption	100% Ar–H_2O	10	0.35	1.30	0.0719	1.70	2.12
Absorption	100% Ar–C_2H_5OH	20	0.49	1.52	0.427	2.66	3.29
Desorption	(CO_2/H_2O)–N_2	20	1.88	0.97	0	0	−0.303

of the process provoked by small perturbations of the concentration of absorbed gas at the liquid surface. These periodic perturbations with small amplitude grow continuously until a new stable state is established (i.e., self-organizing dissipative process). The parameter γ of this self-organizing dissipative structure is related to the amplitude of the new stable state and can be obtained using the experimental data [55, 63, 64] summarized in Table 7.

The experimental data [55, 63, 64] permit us to obtain the parameter γ in (4.135) and (4.136) using the least-squares method [66]. The average values obtained for every process (γ_{av}) are shown in Table 7.

Obviously the values of γ_{av} depend on the interphase concentration c^*. The suitable correlation developed is

$$\gamma = 3.29 \times 10^{-4}(c^*)^{\frac{1}{4}}. \tag{4.137}$$

The values of γ, calculated from (4.137), are presented in Table 7.

The substitution of (4.137) into (4.135) and (4.136) leads to a definitive relationship [66] expressing the amount of absorbed (desorbed) substance:

$$Q = 2c^*\left(\frac{Dt_0}{\pi}\right)^{\frac{1}{2}} + 3.29 \times 10^{-4} c^{*\frac{5}{4}}\left(\frac{g}{\nu}\right)^{\frac{1}{2}} D^{\frac{3}{4}} t_0^{\frac{5}{4}}, \tag{4.138}$$

$$Q = 2c_0\left(\frac{Dt_0}{\pi}\right)^{\frac{1}{2}}. \tag{4.139}$$

Figures 18, 19, 20, 21, 22 and 23 show experimental data concerning the systems summarized in Table 7. They correlate well with the values of γ_{av} (obtained from the least-squares method) and with the values of γ calculated from (4.137).

The theoretical analysis of the absorption kinetics complicated by a chemical reaction in the liquid phase done here continues the investigations performed in [28, 43, 58, 65, 66, 69]. The main results show that the deviations from the linear mass transfer theory cannot be explained by the Marangoni effect. The latter assumption is based on the fact that the thermal effect of the absorption is not sufficient to create a gradient of the liquid surface tension and a consequent loss of stability of the system.

The differences observed may be explained directly by the nonlinear mass transfer effects as a result of the large concentration gradients which induce secondary flows oriented along the normal with respect to the interphase boundary. The result of these secondary flows is the loss of stability of the system with respect of the small perturbations of concentration at the interface. The final result is that the perturbations grow until to a new stable (self-organizing) dissipative structure with a mass transfer rate significantly greater than one predicted by the linear mass transfer theory.

In the cases of a non-stationary absorption of a gas in a stagnant liquid, these secondary flows are induced as a result of the natural convection and the large concentration gradients.

4.3 Liquid Evaporation

The studies described in the previous section addressed the absorption (desorption) of low-solubility gases when the liquid mass transfer resistance limits the mass transfer rates. Obviously, the situation where the mass transfer is limited by the gas phase is interesting. An adequate example is the case of *nonstationary evaporation* of a stagnant liquid layer in a stagnant gas phase above it [67, 68]. The detailed experimental investigation of such systems was reported in [73].

The nonstationary evaporation of a liquid with a moderate partial pressure (water, methanol, ethanol, and isopropyl alcohol) at 20 °C in an inert gas (nitrogen, argon, and helium) is now investigated. The process occurs in a thermostatic condition, corresponding to the experiments performed reported in [73]. Under

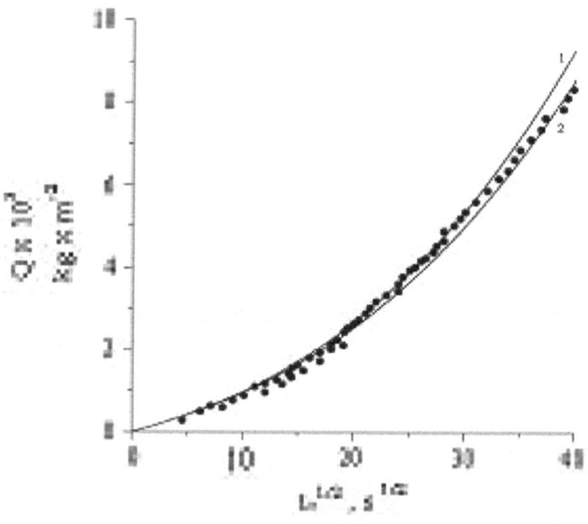

Fig. 18 Absorption of 100% CO_2 in water ($c^* = 1.60$ kg/m^3) at 23 °C. *Circles* experimental data, *1* γ_{av} obtained from the least-squares method, *2* γ calculated from (4.37)

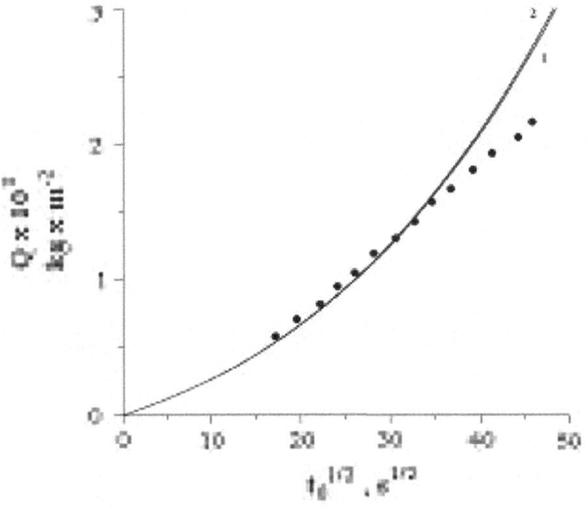

Fig. 19 Absorption of 30% CO_2 (70% N_2) in water ($c^* = 0.471$ kg/m^3) at 23 °C. *Circles* experimental data, *1* γ_{av} obtained from the least-squares method, *2* γ calculated from (4.137)

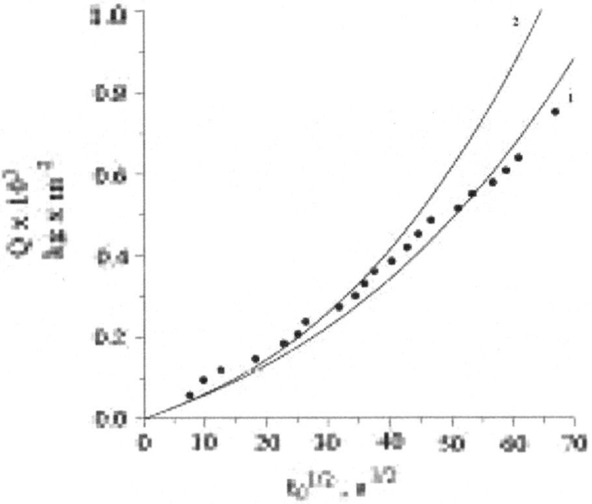

Fig. 20 Absorption of 7.1% CO_2 (92.9% N_2) in water ($c^* = 0.112$ kg/m^3) at 23 °C. *Circles* experimental data, *1* γ_{av} obtained from the least-squares method, *2* γ calculated from (4.137)

such a condition the nonstationary mass transfer of the liquid vapor in the gas phase limits the process rate.

The *mechanism of nonstationary evaporation* may be considered as a nonstationary diffusion complicated with additional effects of a variable temperature at the liquid surface (as a result of the thermal effect of the evaporation phenomenon) and a convection (secondary Stefan flow) as well as a natural convection. The contribution of these phenomena to the evaporation rate will be analyzed next.

The investigations [73] on the evaporation rates of liquids (H_2O, CH_3OH, C_2H_5OH, *i*-C_3H_7OH) showed a time-dependent average liquid temperature. In fact, the process only depends on the surface temperature (θ^*, °C).

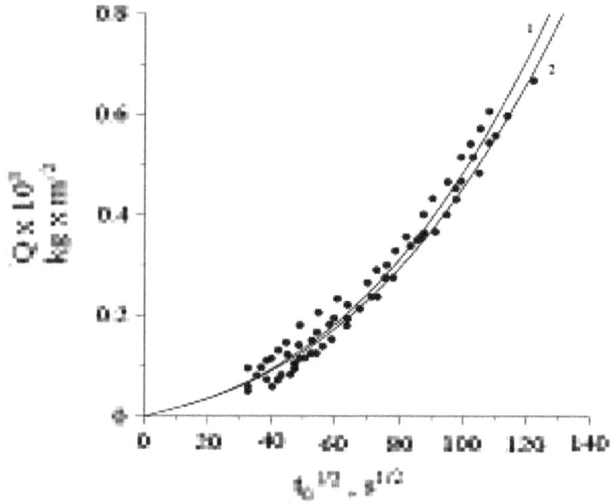

Fig. 21 Absorption of 100% Ar in water ($c^* = 0.0599$ kg/m^3) at 20 C. *Circles* experimental data, *1* γ_{av} obtained from the least-squares method, *2* γ calculated from (4.137)

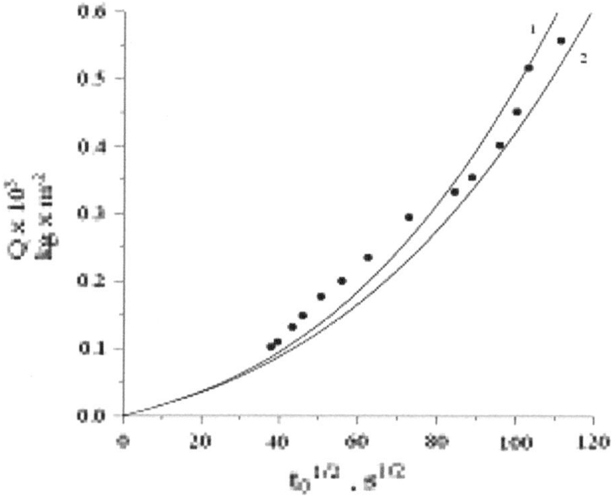

Fig. 22 Absorption of 100% Ar in water ($c^* = 0.0719$ kg/m^3) at 10 °C. *Circles* experimental data, *1* γ_{av} obtained from the least-squares method, *2* γ calculated from (4.137)

The temperature distribution in a layer of an evaporating liquid is described by

$$\frac{\partial \theta}{\partial t} = a\frac{\partial^2 \theta}{\partial z^2}; \quad t = 0, \quad \theta = \theta_0; \quad z = 0, \quad \lambda\frac{\partial \theta}{\partial z} = qJ; \quad z = h, \quad \theta = \theta_0, \tag{4.140}$$

where the coordinate z-axis is oriented normally to the liquid–gas interface ($z = 0$), θ and θ_0 (°C) are the temperatures of the liquid and the initial temperature; t (s) is time, λ (kcal/ms °C) is the thermal conductivity of the liquid, a (m^2/s) is the temperature diffusivity, q (kcal/kg) is the latent heat of the evaporation, J (kg/m^{2}s) is the evaporation rate, and h (m) is the thickness of the evaporating liquid.

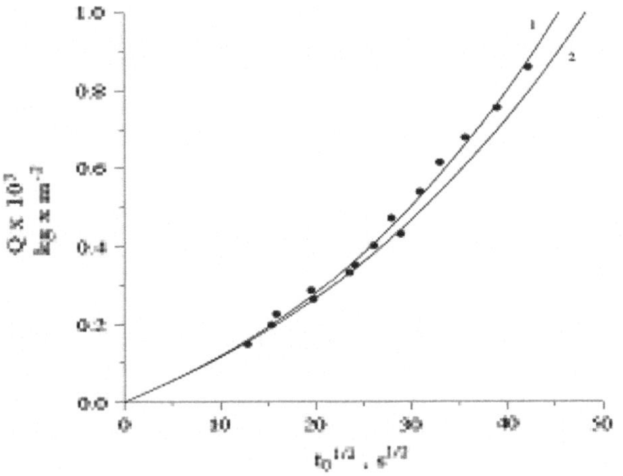

Fig. 23 Absorption of 100% Ar in C_2H_5O ($c^* = 0.427$ kg/m^3) at 20 °C. *Circles* experimental data, *1* γ_{av} obtained from the least-squares method, *2* γ calculated from (4.137)

The evaporation rate J in (4.140) may be determined through experimental data ($J = J_e$) concerning the amount of evaporated liquid Q_e (kg/m^2) at time t (s). Such data are available in [73]. After a sufficiently large initial time interval (where J has its greatest values), the relationship may be expressed as

$$Q_e = A_e\sqrt{t}, \tag{4.141}$$

where A_e (kg/m^{2}s$^{1/2}$) may be determined on the basis of the experimental data reported in [73] for systems such as H_2O–N_2, H_2O–He, H_2O–Ar, CH_3OH–Ar, C_2H_5OH–Ar, and i-C_3H_7OH–Ar (see Table 8). This allows us to define (by means of 4.141) the values of J_e as

$$J_e = \frac{dQ_e}{dt} = \frac{A_e}{2\sqrt{t}}. \tag{4.142}$$

Substitution of (4.142) into (4.140) permits us to define [74] the *temperature distribution* within the evaporating liquid layer with thickness h:

$$\theta - \theta_0 - \frac{qA_e\sqrt{\pi a}}{2\lambda}\sum_{n=0}^{\infty}(-1)^n\left[erfc\frac{2nh + z}{2\sqrt{at}} - erfc\frac{2(n+1)h - z}{2\sqrt{at}}\right]. \tag{4.143}$$

Equation (4.143) allows us to determine the temperature variations at the liquid top surface ($z = 0$) as a result of the evaporation process:

$$\theta_0 - \theta^* = \frac{qA_e\sqrt{\pi a}}{2\lambda}\left[1 + 2\sum_{n=1}^{\infty}(-1)^n erfc\frac{nh}{\sqrt{at}}\right], \tag{4.144}$$

where the liquid thickness was assumed to be 3×10^{-3} m.

It follows from (4.144) that the maximum temperature at the interface may be reached at the limiting situations of $t \to 0$ or $h \to 0$:

Table 8 Characteristic parameters of gas–liquid systems (20 °C)

Parameters	Systems					
	$H_2O–N_2$	$H_2O–He$	$H_2O–Ar$	$CH_3OH–Ar$	$C_2H_5OH–Ar$	i-$C_3H_7OH–Ar$
q (kcal/kg)	584.3	584.3	584.3	280.0	217.9	179.1
$a \times 10^7$ (m²/s)	1.43	1.43	1.43	1.05	0.888	0.752
$\lambda \times 10^5$ (kcal/msC)	1.448	1.448	1.448	4.875	4.015	3.657
$D \times 10^5$, (m²/s)	2.41	2.57	8.86	0.98	1.0	0.846
$\theta_0 - \theta^*$ (°C)	0.3	0.3	0.2	3.4	0.6	0.4
$A_e \times 10^4$ (kg/m²s$^{1/2}$)	2.051	2.190	1.633	19.47	3.963	3.287
$A_D \times 10^4$ (kg/m²s$^{1/2}$)	0.936	0.967	1.79	5.09	3.89	1.51
$A \times 10^4$ (kg/m²s$^{1/2}$)	0.954	0.961	1.94	5.92	4.28	1.57
c^* (kg/m³)	0.0142	0.0143	0.0161	0.142	0.0995	0.0939
c_0^* (kg/m³)	1.13	1.66	0.162	1.48	1.57	1.63
$\vartheta \times 10^5$ (m²/s)	1.441	1.360	12.12	1.360	1.360	1.360
ρ_0 (kg/m³)	1.16	1.66	0.166	1.66	1.66	1.66
$Q_{max} \times 10^2$ (kg/m²)	0.3660	0.3684	0.4129	3.644	2.558	2.414
α	−0.555	0.778	−1.216	−0.246	0.133	0.335
β	0.332	0	0.332	0.332	0	0
γ	1.70	0	1.7	1.7	0	0
ε	2.40	0	2.40	2.40	0	0
$\rho_0/c_0^*c_0^*$	1.02	1.02	1.02	1.12	1.05	1.04

$$\theta_0 - \theta^* = \frac{qA_e\sqrt{\pi a}}{2\lambda}. \qquad (4.145)$$

The result (4.145) can be obtained also in the case when the liquid layer is not subject to thermostatic conditions. In such a situation the last boundary condition in (4.140) becomes $z \to \infty$, $\theta. = \theta_0$ ($z > h_a = \sqrt{at}$), where h_a is the thickness of thermal boundary layer. Thus, the temperature distribution is

$$\theta = \theta_0 - \frac{qA_e\sqrt{\pi a}}{2\lambda} erfc \frac{z}{2\sqrt{at}}. \qquad (4.146)$$

Equation (4.146) shows that the results (4.145) follow directly at $z = 0$.

The results concerning the interface temperature ($\theta_0 - \theta^*$, °C) are listed in Table 8. It is clear that the temperature at the liquid surface practically remains unchanged as a result of the evaporation process. Some deviations from that "rule" are demonstrated by the CH_3OH–Ar system, but they change the partial pressure of the vapors above the liquid in the range of 10%, which should be neglected (the data in [73] have the same experimental error).

The experimental relationship $\theta(\sqrt{t})$ (obtained in [73]) shows that the asymptotic value $Q = Q_{max}$ is reached at large values of $\sqrt{t}$. It allows determination of the exact vapor concentration at the interface $c^* = Q_{max}/l$ $(\mathrm{kg/m^3})$, where $l = 0.257$ m is the gas-phase thickness (depth).

If the evaporation rate is limited by the nonstationary diffusion, the concentration distribution is [58, 65, 66]

$$c = c^* erfc \frac{z}{2\sqrt{Dt}}, \tag{4.147}$$

where c and c^* (kg/m^3) are the concentrations of the vapors in the gas phase and at the interface respectively, and D (m^2/s) is the diffusivity. The solution (4.147) permits us to obtain the rate of the diffusion:

$$J_D = -D\left(\frac{\partial c}{\partial z}\right)_{z=0} = c^*\sqrt{\frac{D}{\pi t}} = \frac{A_D}{2\sqrt{t}}, \quad A_D = 2c^*\sqrt{\frac{D}{\pi}}. \tag{4.148}$$

Comparison of the values of A_D and A_e (see Table 8) indicates significant differences that may be attributed to the occurrence of a Stefan flow [75]. Some strange behaviors are demonstrated by the C_2H_5OH–Ar ($A_e \approx A_D$) and H_2O–He ($A_e < A_D$) systems.

The difference between the evaporation rate and the rate of the nonstationary diffusion indicates that a convective contribution exists. The evaporation of a liquid in an inert gas is a result of a liquid–vapor phase transition, so there is a volumetric effect of a heterogeneous reaction at the interface [75] that creates the Stefan flow. If the process occurs in thermostatic conditions it is limited by both the diffusive and the convective transports in the gas phase. Thus, the evaporation rate can be expressed as

$$J = -D\left(\frac{\partial c}{\partial z}\right)_{z=0} + c^* v_s, \tag{4.149}$$

where v_s (m/s) is the velocity of the Stefan flow.

The mass transfer rate of the inert gas (in the gas phase) in the case of a binary gas–vapor system may be expressed in a similar manner:

$$J_0 = -D\left(\frac{\partial c_0}{\partial z}\right)_{z=0} + c_0^* v_s, \tag{4.150}$$

where c_0 and c_0^* (kg/m^3) are the concentrations of the inert gas in the bulk of the gas phase and at the interface. If the evaporating liquid is saturated by the inert gas (like in the experiments reported in [73]), it follows that

$$J_0 = 0, \quad c_0 + c = \rho = \rho_0 + \alpha c, \quad \frac{\partial c_0}{\partial z} = -(1-\alpha)\frac{\partial c}{\partial z}, \quad \alpha = \frac{\rho_V - \rho_0}{\rho_V}, \tag{4.151}$$

where ρ_V (kg/m^3) is the density of the vapor of the liquid, ρ_0(kg/m^3) is the density of the inert gas, and ρ (kg/m^3) is the density of the gas phase. In this way, we obtain from Eqs. 4.149 and 4.150

$$v_s = -\frac{D(1-\alpha)}{c_0^*}\left(\frac{\partial c}{\partial z}\right)_{z=0}, \quad J = -D\frac{\rho_0}{c_0^*}\left(\frac{\partial c}{\partial z}\right)_{z=0}, \tag{4.152}$$

Comparison of the velocity of the Stefan flow (4.152) and the velocity of the secondary flow induced by the large concentration gradients [69] shows that they are different when there is evaporation of liquid in inert gases under isothermal conditions (a thermostated system) since $\rho_0 \neq \rho^*$. Here

$$\rho^* = \rho_0 + \alpha c^*. \tag{4.153}$$

The *convective mass transfer* upon nonstationary evaporation from a stagnant liquid into a stagnant gas above it (within a large initial time interval) could be attributed to the Stefan flow and the natural convection. In the case of a gaseous layer above a stagnant liquid, the momentum equations of the gas phase and the convection–diffusion equations of the liquid vapors (under the assumption of a one-dimensional approximation) are

$$\frac{\partial v}{\partial t} + v\frac{\partial v}{\partial z} = \vartheta\frac{\partial^2 v}{\partial z^2} - \frac{1}{\rho_0}\frac{\partial p}{\partial z} - \frac{g\alpha c}{\rho_0}, \quad \frac{\partial c}{\partial t} + v\frac{\partial c}{\partial z} = D\frac{\partial^2 c}{\partial z^2};$$

$$t = 0, \quad v = c = 0; \quad z = 0, \quad v = -\frac{D(1-\alpha)}{c_0^*}\left(\frac{\partial c}{\partial z}\right)_{Z=0}, \quad c = c^*; \quad z \to \infty, \quad v = c = 0, \tag{4.154}$$

where ϑ (m^2/s) is the kinematic viscosity .

The coordinate z is oriented vertically upwards and the liquid interface is $z = 0$. In cases when the Stefan flow does not exist (see 4.152), the velocity is zero,

$$v(0,t) \equiv 0, \tag{4.155}$$

which leads to a stable solution of (4.154),

$$\bar{v} \equiv 0, \quad \bar{c} = a_1 z + a_2, \quad \frac{\partial \bar{p}}{\partial z} = g\alpha c, \tag{4.156}$$

i.e., the gas phase is stagnant, the concentration distribution is linear, and the pressure gradient depends on the concentration distribution along the gaseous layer depth [8]. This is a stable state of the system, but small disturbances could lead to a new stable state, where the motion of the gas phase is a result of the natural convection.

It is possible to introduce into (4.154) the dimensionless variables

$$t = t_0 T, \quad z = \delta Z, \quad v = u_0 V, \quad p = \rho_0 u_0^2 P, \quad c = c^* C, \tag{4.157}$$

where t_0 (s) is the characteristic timescale of the process. The length δ denotes the depth of the gaseous layer above the liquid where the principal changes of both the concentration and the velocity occur. The value of u_0 is the characteristic velocity scale. The results are

$$\frac{\partial V}{\partial T} + \frac{u_0 t_0}{\delta}V\frac{\partial V}{\partial Z} = -\frac{u_0 t_0}{\delta}\frac{\partial P}{\partial Z} + \frac{\vartheta t_0}{\delta^2}\frac{\partial^2 V}{\partial Z^2} - \frac{g\alpha t_0 c^*}{\rho_0 u_0}C, \quad \frac{\partial C}{\partial T} + \frac{u_0 t_0}{\delta}V\frac{\partial C}{\partial Z} = \frac{D t_0}{\delta^2}\frac{\partial^2 C}{\partial Z^2};$$

$$T = 0, \quad V = C = 0; \quad Z = 0, \quad V = -\frac{Dc^*(1-\alpha)}{c_0^*\delta u_0}\left(\frac{\partial C}{\partial Z}\right)_{z=0}, \quad C = 1;$$
$$Z \to \infty, \ V = C = 0. \tag{4.158}$$

The existence of the Stefan flow leads to the occurrence of flow inside the gas phase whose characteristic velocity is defined by the boundary condition in (4.158):

$$\frac{Dc^*(1-\alpha)}{c_0^*\delta u_0} \sim 1, \quad u_0 = \frac{Dc^*(1-\alpha)}{c_0^*\delta}, \tag{4.159}$$

i.e., both terms in the boundary conditions of the Stefan flow should have equal orders of magnitude.

If the evaporation rate is limited by the nonstationary diffusion, the parameters of both the nonstationary and the diffusion terms of the diffusion equation of the set (4.158) should have equal orders of magnitude:

$$\frac{Dt_0}{\delta^2} \sim 1, \quad \delta = \sqrt{Dt_0} \sim 10^{-2}\text{m} \quad \text{if} \quad t_0 \sim 10^2\text{s}. \tag{4.160}$$

In this way the characteristic velocity of (4.159) is

$$u_0 = \frac{c^*}{c_0^*}\sqrt{\frac{D}{t_0}}(1-\alpha) \sim 10^{-5}\text{m/s}. \tag{4.161}$$

Including (4.160) in the first equation in (4.158) leads to

$$\frac{\vartheta t_0}{\delta^2} = Sc \sim 1, \quad Sc = \frac{\vartheta}{D}, \tag{4.162}$$

If we assume that the flow is limited by the natural convection, the first equation in (4.158) becomes

$$\frac{\rho_0 u_0}{g\alpha t_0 c^*}\frac{\partial V}{\partial T} + \frac{\rho_0 u_0^2}{g\alpha\delta c^*}V\frac{\partial V}{\partial Z} = -\frac{\rho_0 u_0^2}{g\alpha\delta c^*}\frac{\partial P}{\partial Z} + \frac{\vartheta\rho_0 u_0}{\delta^2 g\alpha c^*}\frac{\partial^2 V}{\partial Z^2} - C. \tag{4.163}$$

In this particular case the parameters of the last two terms of (4.163) should have equal orders of magnitude:

$$\frac{\vartheta\rho_0 u_0}{\delta^2 g\alpha c^*} \sim 1, \quad \delta = \sqrt{\frac{\nu\rho_0}{g\alpha c_0^*}\sqrt{\frac{D}{t_0}}} \sim 10^{-5}\text{m}. \tag{4.164}$$

Conditions (4.160) and (4.164) indicate that the effects of the Stefan flow and the natural convection occur in different zones of the gaseous layer above the liquid surface. This fact permits us to separate these effects if the velocity, pressure, and concentration in the first equation in (4.158) may be expressed in the form

$$v + \bar{v}, \quad \bar{p}, \quad \bar{c}, \tag{4.165}$$

Here $\bar{v}$, $\bar{p}$, and $\bar{c}$ are determined by (4.156), whereas (4.154) with the last two terms of the first equation neglected determines the values of v and c. In this way, the form of (4.158) becomes

$$\begin{gathered}\frac{\partial V}{\partial T} + \beta V \frac{\partial V}{\partial Z} = Sc\frac{\partial^2 V}{\partial Z^2}, \quad \frac{\partial C}{\partial T} + \beta V \frac{\partial C}{\partial Z} = \frac{\partial^2 C}{\partial Z^2};\\ T = 0, \quad V = C = 0; \quad Z = 0, \quad V = -\frac{\partial C}{\partial Z}, \quad C = 1; \quad Z \to \infty, \quad V = C = 0,\end{gathered} \tag{4.166}$$

where β follows directly from (4.158) to (4.160)–(4.162),

$$\beta = (1 - \alpha)\frac{c^*}{c_0^*} \sim 10^{-1}. \tag{4.167}$$

Obviously Eq. 4.166 are valid within a broad initial time interval t_0, when the thickness of the diffusion boundary layer $\delta = \sqrt{Dt_0}$ is less than the depth of the gas phase l (in the cases studied in [73], $l = 0.257$ m).

The solution of (4.166) may be obtained as series of the powers of a small parameter β:

$$V = V_0 + \beta V_1, \quad C = C_0 + \beta C_1. \tag{4.168}$$

Thus, the zeroth-order approximation is

$$\begin{gathered}\frac{\partial V_0}{\partial T} = Sc\frac{\partial^2 V_0}{\partial Z^2}, \quad \frac{\partial C_0}{\partial T} = \frac{\partial^2 C_0}{\partial Z^2}; \quad T = 0, \quad V_0 = C_0 = 0;\\ Z = 0, \quad V_0 = -\frac{\partial C_0}{\partial Z}, \quad C_0 = 1; \quad Z \to \infty, \quad V_0 = C_0 = 0.\end{gathered} \tag{4.169}$$

The solution for C_0 is

$$C_0 = erfc\frac{Z}{2\sqrt{T}}. \tag{4.170}$$

The problem for V_0 is

$$\begin{gathered}\frac{\partial V_0}{\partial T} = Sc\frac{\partial^2 V_0}{\partial Z^2}; \quad T = 0, \quad V_0 = 0; \quad Z = 0, \quad V_0 = \varphi(T) = \frac{1}{\sqrt{\pi T}}; \quad Z \to \infty,\\ V_0 = 0.\end{gathered} \tag{4.171}$$

The solution of (4.171) may be obtained by Green's function [23]:

$$V_0 = \varphi(t)\exp\left(-\frac{Z^2}{4ScT}\right) - \int_0^T \frac{\varphi(\tau)+2\tau\varphi'(\tau)}{\sqrt{\tau}}\left[\int_0^{\frac{\tau Z}{\sqrt{4ScT\tau(T-\tau)}}} \exp(-u^2)du\right]d\tau. \tag{4.172}$$

This permits us to determine V_0 at $\varphi = (\pi T)^{-1/2}$:

$$V_0 = \frac{\exp\left(-\frac{Z^2}{4ScT}\right)}{\sqrt{\pi T}}. \tag{4.173}$$

The problem formulation for C_1 follows from (4.166):

$$\frac{\partial C_1}{\partial T} - \frac{\partial^2 C_1}{\partial Z^2} = V_0\frac{\partial C_0}{\partial Z} = -\frac{\exp\left[-\frac{Z^2}{4T}\left(1+\frac{1}{Sc}\right)\right]}{\pi T}; \tag{4.174}$$
$$T=0,\quad C_1=0;\quad Z=0,\quad C_1=0;\quad Z\to\infty,\quad C_1=0.$$

The solution obtained for (4.174) through Green's functions [23] is

$$C_1 = \frac{\exp\left(-\frac{Z^2}{4T}\right)}{2\pi\sqrt{\pi}}\int_0^T \frac{1}{\tau\sqrt{T-\tau}}\int_0^\infty \left\{\exp\left[-\frac{\xi(T-Z\tau)^2}{4T\tau(T-\tau)}\right] - \exp\left[-\frac{(\xi T+Z\tau)^2}{4T\tau(T-\tau)}\right]\right\} \tag{4.175}$$
$$\times\exp\left(-\frac{\xi^2}{4Sc\tau}\right)d\xi d\tau.$$

From (4.175) it follows that

$$\left(\frac{\partial C_1}{\partial Z}\right)_{Z=0} = -\frac{2}{\pi}\sqrt{\frac{Sc}{\pi T}}\operatorname{arctg}\left(Sc^{-1/2}\right). \tag{4.176}$$

The amount of evaporated liquid Q (kg/m^2) is obtained from (4.152), (4.168), (4.170), and (4.176):

$$Q = \int_0^{t_0} Jdt = -c^*\frac{\rho^*}{c_0^*}\sqrt{Dt_0}\int_0^1\left[\left(\frac{\partial C_0}{\partial Z}\right)_{Z=0} + \frac{c^*(1-\alpha)}{c_0^*}\left(\frac{\partial C_1}{\partial Z}\right)_{Z=0}\right]dT = A\sqrt{t_0}, \tag{4.177}$$

where

$$A = 2c^*\frac{\rho_0}{c_0^*}\left(\sqrt{\frac{D}{\pi}} + \frac{2c^*(1-\alpha)}{\pi c_0^*}\sqrt{\frac{\vartheta}{\pi}}\operatorname{arctg}\left(Sc^{-1/2}\right)\right). \tag{4.178}$$

Expression (4.178) permits us to calculate the parameter A for various systems, summarized in Table 8. The values are close to those of A_D, but in the dominating situations are quite different from the values of A_e. This indicates the existence of an additional convective transport, which could be provoked by a loss of stability of the system. Thus, the small disturbances grow until stable amplitudes are established and the dissipative structures formed have a greater rate of the transport processes.

The results obtained here and their comparisons with the experimental data in [73] indicate that during the evaporation of low-volatility liquids (H_2O, CH_3OH, C_2H_3OH, i-C_3H_7OH) in gaseous media of indifferent gases (N_2, He, Ar) and external thermostatic conditions the temperature of the liquid surface is practically constant. This shows that there is no possibility to create thermocapillary flow as suggested in [73]. Under such conditions the mass transfer in the gas phase limits the evaporation rate, where the convective contribution depends on the Stefan flow.

Comparison of the theoretical and the experimental results shows the existence of an additional convective transport probably induced not only by the Stefan flow but also by the loss of stability of the system.

The experimental data in [73] show that at large characteristic time ($\delta = \sqrt{Dt_0} \geq l = 0.257\text{m}$), $Q = \text{const.}$, which may be explained by the non-stationary diffusion between both surfaces. Thus, the boundary condition in (4.154), $z \to \infty$, $c = 0$, must be replaced by $z = l$, $\frac{\partial c}{\partial z} = 0$. Under conditions imposed by small characteristic times (where $\delta < 1$), the determination of the additional convective transport (induced by the system instability and the onset of a natural convection) is required.

The *stability of the evaporation process* described by model (4.154) will be investigated [68] by means of perturbations of the velocity (v'), pressure (p'), and concentration (c'). Their superposition on the main process leads to

$$v + v', \quad p + p', \quad c + c', \tag{4.179}$$

where (4.179) satisfies model (4.154). The result is two sets of equations:

$$\begin{gathered}\frac{\partial v}{\partial t} + v\frac{\partial v}{\partial z} = \vartheta\frac{\partial^2 v}{\partial z^2}, \quad \frac{\partial c}{\partial t} + v\frac{\partial c}{\partial z} = D\frac{\partial^2 c}{\partial z^2};\\ t = 0, \quad v = c = 0; \quad z = 0, \quad v = -\frac{D(1-\alpha)}{c_0^*}\left(\frac{\partial c}{\partial z}\right)_{z=0}, \quad c = c^*; \quad z = l, \quad v = 0, \quad \frac{\partial c}{\partial z} = 0.\end{gathered} \tag{4.180}$$

$$\frac{\partial v'}{\partial t} + v'\frac{\partial v'}{\partial z} + v\frac{\partial v'}{\partial z} + v'\frac{\partial v}{\partial z} = -\frac{1}{\rho_0}\left(\frac{\partial p}{\partial z} + \frac{\partial p'}{\partial z}\right) + \vartheta\frac{\partial^2 v'}{\partial z^2} - \frac{\alpha g}{\rho_0}(c + c'),$$

$$\begin{gathered}\frac{\partial c'}{\partial t} + v'\frac{\partial c'}{\partial z} + v\frac{\partial c'}{\partial z} + v'\frac{\partial c}{\partial z} = D\frac{\partial^2 c'}{\partial z^2};\\ z = 0, \quad v' = c' = 0; \quad z = 0, \quad v' = -\frac{D(1-\alpha)}{c_0^*}\left(\frac{\partial c'}{\partial z}\right)_{z=0}, \quad c' = 0; \quad z = l, \quad v' = 0.\end{gathered} \tag{4.181}$$

The dimensionless variables

$$t = t_0 T, \quad z = lZ, \quad v = u_0 V, \quad c = c^* C, \tag{4.182}$$

may be introduced into (4.180) and the new form of the sets of equations becomes

$$\frac{\partial V}{\partial T} + \frac{u_0 t_0}{l} V \frac{\partial V}{\partial Z} = \frac{\vartheta t_0}{l^2} \frac{\partial^2 V}{\partial Z^2}, \quad \frac{\partial C}{\partial T} + \frac{u_0 t_0}{l} V \frac{\partial C}{\partial Z} = \frac{D t_0}{l^2} \frac{\partial^2 C}{\partial Z^2};$$
$$T = 0, \quad V = C = 0; \quad Z = 0, \quad V = -\frac{D(1-\alpha)c^*}{c_0^* l u_0} \left(\frac{\partial C}{\partial Z} \right)_{z=0}, \tag{4.183}$$
$$C = 1; \quad Z = 1, \quad V = 0, \quad \frac{\partial C}{\partial Z} = 0.$$

The characteristic velocity of the flow as a result of the Stefan flow is determined from the boundary condition at the interface ($z = 0$):

$$\frac{D(1-\alpha)c^*}{c_0^* l u_0} \sim 1, \quad u_0 = \frac{D(1-\alpha)c^*}{c_0^* l} \sim 10^{-6} \text{m/s}. \tag{4.184}$$

This allows us to determine also the orders of magnitude of the other parameters in (4.183):

$$\frac{u_0 t_0}{l} \sim 10^{-2}, \quad \frac{v t_0}{l} \sim \frac{D t_0}{l^2} \sim 1. \tag{4.185}$$

It is clear that from (4.183) to (4.185) if we neglect the convective terms we may obtain the solution of (4.180):

$$v \frac{\partial v}{\partial z} \sim v \frac{\partial c}{\partial z} \sim 0. \tag{4.186}$$

The solution of (4.180) with respect to the concentration (under condition 4.186) can be obtained by a Laplace transformation [74] in the form

$$c = c^* \sum_{n=0}^{\infty} (-1)^n \left[erfc \frac{2nl + z}{2\sqrt{Dt}} + erfc \frac{(2n+2)l - z}{2\sqrt{Dt}} \right]. \tag{4.187}$$

Thus,

$$\left(\frac{\partial c}{\partial z} \right)_{z=0} = -\frac{c^*}{\sqrt{\pi D t}} \left[1 + 2 \sum_{n=1}^{\infty} (-1)^n \exp\left(-\frac{n^2 l^2}{Dt} \right) \right], \tag{4.188}$$

and the velocity of the Stefan flow is

$$v(0,t) = (1-\alpha) \frac{c^*}{c_0^*} \sqrt{\frac{D}{\pi t}} \left[1 + 2 \sum_{n=1}^{\infty} (-1)^n \exp\left(-\frac{n^2 l^2}{Dt} \right) \right]. \tag{4.189}$$

The solution of (4.181) is derived in the form of "normal" perturbations:

$$v' = v_1(z,t)\exp(-\lambda t), \quad p' = p_1(z,t)\exp(-\lambda t), \quad c' = c_1(z,t)\exp(-\lambda t), \tag{4.190}$$

where v_1, p_1, and c_1 are amplitudes of the perturbations.

The problem concerning the determination of v_1, p_1, and c_1 is an eigenvalue problem, where λ is the eigenvalue and v_1, p_1, and c_1 are the eigenfunctions. The solution is derived in the form of "neutral" perturbations, whose amplitudes neither attenuate nor grow with time, i.e., eigenfunctions at $\lambda = 0$. Thus, from (4.180) to (4.181) we obtain

$$\begin{aligned}
&v_1\frac{\partial v_1}{\partial z} + v\frac{\partial v_1}{\partial z} + v_1\frac{\partial v}{\partial z} = -\frac{1}{\rho_0}\left(\frac{\partial p}{\partial z} + \frac{\partial p_1}{\partial z}\right) + \vartheta\frac{\partial^2 v_1}{\partial z^2} - \frac{g\alpha}{\rho_0}(c + c_1),\\
&v_1\frac{\partial c_1}{\partial z} + v\frac{\partial c_1}{\partial z} + v_1\frac{\partial c}{\partial z} = D\frac{\partial^2 c_1}{\partial z^2};\\
&t = 0, \quad v_1 = c_1 = 0; \quad z = 0, \quad v_1 = -\frac{D(1-\alpha)}{c_0^*}\left(\frac{\partial c_1}{\partial z}\right)_{Z=0}, \quad c_1 = 0; \quad z = l, \quad v_1 = 0.
\end{aligned} \tag{4.191}$$

The solution of (4.191) is achieved by means of dimensionless variables:

$$t = t_0 T,\ z = l_1 Z_1,\ v = u_1 V,\ v_1 = u_1 V_1,\ c = c^* C,\ c_1 = c^* C_1,\ p = \rho_0 u_1^2 P,\ p_1 = \rho_0 u_1^2 P_1, \tag{4.192}$$

The substitution of (4.192) into (4.191) leads to

$$\begin{aligned}
&\frac{u_1^2\rho_0}{l_1 g\alpha c^*}\left(V_1\frac{\partial V_1}{\partial Z} + V\frac{\partial V_1}{\partial Z} + V_1\frac{\partial V}{\partial Z}\right) = -\frac{u_1^2\rho_0}{l_1 g\alpha c^*}\left(\frac{\partial P}{\partial Z} + \frac{\partial P_1}{\partial Z}\right) + \frac{\vartheta u_1\rho_0}{l_1^2 g\alpha c^*}\frac{\partial^2 V_1}{\partial Z^2} - C - C_1,\\
&\frac{u_1 l_1}{D}\left(V_1\frac{\partial C_1}{\partial Z} + V\frac{\partial C_1}{\partial Z} + V_1\frac{\partial C}{\partial Z}\right) = \frac{\partial^2 C_1}{\partial Z^2}; \quad T = 0, \quad V_1 = C_1 = 0;\\
&Z = 0, \quad V_1 = -\frac{D(1-\alpha)c^*}{u_1 l_1 c_0^*}\left(\frac{\partial C_1}{\partial Z}\right)_{z=0}; \quad C_1 = 0; \quad Z = 1, \quad V_1 = C_1 = 0.
\end{aligned} \tag{4.193}$$

It is possible to suggest that in (4.193) the process is limited by the natural convection and the velocity at the interface, i.e.,

$$\frac{\vartheta u_1\rho_0}{l_1^2 g\alpha c^*} = 1, \quad \frac{D(1-\alpha)c^*}{u_1 l_1 c_0^*} = 1. \tag{4.194}$$

This permits us to define the characteristic parameters (the velocity and the thickness of the boundary layer) and their orders of magnitude:

$$u_1 = D(1-\alpha)\frac{c^*}{c_0^*}\sqrt[3]{\frac{g\alpha c_0^*}{D(1-\alpha)\vartheta\rho_0}} \sim 10^{-3}\text{m/s}, \quad l_1 = \sqrt[3]{\frac{(1-\alpha)\vartheta\rho_0 D}{\overset{*}{0}}} \sim 10^{-4}\text{m}. \tag{4.195}$$

Expressions (4.195) permit us to define the order of magnitude of the dimensionless parameters of (4.193) too:

$$\frac{u_1^2\rho_0}{l_1 g c^*} \sim 10^{-2}, \quad \frac{u_1 l_1}{D} \sim 10^{-2}. \tag{4.196}$$

This allows us to express (4.193) in a zeroth-order approximation with respect to the small parameters (4.196):

$$\frac{\partial^2 V_1}{\partial Z_1^2} = C + C_1, \quad \frac{\partial^2 C_1}{\partial Z_1^2} = 0;$$
$$Z_1 = 0, \quad \left(\frac{\partial C_1}{\partial Z}\right)_{Z=0} = -V_1(0,T), \quad C_1 = 0; \quad Z_1 = 1, \quad V_1 = C_1 = 0. \tag{4.197}$$

The concentration $c = c^*C$ in (4.197) varies within a layer with thickness $\delta \sim 10^{-2}$ m (see 4.182), whereas the concentration $c_1 = c_1{}^*C$ varies within the range defined by a layer of depth $l \sim 10^{-4}$. Thus, the form of (4.197) allows us to introduce $C = 1$ for the surface concentration.

The solution of (4.197) concerning C_1 may be obtained with accuracy defined by an arbitrary function $V_1(0,T)$:

$$C_1 = (1 - Z_1)\,V_1(0,T). \tag{4.198}$$

The solution concerning the velocity can be obtained in a similar manner:

$$V_1 = \frac{1}{2}Z_1^2 + V_1(0,T)\left(\frac{Z_1^2}{2} - \frac{Z_1^3}{6}\right) - \left[\frac{1}{2} + \frac{4}{3}V(0,T)\right]Z_1 + V_1(0,T). \tag{4.199}$$

The velocity $V_1(0,\mathrm{T}) = \frac{v_1(0,t)}{u_1}$ cannot be determined theoretically under the basic assumptions (approximations) of the linear stability theory because the normal component of the disturbance velocity $v_1(0,t)$ *is related to the disturbance amplitude*. However, it is possible to establish it from experimental data.

If we suppose [68] that the velocity of the disturbance differs from the velocity of the main flow (4.189), $v_1(0,t)$ can be represented as

$$v_1(0,\, t) = \gamma(1-\alpha)\frac{c^*}{c_0^*}\sqrt{\frac{D}{\pi t}}\left[1 + 2\varepsilon\sum_{n=1}^{\infty}(-1)^n \exp\left(-\frac{n^2 l_1^2}{Dt}\right)\right], \quad l_1 = \beta l, \tag{4.200}$$

where γ is the scale of the disturbance velocity, ε is the scale of the influence of the nonlinear mass transfer effect on the disturbance, and β is the scale of the penetrating depth of the influence of the nonlinear mass transfer effect on the disturbance. The parameters β, γ, and ε should be obtained from experimental data.

Introduction of Eq. 4.200 into the boundary condition in (4.191) at $z = 0$ leads to

$$\left(\frac{\partial c_1}{\partial z}\right)_{z=0} = -\gamma \frac{c^*}{\sqrt{\pi D t}}\left[1 + 2\varepsilon \sum_{n=1}^{\infty} (-1)^n \exp\left(-\frac{n^2 l_1^2}{Dt}\right)\right], \tag{4.201}$$

The mass transfer rate upon evaporation depends on both the diffusive and the convective transports through the liquid–gas interface:

$$J = -D\left(\frac{\partial c}{\partial z} + \frac{\partial c_1}{\partial z}\right)_{z=0} + c^*(v + v_1)_{z=0} = -\frac{D\rho_0}{c_0^*}\left(\frac{\partial c}{\partial z} + \frac{\partial c_1}{\partial z}\right)_{z=0}, \tag{4.202}$$

The introduction of (4.188) and (4.201) into (4.202) leads to

$$J = \frac{\rho_0 c^*}{c_0^*}\sqrt{\frac{D}{\pi t}}\left[1 + 2\sum_{n=1}^{\infty} (-1)^n \exp\left(-\frac{n^2 l^2}{Dt}\right)\right] + \gamma \frac{\rho_0 c^*}{c_0^*}\sqrt{\frac{D}{\pi t}}\left[1 + 2\varepsilon \sum_{n=1}^{\infty} (-1)^n \exp\left(-\frac{n^2 \beta^2 l^2}{Dt}\right)\right]. \tag{4.203}$$

Taking into account that

$$\rho^* = \rho_0 + \alpha c^* = c_0^* + c^*, \tag{4.204}$$

it is possible to find the amount of evaporated liquid through (4.203):

$$Q = \int_0^{t_0} J dt = A\left[(1+\gamma)\sqrt{t} + \sum_{n=1}^{\infty} (-1)^n \int_0^{t_0} \frac{1}{\sqrt{t}} \exp\left(-\frac{n^2 l^2}{Dt}\right) dt + \varepsilon\gamma \sum_{n=1}^{\infty} (-1)^n \int_0^{t_0} \frac{1}{\sqrt{t}} \exp\left(-\frac{n^2 \beta^2 l^2}{Dt}\right) dt,\right] \tag{4.205}$$

where

$$A = 2\frac{\rho_0 c^*}{c_0^*}\sqrt{\frac{D}{\pi}}. \tag{4.206}$$

The experimental data concerning evaporation of various liquids in [73] permits the determination of the values of β, γ and ε. In the calculation of the sums in (4.205) three to four terms were required for the first sum and six to eight terms were required for the second sum.

Figures 24, 25, 26, 27, 28 and 29 show a comparison between the values of Q determined through (4.205) and the experimental data in [73]. The corresponding values of β, γ, and ε are given in Table 8.

Fig. 24 Evaporation in the H_2O–N_2 system

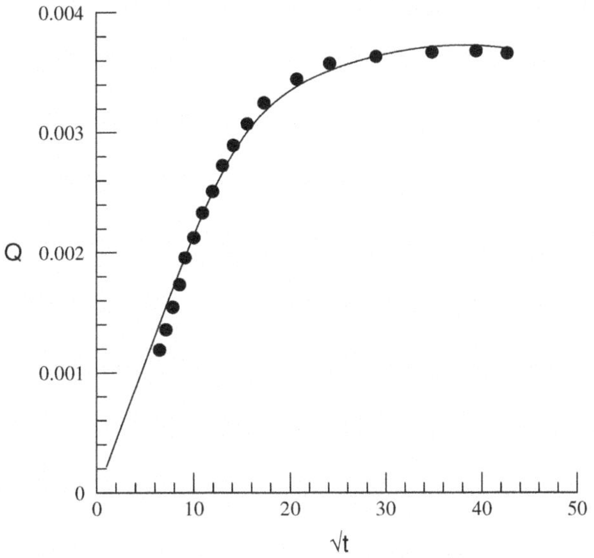

The effect of the Stefan flow is determined by the ratio $\rho_0/c_0{}^*$, whose values are shown in Table 8. The results obtained show that the Stefan flow is greatest for the evaporation of CH_3OH.

The results presented [68] indicate that the rate of evaporation from a stagnant liquid into a gaseous layer having a limited depth is controlled by the rate of the nonstationary diffusion. The concentration of the vapors at the liquid surface is practically constant, whereas the upper boundary of the gaseous layer is

Fig. 25 Evaporation in the H_2O–He system

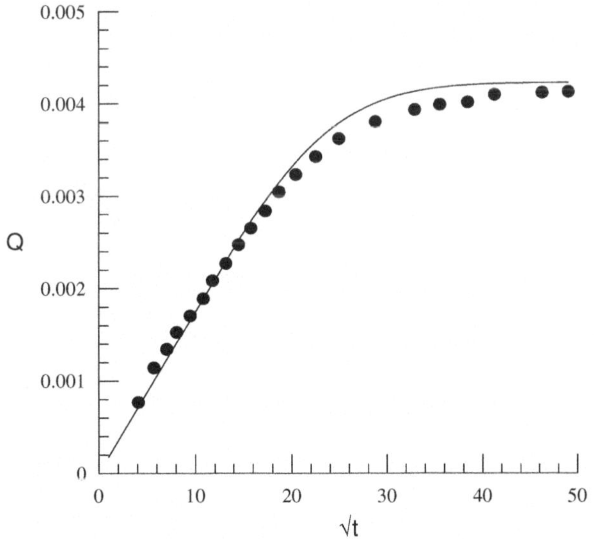

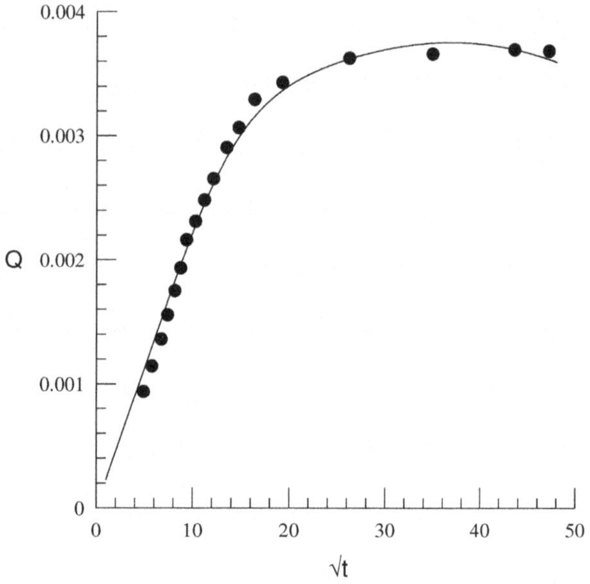

Fig. 26 Evaporation in the H_2O–Ar system

impermeable for the vapors (i.e., diffusive flux is zero). This effect is augmented by the contribution of a convective transport in the gas phase. Both the Stefan flow and the natural convection contribute to this latter effect. The Stefan flow occurs owing to the phase transition at the interface, whereas the natural convection is caused by the loss of stability of the entire process as a result of the growth of

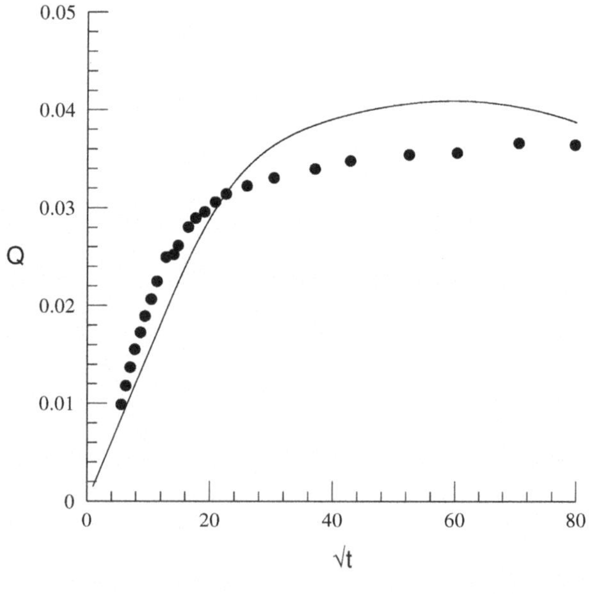

Fig. 27 Evaporation in the CH_3OH–Ar system

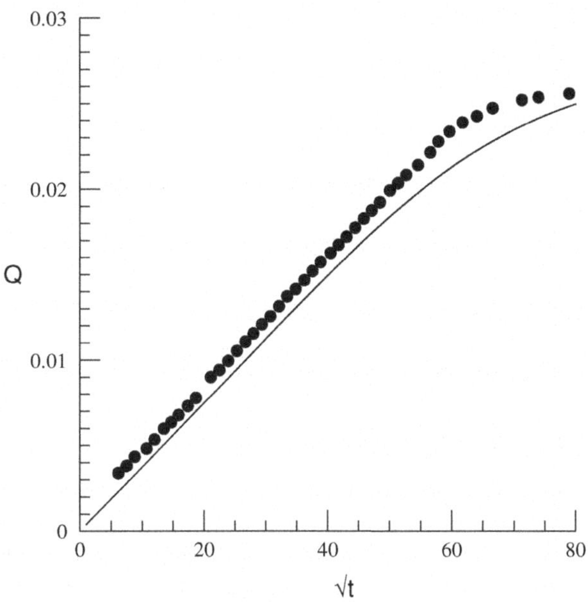

Fig. 28 Evaporation in the C_2H_5OH–Ar system

small perturbations. The latter leads to a self-organization of the system and the formation of dissipative structures.

It can be seen from Table 8 that in the cases when the vapors of the liquid are denser than the inert gas (H_2O–He, C_2H_5OH–Ar, *i*-C_3H_7OH–Ar) the process is stable ($\beta = \gamma = \varepsilon = 0$) and the rate of evaporation can be determined from the nonstationary diffusion rate (increased by 2–5% by the Stefan flow effect, ρ_0/c_0*).

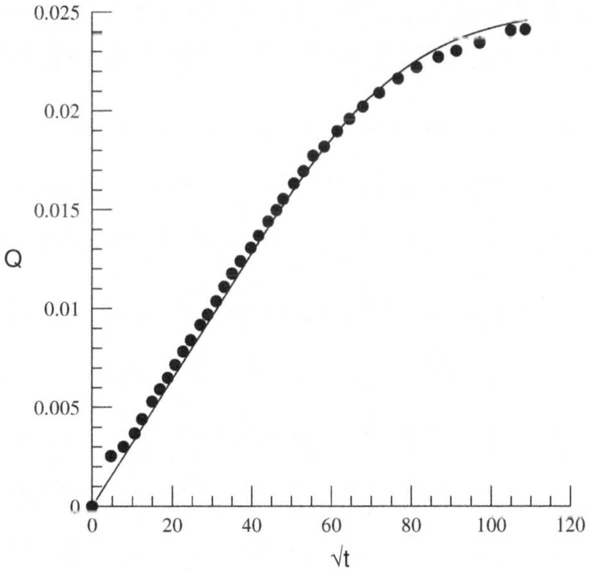

Fig. 29 Evaporation in the *i*-C_3H_7OH–Ar system

The process is unstable when the vapors are less dense than the gas. In these conditions a natural convection appears as a result of the instability. Thus, the evaporation rate ($\gamma = 1.70$) is essentially increased and is 2.7 times larger than the diffusion rate.

It is important to note that the parameters of the dissipative structure (as a result of instability) are equal ($\beta = 0.332$, $\gamma = 1.7$, $\varepsilon = 2.4$) for different liquid–gas systems (H_2O–N_2, H_2O–Ar, CH_3OH–Ar). The deviation of the CH_3OH–Ar system could be a result of experimental errors [68].

5 Examples

5.1 Gas–Liquid System

Significant interaction between flows in gas and liquid will be observed if a movable liquid surface replaces the unmovable solid surface. There will also be the effect of induction of secondary flows as a result of intensive interphase mass transfer in gas–liquid systems, but this effect is superposed with the hydrodynamic interaction between the above-mentioned two phases. The stability under these conditions is not only of theoretical interest, but is also of practical interest in view of the fact that it defines the rate of a number of industrial absorption and desorption processes.

The mathematical model of the nonlinear mass transfer in gas–liquid systems (see Sect. 3.3 in Chap. 3) will be considered in the approximations of the boundary layer theory [29, 30, 41], taking into account that the diffusive resistance is concentrated in the gas phase [31]. It was shown in [32] that the nonlinear effects in the liquid can be neglected. The mathematical description has the following form:

$$u_j\frac{\partial u_j}{\partial x}+v_j\frac{\partial u_j}{\partial y}=\nu_j\frac{\partial^2 u_j}{\partial y^2},\quad \frac{\partial u_j}{\partial x}+\frac{\partial v_j}{\partial y}=0,\quad j=1,2,\quad u_1\frac{\partial c}{\partial x}+v_1\frac{\partial c}{\partial y}=D_1\frac{\partial^2 c}{\partial y^2},\tag{5.1}$$

where the indexes 1 and 2 denote the gas and the liquid, respectively. The boundary conditions of (5.1) are

$$\begin{aligned}&x=0,\quad u=u_{j0},\quad c=c_0,\quad j=1,2;\quad y=0,\quad u_1=u_2,\quad c=c^*,\\&\mu_1\frac{\partial u_1}{\partial y}=\mu_2\frac{\partial u_2}{\partial y},\\&v_1=-\frac{MD_1}{\rho_{0j}^*}\frac{\partial c}{\partial y},\quad v_2=0;\quad y\to\infty,\quad u_1=u_{10},\quad c=c_0;\quad y\to-\infty,\\&u_2=u_{20}.\end{aligned}\tag{5.2}$$

Problem (5.1) and (5.2) was solved numerically and asymptotically [13, 29, 33]. Further, similarity variables will be applied:

$$u_j = 0.5 j u_{j0} \varepsilon_j \Phi_j', \quad v_j = (-1)^{j-1} 0.5 j \left(\frac{u_{j0} v_j}{x}\right)^{0.5} \left(\xi_j \Phi_j' - \Phi_j\right),$$
$$c = c_0 - (c_0 - c^*)\Psi,$$

$$\Phi_j = \Phi_j(\eta_j), \quad \Psi_j = \Psi_j(\eta_j), \quad \eta_j = (-1)^{j-1} y \left(\frac{u_{j0}}{4D_j x}\right)^{0.5}, \quad \varepsilon_j = Sc_j^{0.5},$$
$$Sc_j = \frac{v_j}{D_j}, \quad j = 1, 2. \tag{5.3}$$

The substitution of (5.3) into (5.1, 5.2) leads to

$$\Phi_1''' + \varepsilon_1^{-1} \Phi_1 \Phi_1'' = 0, \quad \Phi_2''' + 2\varepsilon_2^{-1} \Phi_2 \Phi_2'' = 0, \quad \Psi'' + \varepsilon_1 \Phi_1 \Psi' = 0,$$

$$\Phi_1(0) = -\theta_3 \Psi'(0), \quad \Phi_2(0) = 0, \quad \Phi_1'(\infty) = \frac{2}{\varepsilon_1}, \quad \Phi_2'(\infty) = \frac{1}{\varepsilon_2},$$
$$\Phi_1'(0) = 2\theta_1 \frac{\varepsilon_2}{\varepsilon_1} \Phi_2'(0),$$

$$\Phi_2''(0) = -0.5\theta_2 \left(\frac{\varepsilon_1}{\varepsilon_2}\right)^2 \Phi_1''(0), \quad \Psi(0) = 1, \quad \Psi(\infty) = 0. \tag{5.4}$$

The solution of Eq. 5.4 is obtained [13, 33] by determining the initial values of f, which allows us to define the velocity profiles in the gas and the liquid as solutions of a problem with initial conditions

$$2f_j''' + f_j f_j'' = 0, \quad f_j = f_j(\xi_j), \quad \xi_j = \frac{2}{\varepsilon_j} \eta_j,$$

$$f_j(0) = a_j, \quad f_j'(0) = b_j, \quad f_j''(0) = c_j, \quad f_j'(\infty) = 1, \quad j = 1, 2, \tag{5.5}$$

where

$$a_1 = a_{10}, \quad b_1 = \frac{\varepsilon_1}{2} b_{10}, \quad c_1 = \frac{\varepsilon_1^2}{4} c_{10}, \quad a_2 = 0, \quad b_2 = \varepsilon_2 b_{20}, \quad c_2 = -\frac{\varepsilon_2^2}{2} c_{20}, \tag{5.6}$$

where the values of a_{10}, b_{10}, c_{10}, b_{20}, and c_{20} for $\varepsilon_1 = 1$, $\varepsilon_2 = 20$, $\theta_1 = 0.1$, and $\theta_2 = 0.152$ are given in Tables 9 and 10.

It was shown in [34] that the Orr–Sommerfeld equation in the approximations of almost parallel flows has the same form for the gas and the liquid, as follows:

$$(f' - C)(\varphi'' - A^2\varphi) - f'''\varphi$$
$$= -\frac{i}{ARe}\left\{(\varphi^{IV} - 2A^2\varphi'' + A^4\varphi) - \frac{1}{2}(\xi f' - f)\varphi''' + \left[\frac{1}{2}(\xi f''' + f'') + \frac{A^2}{2}(\xi f' - f)\right]\right\}\varphi';$$

Table 9 Liquid flow ($\varepsilon_2 = 20$, $\theta_1 = 0.1$, $\theta_2 = 0.152$, $\theta_3 = \theta$)

θ	b_{20}	c_{20}	k
−0.3	0.0546	0.00033	−0.1
0.0	0.0536	0.00026	−0.086
0.3	0.0527	0.00022	−0.13

Table 10 Gas flow ($\varepsilon_1 = 1$, $\theta_1 = 0.1$, $\theta_2 = 0.152$, $\theta_3 = \theta$)

θ	a_{10}	b_{10}	c_{10}	k
−0.3	0.2797	0.2185	1.662	0.953
−0.2	0.1703	0.2166	1.520	1.133
−0.1	0.07852	0.2152	1.402	1.301
0.0	0.0	0.2138	1.304	1.428
0.1	−0.06822	0.2129	1.220	1.552
0.2	−0.1283	0.2118	1.084	1.665
0.3	−0.1816	0.2107	1.084	1.768

$$\xi = 0, \quad \varphi = 0, \quad \varphi' = 0; \quad \xi = \xi_\infty \geq 6, \quad (\varphi'' - A^2\varphi') - \gamma(\varphi'' - A^2\varphi) = 0,$$
$$k = \lim_{\xi \to \infty} (\xi f' - f),$$

$$(\varphi'' - \gamma^2\varphi') + A(\varphi'' - \gamma^2\varphi) = 0, \quad \gamma = \frac{k}{4} - \frac{\sqrt{k^2 + 16A[A + iRe(1 - C)]}}{4}, \tag{5.7}$$

where

$$f(\xi) = f_j(\xi_j), \quad \xi = \xi_j, \quad \varphi = \varphi_j, \quad \gamma = \gamma_j, \quad k = k_j, \quad j = 1, 2. \tag{5.8}$$

The values of k_j (j = 1, 2) are calculated and are shown in Tables 9 and 10.

The neutral curves (Re,A) and (Re,C_r) for the gas are obtained and the critical Reynolds numbers, corresponding to wave numbers and phase velocities, are presented in Tables 11 and 12.

It can be clearly seen that the direction of the influence of the intensive interphase mass transfer on the hydrodynamic stability of the flow in the gas-phase boundary layer is analogous to the case of the solid interphase surface. Hence, in the case of absorption ($\theta_3 > 0$), increase of stability is observed. In the opposite case of desorption ($\theta_3 < 0$), the stability decreases. The motion of the interface [$f'(0) > 0$] leads to a decrease of the velocity gradients, which is the cause of stabilization of the flow in all cases (increase of Re_{cr}).

The solution of (5.7) for the liquid phase ($f = f_2$) shows that the flow is stable at large Reynolds numbers ($Re \approx$ 25,000), which can be explained by the fact that the velocity gradient in the liquid boundary layer is low and is shaped as the profile of the Couette flow.

The effects of the intensive interphase mass transfer in gas–liquid systems appear as a difference in the rates of absorption and desorption. In the cases where the process is limited by the diffusion resistance in the gas phase, this difference is

Table 11 Values of the critical Reynolds numbers Re_{cr}, wave velocities C_r, wave numbers A, A_{min}, and $C_{r\ min}$

θ	Re_{cr}	A	C_r	A_{min}	$C_{r\ min}$
−0.3	2,511	0.270	0.3863	0.304	0.3878
−0.2	1,605	0.285	0.4095	0.325	0.4108
−0.1	1,078	0.295	0.4264	0.341	0.4281
0.0	795	0.305	0.4469	0.356	0.4493
0.3	397	0.330	0.4866	0.398	0.4902
0.2	483	0.320	0.4749	0.386	0.4786
0.1	605	0.315	0.4620	0.373	0.4645

explained by the Marangoni effect, which manifests itself in the liquid phase. The higher rate of absorption (compared with the rate of desorption) can be explained by the effect of nonlinear mass transfer, i.e., the influence of the induced secondary flow on the kinetics of the mass transfer. Cases where the desorption rate is higher than the absorption rate can be explained by loss of stability and transition to turbulence, since it is possible for the flow in the gas phase to be turbulent for desorption and laminar for absorption at equal Reynolds numbers.

5.2 Liquid–Liquid System

The nonlinear effects in the case of an intensive interphase mass transfer between two liquids can manifest themselves with the same intensity in both phases. In a number of extraction processes where the motion of one of the phases (the dispersion medium) induces motion in the other (the dispersed phase), these effects

Table 12 Computed values of A_j, B_j, C_j, and k; $(m/b = 0, \theta_1 = \theta, \theta_2 = 0)$, $(b/m = 1, \theta_1 = \theta_2 = \theta)$

$\varepsilon = 10$	θ	A_j	B_j	C_j	k
$m/b = 0$	−0.5	0.6652	0.439	0.265	0.673
$m/b = 0$	−0.3	0.0329	0.420	0.265	0.747
$m/b = 0$	−0.1	0.0094	0.405	0.265	0.805
$m/b = 0$	0.0	0.0	0.400	0.265	0.823
$m/b = 0$	0.1	−0.0082	0.394	0.265	0.846
$m/b = 0$	0.3	−0.0221	0.384	0.265	0.883
$m/b = 0$	0.5	−0.0334	0.375	0.265	0.915
$b/m = 1$	−0.5	0.0211	0.413	0.265	0.773
$b/m = 1$	−0.3	0.0128	0.407	0.265	0.800
$b/m = 1$	−0.1	0.0043	0.402	0.265	0.820
$b/m = 1$	0.0	0.0	0.400	0.265	0.823
$b/m = 1$	0.1	−0.0043	0.396	0.265	0.836
$b/m = 1$	0.3	−0.0128	0.390	0.265	0.862
$b/m = 1$	0.5	−0.0211	0.385	0.265	0.880

are of great interest. Further, one could consider [35] the hydrodynamic stability under the condition of an intensive interphase mass transfer between two liquid phases where the velocity in the volume of one of them is zero.

The mathematical model of the nonlinear mass transfer in liquid–liquid systems where the first liquid is in motion over the second one (which is at rest) can be obtained from the "gas–liquid" model (equations 3.3.34, 3.3.35) when $u_{20} = 0$ and $\chi = m$, where m is the distribution coefficient and the indexes 1 and 2 in equations (3.3.34) and (3.3.35) denote liquid 1 and liquid 2, respectively.

The problem formulated has been solved numerically [29, 36] and the boundary values for the velocity and its derivatives have been obtained. This gives us the opportunity to generate the velocity profiles of the following hydrodynamic problem:

$$u_j \frac{\partial u_j}{\partial x} + v_j \frac{\partial v_j}{\partial y} = \nu_j \frac{\partial^2 u_j}{\partial y^2}, \quad \frac{\partial u_j}{\partial x} + \frac{\partial v_j}{\partial y} = 0;$$

$$x = 0, \quad u_1 = u_0, \quad u_2 = 0; \quad y = 0, \quad u_j = u_{j0}, \quad v_j = v_{j0}, \quad \frac{\partial u_j}{\partial y} = R_j, \qquad (5.9)$$
$$j = 1, 2,$$

where u_{j0}, v_{j0} and R_j (j = 1, 2) were determined in Sect. 3.3 in Chap. 3.

The introduction of the following similarity variables

$$u_j = u_0 f_j'(\xi_j), \quad v_j = \left(\frac{u_0 \nu_j}{4x}\right)^{0.5} \left(\xi_j f_j' - f_j\right), \quad \xi_j = (-1)^{j-1} y \left(\frac{u_0}{\nu_j x}\right)^{0.5}, \quad j = 1, 2, \qquad (5.10)$$

leads to a problem which allows us to determine the velocity profiles:

$$2f_j''' + f_j f_j'' = 0,$$

$$f_j(0) = A_j, \quad f_j'(0) = B_j, \quad f_j''(0) = C_j, \quad j = 1, 2, \quad \left(f_1'(\infty) = 1, \quad f_2'(\infty) = 0\right), \qquad (5.11)$$

where A_j, B_j, and C_j are results of the numerical solution [36] and they are displayed in Table 6.

The velocity profiles $f_j'(\xi_j)$(j = 1, 2) depend substantially on the effect of the nonlinear mass transfer (A_j, j = 1, 2), which is characterized by the parameters θ_j(j = 1, 2) [36]:

$$\theta_j = \frac{M(mc_{20} - c_{10})}{\rho_{j0}^* m^{j-1}}, \quad j = 1, 2. \qquad (5.12)$$

This effect is superposed with the effect of the hydrodynamic interaction between the phases (C_j, j = 1, 2). Hence, the interface velocity (B_j, j = 1, 2) takes into account both of the above-mentioned effects.

The linear analysis of the hydrodynamic stability in liquid–liquid systems is made similarly to that in the case of gas–liquid systems. The velocity profiles (5.10) are introduced into the Orr–Sommerfeld equation. The results obtained show that the stability of the profiles depends considerably on the nonlinear effects of the mass transfer $\theta_j (j = 1, 2)$, as well as on the interface velocity (B_j, $j = 1, 2$).

The effect of the nonlinear mass transfer in liquid 1 and the effects of the increase of the interface velocity are superposed and their total influence on the stability of the flow in phase 1 is shown in Table 13 ($m/b = 0$).

Under the conditions of commensurable diffusive resistances in the two liquids, the nonlinear effects are lower (Table 13, $m/b = 1$). The influence of the nonlinear effects (θ) on the stability of the flow decreases.

The linear analysis of the hydrodynamic stability of phase 2 [35] produces results analogous with those for the gas–liquid system. The flow is stable up to large Reynolds numbers ($Re \approx 25{,}000$), which can be explained by the shape of the velocity profile (approximately the same as the Couette one).

Studies on the hydrodynamic stability in systems with intensive interphase mass transfer have shown that the stability increases with the increase of the interface velocity and the increase of concentration gradients in the case of interphase mass transfer directed from the volume to the phase boundary. The decrease of the interface velocity and the change of direction of the interphase mass transfer destabilize the flow in the boundary layer.

Experimental studies [37–39] of mass transfer in systems with intensive interphase mass transfer between two liquids have shown in a number of cases a higher mass transfer rate compared with that predicted by the linear theory of mass transfer. So far this has been explained by the Marangoni effect, i.e., the creation of interfacial tension gradients as a result of temperature or concentration heterogeneity on the phase boundary. The interfacial tension gradient induces

Table 13 Values of the critical Reynolds numbers Re_{cr}, wave velocities C_r, wave numbers A, $C_{r\,min}$, and A_{min} ($m/b = 0$, $\theta_1 = \theta$, $\theta_2 = 0$ and $b/m = 1$, $\theta_1 = \theta_2 = \theta$)

$\varepsilon = 10$	θ	Re_{cr}	A	C_r	A_{min}	$C_{r\,min}$
$m/b = 0$	−0.5	3,145	0.315	0.6235	0.358	0.6246
$m/b = 0$	−0.3	2,663	0.320	0.6155	0.364	0.6163
$m/b = 0$	−0.1	2,343	0.325	0.6092	0.372	0.6101
$m/b = 0$	0.0	2,243	0.330	0.6081	0.372	0.6085
$m/b = 0$	0.1	2,145	0.320	0.6042	0.374	0.6053
$m/b = 0$	0.3	1,983	0.320	0.5997	0.375	0.6009
$m/b = 0$	0.5	1,859	0.330	0.5969	0.377	0.5974
$m/b = 0$	−0.5	2,503	0.325	0.6130	0.367	0.6135
$m/b = 0$	−0.3	2,398	0.325	0.6099	0.370	0.6111
$m/b = 0$	−0.1	2,288	0.325	0.6079	0.371	0.6086
$b/m = 1$	0.0	2,243	0.330	0.6081	0.372	0.6085
$b/m = 1$	0.1	2,170	0.330	0.6064	0.374	0.6066
$b/m = 1$	0.3	2,079	0.320	0.6020	0.375	0.6036
$b/m = 1$	0.5	1,999	0.325	0.6008	0.375	0.6015

secondary flows directed tangentially to the phase boundary. They change the velocity profiles in the boundary layer. Thus, the mass transfer rate is directly affected. In the case of hydrodynamic instability of the new profiles, the flow spontaneously evolves from laminar into turbulent and the mass transfer rate increases drastically.

The results obtained shows that under the condition of intensive interphase mass transfer high mass fluxes induce secondary flows directed normally to the phase boundary. These secondary flows change the velocity profiles, and consequentially they change the kinetics of mass transfer (nonlinear mass transfer) and the hydrodynamic stability of the flow. This is a radically different mechanism for the influence of the intensive interphase mass transfer on the kinetics of the mass transfer and the hydrodynamic stability in liquid–liquid systems.

The theoretical results obtained allow a comparative analysis of the influence of the Marangoni effect and the effect of the nonlinear mass transfer on the mass transfer rate and the hydrodynamic stability of systems with intensive interphase mass transfer.

The results obtained for gas–liquid and liquid–liquid systems show that the stability of the flow in the boundary layer depends considerably on the interface velocity. This velocity is a result of the superposed influence of the flux of momentum (hydrodynamic interaction between the two phases) and the mass flux (inducing of parallel secondary flows) through the phase boundary. On this basis, the influence of the normal and the tangential components of the interface velocity on the hydrodynamic stability of the velocity profiles have been demonstrated [40, 41].

The results obtained allow us to obtain the dependence of the critical Reynolds number Re_{cr} on the tangential interface velocity component $f'(0)$ under the condition of a constant value of the normal velocity component on the interface $f(0)$ (Fig. 30) and the influence of the normal velocity component on the interface $f(0)$ at a constant value of the tangential interface velocity component $f'(0)$ on the critical Reynolds number Re_{cr} (Fig. 31).

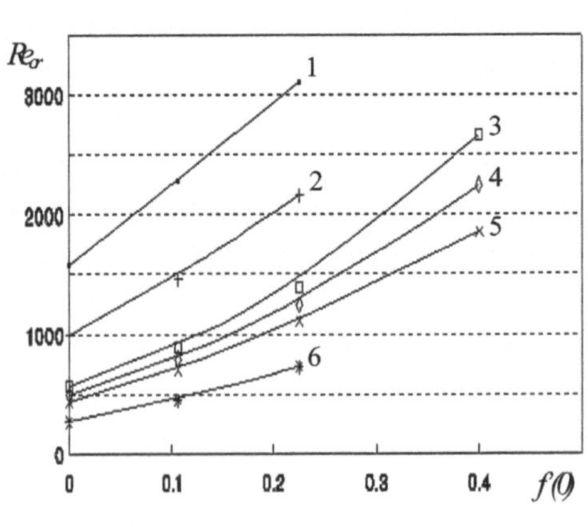

Fig. 30 The dependence of Re_{cr} on $f'(0)$ at $f(0) = \text{const.}$: *1* $f(0) = 0.25$, *2* $f(0) = 0.15$, *3* $f(0) = 0.03$, *4* $f(0) = 0.0$, *5* $f(0) = -0.03$, *6* $f(0) = -0.15$

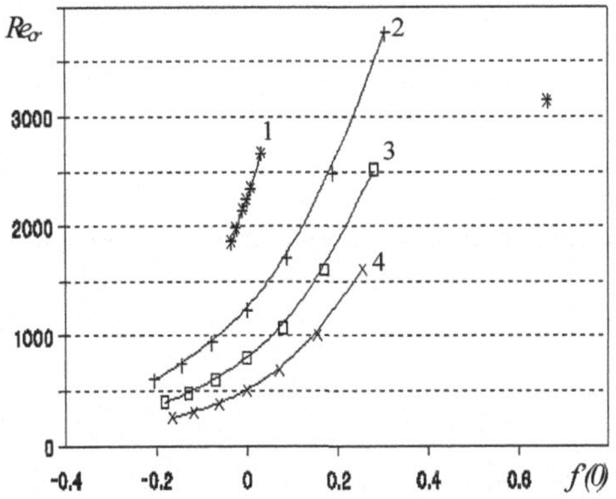

Fig. 31 The dependence of Re_{cr} on $f(0)$ at $f(0) = \text{const.}$: *1* $f(0) = 0.40$, *2* $f(0) = 0.23$, *3* $f(0) = 0.11$, *4* $f(0) = 0.0$

The comparison of the effects of the normal and tangential velocity components on the critical Reynolds number Re_{cr} show a powerful influence of the normal velocity component.

Systems with intensive interphase mass transfer are characterized by the fact that the kinetics of mass transfer does not follow the linear theory of the mass transfer and obvious changes in the hydrodynamic stability are observed. These effects have been explained quite often [37–39, 42] by the Marangoni effect, i.e., the induction of tangential secondary flow on the phase boundary. The investigations of the kinetics of mass transfer in systems with intensive interphase mass transfer [33] and their hydrodynamic stability show that the same effects can be explained by the influence of the nonlinear mass transfer, i.e., the induction of normal secondary flows on the phase boundary. A comparison of the Marangoni effect with the effect of the nonlinear mass transfer will be made.

5.3 Effect of Concentration

In many cases large concentration gradients are realized in large concentration conditions, where density, viscosity, and diffusivity are a function of the concentration, i.e.,

$$\rho = \rho(c), \quad \mu = \mu(c), \quad D = D(c). \tag{5.13}$$

The influence of the concentration and its gradient on the velocity distribution is investigated in the laminar boundary layer approximation:

$$\rho\left(u\frac{\partial u}{\partial x} + v\frac{\partial u}{\partial y}\right) = \frac{\partial}{\partial y}\left(\mu\frac{\partial u}{\partial y}\right), \quad \frac{\partial}{\partial x}(\rho u) + \frac{\partial}{\partial y}(\rho v) = 0,$$

$$\rho\left(u\frac{\partial c}{\partial x} + v\frac{\partial c}{\partial y}\right) = \frac{\partial}{\partial y}\left(\rho D\frac{\partial c}{\partial y}\right); \quad x = 0, \quad u = u_0, \quad c = c_0;$$

$$y = 0, \quad u = 0, \quad v = -\frac{MD\rho^*}{\rho_0^*}\frac{\partial}{\partial y}\left(\frac{c}{\rho}\right), \quad c = c^*; \quad y \to \infty, \quad u = u_0, \quad c = c_0. \tag{5.14}$$

The linear stability analysis [43] considers a nonstationary flow (U, V, P), obtained as a combination of a basic stationary flow (u, v) and two-dimensional periodic disturbances (u_1, v_1, p_1) with small amplitudes ω:

$$U(x,y,t) = u(x,y) + \omega u_1(x,y,t), \quad V(x,y,t) = v(x,y) + \omega v_1(x,y,t),$$

$$P(x,y,t) = \omega p_1(x,y,t), \quad C(x,y,t) = c(x,y) + \omega c_1(x,y,t). \tag{5.15}$$

The nonstationary flow (U, V, P), satisfies the full system of Navier–Stokes equations:

$$\rho\frac{\partial U}{\partial t} + \rho U\frac{\partial U}{\partial x} + \rho V\frac{\partial U}{\partial y} = -\frac{\partial P}{\partial x} + \frac{\partial}{\partial x}\left(\mu\frac{\partial U}{\partial x}\right) + \frac{\partial}{\partial y}\left(\mu\frac{\partial U}{\partial y}\right),$$

$$\rho\frac{\partial V}{\partial t} + \rho U\frac{\partial V}{\partial x} + \rho V\frac{\partial V}{\partial y} = -\frac{\partial P}{\partial y} + \frac{\partial}{\partial x}\left(\mu\frac{\partial V}{\partial x}\right) + \frac{\partial}{\partial y}\left(\mu\frac{\partial V}{\partial y}\right),$$

$$\frac{\partial}{\partial x}(\rho U) + \frac{\partial}{\partial y}(\rho V) = 0, \tag{5.16}$$

with the boundary conditions

$$x = 0, \quad U = u_0, \quad V = 0, \quad P = p_0; \quad y = 0, \quad U = 0,$$
$$V = -\theta_0\frac{\rho_0 D_0}{\Delta c_0}\frac{\partial}{\partial y}\left(\frac{c}{\rho}\right);$$

$$y \to \infty, \quad U = u_0, \quad V = 0, \quad P = p_0, \tag{5.17}$$

where

$$\theta_0 = \left(\frac{D\rho\Delta c_0}{\rho_0^*\rho_0 D_0}\right)_{y=0}, \quad \Delta c_0 = c^* - c_0. \tag{5.18}$$

Linear approximations can be introduced into (5.16)–(5.18) for the dependencies of the density, viscosity, and diffusivity on the concentration:

$$\rho = \rho_0(1 + \bar{\rho}\bar{c}), \quad \mu = \mu_0(1 + \bar{\mu}\bar{c}), \quad D = D_0(1 + \bar{D}\bar{c}), \quad \bar{c} = \frac{c - \theta_1 c_1 - c_0}{\Delta c_0}, \tag{5.19}$$

where the parameters $\bar{\rho}$, $\bar{\mu}$ and $\bar{D}$ are small.

Upon consequential introduction of (5.14), (5.15), and (5.19) into (5.16) and (5.17) and after long transformations, using the linear approximations regarding the small parameters θ_0, ω, $\bar{\rho}$, $\bar{\mu}$ and $\bar{D}$, a set of equations describing the evolution of the superposed periodic flow (disturbance) is obtained:

$$\frac{\partial u_1}{\partial t} + u\frac{\partial u_1}{\partial x} + v\frac{\partial u_1}{\partial y} + u_1\frac{\partial u}{\partial x} + v_1\frac{\partial u}{\partial y} = -\frac{1}{\rho_0}\frac{\partial p_1}{\partial x} + v_0\left(\frac{\partial^2 u_1}{\partial x^2} + \frac{\partial^2 u_1}{\partial y^2}\right),$$

$$\frac{\partial v_1}{\partial t} + u\frac{\partial v_1}{\partial x} + v\frac{\partial v_1}{\partial y} + u_1\frac{\partial v}{\partial x} + v_1\frac{\partial v}{\partial y} = -\frac{1}{\rho_0}\frac{\partial p_1}{\partial y} + v_0\left(\frac{\partial^2 v_1}{\partial x^2} + \frac{\partial^2 v_1}{\partial y^2}\right),$$

$$\frac{\partial u_1}{\partial x} + \frac{\partial v_1}{\partial y} = 0; \quad x = 0, \quad u_1 = 0, \quad v_1 = 0, \quad p_1 = p_0;$$

$$y = 0, \quad u_1 = 0, \quad v_1 = 0: \quad y \to \infty, \quad u_1 = 0, \quad v_1 = 0, \quad p_1 = p_0. \tag{5.20}$$

No equations for the concentration (c_1) are included in (5.20) because for the linear approximation for the small parameters θ_0 and ω the disturbances do not influence the velocity (u_1, v_1).

The periodic disturbances can be considered as a running wave with variable amplitude:

$$u_1 = G'(y)\ \exp i(\alpha x - \beta t), \quad v_1 = -i\alpha G(y)\ \exp i(\alpha x - \beta t), \quad \beta = \beta_r + i\beta_i, \tag{5.21}$$

where $G(y)$ is the amplitude of the disturbance (regarding y) and α and β/α are its wave number and phase velocity respectively.

It can be clearly seen that the amplitude of the disturbance decreases when $\beta_i < 0$ ($c_i < 0$), i.e., the basic flow is stable. At $\beta_i > 0$ ($c_i < 0$) the flow is unstable.

Hence, from (5.20) to (5.21) an equation of the Orr–Sommerfeld type is directly obtained (for almost parallel flow):

$$\left(u - \frac{\beta}{\alpha}\right)(G'' - \alpha^2 G) - \frac{\partial^2 u}{\partial y^2}G = -\frac{iv_0}{\alpha}(G^{IV} - 2\alpha^2 G'' + \alpha^4 G)$$
$$+ \frac{i}{\alpha}\left[vG''' + \left(\frac{\partial^2 u}{\partial x \partial y} - \alpha^2 v\right)G'\right];$$
$$y = 0, \quad G = 0, \quad G' = 0; \quad y \to \infty, \quad G = 0, \quad G' = 0. \tag{5.22}$$

The analysis of stability requires the introduction of the basic flow velocity into (5.22). In the case of gases, one obtains (see Sect 3.4 in Chap. 3):

$$u(x,y) = u_0\frac{\Phi'(\eta)}{\varphi}, \quad v = \frac{u_0\delta}{2x}\frac{\eta\Phi'(\eta) - \Phi(\eta)}{\varphi}, \quad G(y) = \gamma(\eta), \quad \bar{\rho} \ll 1,$$

$$\eta = \frac{y}{\delta}, \quad \delta = \sqrt{\frac{D_0 x}{u_0}}, \quad \varphi = 1 + \bar{\rho}F(\eta), \quad F(\eta) = \frac{c(x,y) - c_0}{c^* - c_0}. \tag{5.23}$$

The introduction of (5.23) into (5.22) leads to

$$\left(\frac{\Phi'}{\varphi} - C\right)(\gamma'' - A_0^2\gamma) - \left(\frac{\Phi'''}{\varphi} - 2\bar{\rho}F'\frac{\Phi''}{\varphi^2} - \bar{\rho}F''\frac{\Phi'}{\varphi^2}\right)\gamma$$
$$= -\frac{i}{A_0 Re_0}(\gamma^{IV} - 2A_0^2\gamma'' + A_0^4\gamma) + \frac{i}{2\varepsilon^2 A_0 Re_0}\frac{\eta\Phi' - \Phi}{\varphi}\gamma'''$$
$$- \frac{i}{2\varepsilon^2 A_0 Re_0}\left(\frac{\eta\Phi''' + \Phi''}{\varphi} - \frac{2\bar{\rho}\eta F'\Phi'' + \bar{\rho}\eta F''\Phi' + \bar{\rho}F'\Phi'}{\varphi^2} + A_0^2\frac{\eta\Phi' - \Phi}{\varphi}\right)\gamma', \tag{5.24}$$

where

$$A = \alpha\delta, \quad C = \frac{\beta}{\alpha u_0} = C_r + iC_i, \quad Re_0 = \frac{u_0\delta}{\nu_0}. \tag{5.25}$$

The solution of (5.25) was in obtained [41] and the results are shown in Table 8 for the dependence on the concentration of the transferred substance ($\bar{\rho}$ and $\bar{\mu}$) and its gradient (θ), where $Re = 1.72Re_0$.

The results obtained show that the effect of the concentration dependencies of the viscosity ($\bar{\mu}$) is analogous to that of the large concentration gradient (θ_0), whereas the change in the density ($\bar{\rho}$) has an insignificant effect and this dependence is not monotonous.

In the case of liquids, the basic flow velocity (see Sect 3.4 in Chap. 3),

$$u(x,y) = u_0\frac{\Phi_1'(\eta_1)}{\varphi}, \quad v = \frac{u_0\delta_1}{2x}\frac{\eta_1\Phi_1'(\eta_1) - \Phi_1(\eta_1)}{\varphi}, \quad G(y) = \gamma_1(\eta_1),$$

$$\eta_1 = y\sqrt{\frac{u_0}{\nu_0 x}}, \quad \delta_1 = \sqrt{\frac{\nu_0 x}{u_0}}, \quad \varphi = 1 + \bar{\rho}F_1(\eta_1), \tag{5.26}$$

is introduced into (5.22) and the Orr–Sommerfeld-type equation can be obtained directly from (5.24) using the substitutions

$$\Phi(\eta) = \Phi_1(\eta_1), \quad F(\eta) = F_1(\eta_1), \quad \gamma(\eta) = \gamma_1(\eta_1),$$

$$\eta = \eta_1, \quad A_0 = A_1 = \alpha\delta_1, \quad Re_0 = Re_1 = \frac{u_0\delta_1}{\nu_0}, \quad \varepsilon = 1, \quad Re = 1.72Re_1. \tag{5.27}$$

The data presented in Tables 14 and 15 permit us to obtain the dependence (Figs. 32, 33) of Re_{cr} on the parameters characterizing the concentration dependencies on the density ($\bar{\rho}$), viscosity ($\bar{\mu}$), diffusivity ($\bar{D}$), and large concentration gradients (θ_0).

The data presented in Tables 14 and 15 and in Figs. 32 and 33 show that in gases and liquids:

- The stability of flows (Re_{cr}) increases if the density depends on concentration ($\bar{\rho} \neq 0$).

Table 14 Effects of high concentrations ($\bar{\rho} \neq 0$, $\bar{\mu} \neq 0$) and large concentration gradients ($\theta_0 \neq 0$) on the critical Reynolds numbers Re_{cr} in gases

θ_0	$\bar{\rho}$	$\bar{\mu}$	Re_{cr}
0.0	0.0	0.0	501
0.0	0.0	0.2	285
0.0	0.0	−0.2	1,135
0.0	0.15	0.0	608
0.0	0.15	0.2	443
0.0	−0.15	0.0	559
0.0	−0.15	−0.2	2,972
0.3	0.0	0.0	1,619
−0.3	0.0	-0.2	2,238
0.3	−0.15	0.0	1,508
0.3	0.15	0.2	547
0.3	0.0	0.0	345
0.3	0.0	0.2	215
0.3	0.15	0.0	491
0.3	0.15	0.2	367

Table 15 Effects of high concentrations ($\bar{\rho} \neq 0$, $\bar{\mu} \neq 0$, $\bar{D} \neq 0$) and large concentration gradients ($\theta_0 \neq 0$) on the critical Reynolds numbers Re_{cr} in liquids

θ	$\bar{\rho}$	$\bar{\mu}$	$\bar{D}$	Re_{cr}
0.0	0.0	0.0	0.0	501
0.3	0.0	0.0	0.0	422
−0.1	0.0	0.0	0.0	564
0.0	0.15	0.0	0.0	556
0.0	−0.15	0.0	0.0	1,073
0.0	0.0	0.2	0.0	373
0.0	0.0	-0.2	0.0	742
0.0	0.0	0.0	0.3	502
0.0	0.0	0.0	-0.3	501

- The decrease of the concentration gradient (θ_0) leads to a decrease of the stability (Re_{cr}).
- In cases where the increase of the concentration leads to an increase (a decrease) of viscosity, i.e., $\bar{\mu} > 0\,(\bar{\mu} < 0)$, one can observe an increase of stability, i.e., high concentrations lead to high (low) mass transfer rates in gases.
- A change in diffusivity ($\bar{D}$) does not influence the stability.
- Additivity of the separated effects is observed.

5.4 Effect of Temperature

A great number of investigations [9, 17, 19, 21, 25, 27, 37–39, 42, 44–47]) have shown that the tangential flows (as a result of interfacial tension gradients) affect considerably the hydrodynamic stability of the interface and the flow in the boundary layer.

Fig. 32 Dependence of the critical Reynolds numbers (Re_{cr}) on high concentrations through the viscosity ($\bar{\mu}$) and density ($\bar{\rho}$), and the influence of large concentration gradients (θ_0) in gases

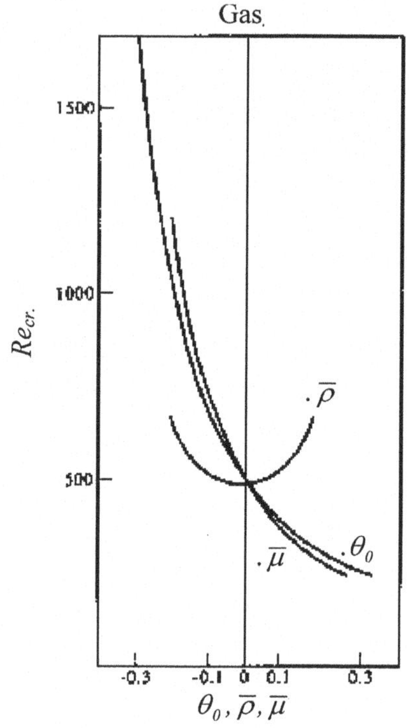

Fig. 33 Dependence of the critical Reynolds numbers (Re_{cr}) on high concentrations through the viscosity ($\bar{\mu}$), density ($\bar{\rho}$), and diffusivity ($\bar{D}$), and large concentration gradients (θ_0) in liquids

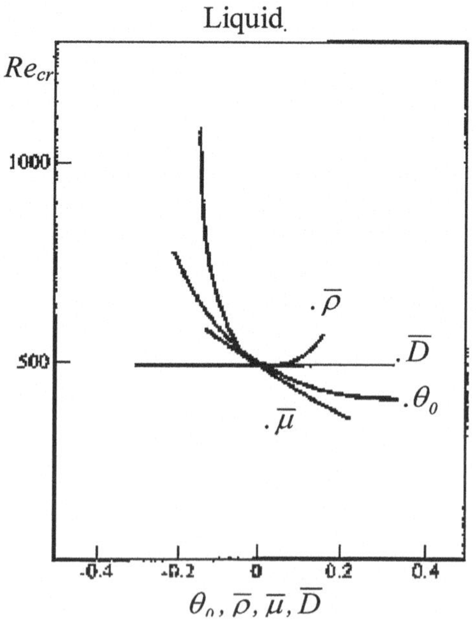

The induction of normal flows (due to large concentration gradients) has an effect of "injection" or "suction" of fluid in the boundary layer, which also changes the hydrodynamic stability in the boundary layer [16, 23, 27, 34, 35, 40]. It has been shown [40] that changes in the normal component of the velocity on the interface influence the hydrodynamic stability more strongly than changes in the tangential component.

The results obtained in Sect. 3.3 in Chap. 3 give us the opportunity to define the influence of the nonlinear mass transfer and the Marangoni effect on the hydrodynamic stability of the flow in the boundary layer.

The numerical analysis [43] of the influence of the effect of nonlinear mass transfer and the Marangoni effect on the hydrodynamic stability in gas–liquid systems leads to some basic conclusions:

1. In the case of absorption, the increase of intensity of the mass transfer directed from the volume of the gas phase towards the phase boundary leads to an increase of the critical Reynolds numbers, i.e., the flow is stabilized.
2. In the case of desorption, the increase of intensity of the mass transfer directed from the phase boundary towards the volume of the gas phase leads to a decrease of the critical Reynolds numbers, i.e., the flow is destabilized.
3. The increase of the temperature gradient along the phase boundary length leads to a decrease of the critical Reynolds numbers, i.e., destabilizes the flow. This Marangoni effect, however, is insignificant in gas–liquid systems with a movable phase boundary.
4. The flow in the liquid phase is globally stable.

References

1. Arnold VI (1984) Ordinary differential equations. Nauka, Moskow (in Russian)
2. Braun M (1978) Differential equations and their applications. Springer, New York
3. Zwillinger D (1957) Handbook of differential equations, 2nd edn. Academic, Boston
4. Marsden JE, McCracken M (1976) The Hopf bifurcation and its application. Springer, New York
5. Kamke E (1959) Differentialgleichungen. Chelsea, New York
6. Joseph DD (1976) Stability of fluid motion. Springer, New York
7. Schlichting H, Gerstein K (2000) Boundary layer theory, 8th revised and enlarged edn. Springer, Berlin
8. Gershuni GZ, Zhuhovitski EM (1972) Convective stability in compressible liquids. Nauka, Moskow (in Russian)
9. Buevich YA (1984) Hydrodynamics of interphase surfaces. Notes of mechanics, vol 34, Nauka, Moscow (in Russian)
10. Betchov R, Crimminale WO (1967) Stability of parallel flows. Academic, New York
11. Zhigulev VU, Tumin AM (1987) Arising of turbulence. Nauka, Novosibirsk (in Russian)
12. Landau LD, Lifshitz EM (1989) Fluid mechanics, 2nd edn. Pergamon, Oxford
13. Boyadjiev Chr (1991) J Phys Eng (Russia) 60:845
14. Goldshtik MA, Shtern VN (1977) Hydrodynamic stability and turbulence. Nauka, Novosibirsk (in Russian)
15. Abramov AA (1961) J Comput Math Math Phys (Russia) 1:542

16. Boyadjiev Chr, Halatchev I, Tchavdarov B (1996) Int J Heat Mass Transf 39:2571
17. Bussman K, Munz H (1942) Absaug Jahrb Dtsch Luftfahrtforsch I:36
18. Chr Boyadjiev, Toschev E (1990) Hung J Ind Chem 18:7
19. Van Stijn ThL (1983) Stability of almost parallel boundary layer flows. PhD thesis, Royal University of Groningen
20. Boyadjiev Chr, Vulchanov N (1987) C R Acad Bulg Sci 40:35
21. Van Stijn ThL, van de Vooren AI (1980) J Eng Math (Netherlands) 14:25
22. Jordinson R (1970) J Fluid Mech 43:801
23. Boyadjiev Chr, Beshkov V (1984) Mass transfer in liquid film flows. Bulgarian Academy of Sciences, Sofia
24. Boyadjiev Chr, Beshkov V (1988) Mass transfer in falling liquid films. Mir, Moscow (in Russian)
25. Brown WB (1961) Boundary Layer Flow Control 2:913
26. Barry MD, Ross MAS (1970) J Fluid Mech 43:813
27. Van Stijn ThL, van de Vooren AI (1983) Comput Fluids 10:223
28. Boyadjiev Chr, Doichinova M (1999) Hung J Ind Chem 27:215
29. Boyadjiev Chr, Vulchanov N (1988) Int J Heat Mass Transf 31:795
30. Vulchanov N, Boyadjiev Chr (1988) Int J Heat Mass Transf 31:801
31. Vulchanov N, Boyadjiev Chr (1990) Int J Heat Mass Transf 33:2045
32. Ostroumov GA (1952) Free convection under internal problem conditions. Technika, Moscow (in Russian)
33. Boyadjiev Chr (1993) Bulg Chem Commun 26:33
34. Boyadjiev Chr, Halatchev I (1996) Int J Heat Mass Transf 39:2581
35. Halatchev I, Boyadjiev Chr (1996) Int J Heat Mass Transf 39:2587
36. Sapundzhiev Ts, Boyadjiev Chr (1993) Russ J Eng Thermophys 3:185
37. Hennenberg M, Bisch PM, Vignes-Adler M, Sanfeld A (1979) In: Sorensen TS (ed) Lecture notes in physics, vol 105, Springer, Berlin, p 229
38. Linde H, Schwartz P, Wilke H (1979) In: Sorensen TS (ed) Lecture notes in physics, vol 105. Springer, Berlin, p 75
39. Savistowski H (1981) Ber Bunsenges Phys Chem 85:905
40. Boyadjiev Chr, Halatchev I (1996) Int J Heat Mass Transf 39:2593
41. Boyadjiev Chr (1990) J Phys Eng (Russia) 59:92, 277
42. Velarde MG, Castilo JL (1981) In: Zieper J, Oertel H (eds) Convective transport and instability phenomena. Braun, Karlsruhe, p 235
43. Boyadjiev Chr, Halatchev I (1998) Int J Heat Mass Transf 41:197, 945
44. Buevich JA (1985) J Phys Eng (Russia) 49:230
45. Sanfeld A, Steinchen A, Hennenberg M, Bisch PM, Van Lamswerde-Galle D, Dall-Vedove W (1979) In: Sorensen TS (ed) Lecture notes in physics, vol 105. Springer, Berlin, p 168
46. Scriven LE, Sterling CV (1960) Nature 127(4733):186
47. Sorensen TS, Hennenberg M (1979) In: Sorensen TS (ed) Lecture notes in physics, vol 105. Springer, Berlin, p 276
48. Boyadjiev Chr (1999) In: Proceedings Apollonia'99, 4th workshop on transport phenomena in two-phase flow, Sozopol, Bulgaria, 11–16 September, pp 141, 155
49. Sorensen TS (ed) (1979) Lecture notes in physics, vol 105. Springer, Berlin
50. Benard H (1990) Rev Gen Sci Pures Appl 12:1261
51. Benard H (1901) Ann Chim Phys 23(7):62
52. Pearson JKA (1958) J Fluid Mech 4:489
53. Nield DA (1964) J Fluid Mech 19:341
54. Nield DA (1996) Z Angew Math Phys 17:226
55. Dilman VV, Kulov NN, Lothov VA, Kaminski VA, Najdenov VI (1998) Theor Fundam Chem Technol 32:377
56. Krylov VS (1980) Success Chem (Russia) 49(11):118
57. Sternling CV, Scriven LE (1959) AIChE J 5:514
58. Boyadjiev Chr (2000) Int J Heat Mass Transf 43:2749, 2759

59. Warmuzinski K, Buzec I (1990) Chem Eng Sci 45:243
60. Brian PLT, Ross JR (1972) AIChE J 18:582
61. Vidal A, Acrivos A (1968) Eng Chem Eng Fundam 7:894
62. Muenz K, Marchello JM (1966) AIChE J 12:249
63. Plevan RE, Quinn JA (1966) AIChE J 12:894
64. Dilman VV, Najdenov VI, Olevskii VV (1992) Chem Ind (Russia) 8:465
65. Boyadjiev Chr (2001) Int J Heat Mass Transf 44:1119
66. Boyadjiev B, Boyadjiev Chr (2001) Int J Heat Mass Transf 44:2505
67. Boyadjiev Chr, Boyadjiev B (2003) Int J Heat Mass Transf 46:1679
68. Boyadjiev B, Boyadjiev Chr (2003) Int J Heat Mass Transf 46:1687
69. Boyadjiev Chr, Babak VN (2000) Non-linear mass transfer and hydrodynamic stability. Elsevier, Amsterdam
70. Boyadjiev Chr, Vulchanov N (1990) Int J Heat Mass Transf 33:2039
71. Boyadjiev Chr, Doichinova M (2004) Therm Sci 8:95
72. Boyadjiev Chr, Doichinova M (2008) J Eng Thermophys 17:142
73. Dilman VV, Lothor VA, Kulov NN, Najdenov VI (2000) Theor Found Chem Eng 34:227
74. Carslaw HS, Jaeger JC (1959) Conduction of heat in solids, 2nd edn. Oxford University Press, London
75. Boyadjiev Chr (2002) Hung J Ind Chem 30:13

Part III
Calculation Problems

The calculation problems in chemical engineering arise with every step of the modeling and simulation procedures. Most chemical engineering models are built on the basis of differential equations and the process simulations are the results of their solutions.

The calculation problems of the modeling are related to parameter identification, where the inverse problem solutions use the solutions of differential equations and function minimization methods.

All these methods permit us to solve the process optimization problems using mathematical programming methods

Solution of Differential Equations

The basic part of chemical engineering processes occurs in thin layers near the phase interface. That is why the mathematical models for the hydrodynamics, heat transfer, and mass transfer are used in the boundary layer approximation, i.e., in the form of parabolic partial differential equations.

1 Analytical Methods

Analytical solutions of the parabolic partial differential equations in the general case use the canonical form:

$$\frac{\partial U}{\partial t} = \frac{\partial^2 U}{\partial x^2} + F(x, t). \tag{1.1}$$

The methods for solution of (1.1) are different depending on the function F and the kind of boundary conditions.

1.1 Green's Functions

The solution of (1.1) can be obtained with the help of Green's functions in the domain $(0 \le x < \infty, 0 \le t < \infty)$ and initial and first kind of boundary conditions:

$$U(0, t) = 0, \quad U(\infty, t) = 0, \quad U(x, 0) = \varphi(x). \tag{1.2}$$

The solution of (1.1, 1.2) is

$$U(x, t) = \int_0^t \int_0^\infty G(x, \xi, t - \tau) F(\xi, \tau) d\xi d\tau + \int_0^\infty G(x, \xi, t) \varphi(\xi) d\xi, \tag{1.3}$$

C. Boyadjiev, *Theoretical Chemical Engineering*,
DOI: 10.1007/978-3-642-10778-8_7, © Springer-Verlag Berlin Heidelberg 2010

where G is Green's function:

$$G(x,\xi,t)=\frac{1}{\sqrt{4\pi t}}\left[\mathrm{e}^{-\frac{(x-\xi)^2}{4t}}-\mathrm{e}^{-\frac{(x+\xi)^2}{4t}}\right]. \tag{1.4}$$

If the boundary conditions are of the second kind

$$\frac{\partial U}{\partial x}(0,t)=0,\quad \frac{\partial U}{\partial x}U(\infty,t)=0,\quad U(x,0)=\varphi(x). \tag{1.5}$$

Green's function has the form:

$$G(x,\xi,t)=\frac{1}{\sqrt{4\pi t}}\left[\mathrm{e}^{-\frac{(x-\xi)^2}{4t}}+\mathrm{e}^{-\frac{(x+\xi)^2}{4t}}\right]. \tag{1.6}$$

In the domain $(0\le x\le l,\ 0\le t<\infty)$, the solution is

$$U(x,t)=\int_0^t\int_0^l G(x,\xi,t-\tau)F(\xi,\tau)d\xi\,d\tau+\int_0^l G(x,\xi,t)\varphi(\xi)d\xi \tag{1.7}$$

and Green's function has the form

$$G(x,\xi,t)=\frac{2}{l}\sum_{\mathrm{n}=1}^{\infty}\sin\frac{\pi \mathrm{n}x}{l}\,\sin\frac{\pi \mathrm{n}\xi}{l}\mathrm{e}^{-\lambda_{\mathrm{n}}^2 t},\quad \lambda_{\mathrm{n}}=\frac{\pi \mathrm{n}}{l} \tag{1.8}$$

if the boundary conditions are of the first or the second kind:

$$U(0,t)=0,\quad U(l,t)=0,\quad U(x,0)=\varphi(x);$$

$$\frac{\partial U}{\partial x}(0,t)=0,\quad \frac{\partial U}{\partial x}U(l,t)=0,\quad U(x,0)=\varphi(x). \tag{1.9}$$

In the cases when

$$U(0,t)=\mu_1(t),\quad U(l,t)=\mu_2(t), \tag{1.10}$$

the solution is

$$U(x,t)=W(x,t)+\frac{x}{l}\mu_2(t)+\frac{l-x}{l}\mu_1(t) \tag{1.11}$$

and the solution for W is like (1.3).

In many cases [1, 2] the function F in (1.1) (volume source or sink) has the form

$$F(x,t) = \psi(t)x^k e^{-\frac{x^2}{4t}}, \tag{1.12}$$

i.e. in (1.7) the integral has to be solved:

$$J_k = \int_0^\infty \xi^k e^{-\frac{\xi^2}{4\tau}} \left[e^{-\frac{(x-\xi)^2}{4(t-\tau)}} - e^{-\frac{(x+\xi)^2}{4(t-\tau)}} \right] d\xi = e^{-\frac{x^2}{4t}} \int_0^\infty \xi^k \left[e^{-\frac{(\xi t - x\tau)^2}{4t\tau(t-\tau)}} - e^{-\frac{(\xi t + x\tau)^2}{4t\tau(t-\tau)}} \right] d\xi = J_k^{(0)}(x,t,\tau) e^{-\frac{x^2}{4t}},$$

$$J_k^{(0)} = \int_0^\infty \xi^k \left[e^{-\frac{(\xi t - x\tau)^2}{4t\tau(t-\tau)}} - e^{-\frac{(\xi t + x\tau)^2}{4t\tau(t-\tau)}} \right] d\xi = a \int_{-c}^\infty (au+b)^k e^{-u^2} du - a \int_c^\infty (au-b)^k e^{-u^2} du, \tag{1.13}$$

where

$$a = \frac{\sqrt{4t\tau(t-\tau)}}{t}, \quad b = \frac{x\tau}{t}, \quad c = \frac{x\tau}{\sqrt{4t\tau(t-\tau)}}, \quad b = ac. \tag{1.14}$$

The solutions of (1.13) for $k = 0, 1, 2, \ldots$ lead to

$$\begin{aligned}
&J_0^{(0)} = \sqrt{\pi} a\, erfc, \quad erfc = \frac{2}{\sqrt{\pi}} \int_0^c e^{-u^2} du, \quad J_1^{(0)} = \sqrt{\pi} ab, \\
&J_2^{(0)} = a^2 b e^{-c^2} + \sqrt{\pi}\left(\frac{a^3}{2} + ab^2 \right) erfc, \quad J_3^{(0)} = \sqrt{\pi}\left(\frac{3}{2} a^3 b + ab^3 \right), \\
&J_4^{(0)} = \left(a^2 b^3 + \frac{5}{2} a^4 b \right) e^{-c^2} + \sqrt{\pi}\left(\frac{3}{4} a^5 + 3a^3 b^2 + ab^4 \right) erfc, \\
&J_5^{(0)} = \sqrt{\pi}\left(\frac{15}{4} a^5 b + 5a^3 b^3 + ab^5 \right).
\end{aligned} \tag{1.15}$$

The cases of $F(x,t) = \psi(t)x^{2k}$ lead to the integral

$$\begin{aligned}
J_k &= \int_0^\infty \xi^{2k} \left[e^{-\frac{(x-\xi)^2}{4(t-\tau)}} + e^{-\frac{(x+\xi)^2}{4(t-\tau)}} \right] d\xi = b \times \int_{-\infty}^\infty (x - bu)^{2k} e^{-u^2} du - 2b \\
&\times \int_0^\infty (x - bu)^{2k} e^{-u^2} du + 2b \int_0^a (x - bu)^{2k} e^{-u^2} du,
\end{aligned} \tag{1.16}$$

$$a = \frac{x}{2\sqrt{t-\tau}}, \quad b = \sqrt{t-\tau}.$$

The solutions of (1.16) for $k = 0, 1, 2, \ldots$ lead to

$$J_0 = \sqrt{\pi} b erfa, \quad J_1 = \sqrt{\pi}\left(bx^2 + \frac{1}{2}b^3\right) erfa + b^2 x e^{-a^2},$$

$$J_2 = \sqrt{\pi}\left(bx^4 + 3x^2b^3 + \frac{3}{4}b^5\right) erfa + \left(x^3b^2 + \frac{5}{2}xb^4\right) e^{-a^2},$$

$$J_3 = \sqrt{\pi}\left(x^6 b + \frac{15}{2}x^4b^3 + \frac{45}{4}x^2b^5 + \frac{15}{8}b^7\right) erfa + \left(x^5b^2 + 7x^3b^4 + \frac{33}{4}xb^6\right) e^{-a^2},$$

$$J_4 = \sqrt{\pi}\left(bx^8 + 14b^3x^6 + \frac{105}{2}b^5x^4 + \frac{105}{2}b^7x^2 + \frac{105}{16}b^9\right) erfa$$
$$+ \left(x^7b^2 + \frac{27}{2}x^5b^4 + \frac{185}{4}x^3b^6 + \frac{279}{8}xb^8\right) e^{-a^2}. \tag{1.17}$$

In many cases the problem is presented in noncanonical form (1.1) and for it to be used Green's functions are necessary to introduce new functions. A typical example is the interphase mass transfer [1, 2]:

$$V_i \frac{\partial c_i}{\partial z} = D_i \frac{\partial^2 c_i}{\partial y^2}; \quad z = 0, \quad c_i = c_i^{(0)}; \quad y = (-1)^{i+1}\infty, \quad z = 0, \quad c_i = c_i^{(0)};$$
$$i = 1, 2, \tag{1.18}$$

where the boundary conditions at the interphase surface are

$$y = 0, \quad c_1 = \chi c_2, \quad D_1 \frac{\partial c_1}{\partial y} = D_2 \frac{\partial c_2}{\partial y}. \tag{1.19}$$

Problem (1.18) and (1.19) can be presented as two separate problems if we introduce an arbitrary function $\psi(z)$ in the first boundary condition of (1.19):

$$y = 0, \quad c_1 = \psi(z), \quad c_2 = \frac{1}{\chi}\psi(z). \tag{1.20}$$

After the solution of the two separate problems (1.18) and (1.20), the functions obtained $c_i(y, z)$, i = 1, 2, must be replaced in the second boundary condition of (1.19) to obtain the arbitrary function $\psi(z)$. In this way the solution of the interphase mass transfer problem [1, 2] reduces to the next problem:

$$V \frac{\partial c}{\partial z} = D \frac{\partial^2 c}{\partial y^2}, \quad c = c(y, z), (y, 0) = c^{(0)}, \quad c(0, z) = \psi(z), \quad c(\infty, z) = c^{(0)}. \tag{1.21}$$

Green's function (1.4) can be used to solve (1.21) if we introduce the new function $U(x, t)$:

$$U(x,t) = c(y,z) - c^{(0)} - \left[\psi(z) - c^{(0)}\right] \mathrm{e}^{-\frac{y^2}{z}}, \tag{1.22}$$

where

$$t = \frac{D}{V} z, \quad x = y, \quad \psi(z) = \psi\left(\frac{V}{D} t\right) = \varphi(t). \tag{1.23}$$

The introduction of (1.22) and (1.23) leads to

$$\frac{\partial U}{\partial t} = \frac{\partial^2 U}{\partial x^2} - \left\{ \varphi'(t) + \frac{D}{V}\left[\varphi(t) - c^{(0)}\right] \left[\frac{x^2}{t^2}\left(1 - \frac{4D}{v}\right) + \frac{2}{t}\right] \right\}, \tag{1.24}$$
$$U(x,0) = 0, \quad U(0,t) = 0, \quad U(\infty,t) = 0.$$

The solution of (1.24) is obtained [1, 2] by (1.3), using Green's function (1.4) and after replacement of the old variables the solution is

$$c(z,y) = \psi(z)\mathrm{e}^{-\frac{Vy^2}{4Dx}} - \frac{1}{\sqrt{\pi x}} \mathrm{e}^{-\frac{Vy^2}{4Dx}} \int_0^x \frac{\psi(\xi) + 2\xi\psi'(\xi)}{\sqrt{\xi}} \left[\int_0^{\frac{y\xi}{\sqrt{\frac{4Dx}{V}\xi(x-\xi)}}} \mathrm{e}^{-s^2}\, ds \right] d\xi. \tag{1.25}$$

1.2 Similarity Variables Method

The similarity variables method transforms the parabolic parcel differential equations into ordinary differential equations. The two-variable problem (1.1) and (1.2) can be solved as a one-variable problem

$$U(x,t) = f(\eta), \quad \eta = \frac{x}{\sqrt{t}}, \tag{1.26}$$

if

$$F(x,t) = t\psi(\eta), \quad \varphi(x) \equiv 0. \tag{1.27}$$

The introduction of (1.26) and (1.27) into (1.1) and (1.2) leads to

$$2f'' + \eta f' + 2\psi = 0, \quad f(0) = 0, \quad f(\infty) = 0, \tag{1.28}$$

where problem (1.28) must be solved numerically.

In the case when $\psi \equiv 0$, the solution is $U(x, t) = f(\eta) \equiv 0$, but for another boundary conditions the solutions are

$$U(0,t)=0,\quad U(\infty,t)=1,\quad U(x,0)=1,\quad U=\frac{2}{\sqrt{\pi}}\int\limits_0^{\frac{x}{2\sqrt{t}}}\mathrm{e}^{-s^2}ds=erf\frac{x}{2\sqrt{t}};$$

$$U(0,t)=1,\quad U(\infty,t)=0,\quad U(x,0)=0,\quad U=1-erf\frac{x}{2\sqrt{t}}=erfc\frac{x}{2\sqrt{t}}. \tag{1.29}$$

The similarity variables are used (see Chap. 3.1) to solve the boundary layer problems (3.1.15). The similarity variables model (3.1.17) shows that all processes in the boundary layer approximation are similar and the model parameter is only the Schmidt number.

1.3 Eigenvalue Problem

The transformation of the parabolic parcel differential equations into ordinary differential equations can be done using the *separated variables method.*

Let us consider the problem

$$\frac{\partial U}{\partial t}=a^2\frac{\partial^2 U}{\partial x^2},\quad U(0,t)=0,\quad U(l,t)=0,\quad U(x,0)=\varphi(x). \tag{1.30}$$

The function U can be represented as

$$U(x,t)=X(x)T(t). \tag{1.31}$$

Substitution of (1.31) into (1.30) leads to

$$\frac{1}{a^2}\frac{T'}{T}=\frac{X''}{X}=-\lambda^2, \tag{1.32}$$

i.e.,

$$X''+\lambda^2X=0,\quad T'+\lambda^2a^2T=0. \tag{1.33}$$

From (1.30) and (1.31) it follows that

$$U(0,t)=X(0)T(t)=0,\quad X(0)=0, \tag{1.34}$$

$$U(l,t)=X(l)T(t)=0,\quad X(l)=0. \tag{1.35}$$

The first equation in (1.33) leads to

$$X(x)=A\cos\lambda x+B\sin\lambda x, \tag{1.36}$$

but from the boundary condition $X(0)=0$ it follows that $A=0$, i.e.,

$$X(x)=B\sin\lambda x. \tag{1.37}$$

Solution (1.37) satisfies the boundary condition $X(l) = 0$,

$$X(l) = B \sin \lambda l = 0, \tag{1.38}$$

if $\sin \lambda l = 0$, i.e.,

$$\lambda = \lambda_n = n\frac{\pi}{l}, \quad n = 1, 2, \ldots. \tag{1.39}$$

The numbers λ_n, n = 1, 2,... are *eigenvalues* of the first equation in (1.33) with boundary conditions $X(0) = 0$ and $X(l) = 0$ (spectrum of the equation).

The introduction of (1.39) into (1.37) leads to

$$X_n(x) = B_n \sin \frac{n\pi}{l} x, \quad n = 0, 1, 2, \ldots. \tag{1.40}$$

The introduction of (1.39) into the second equation in (1.33) leads to

$$T' + (\lambda_n a)^2 T = 0, \quad T_n = e^{-\left(\frac{n\pi}{l}\right)^2 t}, \quad n = 0, 1, 2, \ldots, \tag{1.41}$$

i.e.,

$$U_n(x, t) = B_n e^{-\left(\frac{n\pi}{l}\right)^2 t} \sin \frac{n\pi}{l} x, \quad n = 0, 1, 2, \ldots. \tag{1.42}$$

If function (1.42) satisfies (1.30), its sum must satisfy (1.30) too, i.e.,

$$U(x, t) = \sum_{n=1}^{\infty} B_n e^{-\left(\frac{n\pi}{l}\right)^2 t} \sin \frac{n\pi}{l} x, \tag{1.43}$$

where the coefficients B_n, n = 0, 1, 2,..., must be obtained from the initial condition in (1.30) and (1.43),

$$U(x, 0) = \varphi(x) = \sum_{n=1}^{\infty} B_n \sin \frac{n\pi}{l} x, \tag{1.44}$$

i.e., $\varphi(x)$ allows a representation as a sine trigonometric series. This condition permits us to obtain the coefficients B_n, n = 0, 1, 2,... if we use the new variable $\xi = \frac{\pi x}{l}$:

$$\varphi\left(\frac{l}{\pi}\xi\right) = \sum_{n=1}^{\infty} B_n \sin n\xi, \tag{1.45}$$

i.e., B_n, $n = 0, 1, 2, \ldots$ are Fourier coefficients of the function $\psi(\xi) = \varphi\left(\frac{l}{\pi}\xi\right)$. According to the Euler–Fourier formulas

$$B_n = \frac{2}{\pi} \int_0^{\pi} \psi(\xi) \sin n\xi d\xi = \frac{2}{\pi} \varphi \int_0^{\pi} \varphi\left(\frac{l}{\pi}\xi\right) \sin n\xi d\xi, \tag{1.46}$$

i.e.

$$B_n = \frac{2}{l}\int_0^l \varphi(x)\sin\frac{n\pi}{l}x\,dx. \tag{1.47}$$

1.4 Laplace Transformation

Linear differential equations can be solved by making use of the Laplace transformation. This method transforms ordinary differential equations into algebraic equations, whereas partial differential equations lead to ordinary differential equations [3].

Let us consider two functions $f(t)$ and $F(p)$:

$$F(p) = p\int_0^\infty e^{-pt} f(t)dt, \tag{1.48}$$

where p is a complex-valued variable, $f(t) = 0$ at $t < 0$. Expression (1.48) is the *Laplace transformation of the function* $f(t)$ and will be symbolized as

$$F(p) = Lf(t), \tag{1.49}$$

where $f(t)$ is the original and $F(p)$ is the image. The inverse operator is L^{-1}:

$$f(t) = L^{-1}F(p). \tag{1.50}$$

The *Laplace transformation of the derivative* of the *function* $f(t)$ is expressed through $F(p)$:

$$L\left(\frac{dt}{dt}\right) = p\int_0^\infty e^{-pt}\left(\frac{dt}{dt}\right)dt = pe^{-pt}f(t)\big|_{t=0}^{t=\infty} + p^2\int_0^\infty e^{-pt}f(t)dt, \tag{1.51}$$

where

$$\lim_{t\to\infty}[e^{-pt}f(t)] = 0, \tag{1.52}$$

i.e.

$$L\left(\frac{dt}{dt}\right) = -pf_0 + pF, \quad f_0 = f(0). \tag{1.53}$$

To obtain the second-order derivative we must use $\psi(t) = df/dt$:

$$L\left(\frac{d^2f}{dt^2}\right) = L\left(\frac{d\psi}{dt}\right) = -p\psi_0 + pL\left(\frac{dt}{dt}\right) = -pf_1 - p^2f_0 + p^2F, \quad f_1 = \left.\frac{df}{dt}\right|_{t=0}. \tag{1.54}$$

Let us consider the problem

$$\frac{df}{dt} + af = 0, \quad f(0) = f_0. \tag{1.55}$$

The Laplace transformation (1.48) of problem (1.55) leads to

$$pF - pf_0 + aF = 0, \tag{1.56}$$

i.e.

$$F = \frac{pf_0}{p+a} = L(f). \tag{1.57}$$

From the table of the originals and images [3] it is possible to obtain

$$f = L^{-1}\frac{pf_0}{p+a} = f_0 e^{-at}. \tag{1.58}$$

As an example of a partial differential equation we will solve (1.1) and (1.2) at $F \equiv 0$, $\varphi = \varphi_0$, and boundary conditions

$$U(0,t) = 0, \quad \frac{\partial U}{\partial x}(\infty, t) = 0. \tag{1.59}$$

The image of U will be V:

$$V = p\int_0^\infty e^{-pt}U dt. \tag{1.60}$$

The multiplication of (1.1) and (1.2) with e^{-pt} and integration over t in the interval $[0,\infty)$ leads to

$$\begin{gathered} p\int_0^\infty \frac{\partial U}{\partial t}e^{-pt}dt = p\int_0^\infty \frac{\partial^2 U}{\partial x^2}e^{-pt}dt; \\ x = 0, \quad \int_0^\infty Ue^{-pt}dt = 0; \quad x \to \infty, \quad \int_0^\infty \frac{\partial U}{\partial x}e^{-pt}dt \to 0. \end{gathered} \tag{1.61}$$

After the integration in (1.61) we obtain

$$\frac{d^2V}{dx^2} - pV + p\varphi_0; \quad x = 0, \quad V = 0; \quad x \to \infty, \quad \frac{dV}{dx} \to 0. \tag{1.62}$$

The solution of (1.62) is

$$V = \varphi_0\left(1 - e^{-x\sqrt{p}}\right). \tag{1.63}$$

From the table of the originals and images [3] we find that the original of (1.63) is

$$U = \frac{2}{\sqrt{\pi}} \int_0^{\frac{x}{2\sqrt{t}}} e^{-s^2} ds. \tag{1.64}$$

The analytical methods presented lead to exact solutions of the differential equations. The check of the functions obtained is whether they satisfy the equations and boundary conditions.

2 Perturbation Methods

Many complicated problems can be solved using an approximate method. For that purpose asymptotic methods may be applied. The perturbation method is an asymptotic method of the small parameter for the solution of differential equations. The small parameter methods are the most vigorous means in contemporary applied mathematics [4–6].

The asymptotic methods of the perturbations use as a solution the first two or three terms of the asymptotic expansion. The expansions can be with respect to small or big parameters (*expansions with respect to a parameter*). Another possibility is *expansions with respect to the a coordinate* (for small or big values). *Nonuniform expansions* are possible too. All these methods find wide application in chemical engineering theoretical investigations [1, 2].

2.1 Expansions with Respect to a Parameter

Let us consider a solution of the differential equation

$$y = f(x, a), \tag{2.1}$$

where a is small parameter in the differential equation or in the boundary conditions. The Taylor series presentation of function (2.1) for small values of the parameter a is

$$f(x,a) = f(x,0) + \frac{f'(x,0)}{1!}a + \frac{f''(x,0)}{2!}a^2 + \cdots \tag{2.2}$$

It is obvious that the solution of the differential equation can be searched for as a power series with respect to the small parameter a:

$$y = y^{(0)} + y^{(1)}a + y^{(2)}a^2 + \cdots \tag{2.3}$$

As an example we will use the solution of the velocity distribution problem in gas–liquid boundary layers with a flat interphase surface [1, 2, 7]. If the characteristic velocities are constants, from (1.2.46) it follows that

$$\begin{gathered}
U_i\frac{\partial U_i}{\partial X} + V_i\frac{\partial U_i}{\partial Y_i} = \frac{\partial^2 U_i}{\partial Y_i^2}, \quad \frac{\partial U_i}{\partial X} + \frac{\partial V_i}{\partial Y_i} = 0; \\
X = 0, \quad U_i = 1; \quad Y_1 \to \infty, \quad U_1 = 1; \quad Y_2 \to -\infty, \quad U_2 = 1; \\
Y_1 = Y_2 = 0, \quad U_1 = \theta_1 U_2, \quad \frac{\partial U_1}{\partial Y_1} = \theta_2\frac{\partial U_2}{\partial Y_2}, \quad i = 1, 2,
\end{gathered} \tag{2.4}$$

where

$$\theta_2 = \theta_1^{3/2}\left(\frac{\upsilon_2^{\infty}}{\upsilon_1^{\infty}}\right)^{1/2}. \tag{2.5}$$

In (2.4) the orders of magnitude of all the functions and their derivatives are equal to 1 and the orders of magnitude of the different terms in (2.4) are equal to the orders of magnitude of the parameters in these terms, i.e. the orders of magnitude of the physical effects (mathematical operators) in the model are equal to the orders of magnitude of the parameters. In a gas–liquid system the parameters θ_1 and θ_2 are small ($\theta_1 \sim 10^{-1}$, $\theta_2 \sim 10^{-1}$) and second-order approximations lead to 1% error of the solution (the error of the velocity measurements is greater).

Problem (2.4) can be solved in similarity variables:

$$\begin{gathered}
U_1 = f', \quad V_1 = \frac{1}{2\sqrt{X}}(\eta f' - f), \quad f = f(\eta), \quad \eta = \frac{Y_1}{\sqrt{X}}; \\
U_2 = \varphi', \quad V_2 = -\frac{1}{\sqrt{X}}(\xi\varphi' - \varphi), \quad \varphi = \varphi(\xi), \quad \xi = -\frac{Y_2}{2\sqrt{X}}.
\end{gathered} \tag{2.6}$$

In similarity variables problem (2.4) has the form

$$\begin{gathered}
2f''' + f''f = 0, \quad \varphi''' + 2\varphi''\varphi = 0, \quad f(0) = \varphi(0) = 0, \quad f'(\infty) = \varphi'(\infty) = 1, \\
f'(0) = \theta_1\varphi'(0), \quad \varphi''(0) = -2\theta_2 f''(0).
\end{gathered} \tag{2.7}$$

The solution of (2.7) will be searched for in the form:

$$\begin{gathered}
f = f_0 + \theta_1 f_1 + \theta_2 f_2 + \theta_1^2 f_{11} + \theta_2^2 f_{22} + \theta_1\theta_2 f_{12}, \\
\varphi = \varphi_0 + \theta_1\varphi_1 + \theta_2\varphi_2 + \theta_1^2\varphi_{11} + \theta_2^2\varphi_{22} + \theta_1\theta_2\varphi_{12}.
\end{gathered} \tag{2.8}$$

The introduction of (2.8) into (2.7) and the unification of the terms with the same order of magnitude leads to

$$
\begin{aligned}
&(2f_0''' + f_0''f_0) + \theta_1(2f_1''' + f_0''f_1 + f_1''f_0) + \theta_2(2f_2''' + f_0''f_2 + f_2''f_0) \\
&\quad + \theta_1^2(2f_{11}''' + f_0''f_{11} + f_{11}''f_0 + f_1''f_1) + \theta_2^2(2f_{22}''' + f_0''f_{22} + f_{22}''f_0 + f_2''f_2) \\
&\quad + \theta_1\theta_2(2f_{12}''' + f_0''f_{12} + f_{12}''f_0 + f_1''f_2 + f_2''f_1) + \cdots = 0, \\
&(2\varphi_0''' + \varphi_0''\varphi_0) + \theta_1(2\varphi_1''' + \varphi_0''\varphi_1 + \varphi_1''\varphi_0) + \theta_2(2\varphi_2''' + \varphi_0''\varphi_2 + \varphi_2''\varphi_0) \\
&\quad + \theta_1^2(2\varphi_{11}''' + \varphi_0''\varphi_{11} + \varphi_{11}''\varphi_0 + \varphi_1''\varphi_1) + \theta_2^2(2\varphi_{22}''' + \varphi_0''\varphi_{22} + \varphi_{22}''\varphi_0 + \varphi_2''\varphi_2) \\
&\quad + \theta_1\theta_2(2\varphi_{12}''' + \varphi_0''\varphi_{12} + \varphi_{12}''\varphi_0 + \varphi_1''\varphi_2 + \varphi_2''\varphi_1) + \cdots = 0, \\
&f_0(0) + \theta_1 f_1(0) + \theta_2 f_2(0) + \theta_1^2 f_{11}(0) + \theta_2^2 f_{22}(0) + \theta_1\theta_2 f_{12}(0) = 0, \\
&\varphi_0(0) + \theta_1\varphi_1(0) + \theta_2\varphi_2(0) + \theta_1^2\varphi_{11}(0) + \theta_2^2\varphi_{22}(0) + \theta_1\theta_2\varphi_{12}(0) = 0, \\
&f_0'(\infty) - 1 + \theta_1 f_1'(\infty) + \theta_2 f_2'(\infty) + \theta_1^2 f_{11}'(\infty) + \theta_2^2 f_{22}'(\infty) + \theta_1\theta_2 f_{12}'(\infty) = 0, \\
&\varphi'(\infty) - 1 + \theta_1\varphi_1'(\infty) + \theta_2\varphi_2'(\infty) + \theta_1^2\varphi_{11}'(\infty) + \theta_2^2\varphi_{22}'(\infty) + \theta_1\theta_2\varphi_{12}'(\infty) = 0, \\
&f_0'(0) + \theta_1[f_1'(0) - \varphi_0'(0)] + \theta_2 f_2'(0) + \theta_1^2[f_{11}'(0) - \varphi_1'(0)] + \theta_2^2 f_{22}'(0) \\
&\quad + \theta_1\theta_2[f_{12}'(0) - \varphi_2'(0)] = 0, \\
&\varphi_0''(0) + \theta_1\varphi_1''(0) + \theta_2[\varphi_2''(0) + 2f_0''(0)] + \theta_1^2\varphi_{11}''(0) \\
&\quad + \theta_2^2[\varphi_{22}''(0) + 2f_2''(0)] + \theta_1\theta_2[\varphi_{12}''(0) + 2f_1''(0)] = 0. \qquad (2.9)
\end{aligned}
$$

Solution (2.8) satisfies equations (2.7) if the zeroth-, first-, and second-order approximations are equal to zero. As a result, the next problems are obtained:

$$
\begin{aligned}
&(2f_0''' + f_0''f_0) = 0, \quad f_0(0) = 0, \quad f_0'(0) = 0, \quad f_0'(\infty) - 1 = 0; \\
&(2f_1''' + f_0''f_1 + f_1''f_0) = 0, \quad f_1(0) = 0, \quad f_1'(0) - \varphi_0'(0) = 0, \quad f_1'(\infty) = 0; \\
&(2f_2''' + f_0''f_2 + f_2''f_0) = 0, \quad f_2(0) = 0, \quad f_2'(0) = 0, \quad f_2'(\infty) = 0; \\
&(2f_{11}''' + f_0''f_{11} + f_{11}''f_0 + f_1''f_1) = 0, \quad f_{11}(0) = 0, \quad f_{11}'(0) - \varphi_1'(0) = 0, \quad f_{11}'(\infty) = 0; \\
&(2f_{22}''' + f_0''f_{22} + f_{22}''f_0 + f_2''f_2) = 0, \quad f_{22}(0) = 0, \quad f_{22}'(0) = 0, \quad f_{22}'(\infty) = 0; \\
&(2f_{12}''' + f_0''f_{12} + f_{12}''f_0 + f_1''f_2 + f_2''f_1) = 0, \quad f_{12}(0) = 0, \quad f_{12}'(0) - \varphi_2'(0) = 0; \\
&(2\varphi_0''' + \varphi_0''\varphi_0) = 0, \quad \varphi_0(0) = 0, \quad \varphi_0''(0) = 0, \quad \varphi_0'(\infty) - 1 = 0; \\
&(2\varphi_1''' + \varphi_0''\varphi_1 + \varphi_1''\varphi_0) = 0, \quad \varphi_1(0) = 0, \quad \varphi_1''(0) = 0, \quad \varphi_1'(\infty) = 0; \\
&(2\varphi_2''' + \varphi_0''\varphi_2 + \varphi_2''\varphi_0) = 0, \quad \varphi_2(0) = 0, \quad \varphi_2''(0) + 2f_0''(0) = 0, \quad \varphi_2'(\infty) = 0; \\
&(2\varphi_{11}''' + \varphi_0''\varphi_{11} + \varphi_{11}''\varphi_0 + \varphi_1''\varphi_1) = 0, \quad \varphi_{11}(0) = 0, \quad \varphi_{11}''(0) = 0, \quad \varphi_{11}'(\infty) = 0; \\
&(2\varphi_{22}''' + \varphi_0''\varphi_{22} + \varphi_{22}''\varphi_0 + \varphi_2''\varphi_2) = 0, \quad \varphi_{22}(0) = 0, \quad \varphi_{22}''(0) + 2f_2''(0) = 0, \\
&\quad \varphi_{22}'(\infty) = 0; \\
&(2\varphi_{12}''' + \varphi_0''\varphi_{12} + \varphi_{12}''\varphi_0 + \varphi_1''\varphi_2 + \varphi_2''\varphi_1) = 0, \quad \varphi_{12}(0) = 0, \\
&\quad \varphi_{12}''(0) + 2f_1''(0) = 0, \quad \varphi_{12}'(\infty) = 0.
\end{aligned}
$$

(2.10)

The solutions of the separate equations in (2.10) can be made in determinate sequence. The solution of the first equation in (2.10) is the Blasius function f_0, which is tabulated in [8], and the value $\alpha = f''(0) = 0.3320$ permits us to solve this problem as a Cauchy problem. The solutions for $\varphi_0, f_1, \varphi_1, f_2$ can be obtained [1, 2, 7] immediately:

$$\varphi_0 = \xi; \quad f_1 = \frac{1}{\alpha} f_0'; \quad \varphi_1 \equiv 0, \quad f_2 \equiv 0. \tag{2.11}$$

The values of f_{11} are tabulated in [8] after the numerical solution of the next set:

$$2f_0''' + f_0'' f_0 = 0, \quad f_0(0) = 0, \quad f_0'(0) = 0, \quad f_0'(\infty) = 1,$$

$$2f_{11}''' + f_0'' f_{11} + f_{11}'' f_0 = -\alpha^{-2} f_0'' f_0', \quad f_{11}(0) = 0, \quad f_{11}'(0) = 0, \quad f_{11}'(\infty) = 0, \tag{2.12}$$

where the value $\beta = f_{11}''(0) = -0.5447$ permits us to solve (2.12) as a Cauchy problem.

The solutions of the rest of the problems are

$$\varphi_2 = \alpha\sqrt{\pi} \int_0^{\xi} erfc\, z dz, \quad f_{12} = \sqrt{\pi} f_0', \quad \varphi_{12} \equiv 0,$$

$$\varphi_{22} = \alpha^2 \sqrt{\pi} \int_0^{\xi} \left[-z\,e^{-z^2}\, erfc\, z + \frac{\sqrt{\pi}}{2} erf\, z\, erfc\, z - \frac{2}{\sqrt{\pi}} e^{-z^2} + \frac{1}{\sqrt{\pi}} e^{-2z^2} \right]. \tag{2.13}$$

The expansions with respect to a parameter are correct if the solution of the differential equation has a Taylor series representation. It is obvious that this check is not possible and the check of the correctness of solution (2.8) can only be made by means of comparison with the numerical solution of (2.7).

2.2 Expansions with Respect to a Coordinate

Let us consider the influence of surfactants on the vertical laminar film flows [1]. The velocity distribution in the film $u(x, y)$ is the solution of the problem

$$u\frac{\partial u}{\partial x} - \frac{\partial u}{\partial y} \int \frac{\partial u}{\partial x} dy = \nu \frac{\partial^2 u}{\partial y^2} + g; \quad y = 0, \quad u = 0; \quad y = h(x), \quad -\mu \frac{\partial u}{\partial y} + \frac{\partial \sigma}{\partial x} = 0, \tag{2.14}$$

where ν and μ are the kinematic and the dynamic viscosity, g is gravity, σ is the surface tension at the gas–liquid interface $y = h(x)$, and $y = 0$ is the solid interface.

The vertical laminar film flows are very thin $(y \sim h \ll 1)$ and the solution of problem (2.14) can be searched for as a power series with respect to the small coordinate y:

$$u = \sum_{\mathrm{k}=0}^{\infty} f_{\mathrm{k}}(x) y^{\mathrm{k}}. \tag{2.15}$$

The introduction of (2.15) into (2.14) permits us to obtain the functions f_{k}, $\mathrm{k} = 0, 1, 2, \ldots$:

$$\begin{aligned} &f_0 = 0, \quad f_1 = f, \quad f_2 = -\frac{g}{2v}, \quad f_3 = 0, \quad f_4 = \frac{ff'}{24v}, \quad f_5 = 0, \quad f_6 = 0, \\ &f_7 = \frac{4f'^2 f + 4f^2 f'' - 5ff'}{5040v^2}, \quad f_8 = -\frac{g(ff'' + f'^2)}{4480v^3}, \quad f_9 = 0, \\ &f_{10} = \frac{1}{90v}\left(\frac{7}{8} f_7' f + \frac{1}{5} f_4 f_4' - \frac{5}{2} f_7 f'\right), \end{aligned} \tag{2.16}$$

where f_0 and f_1 must be obtained from the boundary conditions in (2.14).

2.3 Nonuniform Expansions (Poincaret–Lighthill–Ho Method)

The perturbation method often to leads considerable difficulties ensuing from the inappropriate choice of the scale for the different coordinate axes [4]. As a result, a “contraction” of one of them appears to lead to divergence of the solution. This divergence vanishes after an appropriate “stretching” of the scale. This stretching is known as the *Poincaret–Lighthill–Ho method* [1] and it will be demonstrated in the case of mass transfer in liquid film flow [9], where the film thickness h_0 is much greater than the diffusion boundary layer thickness δ. The mathematical description of the process (see Chap. 2.2) is

$$\frac{g}{2v}(2h_0 y - y^2)\frac{\partial c}{\partial x} = D\left(\frac{\partial^2 c}{\partial x^2} + \frac{\partial^2 c}{\partial y^2}\right), \tag{2.17}$$

with boundary conditions

$$x = 0, \quad c = c_0; \quad x \to \infty, \quad c = c^*; \quad y = 0, \quad \frac{\partial c}{\partial y} = 0; \quad y = h_0, \quad c = c^*. \tag{2.18}$$

The film thickness h_0 and the diffusion boundary layer thickness δ will be used as linear scales for the velocity and concentration distributions:

$$x = h_0 X, \quad y = h_0 - \delta\eta, \quad c = c^* + (c_0 - c^*)C, \quad \delta = h_0\sqrt{\frac{2}{3Pe}}, \quad Pe = \frac{u_{av}h_0}{D},$$
$$u_{av} = \frac{gh_0^2}{3\nu}. \tag{2.19}$$

The mass transfer rate J can be represented by the mass transfer coefficient k and the average mass flux for a liquid film with length l:

$$J = k(c_0 - c^*) = -\frac{D}{l}\int_0^l \left(\frac{\partial c}{\partial y}\right)_{y=h_0} dx. \tag{2.20}$$

The introduction of (2.19) into (2.17), (2.18), and (2.20) leads to

$$(1 - \alpha\eta^2)\frac{\partial C}{\partial X} = \alpha\frac{\partial^2 C}{\partial X^2} + \frac{\partial^2 C}{\partial \eta^2}; \quad X = 0, = 1; \quad X \to \infty, \quad C \to 0;$$
$$\eta = 0, = 0; \quad \eta \to \infty, \quad C \to 1. \tag{2.21}$$

$$Sh = \frac{kl}{D} = \alpha^{-1/2}\int_0^{1/\varepsilon} \left(\frac{\partial C}{\partial \eta}\right)_{\eta=0} dX, \tag{2.22}$$

where

$$\alpha = Fo = \frac{2}{3Pe} \ll 1. \tag{2.23}$$

The solution of (2.21) can be sought as a power series expansion:

$$C(X,\eta) = C_0(X,\eta) + \alpha C_1(X,\eta) + \alpha^2 C_2(X,\eta) + \cdots. \tag{2.24}$$

The use of the traditional perturbation method is impossible, because if we put (2.24) in (2.21) and replace the different approximations obtained in (2.22),

$$\left(\frac{\partial C_1}{\partial \eta}\right)_{\eta=0} \sim X^{-3/2}. \tag{2.25}$$

The singularity at $X = 0$ is amplified in the further corrections obtained from (2.25). To avoid the singularity in the local mass flux at $X = 0$, new coordinates are necessary:

$$X = \xi + \alpha\varphi(\xi,\eta), \quad \eta = \eta, \tag{2.26}$$

where φ is an arbitrary function to be determined in such a way that the singularity in $C_1(X,\eta)$ at $X = 0$ will be dominated by a singularity in $C_0(X,\eta)$. In new coordinates $C(X,\eta) = \hat{C}(\xi,\eta)$

$$\begin{aligned}&\frac{\partial C}{\partial X}=\frac{\partial \hat{C}}{\partial \xi}\frac{\partial \xi}{\partial X}, \quad \frac{\partial^2 C}{\partial X^2}=\frac{\partial^2 \hat{C}}{\partial \xi^2}\left(\frac{\partial \xi}{\partial X}\right)^2+\frac{\partial \hat{C}}{\partial \xi}\frac{\partial^2 \xi}{\partial X^2},\\ &\frac{\partial C}{\partial \eta}=\frac{\partial \hat{C}}{\partial \xi}\frac{\partial \xi}{\partial \eta}+\frac{\partial \hat{C}}{\partial \eta}, \quad \frac{\partial^2 C}{\partial \eta^2}=\frac{\partial^2 \hat{C}}{\partial \xi^2}\left(\frac{\partial \xi}{\partial \eta}\right)^2+\frac{\partial \hat{C}}{\partial \xi}\frac{\partial^2 \xi}{\partial \eta^2}+\frac{\partial^2 \hat{C}}{\partial \eta^2},\end{aligned} \tag{2.27}$$

where $\frac{\partial \xi}{\partial X}$, $\frac{\partial^2 \xi}{\partial X^2}$, $\frac{\partial \xi}{\partial \eta}$ and $\frac{\partial^2 \xi}{\partial \eta^2}$, must be obtained after differentiation of (2.26),

$$\begin{aligned}&1=\frac{\partial \xi}{\partial X}+\alpha\frac{\partial \varphi}{\partial \xi}\frac{\partial \xi}{\partial X}, \quad \left(\frac{\partial \eta}{\partial X}=0\right), \quad 0=\frac{\partial^2 \xi}{\partial X^2}+\alpha\left[\frac{\partial^2 \varphi}{\partial \xi^2}\left(\frac{\partial \xi}{\partial X}\right)^2+\frac{\partial \varphi}{\partial \xi}\frac{\partial^2 \xi}{\partial X^2}\right],\\ &0=\frac{\partial \xi}{\partial \eta}+\alpha\frac{\partial \varphi}{\partial \eta}, \quad 0=\frac{\partial^2 \xi}{\partial \eta^2}+\alpha\frac{\partial^2 \varphi}{\partial \eta^2}.\end{aligned} \tag{2.28}$$

The new function $C(\xi,\eta)$ can be represented as a power series expansion (2.24):

$$\hat{C}(\xi,\eta)=\hat{C}_0(\xi,\eta)+\alpha\hat{C}_1(\xi,\eta)+\cdots \tag{2.29}$$

The Taylor series representation of the function $C(X,\eta)=C[\xi+\alpha\varphi(\xi,\eta),\eta]$ for small values of the parameter α is

$$C(X,\eta)=[C(X,\eta)]_{\alpha=0}+\alpha\left(\frac{\partial C}{\partial X}\right)_{\alpha=0}\frac{\partial X}{\partial \alpha}, \tag{2.30}$$

where

$$[C(X,\eta)]_{\alpha=0}=\hat{C}_0(\xi,\eta), \quad \left(\frac{\partial C}{\partial X}\right)_{\alpha=0}\frac{\partial X}{\partial \alpha}=\frac{\partial \hat{C}_0}{\partial \xi}\varphi(\xi,\eta)=\hat{C}_1(\xi,\eta), \tag{2.31}$$

i.e.,

$$C(X,\eta)=\hat{C}_0(\xi,\eta)+\alpha\hat{C}_1(\xi,\eta). \tag{2.32}$$

From the boundary conditions in (2.21) and (2.31) and (2.32) the boundary conditions of the new functions follow:

$$\hat{C}_0(0,\eta)=1, \quad \hat{C}_0(\xi,0)=0, \quad \hat{C}_0(\xi,\infty)=1. \tag{2.33}$$

$$\begin{aligned}&\hat{C}_1(\xi,0)=\lim_{\eta\to 0}\left[\varphi(\xi,\eta)\frac{\partial \hat{C}_0}{\partial \xi}\right], \quad \xi>0;\\ &\hat{C}_1(\xi,\infty)=\lim_{\eta\to \infty}\left[\varphi(\xi,\eta)\frac{\partial \hat{C}_0}{\partial \xi}\right], \quad \xi>0;\\ &\hat{C}_1(0,\eta)=\lim_{\xi\to 0}\left[\varphi(\xi,\eta)\frac{\partial \hat{C}_0}{\partial \xi}\right], \quad \eta>0.\end{aligned} \tag{2.34}$$

The results presented permit us to obtain the zeroth approximation ($\alpha = 0$) in (2.32) and its solution:

$$\frac{\partial \hat{C}_0}{\partial \xi} = \frac{\partial^2 \hat{C}_0}{\partial \eta^2}, \quad \hat{C}_0(\xi, \eta) = erf\left(\frac{\eta}{2\sqrt{\xi}}\right). \tag{2.35}$$

The first approximation of the problem has the form

$$\frac{\partial \hat{C}_1}{\partial \xi} = \frac{\partial^2 \hat{C}_1}{\partial \eta^2} + \frac{\eta \exp(-\eta^2/4\xi)}{\sqrt{4\pi\xi^3}}\left\{\eta^2 + \left[\frac{\partial \varphi}{\partial \xi} - \frac{\partial^2 \varphi}{\partial \eta^2} + \frac{\eta^2}{4\xi^2} - \frac{3}{2\xi} - 2\frac{\partial \varphi}{\partial \eta}\left(\frac{1}{\eta} - \frac{\eta}{2\xi}\right)\right]\right\}. \tag{2.36}$$

The strongest singularity of the local mass flux (2.25) is caused by the terms

$$\frac{\eta^2}{4\xi^2} - \frac{3}{2\xi} \tag{2.37}$$

on the right-hand side of (2.36). Obviously $\phi(\xi, \eta)$ has to be chosen so that the terms in the square brackets in (2.36) will vanish, i.e.,

$$\psi'' + \left(\frac{2}{\zeta} - \frac{\zeta}{2}\right)\psi' = \frac{\zeta^2}{4} - \frac{3}{2}, \tag{2.38}$$

where

$$\psi(\zeta) = \varphi(\xi, \eta), \quad \zeta = \frac{\eta}{\sqrt{\xi}}. \tag{2.39}$$

The boundary conditions of (2.38) are specified with respect to physical considerations, according to the limitedness of the concentration and the local mass flux.

Finally, we can write

$$\varphi(\xi, \eta) = -\frac{\eta^2}{4\xi^2} \tag{2.40}$$

and for $\hat{C}_1(\xi, \eta)$ we obtain

$$\frac{\partial \hat{C}_1}{\partial \xi} = \frac{\partial^2 \hat{C}_1}{\partial \eta^2} - \frac{\eta^3 \exp(-\eta^2/4\xi)}{\sqrt{4\pi\xi^3}}, \quad \hat{C}_1(0, \eta) = 0, \quad \hat{C}_1(\xi, 0) = 0, \quad \hat{C}_1(\xi, \infty) = 0,$$
$$\hat{C}_1 = \frac{\eta}{4}\sqrt{\xi\left(1 - \frac{\eta^2}{3\xi}\right)} \exp\left(-\frac{\eta^2}{4\xi}\right) \tag{2.41}$$

and for the Sherwood number we obtain

$$Sh = \sqrt{\frac{6Pe}{\varepsilon\pi}}\left(1 - \frac{Fo}{6}\right), \quad \varepsilon = \frac{h_0}{l}. \tag{2.42}$$

The function obtained (2.40) cannot avoid the singularity in the next approximations ($\hat{C}_2$, etc.). The solution of this problem requires the introduction of another function into (2.26) and its appropriate determination:

$$X = \xi + \alpha\varphi(\xi,\eta) + \alpha^2\varphi_1(\xi,\eta). \tag{2.43}$$

The necessity of use of the Poincaret–Lighthill–Ho method arises in cases of a bad choice of the linear scale. The use of h_0 as a linear scale in (2.19) leads to the singularity at $X = 0$ because $X \gg 1$. Replacement of h_0 with the film length l leads to $0 \leq X \leq 1$ and the singularity problem is avoided [9].

The analytical methods presented can be combined very usefully with numerical methods (see 2.12).

3 Numerical Methods

The analytical methods permit us to solve linear ordinary differential equations. In cases of weak nonlinearity, when the nonlinear effects are related to small parameters, the use of the perturbation method linearizes the problem. Multidimensional or strong nonlinear problems need *numerical methods.*

The different types of the partial differential equations of mathematical physics use a variety of numerical methods, such *finite differences*, *finite elements*, *integral relations*, and *characteristics*, which have different degrees of universality [10]. The most applicable methods in chemical engineering for solving convection–diffusion problems [11–14] are the methods of finite differences and finite elements.

3.1 Finite Differences Method

Let us consider a stationary one-dimensional convection–diffusion equation:

$$u(x)\frac{dc}{dx} = \frac{d}{dx}\left[D(x)\frac{dc}{dx}\right] + f(x), \quad 0<x<l, \quad c(0) = 0, \quad c(l) = 0. \tag{3.1}$$

The values of x and $c(x)$ at the knots of a lattice in the interval $[0, l]$ are

$$x = x_i = ih, \quad i = 0, 1, \ldots, N, \quad Nh = l, \quad c_i = c(x_i). \tag{3.2}$$

The formula of the Taylor expansion in the neighborhood of an arbitrary point x_i is

$$c_{i\pm1} = c_i \pm h\left(\frac{dc}{dx}\right)_{x=x_i} + \frac{h^2}{2}\left(\frac{d^2c}{dx^2}\right)_{x=x_i} \pm \frac{h^3}{6}\left(\frac{d^3c}{dx^3}\right)_{x=x_i} + O(h^4). \tag{3.3}$$

From (4.3) it is possible to obtain the finite differences derivatives (c_x, c_{xx}), where for the sake of convenience the index i is omitted. The first derivatives are on the left $(c_{\bar{x}})$ or on the right (c_x):

$$\begin{aligned} c_{\bar{x}} &\equiv \frac{c_i - c_{i-1}}{h} = \left(\frac{dc}{dx}\right)_{x=x_i} - \frac{h}{2}\left(\frac{d^2c}{dx^2}\right)_{x=x_i} + O(h^2), \\ c_x &\equiv \frac{c_{i+1} - c_i}{h} = \left(\frac{dc}{dx}\right)_{x=x_i} + \frac{h}{2}\left(\frac{d^2c}{dx^2}\right)_{x=x_i} + O(h^2). \end{aligned} \tag{3.4}$$

Obviously, the finite differences derivatives (4.4) can be used to approximate the first derivate in (4.3) and the approximation error is $O(h)$. By analogy, is possible to obtain the central finite differences derivative $(c_{\dot{x}})$ and the second derivative $c_{\bar{x}x}$:

$$\begin{aligned} c_{\dot{x}} &\equiv \frac{c_{i+1} - c_{i-1}}{2h} = \left(\frac{dc}{dx}\right)_{x=x_i} + \frac{h^2}{3}\left(\frac{d^3c}{dx^3}\right)_{x=x_i} + O(h^3), \\ c_{\bar{x}x} &= \frac{c_x - c_{\bar{x}}}{h} = \frac{c_{i+1} - 2c_i + c_{i-1}}{h^2} = \left(\frac{d^2c}{dx^2}\right)_{x=x_i} + O(h^2). \end{aligned} \tag{3.5}$$

The differential operator of diffusive transfer can be represented as

$$\frac{d}{dx}\left[D(x)\frac{dc}{dx}\right] = \frac{dD}{dx}\frac{dc}{dx} + D(x)\frac{d^2c}{dx^2}, \tag{3.6}$$

where the next approximation is used:

$$D_i = D(x_i), \quad i = 0, 1, \ldots, N, \quad \frac{D_{i+1} + D_i}{2} = D(x_i) + O(h^2), \tag{3.7}$$

$$D_x = \frac{D_{i+1} - D_i}{h} = \left(\frac{dD}{dx}\right)_{x=x_i} + O(h^2), \tag{3.8}$$

The approximations of $u(x)$ and $f(x)$ are similar to (4.7). The introduction of finite differences derivatives and functions into (3.1) leads to a set of linear equations for $c(x_i)$, $i = 0, 1, \ldots, N$.

3.2 Finite Elements Method

The finite elements scheme for problem (3.1) will be made using Galerkin's method [15] and the function $c(x)$ searched for will be represented as a sum:

$$c(x) = \sum_{i=1}^{N-1} c_i w_i(x), \tag{3.9}$$

where $w_i(x)$ are linear elements,

$$w_i(x) = \begin{Bmatrix} 0, & x < x_{i-1}; \\ \frac{x - x_{i-1}}{h}, & x_{i-1} \leq x \leq x_i; \\ \frac{x_{i+1} - x}{h}, & x_i \leq x \leq x_{i+1}; \\ 0, & x > x_{i+1}. \end{Bmatrix}. \tag{3.10}$$

The finite elements scheme for problem (3.1) may be obtained if we multiply (3.1) by $w_i(x)$ and integrate over x in the interval $[x_{i-1}, x_{i+1}]$. As a result, we obtain

$$\int_{x_{i-1}}^{x_{i+1}} u(x)\frac{dc}{dx}w_i(x)dx + \int_{x_{i-1}}^{x_{i+1}} k(x)\frac{dc}{dx}\frac{dw_i}{dx}dx = \int_{x_{i-1}}^{x_{i+1}} f(x)w_i(x)dx, \tag{3.11}$$

having in mind that $w_i(x) = 0$ for $x \leq x_{i-1}$ and $x \geq x_{i+1}$. The introduction of (3.10) into (3.11) leads to a set of linear equations for $c(x_i)$, $i = 1, \ldots, N-1$.

An increase of the exactness of the approximation is possible by the method of Petrov and Galerkin [13], which uses quadratic or cubic polynomials in (3.10).

4 Examples

4.1 Application of Green's Functions

Many problems in the modeling of mass transfer kinetics [1, 2] lead to the necessity to solve (1.1) for different functions $F(x,t)$ and boundary conditions:

$$F(x,t) = \frac{x}{\sqrt{4\pi t^3}}[f(t) + 2tf'(t)]e^{-\frac{x^2}{4t}}, \quad U(0,t) = 0, \quad U(\infty,t) = 0, \quad U(x,0) = 0,$$

$$U(x,t) = \frac{x}{\sqrt{4\pi t^3}}\left[2tf(t) - \int_0^t f(\tau)d\tau\right]e^{-\frac{x^2}{4t}}. \tag{4.1}$$

$$F(x,t) = \frac{x}{\sqrt{\pi t^3}}[f(t) + tf'(t)], \quad U(0,t) = 0, \quad U(\infty,t) = 0, \quad U(x,0) = 0,$$

$$U(x,t) = \frac{x}{\sqrt{\pi t}}f(t)e^{-\frac{x^2}{4t}}. \tag{4.2}$$

$$F(x,t)=-\frac{x^3}{4\sqrt{\pi t}}\left(1+\frac{x^2}{6t}+\frac{x^4}{6t^2}\right)e^{-\frac{x^2}{4t}},\ U(0,t)=0,\ U(\infty,t)=0,\ U(x,0)=0,$$
$$U(x,t)=-\frac{1}{24\sqrt{\pi}}\left(13xt\sqrt{t}+\frac{13}{2}x^3\sqrt{t}+\frac{8}{5}\frac{x^5}{\sqrt{t}}+\frac{1}{6}\frac{x^7}{t\sqrt{t}}\right)e^{-\frac{x^2}{4t}}. \tag{4.3}$$

$$F(x,t)=-\frac{f(t)+2tf'(t)}{2t}e^{-\frac{x^2}{4t}},\quad U(0,t)=0,\quad U(\infty,t)=0,\quad U(x,0)=0,$$
$$U(x,t)=-\frac{1}{\sqrt{\pi t}}e^{-\frac{x^2}{4t}}\int_0^t\frac{f(\tau)+2\tau f'(\tau)}{\sqrt{\tau}}\left[\int_0^{\frac{\tau x}{\sqrt{4t\tau(t-\tau)}}}e^{-s^2}ds\right]d\tau. \tag{4.4}$$

$$F(x,t)=\frac{x^3}{\sqrt{4\pi t^3}}e^{-\frac{x^2}{4t}},\quad U(0,t)=0,\quad U(\infty,t)=0,\quad U(x,0)=0,$$
$$U(x,t)=\frac{x}{2}\sqrt{\frac{t}{\pi}}\left(1+\frac{x^2}{3t}\right)e^{-\frac{x^2}{4t}}. \tag{4.5}$$

4.2 Sturm–Liouville Problem

Theoretical analysis of the mass transfer in liquid film flow [1, 16–20] leads to the problem

$$(1-Y^2)\frac{\partial C}{\partial X}=\frac{\partial^2 C}{\partial Y^2},\quad C(0,Y)=1,\quad C(X,1)=0,\quad \frac{\partial C}{\partial Y}(X,0)=1. \tag{4.6}$$

The *separated variables* approach leads to next solution of (4.6):

$$C(X,Y)=\sum_{n=1}^{\infty}A_nH_n(Y)\exp\left(-\lambda_n^2X\right). \tag{4.7}$$

Substitution of (4.7) into (4.6) leads to the Sturm–Liouville problem:

$$H_n''+\lambda_n^2(1-Y^2)H_n=0,\quad H(1)=0,\quad H'(0)=0, \tag{4.8}$$

where λ_n and H_n ($n=1,2,\ldots$) are eigenvalues and eigenfunctions. The solution of (4.8) is obtained [21] in terms of confluent hypergeometric functions:

$$H_n(Y)=\exp\left(-\lambda_n\frac{Y^2}{2}\right)\varphi\left(a,\frac{1}{2},z\right),\quad a=\frac{1}{4}(1-\lambda_n),\quad z=\lambda_nY^2,$$
$$\varphi(a,c,z)=\sum_{k=0}^{\infty}\frac{a(a+1)\ldots(a+k-1)z^k}{c(c+1)\ldots(c+k-1)k!}. \tag{4.9}$$

Table 1 The first ten values of $\lambda_n, A_n, H'_n(1)$ at $Pe \to \infty$

N	λ_n	A_n	$N'_n(1)$	n	λ_n	A_n	$N'_n(1)$
1	1.68159	1.20083	−1.4291	6	21.66721	−0.06277	11.5797
2	5.66986	−0.29916	3.8071	7	25.66710	0.05152	−13.3337
3	9.66824	0.16083	−5.9202	8	29.66702	−0.04351	15.0429
4	13.66769	−0.10744	7.8932	9	33.66696	0.03753	−16.7136
5	17.66737	0.07964	9.7709	10	37.66692	−0.03283	18.3516

From the first boundary condition in (4.6) it follows that

$$\sum_{n=1}^{\infty} A_n H_n(Y) = 1 \tag{4.10}$$

and the orthogonality of the eigenfunctions is easily proved:

$$A_n = \frac{\int_0^1 (1 - Y^2) H_n(Y) dY}{\int_0^1 (1 - Y^2) H_n^2(Y) dY}. \tag{4.11}$$

The results obtained permit us to calculate the Sherwood number [19]:

$$Sh = -Pe \sum_{n=1}^{\infty} \frac{A_n H'_n(1)}{\lambda_n^2}, \tag{4.12}$$

where $\lambda_n, A_n, H'_n(1)$, $n = 1, 2, \ldots, 10$ are presented [19] in Table 1.

References

1. Boyadjiev Chr, Beschkov V (1984) Mass transfer in liquid film flows. Bulgarian Academy of Sciences, Sofia
2. Boyadjiev ChrB, Babak VN (2000) Non-linear mass transfer and hydrodynamic stability. Elsevier, Amsterdam
3. Carslaw HS, Jaeger JC (1959) Conduction of heat in solids. Oxford University Press, London
4. Nayfeh AH (1973) Perturbation methods. Wiley, New York
5. Van Dyke MD (1964) Perturbation methods in fluid mechanics. Academic, New York
6. Bellman R (1964) Perturbation techniques in mathematics, physics and engineering. Holt, New York
7. Boyadjiev Chr (1971) Int Chem Eng 11:459
8. Boyadjiev Chr, Piperova M (1971) Int Chem Eng 11:479
9. Boyadjiev Chr, Levich VG, Krylov VS (1968) Int Chem Eng 8:393
10. Belotserkovskii OM (1984) Numerical modeling in continuum mechanics. Science, Moscow (in Russian)
11. Samarskii AA, Vabishchevich PN (1999) Numerical methods for solving convection–diffusion problems. Editorial URSS, Moscow (in Russian)
12. Miller JJH, O'Riordan E, Shishkin GI (1996) Fitted numerical methods for singular perturbation problems. World Scientific, Singapore

13. Morton KW (1996) Numerical solution of convection–diffusion problems. Chapman & Hall, London
14. Roos H-G, Stynes M, Tobiska L (1995) Numerical methods for singular perturbed differential equations. Convection–diffusion and flow problems. Springer, Berlin
15. Fletcher CAJ (1984) Computational Galerkin methods. Springer, New York
16. Rotem Z, Neilson JE (1969) Can J Chem Eng 47:341
17. Pirkle JC, Sigillito VG (1972) Appl Sci Res 26:105
18. Hsu CJ (1967) Appl Sci Res 17:359
19. Beschkov V, Boyadjiev Chr, Peev G (1978) Chem Eng Sci 33:65
20. Davis EJ, Bonano EJ (1979) Chem Eng Sci 34:439
21. Davis EJ (1973) Can J Chem Eng 51:562

Parameter Identification (Estimation)

The separate stages of the modeling and simulation lead to specific calculation problems.

The modeling at the stage of parameter estimation must solve incorrect inverse problems and minimize multiextremal functions, where the calculation of the least-squares function needs the solution of differential equations. The simulation at the stages of optimal control and design uses the methods of solution of differential equations and minimization of functions. Model parameter identification (estimation) is an inverse problem solution [1], i.e., a mathematical procedure to calculate parameter values using experimental data.

1 Inverse Problems

The determination (identification) of the model parameters (except the cases of theoretical models and the model theories) is a procedure utilizing experimental data only. In the common case, the mathematical description is a set of equations (algebraic, differential, integral, or integral–differential equations) and the corresponding initial and boundary conditions. These equations link the dependent variables (the objective functions) y, the independent variables (the factors) $\boldsymbol{x}$, and the constants (parameters) $\boldsymbol{b}$:

$$\boldsymbol{f}(\boldsymbol{x}, \boldsymbol{y}, \boldsymbol{b}) = 0, \tag{1.1}$$

where $\boldsymbol{f}$, $\boldsymbol{x}$, $\boldsymbol{y}$, and $\boldsymbol{b}$ can be considered as vectors:

$$f = (f_1, \ldots, f_p), \quad \boldsymbol{x} = (x_1, \ldots, x_m), \quad \boldsymbol{y} = (y_p, \ldots, y_n), \quad \boldsymbol{b} = (b_1, \ldots, b_k). \tag{1.2}$$

When the objective function is time-independent and located at a point in space, the models have nondistributed parameters. This corresponds to a mathematical structure of ordinary differential equations. The models with distributed

C. Boyadjiev, *Theoretical Chemical Engineering*,
DOI: 10.1007/978-3-642-10778-8_8, © Springer-Verlag Berlin Heidelberg 2010

parameters have space-distributed objective functions (and time-dependent too) and consequently their mathematical structures contain partial differential equations.

1.1 Direct and Inverse Problems

Mathematical description (1.1) allows us to solve two main problems:

1. *To find the objective function* through algorithms for given values of the factors and the parameters:

$$y = \varphi(x, b). \tag{1.3}$$

Thus, the *forward (direct) problem* is usually solved through a process of simulations. A characteristic feature of the forward problem is that the objective function can be determined via both experiments and simulations.

2. *The inverse problem* is the process of *parameter identification*:

$$b = \psi(x, y). \tag{1.4}$$

It allows us to calculate the parameter values for known experimentally determined values of both the factors and the objective function. In contrast to the forward problems, the inverse problems can be solved through calculation only. Obviously, the inverse problem solution (1.4) strongly depends on the mathematical structure of the model considered.

1.2 Types of Inverse Problems

The inverse problems [8–15] have various natures. They can be:

- *Retrospective*, i.e., considering nonstationary problems with the so-called *inverse time.*
- *Coefficient problems* or *parameter identification.*
- *Structural problems* looking for the structure of the model.
- *Geometrical problems* identifying the geometrical boundaries.
- *Boundary problems* identifying the boundary conditions and values.

Moreover, the inverse problems can occur under various conditions, such as:

- *Inverse problems on control problem solutions* considering the identification of the control parameters or functions allowing the achievement of desired values of the objective function.
- *Design problems* identifying the parameters of the equipment allowing the establishment of the desired output values as a function of the input variables.

The next comments are concerned mainly with the inverse problems considering parameter identification and particularly some structural problems (related to the model adequacy).

The identification of the model parameters depends on the mathematical structure. Some simple models allow that to be done easily as shown in (1.2.67) after the *analytical solution of the direct problem*. In this case substitution of the objective function with its experimental value leads to an algebraic equation for the model parameter.

In most cases the parameter identification utilizes *integral methods*. As an example, we will use a model with one parameter:

$$Pe\frac{dC}{dX} = \frac{d^2C}{dX^2}; \quad X = 0, \quad Pe = PeC_0 + C_0'; \qquad X = 1, \quad C_1' = 0, \tag{1.5}$$

where

$$C_0 = C(0), \quad C_0' = \left(\frac{dC}{dX}\right)_{X=0}; \quad C_1' = \left(\frac{dC}{dX}\right)_{X=1}. \tag{1.6}$$

The double integration of EQUATION (1.5) with respect to X within the ranges $[0, X)$ and $[0.1]$ yields

$$J.Pe - C_0 Pe = C_1 - C_0 - C_0', \tag{1.7}$$

where

$$J = \int_0^1 C(X)dX, \quad C_1 = C(1). \tag{1.8}$$

Expression (1.7) and the first boundary condition of (1.5) permit us to evaluate Pe:

$$Pe = \frac{C_1 - C_0}{J - 2C_0 + 1}. \tag{1.9}$$

Thus, the determination of the parameter Pe requires experimental data $C(X)$ within the range $0 \leq X \leq 1$.

In the *general case* the values of the parameters can be determined through a *minimization of the least-squares function*:

$$Q(b) = \sum_{\mathrm{n}=1}^{\mathrm{N}} (y_\mathrm{n} - \hat{y}_\mathrm{n})^2, \tag{1.10}$$

where $\hat{y}_\mathrm{n}(\mathrm{n} = 1, \ldots, \mathrm{N})$ are calculated values of the objective function through (1.3):

$$\hat{y}_\mathrm{n} = \varphi(x_\mathrm{n}, b), \quad \mathrm{n} = 1, \ldots, \mathrm{N}. \tag{1.11}$$

Here, x_n and $y_\mathrm{n}(\mathrm{n} = 1, \ldots, \mathrm{N})$ are experimentally determined values of both the factors and the objective function from N experiments.

1.3 Incorrectness of the Inverse Problems

The explanation could create the illusion that the inverse problem solution of (1.4) cannot encounter any difficulties except those of the nonlinear programming for the determination of the minimum of $Q(b)$ (see 1.10). However, the reality for this type of problem is rather different. For example, if $J - 2C_0 + 1 \approx 0$ in (1.9), the value of Pe cannot be determined since under these conditions $C_1 \approx C_2$ Moreover, when the values of the denominator of (1.9) are very small, the experimental errors during the determination of the numerator grow significantly.

These specific characteristics of the inverse problems will be considered below through an example of diffusion in a flowing liquid film [4]. The distribution of the concentration (c) depends on both the coordinates (x, y) and the film surface velocity (v) The diffusivity D is the model parameter:

$$v\frac{\partial c}{\partial x} = D\frac{\partial^2 c}{\partial y^2}; \quad x = 0, \quad c = 0; \quad y = 0, \quad c = c^*; \qquad y \to \infty, \quad c = 0. \tag{1.12}$$

The solution of (1.12) using similarity variables is

$$\frac{c}{c^*} = erfc\frac{\eta}{\sqrt{D}}, \quad \eta = \frac{y}{\sqrt{4x/v}}. \tag{1.13}$$

The plots in Fig. 1 illustrate the variations of the dimensionless concentration with the diffusion coefficient for various values of η. Thus, they represent the forward (direct) problem solution.

Let suppose that the inverse problem (1.4) must be solved, i.e., the diffusion coefficient of CO_2 ($D = 1.8 \times 10^{-9}$ m^2 s^{-1}) in water must be determined on the basis of experimental data (sets of experimentally determined dimensionless concentrations c/c^* as functions of x, y and $x/y = 1$ s). The solution uses the plots in Fig. 1. Obviously, only one experiment is needed to determine D through a graphical treatment of a particular curve. However, if we assume a constant accuracy of the measurement of c/c^*, the final accuracy of the result depends on the curve used. For example, if the experimental error of c/c^* is about 5%, the consequent errors for D through the solution of the inverse problem are 44% (line 1), 18% (line 2), 8% (line 3), and 3% (line 4). Unfortunately, line 5 practically does not allow the determination of D.

Let focus our attention on the physical sense of the problem considering different sections of the curves potted in Fig. 1. The value of D determined through line 1 (accuracy of 44%) can be employed to determine c/c^* via line 3 as a solution of the forward problem (accuracy of 28%). The value of D determined through line 3 (accuracy of 8%) can be employed to find c/c^* in the area of line 1 as a solution of the forward problem (accuracy of 1%). The experimental data defined by line 5 cannot be employed to determine D since the inverse problem (1.4) has a discontinuity point.

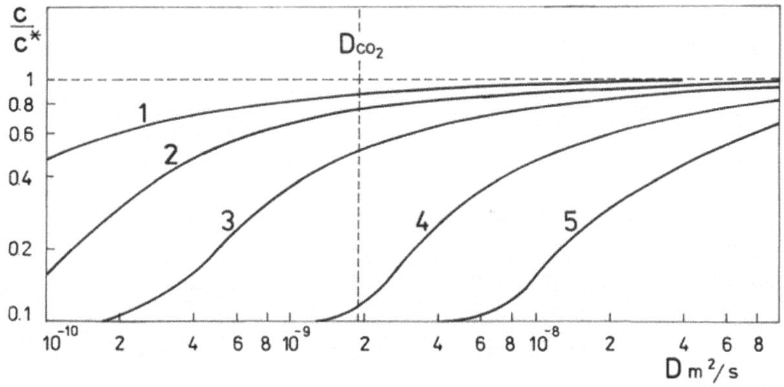

Fig. 1 Graphical explanation of the inverse problem solution concerning the determination of the diffusivity of CO_2 in water based on experimental data with flowing liquid films

The examples discussed above show *how erroneous results could be obtained if the specific features of the inverse problems are not taken into account.* For example, the experimental data c/c^* (presented by curve 2) permit us to determine the values of D within the range $10^{-10} < D < 10^{-9}$(area of stability). From this point of view, the range $10^{-9} < D < 10^{-8}$ is an area of instability, whereas $10^{8} < D < 10^{-7}$is an area of impossible solutions since near the point $c/c^* = 1$ the function has discontinuities.

The data in Fig. 1 indicate that the diffusion coefficient can be determined successfully if a tentative value is known. This allows us to choose a suitable curve for further solution of the problem. For example, if $D \sim 10^{-8}$, the corresponding data are from curve 4, whereas at $D \sim 10^{-9}$ curve 3 must be used and if $D \sim 10^{-10}$, curve 2 must be employed, etc.

The plots in Fig. 1 show that the minimization of $Q(b)$ (see 1.10) cannot allow determination of the parameter with the desired accuracy if the objective function error is high.

2 Sets and Metric Spaces

The specific features of the inverse problems are relevant to the theory of functions and the functional analysis considered next.

2.1 Metrics

The population of elements x forms a set X, i.e., $x \in X$ [5]. The metric space $R(X, \rho)$ is a population of elements of X and the distances ρ between them. The distance ρ between the elements x and y of the set $X(x \in X, y \in X)$, $\rho(x, y) \geq 0$, is termed the *metric* of the space. It satisfies the following three conditions for every x, $y \in X$:

$$\rho(x,y) = 0 \quad at\ x = y, \quad \rho(x,y) = \rho(y,x)$$

$$\rho(x,y) + \rho(y,z) \geq \rho(x,z), \quad z \in X(\text{triangle inequality}). \tag{2.1}$$

The series of points of the metric space $\{x_n\} = x_1, \ldots, x_n$ has a limit if

$$\lim_{n \to \infty} \rho(x, x_n) = 0, \tag{2.2}$$

which is valid for any series (subset). If two arbitrary points x_{n_1} and x_{n_2} of a series satisfy the condition

$$\rho[x_{n_1}, x_{n_2}] < \varepsilon, \quad \varepsilon > 0, \quad n_1 \geq N_\varepsilon, \quad n_2 \geq N_\varepsilon, \tag{2.3}$$

it can be considered as a fundamental series. Obviously the convergent series

$$\lim_{n \to \infty} \rho(x, x_n) = 0 \tag{2.4}$$

is a fundamental one.

If any series of the space converges, it is a complete space. Therefore, it is possible to find arbitrary close points converging to a certain unique limit.

The set A of the metric space R is an ε network with respect to another set (M) from the same space if for $x \in M$ and $a \in M$

$$\rho(a,x) \leq \varepsilon, \tag{2.5}$$

i.e., the elements of the space are contained by a ball $\rho(a,x)$ depending on the value of ε. This allows us to define a completely bounded set M if for any $\varepsilon>0$ there is a finite ε network. This means that if ε is decreased in any case, the set remains finite (the set is bounded by the network). Thus, the set M is bounded not only with respect to the number of the elements, but it is also bounded with respect to the distances between the elements and the finite network (mesh).

The set M is a *compact set* if it is completely bounded in a complete metric space R. It is characterized by elements that are densely located everywhere, but is completely bounded by the finite mesh. Obviously, the ε network is also compact. The compact metric space is usually termed *compact.*

2.2 Linear Spaces

The set of R elements x, y, z, ... forms a linear space [5] if the following conditions are satisfied:

$$x+y=y+x, \quad x+(y+z)=(x+y)+z, \quad x+0=x, \quad x+(-x)=0,$$
$$\alpha(\beta x)=(\alpha\beta)x, \quad 1.x=x, \quad (\alpha+\beta)x=\alpha x+\beta x, \quad \alpha(x+y)=\alpha x+\beta x. \tag{2.6}$$

The linear spaces are *normalized spaces* if for any element $x \in R$ there is a positive value $\|x\|$-a *norm* of x, where

$$\|x\|=0 \quad at\ x=0, \quad \|\alpha x\|=|\alpha|\cdot\|x\|, \quad \|x+y\|=\|x\|+\|y\|. \tag{2.7}$$

The straight line, with the conventional arithmetic operations, is the simplest normalized space with a norm equal to the modulus of the real number:

$$\|x\|=|x|. \tag{2.8}$$

The linear Euclidean space has a norm equal to the length of the vector, i.e., in an n-dimensional case $x = x_1,\ldots,x_n$, and

$$\|x\|=\left[\sum_{i=1}^{n} x_i^2\right]^{\frac{1}{2}}. \tag{2.9}$$

Let consider two vectors of the linear Euclidean space $x = x_1, \ldots, x_n$, $y = y_1, \ldots, y_n$ and their sum $x + y$. Obviously, from (2.9) and the equality

$$\sum_{i=1}^{n}[x_1+y_1]^2=\sum_{i=1}^{n}x_1^2+\sum_{i=1}^{n}y_1^2+\sum_{i=1}^{n}x_1y_1 \tag{2.10}$$

it follows that the norm of the sum may be expressed as

$$\|x+y\|^2=\|x\|^2+\|y\|^2+2(x,y). \tag{2.11}$$

Here, (x, y) denotes the scalar product of the vectors x and y in a Euclidean space, i.e.,

$$(x,y)=\|x\|\cdot\|y\|\cdot\cos\varphi=\sum_{i=1}^{n}x_1y_1. \tag{2.12}$$

and φ is the angle between x and y.

Any normalized space is a *metric space* if

$$\rho(x,y)=\|x-y\|. \tag{2.13}$$

The complete normalized space is termed a *Banach space* (B-space). If x and y are two points in a linear normalized space, the *piece* which connect points x and y is all points z satisfying the conditions

$$z=\alpha x+\beta y, \quad \alpha\geq 0, \quad \beta\geq 0, \quad \alpha+\beta=1. \tag{2.14}$$

The set M of a linear space R is a *convex set* if the piece z which connect the points $x \in M$ and $y \in M$ are elements of the same set, i.e., $z \in M$.

2.3 Functional

If the elements of the set R are functions f, we can consider a set F whose elements are numbers a, so that for every $f \in R$ there is $a \in F$. Thus, the set F is a *functional*. For example, the functional

$$F = \int_0^1 \varphi[x, f(x)]dx \tag{2.15}$$

takes a specific form for different functions $f(x)$ if

$$\varphi[x, f(x)] = xf(x), \tag{2.16}$$

i.e.,

$$f(x) = 1, \quad F(f) = \int_0^1 xdx = \frac{1}{2}; \qquad f(x) = x, \quad F(f) = \int_0^1 x.xdx = \frac{1}{3};$$

$$f(x) = \mathrm{e}^x, \quad F(f) = \int_0^1 x\mathrm{e}^x dx = 1. \tag{2.17}$$

2.4 Operator

Let $x \in X$ and $y \in Y$ be elements of two Banach spaces R and R', and a law (algorithm) A is defined so that for every element $x \in X \subset R$ it is possible to obtain element $y \in Y \subset R'$. This defines an *operator* on the set X,

$$y = Ax, \tag{2.18}$$

with an area of values in the space R'.

Equation (2.18) is an operator equation of the first order. The operators (like functionals) could be linear or nonlinear, continuous or discontinuous, etc. It follows from (2.18) that over $x \in X$ it is possible to define an (operator) algorithm that gives $y \in Y \subset R'$. The inverse problem (to find x if y is known) can be solved through the *inverse operator*:

$$x = A^{-1}y. \tag{2.19}$$

The operator A is fully continuous if it converts every bounded set into a compact set. In this case the corresponding inverse operator A^{-1} is not continuous.

Consider the bounded linear operator (2.18) converting (mapping) $x \in X$ to $R^{'}(y \in R')$. Suppose we have the functional

$$F(x) = (Ax, y), \tag{2.20}$$

where $x \in X$ and $y \in Y$ are fixed elements in R and $R^{'}$, and (Ax, y) is a scalar product. From the boundedness of the operator (2.18) it follows that $y = y^{*} \in R$ exists, so

$$F(x) = (x, y^{*}), \tag{2.21}$$

i.e.,

$$(Ax, y) = (x, y^{*}). \tag{2.22}$$

The elements y^* are defined by $F(x)$. Moreover, y^* are defined by y through (2.20), i.e., there is an operator A^* which converts y to Y^*:

$$y^{*} = A^{*}y. \tag{2.23}$$

The last two definitions (2.22, 2.23) give

$$(Ax, y) = (x, A^{*}y), \tag{2.24}$$

which allows us to define the operator A^* *conjugated* to A. For both operators

$$\|A^{*}\| = \|A\|. \tag{2.25}$$

2.5 Functional of the Misfit

If both sides of (2.18) are not equal (owing to some errors), the norm of the misfit is the functional

$$\Delta(x) = \|Ax - y\|, \tag{2.26}$$

which will be termed hereafter as *misfit* or *error function.*

Similarly, it is possible to define the functional

$$J(x) = \frac{1}{2}\|Ax - y\|^{2} = \frac{1}{2}\Delta^{2}(x). \tag{2.27}$$

If the operator A is defined close to x_0 and Δx is the increment of x close to x_0,

$$A'_{x_0} = \lim_{\Delta x \to 0} \frac{A(x_0 + \Delta x) - Ax_0}{\Delta x} \tag{2.28}$$

is the operator derivative (Freshet's derivative). If the operator is linear,

$$A'_x = A. \tag{2.29}$$

The last expression (2.29) allows us to define the strong [8] differential (Freshet's differential) of the operator A:

$$A(x + \Delta x) - Ax = A'_x \Delta x + \alpha(x, \Delta x), \tag{2.30}$$

where

$$\lim_{\|\Delta x\| \to 0} \frac{\|\alpha(x, \Delta x)\|}{\Delta x} = 0. \tag{2.31}$$

In a similar way it is possible to derive (through 2.30) the differential of the functional $J(x)$:

$$\begin{aligned} J(x + \Delta x) - J(x) &= \frac{1}{2}\|A(x + \Delta x) - y\|^2 - \frac{1}{2}\|Ax - y\|^2 \\ &= \frac{1}{2}\|Ax - y + A'_x \Delta x + \alpha(x, \Delta x)\|^2 - \frac{1}{2}\|Ax - y\|^2. \end{aligned} \tag{2.32}$$

With the help of (2.11), we can transform expression (2.32) into

$$\begin{aligned} &J(x + \Delta x) - J(x) = (Ax - y, A'_x \Delta x) + \frac{1}{2}\|A'_x \Delta x + \alpha\|^2 + (Ax - y, \alpha), \\ &\alpha = \alpha(x, \Delta x). \end{aligned} \tag{2.33}$$

Substitution of (2.25) into (2.33) leads to

$$J(x + \Delta x) - J(x) = \left[A'^{*}_x (Ax - y), \Delta x\right] + \varepsilon(x, \Delta x), \tag{2.34}$$

where $\varepsilon(x, \Delta x)$ denotes all small terms and

$$\lim_{\|\Delta x\| \to 0} \frac{\|\varepsilon(x, \Delta x)\|}{\|\Delta x\|} = 0. \tag{2.35}$$

The above expressions allow the gradient of the functional $J(x)$ to be represented as

$$J'_x = A'^{*}_x (Ax - y). \tag{2.36}$$

If the operator is linear, i.e., $A'_x = A$,

$$J'_x = A^{*}(Ax - y). \tag{2.37}$$

The normalization of the operators employs the inequalities

$$\|A_1 + A_2\| \le \|A_1\| + \|A_2\|, \quad \|Ax\| \le \|A\| \cdot \|x\|. \tag{2.38}$$

2.6 Some Properties of the Direct and Inverse Operators

The solution of the inverse identification problems needs to consider some properties of both the direct (forward) and the inverse operators.

Let us consider a model of mass transfer with a volume reaction [1]:

$$\frac{d^2C}{dX^2} - Pe\frac{dC}{dX} - PeNC = 0; \qquad X = 0, \quad \frac{dC}{dX} - PeC + Pe = 0;$$
$$X = 1, \quad \frac{dC}{dX} = 0. \tag{2.39}$$

Equations (2.39) can be considered as operator equations, i.e., $Ax = y$, which allows us to solve both the direct problem ($y = Ax$) and the inverse one ($x = A^{-1}y$). The direct problem solution is

$$C = a_1 \exp A_1X + a_2 \exp A_2X, \tag{2.40}$$

where

$$A_1 = \frac{Pe + \lambda}{2}, \quad A_2 = \frac{Pe - \lambda}{2}, \quad \lambda = \sqrt{Pe^2 + 4NPe}, \tag{2.41}$$

$$a_1 = \frac{PeA_2 \exp A_2}{-A_1^2 \exp A_1 + A_2^2 \exp A_2}, \quad a_2 = \frac{PeA_1 \exp A_1}{-A_2^2 \exp A_2 + A_1^2 \exp A_1}. \tag{2.42}$$

The solution of (5.2.40–5.2.42) is valid for every value of $Pe > 0$ and for $Pe \to \infty$, i.e.,

$$Pe \to \infty, \quad C = \exp(-NX). \tag{2.43}$$

The inverse problem solution comes through a double integration of (2.39) from 0 up to X and from 0 up to 1. The results is an equation containing two unknown values of Pe and N. The second equation needed can be obtained in a similar way through multiplication of the first equation in (2.39) by X. By the employment of the boundary condition of (2.39), the inverse problem solution is

$$Pe = \frac{J_1(2J_0 - 2C_1) + J_2(C_1 - C_0)}{J_1(C_1 - 2J_1) + J_2(J_0 - C_1)}, \quad N = \frac{2(J_0 - C_1)^2 + (C_0 - C_1)(C_1 - 2J_1)}{J_1(2J_0 - 2C_1) + J_2(C_1 - C_0)}, \tag{2.44}$$

where

$$C_0 = C(0), \quad C_1 = C(1), \quad J_0 = \int_0^1 C(X)dX, \quad J_1 = \int_0^1 XC(X)dX,$$
$$J_2 = \int_0^1 X^2C(X)dX. \tag{2.45}$$

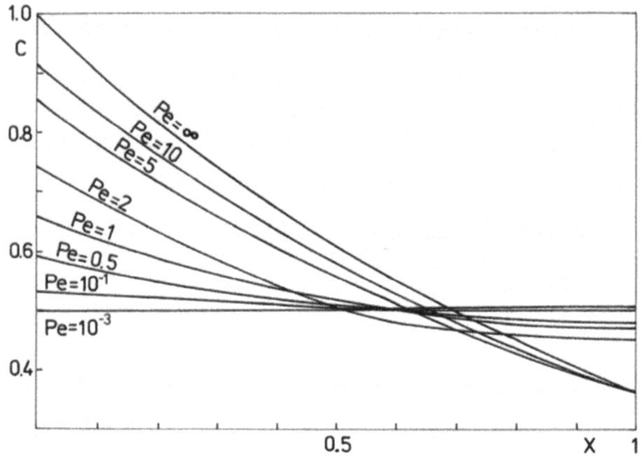

Fig. 2 The variations of the function *C(X)* with *X* for various values of the parameter *Pe* at N = 1

Figure 2 shows the function *C*(*X*) for various values of *Pe* at N $=$ 1. The plot indicates that despite the infinite variations of *Pe* the direct operator (2.40–2.42) always defines a limited function. Thus, the forward operator is a completely continuous (fully bounded) operator. At high values of *Pe*, however, the variations of *Pe* cause insignificant variations of *C*(*X*). Therefore, during the inverse problem solution, small errors in the determination of *C*(*X*) may cause strong errors (several orders of magnitude) of the values of *Pe*. This effect is a characteristic feature of completely limited operators, since experimental data errors often cause instability of their inverse operators.

The graphs in Fig. 2 demonstrate that every function outside the area bounded by the curves $Pe = 10^{-3}$ and $Pe = \infty$ cannot lead to the inverse problem solution. Moreover, the use of experimental values of *C*(*X*) at $X > 0.4$ does not give unique solution of the inverse problem.

3 Incorrectness of the Inverse Problems

The models of the processes (1.1) can be considered as operator equations:

$$Ab = y. \tag{3.1}$$

Here, *A* is the operator (algorithm) allowing calculation of the objective function **y** if the model parameters **b** are known and the values of the independent model variables **x** are available. This direct problem is characterized by the fact that it has a physical analog, i.e., it is possible to determine the objective function *y* experimentally.

The inverse identification problem is

$$b = A^{-1}y. \tag{3.2}$$

The solution of the inverse identification problem is physically impossible since it is not possible to measure experimentally the parameter values. Thus, the only available way is to find them through mathematical methods.

In the common case, it will be assumed that in (3.1) both the objective function y (due to the experiments) and the operator A (due to the mathematical model) could be inaccurate (determined with some errors). Thus, following (3.1) and (3.2), we may write

$$A\tilde{b} = \tilde{y}, \quad \tilde{b} = A^{-1}\tilde{y}, \quad \tilde{A}\tilde{b} = \hat{y}, \tag{3.3}$$

where $\tilde{y}$ and $\hat{y}$ are the experimental and the calculated values, respectively, $\tilde{A}$ is the inaccurate operator, and $\tilde{b}$ are the inaccurate parameters.

Let $\rho(y_1, y_2)$ be a metric describing the difference between both functions. The triangle inequality expresses the difference between y and $\tilde{y}$ as (see 2.1)

$$\rho(y, \tilde{y}) = \rho\left(Ab, A\tilde{b}\right) \leq \rho\left(Ab, \tilde{A}\tilde{b}\right) + \rho\left(\tilde{A}\tilde{b}, A\tilde{b}\right), \tag{3.4}$$

Here, $\rho\left(Ab, \tilde{A}\tilde{b}\right)$ comes from the systematic error of the model, whereas $\rho\left(\tilde{A}\tilde{b}, A\tilde{b}\right)$ is the statistical (random) error of the experiments.

The quality of the model depends on the minimization of (3.4), i.e., the minimization of the errors of both the model and the experiments. This can be achieved though a more complete mathematical description and more precise experiments. Both types of errors depend on $\tilde{b}$, i.e., on the accuracy of the determination of b through the inverse problem (3.2) solution. The main difficulties arise very often from the fact that inverse problem is incorrect (ill posed).

3.1 Correctness After Hadamard

According to Hadamard [2, 3], the correctly posed problems must satisfy three conditions:

1. The solution must exist.
2. The solution must be unique.
3. The solution must be a continuous function of the input data.

The first two conditions concern the deterministic nature of the problem, whereas the third one is relevant to its physical sense.

The inverse problems (employed for parameter identification) presented by (3.2) are incorrect if some of the above conditions are not satisfied. Usually, the incorrectness is due to the nonexistence of the third condition. The operator A^{-1} is not usually continuous and small errors of the experimental values of y cause large errors of the calculated values of b.

The main reason leading to the incorrectness of the inverse problems is the "integral character" of the operator A in (3.1). This means that a large difference between the values of b results in small errors of the values of y. The result of application of the inverse operator A^{-1} is the same: small errors of y cause large errors of the values of b.

3.2 Correctness After Tikhonov

Tikhonov developed some possible ways to solve the inverse problem defined by Hadamard. Following him, the conditions of the inverse problem correctness were formulated by Lavrentiev [2]:

1. The solution is known a priori and it belongs to a set B.
2. The solution is unique.
3. Infinitesimal variations of the input data y, for which the solution for b belongs to B, cause infinitesimal variations of b.

The above conditions permit the problems that were incorrect after Hadamard to be solved through conditionally correct methods (correctness after Tikhonov).

There are different methods for the inverse problem solution, but very often only two approaches are employed—*the selection method and the regularization method*.

4 Methods for Solving Incorrect (Ill-Posed) Problems

Consider two metric spaces B and Y (Fig. 3) and a correctly defined operator A mapping B to Y:

$$AB = Y. \tag{4.1}$$

The elements of both sets $b \in B$ and $y \in Y$ must satisfy the equation

$$Ab = y. \tag{4.2}$$

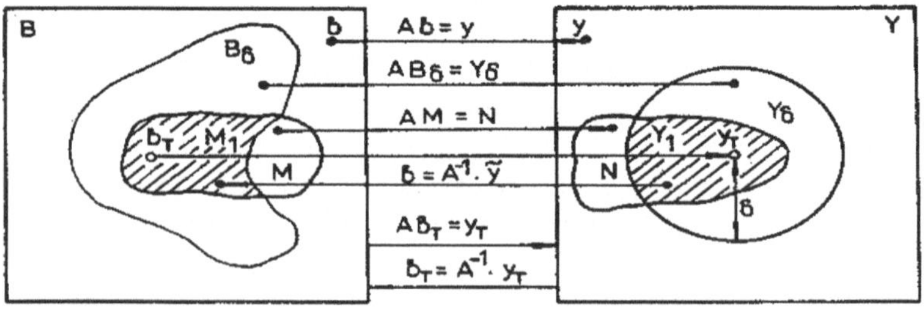

Fig. 3 Graphical representation of the mapping procedures of the inverse problem solution

The explanation developed below supposes that the metrics of both spaces B and Y are uniform and quadratic metrics, so

$$\rho_{\mathrm{B}}(b_1, b_2) = \max_{t \in [t_1, t_2]} \|b_1(t) - b_2(t)\|, \quad \rho_{\mathrm{Y}}(y_1, y_2) = \left\{ \int_{X_1}^{X_2} [y_1(x) - y_2(x)]^2 dx \right\}^{\frac{1}{2}}. \tag{4.3}$$

The above relationships supposes that parameters b depend on the variable t (e.g., assumed as a temperature), whereas the objective function depends on the independent variables x, which allows the integral to be replaced by finite sums. Suppose $y = y_{\mathrm{T}}$ is the exact objective function
(Fig. 3). It permits us to determine b_{T}, i.e., the exact solution of (4.2):

$$Ab_{\mathrm{T}} = y_{\mathrm{T}}. \tag{4.4}$$

In fact the known data are the experimental values of the objective function $(\tilde{y})$ only. This is the reason to look for an approximate solution of the equation

$$Ab = \tilde{y}. \tag{4.5}$$

Very often A is a fully continuous operator, so A^{-1} is not discontinuous. This is the reason why the approximate solution of (4.2) cannot be found as an exact solution of (4.5).

Hence,

$$\tilde{b} = A^{-1}\tilde{y} \tag{4.6}$$

owing to the fact that if $\tilde{y} \in AB$ (the set AB contains all the elements Ab, $b \in B$) such a solution could not exist $(\tilde{b} \notin B)$. If such a solution were possible, it would be unstable owing to the discontinuous character of the operator A^{-1}. These features of the ill-posed inverse problems are discussed with the example illustrated in Fig. 1. If in this particular case the order of the parameter D is about 10^{-9} m^2/s, the experimental values of the objective functions (line 5) cannot be employed to determine the value of D. On the other hand, the data for line 1 lead to an unstable (incorrect) determination of the values of D. The plots in Fig. 1 demonstrate that a stable solution exists if suitable input (experimental) data are chosen (e.g., those for line 3) and the interval of variations of the exact solution is known preliminarily. For example, if the diffusion coefficient is 1 order lower (approximately 10^{-10} m^2/s), it must be determined by line 2.

Practically the experimental values of the objective functions contain some error δ, i.e.,

$$\rho_{\mathrm{Y}}(y_{\mathrm{T}}, \tilde{y}) \leq \delta. \tag{4.7}$$

The elements $\tilde{y}$, satisfying the inequality (4.7), form a subset Y_δ (see Fig. 3). Every $\tilde{y}$ corresponds to an element $b \in B_\delta$ that satisfies the equation

$$AB_\delta = Y_\delta. \tag{4.8}$$

The suggestion from Fig. 3 is that all the elements $b_\delta \in B_\delta$ could be considered as an approximate solution. In fact, however, owing to the instability of the inverse operator A^{-1} the distance between the elements of B_δ could be large. The consequent effect is they differ significantly from the true values b_T.

The difficulties could be avoided if additional information concerning y (or b) were available. In many cases this information could be quantitative. Such an example is the problem looking for a narrow class of possible solutions $\tilde{\delta}_T \in M$ (remember $b_T \in M$). In other cases the information could be qualitative concerning, for example, the smoothness of the function $\tilde{y}$. This allows the development of specific methods [2, 3, 15, 16, 22] for solution of (3.4). Hereafter, four methods will be considered:

1. The method of selections.
2. The method of quasi-solutions.
3. The method of substitution of equations.
4. The method of the quasi-reverse.

4.1 Method of Selections

The method employs a preliminarily known set M containing the exact solution $b_T \in M$. The first step is to find the set N as a solution of the forward problem:

$$AN = N. \tag{4.9}$$

After that the set Y_1 formed by elements satisfying simultaneously (4.8) and (4.9) must be determined. This set contains common elements of N and Y_δ (the dashed area in Fig. 3). It is easy to prove [2] that the minimization of

$$\rho_{Y_l}\left(A\tilde{b}, \tilde{y}\right) \tag{4.10}$$

leads to $\rho_M\left(\tilde{b}, b_T\right) \to 0$, where $\tilde{b} \in M, \tilde{y} \in Y_1$, and the metrics ρ_M and ρ_{Y_1} are defined by (5.4.3).

The set Y_1 is characterized by the fact that the problem

$$b = A^{-1}\tilde{y}, \quad \tilde{y} \in Y_1 \tag{4.11}$$

is correctly posed. The set Y_1 is termed *a class of correctness*, i.e., problem (4.11) is correct after Tikhonov. Obviously $b \in M_1$, where M_1 contains the common elements of M and B_δ (the dashed area in Fig. 3).

4.2 Method of Quasi-Solutions

In many cases experimental errors of the values $\tilde{y}$ lead to the fact that $\tilde{y} \notin N$. Moreover, there are no unique criteria proving that $\tilde{y} \in N = A$. In such cases it is possible to show that $A^{-1}\tilde{y}$ is nonsense and a quasi-solution $\tilde{b} \in M$ must be obtained [2] that minimizes the functional

$$\rho_{\mathrm{Y}}(A\tilde{b}, y) = \inf_{\mathrm{b} \in \mathrm{M}} \rho_{\mathrm{Y}}(Ab, y). \tag{4.12}$$

The quasi-solution $\tilde{b}$is a unique solution if the projection of $\tilde{y} \in Y$ on the set $N = AM \in Y$ is unique. This means that N contains only one element $h \in N$minimizing $\rho_Y(h, \tilde{y})$. In other words, the idea is to find $\tilde{y}$ that is very close to the unique objective function $h \in N$.

4.3 Method of Substitution of Equations

The problems where $\tilde{y} \notin N = AM$ can be solved [2] if the operator equation (4.2) is substituted by

$$(A + \alpha E)b \equiv Ab + \alpha b = y, \tag{4.13}$$

where E is a unit operator. The parameter $\alpha > 0$ must be specified in a manner ensuring the continuity of the inverse operator defining the solution b_α:

$$b_\alpha = (A + \alpha E)^{-1} y. \tag{4.14}$$

4.4 Method of the Quasi-Reverse

In some cases the change of the operator (4.13) is not enough to satisfy the condition of the continuity of the inverse operator (4.14). Under such a condition the method of the quasi-reverse [2] avoids the difficulties through the formation of a significantly new operator.

4.5 Summary

The methods mentioned in the previous sections employ some additional information about the parameters b and the objective function y. The plots in Fig. 1 show that this additional information is very important through the problem solution. For example, the information about the order of magnitude of D is enough to define the experimental data (curves) permitting its determination.

5 Methods for Solving Essentially Ill-Posed Problems

The previous point (Sect. 4) considered several methods for solving the operator equation (8.4.2) when the class of the possible solutions is a compact. However, there are situations when this is impossible and there are experimental errors of the objective function (y) that put it outside the metric space ($y \in Y = AB$). This follows from the fact that B is not a compact set. These are the *essentially ill-posed problems* and the methods discussed in Sect. 4 cannot be applied. Despite this, there are possibilities to create approximate solutions of that class of problems. The basis of that approach is *the regularization operator* [2, 3].

5.1 Regularization Operator

Let us consider the operator equations $Ab = y$, $b \in B$, $y \in Y$, but B is not a compact set and the experimentally obtained objective function $\tilde{y} \in Y$ does not belong to the set $AB(\tilde{y} \notin AB)$. Moreover, let b_T be a solution of $AB = y_T$ i.e., $Ab_T = y_T$ Really, the objective function y_T is unknown, but there are experimental values y_δ obtained with an error of δ, so

$$\rho_Y(y_\delta, y_T) \leq \delta. \tag{5.1}$$

The next assumption is that A is an exact operator. The problem must be solved with the data concerning y_δ, A, and δ only being available. The target is an approximate solution b_δ that approaches b_T and exhibits a stable behavior with small variation of y_δ. As mentioned earlier, b_T cannot be obtained as an exact solution of

$$b_\delta = A^{-1} y_\delta \tag{5.2}$$

since the essential incorrectness means that this solution could not exist for every $y_\delta \in Y$ and that it could be unstable with small variations of y.

Obviously, the approximate solution b_δ must depend on the number parameter $\delta > 0$ characterizing the accuracy of the experimentally obtained objective function y_δ. Taking into account these assumptions, we cannot define the approximate solution of (5.2) as an exact solution of (5.2), but we can define it through the solution of

$$b_\delta = R(y_\delta, \delta). \tag{5.3}$$

Here, $R(y_\delta, \delta)$ is *the regularization operator* of $Ab = y$ (with respect to the element y_T). The equation $Ab = y$ has an exact solution $Ab_T = y_T$, $b_T \in B$, $y_T \in Y$. The regularizing operator must have several features as follows:

The operator $R(y_\delta, \delta)$ maps Y to B. It is defined for every $\delta (0 \leq \delta \leq \delta_1)$ and every $y_\delta \in Y$ such that $\rho_Y(y_\delta, y_T) \leq \delta$.

For every $\varepsilon > 0$ there is $\delta_0 = \delta_0(\varepsilon, y_\delta) \leq \delta_1$ such that $\rho_B(b_\delta, b_T) \leq \varepsilon$ follows directly from the condition $\rho_Y(y_\delta, y_T) \leq \delta \leq \delta_0$, where b_δ is the exact solution of (5.3).

The above definition of the *regularizing operator*$R(y_\delta, \delta)$ does not suggest its uniqueness, so there are possibilities to create various regularization operators (algorithms) for the solution of one incorrect inverse problem.

Very often a more general formulation is needed that considers a dependence of the regularization operator on a parameter α (*regularization parameter*). The operator $R(y,\alpha)$ is defined for $0 < \alpha < \alpha_1$ and for every $y \in Y$, such that $\rho_Y(y, y_T) \leq \delta_1 > 0$. The regularization parameter is defined as a functional $\alpha = \alpha(y, \delta)$, so $y \in Y$and $\rho_Y(y, y_T) \leq \delta_1$. The creation of this functional follows a procedure such that for every $\varepsilon > 0$ a number $\delta \leq \delta(\varepsilon)$ must be found. If $\tilde{y} \in Y$ and $\rho_Y(y, y_T) \leq \delta \leq \delta(\varepsilon)$, $\rho_B(b_T, b_\alpha) \leq \varepsilon$, where b_α is the exact solution of the equation

$$b_\alpha = R[\tilde{y}, \alpha(\tilde{y}, \delta)]. \tag{5.4}$$

The operator in (5.4) is not unique like in (5.3), but at $\alpha = \delta$ they are equivalent.

Equation (5.4) shows that the approximate solution of (4.5) could be an element $b_\alpha = R(y_\delta, \alpha)$, where $\alpha = \alpha(y_\delta) = \alpha_1(\delta)$ confirms the error of the experimental data. The approximate solution b_α is termed *a regularized solution*. The R operator may not be unique and its particular form depends on the type of the operator A. The parameter α must be chosen in a way that confirms the solutions with the experimental error as well as defining a solution having insensitivity to small variations of the experimental data y.

The above explanation demonstrates that the regularization method is an approach looking for *regularization operators* and *regularization parameters* under conditions imposed by additional information (e.g., the magnitude of the error defined by y_δ).

The determination of the regularization operators may be performed through *variational* or *iterative* approaches.

5.2 Variational Approach

As mentioned in Sect. 4, the approximate solutions b_δ should be found within the class B_δ satisfying condition (4.8). This means that the solution accuracy should confirm the accuracy of the input data $y_\delta \in Y_\delta$. It was demonstrated that the set B_δ is sufficiently large, so the selection of a solution must be performed (from B_δ) so it will be stable with respect to small variations of y_δ. Here the selection will be performed by means of a preliminarily defined functional $\Omega(b)$ following from the problem formulation. It should be noted that in many cases in the problems of parameter identification the solutions (the parameters) are constants, whereas $\Omega(b)$ are functions.

5.2.1 Stabilizing Functional

A *stabilizing functional*$\Omega(b)$ can be used to aid in the selection of possible solutions. The functional $\Omega(b) > 0$ is defined for $b \in M_1 \subset M$, where M_1 is a dense subset. It is characterized by the fact that the exact solution exists in its domain of definition, i.e., $b_T \in M_1$ for every $d > 0$, such that $F_{1,d} \subset F_1$, for which $\Omega(b) < d$ is a compact.

Let consider only those elements of B_δ (the domain) where the functional $\Omega(b)$ is defined. Their subset $F_{1,\delta}$ is defined by

$$F_{1,\delta} = F_1 \cap B_\delta. \tag{5.5}$$

We must find this element $b_\delta \in F_{1,\delta}$ that minimizes $\Omega(b)$. This element could be considered as a solution of the operator equation

$$b_\delta = \tilde{R}(y_\delta, \delta), \tag{5.6}$$

since it is easy to prove [2] that $\tilde{R}(y, \delta)$ is a *regularization operator* of the equation $Ab = y$.

The main problem of this approach is the minimization of the functional $\Omega(b)$ under restrictions expressed as the inequality

$$\rho_Y(Ab, y_\delta) \leq \delta. \tag{5.7}$$

5.2.2 Smoothing Functional

An approach that transforms that variational problem into a classic one with the help of (5.7) in the form of an equality employs the Lagrange method [2], i.e., the use of undefined coefficients (multipliers). Let the set M_0 contain the elements $b \in Y_1$ corresponding to the exact lower limit of the functional $\Omega(b)$. Suppose, for simplicity of explanation, that M_0 contains one element. Under such conditions two situations are possible:

The sets M_0 and $Y_{1,\delta}$ have a common domain, i.e., $\rho_Y(Ab_0, y_\delta) \leq \delta$.

The opposite case exists when $\rho_Y(Ab_0, y_\delta) > \delta$.

The first case means that b_0 minimizes the functional $\Omega(b)$, i.e., the desired solution is insensitive to small variations of y_δ.

In the second case (it is possible to prove) the approximate solution b_δ can be found among the elements satisfying the condition

$$\delta_Y(Ab_\delta, y_\delta) = \delta, \quad b_\delta \in M_0 \tag{5.8}$$

and the approximate solution b_δ minimizes the functional

$$M^\alpha(b, y_\delta) = \rho_Y^2(Ab, y_\delta) + \alpha\Omega(b), \tag{5.9}$$

where the parameter α is defined by (5.5.8).

The functional $M^\alpha(b, y)$ is *a smoothing functional* and its minimization is related to the minimization of $\Omega(b)$, since (5.9) indicates that at $\Omega(b) = 0$ the minimum of $\rho_Y^2(Ab, y_\delta)$ with respect to b exists.

The explanation above elucidates the fact that the principal element of the *variational regularization method* is *the choice of the stabilizing functional*$\Omega(b)$. Usually its form follows from the type of the problem, but in many cases its choice is not unique, so various stabilizing functionals through the inverse problem solution can be employed. These problems require specific comments that will be given further when the types of the models are discussed.

5.3 Iterative Approach

An alternative approach to the inverse problem solution is *iterative regularization.* The method concerns the building of regularization algorithms based on various iterative methods where *the regularization parameter* is *the number of the iteration* [8].

Many iterative methods looking for a minimum of a function (the gradient methods too) are insensitive with respect to the input data, but the errors of the calculations can start grow after a certain number of iterations. This requires stopping the iteration procedure at a certain $n = N$ that results in a stable approximation. Thus, N has the sense of a *regularization parameter.*

The convergence of the iterative methods employed for inverse problem solutions depend on their convergence for correct formulation of the problem. As a first step, let us consider the equation

$$Ab = y, \quad b \in B, \quad y \in Y \tag{5.10}$$

with an exact solution $\bar{b}$.

5.3.1 Gradient Methods

There are gradient methods [8] allowing the solution of (5.10). They tend to minimize the functional representing the misfit $\Delta(b)$ of (2.26) via movement in a direction defined by the antigradient of the functional $J(b)$ (see 2.27). The movement employs a step $\beta > 0$ that depends on the strategy of the particular method chosen for the problem solution. The above explanation concerns the choice of an iterative procedure:

$$b_{n+1} = b_n - \beta_n J' b_n, \quad n = 0, 1, 2, \ldots, \tag{5.11}$$

where (through 2.27)

$$J' b_n = A^*(A b_n - y), \quad n = 0, 1, 2, \ldots. \tag{5.12}$$

The first problem that occurs when a particular gradient method is applied is the determination of the step of each iteration as a function of the minimization strategy. The case of a simple iteration usually supposes $\beta_n = \beta > 0$, n = 0, 1, 2, …. The minimization strategies of the more effective methods consider optimizations of each iteration step to enhance the convergence of the procedures. These strategies usually use the minimum condition of a certain functional as a measure of the iteration quality.

The *method of minimum errors* employs the minimum of the error at each iteration as a criterion for the step determination. Thus, β_n follows from the condition ensuring the minimum of $\|b_{n+1} - \bar{b}\|^2$. Through (2.11) and (5.11) it follows that

$$\|b_{n+1} - \bar{b}\|^2 = \|b_n - \beta_n J' b_n - \bar{b}\|^2 = \|b_n - \bar{b}\|^2 - 2\beta_n (b_n - \bar{b}, J' b_n) + \beta_n^2 \|J' b_n\|^2. \tag{5.13}$$

The condition ensuring the minimum of (5.13) with respect to β_n (it follows from 5.12, 2.25) is

$$\beta_n = \frac{(b_n - \bar{b}, J' b_n)}{\|J' b_n\|^2} = \frac{[b_n - \bar{b}, (A b_n - y)]}{\|J' b_n\|^2} = \frac{(A b_n - y, A b_n - y)}{\|J' b_n\|^2} = \frac{\Delta_n^2}{\|J' b_n\|^2}, \tag{5.14}$$

where

$$\Delta_n = \Delta(b_n) = \|A b_n - y\|^2, \quad n = 0, 1, 2, \ldots. \tag{5.15}$$

The *method of the faster slope* [8] uses a step that minimizes the misfit functional with respect to β_n:

$$\Delta_{n+1}^2 = \|A b_{n+1} - y\|^2 = \|A b_n - y\|^2 - 2\beta_n (A b_n - y, J' b_n) + \beta_n^2 \|A J' b_n\|^2, \tag{5.16}$$

i.e.,

$$\beta_n = \frac{(A b_n - y, A J' b_n)}{\|A J' b_n\|^2} = \frac{\|J' b_n\|^2}{\|A J' b_n\|^2}. \tag{5.17}$$

It is possible to use methods having a greater rate of convergence and especially some versions of the conjugated gradients [8].

The result presented above permit us to develop further three sequences of approximations towards the exact solution—$\{u_n\}_1$, $\{u_n\}_2$, and $\{u_n\}_3$—that follow from (5.11) via a substitution $\beta_n = \beta = \text{const.}$ in (5.14) and (5.17).

These conditions make the iterative procedure convergent for every operator A whose inverse operator A^{-1} is continuous.

5.3.2 Uniqueness of the Solution

Let us consider the equation

$$\tilde{A}b = y, \tag{5.18}$$

where $\tilde{A}$ is not continuous, i.e., problem (5.18) does not have a unique solution. Let R_y be a set of solutions and $b^0 \in R_y$ be one of them. This solution is termed *normal with respect to the elementb*$_0 \in B$ if it can be obtained through a minimization of $\|b - b_0\|$. Thus, b^0 is a solution that may be obtained via gradient methods if an initial approximation u_0 is the starting point since the distance between them is minimal. However, a problem concerning the iteration step occurs. The step must be defined in a way that allows the iterative pass $b_0 \to b^0$ to coincide with the approach of b_n to the exact solution $\bar{b}$. To do that, a step β_n satisfying an additional condition is required. This additional condition must ensure the convergence of the iterative procedure. This means the following inequalities must be satisfied simultaneously:

$$\Delta_n^2 - \Delta_{n+1}^2 \geq 0, \quad \|b_n - \bar{b}\|^2 - \|b_{n+1} - \bar{b}\|^2 \geq 0. \tag{5.19}$$

In the case of a *simple iteration* EQUATION (5.19) the conditions take the following forms (with the help of 2.24, 2.28, 5.11, 5.12, 5.15):

$$\Delta_n^2 - \Delta_{n+1}^2 = 2\beta\|J'b_n\|^2 - \beta^2\|AJ'b_n\|^2 \geq \beta\|J'b_n\|^2\left(2 - \beta\|A\|^2\right) \geq 0,$$

$$\|b_n - \bar{b}\|^2 - \|b_{n+1} - \bar{b}\|^2 = 2\beta\Delta_n^2 - \beta^2\|J'b_n\|^2 \geq \beta\Delta_n^2\left(2 - \beta\|A\|^2\right) \geq 0. \tag{5.20}$$

These conditions, ensuring the convergence of the iterative procedure, lead directly to the fact that the successive approximations approach the exact solution when

$$2 - \beta\|A\|^2 > 0, \tag{5.21}$$

i.e.,

$$0 < \beta < \frac{2}{\|A\|^2}. \tag{5.22}$$

The summation in (5.20) (n = 0, 1, 2, ..., ∞) indicates that the sum $\sum_{n=0}^{\infty} \Delta_n^2$ is limited, so $\Delta_n \to 0$ at $n \to \infty$. The latter means that the iterative procedure approaches the exact solution.

As already mentioned, b^0 is a solution of (5.18). The use of the *simple iteration method* with a step satisfying (5.22) leads to the solution $b_c = \lim_{n\to\infty} b_n$. It is easy to prove that $b_c = b^0$.

Let us denote $\Delta b = b_c - b^0$. Hence,

$$A\Delta b = Ab_c - Ab^0 = f - f = 0, \tag{5.23}$$

since $b_c$and $b^0$are exact solutions of (5.10). The conditions ensuring the convergence lead (at n $\to \infty$) to

$$\begin{aligned}(b_n, \Delta b) - (b_c, \Delta b) &= (b_n, \Delta b) - \left(b^0 - \Delta b, \Delta b\right)\\ &= (b_n, \Delta b) - \left(b^0, \Delta b\right) - (\Delta b, \Delta b) = \left(b_n - b^0, \Delta b\right) - \|\Delta b\|^2 \to 0.\end{aligned} \tag{5.24}$$

The first term of the final expression in (5.24) can be represented as

$$\begin{aligned}\left(b_n - b^0, \Delta b\right) &= \left(b_0 - b^0, \Delta b\right) - \left[\sum_{i=0}^{n-1} (u_i - u_{i+1}), \Delta b\right]\\ &= \left(b_0 - b^0, \Delta b\right) - \beta\left[\sum_{i=0}^{n-1} (Ab_i - f), A\Delta b\right] = \left(b_0 - b^0, \Delta b\right).\end{aligned} \tag{5.25}$$

The latter result follows directly from (5.11), (5.12), and (5.23). The vectors $(b_0 - b^0, \Delta b)$ and Δb are orthogonal (in accordance with the definition) and their scalar product (5.25) is zero. Thus, in accordance with (5.24), it follows that $\|\Delta b\| = 0$, i.e., $b_c = b^0$.

The conditions defining the steps of both the *minimum errormethod* and *the faster slope method* [8] can be obtained in a similar way:

$$\beta_n \geq \frac{1}{\|A\|^2}, \quad n = 0, 1, 2, \ldots \tag{5.26}$$

5.3.3 Approximate Equations

Practically, the input data (A and f) needed to solve Eq. (5.10) are tentatively known. The operator A is defined by a set of equations that describes approximately the process mechanism. In other cases the differential operator A could be approximated by finite differences assuming an appropriate error. The objective function is also defined with errors depending on the accuracy of the experimental findings of y. Owing to the above reasons, the following explanations discus the solution of the equation

$$A_h b = y_\delta, \quad b \in B, \quad y_\delta \in Y, \tag{5.27}$$

where A_h is a linear continuous operator approximating A in a way that

$$\|A_h - A\| \le h, \quad y_\delta = y + \tilde{y}, \quad \|\tilde{y}\|_Y \le \delta. \tag{5.28}$$

The iterative gradient methods discussed here could be inapplicable to problem (5.27) for three reasons:

The minimization problem has no solution since $A^{-1}f_\delta$ is an element that does not belong to the set of available solutions of (5.27).

The minimization problem has a solution b_σ that depends on h and $\delta(\sigma = \{h, \delta\})$, but the difference $\bar{b} - b_\sigma$ may be unacceptably large.

If the iterative procedure converges and $b_\sigma \to b^0$ at $\sigma \to 0$, but $b^0 \neq \bar{b}$, the approximate solution does not approach the exact solution of (5.10).

In addition, for sufficiently small errors of the input data (h, σ), the difference between the first approximations of the gradient methods applied to solve (5.27) and the approximations found through the solution of (5.10) are insignificant. Thus, they approach the exact solution. Proof is available elsewhere [8].

The increase of the iteration number could lead to an uncontrolled discrepancy from the desired solution. Taking into account that effect, we can consider the *iterative regularization* as a procedure based on a priori *information* (h, σ) that *should find* an approximate solution *as close as possible to the exact solution* of (5.10). The *number of the iteration* at which the desired approximation is achieved is a *regularization parameter* of the iterative method. A necessary condition is the convergence of the iterative method for exact input data given by the solution of (5.10). It was proved [8] that the iterative methods mentioned satisfy both the necessary and the sufficient conditions required to solve problem (5.27). The latter means that these methods create regularization operators (algorithms) with *the iteration number* as a *regularization parameter.*

The regularized approximation b_n approaches the solution b^0, which is *normal* with respect to the initial approximation b_0 when $\sigma(h, \delta) \to 0$. This approach reduces the main problem (for sufficiently small σ) to detection of the iteration number $N(\sigma)$ at which the approximate solution $b_{N(\sigma)}$ is insensitive with respect to the input data errors.

5.3.4 Criteria for Stopping the Iterations

The criterion needed to stop the iterations could be related to the misfit functional (2.26), i.e., the misfit of (5.27) with respect to $b = b^0$:

$$\begin{aligned} \|A_h b^0 - y_\delta\| &= \|(A_h - A)b^0 + Ab^0 - y_\delta\| = \|(A_h - A)b^0 + y - y_\delta\| \\ &= \|(A_h - A)b^0 - \tilde{y}\| \le h\|b^0\| + \delta = \Delta_H, \end{aligned} \tag{5.29}$$

i.e.,

$$\Delta_n \le \Delta_H. \tag{5.30}$$

Condition (5.30) is *the misfit criterion.* Practically, instead of Δ_n in (5.30) it is more convenient to use the functional $\Phi(n)$ depending on Δ_n, whereas Δ_H could be replaced by a monotonically rising estimator $\Psi(\Delta_H) > \Delta_H$. Thus, inequality (5.30) can be expressed as

$$\Phi(n) < \Psi(\Delta_H). \tag{5.31}$$

Here, the form of $\Phi(n)$ depends on the particular gradient method applied. The desired solution depends on the number of the iteration N_H, i.e., the procedures looks for the minimum n at which (5.31) is satisfied. The procedure concerns the difference

$$\|b_n - b\|^2 - \|b_{n+1} - b\|^2 \tag{5.32}$$

and looks for $n > N_H$ at which it is negative. Thus, the iteration does not converge. On the other hand, at $n \le N_H$ the procedure converges and N_H is the number of the last iteration.

When the *simple iteration method* is applied, the difference (5.32) has the form [8]:

$$\begin{aligned}
&\|b_n - b\|^2 - \|b_{n+1} - b\|^2 = \|b_n - b\|^2 - \|b_n - \beta J' b_n - b\|^2 = \beta\left[2(b_n - b, J' b_n) - \beta\|J' b_b\|^2\right] \\
&= \beta\left[2\Delta_n^2 - \beta\|J' b_n\|^2 + 2(A_h b_n - f_\delta, (A - A_h)b + \tilde{y})\right] \\
&\ge \beta\left[2\Delta_n^2 - \beta\|J' b_n\|^2 - 2\Delta_n(h\|b\| + \delta)\right] \ge \beta\left[\left(2 - \beta\|A_h\|^2\right)\Delta_n^2 - 2\Delta_n \Delta_H\right].
\end{aligned} \tag{5.33}$$

Relationship (5.33) is valid since $\|J' b_n\|^2 \le \|A_h\|\Delta_n$. The last iteration number is the minimum n that causes a negative value of (5.33). Thus, the following inequality can be developed:

$$\left(2 - \beta\|A_h\|^2\right)\Delta_n^2 - 2\Delta_n \Delta_H < 0. \tag{5.34}$$

This leads to

$$\frac{\left(2 - \beta\|A_h\|^2\right)\Delta_n}{2} \equiv \Phi_1'(n) < \Delta_H < \Psi(\Delta_H), \tag{5.35}$$

i.e., the desired functional of (5.31) is $\Phi_1'(n)$.

The substitution of

$$\beta\|J' b_n\|^2 \le \frac{\Delta_n^2 - \Delta_{n+1}^2}{2 - \beta\|A_h\|^2} \tag{5.36}$$

into (5.33) yields a second functional [8]:

$$\Phi_1''(\mathrm{n}) \equiv \frac{\left(3 - 2\beta\|A_\mathrm{h}\|^2\right)\Delta_\mathrm{n}^2 + \Delta_{\mathrm{n}+1}^2}{2\Delta_\mathrm{n}\left(2 - \beta\|A_\mathrm{h}\|^2\right)}. \tag{5.37}$$

The regularization algorithm solving Eq. (5.27) can be formulated as a sequence of steps:

1. Determine the *initial approximationb*$_0$.
2. Determine the *misfit functional*

$$\Delta_\mathrm{n} = \|\mathrm{A_h b_n} - \mathrm{y}_\delta\|, \quad \mathrm{n} = 0, 1, 2, \ldots, \mathrm{N_H}. \tag{5.38}$$

3. Check of *condition* (5.31). If it is satisfied, then $\mathrm{n} = \mathrm{N_H}$ and the iterative procedure stops. If it is not satisfied, then *go to point 4*.
4. Determine the *gradient* $J' b_\mathrm{n}$, n = 0, 1, 2, …, $\mathrm{N_H}$ via (5.12).
5. Determine *step* β_n from (5.14) and (5.17) and *check of conditions* (5.22) and (5.26).
6. *Determineb*$_{\mathrm{n}+1}$ from (5.11) and *start at point* 2.

In the method of *the simple iteration* condition (5.31) *must be checked* through the substitution

$$\Phi_1^{(\mathrm{n})} = \max\left\{\Phi_1'(\mathrm{n}), \Phi_1''(\mathrm{n})\right\}, \tag{5.39}$$

i.e., the greater value of the functional must be used.

The method of *minimum errors* and the method of *the faster slope* formulate the functionals in similar ways:

$$\Phi_2(\mathrm{n}) = \frac{\Delta_\mathrm{n}}{2}, \tag{5.40a}$$

$$\Phi_3(\mathrm{n}) = \frac{\Delta_\mathrm{n}^2 + \Delta_{\mathrm{n}+1}^2}{2\Delta_\mathrm{n}}. \tag{5.40b}$$

The fixed step of the *simple iteration method* must satisfy condition (5.22), so

$$0 < \beta < \frac{2}{(\|A_\mathrm{h} + h\|)^2} \le \frac{2}{\|A\|^2}. \tag{5.41}$$

The justification of the solution can start if $\Phi_1(\mathrm{N_H}) < \Psi(\Delta_\mathrm{H})$. This can be done through a decrease of the value of $\Psi(\Delta_\mathrm{H})$. If $\Delta_{\mathrm{N_H}} > \Psi(\Delta_\mathrm{H})$ the step must be reduced, i.e., define a new step $\beta_1 (0 < \beta_1 < \beta)$. Similar justifications of the iteration step are possible [8] also through the procedures of two other methods mentioned above (*theminimum errors* and *the faster slope* methods).

The regularization algorithm described above is based on the *misfit criterion* Δ_H, where $\|b^0\|$ is the solution that is unknown at the steps performed. Because of that the *misfit criterion* $\varDelta_H$ can be employed if $h\|b^0\| \ll \delta$. Under such a condition it is evident that

$$\varDelta_H = \delta. \tag{5.42}$$

This means that $\varDelta_H$ can be defined in a unique way from the experimental error during the measurements of the objective function. In the other cases, the *general misfit criterion* must be applied. This suggests that $b_n(n = 0, 1, 2, \ldots)$ substitutes for b^0, so

$$\varDelta_H = h\|b_n\| + \delta. \tag{5.43}$$

6 Parameter Identification in Different Types of Models

The methods of parameter identification depend on the mathematical structure of the models employed. This requires the solutions to be considered separately in every particular case.

6.1 Regression Models

Let us consider the regression models formulated earlier (see 2.5.5). The data given are N experimentally determined values of the objective function y_n at fixed values of the factors $x_{in}(n = 1, \ldots, N;\ i = 1, \ldots, m)$. This allows us to define the estimators (b) of the model parameters in a way that minimizes the difference between the experimental (y_n) and the calculated ($\hat{y}_n$) values of the objective function ($n = 1, \ldots, N$). Thus, the procedure looks for a minimum of one of the three functions below:

$$\max_{n=1,\ldots,N}\|e_n\|, \quad \sum_{n=1}^{N}\|e_n\|, \quad \sum_{n=1}^{N} e_n^2, \tag{6.1}$$

where

$$e_n = y_n - \hat{y}_n, \quad n = 1, \ldots, N. \tag{6.2}$$

The values of $\hat{y}_n(n = 1, \ldots, N)$ can be calculated as

$$\hat{y}_n = \sum_{j=1}^{k} b_j f_j(x_{1n}, \ldots, x_{mn}). \tag{6.3}$$

Very common practice is the employment of the *least-squares method* that considers the minimization of the function

$$Q(b_1, \ldots, b_k) = \sum_{n=1}^{N} (y_n - \hat{y}_n)^2. \tag{6.4}$$

The conditions ensuring the minimum of the least-squares function Q are

$$\frac{\partial Q}{\partial b_j} = 0, \quad j = 1, \ldots, k. \tag{6.5}$$

This leads to a set of linear algebraic equations [6, 7]:

$$\sum_{n=1}^{n} f_{jn} \sum_{i=1}^{k} b_i f_{ni} = \sum_{n=1}^{N} f_{jn} y_n, \quad j = 1, \ldots, k, \tag{6.6}$$

where

$$f_{in} = f_{jn} = f_i(x_{1n}, \ldots, x_{mn}), \quad i = j = 1, \ldots, k. \tag{6.7}$$

The set of equations (6.6) allows us to change the order of summation of the left part and the following symbols are introduced:

$$a_{ij} = \sum_{n=1}^{N} f_{in} f_{jn}, \quad z_j = \sum_{n=1}^{N} f_{jn} y_n. \tag{6.8}$$

The manipulations allow us to arrange a *normal set of equations* (with the number of equations equal to the number of unknown variables):

$$\sum_{i=1}^{k} a_{ij} b_i = z_j, \quad j = 1, \ldots, k. \tag{6.9}$$

If $A = \{a_{ij}\}$ denotes the matrix of the coefficients of the unknown variables b_i and A_i is the matrix A formed by substitution of the ith columns by the vector z_j [i, j = 1, ..., k] the model parameters are the solutions of (5.9):

$$b_i = \det A_i / \det A, \quad i = 1, \ldots, k, \tag{6.10}$$

where **det**A and **det**A_i are the determinants of the matrixes AandA_i(i = 1, ..., k).

Solution (6.10) allows us to formulate: that

The error of the experiments is concentrated in detA_i(i = 1, ..., k) if the main assumption that *the model and the independent variables do not introduce errors* in the inverse problem solution is satisfied. If this is not the case, the modifications of the methods discussed above must be applied.

If det$A$ is close to zero, the set of equations (6.9) is not well defined. This leads to a strong effect of the experimental errors (detA_i) on the solution (i.e., the parameter value estimated) b_i, i = 1, ..., k. Therefore, *the inverse problem is incorrect (ill-posed).*

There is a possibility to have det$A$ = 0. This means that equations (6.9) are not independent and the set has a singularity, so no unique solution exists.

6.2 Selection Methods

The last comments on *the least-squares method* indicate that the incorrectness of the inverse problem is accompanied by difficulties concerning the singularity of det$A$ taking into account the preliminarily defined accuracy of the calculation of it. The solutions of such problems need the set (6.9) to be considered as an operator equation:

$$Ab = y \tag{6.11}$$

where $A = \{a_{\mathrm{ij}}\}, \quad b = \{b_{\mathrm{i}}\}, \quad y = \{z_{\mathrm{j}}\}, \quad (\mathrm{i,j} = 1,\ldots,\mathrm{k}).$

Characteristic features of the equation are that not only the right-hand side $\tilde{y}$, but also the operator $\tilde{A}$ are tentatively known. The deficiencies are due to the restrictions coming from the preliminarily assumed accuracy of the calculations despite the suggestion that model structure does not introduce errors. Thus, (6.11) can be expressed as

$$\tilde{A}b = \tilde{y}, \tag{6.12}$$

where

$$\|\tilde{A} - A\| \leq \mathrm{h}, \quad \|\tilde{y} - y\| \leq \delta. \tag{6.13}$$

The sense of the norms used in (6.13) depends on the type of the problem to be solved. Practically, the often-used norms are

$$\|A\| = \left\{\sum_{\mathrm{i,j}} a_{\mathrm{i,j}}^2\right\}^{\frac{1}{2}}, \quad \|y\| = \left\{\sum_{\mathrm{i}=1}^{\mathrm{k}} y_{\mathrm{i}}^2\right\}^{\frac{1}{2}}. \tag{6.14}$$

The exact set of equations (6.11), whose solution is looked for (under the conditions imposed by 6.13), yields an unrestricted number of sets defined by the preliminarily suggested accuracy of both the calculations and the experiments. The tentative solution considered as an approximation of the exact one (6.12) usually does not give a satisfactory result owing to the fact that the problem cannot be solved. The latter point means that the problem is unstable owing to small variations of the right-hand sides of the equations. We therefore need to consider a new class of practically undistinguishable sets of equations (both singular and those without solutions). The development of the problem solution employs the idea of the *selection* considered earlier (Sect. 4).

If $\tilde{b}$ denotes the pseudo-solution of (6.11), it is possible to minimize $\|Ab - y\|$, which yields

$$\sum_{\mathrm{j}=1}^{\mathrm{k}}\left(\sum_{\mathrm{i}=\mathrm{k}}^{\mathrm{k}} a_{\mathrm{ij}}b_{\mathrm{i}} - z_{\mathrm{j}}\right)^2. \tag{6.15}$$

If the set (6.11) has more than one pseudo-solution, this means that (6.15) also has several minima. Therefore, $b \in F_A$, where F_A is the set of all pseudo-solutions. One of them is the desired pseudo-solution determined by the condition of the minimum of the norm:

$$\|\tilde{b}\| = \left\{ \sum_{i=1}^{k} \tilde{b}_i^2 \right\}^{\frac{1}{2}} \to \min . \qquad (6.16)$$

6.3 Variational Regularization

The solution of the essentially ill-posed (incorrect) problem (6.11) utilizes the *variational regularization method*. Suppose that b_0 is the exact solution of *the exact normal set*:

$$Ab = \bar{y}. \qquad (6.17)$$

Really, the approximate values $\tilde{y}$ and the error δ are known:

$$\|\tilde{y} - y\| = \delta, \qquad (6.18)$$

i.e., the approximate solution $\tilde{b}_\delta$ of the approximate set

$$Ab = \tilde{y} \qquad (6.19)$$

could be found.

If the set (6.19) cannot be solved,

$$\inf\|Ab - \bar{y}\| = \mu \geq 0, \qquad (6.20)$$

where $\inf\|Ab - \bar{y}\|$ is the minimum of the norm like (6.15).

Let us express the obvious inequality

$$\|Ab - \tilde{y}\| \leq \|Ab - \bar{y}\| + \|\bar{y} - \tilde{y}\| \leq \mu+\delta \qquad (6.21)$$

and let $\tilde{\mu}$ denote the lower limit of the norm $\|Ab - \tilde{y}\|$. So,

$$\tilde{\mu} = \inf\|Ab - \tilde{y}\|. \qquad (6.22)$$

The obvious inequalities

$$\|Ab - \tilde{y}\| \leq \|Ab - \bar{y}\| + \|\bar{y} - \tilde{y}\|, \quad \|Ab - \bar{y}\| \leq \|Ab - \tilde{y}\| + \|\tilde{y} - \bar{y}\| \qquad (6.23)$$

allow us to derive (with the help of 6.20, 6.22) that

$$\tilde{\mu} \leq \mu + \delta, \quad \mu \leq \tilde{\mu} + \delta, \qquad (6.24)$$

i.e.,

$$|\tilde{\mu} - \mu| \leq \delta. \qquad (6.25)$$

The relationship

$$\|Ab - \tilde{y}\| \le \tilde{\mu} + 2\delta = \delta_1 \tag{6.26}$$

follows directly from (6.21) and (6.25).

If B_δ is the set of solutions b satisfying (6.26), the solutions looked for could be a random element of B_δ, because the solution is unstable with respect to small variations of the right-hand side of (6.19). This imposes conditions on the approximate solutions $\tilde{b}_\delta$ to satisfy (see Sect. 4) not only (6.26), but also to minimize the functional

$$\Omega(b) = \|b\|^2 = \left\{\sum_{i=1}^{k} \tilde{b}_i^2\right\}. \tag{6.27}$$

The regularization parameter R allowing us to determine the approximate solution of (6.19) (i.e., $\tilde{b}$ minimizing 6.26) must do two things simultaneously: (1) satisfy (6.26) and (2) minimize (6.27). Thus,

$$\tilde{b}_\delta = R(\tilde{y}, \delta). \tag{6.28}$$

It is easy to prove [2] that the values of $\tilde{b}_\delta$ which minimize (6.27) are on the border of the area defined by (6.26). This means that $\tilde{b}_\delta$ may be determined through the condition ensuring the minimization of $\|Ab - \tilde{y}\|^2$ if it satisfies simultaneously the relationship $\|Ab - \tilde{y}\|^2 = \delta_1$ and the functional (6.27). The problem is formulated (as shown in Sect. 5.4) in this manner by the *method of Lagrange*. This needs the formulation of an approximate solution $\tilde{b}_\delta$ as a vector b^α which minimizes the functional

$$M^\alpha[b, \tilde{y}] = \|Ab - \tilde{y}\|^2 + \alpha\|b\|^2. \tag{6.29}$$

The parameter $\alpha > 0$ follows from the condition

$$\|Ab^\alpha - \tilde{y}\| = \delta_1. \tag{6.30}$$

From this point of view, the approximate solution b^α can be considered as the action of a certain operator (algorithm) on $\tilde{y}$, i.e.,

$$b^\alpha = R(\tilde{y}, \alpha). \tag{6.31}$$

This approximate solution satisfies the condition ensuring the problem correctness, so R is a regularization functional.

The methods discussed concern the determination of the parameters of regression models. They are built on the basis of different mathematical approaches overcoming the incorrectness of the inverse problems. This incorrectness comes from the ill-posed set of linear algebraic equations (6.9) (i.e., the small values of det A in 6.10 induce the solution instability). There are intrinsic physical methods to solve the problem that concern the optimal planning of the experiment ensuring the maximum of det A.

6.4 Similarity Theory Models

As already mentioned in Sect. 2.4. the similarity theory models (after taking the logarithm) become regression models with respect to parameter identification. They should take into account that the *normal error distribution* of the measurements of the objective function *does not exist* after taking the logarithm of it. Despite this, the identification methods discussed in this chapter are very applicable to the similarity theory models. However, they must take into account the inaccuracy of the determination of b_0 in (2.4.23) due to the error of its of logarithm, which increases after the antilogarithmic operation. In such a situation, a more convenient approach is to determine the parameters of (2.4.23) in two steps. The first step concerns the determination of the power–law coefficients $b_i (i = 1, \ldots, 6)$. The next step involves the substitution of b_i into (2.4.23). The final one-parameter model allows the determination of b_0 as the ratio of the experimental and the calculated values of the objective function, i.e., $b_0 = Sh/\prod_{i=1}^{6} A_i^{b_i}$.

6.5 Diffusion-Type Models

The parameter identification of the diffusion-type models is of special interest owing to their structural variety. Besides that variety, a large class of chemical engineering problems utilize structures of parabolic partial differential equations. This is the basis of the further development of the problems here and especially the iterative methods of parameter identification of parabolic partial differential equations. A typical example of that class is the heat conductivity equation.

6.5.1 Determination of the Heat Conductivity Coefficients in Inverse Heat Transfer Problems

The general formulation of the heat conductivity problem is

$$\begin{aligned}
&c(T)\frac{\partial T}{\partial t} = \frac{\partial}{\partial x}\left[\lambda(T)\frac{\partial T}{\partial x}\right] + W(T)\frac{\partial T}{\partial x}; \qquad t = 0, \quad T = T_0(x);\\
&x = 0, \quad T = g_1(t), \quad \left[-\lambda(g_1(t))\frac{\partial T}{\partial x}|_{x=0} = q_1(t_1)\right];\\
&x = b, \quad T = g_2(t), \quad \left[-\lambda(g_2(t))\frac{\partial T}{\partial x}|_{x=b} = q_2(t_1)\right],
\end{aligned} \tag{6.32}$$

where the second-order boundary conditions expressed in the square brackets can replace the first-order boundary conditions (g_1, g_2).

The parameter identification, i.e., the determination of the coefficients $c(T)$, $\lambda(T)$, and $W(T)$, can be obtained if at points with coordinates $x = d_i (i = 1, \ldots, N)$, so that $0 \le d_i \le b$, there are located temperature pick-ups, recording the temperature variations with time $(0 \le t \le t_m)$:

$$T_e(d_i, t) = f_i(t), \quad i = 1, \ldots, N. \tag{6.33}$$

The problem needs the formulation of the misfit functional in the form

$$J = \frac{1}{2} \sum_{i=1}^{N} \int_0^{t_m} [T(d_i, t) - f_i(t)]^2 dt \le \delta^2, \tag{6.34}$$

where δ is the experimental error of the temperature measurements.

The inverse problem formulated above has solutions [8] only in the cases when at least one boundary condition is second (or third) order and $\frac{\partial T}{\partial x} \neq 0$. For simplicity, the further explanation considers the case of constant coefficients. In this case the problem is

$$\begin{aligned} & c\frac{\partial T}{\partial t} = \lambda \frac{\partial^2 T}{\partial x^2} + W\frac{\partial T}{\partial x}; \quad t = 0, \quad T = T_0(x); \\ & x = 0, \quad -\lambda \frac{\partial T}{\partial x}|_{x=0} = q_1(t); \quad x = b, \quad -\lambda \frac{\partial T}{\partial x}|_{x=b} = q_2(t); \\ & T_e(d_i, t) = f_i(t). \quad i = 1, 2, \ldots, N \ge 3, \end{aligned} \tag{6.35}$$

where the functions $q_1(t)$, $q_2(t)$ and $f_i(t)$, $i = 1, \ldots, N$, are preliminarily known. The problem solution needs the vector of the parameters c, λ, W to be determined, i.e.,

$$p = (c, \lambda, W). \tag{6.36}$$

6.5.2 Iterative Algorithm

The problem solution needs [8] the formulation of a *mean square misfit* (2.27) in the form

$$J(p) = \frac{1}{2} \sum_{i=1}^{3} \int_0^{t_m} [T(d_i, t, p) - f_i(t)]^2 dt \le \delta^2. \tag{6.37}$$

The functional $J(p)$ depends on three variables, i.e., three components of the vector of the gradient must be determined through the conjugate form $J'(p)$ of functional (6.37):

$$J'_c = -\int_0^{t_m}\int_0^{b} \Psi \frac{\partial T}{\partial t} dxdt, \quad J'_W = \int_0^{t_m}\int_0^{b} \Psi \frac{\partial T}{\partial x} dxdt,$$
$$J'_\lambda = \int_0^{t_m}\int_0^{b} \Psi \frac{\partial^2 T}{\partial x^2} dxdt + \int_0^{t_m} \Psi(0,t)\frac{\partial T}{\partial x}|_{x=0}dt - \int_0^{t_m} \Psi(b,t)\frac{\partial T}{\partial x}|_{x=b}dt, \tag{6.38}$$

Here the conjugate function $\Psi(x, t)$ [8, 9] is a solution of the problem

$$-c\frac{\partial \Psi_i}{\partial t} = \lambda\frac{\partial^2 \Psi_i}{\partial x^2} - W\frac{\partial \Psi_i}{\partial x}, \quad i = 1,\ldots,N+1,$$
$$0<t<t_m, \quad d_{i-1}<x<d_i; \qquad d_{i-1}\le x\le d_i, \quad t=t_m, \quad \Psi_i(x,t_m)=0;$$
$$x=0, \quad \lambda\frac{\partial \Psi_1}{\partial x}|_{x=0} = W\Psi_1(0,t); \qquad x=d_i, \quad \Psi_i(d_i,t)=\Psi_{i+1}(d_i,t), \quad i=1,\ldots,N;$$
$$\lambda\left(\frac{\partial \Psi_i}{\partial x}\Big|_{x=d_i} - \frac{\partial \Psi_{i+1}}{\partial x}|_{x=d_i}\right) = T(d_i,t) - f_i(t); \qquad i=1,\ldots,N,$$
$$\lambda\frac{\partial \Psi_{N+1}}{\partial x}|_{x=b} = W\Psi_{N+1}(b,t). \tag{6.39}$$

The iterative procedure [8] could be created on the basis of the *method of the faster slope*. The procedure depends strongly on the initial approximations of c, λ, and W as well as on the iteration steps of each variable. This complexity requires the vector determination of the step

$$p_{n+1} = p_n - \left[\beta_k^n J'_{k,n}\right], \tag{6.40}$$

where $\beta^n = \left[\beta_c^n, \beta_\lambda^n, \beta_W^n\right]$, $J'_{k,n} = \left[J'_{c,n}, J'_{\lambda,n}, J'_{W,n}\right]$, $k = c, \lambda, W$, and n is the iteration number.

The above formulation of the procedure needs the following linear set to be solved:

$$\sum_{j=1}^{3} \beta_j^n a_{jk} = g_k, \quad k = 1,2,3, \tag{6.41}$$

where

$$a_{jk} = \sum_{i=1}^{3}\int_0^{t_m} \vartheta_j(d_i,t,\beta_j=1)\vartheta_k(d_i,t,\beta_k=1)dt,$$
$$g_k = \sum_{i=1}^{3}\int_0^{t_m} [T(d_i,t) - f_i(t)]\vartheta(d_i,t,\beta_k=1)dt. \tag{6.42}$$

The temperature variation $\vartheta(x, t)$ for (k = 1, 2, 3) can be determined from

$$c\frac{\partial \vartheta_k}{\partial t} = \lambda\frac{\partial^2 \vartheta_k}{\partial x^2} + W\frac{\partial \vartheta_k}{\partial x} + \beta_k R_k; \qquad t = 0, \quad \vartheta_k = 0;$$
$$x = 0, \quad \lambda\frac{\partial \vartheta_k}{\partial x}|_{x=0} + \beta_k G_k\frac{\partial T}{\partial x}|_{x=0} = 0; \qquad x = b, \quad \vartheta_k = 0, \tag{6.43}$$

where the functions R_k and G_k are calculated from

$$R_k = \begin{Bmatrix} \frac{\partial T}{\partial t}J'_c, & k = 1; \\ -\frac{\partial^2 T}{\partial x^2}J'_\lambda, & k = 2; \\ -\frac{\partial T}{\partial x}J'_W, & k = 3; \end{Bmatrix}, \quad G_k = \begin{Bmatrix} 0 & k = 1; \\ -J'_\lambda, & k = 2; \\ 0 & k = 3. \end{Bmatrix}. \tag{6.44}$$

The equations developed permit us to suggest the following algorithm [8] for the iterative procedure:

1. Suppose the initial approximations are c_0, λ_0, W_0.
2. Determine the temperature distribution$T_n(x, t)$ by solving (6.25).
3. Solve the conjugated problem (8.6.39) and determine $\Psi_{in}(x, t)$, i = 1, ..., N.
4. Determine the vector components of the gradient of the misfit functional $J'_k(k = c, \lambda, W)$ through (6.38).
5. Solve problems (6.41–6.43).
6. Calculate the new approximations of c, λ, and W through (6.40) and go to point 2.

If the temperature curves $f_i(t)$, $i = 1, \ldots$, N are obtained with a certain experimental error, the misfit criterion (5.31) must be employed to stop the iteration procedure. The value of δ in (6.37) is an estimator of the general error of the input data $f_i(t)$, i = 1, ..., N, so

$$\delta^2 = \sum_{i=1}^{N} \int_0^{t_m} [\sigma_i(t)]^2 dt. \tag{6.45}$$

Here, $\sigma_i(t)$, i = 1, ..., N. is the quadratic mean deviation of the temperature measured at points $x = d_i$, i = 1, ..., N, and the moment t.

6.6 Theoretical Models and Model Theories

The theoretical models and the model theories are characterized by the fact that their parameters are not defined on the basis of experimental data. A more important feature is that these parameters are in fact parameters of elementary processes incorporated into the model of the complex process.

The parameter identification of the models describing elementary processes utilizes experimental data and inverse problem solutions. The common approaches are the variational and the iterative methods.

7 Minimum of the Least-Squares Function

In this chapter it was shown [1, 3, 13, 14, 17–20] that the least-squares method is the general method for solution of the parameter identification problem, i.e., *the minimum of the least-squares function is the inverse problem solution.* Some properties of the minimum of the least-squares function are the cause of the incorrectness of the parameter identification problem [24–33].

Let us consider the expression

$$y = f(x, b), \tag{7.1}$$

where y is the objective function, x is an independent (regime) variable, b is a parameter, and f is a function (operator, algorithm), which permit us to calculate y if x and b are known.

The inverse problem solution uses the expression

$$b = \varphi(x, y, \hat{y}), \tag{7.2}$$

where $\hat{y}$ are experimental values of the objective function. Here φ is an algorithm for minimization of the least-squares function:

$$Q(b) = \sum_{\mathrm{n}=1}^{\mathrm{N}} (y_{\mathrm{n}} - \hat{y}_{\mathrm{n}})^2, \quad y_{\mathrm{n}} = f(x_{\mathrm{n}}, b), \quad \mathrm{n} = 1, \ldots, \mathrm{N}, \tag{7.3}$$

where n is the experiment number. Obviously the properties of the least-squares function are related to the type of objective function.

7.1 *Incorrectness of the Inverse Problem*

Most chemical engineering models (hydrodynamic equations, convection–diffusion equation, convection–conduction equation) are used in the boundary layer approximation, i.e., in the form of parabolic partial differential equations. A characteristic peculiarity of these equations is the presence of a small parameter (viscosity, diffusivity, conductivity) at the highest derivate [23]. In these conditions the direct operator is fully continuous (fully bounded), whereas the corresponding inverse operator is not continuous.

Let us consider a one-parameter model:

$$\varepsilon y'' + y' = 0; \qquad x = 0, \quad y = 1; \qquad y' = b = \varepsilon^{-1}. \tag{7.4}$$

The solution of (7.4.) is

$$y = 1 - \exp(-bx), \tag{7.5}$$

i.e., the dependence of y on ε (for big values of b) is similar to the cases of the diffusion-type models (see Fig. 1, Eqs. 1.13).

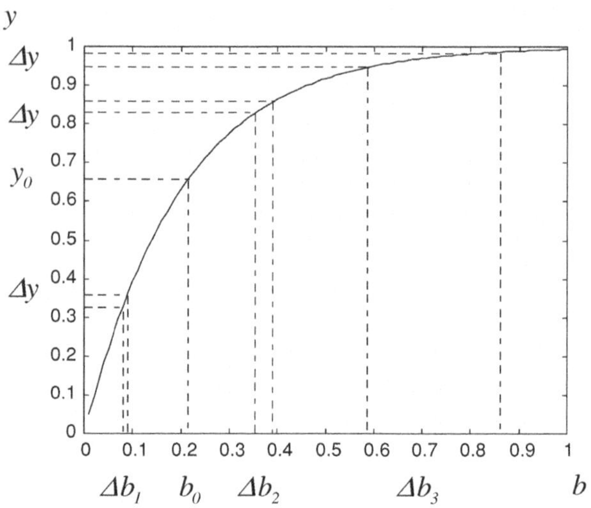

Fig. 4 Objective function y for different values of the model parameter b at $x = x_0 = \text{const}$

Figure 4 shows the relationship between the objective function y and the parameter b for a constant value of the independent variable $x = x_0 = 5$. Such a type of relationship is typical for a number of models of heat or mass transfer processes.

The plot in Fig. 4 permits us to obtain the objective function y_0 if the value of the parameter b_0 is known, which is a *direct problem solution*. However, an *inverse problem* looks for the value of the parameter b_0 if experimental values of the objective function y_0 are known.

Consider Δy as the experimental error of the objective function. Figure 4 shows that the error of the parameter identification depends on the magnitude of the objective function y. For small values of the objective function, there are small errors Δb_1, which shows the inverse identification problem is the *correct* one. However, if the objective function values are large, the corresponding values of Δb_2 are large too and the inverse problem is *incorrect*. In cases of the parameter identification of the models describing elementary processes, extremely large objective function values are utilized and enormous errors Δb_3 occur, which classify the inverse identification problem as *essentially incorrect*.

The results shown in Fig. 4 indicate that the incorrectness of the inverse problem is not a result of the size of the error of y and the cause is the parameter sensitivity with respect to the experimental errors of the objective function.

7.2 Incorrectness of the Least Squares Function Method

Consider the two-parameter model

$$\varepsilon y'' + y' = 0; \qquad x = 0, \quad y = 1 - b_1, \quad y' = -b_1 b_2^2, \quad b_2 = \varepsilon^{-1}. \tag{7.6}$$

The solution of (7.6) is

$$y = 1 - b_1 \exp(-b_2 x), \tag{7.7}$$

where $\bar{b}_1 = 1$ and $\bar{b}_2 = 5$ are the exact values of the parameter.

The parameter identification problem will be solved [24] with help of artificial *experimentaldata* provided by a random number generator:

$$\hat{y}_{\rm n}^{(1)} = (0.95 + 0.1A_{\rm n})y_{\rm n}, \quad \hat{y}_{\rm n}^{(2)} = (0.9 + 0.2A_{\rm n})y_{\rm n}, \tag{7.8}$$

where $A_{\rm n}$ are random numbers within the interval [0, 1]. The values of $y_{\rm n}$ are obtained from the model (7.7) for $x = 0.01$n (n = 1,...,100). The maximum relative errors of these "experimental" data ($\Delta\,\hat{y}$) are ±5 and ±10%. The values of $y_{\rm n}$, $\hat{y}_{\rm n}^{(1)}$, and $\hat{y}_{\rm n}^{(2)}$ are shown in Fig. 5. Comparison of Figs. 4 and 5 shows that when $0 < x < 0.30$ the inverse identification problem is *correct*, whereas in the case of $0.30 < x < 0.65$ it is *incorrect*. The problem becomes *essentially incorrect* when $0.66 < x < 1.00$.

If the experimental data in the separate intervals $0 < x < 0.30$ and $0.30 < x < 0.65$ and $0.66 < x < 1.00$ (in the cases of ±5% relative experimental errors) are used, the least-squares function (7.3) yields horizontal lines (see Figs. 6, 7, 8) when the inverse problem is correct (Fig. 6), incorrect (Fig. 7), or essentially incorrect (Fig. 8). These results show that when the difference between the exact parameter values and the determined value (the point of the function minimum) is very small, the least-squares method is correct (see Fig. 6). In cases of remarkably large differences between the exact parameter values and the minimum of the least-squares function, the inverse problem is incorrect (see Fig. 7). In the extreme case when the least-squares function has no minimum, the inverse problem is essentially incorrect (see Fig. 8).

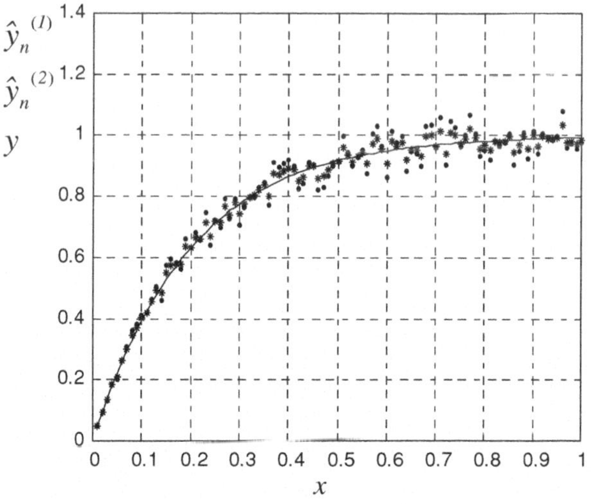

Fig. 5 Mathematical model and "experimental" data: *asterisks* $\hat{y}_{\rm n}^{(1)}$, values of y with a maximum "experimental" error of ±5%; *circles* $\hat{y}_{\rm n}^{(2)}$, values of y with a maximum "experimental" error of ±10%; *line* $y = 1 - \exp(-5x)$

Fig. 6 The horizontals of the least-squares function Q (n = 1–30; $\Delta\hat{y}(\%) = \pm 5$); *circleb* = [1, 5]

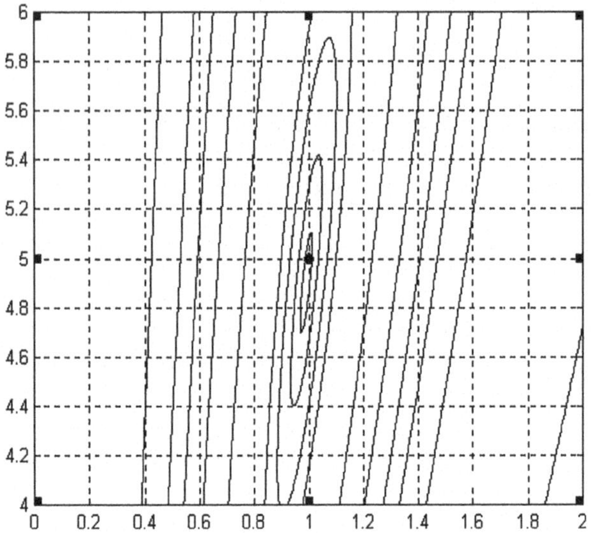

Fig. 7 The horizontals of the least-squares function Q (n = 31–65; $\Delta\hat{y}(\%) = \pm 5$); *circleb* = [1, 5]

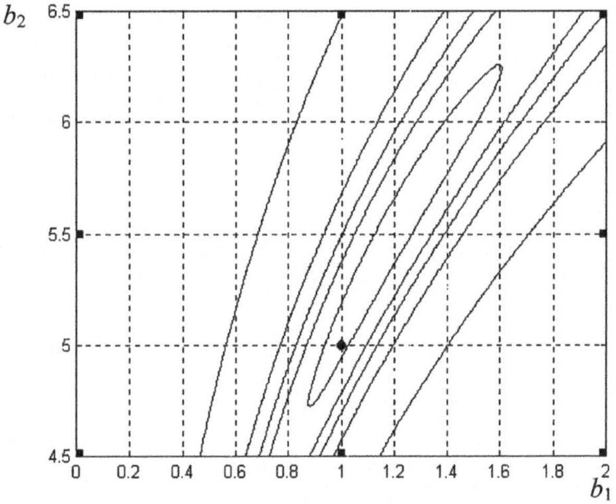

The results obtained show that in the cases of incorrect inverse problems minimization procedures of least-squares functions do not provide solutions, i.e., *the minimum of the least-squares function is not a solution of the incorrect inverse problem.*

Consider a gradient method for a minimum search. If the iterative procedure is convergent, at each step the difference between the iterative solution and the exact one will decrease towards the minimum of the least-squares function. However, there is a step after some iteration where this difference begins to increase. *The number of this last iteration is an additional condition, the iterative procedure must stop, and the last point is the solution of the incorrect inverse problem.*

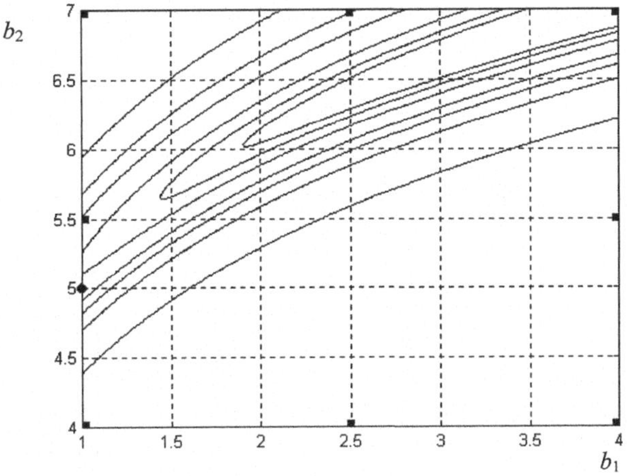

Fig. 8 The horizontals of the least-squares function Q ($n = 66$–100; $\Delta\hat{y}(\%) = \pm 5$); *circleb* = [1, 5]

7.3 Regularization of the Iterative Method for Parameter Identification

Let us assume that the iteration procedure starts with an initial approximation $\mathbf{b}^{(0)} = (b_1^{(0)}, \ldots, b_J^{(0)})$. The values of $\mathbf{b}_i = (b_{1i}, \ldots, b_{Ji})$, where i is the iteration number, are the result of conditions imposed by a movement towards the antigradient of the function $Q(\mathbf{b})$:

$$b_{ji} = b_{j(i-1)} - \beta_{(i-1)} R_{j(i-1)}, \quad j = 1, \ldots, J, \tag{7.9}$$

where

$$R_{j(i-1)} = \frac{\left(\frac{\partial Q}{\partial b_j}\right)_{(i-1)}}{\left[\sum_{j=1}^{J} \left(\frac{\partial Q}{\partial b_j}\right)^2_{(i-1)}\right]^{1/2}}, \quad j = 1, \ldots, J. \tag{7.10}$$

Here β_i is the iteration step and $\beta_0 = 10^{-2}$ (arbitrary small step value). The gradient of $Q(\mathbf{b})$ gives

$$\left(\frac{\partial Q}{\partial b_j}\right)_{(i-1)} = 2 \sum_{n=1}^{N} \left[f\left(\mathbf{x}_n, \mathbf{b}_{(i-1)}\right) - \hat{y}_n\right] \left[\frac{\partial f(\mathbf{x}_n, \mathbf{b})}{\partial b_j}\right]_{(i-1)}, \quad j = 1, \ldots, J, \tag{7.11}$$

where $\frac{\partial f}{\partial b_i}$ have to be calculated analytically or numerically.

Each iteration step is successful if two conditions are satisfied:

$$Q_{i-1} - Q_i = \sum_{n=1}^{N} \beta_{(i-1)} \left\{ \left[2f\left(\mathbf{x}_n, \mathbf{b}_{(i-1)}\right) - 2\hat{y}_n - \beta_{(i-1)} \sum_{j=1}^{J} \left(R_j \frac{\partial f}{\partial b_j} \right)_{(i-1)} \right] \sum_{j=1}^{J} \left(R_j \frac{\partial f}{\partial b_j} \right)_{i-1} \right\} \geq 0, \ \left(b_{j(i-1)} - \bar{b}_j\right)^2 - \left(b_{ji} - \bar{b}_j\right)^2$$

$$= \beta_{(i-1)} \left[2\left(b_{j(i-1)} - \bar{b}_j\right) - \beta_{(i-1)} R_{j(i-1)} \right] R_{j(i-1)} \geq 0, \quad j = 1, \ldots, J.$$

The first condition indicates that the iterative solution ($\mathbf{b}_i$) approaches the solution at the minimum ($\mathbf{b}^*$).On the other hand, the second condition controls the difference between the iterative solution ($\mathbf{b}_i$) and the exact one ($\bar{\mathbf{b}}$). The divergence is due to the effect of the problem incorrectness $\bar{\mathbf{b}} \neq \mathbf{b}^*$(see Figs. 6, 7, 8).

The second condition (7.12) leads to

$$2\left|b_{j(i-1)} - \bar{b}_j\right| \geq \beta_i \left|R_{j(i-1)}\right|, \quad j = 1, \ldots, J, \tag{7.13}$$

where the values of $\bar{b}_j$ (j = 1,. . ., J) are unknown. They can be replaced by

$$\left|b_{j(i-1)} - \bar{b}_j\right| = \Delta_{j(i-1)}^{(0)} = \left|\Delta_j^{(0)} b_{j(i-1)}\right|, \quad j = 1, \ldots, J, \tag{7.14}$$

if the accuracy of the parameter identification is preliminarily defined (desired accuracy). The desired accuracy $\Delta_{j(i-1)}^{(0)}$ should be obtained in each step through the use of the initial value$\Delta_j^{(0)}$:

$$\Delta_j^{(0)} = \gamma \frac{b_{j1} - b_{j0}}{b_{j1}}, \quad j = 1, \ldots, J. \tag{7.15}$$

The parameter γ is related to the desired accuracy (e.g.,$\gamma = 0.9$) and it plays the role of a regularization parameter.

From Eq. (7.9) it follows that

$$\beta_i \left|R_{j(i-1)}\right| = \left|b_{j(i-1)} - b_{ji}\right|, \quad j = 1, \ldots, J. \tag{7.16}$$

Thus, the second condition (7.12) can be expressed as

$$2\Delta_{j(i-1)}^{(0)} \geq \left|b_{j(i-1)} - b_{ji}\right|, \quad j = 1, \ldots, J. \tag{7.17}$$

The condition indicating the point where the iterative solution moves away from the exact one is

$$\left|b_{j(i-1)} - b_{ji}\right| > 2\Delta_{j(i-1)}^{(0)}, \quad j = 1, \ldots, J. \tag{7.18}$$

Hence, condition (7.18) permits a regularization of the parameter identification problem that leads to sufficiently exact values of the model parameters.

7.4 Iteration Step Determination and Iteration Stop Criterion

The step can be modified at each iteration point. In cases of two (or three) successful iterations ($Q_i - Q_{i-1} < 0$, $Q_{i-1} - Q_{ii-2} < 0$), the step should be enlarged twice:

$$\beta_{i+1} = 2\beta_i. \tag{7.19}$$

However, if $Q_i - Q_{i-1} < 0$, $Q_{i-1} - Q_{i-2} \geq 0$, the step should be kept unchanged:

$$\beta_{i+1} = \beta_i. \tag{7.20}$$

If the step is unsuccessful ($Q_i - Q_{i-1} \geq 0$), it should be reduced twice:

$$\beta_{i+1} = \frac{1}{2}\beta_i, \tag{7.21}$$

The steps should be reduced also when the iteration is unsuccessful and there is nonconvergence towards exact parameter values, i.e., when condition (7.18) is satisfied.

The procedure stops after unsuccessful iterations if the last step is smaller than the predefined accuracy:

$$\left|b_{j(i-1)} - b_{ji}\right| < \Delta_{j(i-1)}. \tag{7.22}$$

In cases when the iterative procedure convergences slowly, increase of γ according to (7.17) improves the convergence.

7.5 Iterative Algorithm

The results obtained permit us to build an algorithm for the solution of an inverse identification problem:

1. Put $\beta_0 = 10^{-2}$, $\gamma = 0.9$, $b_{j0} = b_j^{(0)}$ (initial parameter values), j = 1, ..., J.
2. Put i = 1.
3. Calculate $y_{n(i-1)} = f(\mathbf{x}_n, \mathbf{b}_{i-1}), \quad n = 1, \ldots, N$.
4. Calculate $\left(\frac{\partial f}{\partial b_j}\right)_{i-1}, \quad j = 1, \ldots, J$.
5. Calculate $Q_{i-1} = \sum_{n=1}^{N} \left(y_{n(i-1)} - \hat{y}_n\right)^2$.
6. Calculate $\left(\frac{\partial Q}{\partial b_j}\right)_{(i-1)} = 2\sum_{n=1}^{N} \left[y_{n(i-1)} - \hat{y}_n\right]\left[\frac{\partial f}{\partial b_j}\right]_{(i-1)}, \quad j = 1, \ldots, J$.
7. Check if i = 1

   (a) If it does, then go to step 8.
   (b) If it does not, then go to step 10.

8. Calculate the parameters and the accuracy

$$b_{ji} = b_{j(i-1)} - \beta_{(i-1)} R_{j(i-1)}, \quad \Delta_{ji}^{(0)} = \gamma \frac{|b_{j1} - b_{jo}|}{|b_{j1}|} |b_{ji}|, \quad \Delta_{ji} = |b_{ji} - b_{j(i-1)}|,$$

$j = 1, \ldots, J.$

9. Put i = i + 1 and then go back to step 3.
10. Check if $Q_{i-1} - Q_{i-2} > 0$
    (a) If it does, then go to step 11.
    (b) If it does not, then go to step 13.
11. Check if $\Delta_{j(i-1)} < \Delta_{j(i-1)}^{(0)}$
    (a) If it does, then go to step 17.
    (b) If it does not, then go to step 12.
12. Put $\beta_{(i-1)} = \frac{1}{2}\beta_{(i-1)}$ and go back to step 8.
13. Check if $\Delta_{j(i-1)} > 2\Delta_{j(i-1)}^{(0)}$
    (a) If it does, then go to step 12.
    (b) If it does not, then go to step 14.
14. Check if $Q_{(i-2)} - Q_{(i-3)} > 0$
    (a) If it does, then go to step 15.
    (b) If it does not, then go to step 16.
15. Put $\beta_{i-1} = \beta_{i-2}$ and then go back to step 9.
16. Put $\beta_{i-1} = 2\beta_{i-2}$ and then go back to step 9.
17. Stop.

7.6 Correct Problem Solution

Literature sources [1, 3, 8, 10] teach us that every method for solving incorrect problems should also solve correct ones. Therefore, the first solution of the inverse problem considered here corresponds to the interval $0 < x < 0.3$.

Consider one- and two-parameter models (8.7.5, 7.7). Figure 5 shows the models (7.5) and (7.7) with exact parameter values $b^* = 5$, $b_1^* = 1$, $b_2^* = 5$ together with "experimental" data (7.8). The proposed algorithm was used for solution of the identification problem and the results are summarized in Table 1.

The efficiency of every iterative method for function minimization depends on the initial approximation. Parameter values obtained under conditions imposed by different initial approximations are summarized in Table 2.

Table 1 Solutions of one- and two-parameter models ($0 \le x \le 0.3$)

$\Delta\hat{y}(\%)$	$b*$	i	b_1^*	b_2^*	i
±5	4.9678	337	1.0025	5.0674	128
±10	4.9351	339	0.9940	4.9218	172

Table 2 Effect of the initial approximation ($0 \le x \le 0.3$, $\gamma = 0.9$)

$b^{(0)}$	b^*	i	$b_1^{(0)}$	$b_2^{(0)}$	b_1^*	b_2^*	i
1.0	4.9364	33	0.5	6.0	0.9289	4.9372	296
2.0	4.9543	63	0.7	6.0	0.9944	4.9611	240
4.0	4.9672	161	0.9	6.0	0.9955	4.9737	216
6.0	4.9678	337	1.1	6.0	1.0025	5.0674	128
0	4.9678	679	1.3	6.0	0.9934	4.9433	257
10.0	4.9678	1173	1.5	6.0	0.9955	4.9735	224

Table 3 Effect of γ ($0 \le x \le 0.3$)

γ	b^*	i	γ	b_1^*	b_2^*	i
0.9	4.9678	337	0.9	1.0025	5.0674	128
0.1	4.9676	3,044	1.2	1.0108	5.1922	104

7.7 Effect of the Regularization Parameter

The iteration number depends on the value of regularization parameter γ and the efficiency of the minimization increases when the value of γ is increased. This effect is demonstrated through the data summarized in Table 3.

7.8 Incorrect Problem Solution

As mentioned before, if the "experimental" data are captured under conditions (regimes) corresponding to the interval $0.30 < x < 0.65$, the parameter identification problem will be ill-posed. The problem incorrectness is due to solution sensitivity with respect to "experimental" errors associated with determination of the objective function $\hat{y}$.

Consider a solution of the parameter identification problem through minimization of the least-squares function (7.3), with $x_n = 0.01\text{n}$, n = 31, …, 65, i.e., $0.31 \le x \le 0.65$. The solutions of the one-parameter model ($b^{(0)} = 6$, $\gamma = 0.5$) and the two-parameter model ($b_1^{(0)} = 1.1$, $b_2^{(0)} = 6$, $\gamma = 0.05$) are summarized in Table 4. Comparisons between model predictions and "experimental" data are illustrated by the plots in Figs. 9 and 10. These plots indicate permits an inverse problem to be solved in the cases when some of the "experimental" data are not

Table 4 Incorrect problem solution $0.31 \leq x \leq 0.65$

$\Delta\hat{y}$ (%)	b^*	i	b_1^*	b_2^*	i
±5	5.0614	1,213	1.1797	5.4666	642
±10	5.1232	1,217	1.3778	5.9106	416

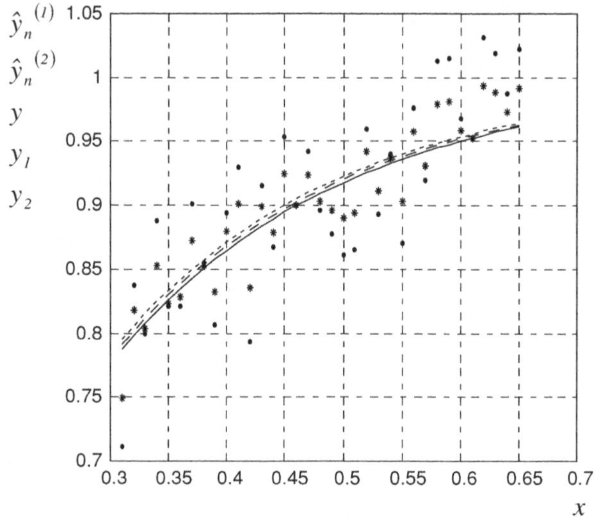

Fig. 9 One-parameter model and "experimental" data: *asterisks* $\hat{y}_n^{(1)}$, "experimental" data with a maximum error of ±5%; *circles* $\hat{y}_n^{(2)}$, "experimental" data with a maximum error of ±10%; *solid line* $y = 1 - \exp(-bx)$; $b = 5$; *dash-dotted line* $y_1 = 1 - \exp(-b^*x)$; $b^* = 5.0614$; *dashed line* $y_2 = 1 - \exp(-b^*x)$; $b^* = 5.1232$

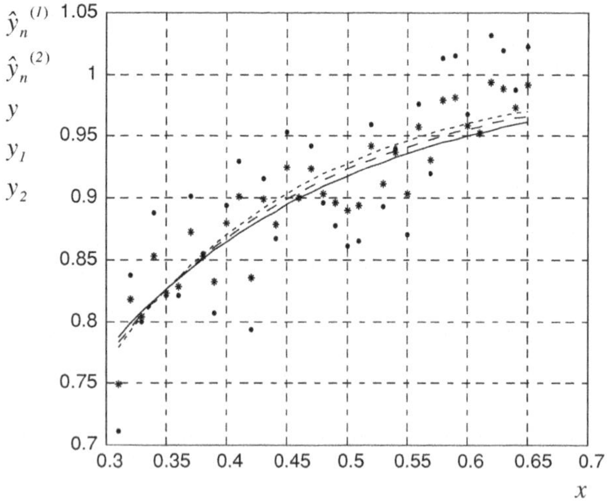

Fig. 10 Two-parameter model and "experimental" data: *asterisks* $\hat{y}_n^{(1)}$, "experimental" data with a maximum error of ±5%; *circles* $\hat{y}_n^{(2)}$, "experimental" data with a maximum error of ±10%; *sold line* $y = 1 - b_1 \exp(-b_2x)$; $b_1 = 1$; $b_2 = 5$; *dash-dotted line* $y_1 = 1 - b_1^*\exp(-b_2^*x)$; r$b_1^* = 1.180$; $b_2^* = 5.467$; *dashed line* $y_2 = 1 - b_1^* \exp(-b_2^*x)$; $b_1^* = 1.3778$; $b_2^* = 5.9106$

Table 5 Effect of the initial approximation ($0.31 \le x \le 0.65$, $\gamma = 0.5$)

$b^{(0)}$	b^*	i	$b_1^{(0)}$	$b_2^{(0)}$	b_1^*	b_2^*	i
1.0	5.0280	25	0.5	6.0	1.1815	5.4708	704
2.0	5.0472	59	0.7	6.0	1.1853	5.4794	619
3.0	5.0562	133	0.9	6.0	1.1797	5.4666	668
4.0	5.0602	301	1.1	6.0	1.1797	5.4666	642
6.0	5.0614	1,213	1.5	6.0	1.1815	5.4709	439
8.0	5.0614	5,171	2.0	6.0	1.1798	5.4664	1,479
10.0	5.0614	15,829	3.0	6.0	1.1798	5.4665	2,944

Table 6 Effect of γ ($0.31 \le x \le 0.65$)

γ	b^*	i	γ	b_1^*	b_2^*	i
0.05	5.0614	13,823	0.05	1.1797	5.4666	642
0.5	5.0614	1,213	0.5	1.2052	5.5246	139
1.2	5.0614	552	1.2	1.2375	5.5951	87

sensible "physically." The latter implies that these data do not have a physical sense since $\hat{y}_n^{(2)} > 1$.

The iteration numbers depend on the initial approximations $b^{(0)}$, $b_1^{(0)}$, $b_2^{(0)}$ of the iterative procedure. The results for $0.31 \le x \le 0.65$ are summarized in Table 5.

The effect of γ (initial accuracy value) on the iteration numbers is summarized in Table 6. The results presented in Figs. 11 and 12 demonstrate that the differences between the exact model and the models derived through parameter identification are very small. On the other hand, the results in Table 5 show that the differences between the exact values of the parameters and the values obtained are significant. The correctness of the parameter identification will be tested next through a criterion of model adequacy [6].

7.9 Essentially Incorrect Problem Solution

The parameter identification problem when the inverse problem is essentially incorrect [28, 30] will be solved by minimization of the least-squares function

$$Q(\mathbf{b}) = \sum_{n=66}^{100} (y_n - \hat{y}_n)^2, \tag{7.23}$$

where $y_n = f(x_n, \mathbf{b})$, $x_n = 0.01n$, $n = 66, \ldots, 100$, $\mathbf{b} = (b_1, b_2)$.

The one-parameter model ($b_1 = 1$ and $b_2 = b$) solution was obtained for different sets of "experimental" data (with errors of about ± 5 and $\pm 10\%$) and the results are shown in Table 7, where $b^{(0)} = 6$, $\gamma = 5$, $0.66 \le x \le 1$.

Fig. 11 One-parameter model and "experimental" data ($0.66 \leq x \leq 1$); *asterisks* $\hat{y}_n^{(1)}$, "experimental" data with a maximum error of ±5%; *circles* $\hat{y}_n^{(2)}$, "experimental" data with a maximum error of ±10%; *solid line* $y = 1 - \exp(-bx)$; $b = 5$; *dash-dotted line* $y_1 = 1 - \exp(-b^* x)$; $b^* = 5.1828$; *dashed line* $y_2 = 1 - \exp(-b^* x)$; $b^* = 5.3816$

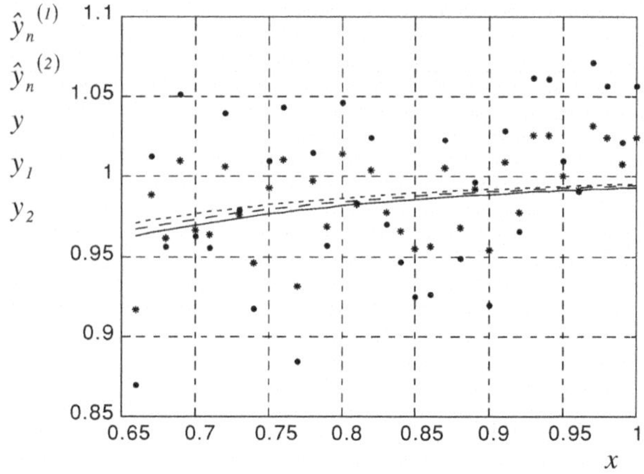

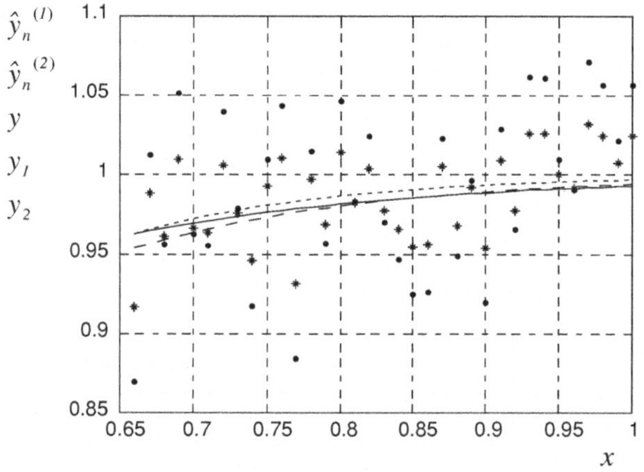

Fig. 12 Two-parameter model and "experimental" data ($0.66 \leq x \leq 1$); *asterisks* $\hat{y}_n^{(1)}$, "experimental" data with a maximum error of ±5%; *circles* $\hat{y}_n^{(2)}$, "experimental" data with a maximum error of ±10%; *solid line* $y = 1 - b_1 \exp(-b_2 x)$; $b_1 = 1$; $b_2 = 5$; *dash-dotted line* $y_1 = 1 - b_1^* \exp(-b_2^* x)$; $b_1^* = 2.1720$; $b_2^* = 6.1731$; *dashed line* $y_2 = 1 - b_1^* \exp(-b_2^* x)$; $b_1^* = 4.9003$; $b_2^* = 7.4004$

Table 7 Solutions of one- and two-parameter models ($0.66 \leq x \leq 1$)

$\Delta\hat{y}$ (%)	b^*	i	b_1^*	b_2^*	i
±5	5.1828	2,066	2.1720	6.1731	54
±10	5.3816	2,156	4.9003	7.4004	128

The results in Fig. 11 show that the difference between the models obtained y_1 and y_2 and the exact model y is very small. The proposed method [28, 30] permits us to obtain a solution of the inverse problem when most of the "experimental" data are not "physically" correct $\left(\hat{y}_n^{(2)} > 1\right)$.

The iteration numbers depend on the initial approximation $b^{(0)}$ of the iterative procedure. The results for $0.66 \leq x \leq 1$ are shown in Table 8. The iteration

Table 8 Effect of the initial approximation ($0.66 \leq x \leq 1$)

$b^{(0)}$	b^*	i	$b_1^{(0)}$	$b_2^{(0)}$	b_1^*	b_2^*	i
1.0	5.2089	18	0.5	6.0	3.2623	6.7358	177
2.0	5.2089	20	0.7	6.0	2.0153	6.0536	43
3.0	5.1883	54	0.9	6.0	2.1252	6.1230	47
4.0	5.1828	186	1.1	6.0	2.1720	6.1731	54
6.0	5.1828	2,066	1.3	6.0	2.1837	6.1803	40
7.0	5.1828	6,912	1.5	6.0	2.1850	6.1811	44
9.0	5.1828	52,993	3.0	6.0	3.1897	6.7284	56

Table 9 Effect of γ ($0.66 \leq x \leq 1$)

γ	b^*	i	γ	b_1^*	b_2^*	i
0.5	5.1828	18,731	0.5	4.5531	7.1961	470
2	5.1828	4,694	1	4.5787	7.2032	433
3	5.1828	3,692	3	2.5261	6.3797	88
5	5.1828	2,066	5	2.1720	6.1731	54
10	5.1828	1,040	10	2.3818	6.2782	53

numbers also depend on γ (the initial desired accuracy) (see Table 9). The results of the parameter identification of the two-parameter model for initial approximations $b_1^{(0)} = 1.1$ and $b_2^{(0)} = 6(\gamma = 5)$ are shown in Table 8 and are compared with "experimental" data in Fig. 12.

The effects of the initial approximation $b_1^{(0)}$, $b_2^{(0)}$ and the value of γ were also investigated (see Tables 8, 9).

The results in Table 8 and those in Figs. 11 and 12 show that the differences between the values obtained and exact parameter values are very large, but the differences between the models obtained and the exact models exhibit the opposite behavior.

7.10 General Case

In practice, it is very often possible to have experimental data in very large regime interval (e.g., $0 \leq x \leq 1$). However, it is unknown which of the experimental data lead to a correct or an incorrect problem. That is why the parameter identification problem will be solved by minimization

where $y_n = f(x_n, \mathbf{b})$, $x_n = 0.01n$, $n = 1, \ldots, 100$, $\mathbf{b} = (b_1, b_2)$.

In Table 10 the results of the parameter identification for one- and two-parameter models are shown. This results were obtained for initial approximation $b^{(0)} = 6$ ($\gamma = 5$) and $b_1^{(0)} = 1.1$, $b_2^{(0)} = 6(\gamma = 2)$. The iteration numbers depend on the initial approximations of the iterative procedure (Table 11) and γ (Table 12). The model adequacy [24] is a criterion for the correctness of the results of the parameter identification only.

Table 10 Solutions of one- and two-parameter models ($0 \le x \le 1$)

$\Delta\hat{y}$ (%)	b^*	i	b_1^*	b_2^*	i
±5	5.0117	50	1.0106	5.1717	65
±10	5.0231	50	1.0196	5.1721	66

Table 11 Effect of the initial approximation ($0 \le x \le 1$)

$b^{(0)}$	b^*	i	$b_1^{(0)}$	$b_2^{(0)}$	b_1^*	b_2^*	i
1.0	4.9876	12	0.5	6.0	0.9987	5.0180	178
2.0	4.9877	12	0.7	6.0	1.0018	5.0475	136
3.0	4.9930	14	0.9	6.0	0.9987	5.0741	106
4.0	5.0106	22	1.1	6.0	1.0106	5.1717	65
6.0	5.0117	50	1.3	6.0	1.0018	5.0474	123
8.0	5.0117	92	2.0	6.0	1.0072	5.1383	92
10.0	5.0117	188	3.0	6.0	1.0040	5.0863	116

Table 12 Effect of γ ($0 \le x \le 1$)

γ	b^*	i	γ	b_1^*	b_2^*	i
0.05	5.0115	4,093	0.5	1.0016	5.0462	106
0.5	5.0116	375	1.0	1.0098	5.0859	94
1.2	5.0116	172	1.2	1.0098	5.0859	94
10	5.0118	28	2.0	1.0106	5.1717	65

7.11 Statistical Analysis of Model Adequacy

A test of the model adequacy was performed through a statistical analysis. The parameters $\mathbf{b}^*$ are derived through calculations of experimental data, so they can also be assumed to be random numbers. The same suggestion is valid for the objective function values calculated with random parameter numbers. Moreover, both the parameter and the objective function also incorporate effects of the model building that implies a lack of knowledge concerning the mathematical structure employed [1].

The model is assumed as an adequate one if the variance of the experimental data error (S_ε) equals the variance of the model error (S). The test was performed with the experimental values of the objective function $\hat{y}_k$, $(k = 1, \ldots, K)$ obtained under identical technical conditions (regimes) $x = \boldsymbol{x}^{(0)} = (x_1^{(0)}, \ldots, x_K^{(0)})$, where κ = 5–10. The experimental data variance requires the mathematical expectation of y $(\tilde{m}_y)$ to be estimated [1, 21]:

$$\tilde{m}_y = \frac{1}{K}\sum_{k=1}^{K} \hat{y}_k, \quad S_\varepsilon^2 = \frac{1}{K-1}\sum_{k=1}^{K} \left(\hat{y}_k - \tilde{m}_y\right)^2. \tag{7.25}$$

Thus, the variance of the model error [1] is

$$S^2 = \frac{1}{N-J}\sum_{n=1}^{N}(y_n - \hat{y}_n)^2 = \frac{Q}{N-J}, \tag{7.26}$$

where N is the number of experimental data and J is the number of parameters.

The model adequacy is defined by the variance ratio

$$F = \frac{S^2}{S_\varepsilon^2}, \tag{7.27}$$

where $S^2 > S_\varepsilon^2$ if S contains the error effect of the both model and experimental data. The value of F is compared with the tabulated values (F_J) of the Fisher distribution [21]. The condition for the model adequacy is

$$F \leq F_J(\alpha, \nu, \nu_\varepsilon), \tag{7.28}$$

where $\nu = N - J$, $\nu_\varepsilon = K - 1$, $\alpha = 0.01\text{-}0.1$.

The statistical analysis of the model adequacy was performed with $0 \leq x \leq 0.30$ and the results are presented in Table 11. For the tests performed, $N = 30$, $J = 1(2)$, $K = 10$, $x^{(0)} = 0.2$, and $\alpha = 0.05$. The results confirm the adequacy of the model.

The statistical analysis of cases corresponding to an incorrect inverse problem ($0.31 \leq x \leq 0.65$) was performed with $N = 35$, $J = 1(2)$, $K = 10$, $x^{(0)} = 0.5$, and $\alpha = 0.05$ (see Table 14). The models are adequate despite the large differences between the calculated and the exact values of the model parameters (see Table 4).

The statistical analysis of the model adequacy in the case of essential incorrectness of the inverse problem ($0.66 \leq x \leq 1$) was done for $N = 35$, $J = 1(2)$, $K = 10$, $x^{(0)} = 0.8$, and $\alpha = 0.05$. The results are collected in Table 15. The models are adequate independently despite the large differences between the calculated and the exact values of the model parameters.

Statistical analysis of the model adequacy in the case of ($0 \leq x \leq 1$) was made for $N = 100$ and the results are shown in [30]. Tables 13, 14 and 15 show that all of models are adequate owing to $F < F_J$ (the general case too).

Table 13 Statistical analysis of the model adequacy ($0 \leq x \leq 0.3$)

J	$\Delta\hat{y}$ (%)	b_1^*	b_2^*	γ	$S_\varepsilon \times^{-2}$	$S \times^{-2}$	F	F_J
1	±5	–	4.9678	0.9	1.7933	1.7071	0.9061	2.24
1	±10	–	4.9351	0.9	3.5867	3.4139	0.9059	2.24
2	±5	1.0025	5.0674	0.9	1.7933	1.8354	1.0475	2.25
2	±10	0.99401	4.9218	0.9	3.5867	3.4434	0.9217	2.25

Table 14 Statistical analysis of the model adequacy ($0.31 \leq x \leq 0.65$)

J	$\Delta\hat{y}\%$	b_1^*	b_2^*	γ	$S_\varepsilon \times ^{-2}$	$S \times ^{-2}$	F	F_J
1	±5	–	5.0614	0.5	2.6042	2.3588	0.8205	2.19
1	±10	–	5.1232	0.5	5.2083	4.7328	0.8257	2.19
2	±5	1.1797	5.4666	0.05	2.6042	2.3656	0.8252	2.20
2	±10	1.3778	5.9106	0.05	5.2083	4.7349	0.8265	2.20

Table 15 Statistical analysis of the model adequacy ($0.66 \leq x \leq 1$)

J	$\Delta\hat{y}\,(\%)$	b_1^*	b_2^*	γ	$S_\varepsilon \times 10^{-2}$	$S \times 10^{-2}$	F	F_J
1	±5	–	5.1828	5	2.7850	2.5988	0.8707	2.19
1	±10	–	5.3816	5	5.5701	5.2482	0.8723	2.19
2	±5	2.1720	6.1731	5	2.7851	2.6221	0.8855	2.20
2	±10	4.9003	7.4004	5	5.5701	5.2482	0.8877	2.20

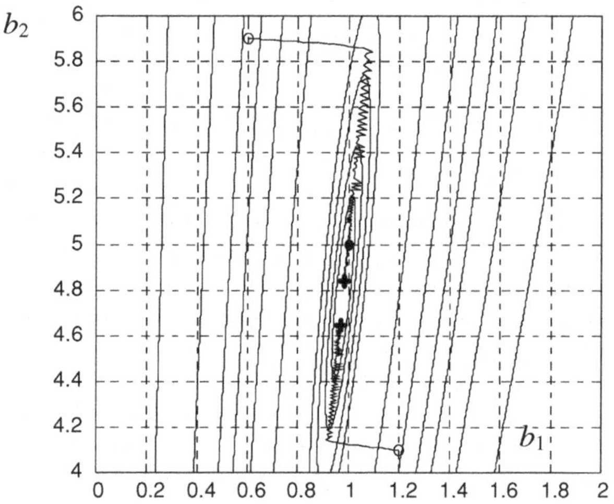

Fig. 13 Road of the iterative procedures ($n = 1$–30; $\Delta\hat{y}(\%) = \pm 10$); *open circles* $b_0 = [0.6, 5.9]$, $b_0 = [1.2, 4.1]$; *filled circles* b = [1, 5], b = [1, 5]; crosses $b^* = [0.9833, 4.8390]$, $b^* = [0.9691, 4.6444]$

7.12 Comparison between Correct and Incorrect Problems

The results obtained for inverse identification problem solutions show a large difference between the least-squares functions in the correct and incorrect problem cases. Horizontals (contour lines) of the least-squares function exist in Figs. 13 and 14. In all cases Q is a ridge-type function. If the inverse problem is correct, the distance between the exact solution points and the minimum defined by the least-squares function should be very small (see Fig. 13). In cases of incorrect problems this distance is very large (see Fig. 14). Figures 13 and 14 show the "road" of the iterative procedures from initial parameter values ($b_1^{(0)}$, $b_2^{(0)}$) towards the parameter values at the last iterations (b_1^*, b_2^*). All these trajectories of the iterative solutions demonstrate the role of the second condition of (7.12).

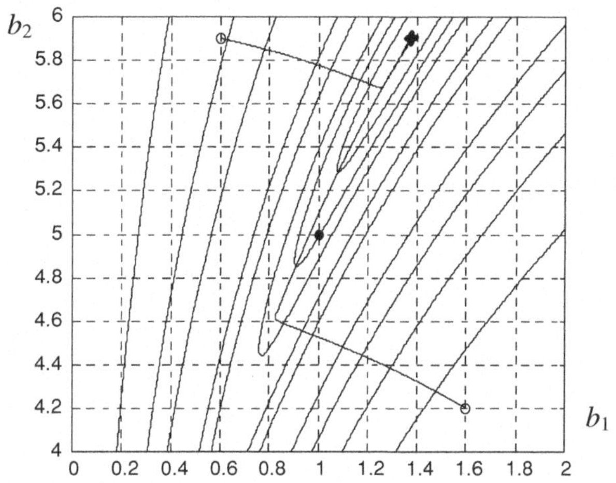

Fig. 14 Road of the iterative procedures (n = 31–65; $\Delta\hat{y}(\%) = \pm 10$); *open circles* $b_0 = (0.6, 5.9)$, $b_0 = (1.6, 4.2)$; *filled circles* $b = [1, 5]$, $b = [1, 5]$; *crosses* $b^* = [1.372, 5.900]$, $b^* = [1.376, 5.907]$

Table 16 Solutions of correct and incorrect problems, using different "experimental" data sets $b_1^{(0)} = 1.1, b_2^{(0)} = 6$

Different "experimental" data		b_1^*	b_1^*	γ	i
$0 \le x \le 0.3$	1	1.0025	5.0674	0.9	128
	2	1.0115	5.1706	0.9	120
	3	1.0068	5.1881	0.9	179
$0.31 \le x \le 0.65$	1	1.1564	5.2675	0.05	798
	2	0.5789	3.7056	0.05	1,803
	3	1.1723	5.2624	0.05	776

In all cases discussed above, the difference between correct and incorrect inverse identification problems is based on the distance between the points of the exact solutions and the minimum of the least-squares function. Actually, exact parameter values practically do not exist, so a useful criterion for the inverse problem "diagnostics" is required. Table 16 summarizes solutions of correct and incorrect inverse problems based on different experimental data sets. It is clear that large differences between solutions can be used as a criterion of inverse problem incorrectness (see the "road" of the iterative procedures in Figs. 15, 16, 17, 18).

Tables 1, 2, 3, 4, 5, 6, 7, 8, 9 and 10 and Figs. 13 and 18 show a very interesting influence of the "correct" experimental data ($0 \le x \le 0.3$) on the correctness of the inverse problem solution in the general case ($0 \le x \le 1$), where the behavior of the least-squares function is like that in the case ($0 \le x \le 0.3$), when the inverse problem solution is correct, i.e., the difference between the exact parameter values and determined value (the minimum of the function) is very small.

Figures 15, 16, 17 and 18 show that the main reason for the inverse problem incorrectness is the sensitivity of the inverse problem solution with respect to the experimental data errors, but not the size of the experimental data error.

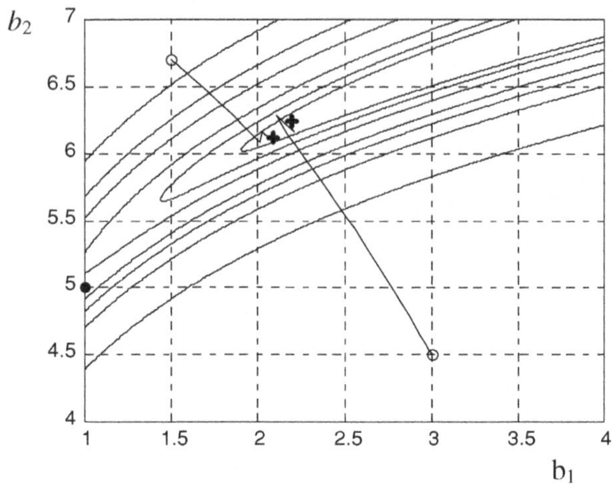

Fig. 15 The horizontals of the least-squares function Q (n = 66–100; $\Delta\hat{y}(\%) = \pm 5$); *open circles* $b_0 = [1.5, 6.7]$, $b_0 = [3, 4.5]$; *filled circles* b = [1, 5], b = [1, 5]; *crosses* $b^* = [2.0848, 6.1160]$, $b^* = [2.1999, 6.2420]$

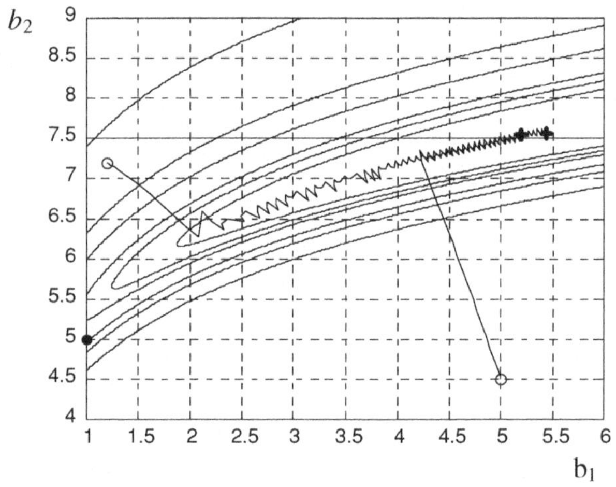

Fig. 16 The horizontals of the least-squares function Q (n = 66–100; $\Delta\hat{y}(\%) = \pm 10$); *open circles* $b_0 = [1.2, 7.2]$, $b_0 = [5, 4.5]$; *filled circles* b = [1, 5], b = [1, 5]; *crosses* $b^* = [5.4289, 7.5596]$, $b^* = [5.1861, 7.542]$

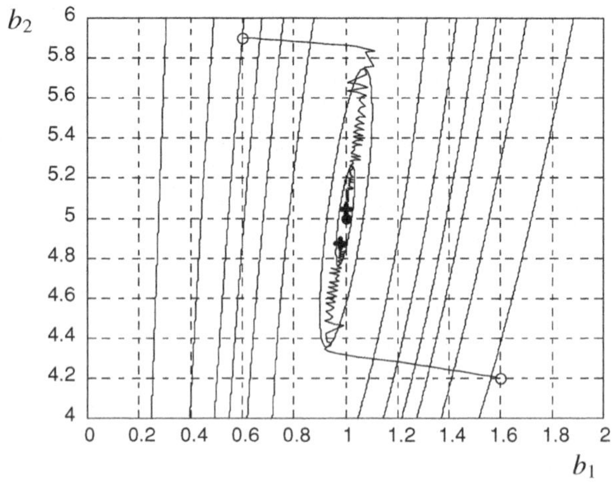

Fig. 17 The horizontals of the least-squares function Q (n = 1–100; $\Delta\hat{y}(\%) = \pm 5$); *open circles* $b_0 = [0.6, 5.9]$, $b_0 = [1.6, 4.2]$; *filled circles* b = [1, 5], b = [1, 5]; *crosses* $b^* = [1.0015, 5.0450]$, $b^* = [0.9793, 4.8743]$

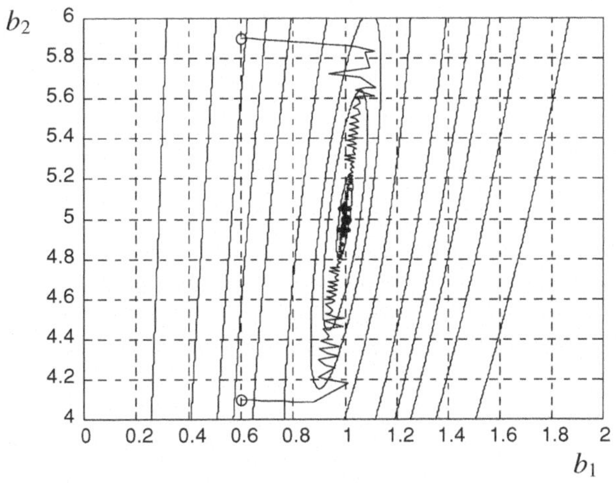

Fig. 18 The horizontals of the least-squares function Q (n = 1–100; $\Delta\hat{y}[\%] = \pm10$); *open circles* $b_0 = [0.6, 5.9]$, $b_0 = [0.6, 4.1]$; *filled circles* b = [1, 5], b = [1, 5]; *crosses* $b^* = [0.9999, 5.0481]$, $b^* = [0.9928, 4.9431]$

Obviously, this peculiarity of the inverse incorrect problems manifests itself not only in the iterative methods, but also in the variational methods for the search of the minimum of the least-squares function (see 4.6).

All of these methods use an additional condition which does not permit a large difference between exact and approximate solutions to be obtained. This additional condition regularizes the problem (the problem becomes conditionally correct) and in the method presented [24] γ is the regularizing parameter.

8 Multiequation Models

The kinetics of many chemical, biochemical, photochemical, and catalytic reactions is very complex, i.e., the kinetic model consists of many equations. The number of parameters of the separate equations is not large, but the total number of parameters is very large [34].

The parameter identification problem for complex kinetic models is described [35–37] on the basis of minimization of the least-squares function using experimental data or their spline approximations.

Model parameter identification in these cases is very difficult because of the multiextremal least-squares function or because of the fact that some minima are of ravine type. The solution of this problem needs very good initial value approximations for the parameters (in the attraction area of the global minimum) for the minimum searching procedure. This main problem in the multiextremal function minimization is solved on the basis of a hierarchical approach [34].

8.1 Problem Formulation

Let us consider the following *multiequation model*:

$$\frac{dc_i}{dt} = F_i(c_1,\ldots,c_M;k_{i1},\ldots,k_{i\alpha_i}),\ c_i(0) = c_{(0)i},\ i = 1,\ldots,M, \tag{8.1}$$

where t is time, $c_i(t_n)$ and k_{ij} are objective functions (concentrations of the reagents) and parameters in the model for i = 1, ..., M, j = 1, ..., α_i, where α_i is the number of parameters of the *i*th equation.

For solution of parameter identification problem we will be use experimental data for the objective functions:

$$c_i^{(e)} = c_i^{(e)}(t_n),\ n = 1,\ldots,N, \tag{8.2}$$

where N is the number of experimental data measurements.

The least-squares functions for the separate model equations are

$$Q_i(k_{i1},\ldots,k_{i\alpha_i}) = \sum_{n=1}^{N}\left[c_i(t_n) - c_i^{(e)}(t_n)\right]^2,\quad i = 1,\ldots,M, \tag{8.3}$$

where $c_i(t_n)$, i = 1, ..., M are obtained after the solution of (8.1).

The least-squares function of the parameter identification in model (8.1) is

$$Q = \sum_{i=1}^{M} Q_i. \tag{8.4}$$

The total number of parameters in (4) is very large,

$$I = \sum_{i=1}^{M}\alpha_i, \tag{8.5}$$

and in many cases it is not possible to minimize function Q, given in equation (8.4), because Q is multiextremal and some minima are of ravine type.

Equation (8.4) shows that the minimization of Q is a multicriteria optimization problem with equal specific weight coefficients of the separate criteria. The obtaining of the global minimum point needs a very good initial approximation, i.e., the initial point of the minimization procedure has to be in the attraction area of the global minimum.

The experimental data for the objective functions (concentrations) can be represented using polynomial approximations:

$$c_i^{(e)}(t_n) \to P_i(t),\quad n = 1,\ldots,N,\quad i = 1,\ldots,M, \tag{8.6}$$

where $P_i(t)$ are polynomials of fifth or sixth power.

Let us consider the first equation in (8.1), where all functions, including the first one, are replaced by the polynomial approximations of the experimental data:

$$c_i(t) \equiv P_i(t), \quad 2 \leq i \leq M, \tag{8.7}$$

i.e.,

$$\frac{dc_1}{dt} = F_1(c_1, P_2, \ldots, P_M; k_{11}, \ldots, k_{1\alpha_1}), \; c_1(0) = c_{(0)1}. \tag{8.8}$$

The minimization of the least-squares function,

$$Q_1(k_{11}, \ldots, k_{1\alpha_1}) = \sum_{n=1}^{N} \left[c_1(t_n) - c_1^{(e)}(t_n) \right]^2, \tag{8.9}$$

where $c_1(t_n)$ is the solution of Eq. (8.8), permits us to calculate the parameters values $(\tilde{k}_{11}, \ldots, \tilde{k}_{1\alpha_1})$ of the first equation in model (8.1). The parameter identification errors for equation (8.8) are the result of the experimental data errors and polynomial approximations errors only, i.e., there is no influence from any other model equation ($2 \leq i \leq M$) errors (as a result of the parameter identification errors of the other parameters in the model). In this way the parameter identification errors are minimal and the small number of parameters in one equation excludes the possibility of many minimum points.

This procedure can be repeated for all equations in (8.1) (step by step). The parameter values obtained $\tilde{k}_{ij}$; $i = 1, \ldots, M$; $j = 1, \ldots, \alpha_i$, can be used as a zeroth-order approximation for the model parameter identification and they are the zeroth hierarchical level in the parameter identification procedure.

Very often the exactness of the zeroth-order approximation of the parameters is not sufficient for the minimization of Q in (8.4), because the parameters in one model do not give an account of the errors of another model. The first-order approximation may be obtained using the zeroth-order one on the first hierarchical level (step by step).

The first step is to solve the first two equations,

$$\begin{aligned} \frac{dc_1}{dt} &= F_1\left(c_1, c_2, P_3, \ldots, P_M; \tilde{k}_{11}, \ldots, \tilde{k}_{1\alpha_1}\right), \; c_1(0) = c_{(0)1}; \\ \frac{dc_2}{dt} &= F_2(c_1, c_2, P_3, \ldots, P_M; k_{21}, \ldots, k_{2\alpha_2}), \; c_2(0) = c_{(0)2}, \end{aligned} \tag{8.10}$$

and to obtain the first-order approximation of the parameters in the second equation $(\bar{k}_{21}, \ldots, \bar{k}_{2\alpha_2})$ by minimization of the function Q_2 (see 8.3).

The next step is the solution of the set of three equations

$$\begin{aligned} \frac{dc_1}{dt} &= F_1\left(c_1, c_2, c_3, P_4, \ldots, P_M; \tilde{k}_{11}, \ldots, \tilde{k}_{1\alpha_1}\right), \; c_1(0) = c_{(0)1}; \\ \frac{dc_2}{dt} &= F_2(c_1, c_2, c_3, P_4, \ldots, P_M; \bar{k}_{21}, \ldots, \bar{k}_{2\alpha_2}), \; c_2(0) = c_{(0)2}; \\ \frac{dc_3}{dt} &= F_3(c_1, c_2, c_3, P_4, \ldots, P_M; k_{31}, \ldots, k_{3\alpha_3}), \; c_3(0) = c_{(0)3}, \end{aligned} \tag{8.11}$$

and determination of $(\bar{k}_{31}, \ldots, \bar{k}_{3\alpha 3})$ after the minimization of Q_3.

The last step is the solution of a set of i equations,

$$\frac{dc_1}{dt} = F_1(c_1, \ldots, c_i; k_{11}, \ldots, k_1\alpha_1),$$
$$c_1(0) = c_{(0)1}; \frac{dc_i}{dt} = F_i(c_1, \ldots, c_i; \bar{k}_{i1}, \ldots, \bar{k}_{i\alpha_i}),\ c_i(0) = c_{(0)i};\ i = 2, \ldots, M, \tag{8.12}$$

and determination of $(\bar{k}_{i1}, \ldots, \bar{k}_{i\alpha_i})$ after minimization of Q_1.

The last hierarchical level is to solve equations (8.8.1) and to minimize Q using the first-order-approximation values $\bar{k}_{ij}(i = 1, \ldots, M; j = 1, \ldots, \alpha_i)$ as initial approximations of the minimization procedure.

The hierarchical approach for parameter identification of the multiequation models is used next for fermentation system modeling.

8.2 Fermentation System Modeling

The mathematical models of fermentation systems contain biomass, product, and substrate material balances. The models obtained [38, 39] consist of three to four equations with six to ten parameters, which have to be obtained using experimental data. The modeling of glucose fermentation was presented in [34]. The fermentation process of gluconic acid production by *Gluconobacter oxydans* [38] (or by *Aspergillus niger* [39]) is oxidation of glucose to gluconic acid and ketogluconic acids [40]. The mathematical model of the fermentation kinetics consists of four equations for biomass, product (gluconic acid), and substrates (glucose and oxygen).

The dependence of the specific growth rate (μ) on glucose and oxygen substrates was assumed to follow the Monod kinetic model [34], which considers substrate limitation. The biomass growth can be described as

$$\frac{dx}{dt} = \mu x, \tag{8.13}$$

where the specific growth rate μ is given by the Monod-type model as

$$\mu = \mu_{max} \frac{c_G}{k_G + c_G} \frac{c_{O_2}}{k_{O_2} + c_{O_2}}. \tag{8.14}$$

The kinetics of gluconic acid formation was based on the Luedeking–Piret equation, originally developed for the fermentation of lactic acid [41]. It is a model which combines growth-associated and non-growth-associated contributions to product formation, i.e., the growth (*dx*/*dt*) and instantaneous biomass concentration (*x*):

$$\frac{dc_{Ga}}{dt} = k_1 \frac{dx}{dt} + k_2 x, \tag{8.15}$$

where k_1 and k_2 are Luedeking–Piret equation parameters for growth-associated and non-growth-associated product formation, respectively.

The rate of glucose utilization is represented by a mass balance equation [42], i.e., the glucose from cell material (dx/dt), metabolic product (dc_{Ga}/dt), and cell activity($k_4 x$):

$$\frac{dc_G}{dt} = -k_3\frac{dx}{dt} - \frac{dc_{Ga}}{dt} - k_4 x. \tag{8.16}$$

The dependence of biomass growth (dx/dt), product formation (dc_{Ga}/dt), and cell activity (x) on the oxygen consumption rate is given by

$$\frac{dc_{O_2}}{dt} = k_L a\left(c^*_{O_2} - c_{O_2}\right) - k_5\frac{dx}{dt} - \frac{1}{2}\frac{dc_{Ga}}{dt} - k_6 x. \tag{8.17}$$

The initial conditions of the model equations (8.13–8.17) are

$$t = 0, \quad x = x_{(0)}, \quad c_{Ga} = c_{(0)Ga}, \quad c_G = c_{(0)G}, \quad c_{O_2} = c_{(0)O_2}. \tag{8.18}$$

The model equations (8.13–8.18) are solved in the time interval $[0 \le t \le t_n]$, where the biomass concentration increases.

The minimization of the least-squares function is used for parameter identification. The problem is solved in dimensionless form, where the characteristic scales are the maximum experimental values of the concentrations $(x^{max}, c_G^{max}, c^*_{O_2})$ in the interval $0 \le t \le t_n$:

$$X = \frac{x}{x^{max}}, \quad C_{Ga} = \frac{c_{Ga}}{c_G^{max}}, \quad C_G = \frac{c_G}{c_G^{max}}, \quad C_{O_2} = \frac{c_{O_2}}{c^*_{O_2}}. \tag{8.19}$$

As a result, the model equations (8.8.13–8.8.18) have the form

$$\frac{dX}{dt} = \mu_{max}\frac{C_G}{K_G + C_G}\frac{C_{O_2}}{K_{O_2} + C_{O_2}}X, \tag{8.20}$$

$$\frac{dC_{Ga}}{dt} = K_1\frac{dX}{dt} + K_2 X, \tag{8.21}$$

$$\frac{dC_G}{dt} = -K_3\frac{dX}{dt} - \frac{dC_{Ga}}{dt} - K_4 X, \tag{8.22}$$

$$\frac{dC_{O_2}}{dt} = k_L a(1 - C_{O_2}) - K_5\frac{dX}{dt} - \frac{\bar{c}_G}{2c^*_{O_2}}\frac{dC_{Ga}}{dt} - K_6 X, \tag{8.23}$$

where

$$K_G = \frac{k_G}{c_G^{max}}, \quad K_{O_2} = \frac{k_{O_2}}{c^*_{O_2}}, \quad K_1 = \frac{k_1 x^{max}}{c_G^{max}}, \quad K_2 = \frac{k_2 x^{max}}{c_G^{max}},$$
$$K_3 = \frac{k_3 x^{max}}{c_G^{max}}, \quad K_4 = \frac{k_4 x^{max}}{c_G^{max}}, \quad K_5 = \frac{k_5 x^{max}}{c^*_{O_2}}, \quad K_6 = \frac{k_6 x^{max}}{c^*_{O_2}}. \tag{8.24}$$

Table 17 Initial and maximum experimental data values

Substance	Initial concentrations	Maximum concentrations	Dimensionless initial conditions
Biomass	$x_{(0)} = 0.2040\left[\frac{\text{kg}}{\text{m}^3}\right]$	$x^{\max} = 2.9238\left[\frac{\text{kg}}{\text{m}^3}\right]$	$X_{(0)} = \frac{x_{(0)}}{x^{\max}} = 0.0698$
Gluconic acid	$c_{(0)\text{Ga}} = 0\left[\frac{\text{kmol}}{\text{m}^3}\right]$	$c_{\text{Ga}}^{\max} = c_{\text{G}}^{\max} = 1.1667\left[\frac{\text{kmol}}{\text{m}^3}\right]$	$C_{(0)\text{Ga}} = \frac{c_{(0)\text{Ga}}}{c_{\text{G}}^{\max}} = 0$
Glucose	$c_{(0)\text{G}} = 1.16667\left[\frac{\text{kmol}}{\text{m}^3}\right]$	$c_{\text{G}}^{\max} = 1.1667\left[\frac{\text{kmol}}{\text{m}^3}\right]$	$C_{(0)\text{G}} = \frac{c_{(0)\text{G}}}{c_{\text{G}}^{\max}} = 1$
Oxygen	$c_{(0)\text{O}_2} = 2.41.10^{-4}\left[\frac{\text{kmol}}{\text{m}^3}\right]$	$c_{\text{O}_2}^{*} = 2.41.10^{-4}\left[\frac{\text{kmol}}{\text{m}^3}\right]$	$C_{(0)\text{O}_2} = \frac{c_{(0)\text{O}_2}}{c_{\text{O}_2}^{*}} = 1$

The initial conditions of equations (8.8.20–8.8.23) are

$$X = X_{(0)},\ C_{\text{Ga}} = C_{(0)\text{Ga}} = 0,\ C_{\text{G}} = C_{(0)\text{G}},\ C_{\text{O}_2} = C_{(0)\text{O}_2}. \tag{8.25}$$

8.2.1 Experimental Data

The parameter identification problem for the fermentation system (8.8.20–8.8.23, 8.8.25) will be solved using the real experimental data [38] (see Table 17).

The concentrations of biomass, gluconic acid, glucose, and oxygen as time functions will be represented in dimensionless forms $\left(X^{(e)}, C_{\text{Ga}}^{(e)}, C_{\text{G}}^{(e)} C_{\text{O}_2}^{(e)}\right)$ using (8.8.19).

The dimensionless experimental data for the concentrations permit us to obtain their polynomial approximations $(P_{\text{X}}, P_{\text{Ga}}, P_{\text{G}}, P_{0_2})$ and to calculate the polynomial approximation error variances $(S_{\text{X}}, S_{\text{Ga}}, S_{\text{G}}, S_{\text{O}_2})$:

$$\begin{aligned} P_{\text{X}}(t) &= -2.7125 \times 10^{-8}t^6 + 2.5023 \times 10^{-6}t^5 - 8.7499 \times 10^{-5}t^4 + 0.00141t^3 \\ &\quad - 0.0090543t^2 + 0.030894t + 0.069996, \\ S_{\text{X}} &= 0.00164; \end{aligned} \tag{8.26}$$

$$\begin{aligned} P_{\text{Ga}}(t) &= -7.3643 \times 10^{-8}t^5 + 3.5119 \times 10^{-6}t^4 - 3.0441 \\ &\quad \times 10^{-5}t^3 + 0.0002176t^2 + 0.00013591t + 9.9851.10^{-5}, \\ S_{\text{Ga}} &= 0.00167; \end{aligned} \tag{8.27}$$

$$\begin{aligned} P_{\text{G}}(t) &= -3.0467 \times 10^{-9}t^6 + 2.4717 \times 10^{-7}t^5 - 6.571.10^{-6}t^4 + 4.8921 \times 10^{-5}t^3 \\ &\quad - 0.0003017t^2 - 0.00058582t + 0.99979, \\ S_{\text{G}} &= 0.00155; \end{aligned} \tag{8.28}$$

$$P_{O_2}(t) = -1.7037 \times 10^{-8}t^6 + 1.3205 \times 10^{-6}t^5 - 4.1355 \times 10^{-5}t^4 + 0.00058519t^3 - 0.0038151t^2 + 0.0037716t + 0.99976, \quad (8.29)$$
$$S_{O_2} = 0.00177$$

It could be supposed that the differences between the polynomial approximation error variances and the experimental data error variances are negligible, i.e., we could use the polynomial approximation instead of concentration as a time function.

8.2.2 Zeroth-Order Approximations of the Model Parameters

The initial approximation of the model parameters can be obtained by solving model equations (8.8.20–8.8.23), where the concentration time distribution has to be replaced by polynomial approximations (8.8.26–8.8.29).

For the parameter identification in the biomass model the following problem has to be solved:

$$\frac{dX}{dt} = \mu_{max} \frac{P_G}{K_G + P_G} \frac{P_{O_2}}{K_{O_2} + P_{O_2}} X, \quad t = 0,\ X = X_{(0)}. \quad (8.30)$$

The zeroth-order approximation of the parameter values $\tilde{\mu}_{max}, \tilde{K}_G$, and $\tilde{K}_{O_2}$ is obtained after minimization of the least-squares function:

$$Q_X\left(\tilde{\mu}_{max}, \tilde{K}_G, \tilde{K}_{O_2}\right) = \min \int_0^{t_N} (X - P_X)^2 dt, \quad (8.31)$$

where the sum in (8.8.9) is replaced with integral and experimental data by polynomial approximation (8.8.26).

The next steps are the consecutive solutions of the problems for gluconic acid production and glucose and oxygen consumption,

$$\frac{dC_{Ga}}{dt} = K_1 \frac{dP_X}{dt} + K_2 P_X, \quad t = 0, \quad C_{Ga} = 0; \quad (8.32)$$

$$\frac{dC_G}{dt} = -K_3 \frac{dP_X}{dt} - \frac{dP_{Ga}}{dt} - K_4 P_X, \quad t = 0, \quad C_G = 1; \quad (8.33)$$

$$\frac{dC_{O_2}}{dt} = k_L a(1 - C_{O_2}) - K_5 \frac{dP_X}{dt} - \frac{c_G^{max}}{2c_{O_2}^*} \frac{dP_{Ga}}{dt} - K_6 P_X, \quad t = 0, \quad C_{O_2} = 1; \quad (8.34)$$

Table 18 Parameter identification

$\hat{a}$	$\tilde{a}$	$\bar{\bar{a}}$	$\bar{a}$	a
1	2	3	4	5
$\hat{\mu}_{max} = 2.427 \times 10^{-5}$	$\tilde{\mu}_{max} = 0.5843$	$\bar{\bar{\mu}}_{max} = 0.5999$	$\bar{\mu}_{max} = 0.5064$	$\mu_{max} = 0.4345$
$\hat{K}_G = 3.108 \times 10^{-5}$	$\tilde{K}_G = 1.505$	$\bar{\bar{K}}_G = 1.505$	$\bar{K}_G = 1.541$	$K_G = 0.9914$
$\hat{K}_{O_2} = 3.412 \times 10^{-5}$	$\tilde{K}_{O_2} = 0.9894$	$\bar{\bar{K}}_{O_2} = 0.9753$	$\bar{K}_{O_2} = 0.6308$	$K_{O_2} = 0.8053$
$\hat{K}_1 = 8.067 \times 10^{-6}$	$\tilde{K}_1 = -0.2120$	$\bar{\bar{K}}_1 = -0.2179$	$\bar{K}_1 = -0.2016$	$\bar{K}_1 = -0.2016$
$\hat{K}_2 = 1.555 \times 10^{-4}$	$\tilde{K}_2 = 5.051 \times 10^{-2}$	$\bar{\bar{K}}_2 = 4.998 \times 10^{-2}$	$\bar{K}_2 = 5.047 \times 10^{-2}$	$\bar{K}_2 = 5.047 \times 10^{-2}$
$\hat{K}_3 = 2.304 \times 10^{-5}$	$\tilde{K}_3 = 2.08 \times 10^{-4}$	$\bar{\bar{K}}_3 = 2.096 \times 10^{-4}$	$\bar{K}_3 = 0$	$K_3 = 1.015 \times 10^{-4}$
$\hat{K}_4 = 7 \times 035.10^{-6}$	$\tilde{K}_4 = 3.587 \times 10^{-2}$	$\bar{\bar{K}}_4 = 3.516 \times 10^{-2}$	$\bar{K}_4 = 3.692 \times 10^{-2}$	$K_4 = 3.686 \times 10^{-2}$
$\hat{k}_L a = 1.006 \times 10^2$	$\tilde{k}_L a = 1.333 \times 10^2$	$\bar{\bar{k}}_L a = 1.344 \times 10^2$	$\bar{k}_L a = 1.244 \times 10^2$	$k_L a = 1.309 \times 10^2$
$\hat{K}_5 = -7.896 \times 10^{-6}$	$\tilde{K}_5 = 6.69 \times 10^{-4}$	$\bar{\bar{K}}_5 = 6.868 \times 10^{-4}$	$\bar{K}_5 = 7.667 \times 10^{-4}$	$K_5 = 6.632 \times 10^{-4}$
$\hat{K}_6 = -8.119 \times 10^{-5}$	$\tilde{K}_6 = 1.129 \times 10^{-3}$	$\bar{\bar{K}}_6 = 1.159 \times 10^{-3}$	$\bar{K}_6 = 1.261 \times 10^{-3}$	$K_6 = 1.126 \times 10^{-3}$
$\hat{S} = 2.7831$	$\tilde{S} = 0.8657$	$\bar{\bar{S}} = 0.6586$	$\bar{S} = 0.5199$	$S = 0.3280$

after the minimization of the least-squares functions:

$$Q_{Ga}\left(\tilde{K}_1, \tilde{K}_2\right) = \min \int_0^{t_N} \left(C_{Ga} - P_{Ga}\right)^2 dt, \qquad (8.35)$$

$$Q_{G}\left(\tilde{K}_3, \tilde{K}_4\right) = \min \int_0^{t_N} \left(C_{G} - P_{G}\right)^2 dt, \qquad (8.36)$$

$$Q_{O_2}\left(\tilde{k}_L a, \tilde{K}_5, \tilde{K}_6\right) = \min \int_0^{t_N} \left(C_{O_2} - P_{O2}\right)^2 dt. \qquad (8.37)$$

In Eqs. (8.8.35–8.8.37) the experimental data values are replaced by their polynomial approximations. The zeroth-order approximations obtained for the parameter values are shown in Table 18 (third column).

8.2.3 First-Order Approximations of the Parameters

The first step is to solve the equations for the biomass growth and gluconic acid production,

$$\frac{dX}{dt} = \tilde{\mu}_{max} \frac{P_G}{\tilde{K}_G + P_G} \frac{P_{O_2}}{\tilde{K}_{O_2} + P_{O_2}} X;$$

$$\frac{dC_{Ga}}{dt} = K_1 \frac{dX}{dt} + K_2 X; \quad t = 0, \quad X = X_{(0)}, \quad C_{Ga} = 0, \tag{8.38}$$

and to minimize (8.8.35):

$$Q_{Ga}(\bar{K}_1, \bar{K}_2) \rightarrow \min. \tag{8.39}$$

The parameter values obtained $\bar{K}_1, \bar{K}_2$ are the first-order approximation in the identification problem solution.

The next steps are consecutive solutions of the problems:

$$\begin{aligned}
&\frac{dX}{dt} = \tilde{\mu}\max \frac{C_G}{\tilde{K}_G + C_G} \frac{P_{O_2}}{\tilde{K}_{O_2} + P_{O_2}} X, \quad \frac{dC_{Ga}}{dt} = \bar{K}_1 \frac{dX}{dt} + \bar{K}_2 X, \\
&\frac{dC_G}{dt} = -K_3 \frac{dX}{dt} - \frac{dC_{Ga}}{dt} - K_4 X; \quad t = 0,\ X = X_{(0)},\ C_{Ga} = 0,\ C_G = 1; \\
&Q_G(\bar{K}_3, \bar{K}_4) \rightarrow \min,
\end{aligned} \tag{8.40}$$

$$\begin{aligned}
&\frac{dC_G}{dt} = -\bar{K}_3 \frac{dX}{dt} - \frac{dC_{Ga}}{dt} - \bar{K}_4 X, \quad \frac{dC_{O_2}}{dt} = k_L a(1 - C_{O_2}) \\
&- K_5 \frac{dX}{dt} - \frac{c_G^{\max}}{2c_{O_2}^*} \frac{dC_{Ga}}{dt} - K_6 X; \\
&t = 0,\ X = X^{(0)},\ C_{Ga} = 0,\ C_G = 1,\ C_{O_2} = 1, \quad Q_{O_2}(\bar{k}_L a, \bar{K}_5, \bar{K}_6) \rightarrow \min.
\end{aligned} \tag{8.41}$$

The last step is to solve model equations (8.8.41), where the parameter values obtained for the oxygen consumption are replaced,

$$k_L a = \bar{k}_L a, \quad K_5 = \bar{K}_5, \quad K_6 = \bar{K}_6, \quad \tilde{\mu}_{\max} = \mu_{\max}, \quad \tilde{K}_G = K_G, \quad K_{O_2} = K_{O_2}, \tag{8.42}$$

and the biomass parameter values are calculated by minimization of the least-squares function:

$$Q_X(\bar{\mu}_{\max}, \bar{K}_G, \bar{K}_{O_2}) \rightarrow \min. \tag{8.43}$$

The values of the parameters obtained are the first-order approximation (see Table 18, fourth column).

The first-order approximation of the parameter values permits us to obtain the exact parameter values. For this purpose, model equations (8.8.41) have to be solved and the exact parameter values will be obtained using the minimization of the least-squares function (see Table 18, fifth column):

$$
\begin{aligned}
&Q(\mu_{\max}, \mathrm{K}_G, \mathrm{K}_{\mathrm{O}_2}, \mathrm{K}_1, \mathrm{K}_2, \mathrm{K}_3, \mathrm{K}_4, \mathrm{k}_\mathrm{L}a, \mathrm{K}_5, \mathrm{K}_6) \\
&\quad = \sum_{n=1}^{N}\left(X_n - X_n^{(e)}\right)^2 + \sum_{n=1}^{N}\left(C_{G_n} - C_{G_n}^{(e)}\right)^2 + \sum_{n=1}^{N}\left(C_{Ga_n} - C_{Ga_n}^{(e)}\right)^2 \\
&\quad + \sum_{i=1}^{N}\left(C_{O_{2n}} - C_{O_{2n}}^{(e)}\right)^2,
\end{aligned} \tag{8.44}
$$

where N = 8 is the experimental number of data.

The first-order approximations of the parameter values are used as initial approximations in the minimization procedure (8.8.44). In Table 18 consecutive approximations of the model parameters values in (8.8.30) and (8.8.32–8.8.34) and model error variances (S) are presented.

Arbitrary parameters values of the model equations (8.8.30, 8.8.32–8.8.34) are used as initial values of the parameters for the model parameter identification in (8.8.30) and (8.8.32–8.8.34) and the solution is shown in the first column in Table 18. Using the same initial values of the parameters to obtain the zeroth-order approximation (the zeroth hierarchical level) leads to the parameters value in the second column.

The parameter values in the second column are used as initial values to obtain the parameters in model (8.8.30, 8.8.32–8.8.34) (see the third column), or to obtain the first-order approximation (see column 4). The parameters values in column 4 are initial values to obtain the exact parameters values (fifth column).

The comparison of the model error variances (see the last row in Table 18) shows a consecutive variance decrease which is a result of the proposed hierarchical approach. A comparison of the calculated biomass, gluconic acid, glucose, and oxygen concentrations (using the exact parameter values) with the experimental data is given in Figs. 19, 20, 21 and 22.

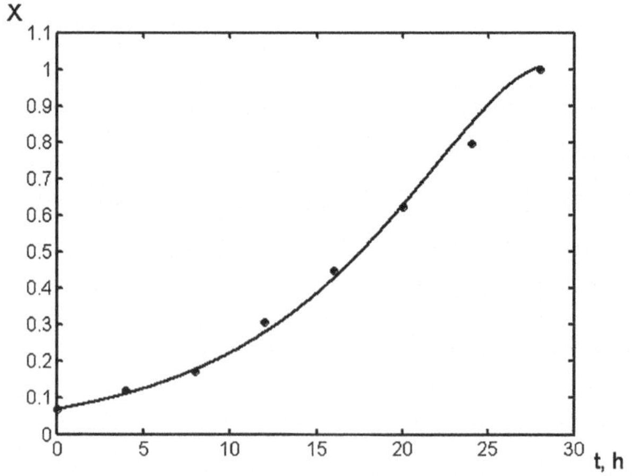

Fig. 19 Comparison of the calculated values and experimental data for biomass dimensionless concentration

Fig. 20 Comparison of the calculated values and experimental data for gluconic acid dimensionless concentration

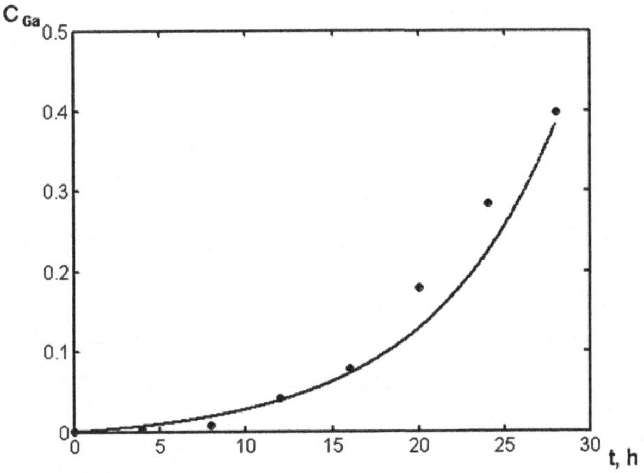

Fig. 21 Comparison of the calculated values and experimental data for glucose dimensionless concentration

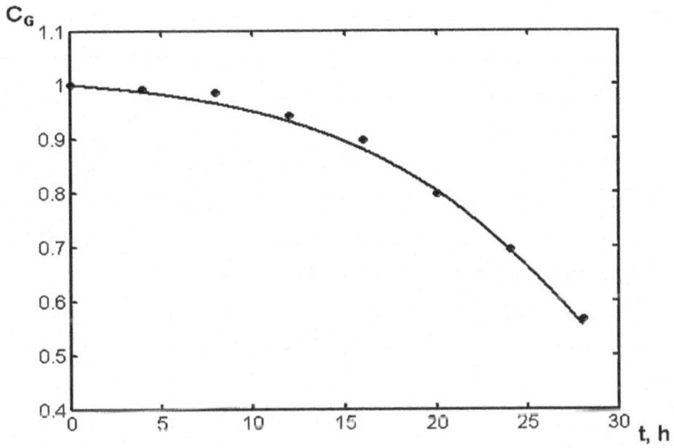

Fig. 22 Comparison of the calculated values and experimental data for oxygen dimensionless concentration

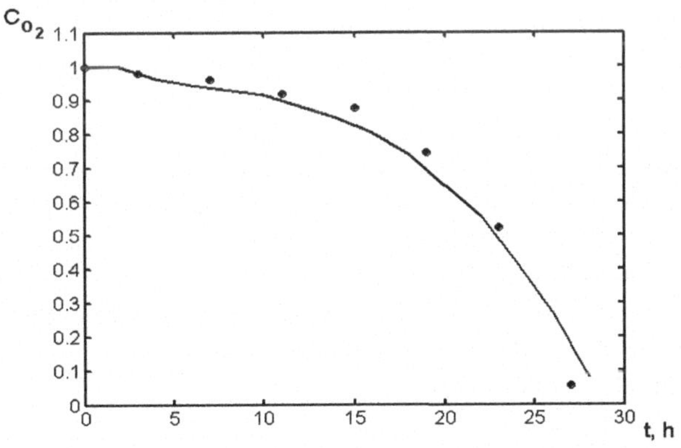

9 Experiment Design

As was shown for modeling of a series of stages, the use of experimental data is required. Most of these data are connected to the identification of the parameters of the model and represent data of the objective function at different regimes (combinations of factors values).

The analysis of significance of the parameters requires the determination of the error of the current experiment, i.e., several measurements of the objective function (five to ten times) per regime.

The model adequacy is determined on the basis of the experiments, as used for determination of the parameter assessments as well the analysis of their significance. In cases where there is a lack of experimental data for several measurements for the same regime, the suitability of the model can be determined by single measurements of the objective function in different regimes if it depends linearly on the model parameters.

One of the fundamental problems of modeling related to the required experimental data is that generally the accuracy of the data obtained increases with increasing number of experiments and that raises the cost of modeling. In this sense a problem emerges for the determination of the admissible minimum number of experiments.

Another underlying problem of modeling is the condition for the selection of the regimes in which the objective function is measured so that the influence of the experimental data error on the accuracy of the results obtained in the calculation of the parameter values is decreased to a minimum.

As a whole, the modeling needs the development of optimal experimental plans, which permit the maximum accuracy to be achieved for the minimum number of experiments.

9.1 Experimental Plans of Modeling

The experimental plans can be considered as matrixes:

$$X = \left\| x_{i,j} \right\|, \quad i = 1, \ldots, m, \quad j = 1, \ldots, k. \tag{9.1}$$

where m is the number of factors (independent variables) and k is number of the levels which each of the factors can take. If we suppose that all of factors can take an equal number of levels, each row of the matrix (8.9.1) represents the experimental plan of one experiment, and the maximum number of rows N is obtained for full combination between different factors and their levels. So the experimental plan obtained represents [6] full classification. For three factors of two levels it directly follows that

$$\begin{vmatrix} x_{11} & x_{21} & x_{31} \\ x_{11} & x_{21} & x_{32} \\ x_{11} & x_{22} & x_{31} \\ x_{11} & x_{22} & x_{32} \\ x_{12} & x_{21} & x_{31} \\ x_{12} & x_{21} & x_{32} \\ x_{12} & x_{22} & x_{31} \\ x_{12} & x_{22} & x_{32} \end{vmatrix} \tag{9.2}$$

The order of the matrix (8.9.1) rises as fast as the increase of the number of factors and their levels. The minimum number of rows N_{min} needs to be determined for different tasks which will be included in the current experimental plan. For this purpose, such a combination of N_{min} rows should be selected for a given maximum number of rows in (8.9.1) for which the accuracy of the results obtained will be a maximum.

9.2 Parameter Identification

The solution of the inverse identification problem places very important requirements on the optimal experiment design. The theoretical models and the model theories use preliminary and exactly identified parameters; because of that they do not need experimental plans.

The diffusion-type models (as the models of the elementary processes) have parameters which are defined on the basis of experimental data for the objective function by solution of inverse incorrect (most frequently) problems. In these models the development of optimal experimental plans can be done so that the maximum accuracy in the identification of the model parameters by means of the minimum number of iterative solutions of the inverse identification problem is obtained. In these cases the optimal plans represent an aggregation of the following quantities: number of points at which the objective function is measured; number of repeated measurements; the coordinates of the points. The determination of the optimal experimental plans depends materially on the mathematical structure of the model through which they will be considered.

Let us consider [8] the problem of determination of the heat conductivity in the following model:

$$\frac{\partial T}{\partial \tau} = a\frac{\partial^2 T}{\partial x^2}; \quad T(x,0) = T_0(x), \quad 0 \le x \le \mathrm{b};$$
$$T_{\mathrm{x}}(0,\tau) = g_1(\tau), \quad 0 \le \tau \le \tau_{\mathrm{m}}; \quad T_{\mathrm{x}}(\mathrm{b},\tau) = g_2(\tau), \quad 0 \le \tau \le \tau_{\mathrm{m}}, \quad T_{\mathrm{x}} = \frac{\partial T}{\partial x}. \tag{9.3}$$

For determination of a in (8.9.3) it is necessary for the temperature variation to be known (e.g., from experimental data) at a given point:

$$T(\mathrm{d},\tau) = f(\tau). \tag{9.4}$$

The determination of a can be realized by means of the functional of the misfit minimization:

$$J = \frac{1}{2}\int_0^{\tau_\mathrm{m}} [T(\mathrm{d},\tau;a) - f(\tau)]^2 d\tau. \tag{9.5}$$

For this purpose the iterative procedure [8] is used:

$$a_{\mathrm{n}+1} = a_\mathrm{n} + \Delta a_\mathrm{n}, \quad \mathrm{n} = 0, 1, \ldots, \tag{9.6}$$

where

$$\Delta a_\mathrm{n} = -\frac{\int_0^{\tau_\mathrm{m}} [T(\mathrm{d},\tau;a_\mathrm{n}) - f(\tau)]\vartheta(\mathrm{d},\tau;a_\mathrm{n})d\tau}{\int_0^{\tau_\mathrm{m}} [\vartheta(d,\tau;a_\mathrm{n})]^2 d\tau}. \tag{9.7}$$

and ϑ is the sensitivity function,

$$\vartheta(\mathrm{d},\tau,a) = \frac{\partial T(x,\tau,a)}{\partial a}\Big|_{\mathrm{x=d}}. \tag{9.8}$$

It is determined from the problem:

$$\frac{\partial \vartheta}{\partial \tau} = a\frac{\partial^2 \vartheta}{\partial x^2} + \frac{\partial^2 T}{\partial x^2}; \quad \vartheta(x,0) = 0, 0 \le x \le \mathrm{b};$$
$$T_\mathrm{x}(0,\tau) + a\vartheta_x(0,\tau) = 0, \quad 0 \le \tau \le \tau_\mathrm{m};$$
$$T_\mathrm{x}(\mathrm{b},\tau) + a\vartheta_\mathrm{x}(\mathrm{b},\tau) = 0, \quad 0 \le \tau \le \tau_\mathrm{m}; \quad \vartheta_\mathrm{x} = \frac{\partial \vartheta}{\partial x}. \tag{9.9}$$

Problem (8.9.3) can be a solution for a given case [8]:

$$\bar{a} = 1.25.10^{-7} m^2/s, \quad \overline{g_1} = 3.10^5 \text{ K/m}, \quad \overline{g_2} = 0, \quad T_0 = 0, \quad \mathrm{b} = 0.008 \text{ m}, \tag{9.10}$$

where $\bar{a}$ is the "exact" value of a in (8.9.3). From the solution obtained for (8.9.3) the function $f(\tau) = T(d,\tau;\bar{a})$ for values of d=0.001,002,…,0.005 m and $\tau_\mathrm{m} = 60$s can be determined. The functions obtained $f(\tau)$ for different **d** will be used as "experimental data" for iterative solution of the inverse problem. For an initial approximation in (8.9.6) and (8.9.7) we use $a_0 = 0.25.10^{-7}\mathrm{m}^2/\mathrm{s}$ and $f(\tau) = T(0, 001, \tau, a_\mathrm{n})$, respectively. The iterations continue while

$$\varepsilon = \frac{\bar{\bar{a}} - a_{n+1}}{\bar{a}} \leq 10^{-4}. \tag{9.11}$$

The number of iterations in this case does not depend on errors in model (8.9.3) or the errors in the "experimental" determination of $f(\tau)$ (8.9.4) and it determines only the structure of the operator (algorithm) giving the opposite problem solution.

If the opposite problem is solved for more than several values of d (0.002, 0.003, 0.004, 0.005) it can be seen that condition (8.9.11) is achieved in a different number of iterations [8] and for d=0.003 m the number of iterations is minimal.

In the example considered the optimal experimental plan represents two numbers (1, 0.003), i.e., a measurement of the temperature (as a function of time) for x=0.003 m. The number of iterative procedures depends on the value of $\bar{a}$. The reason for that is the nonlinearity of the dependency of the objective function on the parameters. That means that above-mentioned algorithm does not give us the possibility to determine the optimal plan, as $\bar{a}$ is unknown. A possible way to cope with this situation is the approach of consecutive design. For this purpose, first, the optimal plan as described above is determined, and $\bar{a}$ is assigned some value on the basis of preliminary information. As a result, an optimal plan d_1 is obtained. This plan is locally optimal and it will be used as a first approximation to find the final optimal plan. In plan d_1 (x=d_1) an experiment is carried out and the temperature is determined as a function of time—$f_1(\tau)$. With this function the inverse problem (8.9.3) is solved and $\overline{a_1}$ is determined. The new value of $a = \bar{a}$ is used for an iterative determination of the new local-optimal plan d_2, etc. while the following condition is satisfied:

$$\frac{\overline{a_{k+1}} - \overline{a_k}}{\overline{a_k}} \leq \varepsilon. \tag{9.12}$$

The problem described above becomes considerably more complicated [8, 44] if there are more parameters in the model and especially when the coefficients of the heat conductivity and the specific heat depend on the temperature. In these cases, the optimality of the experimental plan is determined not only by the solution accuracy and the number of iterative procedures, but also by the condition for unity solution of the inverse problem.

Parameter identification in the similarity criteria models (after the logarithm procedure has been accomplished) and regression models is realized in the same way. In these cases optimal plans are used, which aim to minimize the influence of the experimental error of the objective function on the accuracy of parameter identification. Different criteria for optimality are used.

Most frequently the criterion *D*-optimality is used, in which the experimental plan is determined so as to maximize the determinant of the information matrix of the plan *A*. It can be seen from (8.6.10) that it leads to the minimization of parameter sensitivity as regards the experimental errors in the objective function, i.e., the *D*-optimal experimental plans have a "regularization" effect on the inverse identification problem solution in the regression models.

Other criteria for optimality are G-optimality (minimization of the maximum dispersion of the objective function, calculated through the model), A-optimality (minimization of the average dispersion of the parameters), and orthogonality (nondiagonal elements in matrixes A and A^{-1} are zero, i.e., the model parameters are not correlated).

9.3 Significance of Parameters

The significance analysis of the parameters is realized in two different ways in the different types of models, which is related to different experimental plans being used. In the theoretical models and the model theories, the parameters are not determined by means of experimental data, and significance analysis of the dimension model parameters is not required. Here it is essential to determine the significance of the dimensionless model parameters in the models with generalized variables, which was considered in the quantity analysis of the models in Chap. 4). A similar possibility exists in the case of diffusion-type models.

In the regression models and similarity criteria models for significance analysis of the parameters, a random quantity (5.3.98) is used, from which it can be seen that the experimental data from several measurements of the objective function in same regime are required. Moreover, the analysis method using the Student criterion requires all the nondiagonal elements of the matrix $C = A^{-1}$ to be zero, i.e., the usage of orthogonal experimental plans is needed.

9.4 Adequacy of Models

The difference in the adequacy analysis methods for different types of models leads to different requirements for the experimental plans.

The theoretical models and the model theories need different types of experimental data for determination of the dispersion error of the model and the experimental data. The data correctness for error determination of the model requires the data to be obtained at a sufficient number of the factor levels so that the area of variation of the independent variables is captured maximally uniformly. This results in a great number of combinations between the levels of all the factors, and because of that random combinations should be selected, i.e., randomized plans should be used.

The diffusion-type models can be used in the adequacy analysis of the same experimental data obtained for the purposes of parameter identification.

A general property of theoretical models, model theories, and diffusion-type models is that their objective function, as a rule, depends nonlinearly on their parameters. For this reason it is not possible to prove the suitability of these models through the coefficient of multiple correlations. In all cases the experimental data of multiple measurements in one regime are needed for the dispersion of the experimental error to be obtained and through it the model adequacy to be verified.

Regressions and similarity criteria models are most frequently linear ones regarding their parameters, and because of that the experimental data used for parameter identification are sufficient for the model adequacy analysis. If there is a lack of data of multiple measurements for one regime, the adequacy verification could be replaced by suitability verification.

9.5 Randomized Plans

Randomized plans are used for solution of a wide range of statistical analysis problems and especially in dispersion analysis, where different factors influencing the dispersion of a given random quantity are analyzed. In the modeling it is very important at a previous stage to reject those factors which do not influence the objective function, i.e., the effect is similar to the effect of the random experimental errors. For this purpose sufficient equality between the dispersion of the objective function caused by the variation of the analyzed factor on different levels and the dispersion caused by random disturbances of the objective function must be proven. The quality of this analysis increases as the interval in which the factor levels analyzed is located is enlarged and the number of these levels increases. It can be seen directly from (8.9.2) that an increase in the number of factors and levels leads to a large increase in the number of experiments needed.

A similar problem emerges in the analysis of the dispersion in the regression analysis, where the equality of the error dispersion of the model and the experimental error sets the condition of model adequacy. Here also experimental values of the objective function in the variation of a great number of factors of great number of levels for each of them are needed.

A possible way to solve the problems described above is the usage of *fully randomized plans*. For this purpose the full classification (8.9.2) is used and from it a reasonable number of experimental plans (rows of matrix 8.9.2) are taken quite randomly. So, for example, if the row numbers in (8.9.2) are ordered randomly in a generator of random values in the sequence (4, 1, 5, 2, 6, 3, 8, 7) and we decide to use a plan of three of components, from (8.9.2) we obtain a fully randomized plan, using rows 4, 1, and 5:

$$\begin{matrix} x_{11} & x_{21} & x_{32} \\ x_{11} & x_{22} & x_{32} \\ x_{12} & x_{21} & x_{32} \end{matrix} \qquad (9.13)$$

In many cases it is imposed for some considerations that the plan is responsible for some conditions (e.g., to contain elements x_{11}). Then the randomization accomplished between these rows, which contain x_{11}, i.e., between the first four plans (rows) in (8.9.2) and the *randomized block plan* obtained, is as follows:

$$\begin{matrix} x_{11} & x_{21} & x_{31} \\ x_{11} & x_{22} & x_{31} \\ x_{11} & x_{22} & x_{32} \end{matrix} \tag{9.14}$$

In three-factor experiments with same number of levels for all of them, *Latin squares* are used. This is a square table $n \times n$ of n Latin letters, which are not repeated in a row and a column.

Let us consider three factors with four levels for each of them:

First factor of levels 1, 2, 3, 4

Second factor of levels I II, III, IV

Third factor of levels A, B, C, D

From the data presented above we can compose a standard (canonical) Latin square (Table 19). Each cell of this square is an experimental plan and it is related to an experimental value of the objective function for a given combination of factors and levels. So, for example, 2, III, D corresponds to an objective function obtained in an experiment, in which the first factor has level x_{12}, the second one has level x_{23}, and the third one has level x_{34}. The full classification is represented by a matrix of $4^3 = 64$ rows, whereas the Latin square offers a plan for 20 experiments. This experimental plan for 20 experiments can be obtained quite randomly if two random series of numbers are taken:

$$\begin{matrix} 3, & 2, & 4, & 1 \\ 1, & 3, & 4, & 2, \end{matrix} \tag{9.15}$$

The rows in the Table 19 of the standard Latin square can be rearranged to conform with the first row of the numbers in (8.9.15), i.e., the first row gets the values of the third row, the second row gets the values of the second row, the third row gets the values of fourth row, and the fourth row gets the values of the first row:

$$\begin{matrix} C & D & A & B \\ B & C & D & A \\ D & A & B & C \\ A & B & C & D \end{matrix} \tag{9.16}$$

In the same way the columns of (8.9.16) are rearranged in accordance with the numbers in the second row in (8.9.15):

Table 19 Canonical Latin square

	I	II	III	IV
1	A	B	C	D
2	B	C	D	A
3	C	D	A	B
4	D	A	B	C

$$\begin{matrix} C & A & B & D \\ B & D & A & C \\ D & B & C & A \\ A & C & D & B \end{matrix} \quad (9.17)$$

The Latin square (8.9.17) offers an experimental plan for 20 experiments, chosen quite randomly among 64 feasible ones.

Other similar plans [6] such as Greek–Latin squares (for four-factor experiments), Latin rectangles, and parallelepipeds are known.

9.6 Full and Fractional Factor Experiment

The experimental data obtained for the purposes of the inverse identification problems need the use of optimal experimental plans, which permit the parameter assessments to be obtained with the minimum error and the maximum sustainability regarding the experimental errors of the input data (the experimental data for the objective function). In the theoretical models, model theories, and diffusion-type models it was shown that these optimal experimental plans are made specifically for each of the models. In the regression models (and similarity criteria models after the logarithmic procedure has been accomplished) plans for different types of models can be used [6, 43] such as full factor experiment, fractional replicas, and compositional plans of second and third rows. In these cases the use of catalogues of consequently generated plans [45] is suitable.

The largest group of regression models are characterized by only the first powers of the factors and multiplication between them being involved. For m factors the regression model can have 2^m parameters. For example, for $m = 2$ the model is as follows:

$$y = \mathrm{b}_0 + \mathrm{b}_1 x_1 + \mathrm{b}_2 x_2 + \mathrm{b}_{12} x_1 x_2. \quad (9.18)$$

Let us consider the most frequently occurring cases when for the parameter identification in (8.9.18) two levels of the factors are used: +1 and −1.

A full factor experiment of two levels of the factors represents an experimental plan obtained through a full combination of the levels of all the factors (full classification). For model (9.18) the full factor experiment is represented by Table 20.

Table 20 Full factor experiment for (9.18)

No.	x_1	x_2	y
1	1	1	y_1
2	−1	1	y_2
3	1	−1	y_3
4	−1	−1	y_4

Table 21 Enlarged matrix of the plan for (9.18)

No.	x_0	x_1	x_2	x_1x_2	y
1	1	1	1	1	y_1
2	1	−1	1	−1	y_2
3	1	1	−1	−1	y_3
4	1	−1	−1	1	y_4

Table 22 Enlarged matrix of the plan for (9.19)

No.	x_0	x_1	x_2	x_3	x_1x_2	x_1x_3	x_2x_3	$x_1x_2x_3$	y
1	1	1	1	1	1	1	1	1	y_1
2	1	−1	1	1	−1	−1	1	−1	y_2
3	1	1	−1	1	−1	1	−1	−1	y_3
4	1	−1	−1	1	1	−1	−1	1	y_4
5	1	1	1	−1	1	−1	−1	−1	y_5
6	1	−1	1	−1	−1	1	−1	1	y_6
7	1	1	−1	−1	−1	−1	1	1	y_7
8	1	−1	−1	−1	1	1	1	−1	y_8

The four experiments in Table 20 permit through the least-squares method four coefficients to be determined in (9.18). From Table 20 the enlarged matrix of the plan can be obtained (Table 21).

It can be seen from Table 21 that if we substitute column x_1x_2 by x_1^2 the matrix plan degenerates, as the first column and the second one become the same. An analogous result is obtained if we introduce x_1^3, x_1^4, etc., i.e., the full factor experiment of two levels of the factors can be used only for models which do not contain powers of higher factors.

In the literature it has been proven [6, 43] that the full factor experiment is an experimental plan which is orthogonal, *D*-optimal, *G*-optimal, and *A*-optimal.

The matrix for the plan in the full factor experiment of two levels of the factors for different numbers of the factors *m* can be obtained by the last rule. The first column contains only ones, owing to $x_0 = 1$. In the second column the change of sign is done for each of 1 (2^0) row, in the third column for each of 2 (2^1) rows, in the fourth one for each of 4 (2^2) rows, etc., i.e., in the *k*th column for each of (2^{k-2}) rows. This rule holds for the $(m+1)$th column, and the following columns are obtain directly through proper multiplication of the first $(m+1)$ columns.

For example, for the model

$$y = \mathrm{b}_0 + \mathrm{b}_1x_1 + \mathrm{b}_2x_2 + \mathrm{b}_3x_3 + \mathrm{b}_{12}x_1x_2 + \mathrm{b}_{13}x_1x_3 + \mathrm{b}_{23}x_2x_3 + \mathrm{b}_{123}x_1x_2x_3 \quad (9.19)$$

the enlarged matrix of the plan is shown in Table 22. In the general case, the number of experiments N in the full factor experiment of m factors of two levels is $\mathrm{N} = 2^{\mathrm{m}}$.

It can be seen from (8.9.19) that the number of experiments in the full factor experiment can be larger than the number of model parameters, i.e., from the number of experiments needed, sufficient for determination of the unknown

Table 23 Matrix of the plan for (9.20)

No.	x_0	x_1	x_2	x_3	y
1	1	1	1	1	y_1
2	1	−1	1	−1	y_2
3	1	1	−1	−1	y_3
4	1	−1	−1	1	y_4

parameters by means of the least-squares method. In these cases is possible to use *fractional replicas*. So, for example, the model

$$y = b_0 + b_1x_1 + b_2x_2 + b_3x_3 \tag{9.20}$$

has four parameters, which can be identified from a system of four equations, i.e., of four experimental values of y, but the full factor experiment contains eight values.

Let us consider the matrix of the plan in Table 23. It can be seen directly from Table 23 that it contains half of the experiments in Table 22 (rows 1, 6, 7, 4). On the other hand, Table 23 coincides completely with Table 21, i.e., the experimental plan in Table 23 has the same properties (*A*-, *D*-, and *G*-optimality, orthogonality) as the experimental plan in Table 21. In this sense the experimental plan in Table 23 represents a *semireplica* of the experimental plan in Table 22 and contains $N = 2^{m-1}$ experiments.

For a great number of factors one-fourth replica ($N = 2^m/4 = 2^{m-2}$), one-eighth replica ($N = 2^m/8 = 2^{m-3}$), etc. can be used. In the general case of the experiments in the fractional replica (fractional factor experiment) we have

$$N = 2^{m-t}, \tag{9.21}$$

where t is a power of replica fractionality.

The composition of the experimental plan in a fractional factor experiment will be shown by the following example.

Let us suppose that the parameters are searched for in the model

$$y = b_0 + \sum_{i=1}^{6} b_1x_1. \tag{9.22}$$

Apparently the minimum number of experiments required is $N_{min} = 7$, and the full factor experiment contains 64 experiments. A fractional replica would be searched for which does not exceed in the number of the experiments a given maximum number, for example, $N_{max} = 10$.

From the condition

$$7 \le 2^{m-t} \le 10 \tag{9.23}$$

we obtains $t = 3$ because $2^{6-3} = 8$.

The factors are divided into main ones, whose number is $m - t = 6 - 3 = 3$, and additional ones. Let us suppose that for the main factors we have accepted the

Table 24 Fractional factor experiment for (9.22)

No.	x_0	x_1	x_2	x_3	x_1x_2	x_1x_3	$x_1x_2x_3$	y
1	1	1	1	1	1	1	1	y_1
2	1	−1	1	1	−1	−1	−1	y_2
3	1	1	−1	1	−1	1	−1	y_3
4	1	−1	−1	1	1	−1	1	y_4
5	1	1	1	−1	1	−1	−1	y_5
6	1	−1	1	−1	−1	1	1	y_6
7	1	1	−1	−1	−1	−1	1	y_7
8	1	−1	−1	−1	1	1	−1	y_8

first three. For them the full factor experiment can be composed (the first three columns in Table 24). The additional three factors equalize the multiplications of the highest power of the main ones:

$$x_7 = x_1x_2x_3, \quad x_6 = x_1x_3, \quad x_5 = x_1x_2. \tag{9.24}$$

Expressions (8.9.24) represent generating relations. Table 24 represents the fractional factor experiment, which contains eight times fewer experiments than the full factor experiment, but has kept the properties of the it. In this way the fractional factor experiment with many factors leads to a full factor experiment with a few ones, keeping the optimality of the full factor experiment.

9.7 Compositional Plans

In the cases when the process depends nonlinearly on the factors, in the regression models higher powers of the factors appear and the matrix of the full factor experiment degenerates owing to the appearance of equal columns. The same also holds for all fractional replicas which contain some of the rows of the matrix of the full factor experiment.

Degeneration of the matrix of the full factor experiment (in the second, third, etc., powers of the factors appearing in the model) can apparently to avoided if only one row is added, which can make the equal columns different ones. In this way the compositional plan is made. Apparently from the fractional replicas also the compositional plans can be made.

In the general case, the number of experiments in the compositional plan is

$$\mathrm{N} = 2^{\mathrm{m-t}} + 2\mathrm{m} + \mathrm{N}_0 \tag{9.25}$$

where 2m is the number of additional experiments noted with "starry" points, for which each of the factors are assigned two values ($\pm\alpha$) for zero values of the other factors. The experiments N_0 are observations in the center of the plan ($x_1 = x_2 = \cdots = x_m = 0$).

Table 25 Optimal compositional plan of second order

No.	x_1	x_2	x_3	No.	x_1	x_2	x_3
1	1	1	1	8	−1	−1	−1
2	−1	1	1	9	1	0	0
3	1	−1	1	10	1	0	0
4	−1	−1	1	11	0	1	0
5	1	1	−1	12	0	−1	0
6	−1	1	−1	13	0	0	1
7	1	−1	−1	14	0	0	−1

Optimal compositional plans are developed [6] through following rule:

1. The plan of the factor experiment of is composed of a number experiments $N = 2^{m-t}$, where $t = 0$ at $m \leq 4$ and $t = 1$ at $m \succ 4$.
2. $2m$ "starry" points are added at $\alpha = 1$.
3. Point $x_1 = x_2 = 0$ is added at $m = 2$. At $m \triangleright 2$, $N_0 = 0$.

In this way the optimal compositional plan for $m = 3$ is shown [6] in Table 25. This is the optimal compositional plan of second order, which has the useful property that if in the verification of the model adequacy with a full and a fractional factor experiment inadequacy is proven, then the addition of $2m$ new experiments permits the verification of the adequacy of the new model, where nonlinear effects are accounted for. When the third powers of the factors are used, the composition of the plans of the third row is needed, which can be found in proper catalogues [45].

10 Examples

10.1 Regression Models

In many cases, the use of regression models is very convenient. In the general case (see 2.2.5) the regression models have the form

$$y = b_0 + \sum_{i=1}^{I} b_i f_i(x), \tag{10.1}$$

where y is an objective function, $b_i(i = 0, \ldots, I)$ are model parameters, $f_i(x)(i = 0, \ldots, I)$ are linear or nonlinear functions, and $x = x_k(k = 0, \ldots, K)$ are independent variables (regime parameters).

The parameter identification uses experimental data for objective function $\hat{y}_n$ for different regimes x_n, $n = 1, \ldots, N$, where N is the number of the experiments. The least-squares function has the form

$$Q(\mathrm{b}_0, \mathrm{b}_1, \ldots, \mathrm{b}_\mathrm{I}) = \sum_{\mathrm{n}=1}^{\mathrm{N}} \left[\hat{y}_\mathrm{n} - \mathrm{b}_0 - \sum_{\mathrm{i}=1}^{\mathrm{I}} \mathrm{b}_\mathrm{i} f_\mathrm{i}(x_\mathrm{n}) \right]^2. \tag{10.2}$$

The iterative method [24] is used [25, 26] for parameter identification of two- and three-parameter regression models.

Let us consider the *strong nonlinear regression model*

$$y = \mathrm{b}_1 x - \mathrm{b}_2 \exp(-5x), \tag{10.3}$$

where $\bar{b}_1 = 1, \bar{b}_2 = -1$ are exact parameter values.

The parameter identification problem is solved with the help of the "experimental" data obtained by a generator for random numbers:

$$\hat{y}_\mathrm{n}^{(1)} = (0.95 + 0.1 A_\mathrm{n}) y_\mathrm{n}, \quad \mathrm{n} = 1, \ldots, \mathrm{N}, \tag{10.4}$$

where A_n are random numbers in the interval [0,1] and y_n are obtained from model (8.10.3) for $\mathrm{b}_1 = \bar{\mathrm{b}}_1 = 1,\ \mathrm{b}_2 = \bar{\mathrm{b}}_2 = -1$,

$$y_\mathrm{n} = 1 - \exp(-5 x_\mathrm{n}),\ x_\mathrm{n} = 0.01\mathrm{n}, \quad \mathrm{n} = 1, \ldots, \mathrm{N}. \tag{10.5}$$

Obviously the maximum error of the "experimental" data (6) is $\pm 5\%$.

The conventional least-squares method for parameter identification in the regression models uses the conditions for the minimum of the least-squares function $\frac{\partial Q}{\partial b_\mathrm{i}} = 0, \quad \mathrm{i} = 0, 1, \ldots, \mathrm{I}$. This is the normal set of equations and its solution is the model parameter identification of model (8.10.1). The least-squares method for parameter identification uses the determinant A of the matrix of the normal set of equations (see Sect. 2.5). For model (8.10.3) we obtain

$$\mathrm{A} = \sum_{\mathrm{n}=1}^{\mathrm{N}} x_\mathrm{n}^2 \sum_{\mathrm{n}=1}^{\mathrm{N}} \exp(-10 x_\mathrm{n}) - \left[\sum_{\mathrm{n}=1}^{\mathrm{N}} \exp(-10 x_\mathrm{n}) \right]^2. \tag{10.6}$$

For large values of A the inverse problem is correct, for small values it is incorrect, and for very small values it is essentially incorrect. If the normal set of equations has the form (for a two-parameter model):

$$\begin{aligned} \mathrm{a}_{11}\mathrm{b}_1 + \mathrm{a}_{12}\mathrm{b}_2 &= \mathrm{B}_1 \\ \mathrm{a}_{21}\mathrm{b}_1 + \mathrm{a}_{22}\mathrm{b}_2 &= \mathrm{B}_2 \end{aligned} \tag{10.7}$$

and

$$\mathrm{a}_1 = \max\{\mathrm{a}_{11}, \mathrm{a}_{12}\}, \quad \mathrm{a}_2 = \max\{\mathrm{a}_{21}, \mathrm{a}_{22}\}, \tag{10.8}$$

the scale of the determinant A leads to

$$\overline{\mathrm{A}} = \mathrm{A}_{11}\mathrm{A}_{22} - \mathrm{A}_{12}\mathrm{A}_{21}, \tag{10.9}$$

Table 26 Correct problem solutions

$N = 50$, $b_1^0 = 0.85$, $b_2^0 = -1.4$, $\bar{b}_1 = 1, \bar{b}_2 = -1$								
n	Data no.	b_1^{min}	b_2^{min}	b_1^*	b_2^*	γ	i	$\bar{A}$
1–60	1	0.9986	−0.9950	0.9989	−0.9945	0.2	90	0.8531
	2	0.9935	−1.0035	0.9932	−1.0038	0.1	164	

Table 27 Incorrect problem solutions

$\beta_0 = 0.01$; $N = 50$, $b_1^0 = 0.85$, $b_2^0 = -1.4$, $\bar{b}_1 = 1, \bar{b}_2 = -1$								
n	Data no.	b_1^{min}	b_2^{min}	b_1^*	b_2^*	γ	i	$\bar{A}$
61–80	1	1.0052	−1.2008	1.0109	−1.3441	0.2	543	0.0056
	2	0.9781	−0.6099	0.9820	−0.7050	0.2	3,830	
81–100	1	1.0060	−1.7821	1.0050	−1.3982	0.5	17	0.0014
	2	0.9767	−0.4769	0.9874	−0.5045	0.5	15,652	
101–150				$\beta_0 = 0.1$				
	2	1.0003	−3.0484	1.0500	−1.3997	1	3	0.0013
	1	0.9872	−2.8910	0.9500	−1.4000	1	5	

where

$$A_{11} = \frac{a_{11}}{a_1}, \quad A_{12} = \frac{a_{21}}{a_1} \quad A_{21} = \frac{a_{21}}{a_2}, \quad A_{22} = \frac{a_{22}}{a_2}. \tag{10.10}$$

The values of $\bar{A}$ for different intervals of x_n are shown in Table 27.

The conditions of correctness are:

$10^{-2} < \bar{A} < 1$—correct inverse problem

$10^{-3} < \bar{A} < 10^{-2}$—incorrect inverse problem

$\bar{A} < 10^{-4}$—essentially incorrect problem

In the cases $0 < x_n < 0.6$ the inverse problem is correct and a solution can be obtained as a solution of the normal set (b_1^{min}, b_2^{min}) or by minimization of the least-squares function (b_1^*, b_2^*).

The results of the solutions are shown in Table 26 for different experimental data ($0 < x_n < 0.6$).

From Table 26 it is seen that in the cases of correct inverse problems, the differences between the exact and calculated parameter values are very small. The differences between calculated values for different experimental data sets (data nos. 1 and 2 are obtained using different random number sets in 10.4) are very small too and this is a criterion for the problem correctness. In Fig. 23 $b_0 = [b_1^0, b]$ is the initial step, $b^* = [b_1^*, b_2^*]$ is the end step, and $\bar{b} = [\bar{b}_1, \bar{b}_2]$ is the exact solution. In Fig. 24 a comparison between the mathematical model and the"experimental" data for the correct inverse problem ($0 < x_n < 0.6$) is shown.

In the cases ($0.6 < x_n < 0.8$), ($0.8 < x_n < 1$), and ($1 < x_n < 1.5$) the inverse problem is incorrect, and the results of the solutions are shown in Table 27.

Fig. 23 Horizontals of the least-squares function Q and the iteration steps for $0 < x_n < 0.6dx$: *open circle* $b_0 = [0.85, -1.4]$; *filled circle* $\bar{b} = [1, -1]$; *asterisk* $b^* = [0.9989, -0.9945]$

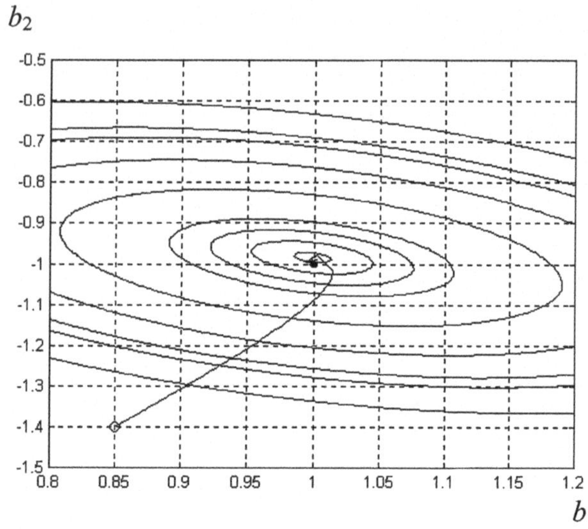

Fig. 24 Mathematical model and "experimental" data ($0 < x_n < 0.6$): *circles* $\hat{y}_n$, values of y with a maximum "experimental" error of $\pm 5\%$; *solid line* $\bar{y} = 1 - \exp(-5x)$ and *dashed line* $y = 0.9989 - 0.9945 \exp(-5x)$ coincide practically

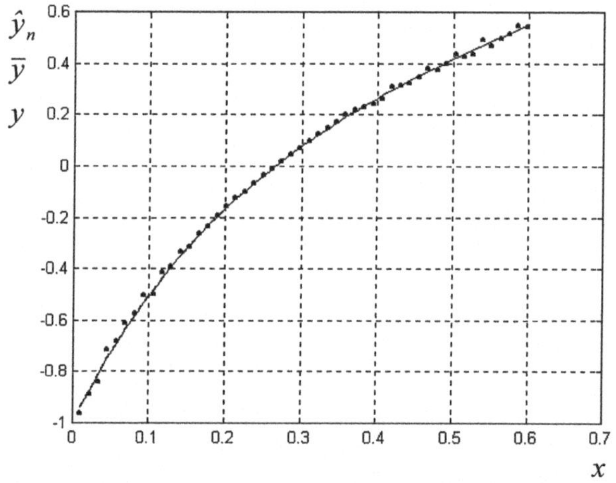

From Table 27 it is obvious that in the cases of incorrect inverse problems, the differences between the exact and the calculated parameter values are large. The differences between the calculated parameters values for different experimental data sets (data no. 1, 2, obtained with the help of different sets of random numbers) are large too and this is a criterion for the problem incorrectness.

In Figs. 25, 26 and 27 the horizontals of the least-squares function Q and the iteration steps are shown. Figures 28, 29 and 30 show comparisons between the mathematical model and the "experimental" data for the incorrect inverse problem ($0.6 < x_n < 1.5$).

In the case ($1.5 < x_n < 2$) the inverse problem is essentially incorrect and the results of the solutions are shown in Table 28.

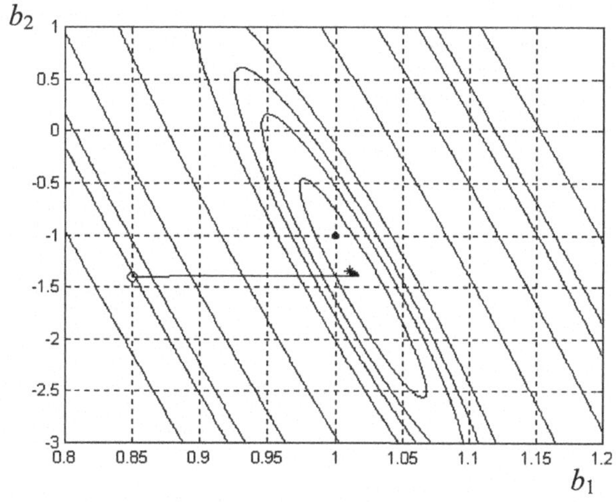

Fig. 25 The horizontals of the least-squares function Q and the iteration steps for $0.6 < x_n < 0.8$: *open circle* $b_0 = [0.85,-1.4]$; *filled circle* $\bar{b} = [1, -1]$; *asterisk* $b^* = [1.0109, -1.3441]$

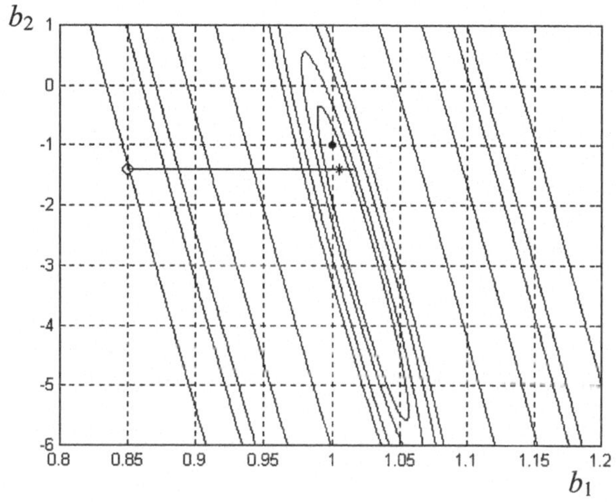

Fig. 26 The horizontals of the least-squares function Q and the iteration steps for $0.8 < x_n < 1$: *open circle* $b_0 = [0.85, -1.4]$; *filled circle* $\bar{b} = [1, -1]$; *asterisk* $b^* = [1.0050, -1.3982]$

In Table 28 it is seen that the differences between the exact and the calculated parameter values are very large. The differences between the calculated parameter values for different experimental data sets ($1.5 < x_n < 2$) are very large too and show that the inverse problem is essentially incorrect. Figure 31 shows that in this case the least-squares function does not have a minimum. A comparison between the mathematical model and the "experimental" data in this case is shown in Fig. 32.

Very often the regression models have many parameters. As an example of a three-parameter model we can consider

$$y = b_1 + b_2 x - b_3 \exp(-5x), \tag{10.11}$$

where the exact parameter values are $b_1 = 0$, $b_2 = 1$, $b_3 = -1$. The "experimental" data were obtained from (8.10.4), where y_n are obtained from (8.10.5).

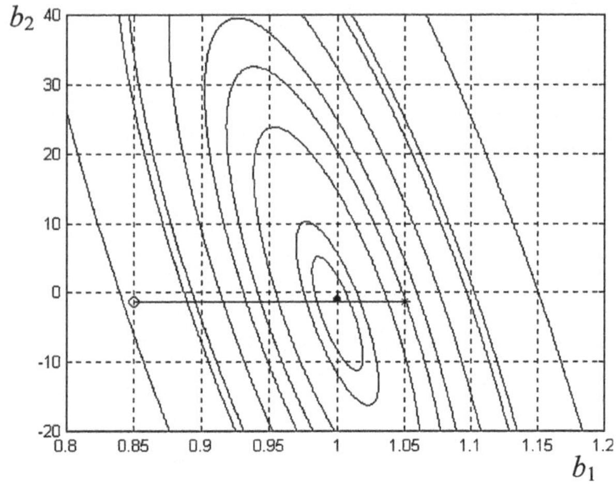

Fig. 27 The horizontals of the least-squares function Q and the iteration steps for $1 < x_n < 1.5$: *open circle* $b_0 = [0.85, -1.4]$; *filled circle* $b = [1, -1]$; *asterisk* $b^* = [1.0500, -1.3997]$

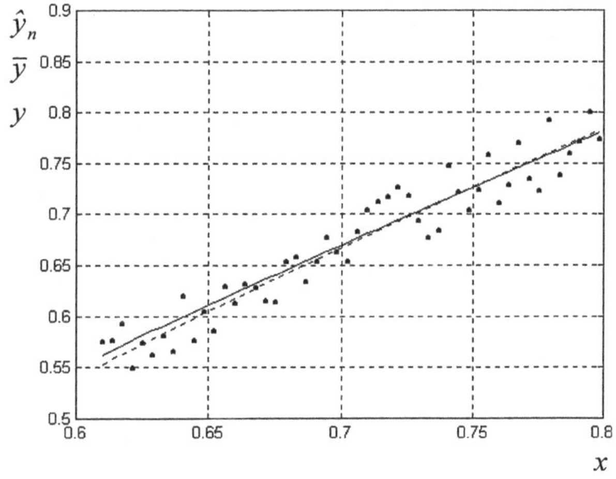

Fig. 28 Mathematical model and "experimental" data ($0.6 < x_n < 0.8$): *circles* $\hat{y}_n$, values of y with a maximum "experimental" error of $\pm 5\%$; *solid line* $\bar{y} = 1 - \exp(-5x)$;*dashed line* $y = 1.0109 - 1.3441\exp(-5x)$

The solutions of the inverse problem for different experimental data sets are shown in Table 29.

In this case we must analyze the parameter significance and the model adequacy.

10.2 Statistical Analysis of the Parameter Significance and Model Adequacy of the Regression Models

Statistical analysis of the parameter significance [21] in the case when the inverse matrix is not diagonal is very hard. That is why the SROV procedure [47] is used

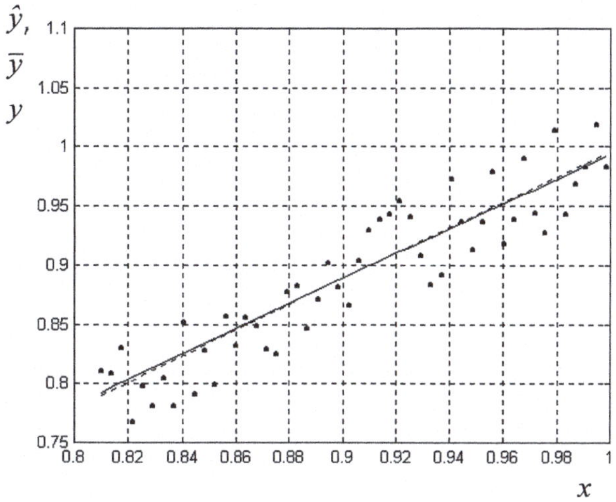

Fig. 29 Mathematical model and "experimental" data ($0.8 < x_n < 1$): *circles* $\hat{y}_n$, values of y with a maximum "experimental" error of ±5%; *solid line* $\bar{y} = 1 - \exp(-5x)$;*dashed line*$y = 1.0050$ - 1.3982exp$(-5x)$

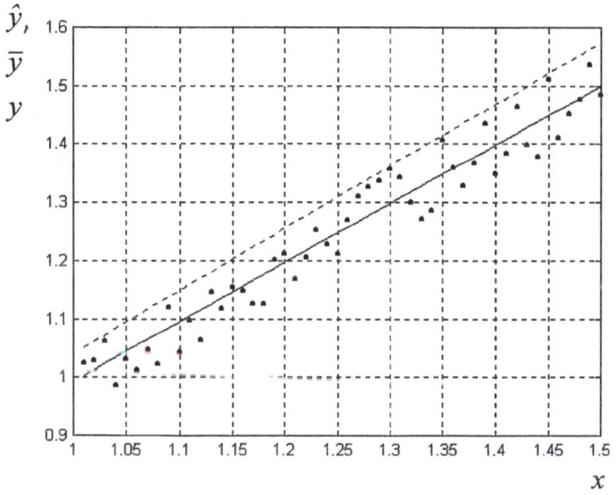

Fig. 30 Mathematical model and "experimental" data ($1 < x_n < 1.5$): *circles* $\hat{y}_n$, values of y with a maximum "experimental" error of ±5%; *solid line* $\bar{y} = 1 - \exp(-5x)$;*dashed line*$y = 1.0500$ - 1.3997exp$(-5x)$

[25, 26], where the relative errors are used because the changes of the objective function values are in a very large interval.

The results of the statistical analysis of the adequacy of two- and three-parameter models are shown in Tables 30 and 31.

The results in Figs. 30 and 32 and in Table 30 show that some of the models ($1 < x_n < 1.5$ and $1.5 < x_n < 2$) are not adequate because the experimental error is very small .

The results in the cases when the experimental error is ±10% are shown in Table 32. Comparison of Table 27 and Table 28 shows that the accuracy parameter of identification is the same, but the models are adequate (see Table 32).

The comparison of the variance of model error of both the two-parameter model and the three-parameter model (see Tables 30, 31) does not provide the possibility to analyze the parameter significance.

Table 28 Essentially incorrect problem solutions

$\beta_0 = 0.1$; $N = 50$, $b_1^0 = 0.85$, $b_2^0 = -1.4$, $\bar{b}_1 = 1$, $\bar{b}_2 = -1$								
n	Data no.	b_1^{min}	b_2^{min}	b_1^*	b_2^*	γ	i	$\bar{A}$
151–200	1	1.0004	–37.7863	1.0500	–1.4000	1	3	$6.97.10^{-5}$
	2	0.9870	69.2971	0.9500	–1.4000	1	5	

Fig. 31 The horizontals of the least-squares functions Q and the iteration steps for $1.5 < x_n < 2$: *open circle* $b_0 = [0.85, -1.4]$; *filled circle* $\bar{b} = [1, -1]$; *asterisk* $b^* = [1.0500, -1.4000]$

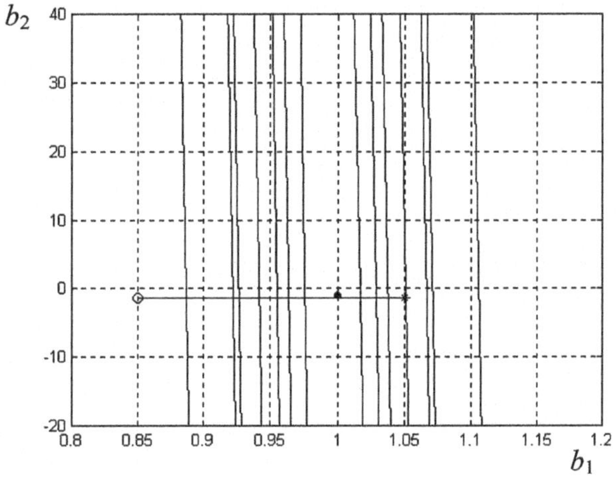

Fig. 32 Mathematical model and "experimental" data ($1.5 < x_n < 2$): *circles* $\hat{y}_n$, values of y with a maximum "experimental" error of $\pm 5\%$; *solid line* $\bar{y} = 1 - \exp(-5x)$ *dashed line* $y = 1.05 - 1.4\exp(-5x)$

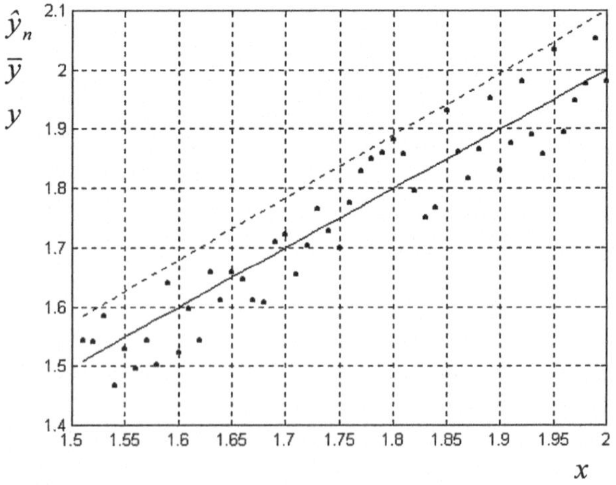

Table 29 Solutions of the three-parameter problem

$N = 50, \beta = 0.01,\ b_1^0 = 0.5,\ b_2^0 = 0.85,\ b_3^0 = -1.4, \bar{b}_1 = 0, \bar{b}_2 = 1, \bar{b}_3 = -1$

n	Data no.	γ	b_1^*	b_2^*	b_3^*	i
1–60	1	0.5	0.0196	0.9607	−1.0203	327
	2	1	0.1459	0.7142	−1.1926	100
61–80	1	0.2	0.0591	0.9334	−1.4731	758
	2	0.2	0.1369	0.8204	−1.4452	594
81–100	1	0.2	0.0044	1.0068	−1.4403	1415
	2	0.2	0.1944	0.7859	−1.4153	454
101–150	1	0.2	−0.0211	1.0141	−1.4103	555
	2	0.2	0.0887	0.9250	−1.4075	439
151–200	1	0.2	−0.0495	1.0246	−1.4011	1135
	2	0.2	0.1694	0.8987	−1.4006	752

Table 30 Adequacy of the two-parameter model

$\Delta\hat{y} = \pm 5\%$; $S_{(2)}^e = 0.0283$; $F_{(2)}^t = 2.8111$

n	x	Data no.	b_1^*	b_2^*	$F_{(2)}$	$S_{(2)}$
60–80	0.7	1	1.0109	−1.3441	1.1361	0.0302
		2	0.9820	−0.7050	1.0361	0.0288
80–100	0.9	1	1.0050	−1.3982	1.1401	0.0302
		2	0.9874	−0.5045	1.0654	0.0292
100–150	1.3	1	1.0500	−1.3997	4.3717	0.0592
		2	0.9500	−1.4000	4.3424	0.0590
150–200	1.8	1	1.0500	−1.4000	4.4571	0.0597
		2	0.9500	−1.4000	4.1995	0.0580

Table 31 Adequacy of the three-parameter model

$\Delta\hat{y} = \pm 5\%$; $S_{(3)}^e = 0.0283$; $F_{(3)}^t = 2.8148$

n	x	Data no.	b_1^*	b_2^*	b_3^*	$F_{(3)}$	$S_{(3)}$
60–80	0.7	1	0.0591	0.9334	−1.4731	1.1571	0.0304
		2	0.1369	0.8204	−1.4452	1.0598	0.0291
80–100	0.9	1	0.0044	1.0068	−1.4403	1.2665	0.0318
		2	0.1944	0.7859	−1.4153	1.0872	0.0295
100–150	1.3	1	−0.0211	1.0141	−1.4103	1.1497	0.0303
		2	0.0887	0.9250	−1.4075	1.0554	0.0291
150–200	1.8	1	−0.0495	1.0246	−1.4011	1.1498	0.0304
		2	0.1694	0.8987	−1.4006	1.0551	0.0291

Table 32 Adequacy of the two-parameter model (error ±10%)

$\Delta\hat{y} = \pm 10\%$; $S^{e}_{(2)} = 0.0564$; $F^{t}_{(2)} = 2.8111$

n	x	Data no.	b_1^*	b_2^*	$F_{(2)}$	$S_{(2)}$
100–150	1.3	1	1.0500	−1.3997	2.0087	0.0800
		2	0.9500	−1.3997	1.8498	0.0768
150–200	1.8	1	1.0500	−1.4000	2.0315	0.0804
		2	0.9500	−1.4000	1.8115	0.0760

Table 33 Linear form of the Clapeyron model

$y = \ln P = b_1 + \frac{b_2}{T}$

β_0	γ	b_1^0	b_2^0	b_1^*	b_2^*	S	R	F
0.01	5	22	−1,915	21.801	−1,915.2	0.058672	0.9995	1.0724×10^5
0.01	10	22	−1,916	21.840	−1,916.0	0.061541	0.9945	9.5366×10^4
				21.815	−1,919.5	0.055669	0.9995	1.0851×10^5

10.3 Clapeyron and Antoan Models

Let us consider experimental data [46] for the temperature dependence of the vapor pressure of ethane. As mathematical models we may use linear or nonlinear forms of the Clapeyron and

Antoan models:

$$y = \ln P = b_1 + \frac{b_2}{T}, \quad P = \exp\left(b_1 + \frac{b_2}{T}\right),$$

$$y = \ln P = b_1 + \frac{b_2}{b_3 + T}, \quad P = \exp\left(b_1 + \frac{b_2}{b_3 + T}\right), \tag{10.12}$$

where P (mmHg) is the vapor pressure and T (K) is the temperature. The parameter identification results, obtained with the help of the method presented [24] and a comparison (the last lines in Tables (33, 34, 35, 36) with the conventional method [47] are shown in Tables 33, 34, 35 and 36. The solutions were obtained for different initial parameter values.

The calculation of the variance S for different cases shows that increase of the nonlinearity (for the model parameters) of the model leads to the inverse problem incorrectness (increase of the model error variance). In the cases when we do not have different experimental data for one temperature, it is not possible to obtain the experimental error variance and to analyze the model adequacy. That is why we must use *model suitability* (see Sect. 5.3) on the basis of the set *correlation coefficientR* (5.3.115). The model is suitable if $R \approx 1$.

Table 34 Nonlinear form of the Clapeyron models

$P = \exp(b_1 + \frac{b_2}{T})$

β_0	γ	b_1^0	b_2^0	b_1^*	b_2^*	S	R	F
0.01	6	21.80	−1,915	21.70	−1,915	0.16530	0.99939	8.5932×10^4
0.1	3	21.84	−1,916	21.84	−1,916	0.11798	0.98066	2.6363×10^3
0.1	10	22.00	−1,920	22.00	−1,920	0.18904	0.92448	6.1746×10^2
		22.00	−1920	21.3205	−1,811	0.19806	0.99980	2.9257×10^6

Table 35 Linear form of the Antoan model

$y = \ln P = b_1 + \frac{b_2}{b_3+T}$

β_0	γ	b_1^0	b_2^0	b_3^0	b_1^*	b_2^*	b_3^*	S	R	F
0.01	5	21.55	−1,935	9	21.885	−1,935.6	0.2763	0.0519	0.9995	5.073×10^4
0.1	5	21.55	−1,935	9	21.530	−1,935.1	7.9571	0.1297	0.9944	4.584×10^3
0.01	5	50	−2,000	0	22.400	−2,000.1	0.0607	0.0295	0.9981	1.353×10^4
		50	−2000	0	20.790	−1,574.1	−14.18	8.5×10^{-3}	0.9999	6.689×10^5

Table 36 Nonlinear form of the Antoan model

$P = \exp\left(b_1 + \frac{b_2}{b_3+T}\right)$

β_0	γ	b_1^0	b_2^0	b_3^0	b_1^*	b_2^*	b_3^*	S	R	F
0.01	0.5	21.88	−1,935.6	0.2763	21.744	−1,935.6	0.5443	0.2001	0.9994	4.517×10^4
0.01	0.8	21.53	−1,935.1	7.9571	21.570	−1,935.6	8.1732	0.2715	0.9999	1.279×10^6
0.01	0.5	22.40	−2,000.1	0.0607	21.975	−2,000.1	0.2073	1.0138	0.6876	46.644
0.01	10	21.55	−1,935.6	9	21.540	−1,935.6	8.9998	0.2874	0.9999	3.88×10^5
		21.55	−1,935.6	9	21.772	−2,061.1	17.728	0.3440	0.9999	3.234×10^6

Another possibility is to use the variance ratio F (5.3.116), where the condition $F > F(\alpha, \nu_0, \nu_1)$ shows that the coefficient R is significant and the model may be used for simulation. In this inequality $F(\alpha, \nu_0, \nu_1)$ is the tabulated value of the Fisher distribution, α—level of significance.

Tables 33, 34, 35 and 36 show values of R and F for all cases. These results show that the models are suitable.

10.4 Incorrectness Criterion

The inverse problem solutions obtained using different experimental data sets permit us to obtain an incorrectness criterion [25]. For this purpose, after solution of the identification problem, the parameter values b* obtained may be put into the model equation:

Table 37 Two-parameter model

$y = 1 - b_1\exp(-b_2x_n);\ \beta_0 = 0.01,\quad b_1^0 = 1.1, b_2^0 = 6, \Delta\hat{y} = \pm5\%$					
$b_1 = 1.0025,\quad b_2 = 5.0674$					
n	Data no.	b_1^*	b_2^*	γ	i
1–30	1	1.0188	5.2191	0.9	136
	2	1.0167	5.1762	0.9	132
	3	1.0032	5.1133	0.9	148
	4	1.0220	5.2217	0.9	120
	5	1.0136	5.1107	0.9	141
$b_1 = 1.1797,\ b_2 = 5.4666$					
31–65	1	1.5104	6.1650	0.05	503
	2	1.2751	5.8012	0.05	456
	3	0.9009	4.8649	0.05	956
	4	1.7406	6.4392	0.05	648
	5	1.2914	5.7424	0.05	482
$b_1 = 2.1720,\ b_2 = 6.1731$					
66–100	1	4.7330	7.3917	5	113
	2	4.3706	7.7028	5	256
	3	0.3071	3.8643	5	175
	4	6.1273	7.3698	5	148
	5	2.9416	6.7350	5	144

$$y = f(b^*, \mathbf{x}), \tag{10.13}$$

where $\boldsymbol{x}$ is vector of the independent variables. If we put (10.13) into (10.4), it is possible to obtain different (four to five) experimental data sets using different sets of random numbers.

Inverse problem solutions for different models using different (five) experimental data sets (different random numbers sets in 10.4) are shown in Tables 37, 38, 39, 40, 41 and 42. The results obtained show that increase of the inverse problem incorrectness leads to increase of the differences between the inverse problem solutions, i.e., these differences are a criterion for the identification problem incorrectness.

10.5 Increase of the Exactness of the Identification Problem Solution

In many cases (see, e.g., 10.12) the parameter values b* obtained are very large. A scaling of the model parameters leads to an increase of the exactness the identification problem solution. Comparison results for models (8.10.12) are shown in Tables 43 and 44. The model error variance values show the scale effect. The last lines in Tables 43 and 44 show that the scaling of the objective function and

Table 38 Two-parameter model

$y = b_1x_n + b_2\exp(-5x_n)$; $\beta_0 = 0.01$, $b_1^0 = 0.85$, $b_2^0 = -1.4$, $\Delta\hat{y} = \pm5\%$					
$b_1 = 0.9989$, $b_2 = -0.9945$					
n	Data no.	b_1^*	b_2^*	γ	i
1–60	1	1.0009	−1.0025	0.2	93
	2	0.9976	−0.9929	0.2	88
	3	1.0053	−0.9960	0.2	90
	4	1.0005	−0.9880	0.2	90
	5	0.9990	−0.9923	0.2	90
$y = b_1x_n + b_2\exp(-5x_n)$ $\beta_0 = 0.1$, $b_1^0 = 0.85$, $b_2^0 = -1.4$, $\Delta\hat{y} = \pm5\%$					
$b_1 = 0.95022$, $b_2 = -1.3308$					
61–150	1	0.8518	−1.3356	1.5	63
	2	1.0496	−1.3554	1.5	267
	3	1.0500	−1.3985	1.5	3
	4	1.0500	−1.3992	1.5	3
	5	1.0487	−1.3348	1.5	210

Table 39 Clapeyron model

$y = LP = b_1 + \frac{b_2}{x_n}$; $b_1^0 = 22$, $b_2^0 = -1915 \Rightarrow b_1 = 21.801$, $b_2 = -1915.2$

$92 \le x_n \le 304$; n = 107

$b_1^0 = 22$, $b_2^0 = -1915$, $\gamma = 10$, $\beta_0 = 0.01$				$b_1^0 = 15$, $b_2^0 = -1500$, $\gamma = 3$, $\beta_0 = 1$			
Data no.	b_1^*	b_2^*	i	Data no.	b_1^*	b_2^*	i
1	21.760	−1,915	11	6	23	−1,500	8
2	21.840	−1,915	9	7	23	−1,500	8
3	21.760	−1,915	11	8	23	−1,500	8
4	21.840	−1,915	9	9	23	1,500	8
5	21.840	−1,915	9	10	23	−1,500	8

Table 40 Clapeyron models

$y = \exp\left(b_1 + \frac{b_2}{x_n}\right)$; $b_1^0 = 21.801$, $b_2^0 = -1915.2 \Rightarrow b_1 = 21.7$, $b_2 = -1915.2$

$92 \le x_n \le 304$; n = 107

$b_1^0 = 21.801$, $b_2^0 = -1915.2$, $\gamma = 6$, $\beta_0 = 0.01$				$b_1^0 = 15$, $b_2^0 = -1500$, $\gamma = 6$, $\beta_0 = 0.01$			
Data no.	b_1^*	b_2^*	i	Data no.	b_1^*	b_2^*	i
1	21.721	−1,915.2	8	6	20.4	−1,500	25
2	21.721	−1,915.2	8	7	20.4	−1,500	25
3	21.721	−1,915.2	8	8	20.4	−1,500	25
4	21.721	−1,915.2	8	9	20.4	−1,500	25
5	21.721	−1,915.2	8	10	20.4	−1,500	25

Table 41 Antoan model

$y = LP = b_1 + \frac{b_2}{x_n + b_3}; b_1^0 = 21.55, b_2^0 = -1935, b_3^0 = 9 \Rightarrow b_1 = 21.725, b_2 = -1935.5,$
$b_3 = 3.007892 \leq x_n \leq 304; n = 107$

$b_1^0 = 21.55, b_2^0 = -1935, b_3^0 = 9,$ $\gamma = 5, \beta_0 = 0.01$					$b_1^0 = 15, b_2^0 = -1500, b_3^0 = 9,$ $\gamma = 5, \beta_0 = 0.01$				
Data no.	b_1^*	b_2^*	b_3^*	i	Data no.	b_1^*	b_2^*	b_3^*	i
1	21.732	-1,935.9	2.9160	581	6	20.362	-1,505.4	-14.499	1814
2	21.814	-1,935.9	2.5890	634	7	20.487	-1,505.6	-15.246	1718
3	21.756	-1,935.8	3.0162	605	8	20.465	-1,505.1	-14.869	924
4	21.753	-1,935.7	2.9115	554	9	20.401	-1,504.7	-14.448	762
5	21.648	-1,935.6	3.7792	608	10	20.388	-1,504.8	-14.266	825

Table 42 Antoan model

$y = \exp\left(b_1 + \frac{b_2}{x_n + b_3}\right); b_1^0 = 21.725, b_2^0 = -1935.5, b_3^0 = 3.0078 \Rightarrow b_1 = 21.681,$
$b_2 = -1935.5, b_3 = 3.195692 \leq x_n \leq 304; \quad n = 107$

$b_1^0 = 21.725, b_2^0 = -1935.5, b_3^0 = 3.0078,$ $\gamma = 0.5, \beta_0 = 0.01$					$b_1^0 = 15, b_2^0 = -1500, b_3^0 = 9,$ $\gamma = 15, \beta_0 = 0.1$				
Data no.	b_1^*	b_2^*	b_3^*	i	Data no.	b_1^*	b_2^*	b_3^*	i
1	21.690	-1,935.5	3.0070	9	6	21.400	-1,500	9.1099	10
2	21.700	-1,935.5	2.9348	1477	7	21.400	-1,500	9.1101	10
3	21.700	-1,935.5	2.8877	2530	8	21.400	-1,500	9.1100	10
4	21.680	-1,935.5	3.0930	2718	9	21.400	-1,500	9.1099	10
5	21.700	-1,935.5	2.9054	2788	10	21.400	-1,500	9.1100	10

independent variable leads to an additional increase of the solution exactness (see variance values). The influence of the initial iterative step value β_0 on the exactness of the identification problem solution is very big. Results where a decrease of the β_0 values leads to a decrease of the model error variance are shown in Tables 43 and 44.

10.6 Incomplete Experimental Data Cases

The use of the hierarchical approach [34] to solve *multiequation models* is impossible in the case of *incomplete experimental data*, i.e., the experimental data for concentration–time dependences of some reagents or reaction products are missing. An example of this case is *modeling ofmicroalgae growth kinetics* [72].

Microalgae are a natural source of high-value compounds for the pharmaceutical and food industries, such as bioactive compounds, vitamins, pigments, and fatty acids [48]. In addition, in the long term, algal culture may be useful for

Table 43 Clapeyron models

Model	β_0	γ	b_1^0	b_2^0	b_1^*	b_2^*	i	S
$y_n = b_1 + \frac{b_2}{x_n}$	0.1	30	–60	–5,000	37.954	–4,916.7	2712	5.6290
$y_n = 10B_1 + \frac{1{,}000B_2}{x_n}$	0.1	44	–6	–5	1.1566	–1.3772	15	2.7860
$y_n = \exp\left(b_1 + \frac{b_2}{x_n}\right)$	0.1	8	19	–2,400	23	–2,400	8	11.039
$y_n = \exp\left(10B_1 + \frac{1{,}000B_2}{x_n}\right)$	0.1	5	1.9	–2.4	2.6537	–2.1317	5	0.97665
$\eta_n = \exp\left(b_1 + \frac{b_2}{\xi_n}\right)$ $\eta_n = y_n/10;\quad \xi_n = x_n/100$	0.1	18	16	–25	22.039	–22.881	10	0.73217

Table 44 Antoan models

β_0	γ	b_1^0	b_2^0	b_3^0	b_1^*	b_2^*	b_3^*	i	S
$y_n = b_1 + \frac{b_2}{x_n + b_3}$									
0.01	2	–60	–5,000	9	28.606	–4,982.1	97.815	6,210	0.12975
$y_n = 10B_1 + \frac{1{,}000B_2}{x_n + b_3}$									
β_0	γ	b_1^0	b_2^0	b_3^0	b_1^*	b_2^*	b_3^*	i	S
0.01	10	–6	–5	9	2.3065	–2.2419	9.4647	4,688	0.058769
$y_n = \exp\left(b_1 + \frac{b_2}{x_n + b_3}\right)$									
β_0	γ	b_1^0	b_2^0	b_3^0	b_1^*	b_2^*	b_3^*	i	S
0.01	1	19	–2400	9	22.907	–2,400	17.001	8,165	0.53226
$y_n = \exp\left(10B_1 + \frac{1{,}000B_2}{x_n + b_3}\right)$									
β_0	γ	b_1^0	b_2^0	b_3^0	b_1^*	b_2^*	b_3^*	i	S
0.01	0.05	1.9	–2.4	9	2.2003	–2.0696	9.0044	1,508	0.20390
$\eta_n = \exp\left(b_1 + \frac{b_2}{\xi_n + b_3}\right);\ \eta_n = y_n/10;\quad \xi_n = x_n/100$									
β_0	γ	b_1^0	b_2^0	b_3^0	b_1^*	b_2^*	b_3^*	i	S
0.01	0.5	16	–25	9	20.197	–25.001	0.46693	1,832	0.43947

production of clean fuels. Photosynthetic algal culture is carried out in photobioreactors that may be illuminated naturally (outdoors) or artificially (indoors). The availability and the intensity of light are the major factors affecting the productivity of photosynthetic cultures [49].

Photobioreactors using algae, plants cells, or photosynthetic bacteria have received considerable attention from biochemical engineers. Industry is presently engaged in developing new products and testing a new generation of algae-derived natural products [50]. The algae near the irradiation source are exposed to a high photon flux density, which enhances their growth rate. The cells at the core of the reactor receive less light as a result of mutual shading and will show a lower growth rate [51–54].

The process of photosynthesis can be divided into light and carbon-fixation reactions because they are physically separated [55, 56]. Photosynthesis is obviously linked to the availability of CO_2.

Although several cell-based models of photosynthesis have been proposed [57–66], they consider only light availability. These models use classic enzyme kinetics and assume slow enzyme-controlled reactions dependent only on light to account for the carbon-fixation reactions [57, 60, 63, 66] or assume that photosynthesis rates are mainly related to light intensity [61, 63, 65]. Other models recognize the CO_2 dependence but ignore it in the model [62] or assume that carbon fixation is proportional to the light intensity and the available carbon [33, 34].

The rate of biomass concentration increase is determined by the photon flux intensity and interphase mass transfer rate of CO_2. If the photon flux is constant, the process rate depends on the CO_2 concentration in the liquid phase, i.e., on the interphase mass transfer rate in the gas–liquid system, and the modeling of the photosynthetic processes in an airlift reactor will be presented.

Let us consider the airlift reactor for photosynthetic processes where the interphase mass transfer of CO_2 is realized in the riser zone and the photochemical reaction is in the downcomer zone. The main particularity in these cases is the low rate of the photosynthetic processes. The theoretical analysis will be made on the basis of the diffusion model of the airlift reactor [68, 69]. In the case of a non-stationary process, the airlift reactor can be considered as a circulation tubular reactor, where the CO_2 distribution in the liquid phase is determined by the convection–diffusion equation with a volume reaction:

$$\frac{\partial c}{\partial t}+u\frac{\partial c}{\partial x}=D\left(\frac{\partial^2 c}{\partial x^2}+\frac{1}{r}\frac{\partial c}{\partial r}+\frac{\partial^2 c}{\partial r^2}\right)-\mathrm{k}c; \quad t=0, c=c_0; \quad r=0, \frac{\partial c}{\partial r}=0;$$

$$r=r_0, \frac{\partial c}{\partial r}=0;$$

$$x=0, \quad c(t,r,0)=\bar{c}(t,l), \bar{\mathrm{u}}\bar{c}(t,l)=u\bar{c}(t,l)-D\frac{\partial c}{\partial x}, \tag{10.14}$$

where $u(r)$ and $c(x, r, t)$ are the velocity and the concentration distribution of CO_2 in the liquid phase in the reactor, $\bar{u}$and $\bar{c}$ are the average velocity and concentration of CO_2 at the entrance (exit) of the reactor, D is diffusivity, k is the coefficient of the reaction rate, l is the height of the liquid column in the reactor, r_0 is the reactor radius, and t is time.

Problem (8.10.14) is analyzed [72] in dimensionless form using the characteristic scale of the process

$$t=t_0T, \quad x=lX, \quad r=r_0R, \quad u=\bar{u}U, \quad c=c_0C, \quad \bar{c}=c_0\bar{C}. \tag{10.15}$$

Substitution of (8.10.15) into (8.10.14) leads to

$$\frac{\partial C}{\partial T}+\frac{\bar{u}t_0}{l}u\frac{\partial C}{\partial X}=\frac{\bar{u}t_0}{l}\frac{1}{Pe}\left[\frac{\partial^2 C}{\partial X^2}+\frac{l^2}{r_0^2}\left(\frac{1}{R}\frac{\partial C}{\partial R}+\frac{\partial^2 C}{\partial R^2}\right)\right]-\mathrm{k}t_0C;$$

$$T=0,\quad C=1;\quad R=0,\quad \frac{\partial C}{\partial R}=0;\quad R=1,\quad \frac{\partial C}{\partial R}=0;$$

$$X=0,\quad C(T,R,0)=\bar{C}(T,1),\quad (\mathrm{U}-1)\bar{C}(T,1)=\frac{1}{Pe}\frac{\partial C}{\partial X},\tag{10.16}$$

where

$$Pe=\frac{\bar{u}\,l}{D}.\tag{10.17}$$

The process is nonstationary as a result of the recirculation (see the boundary conditions at $x=0$), and depends on the coefficient of the reaction rate (k), i.e., for the characteristic time we can use

$$t_0=\frac{1}{\mathrm{k}}.\tag{10.18}$$

Substitution of (8.10.18) into (8.10.16) leads to

$$\delta\left(\frac{\partial C}{\partial T}+C\right)=\frac{1}{Pe}\left[\frac{\partial^2 C}{\partial X^2}+\frac{l^2}{r_0^2}\left(\frac{1}{R}\frac{\partial C}{\partial R}+\frac{\partial^2 C}{\partial R^2}\right)\right]-u\frac{\partial C}{\partial X},\tag{10.19}$$

where $\delta=\frac{kl}{\bar{u}}$ is a small parameter ($\delta \ll 1$) in cases of a small coefficient of the reaction rate (k $\ll$ 1). That gives us the possibility to find the solution of (8.10.19) in the form

$$C(T,R,X)=C_0(R,X)+C_l(T)\tag{10.20}$$

and from (8.10.19) and (8.10.20) we obtain

$$\delta\left(\frac{\partial C_1}{\partial T}+C_0+C_1\right)=\frac{1}{Pe}\left[\frac{\partial^2 C_0}{\partial X^2}+\frac{l^2}{r_0^2}\left(\frac{1}{R}\frac{\partial C_0}{\partial R}+\frac{\partial^2 C_0}{\partial R^2}\right)\right]-u\frac{\partial C_0}{\partial X};$$

$$T=0,\quad C_0=0,\quad C_1=1;\quad R=0,\quad \frac{\partial C_0}{\partial R}=0;\quad R=1,\quad \frac{\partial C_0}{\partial R}=0;$$

$$X=0,\quad C_0(R,0)=C_0(1),\quad (U-1)(\bar{C}_0(1)+\bar{C}_l(1))=\frac{1}{Pe}\frac{\partial C_0}{\partial X}.\tag{10.21}$$

The solution for $C_0(R, X)$ can be obtained from (8.10.21) in the zeroth approximation of parameter δ:

$$u\frac{\partial C_0}{\partial X}=\frac{1}{Pe}\left[\frac{\partial^2 C_0}{\partial X^2}+\frac{l^2}{r_0^2}\left(\frac{1}{R}\frac{\partial C_0}{\partial R}+\frac{\partial^2 C_0}{\partial R^2}\right)\right];$$

$$X=0,\quad C_0(R,0)=0,\quad \frac{\partial C_0}{\partial X}=0;\quad R=0,\quad \frac{\partial C_0}{\partial X}=0;\quad R=1,\quad \frac{\partial C_0}{\partial X}=0.\tag{10.22}$$

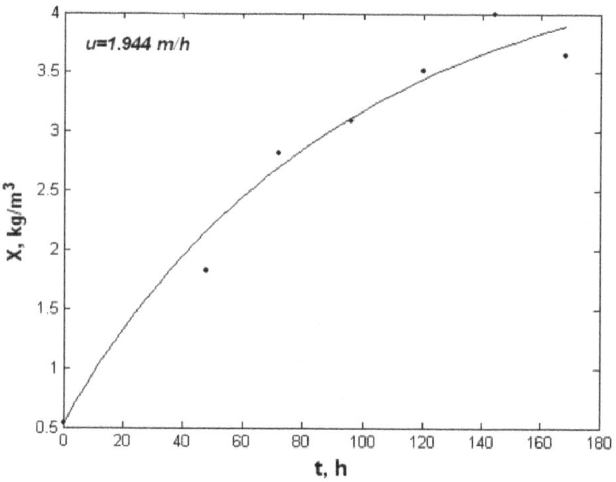

Fig. 33 Comparison of the calculated values and experimental data for biomass concentration

The boundary conditions in (8.10.22) at $X = 0$ follow from the absence of a volume reaction $\left(\frac{\partial C_0}{\partial X} = 0\right)$ and the initial condition at $T = 0 (C_0 = 0)$. Under these conditions the solution of (8.10.22) is $C_0(R, X) \equiv 0$, which permits us to obtain $C_l(T)$:

$$\frac{\partial C_l}{\partial T} = -C_l, \quad T = 0, \quad C_l = 1; \quad C_l = e^{-T}. \tag{10.23}$$

This result shows that *for slow volume reactions the process rate is determined by the kinetics and the column (airlift) reactor exercises the function of an apparatus with an ideal mixing regime. Thus, the photosynthetic process model in an airlift reactor is reduced to a model of photosynthesis kinetics.*

The experimental data for increasing the microalgae (*Pophyridium* sp.) concentration the time (the points in Figs. 33, 35, 37) lead to the hypothesis that the growth mechanism comprises two processes whose rates level off with time, and the concentration of the microalgae becomes steady.

The kinetic equation corresponding to this mechanism has the form

$$\frac{dc_X}{dt} = \mu_{max} \frac{c}{k_1 + c} c_X - k_0 c_X, \tag{10.24}$$

where $c_X = c_X(t), c = c(t)$.

Porphyridium sp. was grown in artificial seawater [70]. Air enriched with $w = 3\%$ CO_2 was sparged into the reactor. A bank of fluorescent lamps was used as an illumination source giving a photon flux density 250 $\mu Em^{-2}s^{-1}$. All experiments were carried out in a room with controlled temperature (23–25°C).

The cultivation of *Porphyridium* sp. experiments were carried out in a laboratory tubular device in an airlift photobioreactor of 13-dm^3 volume [71] in conditions close to the ideal mixing regime in the liquid phase.

The dependence of the change of CO_2 concentration with time $c(t)$ on the balance of the rate of CO_2 consumption for biomass growth and the interphase mass transfer rate in the gas phase is

$$\frac{dc}{dt} = Q - \mathrm{A_x}\mu_{\max}\frac{c}{\mathrm{k_1}+c}c_\mathrm{X}. \qquad (10.25)$$

The volumetric mass transfer rate Q can be determined from the average CO_2 concentration in the gas phase of the input and output from the column:

$$Q = \frac{u}{\mathrm{h}}\left(c_{0,\mathrm{gas}} - c_{\mathrm{h,gas}}\right). \qquad (10.26)$$

The overall mass transfer rate Q depends on the local mass transfer rate q in the column:

$$Q = \frac{1}{\mathrm{h}}\int_0^{\mathrm{h}} \mathrm{q}\,dc_\mathrm{X}, \quad \mathrm{q} = \mathrm{k}\left(c_\mathrm{gas} - \mathrm{k_H}c\right). \qquad (10.27)$$

Assuming that the CO_2 concentration in the gas phase changes linearly in the column from $c_\mathrm{gas}(0, t) = c_{0,\mathrm{gas}}$ to $c_\mathrm{gas}(\mathrm{h}, t) = c_{\mathrm{h,gas}}$,

$$c_\mathrm{gas} = c_{0,\mathrm{gas}} - \frac{\chi}{\mathrm{h}}\left(c_{0,\mathrm{gas}} - c_{\mathrm{h,gas}}\right). \qquad (10.28)$$

Putting c_gas in (8.10.27) leads to

$$Q = \mathrm{k}\left(\frac{c_{0,\mathrm{gas}} + c_{\mathrm{h,gas}}}{2} - \mathrm{k_H}c\right). \qquad (10.29)$$

The expressions for Q (8.10.26, 8.10.29) allow determination of the average CO_2 concentration in the gas outlet:

$$c_{\mathrm{h,gas}} = \frac{\left(\frac{u}{\mathrm{h}} - \frac{\mathrm{k}}{2}\right)c_{0,\mathrm{gas}} + \mathrm{kk_H}c}{\frac{u}{\mathrm{h}} + \frac{\mathrm{k}}{2}}. \qquad (10.30)$$

The process model is represented by (8.10.24–8.10.26) and (8.10.30) with the following boundary conditions:

$$t = 0, \quad c_\mathrm{X} = c_{\mathrm{X}_0}, \quad c = c_0, \qquad (10.31)$$

where c_{X_0} is the initial biomass concentration and $c_0 = c_{0,\mathrm{gas}}/\mathrm{k_H}$ if the process starts with the start of illumination.

This model is characterized by four parameters that can be obtained from the experimental data.

A previous study [33, 34] showed that in cases of models with many equations and parameters the least-squares function is frequently multiextremal or of ravine type. Therefore, a very good initial parameter value for determining the coefficients is needed. For that purpose the parameters were obtained consecutively in the separate equations, where the unknown functions were replaced by for a polynomial approximation of the experimental data. In this case the substitution is difficult, because there are published experimental data for the biomass concentration only.

As already demonstrated [33, 34], the experimental data for the biomass concentration will be represented by the polynomial approximation:

$$c_X(t) = P(t), \quad \frac{dc_X}{dt} = \frac{dP}{dt} = P'(t), \quad A(t) = \frac{P'(t)}{P(t)}. \qquad (10.32)$$

Owing to the lack of experimental data for the concentration, c(t) will be substituted by the "experimental data" $\hat{c}(t)$ that are obtained from (8.10.24) after using (8.10.32):

$$\hat{c}(t) = \frac{k_1[k_0 + A(t)]}{\mu_{max} - k_0 - A(t)}. \qquad (10.33)$$

The "experimental data" $\hat{c}(t)$ are obtained from the experimental data for the biomass (microalgae), but they are conditional because they depends on several quantities μ_{max}, k_0, k_1 that are the subject of determination.

For identification of the model parameters the least-squares function must be used:

$$F = \sum_{i=1}^{N} \left[c_X(t_i) - c_{X_{exp}}(t_i)\right]^2 + a\sum_{i=1}^{N} \left[c(t_i) - \hat{c}(t_i)\right]^2, \qquad (10.34)$$

where t_i(i = 1, ..., N) are the times at which the biomass is quantified and $\alpha =$ 10-100 is a specific weight that compensates for the differences in the orders of magnitude of the two sums.

To determine the function F it is necessary to the solve the model equations (8.10.24), (8.10.25), (8.10.29), and (8.10.30) at the given quantity values μ_{max}, k_0, k_1.

The boundary conditions (8.10.31) must be replaced with

$$t = 0, \quad c_{X_0} = c_{X_{exp,1}}, \quad c_0 = \frac{k_1 A(0)}{\mu_{max} - A(0)}, \qquad (10.35)$$

where it is supposed that at the beginning of the process the effect of the second term in (8.10.24), i.e., k_0, can be ignored.

The experimental data for the biomass show that for $t \geq t_N$, $c_X = c_{X_N}$, $\frac{dc_X}{dt} =$ 0 and from (8.10.24) it follows that

$$c_N = c(t_N) = \frac{k_0 k_1}{\mu_{max} - k_0} = \text{const.} \qquad (10.36)$$

Under this condition $(t \geq t_N) \frac{dc}{dt} = 0$, i.e.,

$$\frac{u}{\mathrm{h}}\left(c_{0,\mathrm{gas}} - c_{\mathrm{h,gas}}(t_N)\right) - A_x \mu_{\max} \frac{c_N}{\mathrm{k}_1 + c_N} c_{X_N} = 0,$$

$$c_{\mathrm{h,gas}}(t_N) = \frac{c_{0,\mathrm{gas}}\left(\frac{u}{\mathrm{h}} - \frac{\mathrm{k}}{2}\right) + \mathrm{k}\mathrm{k}_H c_N}{\frac{u}{\mathrm{h}} + \frac{\mathrm{k}}{2}}. \tag{10.37}$$

Substitution of (10.36) into (10.37) and (10.38) and subsequently of (10.38) into (10.37) leads to an equation that represents the relationship between the parameters $\mu_{\max}$, k, k_0, k_1.

The latter enables the determination of k as a function of the other quantities:

$$\mathrm{k} = \frac{2uA_x c_{X_N} \mathrm{k}_0(\mu_{\max} - \mathrm{k}_0)}{2u\left(c_{0,\mathrm{gas}}\mu_{\max} - c_{0,\mathrm{gas}}\mathrm{k}_0 - \mathrm{k}_H \mathrm{k}_0 \mathrm{k}_1\right) - A_x \mathrm{h}\, c_{X_N} \mathrm{k}_0(\mu_{\max} - \mathrm{k}_0)}. \tag{10.38}$$

Substitution of (8.10.38) into (8.10.29) leads to a model with three parameters and the least-squares function (8.10.34) depends on the quantities $\mu_{\max}$, k_0, k_1.

The determination of the model parameters is made [72] by minimization of (8.10.34) using the procedure *fminsearch* in MATLAB 6.5 and the results obtained are

$$\begin{aligned} \mu_{\max} &= 0.7386\,\mathrm{h}^{-1}, \quad \mathrm{k} = 2.5\,\mathrm{h}^{-1}, \quad \mathrm{k}_0 = 0.01095\,\mathrm{h}^{-1}, \\ \mathrm{k}_1 &= 0.2715\,\mathrm{kg\,m}^{-3}, F = 0.812. \end{aligned} \tag{10.39}$$

A test for the correctness of the inverse identification problem is made [72] using different sets of experimental data.

The coefficient k can be obtained more accurately by minimization of F as a function of four parameters using (8.10.34) as a zeroth approximation. The result is

$$\begin{aligned} \mu_{\max} &= 0.9314\,\mathrm{h}^{-1}, \mathrm{k} = 2.742\,\mathrm{h}^{-1}, \mathrm{k}_0 = 0.0113\,\mathrm{h}^{-1}, \mathrm{k}_1 = 0.0642\,\mathrm{kg\,m}^{-3}, F \\ &= 0.721, \end{aligned} \tag{10.40}$$

i.e., this solution can be regarded as more precise given the value of F being less than in (8.10.39).

The gas velocity is the most important variable in pneumatic reactor operation. The amount of gas supplied to the reactor strongly influences the mixing of the medium, the distribution of cells in the reactor, availability of nutrient to cells, and absorption of CO_2. Increased gas velocity improves mixing and therefore mass transfer [49, 67].

Intensification of growth with increasing gas velocity can be explained by improvement of mass transfer in the reactor. The interphase mass transfer is obviously very important since it is responsible for the provision of the CO_2 required as a building block for the cells' growth. This step is relatively fast owing

Table 45 The model parameter values and the values of the least-squares function for different superficial gas velocities

Model parameter values Superficial gas velocity (m h^{-1})	μ_{max}, h^{-1}	k_0, h^{-1}	k_1, kg m^{-3}	k, h^{-1}	F
$u = 1.944$	1.0185	0.0094	0.0177	3.7282	0.281
$u = 5.76$	0.93139	0.01127	0.06419	2.7417	0.7213
u = 11.88	0.45007	0.00805	0.7758	1.6004	1.304

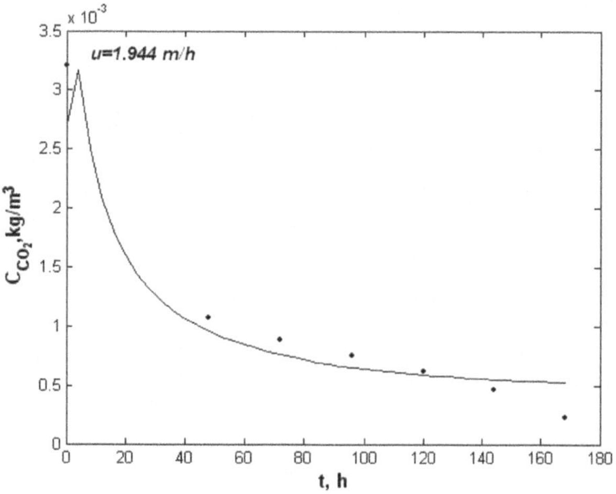

Fig. 34 Comparison of the calculated values and experimental data for CO_2 concentration in the liquid phase

to the high solubility of CO_2 in the gas used. Within the liquid itself, far from the gas–liquid interface, two mechanisms of mass transfer can be distinguished. The first is convective transfer that takes place throughout the reactor and is related to the total liquid circulation and macromixing. This is a function of the reactor design, the physical properties of the medium, and the gas flow rate. The second is the transfer from the bulk of the liquid towards the suspended cells [49]. Liquid–cell mass transfer is influenced by the liquid properties and fluid dynamics, and also depends on cell aggregation.

Comparisons between the model, with calculated parameters (see Table 45), and the experimental data for different superficial gas velocities are shown in Figs. 33, 34, 35, 36, 37 and 38. Table 45 shows the model parameter values and the values of the least-squares function (F).

Assuming that the gas velocity does not essentially influence the coefficients μ_{max}, k_0, k_1, we can calculate their average values $\bar{\mu}_{max}, \bar{k}_0, \bar{k}_1$ from the three given superficial gas velocities. Then with use of the averages we can minimize the least-squares function (F) to determine the mass transfer coefficient ($\bar{k}$):

$$\bar{\mu}_{max} = 0.9749\,h^{-1}; \quad \bar{k}_0 = 0.0104\,h^{-1}; \quad \bar{k}_1 = 0.041\,kg\,m^{-3}. \qquad (10.41)$$

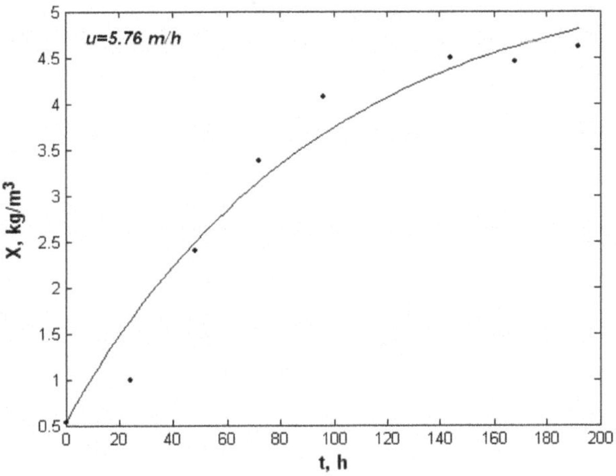

Fig. 35 Comparison of the calculated values and experimental data for biomass concentration

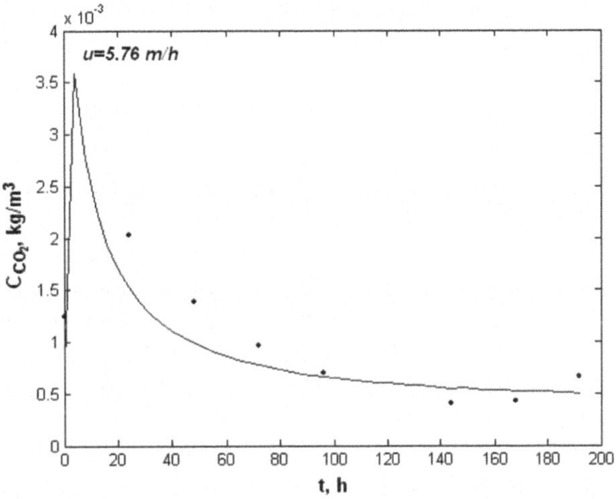

Fig. 36 Comparison of the calculated values and experimental data for CO_2 concentration in the liquid phase

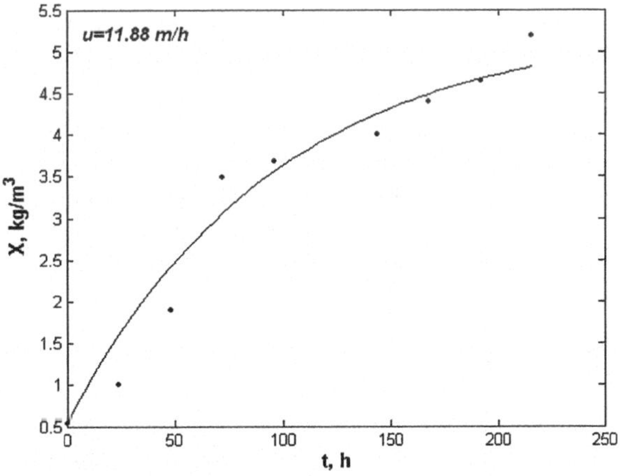

Fig. 37 Comparison of the calculated values and experimental data for biomass concentration

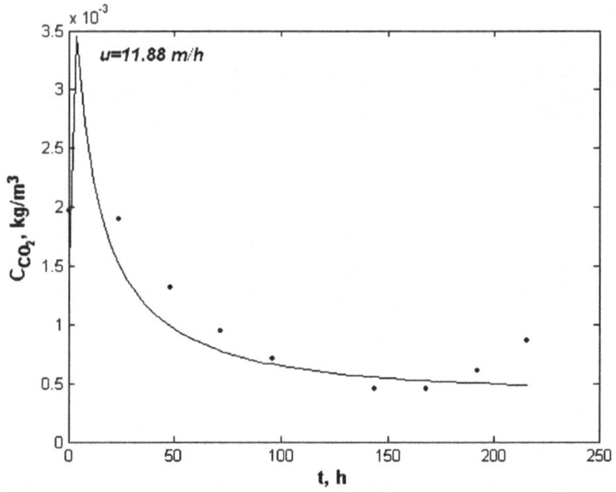

Fig. 38 Comparison of the calculated values and experimental data for CO_2 concentration in the liquid phase

The results obtained show a dependence of the mass transfer coefficient on the superficial gas velocity:

$$k = 5.286u^{-0.3811}. \tag{10.42}$$

The "experimental data" for c(*t*) are obtained from (8.10.33) with the parameter values given in Table 45.

The proposed parameter identification method offers a possibility for finding a solution to a problem connected with insufficient experimental information.

The comparison of the theoretical and experimental data in Figs. 33, 34, 35, 36, 37 and 38 shows that the accuracy of the solution could be increased provided more detailed experimental data for the beginning of the process ($0 \leq t \leq 24$ h) are available.

The good agreement between the model and the experimental data confirms the hypothesis that the growth mechanism comprises two processes whose rates level off with time, and the concentration of biomass becomes steady. On the basis of the existence of a relationship between the concentration of biomass and the CO_2 concentration in the liquid phase, it is demonstrated that the comparison of the theoretical and experimental data in Figs. 33, 34, 35, 36, 37 and 38 shows that the accuracy of the solution could be increased provided more detailed experimental data for the substitution of missing experimental data with a "provisional experimental data set" are available The last set depends on the model parameters.

The method presented is applicable for different photosynthetic processes.

References

1. Boyadjiev Chr (1993) Fundamentals of modeling and simulation in chemical engineering and technology. Bulgarian Academy of Sciences, Sofia (in Bulgarian)

2. Tikhonov AN, Arsenin VI (1986) Methods of inverse problem solution. Nauka, Moscow (in Russian)
3. Tikhonov AN, Kal'ner VD, Glasko VB (1990) Mathematical modelling of technological processes and methods of inverse problem solution in the machine industry. Mashinostroenie, Moscow (in Russian)
4. Chr Boyadjiev, Beschkov V (1988) Mass transfer in liquid film flows. Mir, Moscow (in Russian)
5. Kolmogorov AN, Fomin SV (1954) Topic of theory of functions and functional analysis, 1st edn. Moscow State University, Moscow (in Russian)
6. Vutchkov I, St Stoyanov (1986) Mathematical modelling and optimization of technological objects. Technika, Sofia (in Bulgarian)
7. Ermakov SM, Zhigliavkii NA (1987) Mathematical theory of optimal experiments. Nauka, Moscow (in Russian)
8. Alifanov OM, Artukhin EA, Rumiantsev SV (1988) Extremal methods of incorrect problem solutions. Nauka, Moscow (in Russian)
9. Artukhin EA, Okhapkin AS (1982) J Eng Phys (Russia) 42:1013
10. Alifanov OM (1974) J Eng Phys Russia 26:682
11. Alifanov OM (1977) J Eng Phys Russia 33:972
12. Alifanov OM (1983) J Eng Phys Russia 45:742
13. Alifanov OM, Rumiantsev CB (1980) J Eng Phys Russia 39:253
14. Alifanov OM (1994) Inverse heat transfer problems. Springer, Berlin
15. Beck JV, Arnold KI (1977) Parameter estimation in engineering and science. Wiley, New York
16. Beck JV, Blackwell B, St Clair CR Jr (1985) Inverse heat condition. Ill-posed problems. J. Wiley–Interscience Publications, New York
17. Brakham RL Jr (1989) Scientific data analysis. Springer, Berlin
18. Banks HT, Kunsch K (1989) Estimation techniques for distributed parameters. Birkhäuser, Basel
19. Chavent G (1973) In: Proceedings of the 3rd IFAC symposium. North-Holland, Amsterdam, p 649
20. Chavent G (1980) In: Proceedings of the SIFAC symposium, vol 1. Pergamon, New York
21. Draper NR, Smith H (1966) Applied regression analysis. Wiley, New York
22. Glasko VB (1994) Inverse problem of mathematical physics (trans: Bincer A) American Institute of Physics, Melville
23. Nayfeh AH (1976) Perturbation methods. Wiley, New York
24. Chr Boyadjiev, Dimitrova E (2005) Comput Chem Eng 29:941
25. Dimitrova E, Brauner N, Chr Boyadjiev, Shacham M (2010) Trans Academenergo 1:3
26. Boyadjiev Chr, Dimitrova E (2002) In: Proceedings of the 7th workshop on transport phenomena in two-phase flow (Warna' 2002), p 45
27. Dimitrova E, Boyadjiev Chr (2002) In: Proceedings of the 7th workshop on transport phenomena in two-phase flow (Warna' 2002), p 59
28. Dimitrova E, Boyadjiev Chr (2002) In: Proceedings of the 7th workshop on transport phenomena in two-phase flow (Warna' 2002), p 67
29. Boyadjiev Chr, Dimitrova E (2003) In: Proceedings of the 8th workshop on transport phenomena in two-phase flow (Sunny Beach' 2003), p 105
30. Dimitrova E, Boyadjiev Chr (2003) In: Proceedings of the 8th workshop on transport phenomena in two-phase flow (Sunny Beach'2003), p 123
31. Dimitrova E, Boyadjiev Chr (2003) In: Proceedings of the 8th workshop on transport phenomena in two-phase flow (Sunny Beach'2003), p 133
32. Dimitrova E, Boyadjiev Chr (2004) In: Proceedings of the 9th workshop on transport phenomena in two-phase flow (Borovets'2004), p 137
33. Popova P, Boyadjiev Chr (2004) In: Proceedings of the 9th workshop on transport phenomena in two-phase flow (Borovets'2004), p 145
34. Popova P, Boyadjiev Chr (2008) Biochem Eng J 39:397

35. Madar J, Abonyi J, Roubos H, Szeifert F (2003) Ind Eng Chem Res 42:4043
36. Himmelblau DM, Jones CR, Sichoff KB (1967) Ind Eng Chem Fundam 6:539
37. Tang YP (1971) Ind Eng Chem Fundam 10:321
38. Beschkov V, Velizarov S (1971) C R Acad Bulg Sci 47(8):53
39. Znad H, Bales V, Kawase J (2004) Chem Pap 58:23
40. Olijve W, Kok J (1979) Arch Microbiol 121:283
41. Luedeking R, Piret EL (1959) Biotechnol Bioeng 67:636
42. Velizarov S, Beschkov V (1994) Biotechnol Lett 16:715
43. Vuchkov I (1978) Optimal design of experimental research. Technika, Sofia (in Bulgarian)
44. Artyuhin EA (1985) J Eng Phys Russia 48:490
45. Vuchkov IN et al (1978) Catalogue of consequal generated plans. MHP, Sofia
46. Perry RH, Green DW (1999) Perry's chemical engineering handbook. McGraw-Hill, New York
47. Shacham M, Brauner N (1999) Chem Eng Process 38:477
48. Barbosa M, Hoogakker J, Wijffels R (2003) Biomol Eng 20:115
49. Merchuk J, Gluz M, Mukmenev I (2000) J Chem Technol Biotechnol 75:1119
50. Merchuk J, Wu X (2001) Chem Eng Sci 56:2527
51. Banister T (1979) Limnol Oceanogr 24:76
52. Laws E, Bannister T (1980) Limnol Oceanogr 25:457
53. Ree G, Gotham I (1980) Limnol Oceanogr 26:649
54. Sanchez Perez J, Garcia Camacho F, Garcia Sanchez J, Lopez Alonso D (1993) Appl Microbiol Biotechnol 38:599
55. Papadakis I, Kozabasis K, Lika K (2005) Biochim Biophys Acta 1708:250
56. Prézelin B (1981) Can J Fish Aquat Sci 210:1
57. Fasham M, Platt T (1983) Proc R Soc Lond B Biol Sci 219:355
58. Megard R, Tonkyn D, Senft W (1984) J Plankton Res 6:325
59. Eilers P, Peeters J (1988) Ecol Model 42:199
60. Eilers P, Peeters J (1993) Ecol Model 69:113
61. Zonneveld C (1997) Theor Biol 186:381
62. Zonneveld C, van Den Berg HA, Kooijman S (1997) J Theor Biol 188:215
63. Zonneveld C (1998) J Theor Biol 193:115
64. Zonneveld C (1998) Ecol Model 113:55
65. Rubio C, Camacho F, Sevilla J, Chisti Y, Grima E (2003) Biotechnol Bioeng 81:459
66. Shuter B (1979) J Theor Biol 78:519
67. Prokop A, Erickson L (1994) In: Asenjo JA, Merchuk JC (eds) Bioreactor system design. Dekker, New York
68. Chr Boyadjiev (2006) Int J Heat Mass Transf 49:796
69. Chr Boyadjiev (2006) Int J Heat Mass Transf 4:2053
70. Jones R, Speer H (1963) Physiol Plant 16:636
71. Merchuk J, Ronen M, Giris S, Arad Sh (1998) Biotechnol Bioeng 59:705
72. Popova P, Chr Boyadjiev (2008) Chem Biochem Eng Q 22:491

Optimization

The optimization of chemical engineering processes addresses the task of obtaining the best results under given conditions, i.e., the maximum (minimum) value of the target (object) function of the process. As a rule, the optimization methods are methods for function minimization because in the opposite case a change of the sign of the function is enough. On the other hand, the main problem of the identification of the model parameters is the minimization of the least-squares function. In all these cases, the problem is to minimize a function when the independent variables satisfy different equality and inequality constraints:

$$K = f(x_1, x_2, \ldots, x_N), \tag{1}$$

$$\begin{aligned} &\phi_1(x_1, x_2, \ldots, x_N) = 0, \quad i = 1, 2, \ldots, I; \\ &\psi_j(x_1, x_2, \ldots, x_N) \leq 0, \quad j = 1, 2, \ldots, J. \end{aligned} \tag{2}$$

According to the type of the functions f, ϕ, and ψ, different optimization methods exist.

1 Analytical Methods

The analytical solution of the minimization problem is an exact solution. The basis of every optimization method is minimization of (1) without the constraints (2).

1.1 Unconstraints Minimization

In the case of a one-variable function $f(x)$, the solution of the minimization problem is the solution of the equation

C. Boyadjiev, *Theoretical Chemical Engineering*,
DOI: 10.1007/978-3-642-10778-8_9, © Springer-Verlag Berlin Heidelberg 2010

$$\frac{df}{dx} = 0, \quad x = x_s, \quad s = 1, 2, 3, \ldots, \tag{1.1}$$

where x_s are the roots of (1.1) which satisfy the conditions

$$\left(\frac{d^2 f}{dx^2}\right)_{x=x_s} > 0. \tag{1.2}$$

The minimization of a two-variable function uses the conditions

$$\frac{\partial f}{\partial x_1} = 0, \quad \frac{\partial f}{\partial x_2} = 0, \quad x_1 = x_1^s, \quad x_2 = x_2^s, \quad s = 1, 2, 3, \ldots, \tag{1.3}$$

where the roots of (1.3) satisfy the conditions

$$\frac{\partial^2 f}{\partial x_1^2} > 0, \quad \frac{\partial^2 f}{\partial x_2^2} > 0, \quad \left(\frac{\partial^2 f}{\partial x_1 2}\right)^2 - \frac{\partial^2 f}{\substack{2\\1}} \frac{\partial^2 f}{\partial x_2^2} < 0, \quad x_1 = x_1^s, \quad x_2 = x_2^s, \quad s = 1, 2, 3, \ldots. \tag{1.4}$$

In the case of multivariable functions, conditions (1.4) must be replaced by additional analysis.

1.2 Constraints Minimization

The minimization of function f (1) in the case of equality constraints φ (2) uses Lagrange's method of indeterminate coefficients, introducing the Lagrange function:

$$F = f(x_1, x_2, \ldots, x_N) + \sum_{i=1}^{I} \lambda_i \phi_i(x_1, x_2, \ldots, x_N), \tag{1.5}$$

where λ_i, $i = 1, 2, \ldots, I$, are the Lagrange multipliers (indeterminate coefficients). The minimum of f in the case of equality constraints $\phi_i(x_1, x_2, \ldots, x_N) = 0$, $i = 1, \ldots, I$, is equivalent to the minimum of F without equality constraints, i.e., the roots of the next set of $N + I$ equations are solution of the problem:

$$\frac{dF}{dx_n} + \sum_{i=1}^{I} \lambda_i \frac{d\phi_i}{dx_n} = 0, \quad \phi_i = 0, \quad n = 1, 2, \ldots, N, \quad n = 1, 2, \ldots, N, \tag{1.6}$$

where x_n, $n = 1, 2, \ldots, N$ and λ_i, $i = 1, 2, \ldots, I$, are the independent variables.

The method of Lagrange has a generalization in the case of inequality constraints $\psi_j \leq 0$, $j = 1, 2, \ldots, J$ (see Sect. 2).

1.3 Calculus of Variations

In Sect. 8.2 in Chap. 8 a functional was shown [1] (see 8.2.15 in Chap. 8). Here we will consider the functional

$$z = \int_{x_0}^{x_1} F(x, y, y')\,dx, \quad y = y(x), \tag{1.7}$$

where z has different values for different functions $y = y(x)$. The problem of the calculus of variations is to obtain a continuous function $y = y(x)$ (along with the first derivative $y' = \frac{dy}{dx}$) in the interval $[x_0, x_1]$ which minimize the functional z. The properties of the functionals and the calculus of variations will be considered on the basis of comparison with the functions and minimization of the functions.

Let us consider the function y and the functional z:

$$y = f(x), \quad z = z[y(x)]. \tag{1.8}$$

The increment of the argument (independent variable) is equal to the *differential of the argument dx*. The analogue in the calculus of variations is the *variation of the function* δy:

$$\Delta x = \bar{x} - x = dx, \quad \delta y = \bar{y}(x) - y(x). \tag{1.9}$$

The linearities of the functions and the functionals are similar:

$$l(cx) = cl(x), \quad L[cy(x)] = cL[y(x)]. \tag{1.10}$$

The function increment and the functional increment can be represented in a similar way:

$$\Delta f = f(x + \Delta x) - f(x), \quad \Delta z = z[y(x) + \delta y] - z[y(x)]. \tag{1.11}$$

To obtain the function differential df and the functional variation δz we will use the expressions

$$f(x + \alpha\Delta x) = \left\{ \begin{matrix} f(x), & \alpha = 0 \\ f(x) + \Delta f, & \alpha = 1 \end{matrix} \right\}, \quad z[y(x) + \alpha\delta y] = \left\{ \begin{matrix} z, & \alpha = 0 \\ z + \Delta z, & \alpha = 1 \end{matrix} \right\}. \tag{1.12}$$

The differentiation of (1.12) leads to

$$\begin{aligned} &\frac{\partial}{\partial\alpha}[f(x + \alpha\Delta x)]_{\alpha=0} = [f\prime(x + \alpha\Delta x)\Delta x]_{\alpha=0} = f'(x)\Delta x = f'\,dx = df, \quad f' = \frac{df}{dx}, \\ &\frac{\partial}{\partial\alpha}\{z[y(x) + \alpha\delta y]\}_{\alpha=0} = \delta z. \end{aligned} \tag{1.13}$$

The extremum conditions are $df = 0$ and $\delta z = 0$, i.e.,

$$\frac{\partial}{\partial \alpha}[f(x + \alpha \Delta x)]_{\alpha=0} = 0, \quad \frac{\partial}{\partial \alpha}\{z[y(x) + \alpha \delta y]\}_{\alpha=0} = 0. \tag{1.14}$$

Let us consider functional (1.7):

$$z = \int_{x_0}^{x_1} F[x, y(x), y'(x)]\, dx, \quad y_0 = y(x_0) = \bar{y}(x_0), \quad y_1 = y(x_1) = \bar{y}(x_1). \tag{1.15}$$

The function variation and the variation of the first derivative can be represent as

$$\delta y = \bar{y}(x) - y(x), \quad (\delta y)' = \bar{y}'(x) - y'(x) = \delta y', \tag{1.16}$$

where $\bar{y}(x)$ is used for comparison.

The functional extremum will be searched for using the family of characteristics:

$$y = y(x, \alpha) = y(x) + \alpha \delta y = \begin{Bmatrix} y(x), & \alpha = 0 \\ \bar{y}(x), & \alpha = 1 \end{Bmatrix}. \tag{1.17}$$

Substitution of (1.17) into (1.15) leads to

$$\begin{gathered} z = \int_{x_0}^{x_1} F[x, y(x, \alpha), y'(x, \alpha)]\, dx = \varphi(\alpha), \quad \delta z = [\varphi'(\alpha)]_{\alpha=0}, \\ \phi'(\alpha) = \frac{d\phi}{d\alpha} = \int_{x_0}^{x_1} \left\{ \frac{\partial F}{\partial y} \frac{\partial}{\partial \alpha}[y(x, \alpha)] + \frac{\partial F}{\partial y'} \frac{\partial}{\partial \alpha}[y'(x, \alpha)] \right\} dx, \end{gathered} \tag{1.18}$$

where

$$\frac{\partial}{\partial \alpha}[y(x, \alpha)] = \frac{\partial}{\partial \alpha}[y(x) + \alpha \delta y] = \delta y, \quad \frac{\partial}{\partial \alpha}[y'(x, \alpha)] = \frac{\partial}{\partial \alpha}[y'(x) + \alpha \delta y'] = \delta y', \tag{1.19}$$

i.e.,

$$\begin{gathered} \phi'(\alpha) = \int_{x_0}^{x_1} \left\{ \frac{\partial F}{\partial y} \delta y + \frac{\partial F}{\partial y'} \delta y' \right\} dx, \\ \phi'(0) = \int_{x_0}^{x_1} \left\{ \left(\frac{\partial F}{\partial y} \right)_{\alpha=0} \delta y + \left(\frac{\partial F}{\partial y'} \right)_{\alpha=0} (\delta y)' \right\} dx = \delta z. \end{gathered} \tag{1.20}$$

One of the integrals in (2.20) can be represented as

$$\int_{x_0}^{x_1} \left\{ \left(\frac{\partial F}{\partial y'}\right)(\delta y)' \right\} dx = \int_{x_0}^{x_1} \frac{\partial F}{\partial y'} d(\delta y) = \left(\delta y \frac{\partial F}{\partial y'}\right)_{x=x_0}^{x=x_1} - \int_{x_0}^{x_1} \delta y \frac{\partial}{\partial x}\left(\frac{\partial F}{\partial y'}\right) dx$$
$$= \int_{x_0}^{x_1} \frac{\partial}{\partial x}\left(\frac{\partial F}{\partial y'}\right) \delta y \, dx, \tag{1.21}$$

because $\delta y = 0$ at $x = x_0$ and $x = x_1$ (see 1.15, 1.16). As a result, from (1.20) and (1.21) it follows that

$$\delta z = \int_{x_0}^{x_1} \left[\frac{\partial F}{\partial y} - \frac{\partial}{\partial x}\left(\frac{\partial F}{\partial y'}\right)\right] \delta y \, dx \tag{1.22}$$

and the condition of the functional extremum is

$$\begin{aligned} &\frac{\partial F}{\partial y} - \frac{\partial}{\partial x}\left(\frac{\partial F}{\partial y'}\right) = 0, \\ &\frac{\partial F}{\partial y} - \frac{\partial^2 F}{\partial x \partial y'} - \frac{\partial^2 F}{\partial y \partial y'} y' - \frac{\partial^2 F}{\partial y'^2} y'' = 0, \quad y(x_0) = y_0, \quad y(x_1) = y_1. \end{aligned} \tag{1.23}$$

The result obtained (1.23) is the *Euler–Lagrange equation* and permits one to solve different chemical engineering problems (e.g., optimal temperature distribution in the column apparatuses for chemical reactions).

A classic problem of the calculus of variations is obtaining of the shortest distance between two points in a plane. If $y = y(x)$ is the line which connects the points (x_0, y_0) and (x_1, y_1), the length of the route is a functional,

$$z = \int_{x_0}^{x_1} \sqrt{1 + y'^2} \, dx, \quad y(x_0) = y_0, \quad y(x_1) = y_1 \tag{1.24}$$

and the Euler equation has the form

$$(1 + y'^2)^{-3/1} y'' = 0, \quad y'' = 0, \quad y(x_0) = y_0, \quad y(x_1) = y_1, \tag{1.25}$$

i.e., the straight line representing the shortest distance:

$$y = \frac{y_1 - y_0}{x_1 - x_0} x + \frac{x_1 y_0 - x_0 y_1}{x_1 - x_0}. \tag{1.26}$$

For minimization of the functionals with upper order of the derivatives

$$z = \int_{x_0}^{x_1} F(x, y, y', y'', \ldots, y^{(n)})\, dx, \quad y = y(x), \quad y(x_0) = a_0, \quad y(x_1) = a_1 \quad (1.27)$$

we can use the *Euler–Poisson equation*:

$$\frac{\partial F}{\partial y} - \frac{\partial}{\partial x}\left(\frac{\partial F}{\partial y'}\right) + \frac{\partial^2}{\partial x^2}\left(\frac{\partial F}{\partial y''}\right) - \cdots + (-1)^n \frac{\partial^n}{\partial x^n}\left(\frac{\partial F}{\partial y^{(n)}}\right) = 0,$$
$$y^{(i)}(x_0) = y_i, \quad i = 1, \ldots, n-1. \quad (1.28)$$

In the case of a two-variable function $u = u(x, y)$ and its partial derivatives, the functional has the form

$$z = \iint_D F(x, y, u, p, q)\, dx\, dy, \quad p = \frac{\partial u}{\partial x}, \quad q = \frac{\partial u}{\partial y} \quad (1.29)$$

and the Ostrogradski equation can be used:

$$\frac{\partial F}{\partial u} - \frac{\partial}{\partial x}\left(\frac{\partial F}{\partial p}\right) - \frac{\partial}{\partial y}\left(\frac{\partial F}{\partial q}\right) = 0. \quad (1.30)$$

The minimization of multifunction functionals,

$$z = \int_{x_0}^{x_1} F\left(x, y_1, y_2, \ldots, y_n; y_1', y_2', \ldots, y_n'\right) dx, \quad (1.31)$$

uses a *set of Euler equations.*

The minimization of functional (1.31) in the case of *equality constraints* $g(y_1, y_2) = 0$ uses the Lagrange method, where the indeterminate coefficient is the function $\lambda = \lambda(x)$.

The solutions of the functional minimization problems encounter difficulties in the cases:

- Inequality constraints $g(y_1, y_2) \geq 0$.
- Two-point boundary conditions (uniqueness of the solution).
- Linearity of the function F with respect to y and y' (in the absence of equality or inequality constraints); in the opposite case the extremum is on the border.

1.4 Solution of a Set of Nonlinear Equations

The analytical methods presented for function minimization lead to the solution of sets of nonlinear equations which must be solved numerically. Most of the methods use linearization procedures. Here we will present the *method of Newton.*

Let us consider the set of nonlinear equations

$$f_i(x_1,\ldots,x_N) = 0, \quad i = 1,2,\ldots,I, \tag{1.32}$$

and the coordinates of point $P^*(x_1^*,\ldots,x_N^*)$ are the solution of (1.32), i.e.,

$$f_i\left(x_1^*,\ldots,x_N^*\right) \equiv 0, \quad i = 1,2,\ldots,I. \tag{1.33}$$

If point $P^{(s)}(x_1^{(s)},\ldots,x_N^{(s)})$ is near point $P^*(x_1^*,\ldots,x_N^*)$ the differences between the coordinates are small:

$$x_n^* = x_n^{(s)} + \delta x_n, \quad |\delta x_n| \le \varepsilon_n, \quad n = 1,2,\ldots,N. \tag{1.34}$$

Let us substitute (1.34) into (1.33) and tore present function (1.33) in the neighborhood of point $P^*(x_1^*,\ldots,x_N^*)$ as a Taylor series:

$$\begin{aligned} f_i\left(x_1^*,\ldots,x_N^*\right) &= f_i\left(x_1^{(s)} + \delta x_1,\ldots,x_N^{(s)} + \delta x_N\right) \\ &= f_i\left(x_1^{(s)},\ldots,x_N^{(s)}\right) + \sum_{n=1}^{N} \delta x_n\left(\frac{\partial f_i}{\partial x_n}\right) \equiv 0. \end{aligned} \tag{1.35}$$

From (1.35) the set of linear equations for δx_n, n = 1, 2,..., N, follows:

$$\sum_{n=1}^{N} \delta x_n\left(\frac{\partial f_i}{\partial x_n}\right) = -f_i\left(x_1^{(s)},\ldots,x_N^{(s)}\right), \quad n = 1,2,\ldots,N. \tag{1.36}$$

If f_i, i = 1, 2,..., I, are quadratic functions (1.34) is the direct solution of the set. In all other cases an iterative procedure must be used:

$$x_n^{(s+1)} = x_n^{(s)} + (\delta x_n)^{(s)}, \quad n = 1,\ldots,N, \quad s = 0,1,2,\ldots \tag{1.37}$$

2 Numerical Methods

The numerical methods for function minimization are named *mathematical programming* (linear programming, nonlinear programming, convex programming, integer programming, etc.). Here we will be presented linear and nonlinear programming.

2.1 Linear Programming

In many cases the functions f, φ, and ψ in optimization problem (1) and (2) are linear:

$$y = \sum_{i=1}^{n} c_i x_i, \quad \sum_{i=1}^{n} a_{ij} x_i \{<, =, >\} b_j \geq 0, \quad x_i \geq 0, \quad i = 1, \ldots, n, \quad j = 1, \ldots, m, \tag{2.1}$$

where a, b, c are constants.

As a rule the available software minimizes the linear function y at equality constraints, which imposes the condition of introduction of auxiliary variables x_{n+j}:

$$\sum_{i=1}^{n} a_{ij} x_i \pm x_{n+j} = b_j \geq 0, \quad j = 1, \ldots, m. \tag{2.2}$$

In the case of equality constraints and $n = 2$, the geometrical interpretation of the linear programming show that the constraints form a polygon at the plane $x_1 0 x_2$. The vertical projection of this polygon on the plane $y = c_1 x_1 + c_2 x_2$ generates a new polygon and the distance from an apex (or line) of this new polygon to the plane $x_1 0 x_2$ is minimal, i.e., the coordinates of this apex are the solution of the problem.

For the solution of linear programming problems it is possible to use the simplex method of Dantzig [2]. The modified simplex method is more useful for computer realization with specialized software (GAMS, MATLAB).

2.2 Nonlinear Programming

The nonlinear functions f, φ, and ψ in optimization problem (1) and (2) lead to nonlinear programming problems [3–6]. The solutions of these problems use numerical methods based on the iterative approach. The different types of constraints (2) use different iterative algorithms, but the basis is the unconstraints function minimization algorithms.

The *function minimization without constraints* uses algorithms which search for function minimum step by step in the direction of a decrease of the function values. The search strategies are zeroth, first, and second orders, i.e., using the values of the function, first or second derivative.

The main zeroth-order methods are direct search methods and simplex methods.

The *direct search methods* minimize the function with respect to one variable only (all other variables are constants). At the minimum the same procedure start again with respect to the next variable.

The *simplex methods* minimize the function with the help of regular or irregular simplexes, i.e., polyhedrons. In the case of $n = 2$ the procedure starts from function values at three points y_1, y_2, y_3, which form an equilateral triangle. If $y_1 > y_2 > y_3$, the next point y_4 is symmetric with y_1 with respect to the straight line between points y_2 and y_3. The next simplex is y_2, y_3, y_4 and the same procedure starts again. It is possible to use a scalene triangle.

The first-order methods use the first partial derivatives of the function.

The *relaxation method* starts from an initial point and minimizes function (1) with respect to this variable, whose negative partial derivative is the least (the greatest steepness of descent). The minimum obtained is the next initial point, etc.

The *fastest descent method* is similar to the relaxation method. The difference is that the direction of the function minimization is the antigradient direction.

The *gradient method* is the basis of many first-order methods [4–6]. The minimization of function (1) starts from an initial point $f_0 = f(x_1^{(0)}, x_2^{(0)}, \ldots, x_N^{(0)})$. Every next iteration step must be made in the antigradient direction:

$$x_n^{(s+1)} = x_n^{(s)} - h^{(s)} g_n^{(s)}, \quad g_n^{(s)} = g_n\left(x_1^{(s)}, x_2^{(s)}, \ldots, x_N^{(s)}\right), \quad n = 1, \ldots, N, \tag{2.3}$$

where $h^{(s)}$ is the step of the s iteration and g_n are the components of the gradient vector:

$$g_n = \left(\frac{\partial f}{\partial x_n}\right) \bigg/ \sqrt{\sum_{n=1}^{N} \left(\frac{\partial f}{\partial x_n}\right)^2}, \quad n = 1, \ldots, N. \tag{2.4}$$

The iteration step $f_{s+1} = f(x_1^{(s+1)}, x_2^{(s+1)}, \ldots, x_N^{(s+1)})$ is successful if $f_{s+1} < f_s$. In the opposite case ($f_{s+1} \geq f_s$), the step is unsuccessful. The stopping criterion for the iterative procedure is

$$f_{s+1} \geq f_s, \quad \left|x_n^{(s+1)} - x_n^{(s)}\right| \leq \varepsilon \ll 1. \tag{2.5}$$

The convergence speed depends on the strategy for choosing the step value at every step and the difference between gradient methods [3] is related to this strategy.

A simple algorithm for the choice of the step value uses the distance between the iterative point and the minimum point, i.e., the number of successful iteration steps ($\gamma_s = 1$) before an unsuccessful step ($\gamma_s = 0$):

$$\gamma_s = \begin{Bmatrix} 1 & \textit{if}\ f_s < f_{s-1}; \\ 0 & \textit{if}\ f_s \geq f_{s-1}. \end{Bmatrix}. \tag{2.6}$$

The step values are defined from the conditions

$$h^{(s)} = \begin{Bmatrix} 2h^{(s-1)} & \textit{if}\ \delta_1^{(s)} = 1; \\ h^{(s-1)} & \textit{if}\ \delta_2^{(s)} = 1; \\ 0.5h^{(s-1)} & \textit{if}\ \delta_3^{(s)} = 1. \end{Bmatrix}, \tag{2.7}$$

where the values of γ and δ are presented in Table 1.

Table 1 Values of γ and δ in the iterative algorithm

γ_{s-2}	γ_{s-1}	γ_s	$\delta_1^{(s)}$	$\delta_2^{(s)}$	$\delta_3^{(s)}$	γ_{s-2}	γ_{s-1}	γ_s	$\delta_1^{(s)}$	$\delta_2^{(s)}$	$\delta_3^{(s)}$
1	1	1	1	0	0	1	0	0	0	0	1
1	1	0	0	0	1	0	1	0	0	0	1
0	1	1	0	1	0	0	0	1	0	1	0
1	0	1	0	1	0	0	0	0	0	0	1

The method presented is inapplicable to *multiextremal functions*, where one must search for the *global minimum*. A possibility is to obtain consecutively all local minima. For the transition between two minima it is possible to use the same procedure, but after every minimum one must seek in an arbitrary direction a maximum and after a step in the same direction one *falls into the attraction area of the next minimum*.

Functions with a *ravine-type minimum* are another problem. The procedure presented leads to oscillations of the iteration point at the bottom of the ravine. After that a large step must be made in a direction perpendicular to the oscillations, and then the same procedure is continued.

The basis of the second-order methods [4–6] is the *method of Newton,* which can be used in the case of a small distance between the (initial) iterative point and the minimum point.

Let us represent function (1) in the neighborhood of point $P_0(x_1^{(0)}, x_2^{(0)}, \ldots, x_N^{(0)})$ as a Taylor series:

$$K = f_0 + \sum_{n=1}^{N} \left(\frac{\partial f}{\partial x_n}\right)_{P_0} \left(x_n - x_n^{(0)}\right) + \sum_{m,n} \left(\frac{\partial^2 f}{\partial x_m \partial x_n}\right)_{P_0} \left(x_m - x_m^{(0)}\right)\left(x_n - x_n^{(0)}\right), \tag{2.8}$$

where the values of the function and derivatives are at point P_0:

$$f_0 = f\left(x_1^{(0)}, x_2^{(0)}, \ldots, x_N^{(0)}\right). \tag{2.9}$$

The first partial derivatives of K are

$$\frac{\partial K}{\partial x_n} = \left(\frac{\partial f}{\partial x_n}\right)_{P_0} + 2\sum_{m=1}^{N} \left(\frac{\partial^2 f}{\partial x_m \partial x_n}\right)_{P_0} \left(x_n - x_n^{(0)}\right). \tag{2.10}$$

If $P_1\left(x_1^{(1)}, x_2^{(1)}, \ldots, x_N^{(1)}\right)$ is the point of the minimum, the first partial derivatives of K are equal to zero, i.e.,

$$\left(\frac{\partial f}{\partial x_n}\right)_{P_0} + 2\sum_{m=1}^{N} \left(\frac{\partial^2 f}{\partial x_m \partial x_n}\right)_{P_0} \left(x_n^{(1)} - x_n^{(0)}\right) = 0. \tag{2.11}$$

If f is a quadratic function (2.11) is the direct solution of the minimization problem, i.e., it permits the coordinates of the minimum point $P_1(x_1^{(1)}, x_2^{(1)}, \ldots, x_N^{(1)})$ to be obtained. In all other cases we must use an iterative procedure:

$$x_n^{(s)} = x_n^{(s-1)} + \delta_n^{(s)}, \quad n = 1, \ldots, N, \tag{2.12}$$

where $\delta_n^{(s)}$, $n = 1, \ldots, N$, are the solution of the linear set,

$$\left(\frac{\partial f}{\partial x_n}\right)_{P_{(s-1)}} + 2\sum_{m=1}^{N} \left(\frac{\partial^2 f}{\partial x_m \partial x_n}\right)_{P_{(s-1)}} \delta_n^{(s)} = 0, \quad n = 1, \ldots, N. \tag{2.13}$$

An absence of convergence is possible if the initial point $P_0(x_1^{(0)}, x_2^{(0)}, \ldots, x_N^{(0)})$ and the minimum point $P_1(x_1^{(1)}, x_2^{(1)}, \ldots, x_N^{(1)})$ are widely separated. In these cases preliminary use of a gradient method is very useful.

A numerical iterative algorithm [4–6] for minimization of function f (1) in the case of equality constraints φ (2) uses the *penalty function method*, i.e., the minimization of the function:

$$\Phi = f(x_1, x_2, \ldots, x_N) + \alpha \sum_{i=1}^{I} \phi_i^2 \tag{2.14}$$

without constraints, where α is a big enough number. The function Φ is a *ravine-type function* and an increase of the value of α leads to an increase of the exactness of the solution.

Another possibility for solution of this problem is the *gradient projection method* [4–6]. For this method, in the space of the variables ($x_1, x_2, \ldots, x_N$), we must obtained a point P_0 on the hypersurfaces $\phi_i(x_1, x_2, \ldots, x_N) = 0$, $i = 1, 2, \ldots, I$, by the minimization of the function

$$\phi = \sum_{i=1}^{I} \phi_i^2. \tag{2.15}$$

The next step is to obtain the hyperplanes through point P_0, which are tangents to the hypersurfaces $\phi_i = 0$. The vector gradient of the function $f(x_1, x_2, \ldots, x_N)$ must be projected on the lines of intersection of the hyperplanes. These *gradient projections* are the vector $\boldsymbol{v} = (v_1, v_2, \ldots, v_N)$ and show the direction of the iteration steps, but after several steps it is necessary to correct the coordinates of point P_0.

The iterative procedure starts from point P_0:

$$x_n^{(s+1)} = x_n^{(s)} + h^{(s)} v_n, \quad n = 1, 2, \ldots, N, \quad s = 0, 1, 2, \ldots \tag{2.16}$$

The vector $\boldsymbol{v}$ is tangential with respect to the hypersurfaces $\phi_i = 0$ and lies on the hyperplanes, i.e., it must be orthogonal with respect to the vector $\boldsymbol{grad}\ \phi_i$, $i = 1, 2, \ldots, I$:

$$\sum_{n=1}^{N} v_n \frac{\partial \phi_i}{\partial x_n} = 0, \quad i = 1, 2, \ldots, I, \quad \sum_{n=1}^{N} v_n^2 = 1. \tag{2.17}$$

The directional derivative of the function $f(x_1, x_2, \ldots, x_N)$ regarding the vector $\boldsymbol{v} = (v_1, v_2, \ldots, v_N)$ must be obtained from the scalar product:

$$\frac{df}{dv} = \sum_{n=1}^{N} v_n \frac{\partial f}{\partial x_n}. \tag{2.18}$$

It is possible to find the direction of the fastest decrease of the function $f(x_1, x_2, \ldots, x_N)$ by maximization of function (2.18) with the equality constraints (2.17):

$$\Phi = \sum_{n=1}^{N} v_n \frac{\partial f}{\partial x_n} - \frac{\lambda_0}{2}\left(\sum_{n=1}^{N} v_n^2 - 1\right) + \sum_{i=1}^{I} \lambda_i \left(\sum_{n=1}^{N} v_n \frac{\partial \phi_i}{\partial x_n}\right). \tag{2.19}$$

From the conditions $\partial\Phi/\partial v_n = 0, \ n = 1, 2, \ldots, N$ it follows that

$$v_n = \frac{\frac{\partial f}{\partial x_n} + \sum_{i=1}^{I} \lambda_i \frac{\partial \phi_i}{\partial x_n}}{\lambda_0}, \quad n = 1, 2, \ldots, N, \tag{2.20}$$

where λ_0, λ_i, $i = 1, 2, \ldots, I$ can be obtained from (2.17) and (2.20).

The gradient methods for minimization of function f (1) in the case of inequality constraints ψ (2) use the function

$$\Phi = f(x_1, x_2, \ldots, x_N) + \sum_{j=1}^{J} Q_j(\psi_j), \quad Q_j(\psi_j) = \begin{Bmatrix} 0 & \text{if } \psi_j \le 0 \\ k\psi_j & \text{if } \psi_j > 0 \end{Bmatrix}, \tag{2.21}$$

where k is a big enough number and the problem is similar to (2.14).

In the general case,

$$\Phi = f(x_1, x_2, \ldots, x_N) + \alpha \sum_{i=1}^{I} \phi_i^2 + \sum_{j=1}^{J} Q_j(\psi_j). \tag{2.22}$$

Another possibility is to use Lagrange's method of indeterminate coefficients:

$$\begin{gathered}
K = f(x_1, x_2, \ldots, x_N), \quad \phi_i(x_1, x_2, \ldots, x_N) = 0, \quad i = 1, 2, \ldots, I, \\
\psi_j(x_1, x_2, \ldots, x_N) + \gamma_j = 0, \quad \gamma_j \ge 0, \quad j = 1, 2, \ldots, J, \\
F = f(x_1, x_2, \ldots, x_N) + \sum_{i=1}^{I} \lambda_i \phi_i(x_1, x_2, \ldots, x_N) + \sum_{j=1}^{J} \omega_j \psi_j(x_1, x_2, \ldots, x_N), \\
\gamma_j \omega_j = 0, \quad j = 1, 2, \ldots, J.
\end{gathered} \tag{2.23}$$

For solution of concrete minimization problems it is convenient to use specialized software (GAMS, MATLAB).

3 Dynamic Programming and the Principle of the Maximum

Dynamic programming optimizes the multistage processes in time or in space [7–9]. Typical multistage processes in time are change (regeneration) of equipment, catalyst, etc. The processes in the system of apparatuses, dishes (plates), columns, etc. are examples for multistage processes in space.

3.1 *Functional Equations*

The basis of dynamic programming is so-called functional equations. A typical functional equation has the form

$$f_N(x) = \max_{0 \leq y_N \leq x} [g_N(y_N) + f_{N-1}(x - y_N)], \tag{3.1}$$

where $f_N(x)$ is the target (objective) function of the process (income, profit, quantity, quality, etc.) for the next N stages, starting from state x at stage N. The term $g_N(y_N)$ is the target function value at stage N, whereas $f_{N-1}(x - y_N)$ is the target function value for all the other stages, starting from the state $(x - y_N)$.

The solution of Eq. 3.1 is the optimal value of the control variable (strategy) $y_N^*(0 \leq y_N^* \geq x)$, where y_N^* must be obtained by using some of the optimization methods. Equation (3.1) is a recurrent correlation, i.e., it can be used for all the other stages $f_{N-1}(x - y_N^*)$, starting from the state $(x - y_N^*)$:

$$f_{N-1}\left(x - y_N^*\right) = \max_{0 \leq y_{N-1} \leq x - y_N^*} \left[g_N(y_{N-1}) + f_{N-1}\left(x - y_N^* - y_{N-1}\right)\right]. \tag{3.2}$$

3.2 *Principle of Optimality*

The solution of all functional equations leads to a set of values of the control variable $\{y_1^*, y_2^*, \ldots, y_N^*\}$, which is the optimal strategy for the realization of multistage processes according to Bellman's principle of optimality [7–9], i.e., the optimization of every stage of a multistage process must be made so that the complete process is optimal.

3.3 Principle of the Maximum

Pontryagin's method for solution of nonclassical problems of the calculus of variations is based on the principle of the maximum [10]. This method will be considered in the case of a chemical process in a periodic reactor with ideal mixing, where the mathematical model of chemical kinetics is represented by a set of ordinary differential equations:

$$\frac{dx_i}{dt} = f_i(x_1, x_2, \ldots, x_n; u_1, u_2, \ldots, u_r; t), \quad x_i(t_0) = x_i^{(0)}, \quad i = 1, 2, \ldots, n. \tag{3.3}$$

In (3.3) $x_i(t)$, $x_i^{(0)}$, $i = 1, 2, \ldots, n$, are *phase coordinates* of the process (concentrations of the reagents), its initial values, $u_j(t)$, $j = 1, 2, \ldots, r$, are *control strategies* (temperature, pressure), and t is time in the interval $t_0 \le t \le t_1$.

The optimization problem is to obtain the maximum of the expression

$$S = \sum_{i=1}^{n} c_i x_i, \tag{3.4}$$

where c_i, $i = 1, 2, \ldots, n$, are constants, which permits us to formulate different optimization problems. For example, if we must obtain the conditions ($u_j(t)$, $j = 1, 2, \ldots, r$) for the maximum value of x_1 (maximum concentration of the reagent x_1), $c_1 = 1$, $c_2 = c_3 = \cdots = c_n = 0$. In the case of functional maximization (chemical reaction efficiency),

$$S = \int_0^t F(x_1, x_2, \ldots, x_n; u_1, u_2, \ldots, u_r; t)\, dt, \tag{3.5}$$

a new variable must to be introduced,

$$x_{n+1} = \int_0^t F(x_1, x_2, \ldots, x_n; u_1, u_2, \ldots, u_r; t)\, dt, \tag{3.6}$$

and after differentiation

$$\frac{dx_{n+1}}{dt} = F(x_1, x_2, \ldots, x_n; u_1, u_2, \ldots, u_r; t), \quad x_{n+1}(0) = 0. \tag{3.7}$$

Equation (3.7) must be added to (3.3) and we obtain the maximum of the expression

$$S = \sum_{i=1}^{n+1} c_i x_i, \quad c_1 = c_2 = \cdots = c_n = 0, \quad c_{n+1} = 1. \tag{3.8}$$

Let us consider (3.3) in the case $t_1 = T$, where T is not constant, and the minimization of the functional

$$S = \sum_{i=1}^{n} c_i x_i(T). \tag{3.9}$$

This problem is equivalent to the optimization of a chemical reaction in a periodic reactor with ideal mixing, without time restrictions in one working cycle. It is necessary to obtain the optimal time variations of the temperature and pressure in the reactor.

Let us introduce the new set of functions $p_i(t)$, $i = 1, 2,\ldots, n$, which is full determinate if the control strategies (functions) $u_j(t)$, $j = 1, 2,\ldots, r$, and phase coordinates of the process $x_i(t)$, $i = 1, 2,\ldots, n$ (concentrations of the reagents) are known:

$$\frac{dp_i}{dt} = -\sum_{s=1}^{n} p_s \frac{\partial}{\partial x_i}[f_s(x_1, x_2, \ldots, x_n; u_1, u_2, \ldots, u_r; t)], \quad p_i(T) = -c_i, \\ i = 1, 2, \ldots, n. \tag{3.10}$$

Let us consider a new function:

$$\begin{aligned} H &= H(x_1, x_2, \ldots, x_n; p_1, p_2, \ldots, p_n; u_1, u_2, \ldots, u_r; t) \\ &= \sum_{i=1}^{n} p_i \frac{dx_i}{dt} = \sum_{s=1}^{n} p_s f_s(x_1, x_2, \ldots, x_n; u_1, u_2, \ldots, u_r; t). \end{aligned} \tag{3.11}$$

The function H is very useful because it permits us to obtain $x_i(t)$ and $p_i(t)$, $i = 1, 2,\ldots, n$, as a solution of the set of equations

$$\frac{dx_i}{dt} = \frac{\partial H}{\partial p_i}, \quad \frac{dp_i}{dt} = -\frac{\partial H}{\partial x_i}, \quad x_i(t_0) = x_i^{(0)}, \quad p_i(T) = -c_i, \quad i = 1, 2, \ldots, n. \tag{3.12}$$

Let $u_j^{(0)}(t)$, $j = 1, 2, \ldots, r$, be concrete control strategies, which are introduced into (3.3) and (3.10). As a result, we obtain $x_i^{(u)}(t)$ and $p_i^{(u)}(t)$, $i = 1, 2, \ldots, n$. The substitution of $x_i^{(u)}(t)$, $p_i^{(u)}(t)$, and $u_j^{(0)}(t)$ into (3.11) leads to a new function K:

$$K(u_1, u_2, \ldots, u_r; t) \equiv H\left[x_i^{(u)}(t), p_i^{(u)}(t), u_j^{(0)}(t), t\right]. \tag{3.13}$$

According to *Pontryagin's theorem* the control strategies $u_j^{(0)}(t)$ satisfy the condition of the maximum (minimum), i.e., the optimization criterion S (3.9) is maximal (minimal), if at every moment $t(t_0 \leq t \leq t_1)$ the function K reaches the absolute maximum (minimum) for the values of $u_j^{(0)}(t)$ at the same moment.

4 Examples

4.1 Problem of Optimal Equipment Change

In many cases the efficiency of the equipment (catalyzer activity, thermal resistance, etc.) changes with time. As a rule, the profit (related to this equipment) decreases and a natural question is about the stage when it is better to change the equipment than to continue using the current equipment. The answer to this question leads to the problem of the optimal equipment change [9].

Let us introduce the next symbols:

$r(t)$ is the production value, produced for a year, for t-year-old equipment.
$u(t)$ is the servicing expenditure for 1 year, for t-year-old equipment.
$s(t)$ is the residual value of t-year-old equipment.
p is the price of new equipment.

Let us consider N years as a space of time and we must obtain the optimal cycle for equipment change.

The equipment age reading will start from $t = 0$, and the time stages of the process will be numbered from the end of the process, i.e., $N = 1$ is the last stage (a time stage remains to the end of the process) and $N = N$ is the start of the process.

At every stage of an N-stage process a decision must be taken to retain or change the equipment. The aim of this decision is to obtain the maximum profit from the full N-stage process.

Let $f_N(t)$ be the maximum profit, obtained for t-year-old equipment for all the other N years of the equipment-use cycle. The functional equations obtained on the basis of the principle of optimality have the form

$$f_N(t) = \max\left\{\begin{array}{cc} r(t) - u(t) + f_{N-1}(t+1) & \textit{preservation} \\ s(t) - p + r(0) - u(0) + f_{N-1}(1) & \textit{change} \end{array}\right\}. \tag{4.1}$$

The functional equations (4.1) represent the maximum profit $f_N(t_0)$, where the initial state ($t = t_0$) of the system is obtained from the maximum profit of the first stage (the state is $t = t_0$) and the maximum profit of all the other $N - 1$ stages to the end (the state is $t = t_0 + 1$). By analogy, it is possible to obtain

$$f_{N-1}(t+1) = \max\left\{\begin{array}{cc} r(t+1) - u(t+1) + f_{N-2}(t+2) & \textit{preservation} \\ s(t+1) - p + r(1) - u(1) + f_{N-2}(2) & \textit{change} \end{array}\right\}. \tag{4.2}$$

The functional equations (4.1) are recurrent relations and permit us to obtain the equations for the last stage:

$$f_1(t) == \max\left\{\begin{array}{cc} r(t) - u(t) & \textit{preservation} \\ s(t) - p + r(0) - u(0) & \textit{change} \end{array}\right\}. \tag{4.3}$$

Table 2 Values of the function $n(t)$

t	0	1	2	3	4	5	6	7	8	9	10	11	12
$n(t)$	10	9	8	7	6	5	4	3	2	1	0	0	0

The method presented will be exemplified numerically:

$$r(t) - u(t) = n(t), \quad s(t) = 0, \quad p = 10, \tag{4.4}$$

where the function $n(t)$ is presented in Table 2.

For this example, Eqs. 4.1 and 4.3 have the form

$$\begin{aligned} f_N(t) &= \max\left\{ \begin{array}{cc} n(t) + f_{N-1}(t+1) & \textit{preservation} \\ n(0) - 10 + f_{N-1}(1) & \textit{change} \end{array} \right\}, \\ f_1(t) &= \max\left\{ \begin{array}{cc} n(t) & \textit{preservation} \\ n(0) - 10 & \textit{change} \end{array} \right\}. \end{aligned} \tag{4.5}$$

For a one-stage process and $t = 0$,

$$f_1(t) = \max\left\{ \begin{array}{c} n(0) \\ n(0) - 10 \end{array} \right\} = \max\left\{ \begin{array}{c} 10 \\ 10 - 10 = 0 \end{array} \right\} = 10 \rightarrow \textit{preservation}. \tag{4.6}$$

For a one-stage process and $t = 1$,

$$f_1(t) = \max\left\{ \begin{array}{c} n(1) \\ n(0) - 10 \end{array} \right\} = \max\left\{ \begin{array}{c} 9 \\ 10 - 10 = 0 \end{array} \right\} = 9 \rightarrow \textit{preservation}. \tag{4.7}$$

These calculation procedures continue for $t = 2, 3, \ldots, 12$. After that, calculations are started for a two-stage process using the one-stage-process calculations:

$$f_2(0) = \max\left\{ \begin{array}{c} n(0) + f_1(1) \\ n(0) - 10 + f_1(1) \end{array} \right\} = \max\left\{ \begin{array}{c} 10 + 9 = 19 \\ 10 - 10 + 9 = 9 \end{array} \right\} = 19 \rightarrow \textit{preservation}. \tag{4.8}$$

The calculation results are presented in Table 3.

Let us consider a 15-stage process which starts with new equipment. From Table 3 it is seen that the maximum profit is $f_{15}(0) = 100$ and every subsequent year ($t = 1, 2, \ldots$) the maximum profits (see the diagonals) are 90, 81, 73, 66, and 60ch (where ch is a stage for a change), i.e., the first change must be made (instantly) at the beginning of the sixth year. The next year the maximum profit is $f_9(1) = 60$ and after that the maximum profits (see diagonally) are 51, 43, 36, and 30ch, i.e., the next (second) change must be made (instantly) at the beginning of the 11th year. The next year the maximum profit is $f_4(1) = 30$ and after that the

Table 3 Multistage process optimization

t	0	1	2	3	4	5	6	7	8	9
$f_1(t)$	10	9	8	7	6	5	4	3	2	1
$f_2(t)$	19	17	15	13	11	9	9ch			
$f_3(t)$	27	24	21	18	17ch					
$f_4(t)$	34	30	26	24	24ch					
$f_5(t)$	40	35	32	31	30	30ch				
$f_6(t)$	45	41	39	37	36	35	35ch			
$f_7(t)$	51	48	45	43	41	41ch				
$f_8(t)$	58	54	51	48	48ch					
$f_9(t)$	64	60	56	55	54	54ch				
$f_{10}(t)$	70	65	63	61	60	60ch				
$f_{11}(t)$	75	72	69	67	66	65	65ch			
$f_{12}(t)$	82	78	75	73	72ch					
$f_{13}(t)$	88	84	81	79	78	78ch				
$f_{14}(t)$	94	90	87	85	84	84ch				
$f_{15}(t)$	100	96	93	91	90	90ch				

ch, a stage for a change

maximum profits (see diagonally) are 21, 13, and 6. The maximum profit for 15 years can be obtained after the summation of the separate cycles:

$$f_{15}(0) = [f_{15}(0) - f_{10}(5)] + [f_9(1) - f_5(5)] + f_4(1), \tag{4.9}$$

i.e., 100 = 40 + 30 + 30.

4.2 A Calculus of Variations Problem

The principle of the maximum can be exemplified by the maximization of a functional:

$$F = \frac{1}{2}\int_0^T \left(x^2 + u^2\right) dt, \tag{4.10}$$

where $u(t)$ is the control strategy and the object model is

$$\frac{dx}{dt} = -ax + u, \quad x(0) = x_0. \tag{4.11}$$

The introduction of the new variables

$$x_1(t) \equiv x(t), \quad x_2(t) \equiv \frac{1}{2}\int_0^t \left(x^2 + u^2\right) dt \tag{4.12}$$

leads to the set of equations

$$\frac{dx_1}{dt} = -ax_1 + u, \quad \frac{dx_2}{dt} = \frac{1}{2}x_1^2 + \frac{1}{2}u^2, \quad x_1(0) = x_0, \quad x_2(0) = 0 \tag{4.13}$$

and the functional for minimization is

$$S = x_2(T). \tag{4.14}$$

The use of (3.11) and (3.12) permits us to obtain the function H and equations for the functions $p_1(t)$ and $p_2(t)$:

$$H = -ap_1x_1 + \frac{1}{2}p_2x_1^2 + p_1u + \frac{1}{2}p_2u^2, \tag{4.15}$$

$$\frac{dp_1}{dt} = ap_1 - p_2x_1, \quad \frac{dp_2}{dt} = 0, \quad p_1(T) = 0, \quad p_2(T) = -1. \tag{4.16}$$

From (4.16) it follows that $p_2 \equiv -1$ and for H we obtain

$$H = -ap_1x_1 - \frac{1}{2}x_1^2 + p_1u - \frac{1}{2}u^2. \tag{4.17}$$

For the condition of the maximum we obtain

$$\frac{\partial H}{\partial u} = 0, \quad u = p_1(t) \tag{4.18}$$

and introduction of $u = p_1(t)$ and $p_2 \equiv -1$ into (4.13) and (4.16) leads to

$$\frac{dx_1}{dt} = -ax_1 + p_1, \quad \frac{dp_1}{dt} = ap_1 + x_1, \quad x_1(0) = x_0, \quad p_1(T) = 0. \tag{4.19}$$

The solution of (4.19) is

$$\begin{aligned} x_1 &= A_1 \exp(\lambda_1 t) + A_2 \exp(\lambda_2 t), \\ p_1 &= B_1 \exp(\lambda_1 t) + B_2 \exp(\lambda_2 t), \lambda_{1,2} = \pm\sqrt{a^2+1}, \end{aligned} \tag{4.20}$$

where the constants of the integration depend on x_0 and the control strategy is $u = p_1(t)$.

References

1. Boyadjiev Chr (1993) Fundamentals of modeling and simulation in chemical engineering and technology. Bulgarian Academy of Sciences, Sofia (in Bulgarian)
2. Dantzig GB (1963) Linear programming and extension. Princeton University Press, Princeton
3. Schechter RS (1967) The variational method in engineering. McGraw-Hill, New York
4. Himmelblau DM (1972) Applied nonlinear programming. McGraw-Hill, New York
5. Fiacco AV, McCormick GP (1968) Nonlinear programming: sequential unconstrained minimization techniques. Wiley, New York

6. Ostrovskii GM, Volin YuM, Ziyatdinov NN (2005) Optimization in chemical technology. Fen, Kazan (in Russian)
7. Bellman R (1957) Dynamic programming. Princeton University Press, Princeton
8. Hadley G (1964) Nonlinear and dynamic programming. Addison-Wesley, London
9. Roberts SM (1964) Dynamic programming in chemical engineering and process control. Academic, New York
10. Pontryagin LS, Boltyanskii VG, Gamkrelidze RV, Mischenko EF (1969) Mathematical theory of optimal processes. Science, Moscow (in Russian)

Part IV
Chemical Plant Systems

Systems Analysis

Modeling and simulation of *chemical plant systems* (CPS) is a basic task in chemical technology and requires the quantitative description of industrial systems for the purpose of their optimal design, control, and renovation.

CPS represent a combination of mutually influenced processes in chemical production. Because of this the model of CPS obviously represents a combination of the models of the separate processes, which are supplemented by the equations for the connections (interactions) between them. Straightaway it is clear that during modeling of CPS the problems for the creation of mathematical structures occur, for determining the parameters of the model from experimental data, for statistical analysis of the significance of the parameters, and the adequacy of the model. This means that the problem of the creation of the model of CPS is almost solved since the models of the separate processes exist and the only remaining issue is to add the equations linking the submodels; this is not a complicated task. In this manner the qualitative description of CPS as a set of subproblems is complete mainly regarding their simulations.

1 Simulation of Chemical Plant Systems

CPS represent a combination of technological blocks which are connected in a definite way with technological streams. The streams which do not realize connections between the blocks are the inlet and outlet streams of the CPS.

The technological streams are simultaneously material and thermal. In fact they must be examined as informational because they transfer information (without change) from block to block, containing quantitative data for the quantity and composition of the material stream, for its energy, for its physical constants, and parameters. This means that the pipelines which connect two blocks can be examined as a blocks or streams.

C. Boyadjiev, *Theoretical Chemical Engineering*,
DOI: 10.1007/978-3-642-10778-8_10,

1.1 Model of Chemical Plant Systems

The model of CPS is built [1–5] from mathematical structures, providing a connection between inlet regime variables (***x***), outlet regime variables (***y***), constructive variables (***a***), and variables (***b***) characterizing the state of the equipment. For a block with number *i* they could be represented by vectors (arranged as sequences of numbers):

$$\boldsymbol{x}^{i} = \left|x_{1}^{i},\ldots,x_{m}^{i}\right|,\quad \boldsymbol{y}^{i} = \left|y_{1}^{i},\ldots,y_{n}^{i}\right|,\quad \boldsymbol{a}^{i} = \left|a_{1}^{i},\ldots,a_{q}^{i}\right|,\quad \boldsymbol{b}^{i} = \left|b_{1}^{i},\ldots,b_{r}^{i}\right|. \tag{1.1}$$

The inlet and outlet regime variables in essence represent the parameters of the inlet and outlet streams of the block. The constructive variables are parameters of the apparatus, in which the *i*th process runs. The parameters characterizing the state of the equipment account for some effects from quasi-stationary processes, such as decrease of the activity of the catalysts and the appearance of incrustation over the pipes of the heat exchangers.

The model of the *i*th block represents a system of equations giving the connection between all variables:

$$\boldsymbol{f}\left(\boldsymbol{x}^{i},\boldsymbol{y}^{i},\boldsymbol{a}^{i},\boldsymbol{b}^{i}\right) = 0. \tag{1.2}$$

The combination of models of every block ($i = 1, \ldots, I$, where I is the general number of the blocks in CPS) also becomes model of CPS if to it are added equations for the connections, which have the aspect

$$y_{n_0}^{s} = x_{m_0}^{t}, \tag{1.3}$$

i.e., the inlet regime variable y_{n_0}in block *s* is the outlet regime variable x_{m_0} from the *t*th block.

The inlet and outlet variables in (1.1), which do not participate in the equations for the connections (1.3), form the vectors of the inlet (X) and outlet (Y) regime variables of the CPS:

$$\boldsymbol{X} = |X_1,\ldots,X_M|,\quad \boldsymbol{Y} = |Y_1,\ldots,Y_N|,\quad \boldsymbol{A} = |A_1,\ldots,A_Q|,\quad \boldsymbol{B} = |B_1,\ldots,B_R|, \tag{1.4}$$

where ***A*** and ***B*** are the common combined (vectors) of the constructive variables and variables characterizing the state of the equipment. In this way the model of the CPS is obtained analogously to (1.2) as a system of equations with description

$$\boldsymbol{F}(\boldsymbol{X},\boldsymbol{Y},\boldsymbol{A},\boldsymbol{B}) = 0. \tag{1.5}$$

1.2 Simulation Methods

The simulation of the CPS represents the creation of methods, algorithms, and program systems for the determination of the outlet regime variables:

$$\boldsymbol{Y} = \boldsymbol{G}(\boldsymbol{X}, \boldsymbol{A}, \boldsymbol{B}). \tag{1.6}$$

To solve this problem three basic approaches are used: simultaneous equations, simultaneous modules, and sequential modules [4, 5].

The simultaneous approach solves globally the system from Eqs. 1.2 and 1.3. In practice the CPS contains 20–30 apparatuses, which makes the system of equations examined very large. As an example [5], for the production of carbide 250 variables are used; in the model of the production of sulfuric acid around 1,000 variables are used and 500 equations, half of them, are nonlinear.

The computing problems in the simultaneous equation approach come from three characteristics of the task—the large number of the variables and equations, the nonlinearity of the equations, and the small number of variables in the separate equations. The problem with the nonlinearity is usually solved [4] by preliminary linearization of the equations. A large number of equations remain, which, together with the small number of variables in the separate equations, leads to incomplete Jacobians. To overcome these problems, methods are used for packing the Jacobian matrix, which further essentially increases the effectiveness of the algorithms for simulation.

The simultaneous equation approach overcomes the problems during simulation of CPS entirely by mathematical means, but this obstructs the close scrutiny of the physical relevance of the results obtained in progressing to the solution.

The simultaneous module approach is the opposite of the simultaneous equation approach. In this approach equations are searched for which can be solved separately with respect to one of the outlet variables; when this is impossible groups of equations are determined, and these should be solved as a system. The sequential module approach has quite a wide application range. Here the simulation of the CPS is reduced to sequential simulation of the blocks (modules). This approach in practice traces the physical sequence of the processes and at each step the simulation can be stopped if physically unreasonable results are obtained.

1.3 Sequential Module (Hierarchical) Approach

Simulation of CPS using sequential simulation of the separate blocks is possible if it is built as one hierarchical method, i.e., this is possible by building one hierarchy structure for simulation. The main premise of this method is the obvious fact that one block can be simulated if all its inlet variables are regulated (known). From this it follows that CPS can be simulated sequentially block by block if such a sequence (row) of blocks is found where the inlet variables of each block are the outlet variables of the preceding blocks in a row or are inlet blocks of the CPS. This, of course, is possible only if all inlet regime variables are regulated, which is seen directly in the example in Fig. 1. Obviously, for given inlet variables of streams 1 and 2 blocks in the sequence 1, 4, 5, 2, 3, 6 can be simulated and as a result the outlet regime variables of streams 7 and 9 are determined.

1.4 Acyclic Chemical Plant Systems

The order of simulation of CPS, shown in Fig. 1 is obtained directly from the condition that in this sequence only blocks whose inlet variables are known can be recorded, i.e., given (specified) or calculated as a result of the simulation of the blocks already recorded. This is possible only for acyclic (open) CPS, i.e., when in the scheme there are no recirculation streams. For a small number of blocks obviously this task is solved elementarily, but for a large number of blocks or for repeated solving, utilization of computers and appropriate algorithms is necessary.

The topological structure of CPS can be expressed in different ways. Here we shall use [3, 5] a matrix of the streams

$$\boldsymbol{A} = \|a_{\mathrm{ik}}\|, \quad \mathrm{i} = 1, \ldots, \mathrm{I}, \quad \mathrm{k} = 1, \ldots, \mathrm{K}, \tag{1.7}$$

where I and K are the whole numbers of the blocks and streams, and the values of a_{ik} are

$$a_{\mathrm{ik}} = \begin{cases} 1 & \text{if the kth stream is the inlet in the ith block;} \\ -1 & \text{if the kth stream is the outlet from the ith block;} \\ 0 & \text{if the kth stream is not connected to the ith block.} \end{cases} \tag{1.8}$$

The matrix $\boldsymbol{A}$ contains all the information for the CPS and to it can only be added the vector of the number of the parameters of the streams:

$$\boldsymbol{B} = |b_{\mathrm{k}}|, \quad \mathrm{k} = 1, \ldots, \mathrm{K}, \tag{1.9}$$

where b_{k} is the number of the parameters of the kth stream.

All this information is sufficient for determination of the order of the simulation of the blocks which is recorded in the vector of the order of the calculation:

$$\boldsymbol{C} = |c_{\mathrm{j}}|, \quad \mathrm{j} = 1, \ldots, \mathrm{I}, \tag{1.10}$$

where c_{j} are the numbers of the blocks recorded in a suitable sequence.

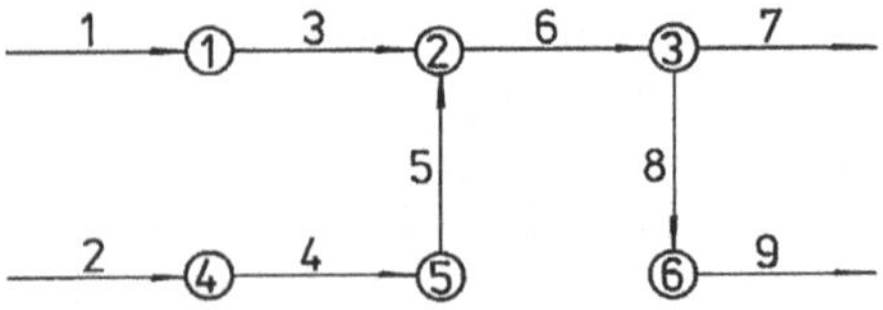

Fig. 1 Order of simulation of chemical plant systems

The vector $\boldsymbol{C}$ can be filled consecutively with the help of two executive vectors $\boldsymbol{D}$ and $\boldsymbol{E}$.

The vector of the inlet blocks contains all the information about the blocks having just one known inlet stream:

$$\boldsymbol{D} = |d_{\mathrm{i}}|, \quad \mathrm{i} = 1, \ldots, \mathrm{I}, \tag{1.11}$$

where

$$d_{\mathrm{i}} = \begin{cases} 1 & \text{if the ith block has just one known inlet stream;} \\ 0 & \text{in the other case.} \end{cases} \tag{1.12}$$

The vector of the particular blocks contains information about the blocks having just one unknown inlet stream:

$$\boldsymbol{E} = |e_{\mathrm{i}}|, \ \mathrm{i} = 1, \ldots, I, \tag{1.13}$$

where

$$e_{\mathrm{i}} = \begin{cases} 1 & \text{if the ith block has just one known inlet stream;} \\ 0 & \text{in the other case.} \end{cases} \tag{1.14}$$

The utilization of vectors $\boldsymbol{D}$ and $\boldsymbol{E}$ enables consecutive filling of $\boldsymbol{C}$ with the help of the following algorithm:

1. Substitute $d_{\mathrm{i}} = 0, \quad e_{\mathrm{i}} = 0, \quad \mathrm{i} = 1, \ldots, \mathrm{I}$.
2. From **D** we determine the numbers of the inlet streams of the CPS (an example from $\sum_{\mathrm{i}=1}^{\mathrm{I}} a_{\mathrm{ik}} = 1$).
3. From the numbers of the inlet streams (through $\boldsymbol{A}$) are determined the numbers of the inlet blocks and we create vector $\boldsymbol{D}$.
4. Check whether there are blocks in $\boldsymbol{D}$ for which all inlet streams are known (determined in step 2).

   (a) If there are, go to step 5.
   (b) If there are not, go to step 6.

5. Blocks for which all inlet streams are known are recorded in $\boldsymbol{C}$. They are deleted from $\boldsymbol{A}$ (their rows) and from $\boldsymbol{D}$. After that put $e_{\mathrm{i}} = d_{\mathrm{i}}, \ \mathrm{i} = 1, \ldots, \mathrm{I}$ and begin from step 1.
6. Check if $e_{\mathrm{i}} = 0, \quad \mathrm{i} = 1, \ldots, \mathrm{I}$.

   (a) If it does, go to step 7.
   (b) If it does not, go to step 8.

7. End (the order of the calculation in acyclic CPS is determined or is an independent contour).
8. Begin determination of the serial independent contour.

1.5 Cyclic Chemical Plant Systems

From the algorithm in the previous section it is obvious that it is not possible in every case to determine to the very end the order of the simulation. In these cases CPS are cyclic (closed), i.e., they have circulation streams. As an example from Fig. 2a, it is clearly seen that the additional circulating stream 10 (in comparison with Fig. 1) does not give us the possibility to apply the algorithm. This problem could be solved if recirculated stream 10 is broken (Fig. 2b). Then the order of the simulation of the CPS in Figs. 1 and 2b could be the same. This would allow us to obtain the values of the outlet regime variables of streams 7, 9, and 10″ for given values of the inlet regime variables of streams 1, 2, and 10′. Obviously breaking stream 10 will not affect the simulation if the variables (the parameters) of streams 10′ and 10″ are equal. In practice this does not happen and that is why after the simulation the values of the parameters of 10″ are assumed for the values of the parameters of 10′ and the simulation is repeated. This iterative procedure ends when the given values of the parameters of 10′ and the calculated values of the parameters of 10″ are equal with the preliminarily required precision. In this way the simulation of one cyclic CPS is reduced to repeated simulation of acyclic CPS obtained after suitable breaking of the circulating streams.

1.6 Independent Contours

From Fig. 2a is clear that the cyclic CPS has an acyclic part (blocks 1 and 4) and contours (blocks 2, 3, 5, and 6). The sequence of blocks (2, 3, 6, 5) is connected with one-way streams (6, 8, 10, 5) and that is why going from a given block by way of the streams we reach the same block; this is called an elementary contour.

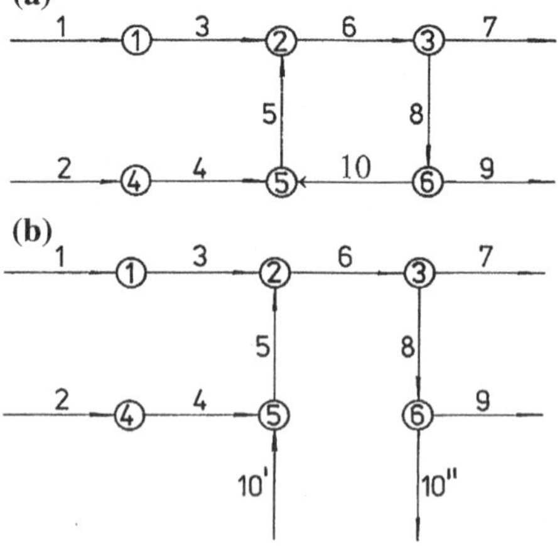

Fig. 2 Cyclic chemical plant systems. **a** Cyclic system; **b** acyclic system if recirculated stream 10 is broken

As already demonstrated, to determine the order of the simulation of the blocks in an elementary contour it is necessary to break one of its streams.

The aggregate of elementary contours which has common blocks represents an independent contour. Determination of the order of the simulation of the blocks in the independent contours is possible only after breaking all elementary contours of which they are composed.

From every mentioned, we see that CPS can be examined as being composed of acyclic parts and independent contours. The blocks in the acyclic parts are simulated only once and the blocks in the separate independent contours (after their breaking) are simulated repeatedly (iteratively). In these iterations blocks from only one independent contour participate. All this is possible if we determine the independent contours in CPS and the elementary contours of which they are composed on the basis of a structural analysis of CPS [1–5].

The algorithm for determination of the elementary contours in an independent contour is shown in Fig. 3.

In the literature different methods for determination of the independent contours have been reported [1–5]. Here will be use the method given in [5], which first determines one block from the independent contour and after that starts from it and traces routes from blocks in a direction opposite that of the streams until the cycle is closed, i.e., finding the elementary contour. Repeating this procedure for all branches from the route leads to the determination of all elementary contours. For this purpose are used the vector of the routes $\boldsymbol{F}$ and the vector of the branchings $\boldsymbol{G}$:

$$\boldsymbol{F} = |f_s|, \quad s = 1,\ldots,S > I, \quad \boldsymbol{G} = |g_t|, \quad t = 1,\ldots,T > I, \quad \boldsymbol{G} = |g_t|, \quad t = 1,\ldots,T > I, \tag{1.15}$$

where in an appropriate way the routes (in a direction opposite that of the streams) and their branches are recorded.

The algorithm for determination of the independent contour [5] starts as a continuation of step 8 in the algorithm in Sect. 1.4 when the vector of the particular blocks is not empty and obviously the numbers of the blocks from the serial independent contour are recorded:

1. First, in $\boldsymbol{F}$ the number of the block corresponding to the first nonzero element in $\boldsymbol{E}$ is recorded. Then, this element in $\boldsymbol{E}$ is equal to zero and in $\boldsymbol{G}$ -1 is recorded (with -1 the branches of the block from the route are delimited). For the concrete example $\boldsymbol{F} = 1, 0, \ldots, \boldsymbol{G} = -1, 0, \ldots$

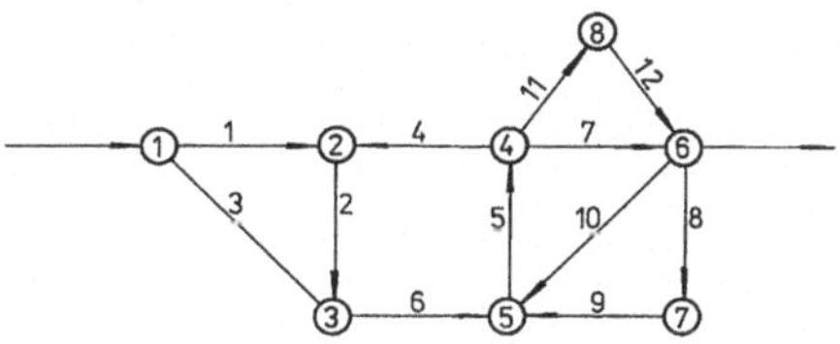

Fig. 3 Independent contours

2. From $\boldsymbol{A}$ are determined the blocks whose outlet streams are the inlet for the block whose number is the last nonzero element in $\boldsymbol{F}$ (branch from the last block in $\boldsymbol{F}$ in a direction opposite that of the streams) and these are recorded consecutively in $\boldsymbol{G}$, i.e., $\boldsymbol{F} = 1, 0,\ldots, \boldsymbol{G} = -1, 3, 0, \ldots$.
3. Check whether there are nonzero elements in $\boldsymbol{G}$.

   (a) If there are, go to step 4.
   (b) If there are not, go to step 13.

4. Find the last nonzero element in $\boldsymbol{G}$ (i.e., 3).
5. Check whether the last nonzero element in $\boldsymbol{G}$ is -1.

   (a) If it is, go to step 6.
   (b) If it is not, go to step 7.

6. Make the last nonzero elements in $\boldsymbol{F}$ and $\boldsymbol{G}$ equal to zero and start from step 3.
7. Check whether the last nonzero element in $\boldsymbol{G}$ is not present in $\boldsymbol{E}$.

   (a) If it is not present, go to step 8.
   (b) If it is present, go to step 9.

8. Make $\boldsymbol{F}$, $\boldsymbol{G}$, and $\boldsymbol{Q}$ equal to zero and start from step 1.
9. Check whether the last nonzero element in $\boldsymbol{G}$ is recorded in $\boldsymbol{F}$.

   (a) If it is, go to step 10.
   (b) If it is not, go to step 12.

10. This means that the serial elementary contour is revealed in $\boldsymbol{F}$. Detect the streams which connect the last blocks in $\boldsymbol{F}$ to the repeated block and record them in an appropriate way in the matrix of the elementary contours $\boldsymbol{Q} = |q_{\mathrm{wk}}|$, $\mathrm{w} = 1, 2, \ldots$, $\mathrm{k} = 1, \ldots, \mathrm{K}$, where $q_{\mathrm{wk}} = 1$ when the kth stream participates in the wth elementary contour and $q_{\mathrm{wk}} = 0$ in the opposite case.
11. The last nonzero element in $\boldsymbol{G}$ is deleted and start from step 3.
12. Make the last nonzero element in $\boldsymbol{G}$ equal to zero and recorded it in $\boldsymbol{F}$; in $\boldsymbol{G}$ record -1 and start from step 2, i.e., $\boldsymbol{F} = 1, 0, \ldots$, $\boldsymbol{G} = -1, 3, 0, \ldots$
13. All elementary contours are discovered in the serial independent contour and start to find the optimal breaking sets (OBS).

In such an algorithm steps 2 and 7 guarantee the determination of the elementary contours only in one independent contour. For the example in Fig. 3, after step 12 we go back to step 2 ($\boldsymbol{F} = 1, 3, 0, \ldots$; $\boldsymbol{G} = -1, 2, 0, \ldots$). After that we again reach step 12 ($\boldsymbol{F} = 1, 3, 2, 0, \ldots$; $\boldsymbol{G} = -1, -1, -1, 0, \ldots$) and go back in step 2 ($\boldsymbol{F} = 1, 3, 2, 0, \ldots$; $\boldsymbol{G} = -1, -1, -1, 4, 1, 0, \ldots$). The process continues through steps 3, 4, 5, 7, 9, and 10, i.e., an elementary contour is discovered which consists of blocks 1, 3, and 2. In this way six elementary contours are determined in the independent contour in Fig. 3: I—1, 2, 3; II—2, 3, 5, 4; III—4, 6, 7, 5; IV—4, 6, 5; V—5, 4, 8, 6; VI—4, 5, 7, 6, 8. They are recorded in the matrix of the elementary contours, which looks like (1.16), where the last row represents the vector of the parameters.

W\K	1	2	3	4	5	6	7	8	9	10	11	12
I	1	1	1									
II		1		1	1	1						
III					1		1	1	1			
IV					1		1			1		
V					1					1	1	1
VI					1			1	1		1	1
b_k	3	2	3	1	3	1	4	1	1	1	1	1

(1.16)

1.7 Breaking Sets

The independent contour is transformed into an open scheme if all its elementary contours are broken. This can be realized from different combinations of a minimum number of streams, which we will call *minimum breaking sets* (MBS). As an example for Fig. 2a, every one of streams 5, 6, 8, and 10 represents an MBS. For the independent contour in Fig. 3, one MBS contains streams 1 and 5.

The presence of more than one MBS for a given independent contour raises the question for selection between them. For this purpose, after the breaking of the streams from the MBS iterative procedures follow till the parameters of the streams are equal (with a given precision) at the places where the breaking occurs. Obviously the number of iterations will depend on the number of parameters of the broken streams. From this it follows that from all the MBS, the one that should be used is the one which has the minimum summary number of parameters of the streams participating in it; we will call this the *optimal breaking set* (OBS).

In the literature different methods for determination of OBS have been described [1–5]. We will consider one of the most effective methods [2], which uses the matrix of the elementary contours (1.16) according to the following algorithm:

1. Those streams in $\boldsymbol{Q}$ which cannot participate in OBS (annulment of the corresponding columns in 1.16), i.e., from two streams one falls away which breaks a smaller (or equal) number elementary contours and has a bigger (or equal) number of parameters. For this purpose it is sufficient to find columns in (1.16) which are contained in other columns and are annulled if they have a greater (equal) number of parameters.
2. As a result, from the procedures in step 1 it is possible for rows to appear in (1.16) with rank 1, i.e., rows which contain only one unit. This means that these are elementary contours which could be broken only with a stream whose number corresponds to this unit. This forces this stream to be memorized and subsequently entered directly in MBS or OBS. All contours (rows) which are broken from these streams could fall away (to be annulled) from matrix $\boldsymbol{Q}$ (1.16).

3. As a result, from the procedures in steps 1 and 2 it is possible for a row (contour) to appear in (1.16) which is entirely contained in another row, i.e., the breaking of the first one leads to automatic breaking of the second one. This gives the opportunity in (1.16) for all elementary contours to fall away which are broken automatically on breaking of the others.
4. After the operations in steps 1–3 in (1.16) there remain the minimum number of streams, from which can be obtained different MBS. For this purpose:
   (a) MBS of the first elementary contour are determined—they contain one stream from it.
   (b) MBS for the first and the second contours are determined through appropriate combinations between MBS for the first contour and the streams from the second contour.
   (c) MBS for the first three contours are determined through appropriate combinations between MBS of the first two contours and the streams of the third contour.

   This recursive procedure continues until all elementary contours in the independent contour are exhausted.
5. To the four MBS obtained are added the streams memorized in step 2. For each of the MBS obtained, the summary number of parameters of the streams are determined. That (those) which has (have) the minimum summary number of parameters represents (represent) the OBS.

This algorithm can be demonstrated simply by matrix (1.16):

1. As a result of step 1 in the previous algorithm columns 1, 3, 4, 7, 8, are 11 are annulled.
2. The first row has rank 1, i.e., the number of stream 2 is memorized and rows I and II are annulled because the first two contours are broken by stream 2.
3. Row III is contained entirely in row IV, i.e., the last one is annulled. Row IV is contained entirely in row V and the last one is annulled.
4. From (1.16) only columns 5, 9, and 10 and rows III and IV remain. We will look further at two contours from the following streams (1.17) 5, 9 and 5, 10. According to the recursive procedure, there are two MBS for the first contour—5 and 9. The MBS for the first two contours are 5 and 9, 10.
5. To the MBS, obtained in step 4, must be added stream 2, i.e., the MBS are 2, 5 and 2, 9, 10. Their summary numbers of parameters are, respectively, 5 and 4, i.e., the OBS is 2, 9, 10.

1.8 Optimal Order

The algorithms examined in this section permits a structural analysis of CPS to be made, as a result the acyclic parts, independent contours, and the streams to be determined have to be broken for the algorithm for simultaneous module simulation to be applied for the independent contours. On this basis we can find the optimal order for simulation of the apparatuses in CPS. This optimum consists of separating the apparatuses into two parts, one of which is simulated once (acyclic part) and the other is simulated repeatedly (iterative simulation of the independent contours). The second substantial peculiarity of the optimal order is that the iterative procedures envelop the apparatuses of only one (serial) independent contour, which decreases the number of iterations. This acceleration of the iterative calculations is increased by OBS, i.e., finding the optimal order of simulation inside the independent contour.

Thus, the optimal order for simulation of the blocks in CPS can be obtained through the following algorithm:

1. Input of the information for the technological structure for CPS, i.e., matrix $\boldsymbol{A}$ (1.7) and vector (1.9).
2. Determination of the order for simulation in acyclic CPS with the help of the algorithm described after (1.14).
3. Check of whether the numbers of all blocks are recorded in the vector of the optimal order $\boldsymbol{C}$ (9.1.10).
   (a) If they are, go to step 7.
   (b) If they are not, go to step 4.
4. Determination of the elementary contours in the serial independent contour with the help of the algorithm described after (1.15).
5. Determination of the OBM of the serial independent contour with the help of algorithm described after (1.16).
6. Breaking of the serial independent contour (in the columns of $\boldsymbol{A}$ corresponding to the OBS -1 is replaced with 0) and start from step 2.
7. End. The vector $\boldsymbol{C}$ is filled up. In parallel, it is convenient to form the vector $\bar{\boldsymbol{C}} = |\bar{c}_j|$, $j = 1, \ldots, I$, where $\bar{c}_j = 0$ for the blocks from the acyclic part, $\bar{c}_j = 1$ for the blocks from the first independent contour, $\bar{c}_j = 2$ for the blocks from the second independent contour, etc.

The optimal order for simulation of apparatuses in CPS obtained in this way is used in cases when all inlet regime variables are regulated and all outlet regime variables are free.

2 Simulation for Specified Outlet Variables

In a lot of cases and especially for the purposes of design, the simulation of CPS when some of the outlet regime variables are specified [5–9] is necessary. Thus, as an example Fig. 4 shows CPS where the variables $\bar{x}$ and $\bar{y}$ are specified. Obviously the simultaneous module approach proposed in Sect. 1 cannot be applied directly because of the existence of *specified outlet variables* (SOV) $(\bar{y}_1, \ldots, \bar{y}_5)$. This problem can be solved if the function is composed as

Fig. 4 Zone of influence

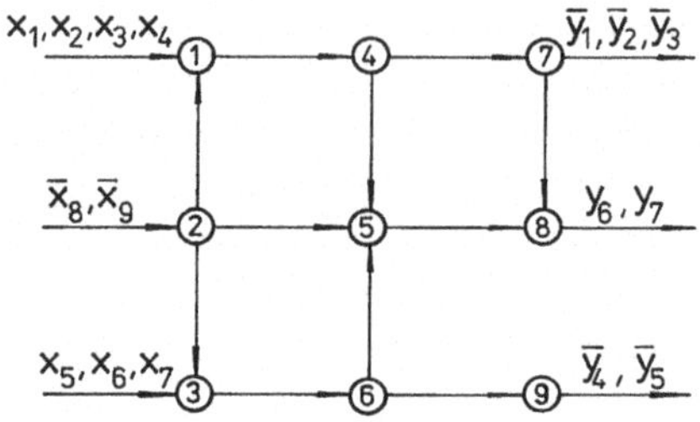

$$F(x_1,\ldots,x_7) = \sum_{i=1}^{5} (y_i - \bar{y}_i)^2, \tag{2.1}$$

which can be minimized by varying the free inlet variables $x_1,\ldots, x_7$, as for each composition of values of $x_1,\ldots,x_7$, are determined $y_1,\ldots,y_5$, using the method for simulation described in Sect. 1 (obviously the order of simulation will be 2, 1, 3, 4, 6, 5, 7, 8, 9). If we denote the coordinates of the minimum of (2.1) as $\bar{x}_1,\ldots,\bar{x}_7$, it is obvious that the simulation of CPS in Fig. 4 can be made through the algorithms in Sect. 1 if we accept $\bar{x}_1,\ldots, \bar{x}_7$, as specified variables and $y_1,\ldots, y_7$, as free variables. Under these conditions the specified values $\bar{y}_1,\ldots, \bar{y}_5$, are obtained automatically as a result of the simulation for such specified variables $\bar{x}_1,\ldots, \bar{x}_9$, [5–9].

The proposed approach for simulation for the existence of SOV is difficult owing to the fact that a minimization problem with large dimension has to be solved. The decrease of the dimension of the problem can be achieved through its decomposition. Thus, as an example instead of the minimum of F from (2.1) we could search for the minimum of two other functions with a smaller number of variables:

$$F_1(x_1,\ldots,x_4) = \sum_{i=1}^{3} (y_i - \bar{y}_i)^2, \quad F_2(x_5,\ldots,x_7) = \sum_{i=4}^{5} (y_i - \bar{y}_i)^2. \tag{2.2}$$

In (2.1) and (2.2) obviously the number of SOV should not exceed the number of free inlet variables, i.e., the CPS should have the necessary number of degrees of freedom. This condition will be further called a "parametric condition."

2.1 Zone of Influence

The decomposition approach is possible because the free inlet variables of block 1 (Fig. 4) influence the SOV of block 7 and do not influence the SOV of block 9. In contrast, the free inlet variables of block 3 influence the SOV of block 9 and have no influence on the SOV of block 7. That is because of the fact that block 1 enters into the zone of influence of block 7 and block 3 enters into the zone of influence of block 9 [5–9].

The zone of influence of a block represents a combination of blocks from which can be reached the block moving towards the direction of the technological streams. Thus, as an example in Fig. 4, the zone of influence of block 8 contains blocks 1–7.

We will further examine zones of influence of blocks with SOV. These blocks will be most often used as the outlet for CPS but they could be the inlet or somewhere in-between. If in CPS there is an independent contour and one block from the contour participates in a zone of influence, then obviously the other blocks from the contour participate in it also. When searching for zones of influence of the independent contours this permits the blocks to be replaced with superblocks, i.e., a CPS is transformed from acyclic into cyclic. If the block with SOV participates in an independent contour, the other blocks from the contour have the same zone. This permits the superblock to replace in these cases the block with SOV. From this it follows that the first step when searching for zones of influence is structure analysis of CPS through the algorithms in Sect. 1 and the independent contours are replaced with superblocks. We will further examine the acyclic CPS thus obtained.

The zone of influence of a block with SOV is obtained if we start from this block and trace all the routes in a direction opposite that of the streams. For this purpose an algorithm is used to search for the elementary contours (Sect. 1), with the difference that we do not start from a particular block but from the block with SOV and stop the searching on a given block.

The route is not complete when the number of the block appears over again (closing of an elementary contour) but when the number of an inlet block in the CPS is reached. The results from these procedures are saved in the matrix of the zones of influence, where in the first column are saved the numbers of the blocks with SOV and the other columns correspond to the numbers of the blocks in CPS. The rows correspond to the blocks with SOV and contain their zones of influence. As an example, the matrix of the zones of influence of blocks 7 and 9 corresponding to Fig. 4 appear like (2.3).

	1	2	3	4	5	6	7	8	9
7	1	1		1					
9		1	1			1			

(2.3)

Having in mind that the SOV will be influenced by the free inlet variables, it is obvious that the inlet blocks should be determined, i.e., the inlet zone of influence should be obtained. In this way from (2.3) we get directly (2.4):

	1	2	3	4	5	6	7	8	9
7	1	1							
9		1	1						

(2.4)

2.2 *Absolutely Independent Influence*

From the matrix of the inlet influence (2.4) we can obtain the zones of absolutely independent inlet influence (2.4). For this purpose all columns in the matrix of the zone of inlet influence are annulled which contain more than one unit. In this way we get the matrix for absolutely independent inlet influence (2.5):

	1	2	3	4	5	6	7	8	9
7	1								
9			1						

(2.5)

From (2.5) it can be seen that in the zone of absolutely independent inlet influence of block 7 (with SOV), block 1 enters and for block 9, block 3 enters. From this it follows that SOV of blocks 7 and 9 can be obtained through minimization of the function in (2.2), i.e., through varying (separately and absolutely independently) the free inlet variables of blocks 1 and 3. In this way we can be solve the problem of the simulation of CPS with SOV as the values of x_j at the points of the minima are accepted as regulated values $\bar{x}_j(j = 1, \ldots, 7)$, and the regulated values $\bar{y}_i$ are for free $y_i(i = 1, \ldots, 5)$. For this purpose obviously two conditions must be fulfilled:

1. The matrix of the zones of absolutely independent inlet influences (2.5) should not have zero rows.
2. The parametric condition should be fulfilled for each row in (2.5)

In cases when one of these conditions is not fulfilled, zones of independent inlet influence are searched for.

2.3 *Independent Influence*

The determination of the zones of independent inlet influence is made in a different way depending on which of the two conditions listed at the end of the previous section is not fulfilled. For this reason, two examples will be scrutinized. Figure 5 shown CPS whose zone of inlet influence looks like (2.6). When searching for the zone of absolutely independent inlet influence, we get matrix (2.7), which has one nonzero row, i.e., there are no zones of absolutely independent inlet influence (but not absolutely independent) and we check whether the parametric condition is fulfilled.

	1	2	3	4	5	6	7	8	9	10	11	12
9	1	1	1									
10		1	1									
12			1	1								

(2.6)

	1	2	3	4	5	6	7	8	9	10	11	12
9	1											
10												
12				1								

(2.7)

If it is fulfilled, minimization of the functions is suggested:

$$F_1(x_1,\ldots,x_3)=\sum_{i=1}^{3}(y_i-\bar{y}_i)^2,\quad F_2(x_6,\ x_7)=\sum_{i=7}^{8}(y_i-\bar{y}_i)^2. \tag{2.8}$$

In (2.6) the rows for which the zones of independent inlet influence are obtained are made equal to zero and to the rest of the rows are applied operations for obtaining the zone for absolutely independent inlet influence. As a result, the matrix (2.9) is obtained. For the nonzero rows thus obtained, the parametric condition is checked and if it is fulfilled, minimization of the function is suggested:

	1	2	3	4	5	6	7	8	9	10	11	12
10		1	1									

(2.9)

$$F_3(x_4,\ x_5)=\sum_{i=4}^{5}(y_i-\bar{y}_i)^2. \tag{2.10}$$

In the example examined (Fig. 5), CPS have no zones of absolutely independent inlet influence but there are zones of independent inlet influence if the minimization of the functions is made in a definite sequence F_3, F_1, F_2, where F_1 and F_2 could arbitrarily change places. In this way, in a first step the specified values $\bar{x}_4$, $\bar{x}_5$, are determined and in the second step $\bar{x}_1,\ldots,\bar{x}_3$ and $\bar{x}_6$, $\bar{x}_7$ or $\bar{x}_6$, x_7 and $\bar{x}_1,\ldots,\bar{x}_3$ are determined.

The other characteristic example is from Fig. 4 if the specified inlet streams are $\bar{x}_3, \bar{x}_4, \bar{x}_7$, and $\bar{x}_9$. From this it follows that the parametric condition for the first row

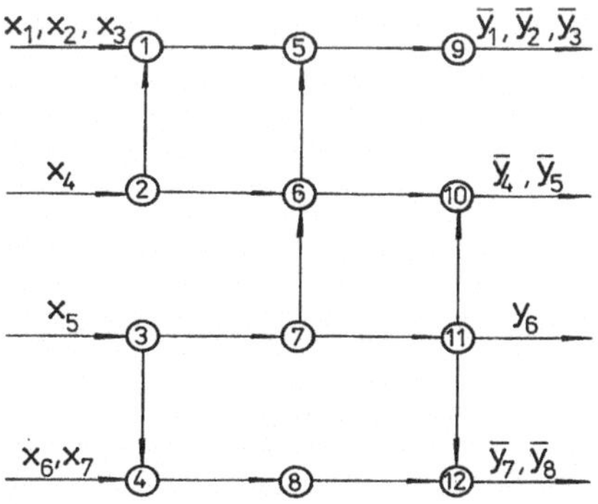

Fig. 5 Independent influence

of matrix (2.5) is not fulfilled. Then the rows (the second) for which the parametric condition is fulfilled require that the function F_1 be minimized:

$$F_1(x_5,\ x_6) = \sum_{i=4}^{5} (y_i - \bar{y}_i)^2. \qquad (2.11)$$

After that these rows are made equal to zero in (2.4) and to it are applied again the operations for obtaining the matrix of absolutely independent inlet influence. As a result, matrix (2.12) is obtained:

	1	2	3	4	5	6	7	8	9	10	11	12
7	1	1										

(2.12)

The parametric condition for the rows in (2.12) is checked and if it is fulfilled, minimization of the function is suggested:

$$F_2(x_1, x_2, x_8) = \sum_{i=1}^{3} (y_i - \bar{y}_i)^2. \qquad (2.13)$$

From what has been described it is seen that CPS in Fig. 4 with specified variables $\bar{x}_3, \bar{x}_4, \bar{x}_7$ and $\bar{x}_9, \bar{y}_1, \bar{y}_2, \bar{y}_3, \bar{y}_4$ and $\bar{y}_5$ could be simulated if functions (1.11) and (2.13) are preliminarily minimized in sequence F_2, F_1. At the points of the minima $\bar{x}_1, \bar{x}_2, \bar{x}_8$ and $\bar{x}_5, \bar{x}_6$ are determined, which are further accepted as specified variables and $\bar{y}_1, \bar{y}_2, \bar{y}_3, \bar{y}_4$, and $\bar{y}_5$ are accepted as free variables. This means this CPS has zones of independent inlet influence but it is not absolutely independent, i.e., it is necessary to respect a definite sequence.

2.4 Combined Zones

The two algorithms examined for simulation of CPS with SOV through determination of the zones of absolutely independent and independent inlet influence cannot cover the cases when the matrix of the zones of absolutely independent inlet influences (2.5) is zero or when in the matrix of independent inlet influence (2.7) there is a row for which the parametric condition is not fulfilled. For these cases two algorithms are created [6] for determination of combined zones for independent inlet influence. For this purpose combinations are found from two (if this is not necessary, from three and more) blocks with SOV and the zone of influence for every combination is determined—combined zone of independent inlet influence.

3 Models of Separate Blocks

The simulation of CPS is connected with the necessity for simulation of a large number of processes (blocks) and one substantial part of them (which participates in the independent contours) is simulated repeatedly. This forces a series of conditions when creating the models of the blocks (calculation modules) in CPS [1, 5].

3.1 Types of Modules

The necessary time for simulation of CPS depends on the time for simulation of the separate blocks, the number of blocks, and the number of iterations (at acyclic CPS). From these, in reality, two factors can be decreased, the first and the third one.

The time for simulation of the separate blocks depends on the effectiveness of the calculation algorithm and the complexity of the mathematical structure of its mathematical description. This means that decrease of this time could be achieved through simplifying the mathematical description. This could be realized in different ways, of which the most used one is not accounting for the temperature dependence of the parameters, replacing the kinetic and balance equations, replacing the kinetics constants with coefficients of effectiveness, and linearization of the equations. This leads to the possibility of composing different kinds of modules depending on their precision (accuracy), as the inaccuracy in creating calculation procedures on this basis is on account of increasing the quickness of their operation.

The number of the iterations (at correctly found OBS of streams) depends on the initial approximations of the values of the parameters of the broken streams, which starts with the iteration procedure. This forces the search for a good initial approximation.

All this gives us the opportunity to create effective algorithms for simulation of two stages. The first stage represents simulation with the help of simple but high-speed modules for not good initial approximations of the parameters. This gives the opportunity to realize quickly a large number of iterations from which are determined not very exact approximate values of the parameters of the broken streams. In the second stage these values are used as good initial approximations for real simulation with precise modules, which leads to one accurate final result with a small number of iterations.

Simulation of CPS for the purposes of the optimal control or design sets also different conditions on the calculation modules. As an example, the modules for control calculate the outlet regime variables for given inlet regime variables of the process and constructive variables of the apparatus. On the other hand, the modules for design calculate the constructive variables (parameters) of the apparatus for given inlet and outlet regime variables of the process. This does not exclude the possibility of using different modules for control or design in a simple or precise variant depending on the accuracy of the initial approximations.

The different kinds of calculation modules described can be applied in different stages when creating CPS:

1. Determination of the outlet regime variables for given inlet regime variables through balance equations with given coefficients of effectiveness and constant physical parameters (example in relation to temperature).
2. Repeat step 1 with good initial approximations and precise balance equations.
3. Determination of the constructive parameters of the apparatuses through modules for design and the values obtained in step 2 for the inlet and outlet regime variables.
4. Determination of the outlet regime variables through kinetic equations and the given inlet regime variables, constructive variables, and constant physical parameters.
5. Simulation of CPS through precise kinetic models and good initial approximations.

The optimal design and control of CPS is achieved through solving the optimization problems at each of the above-mentioned stages.

Some of the considerations so expressed will be illustrated through simplified modules for control of processes in heat exchangers, separators, and chemical reactors [1, 5].

3.2 Heat Transfer

A countercurrent heat transfer apparatus (Fig. 6) will be examined with constructive variables:

$$\boldsymbol{a} = |F,\ a_{\mathrm{q}}|, \tag{3.1}$$

where $\boldsymbol{a}$ is the vector of the constructive variables, F is the heat transfer surface, and a_{q} is the vector of the other constructive variables.

In the heat exchanger cold and hot streams and their vectors of the inlet regime variables are, respectively,

$$\begin{aligned} \boldsymbol{x}_1 &= |x_{1\mathrm{f}}, x_{1\mathrm{t}}, x_{1\mathrm{c}}, x_{1\mathrm{p}}| \quad \text{cold stream,} \\ \boldsymbol{x}_2 &= |x_{2\mathrm{f}}, x_{2\mathrm{t}}, x_{2\mathrm{c}}, x_{2\mathrm{p}}| \quad \text{hot stream,} \end{aligned} \tag{3.2}$$

where x_{f}, x_{t}, x_{c} are the capacity, temperature, and the specific heat of the stream and x_p is the vector of the rest of the parameters of the stream. Analogously we can write the vectors of the outlet regime variables:

$$\boldsymbol{y}_1 = |y_{1\mathrm{f}}, y_{1\mathrm{t}}, y_{1\mathrm{c}}, y_{1\mathrm{p}}|, \quad \boldsymbol{y}_2 = |y_{2\mathrm{f}}, y_{2\mathrm{t}}, y_{2\mathrm{c}}, y_{2\mathrm{p}}|. \tag{3.3}$$

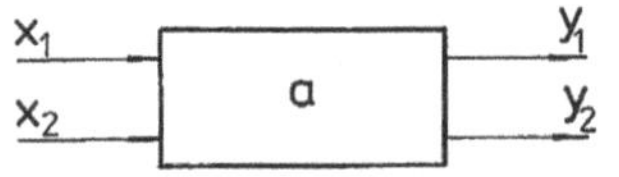

Fig. 6 Heat transfer module

The mathematical description of the process is obtained from the balance equation at countercurrent flow:

$$Q = x_{1f}x_{1c}(x_{1t} - y_{1t}) = x_{2f}x_{2c}(y_{2t} - x_{2t}), \tag{3.4}$$

where Q depends on the effectiveness of the heat transfer and should be expressed through a given coefficient of effectiveness.

The maximum quantity heat which could be exchanged between the cold and the hot stream is determined by their final temperatures for a sufficiently large heat transfer surface:

$$y_{1t} = y_{1t_{max}} = x_{2t}, \quad Q_{max} = Q'_{max}, \quad y_{2t} = y_{2t_{min}} = x_{1t}, \quad Q_{max} = Q''_{max}, \tag{3.5}$$

where

$$Q'_{max} = x_{1f}x_{1c}(y_{1t_{max}} - x_{1t}) = x_{1f}x_{1c}(x_{2t} - x_{1t}),$$

$$Q''_{max} = x_{2f}x_{2c}(x_{2t_{max}} - y_{2t_{min}}) = x_{2f}x_{2c}(x_{2t} - x_{1t}). \tag{3.6}$$

Obviously $Q_{max} = \min\{Q'_{max}, Q''_{max}\}$, i.e.,

$$Q_{max} = \begin{cases} Q'_{max} & \textit{if } x_{1f}x_{1c} < x_{2f}x_{2c}; \\ Q''_{max} & \textit{if } x_{1f}x_{1c} > x_{2f}x_{2c}. \end{cases} \tag{3.7}$$

Now, we can define the coefficient of the effectiveness of the heat transfer:

$$Q = EQ_{max}. \tag{3.8}$$

Thus mathematical description of the heat transfer obtained permits the creation of a simplified (but highly effective) calculation module through the following algorithm:

1. Set x_1, x_2 and E.
2. Determine $(x_f x_c)_{min} = \min\{x_{1f}x_{1c}, x_{2f}x_{2c}\}$.
3. Determine $Q_{max} = (x_f x_c)_{min}(x_{2t} - x_{1t})$.
4. Determine $Q = EQ_{max}$.
5. Determine $y_{1t} = x_{1t} + Q/x_{1f}x_{1c}$.
6. Determine $y_{2t} = x_{2t} - Q/x_{2f}x_{2c}$.
7. Substitute $y_{1f} = x_{1f}$, $y_{2f} = x_{2f}$.
8. Substitute $y_{1c} = x_{1c}$, $y_{2c} = x_{2c}$, $y_{1p} = x_{1p}$, $y_{2p} = x_{2p}$.

3.3 Separation

Different processes could be introduced into the mathematical model as calculation modules "separators" (Fig. 7), where one multicomponent mixture is

Fig. 7 Separation module

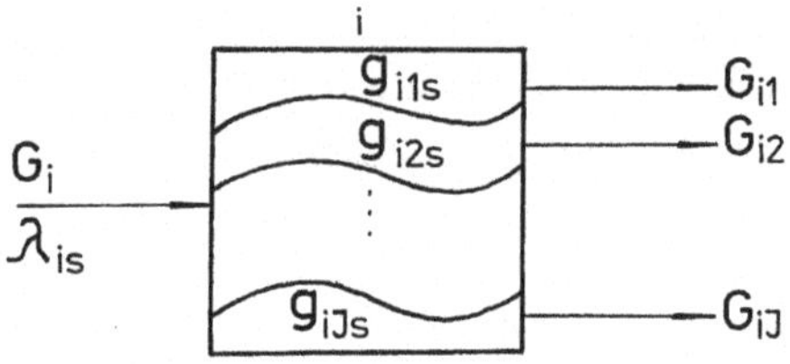

separated into several phases (parts). If with i we denote the number of the block, s = 1,…, S is the number of the component and j = 1,…, J is the number of the phase, then G_i is the quantity of the mixture coming in for separation and λ_{is} is its composition:

$$G_i = \sum_{s=1}^{S} \lambda_{is}. \tag{3.9}$$

In the separator the inlet stream is divided into J phases G_{ij} = (j = 1,…, J), as each phase contains quantity g_{ijs} of the sth component (s = 1,…, S), i.e.,

$$G_i = \sum_{j=1}^{J} G_{ij}, \quad G_{ij} = \sum_{s=1}^{S} g_{ijs}, \quad \lambda_{is} = \sum_{j=1}^{J} g_{ijs}. \tag{3.10}$$

The process in the separator is fully defined if the coefficients of the split are given:

$$\delta_{ijs} = \frac{g_{ijs}}{\lambda_{is}}, \quad j = 1, \ldots, J, \quad s = 1, \ldots, S. \tag{3.11}$$

From (3.10) and (3.11) it follows directly that

$$\sum_{j=1}^{J} \delta_{ijs} = 1. \tag{3.12}$$

Thus mathematical description of the process separation allows the creation of a simple and effective calculation module for dividing multicomponent mixtures through the following algorithm:

1. Set G_i, λ_{is} and δ_{ijs}; $j = 1, \ldots, J$; $s = 1, \ldots, S$.
2. Calculate $g_{ijs} = \lambda_{is}\delta_{ijs}$; $j = 1, \ldots, J$; $s = 1, \ldots, S$..
3. Calculate $G_{ij} = \sum_{s=1}^{S} g_{ijs}$, $j = 1, \ldots, J$.This algorithm allows the parameters of the outlet streams, $y_{ij} = \left|G_{ij}, g_{ij1}, \ldots, g_{ijs}\right|$, to be determined for given values of the parameters of the inlet streams, $x_i = |G_i, \lambda_{i1}, \ldots, \lambda_{is}|$, and the coefficients of separation, δ_{ijs}; $j = 1, \ldots, J$; $s = 1, \ldots, S$.

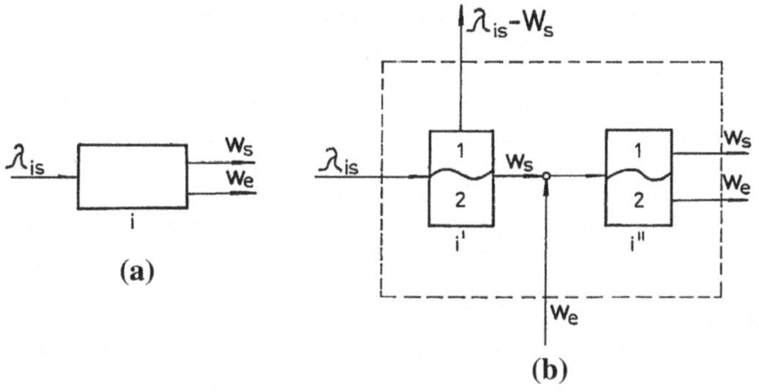

Fig. 8 Chemical processes module (*a*) as a combination of two separators modules (*b*)

3.4 Chemical Processes

The modules of the processes which flow in chemical reactors (Fig. 8a) can be built conditionally from the modules of two separators (Fig. 8b). In reactor i substance λ_{is} enters and nonreacting substance w_s comes out and the product of the reaction is w_e (Fig. 8a). Conditionally it could be accepted that λ_{is} enters separator i' and is divided into a reaction part ($\lambda_{is} - w_s$) and a nonreaction part (w_s). The nonreaction part is mixed with the quantity of component w_e received in the reactor and they are divided in separator i''. For these two separators we can define the coefficients of separation:

$$\delta_{i's1} = \frac{w_s}{\lambda_{is}}, \quad \delta_{i's2} = \frac{\lambda_{is} - w_s}{\lambda_{is}}, \quad \delta_{i''s1} = \frac{w_s}{w_s + w_e}, \quad \delta_{i''s2} = \frac{w_e}{w_s + w_e}. \qquad (3.13)$$

The above equations allows us to build the following simple algorithm for simulation:

1. Set λ_{is} and the coefficients of separation δ.
2. Determine $w_s = \delta_{i's1}\lambda_{is}$.
3. Determine $w_s + w_e = w_s/\delta_{i''s1}$.
4. Determine $w_e = \delta_{i''s2}(w_s + w_e)$.

For solution of concrete problems in the field of the modeling and simulation of CPS it is convenient to use specialized software (ChemCad, Asspen, Pro II, Hysim).

References

1. Ostrovskii GM, Volin YuM (1970) Optimization methods of complex chemical engineering systems. Chemistry, Moscow (in Russian)
2. Ostrovskii GM, Volin YuM (1975) Modeling of complex chemical engineering systems. Chemistry, Moscow (in Russian)

3. Krowe CM et al. (1973) Chemical plant simulation. Prentice Hall, New York
4. Westerberg AW, Hutchison HP, Motard RL, Winter P (1979) Process flowsheeting. Cambridge University Press, London
5. Boyadjiev Chr (1993) Fundamentals of modeling and simulation in chemical engineering and technology. Bulgarian Academy of Sciences, Sofia (in Bulgarian)
6. Botev T, Boyadjiev Chr, Stateva R (1982) Hung J Ind Chem 10:175
7. Botev T, Boyadjiev Chr, Stateva R (1984) Hung J Ind Chem 12:109
8. Botev T, Stateva R, Boyadjiev Chr (1985) Hung J Ind Chem 13:405
9. Stateva R, Boyadjiev Chr (1980) Int Chem Eng 20:306

Synthesis of Systems

1 Optimal Synthesis of Chemical Plants

One of the main tasks of modeling and simulation in chemical engineering and chemical technology is to obtain a quantitative description for the purposes of optimal design or control. This represents the optimization of some parameter (most often economic) in the design of a new operating process or system or the control of operating processes or systems [1].

1.1 Optimization

The optimization of a parameter of the system requires the search for the maxima or minima of one objective function characterizing in some way the system. It depends on the variables of the system (10.1.4) and satisfies the equations of the model in (10.1.5):

$$z = z(X,\ Y,\ A,\ B),\ F(X,\ Y,\ A,\ B) = 0. \tag{1.1}$$

Usually, the optimal design reduces to searching for the maxima (minima) of z in relation to some of the constructive variables A, whereas in the optimal control this role is given to the inlet regime variables X. In both cases the limitation placed upon the variables $F = 0$ must be satisfied and depends on the topological structure of *chemical plant systems* (CPS). Obviously, the attempt to optimize CPS through variation of the structure could lead to new results, but this sets the task for optimal synthesis of CPS.

1.2 Optimal Synthesis

Searching for the optimum of the objective function of CPS for an inconstant topological structure is reduced to determination of

C. Boyadjiev, *Theoretical Chemical Engineering*,
DOI: 10.1007/978-3-642-10778-8_11, © Springer-Verlag Berlin Heidelberg 2010

$$z_{\text{opt}} = \underset{A,\alpha}{\text{opt}}\, z(X,\ Y,\ A,\ \alpha), \tag{1.2}$$

i.e., to minimization (maximization) of z in relation to variables A and α, where α is the vector of the structure variables (parameters). It expresses quantitatively the topological structure of CPS [1].

The mathematical task thus formulated for optimal synthesis of CPS means that for given values of the inlet regime variables X and for desired values of the outlet regime variables Y, there need to be found such a topological structure (α) and apparatuses for the separate processes (A) that CPS are optimal in some meaning (z). Obviously, solving such a global task is impossible not only because of the lack of appropriate methods, but also because of its simple formulation. Having this in mind, we will use a hierarchical approach which will reduce the task for optimal synthesis of CPS several main tasks, where obviously the synthesis of the optimal topological structure is connected with the corresponding synthesis of models of CPS.

1.3 Main Problems

The most general task for optimal synthesis of CPS has the purpose of finding the optimal composition of apparatuses and the technological connections between them in such a way that for a given composition of substances (raw materials, semiproducts) in terms of kind and quantities definite products (of a given kind and quantity) are to be produced through appropriate physical and chemical influences upon the substances entering CPS. This problem can be solved only if it is decomposed in an appropriate way. One possible approach [1] suggests solving different optimization problems at the following hierarchical levels:

1. Selection of routes and conditions for implementing the reactions
2. Determination of optimal chemical reactor systems
3. Determination of optimal systems for separation of mixtures
4. Selection of supporting subsystems
5. Determination of optimal heat exchanger systems
6. Qualitative analysis of the reliability of CPS
7. Analysis of the dynamic properties of CPS
8. Preliminary optimal structure of CPS

The separate tasks of the optimal synthesis thus outlined show that homogeneous CPS which contain monotype processes (chemical, separation, heat transfer, etc., processes) are most often examined. Of no lesser interest is the optimal synthesis of heterogeneous CPS (consisting of different types of processes).

One often-solved problem is the optimal synthesis of a separation system concerning a certain mixture of n substances: the task is to find the optimal separation system by dividing the mixture into pure substances (at minimal annual expense).

Another task of no less importance and especially relevant economically is the synthesis of an optimal system for recuperative heat transfer (OSRHT). There are two given material stream systems. One system represents hot streams which should be cooled to a given temperature and the second system is the cold streams which should be warmed up to a given temperature. A system of heat exchangers is searched for where the cold and hot streams exchange heat and the inadequacies are compensated by heating agents and cooling agents. The optimal system is chosen in such a way that the costs for heat transfer equipment and heating and cooling agents are minimal.

1.4 Methods of Synthesis

The variety of tasks for synthesis of CPS forces the creation of a series of methods, most of which differ in principle.

The method which permits the most mathematical formalization of the task is that of the structural parameters [1]. If the model of CPS is expressed as

$$y^k = f^k\left(x^k, a^k\right),\ k = 1, \ldots, N, \tag{1.3}$$

then the vectors of the outlet and inlet regime variables of the kth block and its constructive variables can be expressed as

$$y^k = \left|y_1^k, \ldots, y_{n_k}^k\right|,\quad x^k = \left|x_1^k, \ldots, x_{m_k}^k\right|,\quad a^k = \left|a_1^k, \ldots, a_{p_k}^k\right|. \tag{1.4}$$

Between the inlet and the outlet regime variables of the different blocks connections exist which can be expressed through structural parameters:

$$x_1^k = \sum_{i=1}^{n} \sum_{j=1}^{n_l} \alpha_{ji}^{ik} y_j^i, \tag{1.5}$$

where

$$\alpha_{ji}^{ik} = \begin{cases} 1 & \text{if there is a connection between the } k\text{th} \\ & \text{and the } i\text{th block in the direction from } l \text{ to } k; \\ 0 & \text{if there is no connection between blocks } k \text{ and } l. \end{cases} \tag{1.6}$$

The quality of the synthesized CPS is determined by the values of a particular objective function z and its extreme value is searched for in relation to parameters α and A:

$$z_{opt} = \underset{A,\alpha}{\text{opt}}\, z(X, Y, A, \alpha) \rightarrow \text{extremum}. \tag{1.7}$$

Evidently the integer variables α create difficulties when searching for the extreme values in (1.7). These inconveniences disappear when α can have arbitrary values in the interval $0 \leq \alpha \leq 1$, but then α expresses the part from the outlet stream from a given block which becomes the inlet stream in other block.

The method of dynamic programming also finds application in the optimal synthesis of CPS. As an example separation of n substances in n stages is examined. In the first stage the costs are determined for the separation of one substance (from all n substances), which are obviously zero. In the second stage the costs are determined for the separation of two substances for all combinations of n substances, i.e., $n(n-1)/2$ combinations from more than two substances. In the third stage the costs are determined for separation of three substances in two parts, one of which contains only one substance. In this way stage n is reached for different variants for separations of n substances, from which the most economical is chosen.

Heuristic methods for optimal synthesis [1] use algorithms which in the first stage are composed of a set of heuristics, i.e., rules for making decisions in given technological situations which usually comprise much engineering experience. The second stage in the heuristic algorithms represents the way to make decisions to choose (at a given stage of the synthesis) one or another heuristic. The last stage of the algorithm is the way to tune and train the heuristic algorithm.

In a series of cases it is especially convenient (example, for reconstruction of CPS) to use evolutionary methods, which start from an initial structure of CPS and then use different algorithms for the gradual optimization. Some tasks for optimal synthesis permit the use of combinatorial methods, through which the full set of variants of CPS is obtained (in the form of a tree of the variants) and the optimal variant is searched for.

The combinatorial method leads to difficulties as a result of the possibility to obtain a large number of variants. It could be combined with the heuristic method, which during the creation of the variants rejects (discards) the futureless variants used for evaluation of the variants of the heuristic rules. Such a combined method [2–4] will be examined more carefully in the next example.

1.5 Optimal Synthesis of a System for Recuperative Heat Transfer

N hot streams which should be cooled from temperature $T_{\mathrm{hi}}^{\mathrm{in}}$ to $T_{\mathrm{hi}}^{\mathrm{out}} (\mathrm{i} = 1, \ldots, \mathrm{m})$ and n cold streams which should be heated from temperature $T_{\mathrm{cj}}^{\mathrm{in}}$ to $T_{\mathrm{cj}}^{\mathrm{out}}$ will be examined. For given thermal/physical parameters of the streams and in particular of the water equivalent of the hot and cold streams W_{hi} and $W_{\mathrm{cj}}, \mathrm{i} = 1, \ldots, \mathrm{m}; \mathrm{j} = 1, \ldots, \mathrm{n}$ (product of the mass consumption and the specific heat), we search for the OSRHT with minimal annual remittance expenses:

$$C = C_k + C_e + C_g, \tag{1.8}$$

where C_k is the capital expenses (for heat exchangers), C_e is exploitation expenses (for heating and cooling agents), and C_g is other expenses (pipelines, pumps, hydraulic losses, etc.).

The solution of the problem will be searched for through consecutive unification of hot and cold streams in appropriate recuperative heat exchangers, calculation of these heat exchangers, and determination of the necessary additional heat exchangers for the necessary additional cooling (heating) of the hot (cold) streams and the necessary cooling agents (heating agents) for this purpose.

The minimum of C is searched for among a *system for recuperative heattransfer* (SRHT) obtained through different combinations of the hot and cold streams. This direct combined approach is difficult because of the large number of possible combinations [2–4]. This forces us to use a heuristic approach for selection of the most likely combinations. For this purpose the combinatorial–heuristic method [2–4] is created in which the procedure for optimal synthesis of CPS is reduced to using the sets of the hot and cold streams:

$$I_h = \{(\mathrm{i} = 1,\ldots,\mathrm{m})\}, I_c = \{\mathrm{j}(\mathrm{j} = 1,\ldots,\mathrm{n})\}, \tag{1.9}$$

where the transfer of heat is realized between streams $i = j$. This could lead to an SRHT if the streams are arranged in a proper (optimal) way in the sets I_h and I_c.

The analysis of a series of particular cases proves [2–4] that an SRHT can be obtained if "the sets of the hot (cold) streams represent sequences at which their initial (final) temperatures decrease and in the heat exchangers are combined sequentially a couple of streams with equal numbers $i = j = 1, 2, \ldots$" This will be used further as a basic heuristic for building the combinatorial–heuristic method.

The heat transfer between the hot and the cold stream ($i = j$) could flow under different conditions. If we use the notation in Fig. 1, where the inlet and outlet temperatures of the streams do not match their initial and final values, we see that the heat transfer depends materially on the admissible minimal temperature difference $\Delta T_{\min}$, which results in the obvious limitations

$$t_h^{in} - t_c^{out} \geq \Delta T_{\min}, \quad t_h^{out} - t_c^{in} \geq \Delta T_{\min}, \quad T_h^{initial} - T_c^{initial} > \Delta T_{\min}, \quad Q \leq Q_{\max}, \tag{1.10}$$

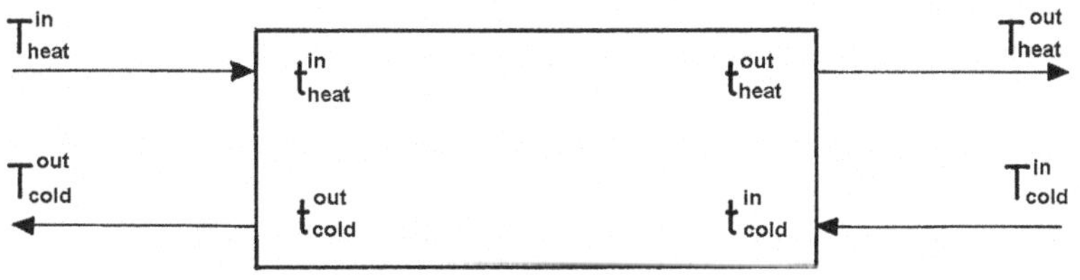

Fig. 1 Inlet and outlet temperatures of the streams

where

$$Q_{\max} = \min(Q_{\mathrm{h}}, Q_{\mathrm{c}}), \quad Q_{\mathrm{h}} = W_{\mathrm{h}}\left(T_{\mathrm{h}}^{\mathrm{initial}} - T_{\mathrm{h}}^{\mathrm{final}}\right), \quad Q_{\mathrm{c}} = W_{\mathrm{c}}\left(T_{\mathrm{c}}^{\mathrm{final}} - T_{\mathrm{c}}^{\mathrm{initial}}\right),$$
$$Q = W_{\mathrm{q}}\left(t_{\mathrm{q}}^{\mathrm{in}} - t_{\mathrm{q}}^{\mathrm{out}}\right) = W_{\mathrm{c}}\left(t_{\mathrm{c}}^{\mathrm{out}} - t_{\mathrm{c}}^{\mathrm{in}}\right). \tag{1.11}$$

Heat transfer between the two streams in Fig. 1 is possible under the following condition

$$Q_{\mathrm{h}} \le Q_{\mathrm{c}}, \quad W_{\mathrm{h}} \le W_{\mathrm{c}}, \quad Q_{\mathrm{h}} > Q_{\mathrm{c}}, \quad W_{\mathrm{h}} > W_{\mathrm{c}}, \quad \left(T_{\mathrm{h}}^{\mathrm{initial}} - T_{\mathrm{c}}^{\mathrm{final}}\right) \le \Delta T_{\min},$$
$$\left(T_{\mathrm{h}}^{\mathrm{final}} - T_{\mathrm{c}}^{\mathrm{initial}}\right) \le \Delta T_{\min}, \quad \left(T_{\mathrm{h}}^{\mathrm{initial}} - T_{\mathrm{c}}^{\mathrm{final}}\right) > \Delta T_{\min}, \quad \left(T_{\mathrm{h}}^{\mathrm{final}} - T_{\mathrm{c}}^{\mathrm{initial}}\right) > \Delta T_{\min}, \tag{1.12}$$

the combination of which leads [2–4] to a large number of variants of the heat transfer between two streams, from which the real ones are those which satisfy conditions (1.10).

The above considerations permit the creation of an algorithm for optimal synthesis of one variant [2–4]:

1. Input of the data for the hot and cold streams.
2. Arranging of the streams in accordance with their final temperatures

$$\left(T_{\mathrm{h1}}^{\mathrm{initial}} > T_{\mathrm{h2}}^{\mathrm{initial}} > \cdots > T_{\mathrm{hm}}^{\mathrm{initial}}, \ T_{\mathrm{c1}}^{\mathrm{initial}} > T_{\mathrm{c2}}^{\mathrm{final}} > \cdots > T_{\mathrm{cn}}^{\mathrm{final}}\right).$$

3. Choosing the serial couple of streams $i = j$.
4. Checking whether the heat transfer is possible $\left(T_{\mathrm{h}}^{\mathrm{initial}} - T_{\mathrm{c}}^{\mathrm{initial}} \ge \Delta T_{\min}\right)$.

   (a) If it is, go to step 7.
   (b) If it is not, go to step 5.

5. An external heat carrier is used to heat the cold stream to specified $T_{\mathrm{c}}^{\mathrm{final}}$. Calculation of the necessary heat exchanger for this heating and the annual cost for its use. This cold stream from the set is excluded and the algorithm starts from step 2.
6. Checking whether there are more cold streams.

   (a) If there are, go to step 7.
   (b) If there are not, go to step 10.

7. Choice of a variant depending on Q_{h}, Q_{c}, W_{h}, W_{c}, and $\Delta T_{\min}$ [2–4].
8. Calculation of the necessary recuperative heat exchanger and its capital expenses.
9. Further the processed parts from the two streams are excluded and the algorithm starts from step 2.
10. The hot parts are cooled with an external cool carrier.

The necessary heat exchanger for this cooling and the annual use expenses are calculated.

This algorithm permits the synthesis of an initial approximation of an OSRHT. It represents one local minimum of (1.8) because such a chosen couple of streams are not always optimal. To search for the optimal couple of streams the main permutation of the set of the cold streams (obtained through the main heuristics) is used and from it another two permutations are obtained in which the second and third elements of the main permutation are exchanged with the first one, i.e., with the above algorithm generally three OSRHT are synthesized. As an example in step 3 besides $i = j = 1$ two more cases are examined: $i = 1$, $j = 2$ and $i = 1$, $j = 3$. The OSRHT are determined, their rendered expenses are compared, and from the minima of these three numbers are determined the optimal couple of streams and the algorithm starts from step 2. This algorithm permits us to get close to the global extreme of the task through the introduction of additional heuristics. The admissible permutations are built from the main one through a limited number (three) of sequential elements. Each next permutation differs from the main one by the number of the element situated in the first place preserving the range of the others.

The optimal couple of streams are excluded from the sets of the hot and cold streams and their places are taken by the remaining hot and cold streams and everything starts from the beginning until all hot streams are exhausted. In this way the approach presented solves the problem of optimal synthesis of an SRHT as a task of optimal arranging [2–4]. Created on this basis, the program SYNTI permits automated creation of OSRHT for the purposes of optimal design and renovation. The test and industrial examples solved show [2–4] good agreement with other methods.

Using the method for optimal synthesis of SRHT leads to serious economy of power. In the literature [4–6] these economies in petrochemical manufacturing are an average 30%, but in separate cases they reach 50% (production of ethylene).

2 Renovation of Chemical Plant Systems

The main task of chemical engineering is the quantitative description of CPS for the purposes of their optimal design and control. The analysis of the solutions of the tasks for optimal design of a new CPS and optimal control of an operating CPS during the last 10–20 years shows that the optimization of the design stage is very advantageous in comparison with the exploitation stage. This advantage obviously arises from the possibility for optimization of the construction of the apparatuses and in some cases of the plant scheme. On the other hand, the solution of the problem for optimal control could have great economical effectiveness owing to the fact that the number of operating CPS is much larger than the number which should be designed in the future and this relation increases with the intensive development of the industry. The advantages of the optimization of the stages of design and exploitation can be combined (summed up) if the operating CPS is driven to a new condition which is more effective (economically) in comparison with the previous one through *renovation* ["renovate" means to bring to a better condition than earlier, or to a more effective (economically) condition], i.e., solving the problem for optimal reconstruction of the operating CPS [7].

The task for renovation of CPS differs from the tasks for optimal design and control through the variables which are used for optimization (free) in the mathematical description of CPS.

2.1 Mathematical Description

A CPS consists of separate blocks connected by streams in correspondence with its topological structure. Each block u can be characterized by an ordered sequence of three numbers

$$u = (r, s, t), \tag{2.1}$$

where r is the number of the apparatus in the register of the type of apparatuses, s is the number of the process in the register of the type of processes, and t is the number of the block in the topological structure of the CPS.

The technological streams between the blocks are characterized by their number q in the topological structure and their direction in the values of its parameters (composition, capacity, temperature, pressure, etc.).

The topological structure of the CPS can be expressed simply with the help of the matrix of the streams:

$$A = \|a_{\mathrm{tq}}\|, \quad \mathrm{t} = 1, \ldots, \mathrm{T}, \quad \mathrm{q} = 1, \ldots, \mathrm{Q}, \tag{2.2}$$

where

$$a_{\mathrm{tq}} = \begin{cases} 1 & \text{if stream q is theinlet in block t;} \\ -1 & \text{if stream q is theoutlet in block t;} \\ 0 & \text{if stream q is not connected with block t,} \end{cases} \tag{2.3}$$

and T and Q are the whole numbers of the blocks and the streams in the CPS.

The set of the inlet regime variables in block $u_0 = (r_0, s_0, t_0)$ is

$$x_{r_0 s_0 t_0} = x_{r_0 s_0 t_0}^{\mathrm{q}_{\mathrm{x}_1}} \cup x_{r_0 s_0 t_0}^{\mathrm{q}_{\mathrm{x}_2}} \cup \ldots, \tag{2.4}$$

where $x_{r_0 s_0 t_0}^{\mathrm{q}}$ are the sets of the variables of the inlet streams in block u_0,

$$x_{r_0 s_0 t_0} = g_q. \tag{2.5}$$

for $q = q_{x_1}, q_{x_2} \dots$ and $a_{t_0 q_{x_1}} = a_{t_0 q_{x_2}} = \dots = 1$, i.e., $q_{x_1}, q_{x_2}, \dots$ are the numbers of the whole inlet streams of block u_0, g_q are the set of the variables of stream q.

The set of the outlet regime variables can be expressed analogously:

$$y_{r_0 s_0 t_0} = y_{r_0 s_0 t_0}^{q_{y_1}} \cup y_{r_0 s_0 t_0}^{q_{y_2}} \cup \dots, \tag{2.6}$$

where $y_{r_0 s_0 t_0}^{q}$ are the sets of the variables of the outlet streams from block u_0,

$$y_{r_0 s_0 t_0}^{q} = g_q. \tag{2.7}$$

for $q = q_{y_1}, q_{y_2}, \dots$ and $a_{t_0 q_{y_1}} = a_{t_0 q_{y_2}} = \dots = -1$, i.e., $q_{y_1}, q_{y_2}, \dots$are the numbers of all outlet streams of block u_0.

The set of constructive variables of apparatus r_0 corresponding to block u_0 can be expressed as

$$z_{r_0 s_0 t_0} = \left\{ z_{r_0 t_0}^{1}, z_{r_0 t_0}^{2}, \dots, z_{r_0 t_0}^{b_{r_0}} \right\}, \tag{2.8}$$

where b_{r_0} is the number of constructive variables of apparatus r_0.

The mathematical description of block u_0 represents a system of equations related to the regime and the constructive variables:

$$f_{rst}^{i}(x_{rst}, y_{rst}, z_{rst}) = 0, \quad i = 1, \dots, I_s, \tag{2.9}$$

where I_s is the number of equations describing the process with number s.

The equations for the connections between the blocks represent all equations of the type

$$y_{r_y s_y t_y}^{q_y} = x_{r_x s_x t_x}^{q_x} \tag{2.10}$$

for $q_y = q_x$ and $r_y \neq r_x, \quad s_y \neq s_x, \quad t_y \neq t_x$.

The combination of Eqs. (2.9) and (2.10) represents the mathematical description of the CPS.

The inlet (x_{rst}^{q}) and outlet (x_{rst}^{q}) regime variables of the separate blocks of the CPS which are not equations of the type (2.10) generate the set of inlet $\left(X_{rst}\right)$ and outlet $\left(Y_{rst}\right)$ variables of the CPS:

$$X_{rst} \subset x_{rst}, \quad Y_{rst} \subset y_{rst}, \tag{2.11}$$

The general number of Eqs. in (2.9) and (2.10) is always less than the general number of the regime and constructive variables. Having this in mind, the use of the mathematical description of CPS for the purposes of their simulation forces the separation of the variables into free (not regulated) and given (regulated) variables in such way that the number of free variables is equal to the numbers of Eqs. in (2.9) and (2.10). The regulated (not regulated) variable could be a regime variable as well as a constructive, variable i.e., several subsets of regulated variables can be differentiated:

$$x^{\mathrm{R}}_{\mathrm{rst}} \subset x_{\mathrm{rst}}, \quad X^{\mathrm{R}}_{\mathrm{rst}} \subset X_{\mathrm{rst}}, \quad y^{\mathrm{R}}_{\mathrm{rst}} \subset y_{\mathrm{rst}}, \quad Y^{\mathrm{R}}_{\mathrm{rst}} \subset Y_{\mathrm{rst}}, \quad z^{\mathrm{R}}_{\mathrm{rst}} \subset z_{\mathrm{rst}}., \tag{2.12}$$

Depending on which are the regulated variables in the system of Eqs. (2.9) and (2.10) the solution requires the creation of different algorithms on the basis of which different mathematical model of CPS are created.

2.2 Mathematical Models

A quantitative description of CPS for the purposes of the optimal control [1] needs the creation of models where the inlet regime variables and the constructive variables are specified:

$$X^{\mathrm{R}}_{\mathrm{rst}} = X_{\mathrm{rst}}, \quad Y^{\mathrm{R}}_{\mathrm{rst}} \neq 0, \quad z^{\mathrm{R}}_{\mathrm{rst}} = z_{\mathrm{rst}}. \tag{2.13}$$

Simulation of CPS can be realized in the cases when some of the outlet regime variables are specified. Obviously some of the inlet regime variables should be free, i.e.,

$$Y^{\mathrm{R}}_{\mathrm{rst}} \neq 0, \quad X^{\mathrm{R}}_{\mathrm{rst}} \neq X_{\mathrm{rst}}. \tag{2.14}$$

Obviously system (2.9) and (2.10) can be solved if it has the necessary degrees of freedom, i.e., if the number of specified outlet variables does not exceed the number of free inlet variables:

$$\left|Y^{\mathrm{R}}_{\mathrm{rst}}\right| \leq \left|X_{\mathrm{rst}} - X^{\mathrm{R}}_{\mathrm{rst}}\right|. \tag{2.15}$$

In these cases one can use the method [8–11] for determination of the zones of independent inlet influence of a block with specified outlet variables.

The quantitative description of CPS for the purposes of the optimal design requires the creation of models where the regime variables are specified:

$$X^{\mathrm{R}}_{\mathrm{rst}} = X_{\mathrm{rst}}, \quad Y^{\mathrm{R}}_{\mathrm{rst}} = Y_{\mathrm{rst}}, \quad z^{\mathrm{R}}_{\mathrm{rst}} \neq 0. \tag{2.16}$$

With this in mind, when solving the design task, it is advisable to use standard equipment. Some of the constructive variables could be given:

$$z^{\mathrm{R}}_{\mathrm{rst}} \neq 0. \tag{2.17}$$

Renovation [2–7] differs from the optimal design or control in that some of the constructive variables should be free, but in the essential part they should be specified. This as a rule requires some of the regime variables to be free also.

In this way a quantitative description of a CPS for the purposes for its renovation should be obtained on the basis of mathematical models where the regime as well as the constructive variables is partially specified:

$$X_{\mathrm{rst}}^{\mathrm{R}} \neq 0, \quad Y_{\mathrm{rst}}^{\mathrm{R}} \neq 0, \quad z_{\mathrm{rst}}^{\mathrm{R}} \neq 0. \tag{2.18}$$

From (2.18) one can see the possibility for a large variety when specifying the variables of CPS and from there the large variety of renovation problems which can be solved. This requires the creation of the necessary variety of methods, algorithms, and mathematical models also. Independently from the large number of models the tasks for renovation can be reduced to several main types.

2.3 Main Problems

The renovation of CPS represents reducing the system to a new, more economically effective condition in comparison with the earlier state. The tasks are more general than the tasks of the optimal reconstruction of CPS since they are realized through introduction of a new apparatus, a new process, or changing the topological structure of CPS. This leads to the introduction of new values of the variables r, s, and t. As far as they simply determine the blocks of CPS this is equivalent to introducing new blocks in CPS. If by $r_{\mathrm{a}}, s_{\mathrm{a}}, t_{\mathrm{a}}$ are denoted the new values of the parameters characterizing the new block, it is possible to introduce the following types of new blocks:

$$(r, s, t_{\mathrm{a}}), (r_{\mathrm{a}}, s, t), (r, s_{\mathrm{a}}, t), (r_{\mathrm{a}}, s, t_{\mathrm{a}}), (r, s_{\mathrm{a}}, t_{\mathrm{a}}), (r_{\mathrm{a}}, s_{\mathrm{a}}, t), (r_{\mathrm{a}}, s_{\mathrm{a}}, t_{\mathrm{a}}). \tag{2.19}$$

In practice, introducing a new process or apparatus in general changes the topological structure of CPS, i.e.,

$$(r_{\mathrm{a}}, s, t) = (r_{\mathrm{a}}, s, t_{\mathrm{a}}), \quad (r, s_{\mathrm{a}}, t) = (r, s_{\mathrm{a}}, t_{\mathrm{a}}), \quad (r_{\mathrm{a}}, s_{\mathrm{a}}, t) = (r_{\mathrm{a}}, s_{\mathrm{a}}, t_{\mathrm{a}}). \tag{2.20}$$

Introducing a new process is practically always connected with introducing new apparatus:

$$(r, s_{\mathrm{a}}, t_{\mathrm{a}}) = (r_{\mathrm{a}}, s_{\mathrm{a}}, t_{\mathrm{a}}). \tag{2.21}$$

From (2.19) to (2.21) it directly follows that all new blocks can be reduced to three main types:

$$(r, s, t_{\mathrm{a}}), \quad (r_{\mathrm{a}}, s, t_{\mathrm{a}}), \quad (r_{\mathrm{a}}, s_{\mathrm{a}}, t_{\mathrm{a}}). \tag{2.22}$$

From (2.22) it can be seen that there are three main tasks of the renovation and they are accomplished on the basis of the optimal synthesis of CPS, creation of highly intensive apparatuses, and introduction of highly effective processes.

2.4 Renovation by Optimal Synthesis of Chemical Plant Systems

The renovation of CPS through optimal synthesis is solved through the introduction of new blocks of type (r, s, t_a), i.e., through finding a new topological structure of CPS for given types of processes and apparatuses. On the other hand, however, from (2.18) it follows that some of the constructive variables are regulated at some of the blocks. In this way, this renovation task is equivalent to the task for optimal synthesis for partially given equipment. In all cases of practical interest, maximal preservation of the present equipment is necessary.

One of the first attempts to solve this problem was realized [5, 6] for renovation of subsystems for recuperative heat transfer in CPS. In this case the solution was obtained in two stages. In the first stage, methods for optimal synthesis of subsystems for recuperative heat transfer are used [2–4]. This part of the subsystem will stay unchanged until the end because of the fact that is too close to the optimal one. In the second stage the problem of the optimal design of subsystems for recuperative heat transfer [6] is solved for the rest of the subsystem.

The problem of renovation on the basis of optimal synthesis can be solved for multirange (multiassortment) CPS [7]. In these cases the blocks of CPS are universal apparatuses with given connections between them where several production activities are realized simultaneously and consecutively. Obviously the time τ should be the fourth parameter, characterizing the blocks of the multirange CPS:

$$u = (r, s, t, \tau), \tag{2.23}$$

i.e., the block with number t at moment τ represents apparatus r where process s is realized.

Multirange CPS are characterized by a set of universal apparatuses connected in a given system for each of them. Very often it should be determined which process will be realized in them in connection with production of different ranges or on changing the production program (changing the stages of the production of the existing assortments, introduction of new ones, etc.). With this in mind, the task of renovation of multiassortment CPS is reduced to finding the distribution of the separate operations of all the production activities over the universal apparatuses. Obviously, here we could also use other criteria for optimization. This problem can be solved by finding the optimal schedules of the apparatuses in CPS [7]. This solution would be significantly more effective if new connections between the apparatuses are added or if the number of some of the universal apparatuses is increased.

For a boundary case, the task of renovation of multiassortment CPS can be reduced to determination of the optimal schedule of the apparatuses of the multiassortment CPS if its multivariant realization is possible. Here the addition of apparatuses and the connections between them could lead to interesting optimal solutions.

2.5 Renovation by Introduction of Highly Intensive Equipment

The optimal reconstruction of CPS is realized most often through exchange of some of the apparatuses with new, more intensive ones. This approach is widely used and that is why it is considered as a main method for renovation of CPS.

The creation of highly intensive apparatuses is realized through intensification of the processes flowing in them. As an example, for the heat and mass transfer apparatuses this is most often achieved through intensification of the hydrodynamic interaction of the phases through new packing (in absorption apparatuses), creation of a boiling layer (in apparatuses for drying and catalytic reactions), etc.

The analysis of the task of creation of apparatuses for realization of intensive processes shows that this is achieved through the introduction of external energy to the system. From an economic point of view, in vapor–liquid and vapor–solid systems the introduction of external energy is most profitable through the vapor phase. In the liquid–liquid systems the small differences in the density do not permit the energy to be introduced in a hydrodynamic way and because of this the use of vibrations, pulsation, etc. is recommended. All these considerations should be used in the creation of intensive apparatuses for the purposes of renovation.

2.6 Renovation by Introduction of Highly Effective Processes

One of the most effective approaches for the renovation of CPS is realized through the introduction of new processes. This provides the opportunity to obtain a greater variety of optimal solutions and, on the other hand, permits the use of the achievements of chemical engineering in the creation of a new, highly effective process. In this approach the new processes replace the old ones or they are introduced additionally with the purpose of solving ecological problems or decreasing of costs of raw materials and energy.

From a economic point of view, the introduction of processes which significantly decrease the energy cost per unit of production is particularly interesting. The first step when solving these problems is the analysis of the thermodynamic effectiveness of the separate processes in CPS, with the purpose of finding those which should be replaced or to which should be added processes to increase the thermodynamic effectiveness of CPS as a whole.

The replacement of one process with another with the purpose of increasing the thermodynamic effectiveness is used in separation processes with a phase transition. The replacement of the distillation processes with extraction or adsorption processes is particularly effective in the cases when that is impossible.

The introduction of additional processes with the purpose of utilization of heat is one of the most important methods for renovation of CPS. This is confirmed by the wide use of heat pumps in distillation, evaporation, drying, etc., installations where huge quantities of low-potential heat are separated. As an example, the use of thermocompressors to increase the pressure and temperature of the steam of the distillate in distillation columns could decrease the energy cost by 40–60%.

The methods for renovation of CPS are created in parallel with the methods of optimal design and control of CPS. In contrast with the optimal control (where the constructive variables are regulated and the free regime variables are changing continuously) and the optimal design (where the regime variables are discreet), in renovation the constructive as well as the regime variables are regulated and the free variables are changing continuously and are discreet.

The renovation of CPS operates with one considerable set of methods which are created for solving different classes of problems. From the description given, it is evident that some of these methods were already known long ago, but for the larger part they have been created comparatively recently. On this basis, a series of interesting practical problems have been solved for renovation of CPS through the introduction of heat pumps, the creation of optimal schemes for recuperative heat transfer, optimization of multiassortment productions, etc.

References

1. Boyadjiev Chr (1993) Fundamentals of modeling and simulation in chemical engineering and technology. Bulgarian Academy of Sciences, Sofia (in Bulgarian)
2. Zhelev T, Boyadjiev Chr (1988a) Int Chem Eng 28:543
3. Zhelev T, Boyadjiev Chr (1988b) Int Chem Eng 28:548
4. Zhelev T, Boyadjiev Chr (1988c) Int Chem Eng 28:554
5. Zhelev T, Boyadjiev Chr, Kantcheva S (1987) Hung J Ind Chem 15:403
6. Zhelev T, Boyadjiev Chr, Hartman K (1984) Hung J Ind Chem 12:461
7. Boyadjiev Chr (1987) Chem Ind 6:368 (in Russian)
8. Botev T, Boyadjiev Chr, Stateva R (1982) Hung J Ind Chem 10:175
9. Botev T, Boyadjiev Chr, Stateva R (1984) Hung J Ind Chem 12:109
10. Botev T, Stateva R, Boyadjiev Chr (1985) Hung J Ind Chem 13:405
11. Stateva R, Boyadjiev Chr (1980) Int Chem Eng 20:306

Conclusion

Chemical engineering is related to *practices* in various industries, such as chemical, food processing, power engineering, biotechnology, and ecology. According to an old sentence "*there is nothing more useful to practice than a good theory*" theoretical chemical engineering provides the background to solve numerous practical problems using modeling and simulation methods.

Most of the theoretical results presented here are related to *chemical macrokinetics*, i.e., a mutual area of chemical, hydrodynamic, mass transfer, and heat transfer processes. This simply means that the reagent concentrations in the chemical kinetics models are related to the equations of hydrodynamics, diffusion, and heat conduction and the models express a *full correspondence between physical effects and mathematical operators.*

The basic difficulty in developing analytical solutions of the models is due to the *nonlinearity of the hydrodynamic equations.* In the first half of the twentieth century, the theory avoided the hydrodynamic problem and *Nernst's film concept* (*Langmuir, Lewis, and Whitman*) was used, where the velocity is assumed to be zero. According to this assumption, the mass transfer taking place is due to a stationary diffusion trough an immovable fluid film with unknown thickness. The basic disadvantages of this theory are (1) the *linear dependence of the mass transfer coefficient on diffusivity*, which is not confirmed experimentally, and (2) the *unknown thickness of the film*, which does not allow theoretical determination of the mass transfer coefficient. However, despite these general drawbacks, the theory is still valid and widely applied to practical problems such as *a thin layer at the phase boundary, the thermodynamic equilibrium at the interphase*, as well as the basic consequence of the theory regarding the *additivity of the diffusion resistances*.

The next step beyond the Nernst concept is *Higbie's penetration theory* and some related versions of it: Higbie's concept refers to an assumed constant fluid velocity and a transient mass transfer in a coordinate system moving with the same velocity. This concept, however, does not take into account the *velocity distribution in the boundary layer.*

Theoretical analyses of various hydrodynamic approximations (zero or constant velocity) without correct physical backgrounds reveal that *the boundary layer approximation is the best concept ever conceived owing to its adequacy in describing the physical conditions in real processes*. The *laminar boundary layer theory* (*Prandtl, Schlichting, Gerstein, and Loitsianskii*) has gained advances in cases such as *turbulent boundary layers and diffusion boundary layers* (*Prandtl, Taylor, Karman, Landau, and Levich*). The diffusion boundary layer theory (*Landau and Levich*) has allowed the modeling of many processes pertinent to *nonlinear mass transfer in electrochemical systems* (*Krylov*) and those of *nonlinear mass transfer and hydrodynamic stability in one- and two-phase systems* (*Boyadjiev*).

Advances in numerical methods and computer hardware and software have permitted quite complicated nonlinear boundary problems to be solved numerically if, however, the differential equations and boundary conditions are well formulated.

Numerous chemical engineering processes take place in two-phase systems and the model boundary conditions have to be formulated at the interphase surfaces. *However, commonly the phase interfaces are unknown, which does not allow the well-developed models to applied and well-developed software codes to be employed.* A way of avoiding some of these problems was the use of *diffusion-type models* of columnar devices, where both the velocity and the concentration distributions are replaced by their *averages over the column cross-sectional area*.

The solutions of many problems of practical interest lead to models where the parameters must be obtained by using experimental data. Two main problems concerning these cases (*incorrectness of the inverse problem and multiextremality of the least-squares function*) can be solved by the methods presented in this book.

The information about the methods applied to model and simulate chemical plant systems allows specially developed software codes to be used.

This book does not address the common topic of *distillation column modeling*. The modeling of distillation devices in plate columns uses the approach "from plate to plate": the model equations relevant to each plate are mass and heat balance equations, where the parameters of mass and heat transfer kinetics are replaced by efficiency coefficients. To this end, the thermodynamic liquid–vapor equilibrium at each plate has to be calculated too. All these parameters depend on the types and concentrations of the components in both the liquid and the vapor phases. The use of iterative methods to solve the model equations at each plate requires very effective thermodynamic methods allowing repeated calculations of the liquid–vapor equilibria. The models of distillation plate columns contain many equations, but every equation contains few variables. This leads to model equations with *scarce matrixes and the use of special mathematical software to solve them* is necessary. Obviously, the modeling of distillation plate columns is more of a thermodynamic and mathematical nature without involving a hydrodynamic background. The solutions of such problems are commonly performed by especially developed codes (e.g., ChemCad).

The basic reason for this book is to provide correct formulations and solutions of theoretical problems in chemical engineering by use of modeling and simulation methods allowing problems in practical cases to be solved.

Index